Grzimek's
Animal Life Encyclopedia

Second Edition

●●●●

Grzimek's
Animal Life Encyclopedia

Second Edition

●●●●

Volume 11
Birds IV

Jerome A. Jackson, Advisory Editor
Walter J. Bock, Taxonomic Editor
Donna Olendorf, Project Editor

Joseph E. Trumpey, Chief Scientific Illustrator

Michael Hutchins, Series Editor
In association with the American Zoo and Aquarium Association

GALE®

THOMSON
™
GALE

Detroit • New York • San Diego • San Francisco • Cleveland • New Haven, Conn. • Waterville, Maine • London • Munich

THOMSON

GALE

Grzimek's Animal Life Encyclopedia, Second Edition
Volume 11: Birds IV

Project Editor
Donna Olendorf

Editorial
Deirdre Blanchfield, Madeline Harris, Christine Jeryan, Kristine M. Krapp, Kate Kretschmann, Melissa C. McDade, Mark Springer

Permissions
Kim Davis

Imaging and Multimedia
Mary K. Grimes, Lezlie Light, Christine O'Bryan, Barbara Yarrow, Robyn V. Young

Product Design
Tracey Rowens, Jennifer Wahi

Manufacturing
Dorothy Maki, Evi Seoud, Mary Beth Trimper

LIBRARY OF CONGRESS CATALOGING-IN-PUBLICATION DATA

Grzimek, Bernhard.
  [Tierleben. English]
  Grzimek's animal life encyclopedia.— 2nd ed.
    v. cm.
  Includes bibliographical references.
  Contents: v. 1. Lower metazoans and lesser deuterosomes / Neil Schlager, editor — v. 2. Protostomes / Neil Schlager, editor — v. 3. Insects / Neil Schlager, editor — v. 4-5. Fishes I-II / Neil Schlager, editor — v. 6. Amphibians / Neil Schlager, editor — v. 7. Reptiles / Neil Schlager, editor — v. 8-11. Birds I-IV / Donna Olendorf, editor — v. 12-16. Mammals I-V / Melissa C. McDade, editor — v. 17. Cumulative index / Melissa C. McDade, editor.
  ISBN 0-7876-5362-4 (set hardcover : alk. paper)
  1. Zoology—Encyclopedias. I. Title: Animal life encyclopedia. II. Schlager, Neil, 1966- III. Olendorf, Donna IV. McDade, Melissa C. V. American Zoo and Aquarium Association. VI. Title.
  QL7 .G7813 2004

          590'.3—dc21
          2002003351

Printed in Canada
10 9 8 7 6 5 4 3 2 1

Recommended citation: *Grzimek's Animal Life Encyclopedia*, 2nd edition. Volumes 8–11, *Birds I–IV*, edited by Michael Hutchins, Jerome A. Jackson, Walter J. Bock, and Donna Olendorf. Farmington Hills, MI: Gale Group, 2002.

Contents

Contents

Volume 11: Birds IV

• • • • •

Foreword

Earth is teeming with life. No one knows exactly how many distinct organisms inhabit our planet, but more than 5 million different species of animals and plants could exist, ranging from microscopic algae and bacteria to gigantic elephants, redwood trees and blue whales. Yet, throughout this wonderful tapestry of living creatures, there runs a single thread: Deoxyribonucleic acid or DNA. The existence of DNA, an elegant, twisted organic molecule that is the building block of all life, is perhaps the best evidence that all living organisms on this planet share a common ancestry. Our ancient connection to the living world may drive our curiosity, and perhaps also explain our seemingly insatiable desire for information about animals and nature. Noted zoologist, E.O. Wilson, recently coined the term "biophilia" to describe this phenomenon. The term is derived from the Greek *bios* meaning "life" and *philos* meaning "love." Wilson argues that we are human because of our innate affinity to and interest in the other organisms with which we share our planet. They are, as he says, "the matrix in which the human mind originated and is permanently rooted." To put it simply and metaphorically, our love for nature flows in our blood and is deeply engrained in both our psyche and cultural traditions.

Our own personal awakenings to the natural world are as diverse as humanity itself. I spent my early childhood in rural Iowa where nature was an integral part of my life. My father and I spent many hours collecting, identifying and studying local insects, amphibians and reptiles. These experiences had a significant impact on my early intellectual and even spiritual development. One event I can recall most vividly. I had collected a cocoon in a field near my home in early spring. The large, silky capsule was attached to a stick. I brought the cocoon back to my room and placed it in a jar on top of my dresser. I remember waking one morning and, there, perched on the tip of the stick was a large moth, slowly moving its delicate, light green wings in the early morning sunlight. It took my breath away. To my inexperienced eyes, it was one of the most beautiful things I had ever seen. I knew it was a moth, but did not know which species. Upon closer examination, I noticed two moon-like markings on the wings and also noted that the wings had long "tails", much like the ubiquitous tiger swallow-tail butterflies that visited the lilac bush in our backyard. Not wanting to suffer my ignorance any longer, I reached immediately for my *Golden Guide to North*

American Insects and searched through the section on moths and butterflies. It was a luna moth! My heart was pounding with the excitement of new knowledge as I ran to share the discovery with my parents.

I consider myself very fortunate to have made a living as a professional biologist and conservationist for the past 20 years. I've traveled to over 30 countries and six continents to study and photograph wildlife or to attend related conferences and meetings. Yet, each time I encounter a new and unusual animal or habitat my heart still races with the same excitement of my youth. If this is biophilia, then I certainly possess it, and it is my hope that others will experience it too. I am therefore extremely proud to have served as the series editor for the Gale Group's rewrite of *Grzimek's Animal Life Encyclopedia*, one of the best known and widely used reference works on the animal world. *Grzimek's* is a celebration of animals, a snapshot of our current knowledge of the Earth's incredible range of biological diversity. Although many other animal encyclopedias exist, *Grzimek's Animal Life Encyclopedia* remains unparalleled in its size and in the breadth of topics and organisms it covers.

The revision of these volumes could not come at a more opportune time. In fact, there is a desperate need for a deeper understanding and appreciation of our natural world. Many species are classified as threatened or endangered, and the situation is expected to get much worse before it gets better. Species extinction has always been part of the evolutionary history of life; some organisms adapt to changing circumstances and some do not. However, the current rate of species loss is now estimated to be 1,000–10,000 times the normal "background" rate of extinction since life began on Earth some 4 billion years ago. The primary factor responsible for this decline in biological diversity is the exponential growth of human populations, combined with peoples' unsustainable appetite for natural resources, such as land, water, minerals, oil, and timber. The world's human population now exceeds 6 billion, and even though the average birth rate has begun to decline, most demographers believe that the global human population will reach 8–10 billion in the next 50 years. Much of this projected growth will occur in developing countries in Central and South America, Asia and Africa—regions that are rich in unique biological diversity.

Finding solutions to conservation challenges will not be easy in today's human-dominated world. A growing number of people live in urban settings and are becoming increasingly isolated from nature. They "hunt" in super markets and malls, live in apartments and houses, spend their time watching television and searching the World Wide Web. Children and adults must be taught to value biological diversity and the habitats that support it. Education is of prime importance now while we still have time to respond to the impending crisis. There still exist in many parts of the world large numbers of biological "hotspots"—places that are relatively unaffected by humans and which still contain a rich store of their original animal and plant life. These living repositories, along with selected populations of animals and plants held in professionally managed zoos, aquariums and botanical gardens, could provide the basis for restoring the planet's biological wealth and ecological health. This encyclopedia and the collective knowledge it represents can assist in educating people about animals and their ecological and cultural significance. Perhaps it will also assist others in making deeper connections to nature and spreading biophilia. Information on the conservation status, threats and efforts to preserve various species have been integrated into this revision. We have also included information on the cultural significance of animals, including their roles in art and religion.

It was over 30 years ago that Dr. Bernhard Grzimek, then director of the Frankfurt Zoo in Frankfurt, Germany, edited the first edition of *Grzimek's Animal Life Encyclopedia*. Dr. Grzimek was among the world's best known zoo directors and conservationists. He was a prolific author, publishing nine books. Among his contributions were: *Serengeti Shall Not Die*, *Rhinos Belong to Everybody* and *He and I and the Elephants*. Dr. Grzimek's career was remarkable. He was one of the first modern zoo or aquarium directors to understand the importance of zoo involvement in *in situ* conservation, that is, of their role in preserving wildlife in nature. During his tenure, Frankfurt Zoo became one of the leading western advocates and supporters of wildlife conservation in East Africa. Dr. Grzimek served as a Trustee of the National Parks Board of Uganda and Tanzania and assisted in the development of several protected areas. The film he made with his son Michael, *Serengeti Shall Not Die*, won the 1959 Oscar for best documentary.

Professor Grzimek has recently been criticized by some for his failure to consider the human element in wildlife conservation. He once wrote: "A national park must remain a primordial wilderness to be effective. No men, not even native ones, should live inside its borders." Such ideas, although considered politically incorrect by many, may in retrospect actually prove to be true. Human populations throughout Africa continue to grow exponentially, forcing wildlife into small islands of natural habitat surrounded by a sea of humanity. The illegal commercial bushmeat trade—the hunting of endangered wild animals for large scale human consumption—is pushing many species, including our closest relatives, the gorillas, bonobos, and chimpanzees, to the brink of extinction. The trade is driven by widespread poverty and lack of economic alternatives. In order for some species to survive it will be necessary, as Grzimek suggested, to establish and enforce a system of protected areas where wildlife can roam free from exploitation of any kind.

While it is clear that modern conservation must take the needs of both wildlife and people into consideration, what will the quality of human life be if the collective impact of short-term economic decisions is allowed to drive wildlife populations into irreversible extinction? Many rural populations living in areas of high biodiversity are dependent on wild animals as their major source of protein. In addition, wildlife tourism is the primary source of foreign currency in many developing countries and is critical to their financial and social stability. When this source of protein and income is gone, what will become of the local people? The loss of species is not only a conservation disaster; it also has the potential to be a human tragedy of immense proportions. Protected areas, such as national parks, and regulated hunting in areas outside of parks are the only solutions. What critics do not realize is that the fate of wildlife and people in developing countries is closely intertwined. Forests and savannas emptied of wildlife will result in hungry, desperate people, and will, in the long-term lead to extreme poverty and social instability. Dr. Grzimek's early contributions to conservation should be recognized, not only as benefiting wildlife, but as benefiting local people as well.

Dr. Grzimek's hope in publishing his *Animal Life Encyclopedia* was that it would "...disseminate knowledge of the animals and love for them", so that future generations would "...have an opportunity to live together with the great diversity of these magnificent creatures." As stated above, our goals in producing this updated and revised edition are similar. However, our challenges in producing this encyclopedia were more formidable. The volume of knowledge to be summarized is certainly much greater in the twenty-first century than it was in the 1970's and 80's. Scientists, both professional and amateur, have learned and published a great deal about the animal kingdom in the past three decades, and our understanding of biological and ecological theory has also progressed. Perhaps our greatest hurdle in producing this revision was to include the new information, while at the same time retaining some of the characteristics that have made *Grzimek's Animal Life Encyclopedia* so popular. We have therefore strived to retain the series' narrative style, while giving the information more organizational structure. Unlike the original *Grzimek's*, this updated version organizes information under specific topic areas, such as reproduction, behavior, ecology and so forth. In addition, the basic organizational structure is generally consistent from one volume to the next, regardless of the animal groups covered. This should make it easier for users to locate information more quickly and efficiently. Like the original Grzimek's, we have done our best to avoid any overly technical language that would make the work difficult to understand by non-biologists. When certain technical expressions were necessary, we have included explanations or clarifications.

Considering the vast array of knowledge that such a work represents, it would be impossible for any one zoologist to have completed these volumes. We have therefore sought specialists from various disciplines to write the sections with which they are most familiar. As with the original *Grzimek's*, we have engaged the best scholars available to serve as topic editors, writers, and consultants. There were some complaints about inaccuracies in the original English version that may have been due to mistakes or misinterpretation during the complicated translation process. However, unlike the original *Grzimek's*, which was translated from German, this revision has been completely re-written by English-speaking scientists. This work was truly a cooperative endeavor, and I thank all of those dedicated individuals who have written, edited, consulted, drawn, photographed, or contributed to its production in any way. The names of the topic editors, authors, and illustrators are presented in the list of contributors in each individual volume.

The overall structure of this reference work is based on the classification of animals into naturally related groups, a discipline known as taxonomy or biosystematics. Taxonomy is the science through which various organisms are discovered, identified, described, named, classified and catalogued. It should be noted that in preparing this volume we adopted what might be termed a conservative approach, relying primarily on traditional animal classification schemes. Taxonomy has always been a volatile field, with frequent arguments over the naming of or evolutionary relationships between various organisms. The advent of DNA fingerprinting and other advanced biochemical techniques has revolutionized the field and, not unexpectedly, has produced both advances and confusion. In producing these volumes, we have consulted with specialists to obtain the most up-to-date information possible, but knowing that new findings may result in changes at any time. When scientific controversy over the classification of a particular animal or group of animals existed, we did our best to point this out in the text.

Readers should note that it was impossible to include as much detail on some animal groups as was provided on others. For example, the marine and freshwater fish, with vast numbers of orders, families, and species, did not receive as detailed a treatment as did the birds and mammals. Due to practical and financial considerations, the publishers could provide only so much space for each animal group. In such cases, it was impossible to provide more than a broad overview and to feature a few selected examples for the purposes of illustration. To help compensate, we have provided a few key bibliographic references in each section to aid those interested in learning more. This is a common limitation in all reference works, but *Grzimek's Encyclopedia of Animal Life* is still the most comprehensive work of its kind.

I am indebted to the Gale Group, Inc. and Senior Editor Donna Olendorf for selecting me as Series Editor for this project. It was an honor to follow in the footsteps of Dr. Grzimek and to play a key role in the revision that still bears his name. *Grzimek's Animal Life Encyclopedia* is being published by the Gale Group, Inc. in affiliation with my employer, the American Zoo and Aquarium Association (AZA), and I would like to thank AZA Executive Director, Sydney J. Butler; AZA Past-President Ted Beattie (John G. Shedd Aquarium, Chicago, IL); and current AZA President, John Lewis (John Ball Zoological Garden, Grand Rapids, MI), for approving my participation. I would also like to thank AZA Conservation and Science Department Program Assistant, Michael Souza, for his assistance during the project. The AZA is a professional membership association, representing 205 accredited zoological parks and aquariums in North America. As Director/William Conway Chair, AZA Department of Conservation and Science, I feel that I am a philosophical descendant of Dr. Grzimek, whose many works I have collected and read. The zoo and aquarium profession has come a long way since the 1970s, due, in part, to innovative thinkers such as Dr. Grzimek. I hope this latest revision of his work will continue his extraordinary legacy.

Silver Spring, Maryland, 2001
Michael Hutchins
Series Editor

• • • • •

How to use this book

Gzimek's Animal Life Encyclopedia is an internationally prominent scientific reference compilation, first published in German in the late 1960s, under the editorship of zoologist Bernhard Grzimek (1909–1987). In a cooperative effort between Gale and the American Zoo and Aquarium Association, the series is being completely revised and updated for the first time in over 30 years. Gale is expanding the series from 13 to 17 volumes, commissioning new color images, and updating the information while also making the set easier to use. The order of revisions is:

Vol 8–11: Birds I–IV
Vol 6: Amphibians
Vol 7: Reptiles
Vol 4–5: Fishes I–II
Vol 12–16: Mammals I–V
Vol 1: Lower Metazoans and Lesser Deuterostomes
Vol 2: Protostomes
Vol 3: Insects
Vol 17: Cumulative Index

Organized by order and family

The overall structure of this reference work is based on the classification of animals into naturally related groups, a discipline known as taxonomy—the science through which various organisms are discovered, identified, described, named, classified, and catalogued. Starting with the simplest life forms, the protostomes, in Vol. 1, the series progresses through the more complex animal classes, culminating with the mammals in Vols. 12–16. Volume 17 is a stand-alone cumulative index.

Organization of chapters within each volume reinforces the taxonomic hierarchy. Opening chapters introduce the class of animal, followed by chapters dedicated to order and family. Species accounts appear at the end of family chapters. To help the reader grasp the scientific arrangement, each type of chapter has a distinctive color and symbol:

▲= Family Chapter (yellow background)

●= Order Chapter (blue background)

◖= Monotypic Order Chapter (green background)

As chapters narrow in focus, they become more tightly formatted. General chapters have a loose structure, reminiscent of the first edition. While not strictly formatted, order chapters are carefully structured to cover basic information about member families. Monotypic orders, comprised of a single family, utilize family chapter organization. Family chapters are most tightly structured, following a prescribed format of standard rubrics that make information easy to find and understand. Family chapters typically include:

Thumbnail introduction
    Common name
    Scientific name
    Class
    Order
    Suborder
    Family
    Thumbnail description
    Size
    Number of genera, species
    Habitat
    Conservation status
Main essay
    Evolution and systematics
    Physical characteristics
    Distribution
    Habitat
    Behavior
    Feeding ecology and diet
    Reproductive biology
    Conservation status
    Significance to humans
Species accounts
    Common name
    Scientific name
    Subfamily
    Taxonomy
    Other common names
    Physical characteristics
    Distribution
    Habitat
    Behavior
    Feeding ecology and diet
    Reproductive biology

Conservation status
Significance to humans
Resources
Books
Periodicals
Organizations
Other

Color graphics enhance understanding

Grzimek's features approximately 3,500 color photos, including approximately 480 in four Birds volumes; 3,500 total color maps, including almost 1,500 in the four Birds volumes; and approximately 5,500 total color illustrations, including 1,385 in four Birds volumes. Each featured species of animal is accompanied by both a distribution map and an illustration.

All maps in *Grzimek's* were created specifically for the project by XNR Productions. Distribution information was provided by expert contributors and, if necessary, further researched at the University of Michigan Zoological Museum library. Maps are intended to show broad distribution, not definitive ranges, and are color coded to show resident, breeding, and nonbreeding locations (where appropriate).

All the color illustrations in *Grzimek's* were created specifically for the project by Michigan Science Art. Expert contributors recommended the species to be illustrated and provided feedback to the artists, who supplemented this information with authoritative references and animal skins from University of Michgan Zoological Museum library. In addition to species illustrations, *Grzimek's* features conceptual drawings that illustrate characteristic traits and behaviors.

About the contributors

The essays were written by expert contributors, including ornithologists, curators, professors, zookeepers, and other reputable professionals. *Grzimek's* subject advisors reviewed the completed essays to insure that they are appropriate, accurate, and up-to-date.

Standards employed

In preparing these volumes, the editors adopted a conservative approach to taxonomy, relying primarily on Peters Checklist (1934–1986)—a traditional classification scheme. Taxonomy has always been a volatile field, with frequent arguments over the naming of or evolutionary relationships between various organisms. The advent of DNA fingerprinting and other advanced biochemical techniques has revolutionized the field and, not unexpectedly, has produced both advances and confusion. In producing these volumes, Gale consulted with noted taxonomist Professor Walter J. Bock as well as other specialists to obtain the most up-to-date information possible. When scientific controversy over the classification of a particular animal or group of animals existed, the text makes this clear.

Grzimek's has been designed with ready reference in mind and the editors have standardized information wherever fea-

sible. For **Conservation status,** *Grzimek's* follows the IUCN Red List system, developed by its Species Survival Commission. The Red List provides the world's most comprehensive inventory of the global conservation status of plants and animals. Using a set of criteria to evaluate extinction risk, the IUCN recognizes the following categories: Extinct, Extinct in the Wild, Critically Endangered, Endangered, Vulnerable, Conservation Dependent, Near Threatened, Least Concern, and Data Deficient. For a complete explanation of each category, visit the IUCN web page at http://www.iucn.org/themes/ssc/redlists/categor.htm

In addition to IUCN ratings, essays may contain other conservation information, such as a species' inclusion on one of three Convention on International Trade in Endangered Species (CITES) appendices. Adopted in 1975, CITES is a global treaty whose focus is the protection of plant and animal species from unregulated international trade.

Grzimek's provides the following standard information on avian lineage in **Taxonomy** rubric of each Species account: [First described as] *Muscicapa rufifrons* [by] Latham, [in] 1801, [based on a specimen from] Sydney, New South Wales, Australia. The person's name and date refer to earliest identification of a species, although the species name may have changed since first identification. However, the organism described is the same.

Other common names in English, French, German, and Spanish are given when an accepted common name is available.

Appendices and index

For further reading directs readers to additional sources of information about birds. Valuable contact information for **Organizations** is also included in an appendix. While the encyclopedia minimizes scientific jargon, it also provides a **Glossary** at the back of the book to define unfamiliar terms. An exhaustive **Aves species list** records all known species of birds, categorized according to Peters Checklist (1934–1986). And a full-color **Geologic time scale** helps readers understand prehistoric time periods. Additionally, each of the four volumes contains a full **Subject index** for the Birds subset.

Acknowledgements

Gale would like to thank several individuals for their important contributions to the series. Michael Souza, Program Assistant, Department of Conservation and Science, American Zoo and Aquarium Association, provided valuable behind-the-scenes research and reliable support at every juncture of the project. Also deserving of recognition are Christine Sheppard, Curator of Ornithology at Bronx Zoo, and Barry Taylor, professor at the University of Natal, in Pietermaritzburg, South Africa, who assisted subject advisors in reviewing manuscripts for accuracy and currency. And, last but not least, Janet Hinshaw, Bird Division Collection Manager at the University of Michigan Museum of Zoology, who opened her collections to *Grzimek's* artists and staff and also compiled the "For Further Reading" bibliography at the back of the book.

·····

Advisory boards

Series advisor

Michael Hutchins, PhD
Director of Conservation and William Conway Chair
American Zoo and Aquarium Association
Silver Spring, Maryland

Subject advisors

Volume 1: Lower Metazoans and Lesser Deuterostomes
Dennis Thoney, PhD
Director, Marine Laboratory & Facilities
Humboldt State University
Arcata, California

Volume 2: Protostomes
Dennis Thoney, PhD
Director, Marine Laboratory & Facilities
Humboldt State University
Arcata, California

Sean F. Craig, PhD
Assistant Professor, Department of Biological Sciences
Humboldt State University
Arcata, California

Volume 3: Insects
Art Evans, PhD
Entomologist
Richmond, Virginia

Rosser W. Garrison, PhD
Systematic Entomologist, Los Angeles County
Los Angeles, California

Volumes 4–5: Fishes I–II
Paul Loiselle, PhD
Curator, Freshwater Fishes
New York Aquarium
Brooklyn, New York

Dennis Thoney, PhD

Director, Marine Laboratory & Facilities
Humboldt State University
Arcata, California

Volume 6: Amphibians
William E. Duellman, PhD
Curator of Herpetology Emeritus
Natural History Museum and Biodiversity Research Center
University of Kansas
Lawrence, Kansas

Volume 7: Reptiles
James B. Murphy, PhD
Smithsonian Research Associate
Department of Herpetology
National Zoological Park
Washington, DC

Volumes 8–11: Birds I–IV
Walter J. Bock, PhD
Permanent secretary, International Ornithological Congress
Professor of Evolutionary Biology
Department of Biological Sciences,
Columbia University
New York, New York

Jerome A. Jackson, PhD
Program Director, Whitaker Center for Science,
Mathematics, and Technology Education
Florida Gulf Coast University
Ft. Myers, Florida

Volumes 12–16: Mammals I–V
Valerius Geist, PhD
Professor Emeritus of Environmental Science
University of Calgary
Calgary, Alberta
Canada

Devra Gail Kleiman, PhD
Smithsonian Research Associate
National Zoological Park
Washington, DC

• • • • •

Contributing writers

Birds I–IV

Michael Abs, Dr. rer. nat.
Berlin, Germany

George William Archibald, PhD
International Crane Foundation
Baraboo, Wisconsin

Helen Baker, PhD
Joint Nature Conservation Committee
Peterborough, Cambridgeshire
United Kingdom

Cynthia Ann Berger, MS
Pennsylvania State University
State College, Pennsylvania

Matthew A. Bille, MSc
Colorado Springs, Colorado

Walter E. Boles, PhD
Australian Museum
Sydney, New South Wales
Australia

Carlos Bosque, PhD
Universidad Simón Bolivar
Caracas, Venezuela

David Brewer, PhD
Research Associate
Royal Ontario Museum
Toronto, Ontario
Canada

Daniel M. Brooks, PhD
Houston Museum of Natural Science
Houston, Texas

Donald F. Bruning, PhD
Wildlife Conservation Society
Bronx, New York

Joanna Burger, PhD
Rutgers University
Piscataway, New Jersey

Carles Carboneras
SEO/BirdLife
Barcelona, Spain

John Patrick Carroll, PhD
University of Georgia
Athens, Georgia

Robert Alexander Cheke, PhD
Natural Resources Institute
University of Greenwich
Chatham, Kent
United Kingdom

Jay Robert Christie, MBA
Racine Zoological Gardens
Racine, Wisconsin

Charles T. Collins, PhD
California State University
Long Beach, California

Malcolm C. Coulter, PhD
IUCN Specialist Group on Storks,
Ibises and Spoonbills
Chocorua, New Hampshire

Adrian Craig, PhD
Rhodes University
Grahamstown, South Africa

Francis Hugh John Crome, BSc
Consultant
Atheron, Queensland
Australia

Timothy Michael Crowe, PhD
University of Cape Town
Rondebosch, South Africa

H. Sydney Curtis, BSc
Queensland National Parks &
Wildlife Service (Retired)
Brisbane, Queensland
Australia

S. J. J. F. Davies, ScD
Curtin University of Technology
Department of Environmental Biology
Perth, Western Australia
Australia

Gregory J. Davis, PhD
University of Wisconsin-Green Bay
Green Bay, Wisconsin

William E. Davis, Jr., PhD
Boston University
Boston, Massachusetts

Stephen Debus, MSc
University of New England
Armidale, New South Wales
Australia

Michael Colin Double, PhD
Australian National University
Canberra, A.C.T.
Australia

Rachel Ehrenberg, MS
University of Michigan
Ann Arbor, Michigan

Eladio M. Fernandez
Santo Domingo
Dominican Republic

Simon Ferrier, PhD
New South Wales National Parks and
Wildlife Service
Armidale, New South Wales
Australia

Kevin F. Fitzgerald, BS
South Windsor, Connecticut

Hugh Alastair Ford, PhD
University of New England
Armidale, New South Wales
Australia

Joseph M. Forshaw
Australian Museum
Sydney, New South Wales
Australia

Bill Freedman, PhD
Department of Biology
Dalhousie University
Halifax, Nova Scotia
Canada

Clifford B. Frith, PhD
Honorary research fellow
Queensland Museum
Brisbane, Australia

Dawn W. Frith, PhD
Honorary research fellow
Queensland Museum
Brisbane, Australia

Peter Jeffery Garson, DPhil
University of Newcastle
Newcastle upon Tyne
United Kingdom

Michael Gochfeld, PhD, MD
UMDNJ-Robert Wood Johnson
Medical School
Piscataway, New Jersey

Michelle L. Hall, PhD
Australian National University
School of Botany and Zoology
Canberra, A.C.T.
Australia

Frank Hawkins, PhD
Conservation International
Antananarivo, Madagascar

David G. Hoccom, BSc
Royal Society for the Protection of
Birds
Sandy, Bedfordshire
United Kingdom

Peter Andrew Hosner
Cornell University
Ithaca, New York

Brian Douglas Hoyle PhD
Bedford, Nova Scotia
Canada

Julian Hughes
Royal Society for the Protection of
Birds
Sandy, Bedfordshire
United Kingdom

Robert Arthur Hume, BA
Royal Society for the Protection of
Birds
Sandy, Bedfordshire
United Kingdom

Gavin Raymond Hunt, PhD
University of Auckland
Auckland, New Zealand

Jerome A. Jackson, PhD
Florida Gulf Coast University
Ft. Myers, Florida

Bette J. S. Jackson, PhD
Florida Gulf Coast University
Ft. Myers, Florida

Darryl N. Jones, PhD
Griffith University
Queensland, Australia

Alan C. Kemp, PhD
Naturalists & Nomads
Pretoria, South Africa

Angela Kay Kepler, PhD
Pan-Pacific Ecological Consulting
Maui, Hawaii

Jiro Kikkawa, DSc
Professor Emeritus
University of Queensland,
Brisbane, Queensland
Australia

Margaret Field Kinnaird, PhD
Wildlife Conservation Society
Bronx, New York

Guy M. Kirwan, BA
Ornithological Society of the Middle
East
Sandy, Bedfordshire
United Kingdom

Melissa Knopper, MS
Denver Colorado

Niels K. Krabbe, PhD
University of Copenhagen
Copenhagen, Denmark

James A. Kushlan, PhD
U.S. Geological Survey
Smithsonian Environmental Research
Center
Edgewater, Maryland

Norbert Lefranc, PhD
Ministère de l'Environnement,
Direction Régionale
Metz, France

P. D. Lewis, BS
Jacksonville Zoological Gardens
Jacksonville, Florida

Josef H. Lindholm III, BA
Cameron Park Zoo
Waco, Texas

Peter E. Lowther, PhD
Field Museum
Chicago, Illinois

Gordon Lindsay Maclean, PhD, DSc
Rosetta, South Africa

Steve Madge
Downderry, Torpoint
Cornwall
United Kingdom

Albrecht Manegold
Institut für Biologie/Zoologie
Berlin, Germany

Jeffrey S. Marks, PhD
University of Montana
Missoula, Montana

Juan Gabriel Martínez, PhD
Universidad de Granada
Departamento de Biologia
Animal y Ecologia
Granada, Spain

Barbara Jean Maynard, PhD
Laporte, Colorado

Cherie A. McCollough, MS
PhD candidate, University of Texas
Austin, Texas

Leslie Ann Mertz, PhD
Fish Lake Biological Program
Wayne State University
Biological Station
Lapeer, Michigan

Derek William Niemann, BA
Royal Society for the Protection of
Birds
Sandy, Bedfordshire
United Kingdom

Malcolm Ogilvie, PhD
Glencairn, Bruichladdich
Isle of Islay
United Kingdom

Penny Olsen, PhD
Australian National University
Canberra, A.C.T.
Australia

Jemima Parry-Jones, MBE
National Birds of Prey Centre
Newent, Gloucestershire
United Kingdom

Colin Pennycuick, PhD, FRS
University of Bristol
Bristol, United Kingdom

James David Rising, PhD
University of Toronto
Department of Zoology
Toronto, Ontario
Canada

Christopher John Rutherford Robertson
Wellington, New Zealand

Peter Martin Sanzenbacher, MS
USGS Forest & Rangeland Ecosystem
Science Center
Corvallis, Oregon

Matthew J. Sarver, BS
Ithaca, New York

Herbert K. Schifter, PhD
Naturhistorisches Museum
Vienna, Austria

Richard Schodde PhD, CFAOU
Australian National Wildlife
Collection, CSIRO
Canberra, A.C.T.
Australia

Karl-L. Schuchmann, PhD
Alexander Koenig Zoological Research
Institute and Zoological Museum
Bonn, Germany

Tamara Schuyler, MA
Santa Cruz, California

Nathaniel E. Seavy, MS
Department of Zoology
University of Florida
Gainesville, Florida

Charles E. Siegel, MS
Dallas Zoo
Dallas, Texas

Julian Smith, MS
Katonah, New York

Joseph Allen Smith
Baton Rouge, Louisiana

Walter Sudhaus, PhD
Institut für Zoologie
Berlin, Germany

J. Denis Summers-Smith, PhD
Cleveland, North England
United Kingdom

Barry Taylor, PhD
University of Natal
Pietermaritzburg, South Africa

Markus Patricio Tellkamp, MS
University of Florida
Gainesville, Florida

Joseph Andrew Tobias, PhD
BirdLife International
Cambridge
United Kingdom

Susan L. Tomlinson, PhD
Texas Tech University
Lubbock, Texas

Donald Arthur Turner, PhD
East African Natural History Society
Nairobi, Kenya

Michael Phillip Wallace, PhD
Zoological Society of San Diego
San Diego, California

John Warham, PhD, DSc
University of Canterbury
Christchurch, New Zealand

Tony Whitehead, BSc
Ipplepen, Devon
United Kingdom

Peter H. Wrege, PhD
Cornell University
Ithaca, New York

· · · · ·

Contributing illustrators

**Drawings by Michigan
Science Art**

Joseph E. Trumpey, Director, AB, MFA
Science Illustration, School of Art and Design, University
of Michigan

Wendy Baker, ADN, BFA

Brian Cressman, BFA, MFA

Emily S. Damstra, BFA, MFA

Maggie Dongvillo, BFA

Barbara Duperron, BFA, MFA

Dan Erickson, BA, MS

Patricia Ferrer, AB, BFA, MFA

Gillian Harris, BA

Jonathan Higgins, BFA, MFA

Amanda Humphrey, BFA

Jacqueline Mahannah, BFA, MFA

John Megahan, BA, BS, MS

Michelle L. Meneghini, BFA, MFA

Bruce D. Worden, BFA

Thanks are due to the University of Michigan, Museum
of Zoology, which provided specimens that served as models
for the images.

Maps by XNR Productions

Paul Exner, Chief cartographer
XNR Productions, Madison, WI

Tanya Buckingham

Jon Daugherity

Laura Exner

Andy Grosvold

Cory Johnson

Paula Robbins

Old World warblers
(Sylviidae)

Class Aves
Order Passeriformes
Suborder Passeri (Oscines)
Family Sylviidae

Thumbnail description
Very small to medium-sized, often dull-colored, songbirds with thin, pointed bills

Size
3.1–9.8 in (8–25 cm); .1–2 oz (4–56 g)

Number of genera, species
60 genera; 350–391 species

Habitat
Highly varied. Largely arboreal, but many species inhabit wetlands, grasslands, thickets, scrub, and riparian zones

Conservation status
Endangered: 10 species; Critically Endangered: 3 species; Vulnerable: 29 species; Near Threatened: 13 species; Data Deficient: 9 species

Distribution
Subfamily Sylviinae: Palearctic region, Africa, Asia, Australasia, Oceania. Two genera, *Regulus* and *Phylloscopus*, reach Nearctic. Subfamily Polioptilinae: New World, from South America to the northern United States.

Evolution and systematics

The taxonomy of the Passeri, the suborder of oscine passerines (the songbirds) that contains the Sylviidae (Old World warblers), has long been debated. Much of the controversy focuses on the delineation of the apparently closely related families Muscicapidae, Turdidae, Timaliidae, and Sylviidae. Currently available molecular data suggest that widely used, traditional family classifications do not represent the evolutionary history of this large and complex group of birds. Sibley and Ahlquist place the Sylviidae within a large superfamily, the Sylvioidea. This superfamily also contains members of the following traditional families: Sittidae (nuthatches), Certhiidae (creepers), Troglodytidae (wrens), Paridae (tits), Aegithalidae (long-tailed tits), Hirundinidae (swallows), Pycnonotidae (bulbuls), Zosteropidae (white-eyes), and Timaliidae (babblers). The molecular data do not support the long-held belief that the Sylviidae are closely related to the Turdidae (thrushes) and the Muscicapidae (Old World flycatchers), for these two families fall into a separate superfamily, the Muscicapoidea. The data do, however, support the inclusion of some of the babblers (Timaliinae) and laughing thrushes (genus *Garrulax*) in the Sylviidae family.

The classification used in the present work is based upon the traditional definition of the Sylviidae. The genera may be divided into six groups, based on molecular evidence of their taxonomic affiliation. The first of these are the gnatcatchers and gnatwrens (*Polioptila*, *Microbates*, and *Ramphocaenus*), which comprise the Sylviid subfamily Polioptilinae, but are placed by Sibley and Ahlquist within an expanded Certhiidae, including both creepers and wrens. The remaining five groups

are considered to be within the sylviid subfamily Sylviinae. These are: (1) the kinglets (*Regulus*), regarded as a separate family by Sibley and Ahlquist and others; (2) the cisticolas and allies (*Cisticola*, *Prinia*, *Apalis*, *Camaroptera*), and other allied genera, a distinct group that Sibley and Ahlquist regard as a separate family, and most authors regard as a subfamily, Cisticolinae, of the Sylviidae; (3) the *Sylvia* warblers (*Sylvia*, *Parisoma*), a group that, according to Sibley and Ahlquist, is more closely related to the timaliine babblers than to traditional sylviids; (4) the grassbirds and allies (*Megalurus*, *Bowdleria*, *Cincloramphus*, *Megalurulus*, *Chaetornis*, *Gramnicola*, *Schoenicola*), and other allied genera, which comprise the subfamily Megalurinae in the Sibley and Ahlquist system; (5) the remaining genera of the traditional Sylviidae, most of which appear to fall within a clade represented by Sibley and Ahlquist's Acrocephalinae subfamily. Groups three, four and five above, plus the *Garrulax* laughing thrushes comprise Sibley and Ahlquist's more restricted Sylviidae, hereafter referred to as Sylviidae *sensu strictu*.

Little is known about the evolutionary history of passerines. The prevailing opinion has been that passerines arose in the Tertiary, specifically in the early Eocene, then underwent a dramatic diversification during the Miocene. The oldest putative passerine fossils are from early Eocene Australia (ca. 54 million years ago), lending support to a Southern Hemisphere origin, since passerine fossils do not appear in the Northern Hemisphere until the Oligocene. By the lower Miocene, passerine fossils greatly outnumber all other taxa in many Northern Hemisphere sites. A recent molecular study suggests that the passerine divergence may be much older and that passerines

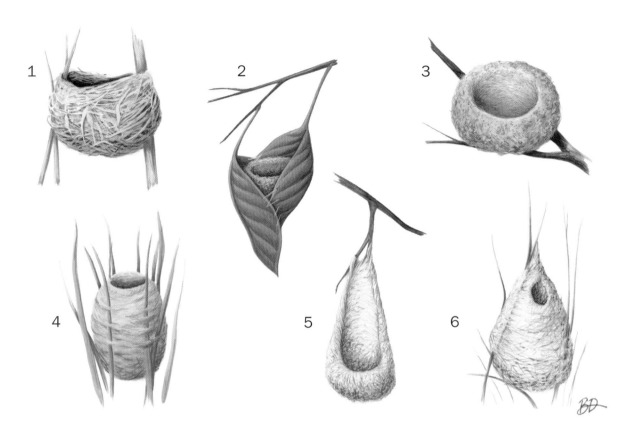

Warbler nests: 1. Great reed warbler (*Acrocephalus arundinaceus*); 2. Common tailorbird (*Orthotomus sutorius*); 3. Blue-gray gnatcatcher (*Polioptila caerulea*); 4. Fan-tailed cisticola (*Cisticola juncidis*); 5. Long-billed crombec (*Sylvietta rufescens*); 6. Tawny-flanked prinia (*Prinia subflava*). (Illustration by Barbara Duperron)

evolved on Gondwana, the Cretaceous supercontinent in the southern hemisphere. If this is the case, oscine passerines may have diverged from suboscine passerines when Australia separated from Gondwana, radiated on Australia, then dispersed throughout the Old World when Australia came in contact with Southeast Asia.

Within the Sylviidae *sensu strictu*, the first divergence was probably between Sibley & Ahlquist's Acrocephaline subfamily (*Acrocephalus, Hippolais, Phylloscopus, Seicercus, Sylvietta, Sphenoeacus*, etc.) and the lineage leading to the other three subfamilies. Next to diverge was *Garrulax*, the laughing-thrushes, which diverged from a lineage leading to the Timaliine babbers and *Sylvia* and *Parisoma* warblers. The genus *Sylvia* is thought to be at least 12-16 million years old. The *Phylloscopus* radiation is about as old as *Sylvia*, while the *Acrocephalus/Hippolais* radiation is only about half as old. These estimates place much of the sylviid radiation during the Miocene and Pliocene epochs of the mid-Tertiary, consistent with the passerine radiation in the fossil record.

Physical characteristics

The traits that have been used to characterize Sylviidae *sensu latu* are: unspotted young (as opposed to the Old World flycatchers and thrushes); rictal bristles at the base of the bill (a widespread adaptation for insectivory); thin, pointed bill; hatchlings naked or only partially downy; ten primaries; scutellate tarsi; and lack of strong sexual dimorphism (*Sylvia* is a notable exception). Most genera have twelve tail feathers, but there are a few exceptions. Wing shape and length ranges from short and rounded in sedentary species, to longer and more pointed in long-distance migrants. Moult timing and pattern are highly variable.

Sylviids are typically dull in color; often in shades of brown, green, yellow, and gray. The family includes some of the tiniest songbirds in the world, the kinglets, as well as some small wren-like birds, many small warblers, and the medium-sized marsh warblers and grassbirds. The smallest sylviids, the kinglets, weigh only a few grams. The largest, the marsh warblers *Acrocephalus* and the grassbirds *Megalurus* can weigh close to 2 oz (60 g). A comprehensive summary of longevity data is unavailable, but many species live at least 8–12 years.

While most species do not have distinctive breeding plumages like those of the New World warblers (Parulidae), a few sylviids show a marked contrast between breeding and nonbreeding plumages. A striking example is the red-winged warbler (*Heliolais erythroptera*), the only member of a genus apparently closely related to *Prinia*. While breeding red-

A grasshopper warbler's (*Locustella naevia*) wing-waving display. (Illustration by Barbara Duperron)

winged warblers of both sexes have dark gray upperparts and tail, they become predominantly tawny-brown in nonbreeding plumage. A somewhat less striking example is the genus *Cisticola*. Members of this genus undergo two moults each season, and most have shorter tails and slightly different plumage coloration during the breeding season. A few members of the genus *Prinia* have similar distinct seasonal plumage variation.

Distribution

The Old World warblers are extraordinarily widespread, occuring on every continent except South America and Antarctica. If the Polioptilinae are included, there are representatives of Sylviidae *sensu latu* in South America as well. Most of the sylviid diversity occurs in the African and Oriental faunal regions, with a less diverse, but widespread group of species in the Palearctic. Nearctic and Australian species represent recent invasions from Siberia and Southeast Asia, respectively.

The centers of distribution for some major sylviid genera are as follows. Sino-Himalayan Region: *Bradypterus*, *Cettia*, *Phylloscopus*, *Seicercus*, *Prinia*; Southeast Asia: *Megalurus* and allies, *Orthotomus*; Temperate Asia (Palearctic): *Locustella*, *Acrocephalus*, *Hippolais*; Africa: *Cisticola* and allies; Mediterranean/Middle Eastern (Palearctic) region: *Sylvia*. Relatively little exchange has

occurred between Northern (Palearctic) and Southern (Oriental and African) Hemisphere faunas in the Old World, perhaps due to the East/West orientation of major barriers, including the Sahara and Gobi deserts, the Himalayas, the Alps, the Atlas, Caucasus, and other mountain ranges, and historically, the Tethys sea, which separated Africa from what is now the Palearctic. Conversely, in the Americas, where the major barriers are oriented North to South, many Neotropical migrant families originated in South America, then spread north to the Nearctic. Voous (1977) is of the opinion that all Old World sylviids arose in either the Indo-African region or the Sino-Himalayan region. While these are areas where the largest radiations have taken place, it is not clear where the common ancestor of the modern Sylviidae arose.

Habitat

The Old World warblers occupy an astonishing variety of habitats, from montane and riparian forests to arid scrublands to marshes and river floodplains to city parks and backyards. Sylviids are found in all extremes of water availability with the exception of open water and harsh deserts. The family is represented at a wide range of altitudes, from lowlands at or near sea level to montane forests and dwarf scrub as high as several thousand meters. Considerable habitat adaptability is

Blue-gray gnatcatcher (*Polioptila caerulea*) chicks in their nest. (Photo by R. & S. Day/VIREO. Reproduced by permission.)

demonstrated by the varied habitats occupied by *Acrocephalus* warblers that have colonized oceanic islands.

Habitat partitioning is widespread among sylviid warblers. An example of foraging height segregation is found in the *Sylvietta* crombecs. The long-billed crombec is restricted to undergrowth in areas of sympatry with red-faced and red-capped crombecs. In other areas, the long-billed crombec is found at higher levels in the forest, demonstrating that competition limits the ability of this species to fully exploit its potential niche. Habitat partitioning also occurs in syntopic species of *Sylvia*, but there is considerable overlap, leading to interspecific territorial interactions.

Exploitation of topological niches within a habitat has been suggested as an important step in ecological and morphological divergence of closely related species. Adam Richman and Trevor Price have shown, in a 1992 study, that such a scenario has apparently occurred among a group of sympatric *Phylloscopus* warblers in the Himalayas. A related phenomenon is replacement, the presence of closely related species in different habitat types, with little overlap. Replacement is essentially habitat partitioning on a larger scale. Many Afrotropical genera contain closely-related species that occupy similar ecological niches in dissimilar habitats.

Behavior

Sylviids range from highly arboreal to almost entirely terrestrial. They are generally very active but often quite secre-

tive and skulking. Most species inhabit dense vegetation, and are most easily located and identified by voice. The carriage of many of the Old World warblers is more or less horizontal. Flight ranges from very weak in some sedentary Afrotropical species, to strong and sustained in long-distance migrants. The gait is typically a hop, but some species run.

Vocalizations

The Old World warblers are one of the most vocally diverse bird families in the world. Most species have well-developed songs, with parameters that vary greatly both within and between genera. Song ranges from the unmusical, repetitive sounds of *Prinia* to the melodious warbles of *Sylvia* and *Phylloscopus*, to the amazingly complex, multi-element songs of some *Acrocephalus* species.

Song is an important component of territorial defense; and the variety of habitats occupied by sylviids have probably contributed to their vocal diversity. Species inhabiting relatively open grasslands, marshes, and brushlands often engage in song-flight displays. Daily and seasonal singing patterns are highly varied as well; *Cettia cetti* sings all day, for most of the year, sometimes at night. Night singing is common in *Acrocephalus*, *Bradypterus baboecala* and *Locustella*.

Most *Sylvia* warblers appear to use subsong, typically a longer, lower, more complex and melodious version of the territorial song, in courtship, but it is also given on the wintering grounds. A few sylviids incorporate mimicry into their songs as well. The marsh warbler *Acrocephalus palustris* learns its song components, from many dozens of other species it hears on both the breeding grounds in Europe, and the wintering grounds in Africa. Some of the *Hippolais* warblers (*H. icterina, H. languida, H. polyglotta*) incorporate mimcry of other species into phrases of their own songs. Mimicry also occurs in *Eremomela* and *Sylvia*.

Duets are thought to be involved in pair-bond maintenance, synchronization of sexual activity, or cooperative territorial defense, and are generally associated with monogamous species that maintain territories year-round. A *Prinia subflava* female occasionally adds her own complementary rattle to the notes of a singing male, creating a duet. The female *Prinia bairdii* is known to duet during territory advertisement. A few *Cisticola* species (*C. hunteri, C. chubbi, C. nigriloris*) also engage in duetting. *Apalis flavida* duets, but not for territory defense. Other genera with species that duet include: *Bathmocercus, Bradypterus, Drymocichla, Schistolais,* and *Spiloptila*.

Migration and dispersal

The Sylviidae exhibit a wide range of movement tendencies. There is even considerable variation within some species, like the fan-tailed warbler (*Cisticola juncidis*), blue-gray gnatcatcher (*Polioptila caerulea*), and Cetti's warbler (*Cettia cetti*). Each of these species includes some populations that are migratory, and others that are mostly sedentary. The true long-distance migrants are typically Palearctic species that escape the northern winter by travelling to sub-Saharan Africa or Southern Asia. These long-distance migrants undertake their movements at night.

Many Asian species are altitudinal migrants, moving between elevations, including the Tesias, several *Phylloscopus* and

Seicercus species, and many *Cettia* species. Many African species undertake local and semi-local movements to moister areas during the dry season. *Cisticola juncidis* and other species of southern affinity tend to erupt northward after mild winters. Philopatry (tendency to remain near the home area) predominates in the great reed warbler, *Acrocephalus arundinaceus*, and is predicted by genetic diversity.

Territorial and social interactions

Almost all Old World warblers are territorial. Typically, the male defends a territory with song, display, and sometimes chasing and fighting. The majority of antagonistic behavior in migratory species occurs in the early spring during initial territory settlement and mate acquisition. In several monogamous species the female shares in territorial defense, sometimes by antagonistic behavior, and occasionally by song or duetting with her mate. Among polygynous species, females sometimes defend feeding and nesting areas within a male's territory against intrusion from neighboring females. Many African species defend territories year-round. Wintering and breeding ground site fidelity is often high in migratory and resident species. Family groups are common in many genera after the breeding season has ended.

Many species are interspecifically territorial. *Hippolais icterina* aggressively defends its territory and nest from many other species. Others, such as *Cisticola anonymus*, are only aggressive toward other species near nest. Some sylviids, such as the barred warbler *Sylvia nisoria*, preferentially associate with certain species of shrikes (*Lanius*) whose vigilance and anti-predator aggression result in higher reproductive success for the barred warbler.

Feeding ecology and diet

The diets of all members of the family consist primarily of insects, other arthropods, and especially spiders. Other prey items include crustaceans, snails, and harvestmen (Opiliones). Sap (golden-crowned kinglet) and nectar (several African and Asian species, especially *Prinia*) are occasionally consumed. *Prinia hodgsonii* and *Orthotomus sutorius* are known to carry pollen attached to the feathers of the throat and forehead; it and other nectar-feeding species may be important pollinators in the tropics. Some of the larger reed-warblers (including *Acrocephalus arundinaceus*, *A. rufesecens*, and *A. stentoreus*) occasionally take small frogs and fish. Young sylviids are fed almost exclusively arthropods, usually soft-bodied larvae and small insects, but in some cases receive berries as well. Variation in prey size and type is found among sympatric foraging guilds. During the pre-migratory period of Palearctic *Sylvia* warblers, individuals shift their diet from largely insects to largely berries and fruits in order to accumulate fat for migration.

Foraging strategies among the Old World warblers are diverse. Some species forage singly or in pairs, while others forage in groups of several family members or other conspecifics. Still others readily join mixed-species foraging parties, especially in the nonbreeding season. *Cisticola nigriloris* forages in groups of 5–8 birds, apparently keeping in contact by group-singing. Many African and nontropical species, join foraging parties in the nonbreeding season.

Garden warbler (*Sylvia borin*) chicks are fed in their nest. (Photo by Jany Sauvanet. Photo Researchers, Inc. Reproduced by permission.)

The typical foraging mode is perched gleaning (also known as standpicking) from the ground or from vegetation. Sallying and hawking of aerial insects is also used by many species, including the flycatcher warblers (genus *Seicercus*). *Bradypterus sylvaticus* forages on the ground by scratching up litter and humus with its feet, or by disturbing the humus with its wings and tail, in a motion reminiscent of dustbathing. Tesias, *Macrosphenus kretschmeri*, and other terrestrial warblers feed in a similar manner. Even some *Phylloscopus* warblers, members of a genus composed mostly of arboreal species, have adapted to ground-feeding. Kemp's longbill *Macrosphenus kempi* climbs about on undergrowth, probing its bill into dead wood.

Reproductive biology

Most Old World warblers are either socially monogamous or polygynous. Monogamous warbler species may have a pair bond that lasts for either a single nesting, a single season, or more frequently the pair may remain monogamous for several years. Polygyny occurs when a single male maintains pair bonds with two or more females, each with her own nest and young. Classical polyandry (sex-role reversal) does not occur in this family. Serial monogamy is fairly common among the sylviids, with pair bonds lasting for only a single nesting. This is the case in *Cisticola juncidis*, a species in which most males mate sequentially with three or four females (as many as eleven) per season. Despite the implications of the term, many socially monogamous species engage in extra-pair copulations in order to enhance their reproductive success. Polygyny has been intensively studied in *Acrocephalus* reed warblers. Other genera with polygynous species include *Cettia*, *Cisticola*, and *Phylloscopus*. The duration of the pair bond is in part determined by the movements of a species.

A black-browed reed-warbler (*Acrocephalus bistrigiceps*) builds its nest. (Photo by A.J. Knystautas/VIREO. Reproduced by permission.)

Courtship

Some species have elaborate courtship displays that may or may not be accompanied by vocal activity. During the courtship dance of *Camaroptera brachyura*, the male repeatedly jumps 12–16 in (30–40 cm) into the air from his perch, while whirring his wings and giving a precisely timed call each second, corresponding with the jumps. Some *Cisticola* species use similar courtship dances, while others attract females and defend their territories with remarkable aerial song-displays. Many species, including the wood warbler (*Phylloscopus sibilatrix*) and the melodious warbler (*Hippolais polyglotta*), use a special flight style, often called flutter-flight or butterfly flight when pursuing a female. A characteristic of *Sylvia* courtship is the building of several cock nests (partial nests decorated with spider silk, flower petals, etc.) by the male. The female apparently uses these partial nests to assess male quality. Courtship displays are diverse, and utilize full song, subsong, calls, posture and flight displays, nest building, and presentation of objects.

Nests

The Old World warblers build some of the most interesting nests in the avian world. They range from simple, open cups to intricately woven domed and rounded structures, to highly camouflaged balls of moss and lichens. Nests may be located on the ground, in low vegetation, or in shrubs and trees as high as 80 ft (26 m) or more. Support may be provided by grass and sedge clumps, tree branches, or other vegetation. Alternatively, the nest may be hanging or nearly

pendulous, attached to living leaves, woven into surrounding grasses, or built as a lining inside an abandoned nest of another species. Nests of the *Orthotomus* tailorbirds are famous for their sewing abilities. The female pierces green leaves with her bill, then draws them together with 'threads' of plant fiber, making several stitches, and sometimes knotting the ends of the fibers. The resulting sac is lined with spider silk and plant fibers. Similar nests are made by *Artisornis*, *Calamonastes*, *Camaroptera*, and *Schistolais*. The nests of most sylviids are built by both sexes, but in some species, especially polygynous ones, the female builds alone.

Care of young

Sylviids generally lay between one and 12 eggs per clutch. The young are usually incubated and brooded by the female, but in some species including *Abroscopus*, *Seicercus*, and *Sylvia*, the male shares in the task. Feeding of the young is usually done by both parents in socially monogamous species, but there is much interspecific and intraspecific variation in the degree of paternal involvement. In socially polygynous species, males tend to take a smaller role in feeding the young. Aquatic warbler (*Acrocephalus paludicola*) males take no role in incubation and rearing of their young. The altricial young, which are usually naked but may be downy, remain in the nest for 10–21 days. In many species, the fledglings are dependent on their parents for an additional 1–4 weeks after leaving the nest. Broods of fledglings often roost in groups to conserve warmth and receive food from the parents. Most northern species are usually single-brooded, but occasionally raise a second brood if conditions permit. Many tropical and sub-tropical species, on the other hand, are double-brooded.

Some sylviid taxa engage in cooperative breeding. Helpers have been reported in *Acrocephalus vaughani*, *Prinia bairdii*, *Cisticola anonymous*, *C. nigriloris*, and most of the Eremomelas. It is not clear, however, whether the helpers in these species are related to the breeding pair, nor is it known whether mating of helpers with breeders occurs. The only known case of true cooperative breeding (a monogamous pair helped by non-breeding kin) among the Sylviidae is the Seychelles warbler (*Acrocephalus sechellensis*). Most of the helpers are females, since males tend to disperse and attempt to set up their own territories. In high quality territories, the presence 1–2 helpers increases the reproductive success of the resident pair, while the presence of 3 or more helpers creates too much competition and decreases reproductive success. Breeding pairs have been shown to facultatively manipulate the sex ratio of their eggs to maximize their future reproductive success.

Conservation status

Several species of Sylviid face serious threats. Many of these species are greatly range-restricted, with small populations inhabiting a single island or tiny patch of isolated habitat. Nearly one third of these threatened species are found in tropical Africa. About as many are endemic to small oceanic islands. The genera with the largest number of threatened or Near Threatened species are: *Acrocephalus* (15), *Apalis* (7), *Bradypterus* (6), and *Cisticola* (5).

The reed-warblers (*Acrocephalus*), are among the most widespread of the Sylviid genera, but their colonization of island habitats makes them vulnerable to anthropogenic and climatic threats. Acrocephaline species that have large ranges are presumably able to survive loss of habitat. However, island populations, and those species with small or already fragmented ranges are susceptible to habitat destruction and introduction of nonindigenous species. The aquatic warbler has probably suffered dramatic population declines due to destruction of its riverine marshland habitat in Eastern Europe and Western Siberia. IUCN lists it as Vulnerable.

The Seychelles warbler is an island species whose population had drooped precipitously to a mere 26 birds occupying the small Cousin Island in the Seychelles. Intensive management of the island resulted in a dramatic increase over the past thirty years, with the island reaching apparent carrying capacity in the early 1980s. Recent translocation efforts have expanded the range of the Seychelles Warbler to include two more islands.

The millerbird (*Acrocephalus familiaris*) is the only Sylviid to colonize the Hawaiian Archipelago. Historically, two populations, which were usually regarded as separate species or subspecies, inhabited the islands, one on Laysan and one on Nihoa. The Laysan form became extinct in the early twentieth century, after much of the native vegetation was destroyed by rabbits. The Nihoan form persists on the steep, rocky slopes of this tiny volcanic island. The Nihoan millerbird is listed as a Federally Endangered species in the United States. A 1996 USFWS population estimate was 155 individuals.

Many of the Endangered African species are found in threatened East African forests. A bird that may once have inhabited natural forest clearings, *Artisornis moreaui*, is limited in its dispersal ability by very weak flight. Forest clearing in its very small range may lead to the extinction of this highly local and poorly known species. Six species of *Apalis*, most found only in the highland forests of one or a few mountain groups of East Africa from Kenya to Mozambique, face serious conservation risks, as do rare species of *Bathmocercus*, *Chloropeta*, *Cisticola* and *Eremomela*.

Several threatened species in the genera *Acrocephalus*, *Bradypterus*, *Cettia*, *Megalurulus*, *Megalurus*, *Locustella*, *Phylloscopus*, and *Orthotomus* occur on islands in Southeast Asia and Australasia.

Among the kinglets, there has been only a single subspecies faced with serious conservation threats. *Regulus calendula obscura* was endemic to a highly restricted area of cypress groves on the island of Guadalupe. The species was last seen in the 1970s and is considered Extinct.

Among the Polioptilinae, only one species, the creamy-bellied gnatcatcher, *Polioptila lactea*, is is Near Threatened.

An arctic warbler (*Phylloscopus borealis*) sings while perched on a branch. (Photo by Doug Wechsler/VIREO. Reproduced by permission.)

This species inhabits lowland coastal forests (to 500m) of Brazil, Argentina, and Paraguay, primarily in the Paraná basin. Its habitat has been severely reduced and deforestation continues to pose a threat to its remaining small habitat.

Significance to humans

Their small size, inconspicuous coloration, and preference for inhabiting vegetation make the Old World warblers unlikely candidates for human attention. However, the songs of many species are complex and well-developed. Since many species inhabit Europe and the Mediterranean, these songs have been familiar to Western civilization for thousands of years.

Sylviids are largely insectivorous, and are thus invaluable agents of pest control, not only in agricultural and suburban areas, but in evergreen and hardwood forests. These small songbirds are part of a natural balance that man repeatedly threatens with pesticide use, introduction of alien species, and planting and maintenance of monocultures in farming and timber production. Along with spiders and other natural enemies, sylviid warblers are part of an already extant system of insect control awaiting utilization by man. In many tropical areas, warblers are major consumers of mosquitos, helping to limit the number of vectors for insect-borne disease.

1. Golden-spectacled warbler (*Seicercus burkii*); 2. Whitethroat (*Sylvia communis*); 3. Golden-crowned kinglet (*Regulus satrapa*); 4. Tawny-flanked prinia (*Prinia subflava*); 5. Long-billed crombec (*Sylvietta rufescens*); 6. Blackcap (*Sylvia atricapilla*); 7. Slaty-bellied tesia (*Tesia olivea*); 8. Arctic warbler (*Phylloscopus borealis*); 9. Long-billed gnatwren (*Ramphocaenus melanurus*); 10. Blue-gray gnatcatcher (*Polioptila caerulea*); 11. Chiffchaff (*Phylloscopus collybita*). (Illustration by Barbara Duperron)

1. Zitting cisticola (*Cisticola juncidis*); 2. Cetti's warbler (*Cettia cetti*); 3. Common tailorbird (*Orthotomus sutorius*); 4. Yellow-breasted apalis (*Apalis flavida*); 5. Yellow-bellied eremomela (*Eremomela icteropygialis*); 6. Icterine warbler (*Hippolais icterina*); 7. Marsh warbler (*Acrocephalus palustris*); 8. Little grassbird (*Megalurus gramineus*); 9. Great reed warbler (*Acrocephalus arundinaceus*); 10. Little rush-warbler (*Bradypterus baboecala*); 11. Golden-headed cisticola (*Cisticola exilis*); 12. Grasshopper warbler (*Locustella naevia*). (Illustration by Barbara Duperron)

Species accounts

Long-billed gnatwren
Ramphocaenus melanurus

SUBFAMILY
Polioptilinae

TAXONOMY
Ramphocaenus melanurus Vieillot, 1819.

OTHER COMMON NAMES
French: Microbate à long bec; German: Schwarzschwanz-Degenschnäbler; Spanish: Chirito Picón.

PHYSICAL CHARACTERISTICS
4.75 in (12 cm); .3–.4 oz (8.5–10 g). A small, distinctive bird with a long, cocked tail and long, pale slender bill. Tawny brown above, whitish below, with buffy wash on face, sides of breast and flanks. Tail dusky black, with white-tipped outer feathers.

DISTRIBUTION
Neotropical. Lowlands (to 4,900 ft [1500 m]) from Yucatan south through Central America and northern South America to central Brazil, also east coast of Brazil, west of Andes to coastal Ecuador.

HABITAT
Undergrowth and thickets in deciduous forest and forest edges, and humid forest interior; vine tangle.

BEHAVIOR
Very active. Solitary or in pairs. Song is a clear musical trill. Male tail fanning and lateral tail movements probably function as territorial or mate-attraction displays.

FEEDING ECOLOGY AND DIET
Forages for insects in low undergrowth, sometimes joins mixed foraging flocks.

REPRODUCTIVE BIOLOGY
Nest is loose, deep cup of leaves, grasses and other plant materials located near the ground. Both parents incubate the two eggs (17 days), and feed the young. Fledging occurs after 12-15 days.

CONSERVATION STATUS
Not threatened.

SIGNIFICANCE TO HUMANS
None known. ◆

Blue-gray gnatcatcher
Polioptila caerulea

SUBFAMILY
Polioptilinae

TAXONOMY
Polioptila caerulea Linnaeus, 1766. Seven to nine subspecies recognized.

Ramphocaenus melanurus
  Resident

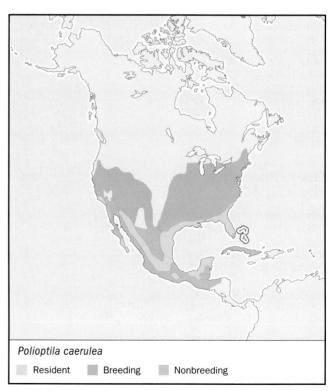

Polioptila caerulea
  Resident   Breeding   Nonbreeding

OTHER COMMON NAMES
English: Cat-bird; French: Gobe-moucherons Gris-bleu; German: Blaumükenfänger; Spanish: Perlita Grisilla.

PHYSICAL CHARACTERISTICS
4–4.5 in (10–11.5 cm); 0.18–0.25 oz (5–7 g). Tiny grayish bird with a distinct white eye ring, and a long tail, often cocked. Upperparts blue-gray, underparts whitish, outer tail feathers white, inner tail feathers black. Breeding male has black 'forehead' stripe extending from base of bill to just above and behind the eye.

DISTRIBUTION
Nearctic. Breeds throughout much of United States and Mexico south to Belize. Winters from extreme southern United States to Baja, Honduras, and Cuba.

HABITAT
Swampy deciduous or pine woods and riparian lowlands in eastern United States. Arid scrub, pinyon-juniper, and open woodland in western United States. Scrub, wood edges, thorn forest, and clearings on wintering grounds.

BEHAVIOR
Usually solitary or in pairs. Call is a thin buzzy whine, the male song, a rather soft series of such notes, interspersed with chips and whistles. In migratory populations, males arrives first and sing to defend their territories. Some populations in Mexico and Bahamas are permanent residents, but most are migratory between April and September.

FEEDING ECOLOGY AND DIET
Will take a wide variety of insects, spiders. Usually gleans while perched, but also hover-gleans and hawks for insects.

REPRODUCTIVE BIOLOGY
Monogamous. During courtship, the male leads the female to potential nest sites. The nest, built by both sexes, is a neat cup of plant fibers often camouflaged and placed high on a branch or fork of a tree or shrub. Four or five eggs are incubated by both parents for 11–15 days. Female broods the nestlings, but later both sexes feed young. Fledging occurs after 10–15 days.

CONSERVATION STATUS
Not threatened.

SIGNIFICANCE TO HUMANS
None known. ◆

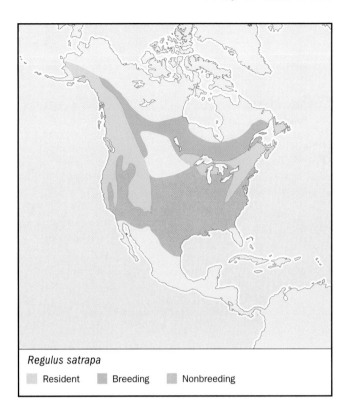

Regulus satrapa
■ Resident ■ Breeding ■ Nonbreeding

Golden-crowned kinglet
Regulus satrapa

SUBFAMILY
Sylviinae

TAXONOMY
Regulus satrapa Lichtenstein, MHK, 1823.

OTHER COMMON NAMES
French: Roitelet à couronne dorée; German: Satrap; Spanish: Reyezuelo Corona Dorada.

PHYSICAL CHARACTERISTICS
3–4 in (8–11 cm); 0.1–0.3 oz (4–7.5 g). Among smallest of all songbirds; olive-green with two whitish wing bars, and a white eye line surmounted by black lateral crown stripes and a yellow crown. Male has orange central crown feathers that are usually concealed.

DISTRIBUTION
Breeds from Nearctic boreal zone south through New England and Appalachians in eastern United States, Rockies, Cascades, Sierra Nevada, Chiricahuas in West. Resident populations in mountains of Mexico, Guatemala. Winters south to Northeastern Mexico.

HABITAT
Dense conifers above 6,560 ft (2,000 m) in Mexico. Sometimes in deciduous forest in winter.

BEHAVIOR
Very active, often hangs upside down. Voice includes a high thin call note and song, given by both sexes, consisting of a series of ascending notes, sometimes followed by a descending warble. Male defends territory with song, and song-displays (crown raised, tail and wing flicking). Often joins mixed-species flocks in fall and winter.

FEEDING ECOLOGY AND DIET
Mostly small insects, spiders, and arthropod eggs; occasionally sap, rarely fruit. Occasionally hover-gleans.

REPRODUCTIVE BIOLOGY
Serially monogamous. Pair bond is maintained through breeding season. Nest, built by both sexes, is a deep hanging cup of moss, lichens, bark, spider webs and other plant material, attached to hanging twigs near trunk, placed high in conifer. Eight to nine (sometimes 5-11) eggs. Incubation by female only (14-15 days). Nestlings are fed by both parents, and leave the nest after 14-19 days.

CONSERVATION STATUS
Not threatened.

SIGNIFICANCE TO HUMANS
None known. ◆

Zitting cisticola
Cisticola juncidis

SUBFAMILY
Sylviinae

TAXONOMY
Sylvia juncidis Rafinesque, 1810.

OTHER COMMON NAMES
English: Fantailed warbler, streaked cisticola; French: Cisticole des joncs; German: Cistensänger; Spanish: Buitrón Común.

PHYSICAL CHARACTERISTICS
3.9–4.7 in (10–12 cm); 0.3–0.4 oz (8–12 g). Small warbler with warm brown upperparts strongly streaked with black, rufous rump and flanks, short, rounded wings, and short, graduated tail, spotted black and white underneath. Bill short, thin, and slightly decurved.

DISTRIBUTION
Widespread. Southern Europe (Iberian Peninsula, Mediterranean rim), sub-Saharan Africa, Indian subcontinent, Southeast Asia, Australasia.

HABITAT
Open tall-grass habitat and grassy wetlands, agricultural lands, primarily in lowlands.

BEHAVIOR
Mostly sedentary, but marked post-breeding dispersal of both adults and juveniles in many populations. Also migratory in Western Mediterranean. Male song is a quick, sharp single note given consistently every 0.5 to 1 seconds. Males are aggressively territorial, especially in vicinity of nest.

FEEDING ECOLOGY AND DIET
Forages mostly on the ground. Takes insects and insect larvae, particularly Lepidoptera, grasshoppers, spiders, and beetles.

REPRODUCTIVE BIOLOGY
Serially monogamous with most males mating with 1–11 females over the course of a season. Occasionally simultaneously polygynous. Pair bond lasts for a single nesting. During courtship, male builds several partially complete nests near the ground, and attracts female with song-flight. Nest is pear-shaped bag made by sewing and weaving grasses together with spider web. Two to six eggs incubated by female for 13 days; young leave nest in 11–15 days. Female feeds young 10–20 days after fledging.

CONSERVATION STATUS
Not threatened.

SIGNIFICANCE TO HUMANS
None known. ◆

Golden-headed cisticola
Cisticola exilis

SUBFAMILY
Sylviinae

TAXONOMY
Cisticola exilis Vigors and Horsfield, 1827.

OTHER COMMON NAMES
English: Bright-headed cisticola, yellow-headed cisticola, tailorbird, corn bird; French: Cisticole à couronne dorée; German: Goldkopg-Cistensänger; Spanish: Buitrón de Capa Dorada.

PHYSICAL CHARACTERISTICS
3.9–4.3 in (10–11 cm). Small warbler, warm brown back, rufous nape, crown, breast and flanks, whitish throat and belly. Strong black streaking on back. Wings and tail short. Bill short, thin, and slightly decurved.

Cisticola exilis
▨ Resident

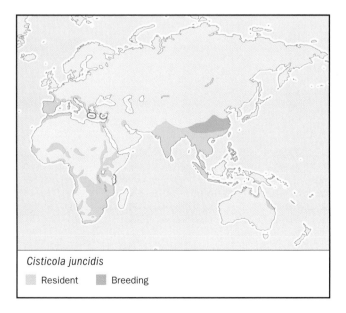

Cisticola juncidis
▨ Resident   ▨ Breeding

DISTRIBUTION
The only cisticola absent from Africa. Southern Asia, from India and China south through Phillipines, Malay Archipelago, and New Guinea, to Australia.

HABITAT
Tall, coarse, open grassland. Often in wet areas.

BEHAVIOR
Usually solitary or in pairs. Scurries through dense grasses in a mouse-like manner. May form loose flocks in nonbreeding season. Male song, given during flight-display, is a nasal wheeze, followed by an explosive, liquid *plook* note.

FEEDING ECOLOGY AND DIET
Insectivorous. Forages on or near ground.

REPRODUCTIVE BIOLOGY
Polygynous. Males with shorter tails have increased reproductive success. Male flight-display consists of characteristic circular flight, singing and a high-speed vertical plummet. Nest is a small, rounded bag of grasses, plant down, and spider silk, attached to grasses or other vegetation. The female builds the nest, sometimes with help from the male, but incubates the 3–4 eggs herself.

CONSERVATION STATUS
Not threatened.

SIGNIFICANCE TO HUMANS
None known. ◆

Tawny-flanked prinia
Prinia subflava

SUBFAMILY
Sylviinae

TAXONOMY
Motacilla subflava Gmelin, 1789. Subspecies *inornata* Sykes, 1832.

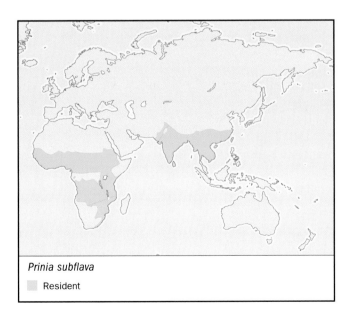

Prinia subflava

☐ Resident

OTHER COMMON NAMES
English: Plain prinia; French: Prinia modeste; German: Braunkopfprinie; Spanish: Prinia de Dorsos Castaños.

PHYSICAL CHARACTERISTICS
5.1–5.5 in (13–14 cm); 0.3–0.4 oz (7–11.5 g). Mmedium to large warbler with short rounded wings and a fairly long, graduated tail. Upperparts uniformly rufous-brown to gray-brown, underparts whitish. Short whitish eye line. Legs long, bill relatively short, slightly decurved.

DISTRIBUTION
Sub-Saharan Africa, southern Asia, from Indian subcontinent to China, Taiwan, most of mainland Southeast Asia.

HABITAT
Lowlands; grassland and savanna, wetlands, mangroves, scrub, forest clearings and edges.

BEHAVIOR
Active, vocal and conspicuous. Often in pairs or family groups. Males territorial, defending year-round by singing. Apparently sedentary, but movement may occur in parts of range during dry season. Duetting occasionally occurs.

FEEDING ECOLOGY AND DIET
Forages in vegetation, gleaning insects and sometimes taking nectar.

REPRODUCTIVE BIOLOGY
Nest is an oval of woven grasses with a side-top entrance. Two to five eggs are incubated for 13–14 days. Young, fed by both parents, fledge after 13–17 days, but remain with parents 2–3 weeks.

CONSERVATION STATUS
Not threatened.

SIGNIFICANCE TO HUMANS
None known. ◆

Yellow-breasted apalis
Apalis flavida

SUBFAMILY
Sylviinae

TAXONOMY
Apalis flavida Strickland, 1852.

OTHER COMMON NAMES
English: Brown-tailed apalis; French: Apalis à gorge jaune; German: Gelbbrust-Feinsänger; Spanish: Apalis de Pecho Amarillo.

PHYSICAL CHARACTERISTICS
4.7–5.1 in (12–13 cm); 0.3 oz (7–9.5 g). Small warbler, olive-green above, whitish throat and belly separated by broad yellow breast band, with a central black marking in some populations. Tail long and graduated, wings short and rounded. Gray on crown and sides of face.

DISTRIBUTION
Endemic to sub-Saharan Africa.

HABITAT
Savanna, forest edge, riparian forest and thickets.

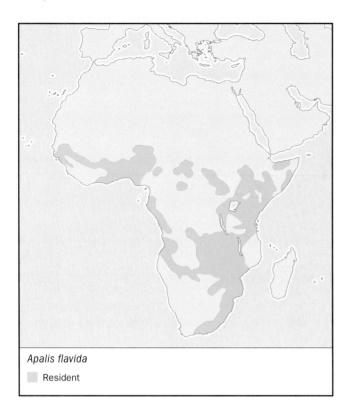

Apalis flavida
■ Resident

Tesia olivea
■ Resident

BEHAVIOR
Often in pairs or family groups. Male song individually distinctive. Pairs sing duets. Maintain year-round feeding territories.

FEEDING ECOLOGY AND DIET
Forages in foliage of trees and thickets. Takes a variety of insects, insect larvae, spiders, and also fruit and nectar. Sometimes joins mixed-species foraging parties.

REPRODUCTIVE BIOLOGY
Nest is a hanging, domed bag with side-top entrance, constructed largely of lichen bound with spider silk, 3–6 ft (1–2 m) above ground. Two to three eggs incubated 12–14 days by the female; the young, fed by both sexes, fledge after 15–16 days.

CONSERVATION STATUS
Not threatened.

SIGNIFICANCE TO HUMANS
None known. ◆

Slaty-bellied tesia
Tesia olivea

SUBFAMILY
Sylviinae

TAXONOMY
Tesia olivea McClelland, 1840.

OTHER COMMON NAMES
English: Slaty-bellied ground warbler; French: Té à ventre ardoise; German: Goldscheiteltesia; Spanish: Trinador de Vientre Pizarro.

PHYSICAL CHARACTERISTICS
3.5–4 in (9–10 cm); 0.2–0.3 oz (6–9 g). Characterized by an exceptionally short tail with 10 rectrices, a large head, rounded wings, long legs, and a more or less upright posture.

DISTRIBUTION
From south Central Asia through lower Himalayas and Nepal to south China, south to northwest Thailand, north Laos, and North Vietnam.

HABITAT
Damp areas and stream courses in evergreen forests; usually 3,280–8,860 ft (1,000–2,700 m).

BEHAVIOR
Keeps to dense cover near the ground. It is skulking, but readily approaches to investigate and scold an intruder. Apparently non-migratory, but descends to lower elevations in winter. Voice a ventriloquial *tchirik-tchirik*.

FEEDING ECOLOGY AND DIET
Forages in litter on the forest floor for insects and spiders.

REPRODUCTIVE BIOLOGY
Little known. Pairs frequently observed.

CONSERVATION STATUS
Not threatened.

SIGNIFICANCE TO HUMANS
None known. ◆

Cetti's warbler
Cettia cetti

SUBFAMILY
Sylviinae

TAXONOMY
Cettia cetti Temminck, 1820.

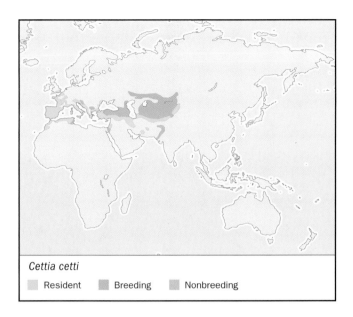

Cettia cetti

■ Resident ■ Breeding ■ Nonbreeding

OTHER COMMON NAMES
French: Bouscarle de Cetti; German: Seidensänger; Spanish:
Ruiseñor Bastardo de Cetti.

PHYSICAL CHARACTERISTICS
5.3–5.7 in (13.5–14.5 cm); 0.4–0.6 oz (10–18g) (males), 0.3–0.6
oz (8–16g) (females). Medium-sized, plump warbler with long,
graduated, rounded tail with 10 rectrices; short, rounded
wings; and a delicate, dark bill. Strongly sexually dimorphic, as
measured by wing length and mass. Upperparts dull chestnut
to rufous, throat white, breast grayish, and belly and flanks
buffy. Characteristic thin, whitish supercilium and eye ring.

DISTRIBUTION
Distribution across Eurasia. In summer, north into Caucasus
region of Russia and Uzbekistan, Kyrgyzstan, Kazakstan. Winters along the Indus River in Pakistan. Eastern race (*albiventris*)
migrates to south to Pakistan and Afghanistan.

HABITAT
Lowlands, usually near water, in dense thickets and reedbeds.

BEHAVIOR
Skulking. Male territorial song is an explosive series of clear
tones. A softer variant is used during courtship. Male song pattern is unique allowing individual recognition. Males aggressively territorial, defending with song, wing-waving display and
fighting. In sedentary populations, territories defended all year.

FEEDING ECOLOGY AND DIET
Forages on or near the ground for insects (especially aquatic
invertebrates), spiders, harvestmen, snails, earthworms, and
some seeds.

REPRODUCTIVE BIOLOGY
Often serially polygynous. Male may mate with the same 1–4 females for several successive years. Nest is loose cup of stems and
leaves, placed low among tangled vegetation. The nest of each
female is placed in her 'range' within the male territory. The
4–5 eggs are incubated by the female. Fledging occurs at 14–16
days, young remain dependent for additional 15 or more days.

CONSERVATION STATUS
Not threatened.

SIGNIFICANCE TO HUMANS
None known. ◆

Little rush warbler
Bradypterus baboecala

SUBFAMILY
Sylviinae

TAXONOMY
Bradypterus baboecala Vieillot, 1817.

OTHER COMMON NAMES
English: African bush-warbler, African sedge warbler; French:
Bouscarle des marais; German: Sumpfbuschsänger; Spanish:
Ruiseñor Africano.

PHYSICAL CHARACTERISTICS
5.9–7.5 in (15–19 cm); 0.4–0.6 oz (11–17 g). Medium-sized
warbler with dark brown upperparts, buffy flanks and breast,
whitish underparts, whitish throat streaked or spotted with
dark brown, and pale eye line. Tail wide and rounded.

DISTRIBUTION
Sub-Saharan Africa. Widespread, but local in the northern
tropics; common throughout much of the south.

HABITAT
Reedbeds and grasses, near marshes, lagoons, sewage ponds,
watercourses.

BEHAVIOR
Found singly or in pairs. Reluctant to fly. Song an accelerating
series of chips, with the tempo of a "bouncing ball." Males defend territory with song and song-flight. Generally sedentary.

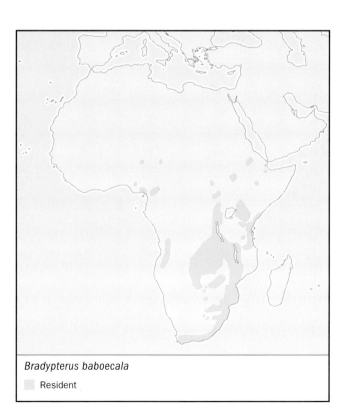

Bradypterus baboecala

■ Resident

FEEDING ECOLOGY AND DIET
Forages for insects. Also known to take ant eggs.

REPRODUCTIVE BIOLOGY
Monogamous. Nest is a tight, bulky cup of grasses and reeds, situated in a tussock and supported by sedge or rush leaves over water. The 2–3 eggs are incubated 12–14 days. Fledging period is 12–13 days, both sexes feed young.

CONSERVATION STATUS
Not threatened.

SIGNIFICANCE TO HUMANS
None known. ◆

Grasshopper warbler
Locustella naevia

SUBFAMILY
Sylviinae

TAXONOMY
Locustella naevia Boddaert, 1783.

OTHER COMMON NAMES
French: Locustelle tachetée; German: Feldschwirl; Spanish: Buscarla Pintoja.

PHYSICAL CHARACTERISTICS
4.7–5.1 in (12–13 cm); 0.3–0.6 oz (9.5–18 g). Small grass warbler with olive-brown upperparts, streaked with black, underparts creamy white, flanks and breast buffy, streaked brown. Short, rounded wings and short graduated tail.

DISTRIBUTION
Breeds from southern Scandinavia, British Isles and France west throughout central Europe and Siberia to Mongolia, Northern China. Winters in Africa and Indian subcontinent.

HABITAT
Grasslands, low scrub, bogs, fens and marshes, with some shrubs and trees.

BEHAVIOR
Very secretive; keeps to dense cover. Gait is a run; flight short and low. Song a high, far-carrying trill, common at night. Female sings during courtship.

FEEDING ECOLOGY AND DIET
Forages in low vegetation and on ground for insects.

REPRODUCTIVE BIOLOGY
Female builds cup nest of grass and plant material on or near ground in thick vegetation. The 5–6 eggs are incubated 12–15 days by both parents. Young leave nest after 10–15 days.

CONSERVATION STATUS
Not threatened.

SIGNIFICANCE TO HUMANS
None known. ◆

Marsh warbler
Acrocephalus palustris

SUBFAMILY
Sylviinae

TAXONOMY
Acrocephalus palustris Bechstein, 1798.

OTHER COMMON NAMES
French: Rousserolle verderolle; German: Sumpfrohrsänger; Spanish: Carricero Poliglota.

PHYSICAL CHARACTERISTICS
5.1 in (13 cm); 0.4–0.7 oz (10–20 g). Heavy, medium-sized warbler with uniform olive-brown upperparts, creamy underparts, buffy wash on sides of breast and flanks, white throat, and light line from eye to base of bill.

DISTRIBUTION
Breeds throughout Central and Western Europe, excluding Iberia and British Isles to southern Scandinavia and western Russia. Winters in Southeast Africa.

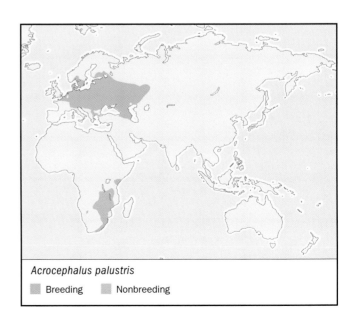

Locustella naevia
■ Breeding   ■ Nonbreeding

Acrocephalus palustris
■ Breeding   ■ Nonbreeding

HABITAT
Breeds in tall herbaceous vegetation and woody cover, often in moist areas, but also on dry slopes, and in parks and open forest.

BEHAVIOR
Solitary and territorial in breeding season, some males polyterritorial. Posture somewhat upright. Song a complex, sweet warble learned by mimicry of other birds. Female sings during courtship.

FEEDING ECOLOGY AND DIET
Gleans insects and spiders from low vegetation. Also takes snails, and rarely berries.

REPRODUCTIVE BIOLOGY
Mostly monogamous, but serial monogamy and opportunistic polygyny occur. Courtship may include aerial dance involving both sexes. Nest is built by female; cylindrical cup of leaves and plant material 3.3–6.6 ft (1–2 m) from ground in tall vegetation, with rim woven around supporting vegetation. Three to six eggs incubated and young, cared for by both parents, leave nest after 11–12 days, remain dependent 15–19 days.

CONSERVATION STATUS
Not threatened.

SIGNIFICANCE TO HUMANS
None known. ◆

Great reed warbler
Acrocephalus arundinaceus

SUBFAMILY
Sylviinae

TAXONOMY
Acrocephalus arundinaceus Linnaeus, 1758. Subspecies *griseldis* Hartlaub, 1891.

OTHER COMMON NAMES
French: Rousserolle turdoïde; German: Drosselrohrsänger; Spanish: Carricero Tordal.

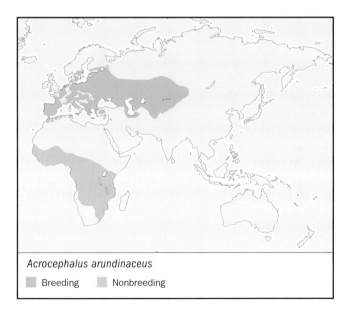

Acrocephalus arundinaceus
■ Breeding   ■ Nonbreeding

PHYSICAL CHARACTERISTICS
7.5–7.9 in (19–20 cm); 0.7–1.8 oz (21–51 g). Large, heavy warbler with strong bill and legs, longish pointed wings. Uniform brown above, whitish below, with buffy wash on flanks, rufous rump, light eyeline.

DISTRIBUTION
Breeds from Iberia and northwest Africa to western Siberia and west China. Winters across sub-Saharan Africa.

HABITAT
Reedbeds and other emergent vegetation, sometimes in thickets bordering water. Swamps, stream courses and savanna in winter.

BEHAVIOR
Movements slow and clumsy. Song deep and slow, full of low croaks and rattles. Male defends large territory.

FEEDING ECOLOGY AND DIET
Insects, spiders, snails, small vertebrates (fish and frogs), some fruits in non-breeding season. Forages in emergent vegetation, trees and shrubs, and water surface.

REPRODUCTIVE BIOLOGY
Many polygynous, but significant portion monogamous. Pair bond lasts only until nestling stage. Nest is built by female; deep cylinder of reeds and leaves, suspended above water. Three to six eggs incubated 14 days by female; young fed by both parents; fledglings dependent additional two weeks.

CONSERVATION STATUS
Not threatened.

SIGNIFICANCE TO HUMANS
None known. ◆

Icterine warbler
Hippolais icterina

SUBFAMILY
Sylviinae

TAXONOMY
Hippolais icterina Viellot, 1817.

OTHER COMMON NAMES
French: Hypolaïs ictérine; German: Gelbspötter; Spanish: Zarcero Icterino.

PHYSICAL CHARACTERISTICS
4.9–5.3 in (12.5–13.5 cm); 0.3–0.8 oz (8–23 g). Small warbler with olive to gray-olive upperparts, yellowish underparts, pale eye line, and long, thin bill.

DISTRIBUTION
France, Balkans, and Scandinavia east to southwest Siberia. Winters in Southern Africa.

HABITAT
Open forests, orchards gardens and edges; often in lowlands and river valleys.

BEHAVIOR
Active and vocal, but usually remains in foliage. Territorial on breeding and wintering grounds. Song rapid, varied, including musical phrases and mimicry.

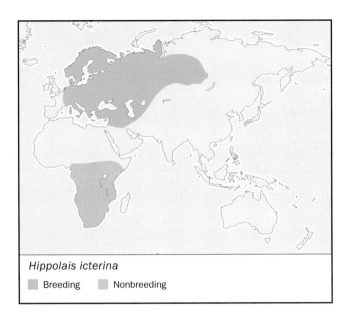

Hippolais icterina

▨ Breeding ▨ Nonbreeding

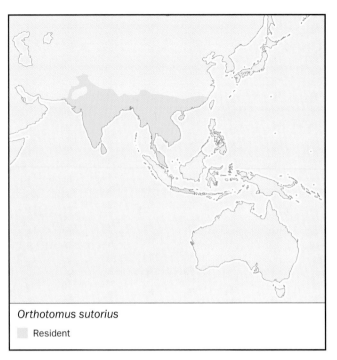

Orthotomus sutorius

▨ Resident

FEEDING ECOLOGY AND DIET
Insects, fruit in late summer. Forages in foliage, gleaning insects and sallying to catch aerial prey.

REPRODUCTIVE BIOLOGY
Monogamous. Nest is neat cup of plant material, decorated on outside with bark, paper, wool, etc., built by female in fork of shrub or tree. Four to five eggs incubated by female for 13–15 days; young, cared for by both parents, leave nest after 12–16 days, independent after 1–2 weeks.

CONSERVATION STATUS
Not threatened.

SIGNIFICANCE TO HUMANS
None known. ◆

Common tailorbird
Orthotomus sutorius

SUBFAMILY
Sylviinae

TAXONOMY
Orthotomus sutoria Pennant, 1769.

OTHER COMMON NAMES
English: Long-tailed tailorbird; French: Couterière à longe queue; German: Rotstirn-Schneidervogel.

PHYSICAL CHARACTERISTICS
5.1 in (13 cm); 0.2–0.4 oz (6–10 g). Small bird with long, thin, decurved bill. Back olive-green, underparts creamy, nape gray, crown rufous, whitish eyeline. Short, spiky tail often held erect.

DISTRIBUTION
Resident throughout India and Southeast Asia to Java and Indonesia.

HABITAT
Deciduous forest, scrubby clearings, gardens, scrubland, and mangroves.

BEHAVIOR
Solitary or in pairs. Skulking but restless and active. Flicks tail side to side often. Song a loud, two-syllable phrase repeated in series of three to four calls.

FEEDING ECOLOGY AND DIET
Insects and insect larvae, flower nectar. Forages on ground, in low vegetation and in trees.

REPRODUCTIVE BIOLOGY
Nest is lined with spider silk, soft plant fibers, etc.; sewn by female from the leaves of a broad-leaved plant, usually well hidden. Incubation of 3–5 eggs for 12 days and feeding nestlings done by both parents.

CONSERVATION STATUS
Not threatened.

SIGNIFICANCE TO HUMANS
None known. ◆

Yellow-bellied eremomela
Eremomela icteropygialis

SUBFAMILY
Sylviinae

TAXONOMY
Eremomela icteropygialis Lafresnaye, 1822.

OTHER COMMON NAMES
English: Salvadori's eremomela; French: Érémomèle à croupion jaune; German: Gelbbauch-Eremomela; Spanish: Eremomela de Vientre Amarillo.

PHYSICAL CHARACTERISTICS
3.9–4.3 in (10–11 cm); 0.26–0.33 oz (7.5–9.3 g). Small bird with gray head, nape, back, and breast, yellow belly and undertail. Dark eyeline surmounted by a lighter one; dark gray wing and tail.

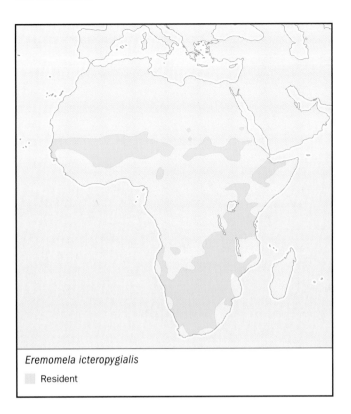

Eremomela icteropygialis

▨ Resident

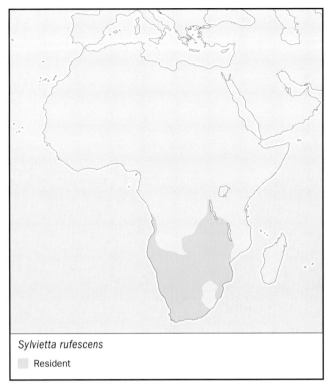

Sylvietta rufescens

▨ Resident

DISTRIBUTION
Widespread throughout non-forested sub-Saharan Africa.

HABITAT
Woodland, forest edge, scrub, gardens, and riparian areas.

BEHAVIOR
Inconspicuous. Usually in pairs or groups of up to eight birds. Territorial. Song is loud, twittery series of 7–8 notes. Mimicry sometimes included.

FEEDING ECOLOGY AND DIET
Foraging mostly in canopy and mid-level for small insects. Sometimes joins mixed-species foraging parties.

REPRODUCTIVE BIOLOGY
Monogamous. Breeds primarily during monsoon season. Nest is deep cup of plant down and spider silk, lined with grass, and suspended from fork of a shrub or tree. One to three eggs are incubated 13–14 days by female; nestlings, fed by both parents, leave after 15–16 days, fed by parents additional two or more weeks.

CONSERVATION STATUS
Not threatened.

SIGNIFICANCE TO HUMANS
None known. ◆

Long-billed crombec
Sylvietta rufescens

SUBFAMILY
Sylviinae

TAXONOMY
Sylvietta rufescens Vieillot, 1817.

OTHER COMMON NAMES
English: Cape crombec; French: Crombec à long bec; German: Langschnabelsylvietta; Spanish: Sylvieta de Pico Largo.

PHYSICAL CHARACTERISTICS
3.9–4.7 in (10–12 cm); 0.3–0.8 oz (8–23 g). Small, nearly tailless bird with brownish gray upperparts, breast and flanks buffy, throat and belly white. Black eyeline surmounted by a light eyeline.

DISTRIBUTION
Endemic resident in Southern Africa.

HABITAT
Thick brush, *Acacia* scrub, open and secondary woodland, often in drier areas.

BEHAVIOR
Solitary or in pairs or family groups. Song a high, variable series of trilled notes. Territorial. Flight bouncy.

FEEDING ECOLOGY AND DIET
Insects, ticks and grass seeds. Forages methodically from bottom to top of bushes and trees. Often joins mixed-species foraging parties.

REPRODUCTIVE BIOLOGY
Monogamous. Nest is large, hanging bag of grasses, spider webs, and plant fibers, close to ground and attached to the lower limbs of a tree, often an acacia. One to three eggs incubated for two weeks; nestlings fed by both parents for two weeks before leaving.

CONSERVATION STATUS
Not threatened.

SIGNIFICANCE TO HUMANS
None known. ◆

Chiffchaff

Phylloscopus collybita

SUBFAMILY
Sylviinae

TAXONOMY
Phylloscopus collybita Vieillot, 1817.

OTHER COMMON NAMES
English: Common, Eurasian chiffchaff; French: Pouillot véloce;
German: Zilpzalp; Spanish: Mosquitero Común.

PHYSICAL CHARACTERISTICS
3.9–4.3 in (10–11 cm); 0.2–0.4 oz (6–12 g). Small warbler with
short, fine bill, olive-green upperparts, white underparts,
brownish flanks, a dark line through the eye, surmounted by a
light line.

DISTRIBUTION
Breeds from Iberia, British Isles and Scandinavia to Caspian
Sea, northern Mongolia, eastern Siberia. Winters in Western
Europe, Mediterranean, northwest Africa, Middle East, India
and sub-Saharan Africa.

HABITAT
Open woodland with tall undergrowth, parks, and scrub, typically in lowlands.

BEHAVIOR
Solitary or in pairs during breeding season. In small groups or
mixed flocks in migration and winter. Territorial. Song highly
variable; in nominate group it is series of two-note alternating
phrases.

FEEDING ECOLOGY AND DIET
Forages from ground to canopy for insects, gleaning from foliage.

REPRODUCTIVE BIOLOGY
Monogamous, sometimes facultatively polygynous. Nest is
built by the female; dome of dry gr–ss and other plant materials with a side entrance, on or near the ground in thick vegetation. Four to seven eggs, incubated 13–15 days by female;

nestlings cared by female; fledge after 14–16 days. Brood independent after 10–19 days.

CONSERVATION STATUS
Not threatened.

SIGNIFICANCE TO HUMANS
None known. ◆

Arctic warbler

Phylloscopus borealis

SUBFAMILY
Sylviinae

TAXONOMY
Phyllopneuste borealis Blasius, 1858.

OTHER COMMON NAMES
French: Pouillot boréal; German: Wanderlaubsänger; Spanish:
Mosquitero Boreal.

PHYSICAL CHARACTERISTICS
4.1–5.1 in (10.5–13 cm); 0.3–0.5 oz (8–15 g). Medium-sized
warbler, olive-green above, yellowish-white below, with a yellow wash in some plumages. Thin, clean whitish eye line; long
wings with two whitish wing bars.

DISTRIBUTION
Breeds in boreal and subalpine zones from Scandinavia
throughout Asia to Japan and Western Alaska. Winters in
Southeast Asia, Wallacea.

HABITAT
Coniferous, deciduous, and mixed forest in taiga zone. Winters
in open woodlands, rainforest, forest edge, gardens, and mangroves.

BEHAVIOR
Arboreal and active, with quick flight and habit of wing- and
tail-flicking. Usually solitary or in pairs or small family groups.
Territorial; male often defends with song and wing-rattling
displays.

FEEDING ECOLOGY AND DIET
Forages in foliage for insects and larvae, usually high, but occasionally close to the ground.

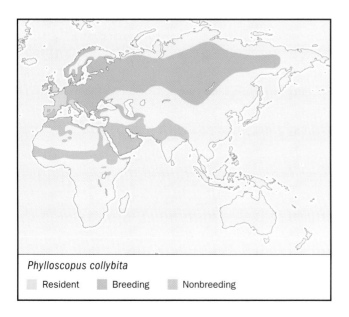

Phylloscopus collybita

▨ Resident   ▨ Breeding   ▨ Nonbreeding

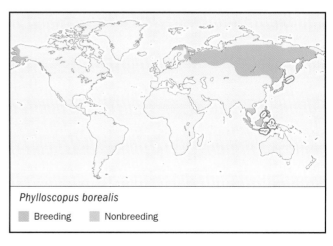

Phylloscopus borealis

▨ Breeding   ▨ Nonbreeding

REPRODUCTIVE BIOLOGY
Monogamous, occasionally polygynous. Courtship involves song, wing-rattling and wing-flapping displays. Nest is built by the female; dome of dry grass and other plant materials with a side entrance on the ground in thick vegetation. Incubation of 5–7 eggs for 11–13 days by female; fledging takes 13–14 days, young fed by both parents, brooded by female; young independent after two weeks.

CONSERVATION STATUS
Not threatened.

SIGNIFICANCE TO HUMANS
None known. ◆

Golden-spectacled warbler
Seicercus burkii

SUBFAMILY
Sylviinae

TAXONOMY
Seicercus burkii Burton, 1836.

OTHER COMMON NAMES
English: Yellow-eyed flycatcher warbler; French: Pouillot de Burke; German: Goldbrillen-Laubsängerl Spanish: Curruca de Burke.

PHYSICAL CHARACTERISTICS
3.9–4.7 in (10–12 cm); 0.2–0.3 oz (6–9 g). Small, plump warbler, with bright green upperparts, bright yellow underparts, a short, broad tail with white undertail feathers. Crown streaked black and gray, eyering yellow; yellow wingbar.

DISTRIBUTION
Southern Asia, from India to south-central China to Thailand, Cambodia and Vietnam.

HABITAT
Mid-level undergrowth of evergreen or mixed forest in highlands. Winters at lower elevations.

BEHAVIOR
Usually solitary or in pairs. Rarely found in canopy. Song of *burkii* is a loud, clear trill.

FEEDING ECOLOGY AND DIET
Insects. Often joins mixed-species flocks in non-breeding season.

REPRODUCTIVE BIOLOGY
Little known. Nest is ball of mosses, grass, and other plant fibers, lined with moss and lichen, and concealed along a bank or slope, often in tree roots. Four eggs are incubated by both parents.

CONSERVATION STATUS
Not threatened.

SIGNIFICANCE TO HUMANS
None known. ◆

Little grassbird
Megalurus gramineus

SUBFAMILY
Sylviinae

TAXONOMY
Megalurus gramineus Gould, 1865.

OTHER COMMON NAMES
English: Little marshbird; striated grassbird; little reedbird; marsh warbler; French: Mégalure menue; German: Zwergschilfsänger.

PHYSICAL CHARACTERISTICS
5.1–5.9 in (13–15 cm). Medium-sized warbler with brownish upperparts, streaked dark; pale grayish underparts finely

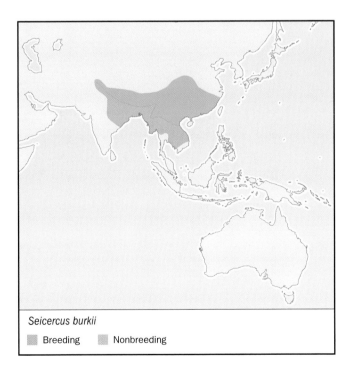

Seicercus burkii
▨ Breeding   ▨ Nonbreeding

Megalurus gramineus
▨ Resident

streaked with dark brown. Flight feathers graduated, edged with white.

DISTRIBUTION
Southern and eastern Australia.

HABITAT
Dense vegetation of marshes, reedbeds, swamps, occasionally mangroves.

BEHAVIOR
Solitary. Skulking. Song is three note, plaintive whistle. Flight is weak. Some movement and nomadism occurs.

FEEDING ECOLOGY AND DIET
Insects, insect larvae, spiders and other arthropods; also takes aquatic mollusks.

REPRODUCTIVE BIOLOGY
Courtship involves chasing and wing-fluttering displays. Nest is suspended above water, a deep cup of grass and plant material, lined with large feathers, and rimmed tops. Three to five eggs.

CONSERVATION STATUS
Not threatened.

SIGNIFICANCE TO HUMANS
None known. ◆

Blackcap
Sylvia atricapilla

SUBFAMILY
Sylviinae

TAXONOMY
Sylvia atricapilla Linnaeus, 1758.

OTHER COMMON NAMES
French: Fauvette à tête noire; German: Mönchsgrasmücke; Spanish: Curruca Capirotada.

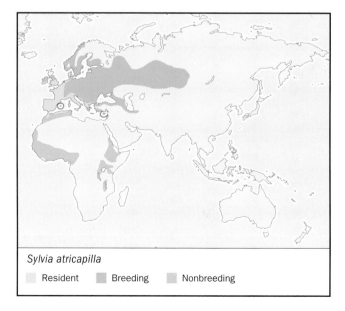

Sylvia atricapilla

☐ Resident ▨ Breeding ▨ Nonbreeding

PHYSICAL CHARACTERISTICS
5.5 in (14 cm); 0.5–0.7 oz (15–21 g). Medium-sized, with plumage ranging from slate gray in adult males to olive or brown in females and juveniles. Crown is distinctive (black in adult males, rufous in females and juveniles). Wings are long and pointed, with long primary projection. Bill is black, relatively long, and pointed. Legs long.

DISTRIBUTION
Breeds from British Isles and southern Scandinavia throughout Western and Central Europe to coastal northwest Africa, Mediterranean, Near East, and west to central Russia and northern Iran.

HABITAT
Forest with tall undergrowth, from riparian areas, parks and gardens to boreal forest and alpine forest to treeline.

BEHAVIOR
Arboreal and very active. Males territorial, defending with song, displays, and agonistic behavior. Mimicry of other birds is occasionally incorporated into song. Mixed partial migrant: individuals in northern range migrate south, while individuals in southern range (the Mediterranean area) are residents or partial migrants.

FEEDING ECOLOGY AND DIET
Feeds in trees and shrubs, gleaning insects and other arthropods from leaves and branches. During migration and on wintering grounds, fruits constitute a large part of the diet.

REPRODUCTIVE BIOLOGY
Pairs solitary and territorial, generally monogamous. Site-fidelity is high in migratory populations. Courtship involves construction, by the male, of several loose 'cock nests'. The female completes the final nest, a fine cup typically located in dense vegetation of a tree or shrub above ground. Both parents incubate 2–6 eggs (typically 5) for 10–16 days. Feeding young in the nest (8–14 days) and after fledging (for about two weeks) is also shared.

CONSERVATION STATUS
Not threatened.

SIGNIFICANCE TO HUMANS
A familiar songbird easily recognized by appearance and voice. It is a model system for the study of the physiology and evolution of bird migration, and for the study of avian diet and energetics, especially as related to movement and seasonal food availability. ◆

Whitethroat
Sylvia communis

SUBFAMILY
Sylviinae

TAXONOMY
Sylvia communis Latham, 1787.

OTHER COMMON NAMES
English: Common whitethroat, greater whitethroat; French: Fauvette grisette; German: Dorngrasmücke; Spanish: Curruca Zarcera.

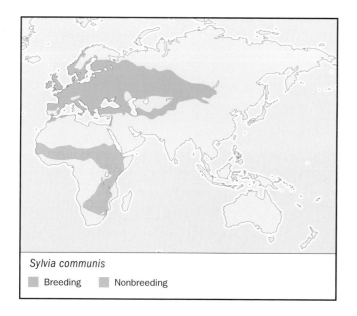

Sylvia communis

■ Breeding ■ Nonbreeding

PHYSICAL CHARACTERISTICS

5.5 in (14 cm); 0.4–0.9 oz (10–24 g). Medium-sized warbler with gray-brown upperparts, whitish underparts, buffy flanks, gray (male) or brownish (female) cap and side of face, and a bright white throat. Folded wing shows significant rufous edging of coverts, secondaries and tertials. Long tail with white outer tail feathers.

DISTRIBUTION

Breeds in Western and Central Europe, southern and coastal Scandinavia, Turkey, North Africa, and Western and Central Siberia. Winters in sub-Saharan Africa.

HABITAT

Open scrubland, farmland and forest edges, with mix of herbaceous and low woody vegetation.

BEHAVIOR

Typically solitary or in pairs. Skulking, but inquisitive. Male song a short, scratchy warble. Territorial. Engages in song-flight display, rising vertically from shrub, then swooping back down.

FEEDING ECOLOGY AND DIET

Forages in low and mid-height vegetation, gleaning arthropods.

REPRODUCTIVE BIOLOGY

Monogamous. Male builds 'cock nests'. Nest a loose, deep cup of grasses in dense, low, tangled vegetation. Three to six eggs incubated 11–13 days, mostly by female; young brooded and fed by both parents, leave nest after 10–12 days, remain with parents 2–3 weeks.

CONSERVATION STATUS

Not threatened.

SIGNIFICANCE TO HUMANS

None known. ◆

Resources

Books

Ali, Salim, and S. Dillon Ripley. *Handbook of the Birds of India and Pakistan*, 2nd ed. Vol. 8. *Warblers to Redstarts*. Delhi: Oxford University Press, 1997.

Baker, Kevin. *Warblers of Europe, Asia and North Africa*. Princeton: Princeton University Press, 1997.

Cramp, Stanley, ed. *The Birds of the Western Palearctic*. Vol. 6. *Warblers*. Oxford: Oxford University Press, 1992.

Kaufman, Kenn. *Lives of North American Birds*. New York: Houghton Mifflin, 1996.

Parmenter, Tim and Clive Byers. *A Guide to the Warblers of the Western Palearctic*. Uxbridge, Middlesex: Bruce Coleman Books, 1991.

Shirihai, Hadoram, Gabriel Gargallo, and Andreas J. Helbig. *Sylvia Warblers*. Princeton: Princeton University Press, 2001.

Sibley, Charles G. and Jon E. Ahlquist. *Phylogeny and Classification of Birds*. New Haven, Yale University Press, 1990.

Urban, Emil K., Hilary C. Fry, and Stuart Keith. *The Birds of Africa*. Vol. 5. San Diego: Academic Press, 1997.

Periodicals

Alström, P. and U. Olsson. "The golden-spectacled warbler: a complex of sibling species, including a previously undescribed species." *Ibis* 151 (1999): 545–568.

Barker, F. Keith, et al. "A phylogenetic hypothesis for passerine birds: taxonomic and biogeographic implications of an analysis of nuclear DNA sequence data." *Proceedings of the Royal Society of London*, Series B, 269 (2002): 295–308.

Catchpole, Clive K. "The evolution of mating systems in *Acrocephalus* warblers." *Japanese Journal of Ornithology* 44 (1995): 195–207.

Helbig, Andreas J. and Ingrid Seibold. "Molecular phylogeny of Palearctic-African *Acrocephalus* and *Hippolais* warblers (Aves: Sylviidae)." *Molecular Phylogenetics and Evolution* 11 (1999): 246–260.

Irwin, Darren E., et al. "Speciation in a ring." *Nature* 409 (2001): 333–337.

Leisler, B., et al. "Taxonomy and phylogeny of reed warblers (genus *Acrocephalus*) based on mtDNA sequences and morphology." *Journal fur Ornithologie* 138 (1997): 469–496.

Matthew J. Sarver, BS

Old World flycatchers
(Muscicapidae)

Class Aves
Order Passeriformes
Suborder Passeri (Oscines)
Family Muscicapidae

Thumbnail description
A large, highly variable group consisting of about 135 species of songbirds. They are small-to medium-sized perching birds, sometimes quite colorful; most feed by making aerial sallies from an exposed perch to catch their prey of flying insects

Size
Body length 3–9 in (7.6–23 cm)

Number of genera, species
16 genera, 135 species, including 9 genera (109 species) of Muscicapinae (typical Old World flycatchers) and 7 genera (26 species) of Platysteirinae (African flycatchers)

Habitat
Species occur in a wide range of habitats, including forest, woodland, savanna, grassland, edges of waterbodies and wetlands, pasture and other agricultural areas, and well-vegetated gardens and residential areas

Conservation status
The World Conservation Union (IUCN) lists 18 species of the Muscicapidae as being at risk, plus another 19 species as being Near Threatened

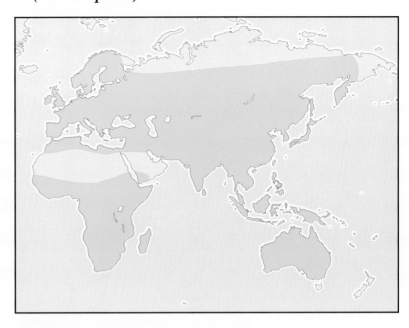

Distribution
Widespread in Eurasia and Africa with the greatest richness of species occurring in tropical and subtropical Africa, India, and Southeast Asia

Evolution and systematics

As treated here, the Old World flycatchers (family Muscicapidae) include 16 genera and about 135 species, including 9 genera and 109 species of the subfamily Muscicapinae (the typical Old World flycatchers) and 7 genera and 26 species of Platysteirinae (the African flycatchers). However, avian systematists are actively studying the composition of this diverse group of birds and related ones. There is ongoing controversy about the relationship of the Muscicapidae to other designated families of passerine birds, and whether to combine some of these with the Old World flycatchers. One classification joins the Muscicapidae with several related families, including the whistlers (Pachycephalidae), the fantails (Rhipiduridae), and the monarchs (Monarchidae). That group would include about 49 genera and 398 species. Some other systematic treatments would include even more families than this, including the Old World warblers (Silviidae) and thrushes (Turdidae). The tyrant flycatchers (Tyrannidae) are generally similar in appearance and behavior to the Muscicapidae, but this is because of convergent evolution as the families are not actually closely related.

Physical characteristics

The Old World flycatchers are a large and variable group consisting of about 135 species of songbirds. The subfamily Muscicapinae includes small- to medium-sized birds, with a body length generally ranging from 3–9 in (7.6–23 cm). The bill is relatively small, short, vertically (dorso-ventrally) flattened, and pointed. There are stout, well-developed bristles (known as rictal bristles) at the gape of the beak, which help increase the effectiveness at catching flying insects. The legs are relatively short and the feet small but well suited to perching. Coloration is highly variable among species, ranging from dull gray or brown to bright blue or vermilion and sometimes occurring in stark patterns. The sexes are colored alike in many species, but are dimorphic in others. Young birds are usually relatively subdued in coloration and spotted on the back and/or breast.

The subfamily Platysteirinae is also a varied group of small- to medium-sized songbirds. The bill is small, short, vertically flattened, and slightly hooked at the tip. There are well-developed rictal bristles at the gape of the beak. The tail

The colorful plumage of the blue-and-white flycatcher (*Ficedula cyanomelana*). (Photo by T. Shimba/VIREO. Reproduced by permission.)

subarctic regions and wintering in subtropical and tropical areas. During their migration they travel at night and feed during the day. Species that breed in subtropical and tropical regions are resident there, although some of them may undertake seasonal movements between low- and high-altitude habitats. Most species are non-gregarious, occurring only singly, as pairs, or in small family groups consisting of parents and their immature progeny. The Old World flycatchers are strong and highly maneuverable fliers. Most species feed by sitting on a conspicuous perch until a flying insect is spotted, which is then pursued and caught in an aerial sally. Some species also glean prey from foliage, bark surfaces, or the ground. Some species have a well-developed song used to proclaim and defend their breeding territory, but in others it is relatively monotonous and weak. All species have calls used to communicate with individuals of their species, or to alert all nearby birds of the presence of a potential predator.

Feeding ecology and diet

Old World flycatchers use a variety of feeding tactics to prey on insects and other arthropods. Most species use an exposed vantage point, such as a high perch in a tree, to spot flying insects, which are then caught in an aerial pursuit. Some species also feed by gleaning prey from bark, branches, foliage, or spider webs, or by swooping down to feed on arthropods spotted on the ground. Some species repeatedly return

is short and the legs and feet are variable in length and stoutness. An area of bare skin around the eye is variable in color among species. The sexes are colored differently, with the females being generally more subdued in pattern and hue. The plumage coloration of the male is mostly glossy black and white, while the female is usually more brown and sometimes light rufous on the belly. In both the Muscicapinae and Platysteirinae, relatively widespread species have been divided into various gepgraphically distinct subspecies and races that are distinct in elements of their size, plumage, behavior, and song.

Distribution

Species of typical Old World flycatchers are found widely through Eurasia and Africa. However, the greatest richness of species occurs in tropical Asia and Africa. Species of the Platysteirinae occur only in Africa.

Habitat

Species of Old World flycatchers occur in a wide range of habitats, including moist and dry forest, woodland, savanna, grassland, edges of waterbodies and wetlands, pasture and other agricultural areas, and well-vegetated gardens and residential areas.

Behavior

The northern species of Old World flycatchers are seasonal migrants, breeding in higher-latitude temperate and

A cape puffback (*Batis capensis*) on its nest. (Photo by W. Tarboton/VIREO. Reproduced by permission.)

to use a favorite perch for hunting, while others move about and frequently change their vantage point.

Reproductive biology

Old World flycatchers are highly territorial during their breeding season, defending a nesting area from others of their species. They do this by proclaiming their territory by frequent renditions of a song, and if this is not sufficient they will fight with intruders. They build a cup-shaped nest of grass, bark, and other plant fibers. The nest is generally placed in the fork of a branch, on a ledge of a bank, or in a cavity in a tree, stump, or cliff. They lay two to seven spotted or mottled eggs. In some species both parents participate in building the nest and incubating the eggs, while in others only the female does this. The incubation period ranges from about 12–22 days. Both parents care for the nestlings and fledglings. In some species, particularly of African flycatchers, immature birds of previous nestings will help their parents raise a new clutch of siblings.

Conservation status

The World Conservation Union IUCN lists 18 species of birds in the Muscicapidae as being at risk, plus another 19 species as being Near Threatened. However, the conservation status of many rare species in this group has not yet been studied. Further research will undoubtedly add additional species to the listings. Examples of listed species at-risk include the Nimba flycatcher (*Melaenornis annamarulae*) of the Ivory Coast and Guinea (Vulnerable), the streaky-breasted jungle-flycatcher (*Rhinomyias addita*) of China (Near Threatened), the white-browed jungle-flycatcher (*Rhinomyias insignis*) of the Philippines (Vulnerable), the Grand Comoro flycatcher (*Humblotia flavirostris*) of the Comoro Islands (Endangered), the Lampobattang flycatcher (*Ficedula bonthaina*) of Indonesia (Endangered), the red-tailed newtonia (*Newtonia fanovanae*) of Madagascar (Vulnerable), and the banded wattle-eye (*Platysteira laticincta*) of Cameroon (Endangered). All of these species are at-risk because of historical and ongoing habitat loss and fragmentation.

Significance to humans

Old World flycatchers are not of direct importance to humans, other than the indirect economic benefits of ecotourism and bird-watching focused on seeing birds in natural habitats. Moreover, it is crucial that research be undertaken to better understand the biology and habitat needs of the rare and endangered species of Old World flycatchers. In addition, their critical habitats must be identified and rigorously protected to prevent the extinction of these birds.

1. Silverbird (*Empidornis semipartitus*); 2. Black-and-white flycatcher (*Bias musicus*); 3. Abyssinian slaty flycatcher (*Melaenornis chocolatinus*); 4. Orange-breasted blue flycatcher (*Cyomis tickelliae*); 5. Fraser's forest-flycatcher (*Fraseria ocreata*); 6. Female shrike-flycatcher (*Megabyas flammulatus*); 7. Black-throated wattle-eye (*Platysteira peltata*); 8. Cape batis (*Batis capensis*); 9. Female chestnut wattle-eye (*Dyaphorophyia castanea*) 10. Large-billed blue-flycatcher (*Cyornis caerulatus*); 11. Red-tailed newtonia (*Newtonia fanovanae*). (Illustration by Barbara Duperron)

1. Male collared flycatcher (*Ficedula albicollis*), spring plumage; 2. Dull-blue flycatcher (*Eumyias sordida*); 3. Male European pied flycatcher (*Ficedula hypoleuca*), spring plumage; 4. Fulvous-chested jungle-flycatcher (*Rhinomyias olivacea*); 5. Female little slaty flycatcher (*Ficedula basilanica*) 6. Brown-chested jungle-flycatcher (*Rhinomyias brunneata*); 7. Spotted flycatcher (*Muscicapa striata*); 8. Large niltava (*Niltava grandis*); 9. Ashy flycatcher (*Muscicapa caerulescens*); 10. Cassin's flycatcher (*Muscicapa cassini*). (Illustration by Barbara Duperron)

Species accounts

Spotted flycatcher
Muscicapa striata

SUBFAMILY
Muscicapinae

TAXONOMY
Muscicapa striata Pallas, 1764.

OTHER COMMON NAMES
French: Gobemouche gris; German: Grauschnäpper; Spanish: Papamoscas Gris.

PHYSICAL CHARACTERISTICS
The body length is about 5 in (13–14 cm). Both sexes are colored alike, having a brownish gray back, head, and tail and a white belly and throat streaked with gray. The juvenile is more spotted. Various subspecies have been described based on plumage and song characters.

DISTRIBUTION
Breeds widely in northern and central Europe and European and central Russia, and winters in southern Africa.

HABITAT
Breeds in temperate forest edges, woods, parks, orchards, and gardens.

BEHAVIOR
A migratory species. Pairs of breeding birds defend a territory. Winters as single birds. Often flicks its wings and tail when perched. The song is delivered from a prominent perch, and is a series of about six squeaky notes.

FEEDING ECOLOGY AND DIET
Spots flying insects from a prominent perch and then pursues the prey by an aerial sally. Usually returns persistently to the same perch.

REPRODUCTIVE BIOLOGY
Builds a cup-shaped nest in a tree crotch, shallow tree-cavity, or behind loose bark, and also uses nest-boxes. Lays four to six greenish eggs with rust-colored spots.

CONSERVATION STATUS
Not threatened. A widespread and locally abundant species.

SIGNIFICANCE TO HUMANS
None known, except for the economic benefits of bird-watching. ◆

Ashy flycatcher
Muscicapa caerulescens

SUBFAMILY
Muscicapinae

TAXONOMY
Muscicapa caerulescens Hartlaub, 1865.

OTHER COMMON NAMES
English: Ashy alseonax, blue-gray flycatcher, cinereus flycatcher; French: Gobemouche à lunettes; German: Schieferschnäpper; Spanish: Papamoscas Ahumado.

PHYSICAL CHARACTERISTICS
The body length is 5.5–6 in (14–16 cm). The sexes are colored similarly, with a light blue-gray back, gray-white underparts, a

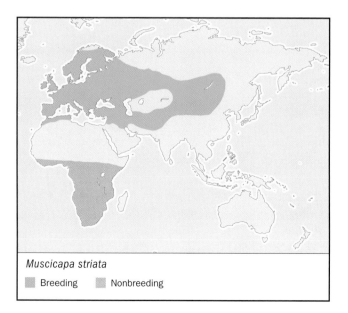

Muscicapa striata

■ Breeding ■ Nonbreeding

Muscicapa caerulescens

■ Resident

white eye-ring, and a black line bordered by a white one extending through the eye. Several subspecies have been identified, which differ somewhat in coloration.

DISTRIBUTION
A resident species of much of southern Africa.

HABITAT
Occurs in a wide range of tropical forest-edges and moist open woodlands and savannas, and also cleared and agricultural areas. Does not occur in the interior of closed forest. Occurs as high as 5,900 ft (1,800 m).

BEHAVIOR
A nonmigratory species. Pairs of breeding birds defend a territory. The song consists of three to five high-pitched notes.

FEEDING ECOLOGY AND DIET
Hunts from a perch for flying insects in the upper canopy. Usually returns to its original perch after an aerial sally.

REPRODUCTIVE BIOLOGY
Builds a bulky cup-shaped nest of moss, grass, and rootlets lined with finer material. Nest is placed in a shallow tree-cavity, a bark crevice, or at a narrow branch-fork. Lays two or three creamy colored, finely speckled eggs.

CONSERVATION STATUS
Not threatened. A widespread and locally abundant species.

SIGNIFICANCE TO HUMANS
None known, except for the economic benefits of bird-watching. ◆

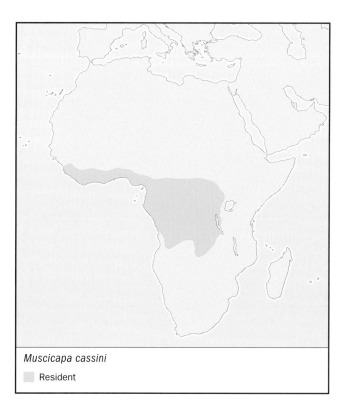

Muscicapa cassini

▨ Resident

Cassin's flycatcher
Muscicapa cassini

SUBFAMILY
Muscicapinae

TAXONOMY
Muscicapa cassini Heine, 1859.

OTHER COMMON NAMES
English: Cassin's alseonax, Cassin's gray flycatcher; French: Gobemouche de Cassin; German: Cassinschnäpper; Spanish: Papamoscas de Cassin.

PHYSICAL CHARACTERISTICS
The body length is about 5.5 in (14 cm). The sexes are colored similarly, with a bluish gray back, black wings and tail, and white underparts with gray flanks and chest.

DISTRIBUTION
A resident species of much of western tropical Africa.

HABITAT
Occurs in the vicinity of rivers, streams, and other surface waters within humid, lowland, tropical forest. Occurs as high as about 5,900 ft (1,800 m).

BEHAVIOR
A nonmigratory species. Pairs of breeding birds defend a linear territory along a watercourse, or a wider one in flooded forest. The song consists of a medley of whistles, buzzes, and chirps sung in bouts of several minutes.

FEEDING ECOLOGY AND DIET
An active hunter that searches for flying insects from an exposed perch, such as a stump or dead tree in the water or from an overhanging branch. Usually returns to its original perch after each sally. Sometimes swoops to take prey from a spider web or the ground.

REPRODUCTIVE BIOLOGY
Builds a cup-shaped nest of grass and other fibers. The nest is placed close to the ground in a shallow cavity in a stump, in other kinds of tree-crevices, or at a narrow branch-fork. Lays two light-green, finely speckled eggs.

CONSERVATION STATUS
Not threatened. A widespread and locally abundant species.

SIGNIFICANCE TO HUMANS
None known, except for the economic benefits of bird-watching. ◆

Pied flycatcher
Ficedula hypoleuca

SUBFAMILY
Muscicapinae

TAXONOMY
Ficedula hypoleuca Pallas, 1764.

OTHER COMMON NAMES
English: European pied flycatcher; French: Gobemouche noir; German: Trauerschnäpper; Spanish: Papamoscas Cerrojillo.

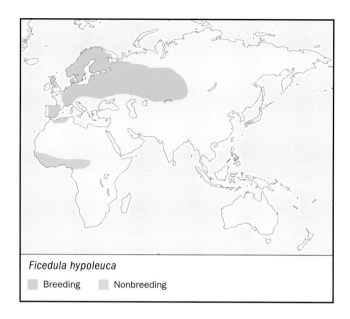

Ficedula hypoleuca

☐ Breeding ☐ Nonbreeding

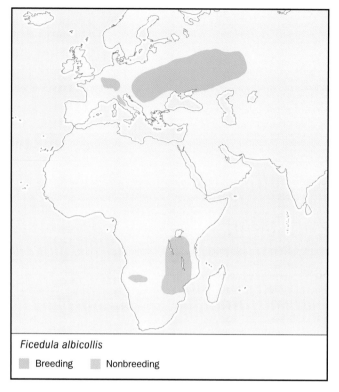

Ficedula albicollis

☐ Breeding ☐ Nonbreeding

PHYSICAL CHARACTERISTICS
The body length is 5 in (13–14 cm). The male is colored strikingly black-and-white, with a black back and head, a white belly and throat, and white wing-flashes. The female and juvenile are gray above, with a white belly and throat, and white wing-flashes. The winter male is colored more grayish.

DISTRIBUTION
Breeds widely in northern and central Europe and European Russia, and winters in western equatorial Africa.

HABITAT
Breeds in temperate forest, woods, parks, orchards, and gardens.

BEHAVIOR
A migratory species. Pairs of breeding birds defend a territory. Winters as single birds. Often raises its fanned tail while perched. The song is delivered from a prominent perch, and is a varied series of loud trills.

FEEDING ECOLOGY AND DIET
Catches flying insects by an aerial sally from a prominent perch. Usually returns persistently to the same perch.

REPRODUCTIVE BIOLOGY
Nests in a tree-cavity, and also uses nest-boxes. Lays five to seven blue eggs.

CONSERVATION STATUS
Not threatened. A widespread and locally abundant species.

SIGNIFICANCE TO HUMANS
None known, except for the economic benefits of bird-watching. ◆

Collared flycatcher
Ficedula albicollis

SUBFAMILY
Muscicapinae

TAXONOMY
Ficedula albicollis Temminck, 1795.

OTHER COMMON NAMES
English: Half-collared flycatcher; semicollared flycatcher; French: Gobemouche à collier; German: Halsbandschnäpper; Spanish: Papamoscas Collarino.

PHYSICAL CHARACTERISTICS
The body length is about 4.5 in (11.5 cm), with the male colored black-and-white, with a black back and head, an intervening white collar at the nape, a white belly and throat, and white wing-flashes. The female and juvenile are gray with a white belly and throat and white wing-flashes. The winter male is colored more grayish. Various subspecies have been described based on plumage and song characters.

DISTRIBUTION
Breeds widely in Europe, and winters in eastern equatorial Africa.

HABITAT
Breeds in temperate forest, woods, parks, orchards, and gardens.

BEHAVIOR
A migratory species. Pairs of breeding birds defend a territory. Winters as single birds. Often raises its fanned tail while perched. The song is delivered from a prominent perch, and is a varied series of soft trills.

FEEDING ECOLOGY AND DIET
Catches flying insects by an aerial sally from a prominent perch. Usually returns persistently to the same perch.

REPRODUCTIVE BIOLOGY
Nests in a tree-cavity or holes in rock walls, and also uses nest-boxes. Lays five to seven blue eggs.

CONSERVATION STATUS
Not threatened. A widespread and locally abundant species.

SIGNIFICANCE TO HUMANS
None known, except for the economic benefits of bird-watching. ◆

Little slaty flycatcher
Ficedula basilanica

SUBFAMILY
Muscicapinae

TAXONOMY
Ficedula basilanica Sharpe, 1877.

OTHER COMMON NAMES
French: Gobemouche de Basilan; German: Schiefergrund-schnäpper; Spanish: Papamoscas Pizarro Chico.

PHYSICAL CHARACTERISTICS
The body length is about 5 in (12 cm). The male is colored slate-gray on the back, head, and tail and has a white belly. The female is more brownish.

DISTRIBUTION
Vulnerable. A rare, endemic, nonmigratory species that only occurs on the islands of Samar, Leyte, Dinagat, Mindanao, and Basilan in the Philippine archipelago.

HABITAT
Occurs in lowland, humid, evergreen, primary and mature secondary tropical forest, including selectively logged areas. It selects microhabitat within the dense forest understorey. It occurs over the altitudinal range of sea level to 3,900 ft (0–1,200 m).

BEHAVIOR
A nonmigratory species. Pairs of breeding birds defend a territory. A quiet, skulking species of the forest understory. The song is a series of warbling phrases.

FEEDING ECOLOGY AND DIET
Searches from a low perch for flying insects.

REPRODUCTIVE BIOLOGY
Builds a cup-shaped nest.

CONSERVATION STATUS
Vulnerable. A rare and declining species. Much of its original natural habitat has been lost to the development of subsistence agriculture, commercial plantations, logging, mining, and in one place, a golf course. Some of its breeding habitat occurs in protected areas, but is not necessarily safe from disturbance there. Areas of its critical breeding habitat must be protected.

SIGNIFICANCE TO HUMANS
None known, except for the economic benefits of bird-watching. ◆

Brown-chested jungle-flycatcher
Rhinomyias brunneata

SUBFAMILY
Muscicapinae

TAXONOMY
Rhinomyias brunneata Slater, 1897.

OTHER COMMON NAMES
English: Brown-chested flycatcher, Chinese olive flycatcher; French: Gobemouche à poitrine brune; German: Weisskehl-Dschungelschnäpper; Spanish: Papamoscas Selvático de Gargantilla Blanca.

Ficedula basilanica
  Resident

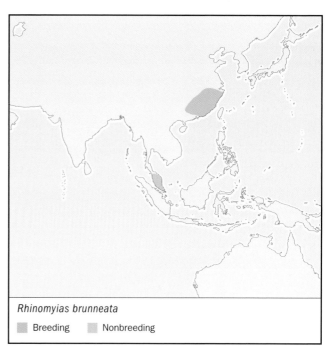

Rhinomyias brunneata
  Breeding   Nonbreeding

PHYSICAL CHARACTERISTICS
The body length is about 6 in (15 cm). The sexes are colored similarly, with a brown back, wings, and tail, a brown-buff breast, white throat, and tan eye-ring.

DISTRIBUTION
A migratory species that breeds in southeastern China and winters on the Malay Peninsula. It migrates through Thailand and perhaps the Nicobar Islands, and winters in parts of Malaysia and Singapore, and possibly parts of the islands of Borneo and Sumatra.

HABITAT
Breeds and winters in dense stands of bamboo and shrubs within subtropical, broadleaf, evergreen forest, and mature second-growth forest. Occurs within an altitudinal range of 2,000–3,900 ft (600–1,200 m).

BEHAVIOR
A migratory species. Pairs of breeding birds defend a territory. The song consists of a series of piping calls.

FEEDING ECOLOGY AND DIET
Searches from a perch for flying insects in the forest canopy.

REPRODUCTIVE BIOLOGY
Builds a cup-shaped nest in a shallow tree-cavity or at a narrow branch-fork.

CONSERVATION STATUS
Vulnerable. A rare and declining species because of widespread loss and fragmentation of its critical habitat. Its critical breeding, migratory, and wintering habitats must be identified and protected.

SIGNIFICANCE TO HUMANS
None known, except for the economic benefits of bird-watching. ◆

Rhinomyias olivacea

▨ Resident

Fulvous-chested jungle-flycatcher
Rhinomyias olivacea

SUBFAMILY
Muscicapinae

TAXONOMY
Rhinomyias olivacea Hume, 1877.

OTHER COMMON NAMES
English: Olive-backed jungle-flycatcher; French: Gobemouche à dos olive; German: Olivrücken-Dschungelschnàpper; Spanish: Papamoscas Selvático de Lomo Olivo.

PHYSICAL CHARACTERISTICS
The body length is about 6 in (15 cm). The sexes are colored similarly, with a brownish gray back, white throat and belly, tawny chest, and rufus tail.

DISTRIBUTION
A widespread species of the Malay Peninsula, parts of the island of Borneo, and the Greater Sunda Islands of Indonesia, including Sumatra, Java, and Bali.

HABITAT
Occurs at forest edges, in secondary forest, and in plantations. Generally occurs below 3,950 ft (1,200 m).

BEHAVIOR
A nonmigratory species. Pairs of breeding birds defend a territory. The song is a series of seven to nine simple notes.

FEEDING ECOLOGY AND DIET
Forages for insects on or near the ground, generally in the lower canopy. Hunts from a perch.

REPRODUCTIVE BIOLOGY
Builds a cup-shaped nest in a shallow cavity or tree-hole.

CONSERVATION STATUS
Not threatened. A widespread and locally abundant species.

SIGNIFICANCE TO HUMANS
None known, except for the economic benefits of bird-watching. ◆

Grand Comoro flycatcher
Humblotia flavirostris

SUBFAMILY
Muscicapinae

TAXONOMY
Humblotia flavirostris Milne-Edwards and Oustalet, 1885.

OTHER COMMON NAMES
English: Humblot's flycatcher; French: Gobemouche des Comores; German: Humblotschnäpper; Spanish: Papamoscas de Humblot.

PHYSICAL CHARACTERISTICS
The body length is 5.5 in (14 cm). The sexes are colored similarly, with a brown back, wings, and tail, a white breast heavily streaked with brown, and a lighter streaked throat.

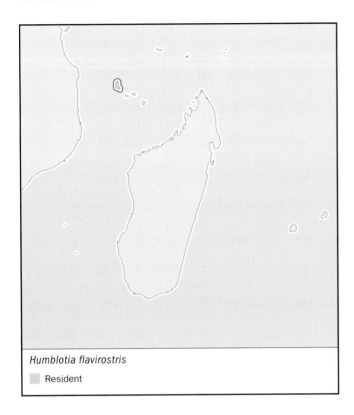

Humblotia flavirostris

■ Resident

DISTRIBUTION
A rare, endemic, nonmigratory species that occurs only in the Comoro Islands, in the Indian Ocean just north of Madagascar.

HABITAT
Only occurs in natural tropical forest on the slopes of Mount Karthala, a periodically erupting volcano on Grand Comoro Island. Occurs within an altitudinal range of 2,600–6,600 ft (800–2,000 m).

BEHAVIOR
A nonmigratory species. Pairs of breeding birds defend a territory. The song is a sharp trill.

FEEDING ECOLOGY AND DIET
Searches from a perch for flying insects in low parts of the forest canopy.

REPRODUCTIVE BIOLOGY
Builds a cup-shaped nest in a relatively tall tree.

CONSERVATION STATUS
Endangered. An extremely local and rare species that only breeds on the slopes of a single, active volcano. Its population is declining because some of its habitat is being lost to the development of subsistence agriculture, commercial plantations, logging, and invasion of disturbed areas by non-native shrubs. Its critical breeding habitat must be protected.

SIGNIFICANCE TO HUMANS
None known, except for the economic benefits of bird-watching. ◆

Dull-blue flycatcher
Eumyias sordida

SUBFAMILY
Muscicapinae

TAXONOMY
Eumyias sordida Walden, 1870.

OTHER COMMON NAMES
English: Dusky-blue flycatcher; French: Gobemouche de Ceylan; German: Ceylonschnäpper; Spanish: Papamoscas de Sri Lanka.

PHYSICAL CHARACTERISTICS
The body length is about 6 in (15 cm). The sexes are colored alike, having an ashy-blue back, head, chest, and tail and a whiter belly. The forehead is a brighter blue color, and a black line runs forward from the eyes. The juvenile is more spotted.

DISTRIBUTION
A resident species occurring only in Sri Lanka.

HABITAT
Occurs at the edges of forest and woods and in shady gardens. Occurs at various altitudes, but most abundant at 3,950–6,000 ft (1,200–1,830 m).

BEHAVIOR
A nonmigratory species. Pairs of breeding birds defend a territory. The song is a series of six to eight clear notes.

FEEDING ECOLOGY AND DIET
Catches flying insects by an aerial sally from a low perch. Often returns to the same perch.

REPRODUCTIVE BIOLOGY
Nests in forest or woods, where a cup-shaped nest made of mosses and other fibers is placed close to the ground on a ledge, in a rocky crevice, or in a shallow tree-cavity.

Eumyias sordida

■ Resident

CONSERVATION STATUS
An endemic species of Sri Lanka, but locally abundant and not threatened according to IUCN criteria. However, its forest habitat is rapidly being lost.

SIGNIFICANCE TO HUMANS
None known, except for the economic benefits of bird-watching. ◆

Large niltava
Niltava grandis

SUBFAMILY
Muscicapinae

TAXONOMY
Niltava grandis Blyth, 1842.

OTHER COMMON NAMES
English: Great Niltava; French: Grand gobemouche; German: Kobaltniltava; Spanish: Niltava grande.

PHYSICAL CHARACTERISTICS
The body length is about 8 in (20 cm). The sexes are colored differently. The male has a dark-blue back and a brilliant-blue top of the head, a black mask through the eyes, a black throat, and a shining orange-rufous belly. The female is rather uniformly olive-brown with a lighter belly, and the juvenile is light-brown with whitish spots.

DISTRIBUTION
A species of the Himalayan region of Nepal, Bhutan, and western India, and extending through the mountains of Myanmar, Thailand, Cambodia, and southwest China and Southeast Asia, including the Malay Peninsula and Sumatra.

Niltava grandis
☐ Resident

HABITAT
Occurs in humid, dense, broad-leafed forest, often near a stream. Mostly breeds at altitudes of 5,000–9,350 ft (1,525–2,850 m).

BEHAVIOR
A nonmigratory species, but undertakes seasonal movements between high- and low-elevation habitats. Pairs of breeding birds defend a territory. Not highly active, and often sits on perch for rather long periods without feeding. The song is a series of three or four simple, rising whistles.

FEEDING ECOLOGY AND DIET
Catches flying insects in the middle part of the canopy by an aerial sally from a perch. Often returns to the same perch. Also eats small fruits.

REPRODUCTIVE BIOLOGY
Nests in forest. Builds a cup-shaped nest of mosses and other fibers in a mossy area among rocks on the ground, against a tree-trunk, or in a shallow cavity in a rotten stump.

CONSERVATION STATUS
A rare, endemic species of the western Himalayas. However, it may be locally abundant and is not threatened according to IUCN criteria. Nevertheless, its forest habitat is rapidly being lost and the species should be closely monitored.

SIGNIFICANCE TO HUMANS
None known, except for the economic benefits of bird-watching. ◆

Orange-breasted blue flycatcher
Cyornis tickelliae

SUBFAMILY
Muscicapinae

TAXONOMY
Cyornis tickelliae Blyth, 1843.

OTHER COMMON NAMES
English: Tickell's blue flycatcher; French: Gobemouche de Tickell; German: Braunbrust-Blauschnäpper; Spanish: Niltava de Tickell.

PHYSICAL CHARACTERISTICS
The body length is about 5.5 in (14 cm). The sexes are dimorphic. The male has a blue back and top of the head, a black mask around the eyes, a rufus chest, and a white belly. The female is colored a more subdued blue above, with a white throat and belly and rufus-washed chest. The juvenile is brown with whitish spots and bluish wings and tail. However, some of these colors and patterns differ among geographic races.

DISTRIBUTION
A widespread species of the Indian subcontinent, occurring in India, southern Nepal, and Sri Lanka.

HABITAT
Occurs in open, dry forest and woodland, and also in well-vegetated gardens.

BEHAVIOR
A nonmigratory species. Pairs of breeding birds defend a territory. Flits actively in the canopy. The song is a metallic trill of six to 10 notes.

Cyornis tickelliae
▢ Resident

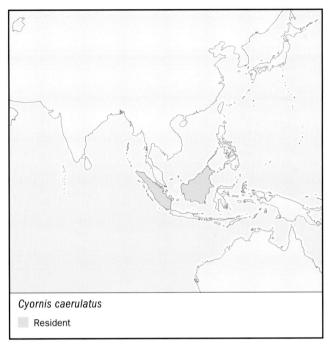

Cyornis caerulatus
▢ Resident

FEEDING ECOLOGY AND DIET
Catches flying insects within the forest canopy, by an aerial sally from a perch.

REPRODUCTIVE BIOLOGY
Builds a cup-shaped nest of mosses on a rocky ledge, in a tree-hole, or among tangles of tree-roots. Nests within about 6 ft (2 m) of the forest floor.

CONSERVATION STATUS
Not threatened. A widespread and locally abundant species.

SIGNIFICANCE TO HUMANS
None known, except for the economic benefits of bird-watching. ◆

Large-billed blue-flycatcher
Cyornis caerulatus

SUBFAMILY
Muscicapinae

TAXONOMY
Cyornis caerulatus Bonaparte, 1857.

OTHER COMMON NAMES
French: Bobemouche à grand bec; German: Breitschnabel-Blauschnäpper; Spanish: Niltava Picuda.

PHYSICAL CHARACTERISTICS
The body length is about 5.5 in (14 cm). The sexes are colored differently. The male has a dark-blue back, tail, and head, and an orange belly and throat. The female is brown above with a blue tail and rump.

DISTRIBUTION
A rare, endemic, nonmigratory species that only occurs on the Indonesian islands of Borneo and Sumatra.

HABITAT
Inhabits humid lowland and mid-slope, evergreen, tropical forest. It occurs in densely vegetated habitats within primary, mature secondary, and selectively logged forest.

BEHAVIOR
A nonmigratory species. Pairs of breeding birds defend a territory.

FEEDING ECOLOGY AND DIET
Searches from a perch for flying insects in middle and upper parts of the canopy.

REPRODUCTIVE BIOLOGY
Builds a cup-shaped nest in a relatively tall tree.

CONSERVATION STATUS
Vulnerable. An increasingly rare and declining species because much of its habitat has been lost to the development of subsistence agriculture, commercial plantations, and logging. Some of its breeding habitat occurs in various protected areas, but these places are still being subjected to commercial logging. Much of its habitat has been degraded by extensive, illegal fires started to clear the natural forest for agricultural use. Areas of its critical breeding habitat must be protected.

SIGNIFICANCE TO HUMANS
None known, except for the economic benefits of bird-watching. ◆

Fraser's forest-flycatcher
Fraseria ocreata

SUBFAMILY
Muscicapinae

TAXONOMY
Fraseria ocreata Strickland, 1844.

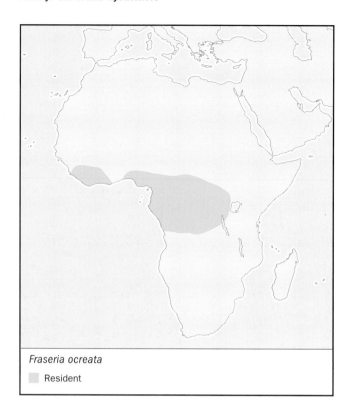

Fraseria ocreata
▪ Resident

OTHER COMMON NAMES
English: African forest-flycatcher; French: Gobemouche forestier; German: Waldschnäpper; Spanish: Papamoscas del Bosque.

PHYSICAL CHARACTERISTICS
The body length is about 6 in (15 cm). The sexes are colored similarly, with upperparts dark-gray, the tail feathers black, and the underparts white with dark, dense, scalloped markings. The juvenile is more brown with some spotting.

DISTRIBUTION
A widespread species of tropical western and central Africa.

HABITAT
Occurs in primary and mature secondary, moist, tropical forest from the lowlands to as high as 5,250 ft (1,600 m). Inhabits stands with a high, thick canopy and often occurs in the vicinity of a forest watercourse.

BEHAVIOR
A nonmigratory species. A relatively social species, which may occur in groups of as many as 30 individuals. It may breed in a semi-colonial manner. It also breeds as isolated, monogamous pairs, but groups may be polygamous. Breeding pairs are often accompanied by their young of previous years, which may help the parents with raising their brood. The family and social groups also react collectively to danger. The song is a melodious series of two to seven notes.

FEEDING ECOLOGY AND DIET
Forages actively for insects, which are gleaned from foliage. Also fly-catches for aerial prey. A noisy, active, gregarious bird that frequently flits its tail and partially spreads its wings while hunting.

REPRODUCTIVE BIOLOGY
Builds a bulky nest with an internal cup. The nest is built by various members of the family group. Lays two or three eggs, which are incubated by the female. The young are raised by the parents plus related helpers.

CONSERVATION STATUS
Not threatened. A widespread and locally abundant species.

SIGNIFICANCE TO HUMANS
None known, except for the economic benefits of birdwatching. ◆

Abyssinian slaty flycatcher
Melaenornis chocolatinus

SUBFAMILY
Muscicapinae

TAXONOMY
Melaenornis chocolatina Rüppell, 1840.

OTHER COMMON NAMES
English: Chocolate flycatcher; slaty flycatcher; French: Gobemouche chocolat; German: Habeschdrongoschnäpper; Spanish: Papamoscas Etíope.

PHYSICAL CHARACTERISTICS
The body length is about 5.5 in (15 cm). The sexes are colored similarly, with a dark-brown back, grayer belly, light under the rump, and buff on breast.

DISTRIBUTION
An endemic (or local) species of the highlands of Ethiopia and Eritrea.

HABITAT
Occurs in a range of types of humid, highland forest and woods and coffee plantations as high as 8,200 ft (2,500 m).

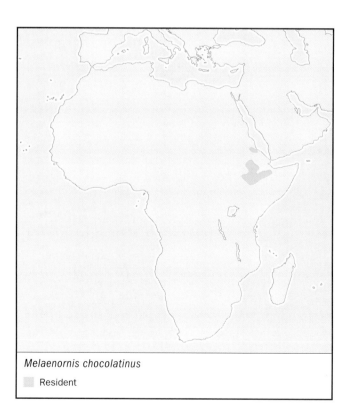

Melaenornis chocolatinus
▪ Resident

BEHAVIOR
A nonmigratory species. Pairs of breeding birds defend a territory. Sometimes wags its tail while perched. The song is a simple, high-pitched phrase of three or four notes.

FEEDING ECOLOGY AND DIET
Hunts from a perch in the forest canopy for flying insects.

REPRODUCTIVE BIOLOGY
Builds a cup-shaped nest at a narrow fork of a horizontal tree branch. Lays three blue-gray, blotched eggs.

CONSERVATION STATUS
Not threatened. A rare endemic species but locally abundant.

SIGNIFICANCE TO HUMANS
None known, except for the economic benefits of birdwatching. ◆

Silverbird
Empidornis semipartitus

SUBFAMILY
Muscicapinae

TAXONOMY
Empidornis semipartita Rüppell, 1840.

OTHER COMMON NAMES
French: Gobemouche argenté; German: Silberschnäpper; Spanish: Papamoscas Plateado.

PHYSICAL CHARACTERISTICS
The body length is about 7.5 in (18 cm). The sexes are colored similarly, with a light silvery grayish blue back, bright rufus

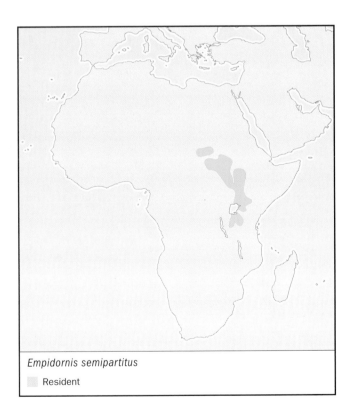

Empidornis semipartitus
☐ Resident

underparts, wings gray above and orange beneath, and silvery markings on the head and tail.

DISTRIBUTION
An endemic (or local) species of the highlands of east-central Africa, including parts of Ethiopia, Sudan, Tanzania, Uganda, and Kenya.

HABITAT
Occurs in dry and semi-arid forest and woods with scattered large trees, especially acacias, in areas as high as 9,050 ft (2,300 m).

BEHAVIOR
A nonmigratory species. Pairs of breeding birds defend a territory. The song is a soft, rich, and warbling.

FEEDING ECOLOGY AND DIET
Hunts from a perch in the canopy for flying insects. Also swoops down to take insects from the ground. Usually returns to its original perch.

REPRODUCTIVE BIOLOGY
Builds a dome-shaped nest of grass and thorny twigs lined with finer fibers, or may used an old nest of a weaver-finch. Lays two or three olive-green eggs.

CONSERVATION STATUS
Not threatened. An endemic species that is locally abundant in its range.

SIGNIFICANCE TO HUMANS
None known, except for the economic benefits of birdwatching. ◆

Red-tailed newtonia
Newtonia fanovanae

SUBFAMILY
Muscicapinae

TAXONOMY
Newtonia fanovanae Gyldenstolpe, 1933.

OTHER COMMON NAMES
English: Fanovana newtonia, fanovana warbler; French: Newtonie à queue rouge; German: Fanovana-Newtonie; Spanish: Papamoscas de Newton.

PHYSICAL CHARACTERISTICS
The body length is about 5 in (12 cm). The back and head are olive-brown, the tail rufous, and the belly white.

DISTRIBUTION
A rare, endemic, nonmigratory species that only occurs on the island of Madagascar.

HABITAT
Occurs in lowland, humid, evergreen tropical forest. It occurs only in large tracts of intact forest, over the altitudinal range of 330–2,950 ft (100–900 m).

BEHAVIOR
A nonmigratory species. Pairs of breeding birds defend a territory. The song is a descending series of notes.

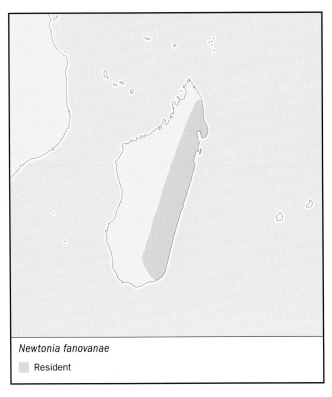

Newtonia fanovanae

Resident

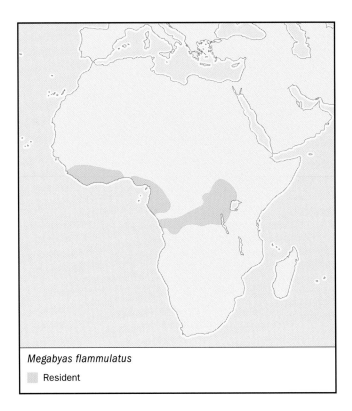

Megabyas flammulatus

Resident

FEEDING ECOLOGY AND DIET
Searches from a perch for flying insects in the middle and up-per parts of the forest canopy. Often associates with mixed-species foraging flocks.

REPRODUCTIVE BIOLOGY
Builds a cup-shaped nest in a relatively tall tree.

CONSERVATION STATUS
Vulnerable. A rare and declining species because much of its habitat has been lost to the development of subsistence agricul-ture, and to logging in some areas. Some of its breeding habi-tat occurs in various protected areas, but it is still at risk of disturbance there. Areas of its critical breeding habitat must be well protected.

SIGNIFICANCE TO HUMANS
None known, except for the economic benefits of bird-watching. ◆

Shrike-flycatcher
Megabyas flammulatus

SUBFAMILY
Platysteirinae

TAXONOMY
Megabyas flammulata Verreaux, 1855.

OTHER COMMON NAMES
English: African shrike-flycatcher; French: Gobemouche écorcheur; German: Schnäpperwürger; Spanish: Atrapamoscas Africano.

PHYSICAL CHARACTERISTICS
The body length is about 6 in (15 cm). The iris is bright or-ange. The male is colored glossy black on the head, back, wings, and tail, and white on the rump, underparts, and under-wings. The female and immature are brownish, and their breast is brown-streaked on white.

DISTRIBUTION
A widespread, nonmigratory species of central tropical Africa.

HABITAT
Occurs in lowland, humid, primary and mature secondary trop-ical forest, as well as the forested edges of clearings. It occurs as high as about 7,000 ft (2,150 m)

BEHAVIOR
A nonmigratory species that occurs in pairs or small groups. Breeding birds defend a territory. Sits quietly on a perch, often swinging its tail slowly sideways. The song is a sustained series of repetitive phrases.

FEEDING ECOLOGY AND DIET
Searches actively or from a perch for flying insects in the lower part of the forest canopy. Insects are also gleaned from foliage. Sometimes joins mixed-species foraging flocks.

REPRODUCTIVE BIOLOGY
Builds a small cup-shaped nest in a narrow fork of a branch. Lays two or three, greenish-gray, mottled eggs that are incu-bated for at least 16 days. Pairs are monogamous but are prob-ably helped with their breeding by their immature progeny.

CONSERVATION STATUS
Not threatened. An endemic species that is locally abundant in parts of its range.

SIGNIFICANCE TO HUMANS
None known, except for the economic benefits of bird-watching. ◆

Black-and-white flycatcher
Bias musicus

SUBFAMILY
Platysteirinae

TAXONOMY
Bias musicus Vieillot, 1818.

OTHER COMMON NAMES
English: Black-and-white shrike-flycatcher; French: Gobe-mouche chanteur; German: Vangaschnäpper; Spanish: Atrapamoscas Blanco y negro.

PHYSICAL CHARACTERISTICS
The body length is about 6 in (15 cm). The iris is bright yellow. Both sexes have a feathered crest on the top of the head. The male is colored glossy black on the head, back, wings, tail, and chest, and white on the rump and lower underparts. The female and immature are mostly brown, with a white throat and tan-colored chest and belly.

DISTRIBUTION
A widespread, nonmigratory species of central tropical Africa.

HABITAT
Occurs in large openings in lowland, humid, primary and secondary tropical and montane forest, including agricultural and village clearings with some tall trees. It occurs as high as about 5,600 ft (1,700 m)

BEHAVIOR
A nonmigratory species that occurs in pairs or as small family groups. Breeding birds defend a territory. Performs slow-flying, noisy, aerial displays. The song is a series of two to four varied notes.

FEEDING ECOLOGY AND DIET
Searches actively or from a perch for insects in the upper part of the tree canopy. Insects are gleaned from foliage, and are also caught in flight.

REPRODUCTIVE BIOLOGY
Builds a small cup-shaped nest in a narrow fork of a branch. Lays two or three, pale blue-green, blotched eggs that are incubated for 18–19 days. Pairs are monogamous but their immature progeny help them with their breeding effort.

CONSERVATION STATUS
Not threatened. An endemic species but locally abundant in parts of its range.

SIGNIFICANCE TO HUMANS
None known, except for the economic benefits of birdwatching. ◆

Chestnut wattle-eye
Dyaphorophyia castanea

SUBFAMILY
Platysteirinae

TAXONOMY
Dyaphorophyia castanea Fraser, 1843.

OTHER COMMON NAMES
French: Gobemouche caronculé châtain; German: Weissbürzel-Lappenschnäpper; Spanish: Ojicarunculado Castaño.

PHYSICAL CHARACTERISTICS
The body length is about 4 in (10–11 cm). The head is relatively large, the tail extremely short, and the wings short and

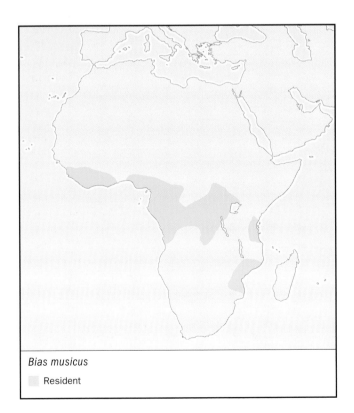

Bias musicus
■ Resident

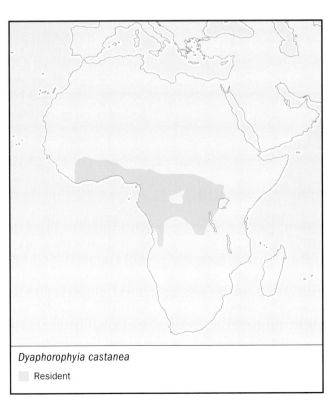

Dyaphorophyia castanea
■ Resident

rounded. The iris is dark brown, and there is a gray patch of bare skin (an eye-wattle) around the eye. The male is colored glossy black above, with a white rump and undersides except for a black band across the breast. The female is a duller brown-black with a gray head, white chin and belly, and sides of head and chest chestnut.

DISTRIBUTION
A widespread, nonmigratory species of tropical, central, western Africa.

HABITAT
Occurs in lowland, humid, primary and mature secondary tropical and montane forest, including flooded forest. It occurs in relatively shrubby and liana-dense habitats. It occurs as high as about 5,900 ft (1,800 m)

BEHAVIOR
A nonmigratory species that occurs in pairs or as small family groups. Breeding birds defend a territory. The song is a series of simple notes.

FEEDING ECOLOGY AND DIET
Searches actively or from a perch for insects in the lower canopy. Insects are gleaned from foliage and also caught in flight.

REPRODUCTIVE BIOLOGY
Builds a small cup-shaped nest in a fork of a branch. Lays one or two, glossy blue-green eggs that are incubated by the female for 17 days. Pairs are monogamous but their immature progeny help them with their breeding effort.

CONSERVATION STATUS
Not threatened. An endemic species that is locally abundant in parts of its range.

SIGNIFICANCE TO HUMANS
None known, except for the economic benefits of bird-watching. ◆

Black-throated wattle-eye
Platysteira peltata

SUBFAMILY
Platysteirinae

TAXONOMY
Platysteira peltata Sundevall, 1850.

OTHER COMMON NAMES
English: Wattle-eyed flycatcher; French: Gobemouche caronculé à gorge noire; German: Schwarzkehl-Lappenschnäpper; Spanish: Ojicarunculado de Garganta Negra.

PHYSICAL CHARACTERISTICS
The body length is about 5.5 in (14 cm). The head is relatively large, the tail short, and the wings rounded. There is a red patch of bare skin (an eye-wattle) over the eye. The male is colored glossy green-black on the head, with a dark gray back, and white belly. The female is a duller color and has a dark ring across the breast.

DISTRIBUTION
A widespread, nonmigratory species of southern subtropical Africa.

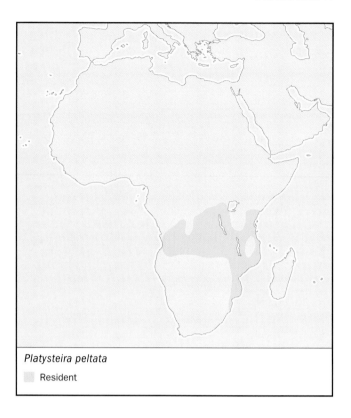

Platysteira peltata

▨ Resident

HABITAT
Occurs in primary and mature secondary lowland and montane forest, often in the vicinity of surface water. It occurs at 5,600–9,900 ft (1,700–3,000 m)

BEHAVIOR
A nonmigratory species that occurs in pairs or as small family groups. Breeding birds defend a territory. The song is a rasping series of notes.

FEEDING ECOLOGY AND DIET
Searches actively or from a perch for insects in the lower canopy and sometimes higher. Insects are gleaned from foliage and also caught in flight.

REPRODUCTIVE BIOLOGY
Builds a small cup-shaped nest in a narrow fork of a branch. Lays one or two, glossy gray-green eggs that are incubated by the female for 16–18 days. Pairs are monogamous but their immature progeny help them with their breeding effort.

CONSERVATION STATUS
Not threatened. An endemic species but locally abundant in parts of its range.

SIGNIFICANCE TO HUMANS
None known, except for the economic benefits of bird-watching. ◆

Cape batis
Batis capensis

SUBFAMILY
Platysteirinae

TAXONOMY
Batis capensis Linnaeus, 1766.

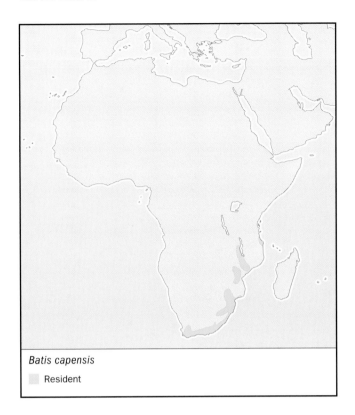

Batis capensis
Resident

black top of head, white throat, black breast-band, white belly, and rufus flanks. The female is a duller color, with no breast-band, and has a brown wash to the breast.

DISTRIBUTION
A local (or endemic) species of coastal southern Africa.

HABITAT
Occurs in primary and mature secondary woodland, montane forest, shrubby scrub, plantations, and gardens with trees. It occurs between sea level and about 7,050 ft (2,150 m).

BEHAVIOR
A nonmigratory species that occurs in pairs or as small family groups. Some populations undertake seasonal altitudinal movements. Breeding birds defend a territory. Sometimes aggregates into a larger, excitable flock of 10–30 birds known as a "batis parliament," and may also join mixed-species foraging flocks. The song is a simple whistle or series of notes.

FEEDING ECOLOGY AND DIET
Searches actively or from a perch for insects at all levels of the canopy. Insects are mostly caught in flight, often after they are scared from a hiding place.

REPRODUCTIVE BIOLOGY
Builds a small cup-shaped nest in a narrow fork of a branch. Lays one to three eggs that are incubated by the female for 17–21 days. Pairs are monogamous but their immature progeny help them with their breeding effort.

CONSERVATION STATUS
Not threatened. An endemic species that is locally abundant in parts of its range.

SIGNIFICANCE TO HUMANS
None known, except for the economic benefits of bird-watching. ◆

OTHER COMMON NAMES
English: Cape puffback; French: Batis du Cap; German: Kap-schnäpper; Spanish: Batis de el Cabo.

PHYSICAL CHARACTERISTICS
The body length is about 6 in (15 cm). The head is relatively large, the tail short, and the wings rounded. The iris is orange. The male is colored dark blue-gray on the back and tail, with a

Resources

Books

BirdLife International. *Threatened Birds of the World.* Barcelona and Cambridge: Lynx Edicions and BirdLife International, 2000.

Cramp, S., and C. M. Perrins, eds. *Handbook of the Birds of Europe, the Middle East, and North Africa: The Birds of the Western Palaearctic: Old World Flycatchers to Shrikes.* New York: Oxford Press, 1993.

Urban, E. K., C. H. Fry, and S. Keith. *The Birds of Africa.* Vol. 5. London: Academic Press.

Organizations

BirdLife International. Wellbrook Court, Girton Road, Cambridge, Cambridgeshire CB3 0NA United Kingdom. Phone: +44 1 223 277 318. Fax: +44-1-223-277-200. E-mail: birdlife@birdlife.org.uk Web site: <http://www.birdlife.net>

IUCN–The World Conservation Union. Rue Mauverney 28, Gland, 1196 Switzerland. Phone: +41-22-999-0001. Fax: +41-22-999-0025. E-mail: mail@hq.iucn.org Web site: <http://www.iucn.org>

Bill Freedman, PhD

Australian fairy-wrens
(Maluridae)

Class Aves
Order Passeriformes
Suborder Passeri (Oscines)
Family Maluridae

Thumbnail description
Small insectivores with long cocked tails; females and young are cryptically colored but breeding-plumage males have gaudy plumage

Size
5.5–8.6in (14–22cm); 0.27–1.2oz (7.6–34.1g)

Number of genera, species
6 genera; 30 species

Habitat
Shrubbery and undergrowth in arid woodlands and forests, though some inhabit rainforests and many species thrive in suburbs, parks, and gardens

Conservation status
No species are currently Endangered but many have localized distributions threatened by habitat loss

Distribution
Found throughout Australia and New Guinea, and associated islands

Evolution and systematics

Fairy-wrens are a distinct and divergent group of passerines, characterized by long, sometimes filamentous, tails, usually with 10 retrices. They are often subdivided into two subfamilies: grasswrens (Amytornithinae) and fairy-wrens (Malurinae), although recent DNA studies are incomplete and details of relationships among species and species groups are still problematic.

Physical characteristics

Grasswrens are cryptically colored shades of tan and brown with black-and-white markings, while fairy-wrens have breeding-plumaged males with bright blues, violets, purples, and russets; some species feature cheek patches of brilliant turquoise that can be extended to form a face fan during agonistic or nuptial displays. Long, cocked tails are displayed while moving and are characteristic of this family. Emu-wrens have long, filamentous tails, and like the other fairy-wrens are dimorphic in plumage. Grasswrens are nearly monomorphic in plumage, with subtle shades distinguishing the sexes: females are usually more russet below.

Distribution

Fairy-wrens are found throughout Australia and New Guinea. Some species have very restricted ranges, while others are distributed continent-wide. Emu-wrens and grasswrens are confined to Australia, while fairy-wrens are found in New Guinea as well. The monotypic genera *Sipodotus* and *Clytomyias* are confined to New Guinea and outlying islands.

Habitat

Grasswrens tend to be birds of spinifex and porcupine grass of the arid interior, and have very limited geographic distribution, reflecting perhaps relic populations isolated by increasingly arid conditions, eventually evolving to become separate species. The emu-wrens occupy a variety of habitats, with the southern emu-wren found in swampy heath and plains thickets of southern Australian coastal belts, while the other two species occupy the arid interior. The fairy-wrens occupy a range of habitats, including tropical grasslands, wet forests and woodlands, and the semi-arid interior. Several species have adapted well to humans and grace the parks and gardens of suburbia.

Behavior

Most malurids are found in family groups and tend to be territorial and sedentary, communicating among group members with a broad spectrum of melodious calls. They are busy foragers, climbing through dense undergrowth and hopping with cocked tails across open patches of ground.

Feeding ecology and diet

Most malurids are ground foragers, gleaning and pecking a wide variety of invertebrates from bare ground, litter, grass, and logs. They glean foliage, twigs, and bark, and occasionally hawk flying insects from the air. Some species are more specialized foragers; for example, the purple-crowned fairy-wren (*Malurus coronatus*), forages largely in pandanus along the edges of tropical streams, rivers, and ponds.

Reproductive biology

Most species are cooperative breeders with surviving progeny from previous years acting as helpers at the nest, or at least having delayed dispersal. Studies indicate high adult survival rates and abundant extra-pair copulations, with resident males often fathering a minority of the offspring produced in their territory. Nests are usually domed balls of woven grass with side entrances, with clutches of two to four red-spotted, white eggs. Incubation, usually by the female, lasts 10–14 days; young are fed for four to six weeks.

Conservation status

No species are currently threatened, but overgrazing and habitat alteration for agriculture and timber production are potential threats.

Significance to humans

None known, though many consider them among the most beautiful and endearing species of birds.

Mating behavior of splendid (black-backed) fairy-wrens (*Malurus splendens melanotus*) in dry gidgee (*Acacia cambagei*) woodland, southwest Queensland, Australia. (Photo by Wayne Lawler. Photo Researchers, Inc. Reproduced by permission.)

1. Variegated fairy-wren (*Malurus lamberti*); 2. Southern emu-wren (*Stipiturus malachurus*); 3. Female Wallace's wren (*Sipodotus wallacii*); 4. Red-backed fairy-wren (*Malurus melanocephalus*); 5. Purple-crowned fairy-wren (*Malurus coronatus*); 6. Orange-crowned wren (*Clytomyias insignis*); 7. Splendid fairy-wren (*Malurus splendens*); 8. Superb fairy-wren (*Malurus cyaneus*); 9. Black grasswren (*Amytornis housei*); 10. Female striated grasswren (*Amytornis striatus*). (Illustration by Joseph E. Trumpey)

Species accounts

Southern emu-wren
Stipiturus malachurus

TAXONOMY
Stipiturus malachurus Shaw, 1798, Sydney and Botany Bay, New South Wales, Australia. Eight subspecies.

OTHER COMMON NAMES
French: Queue-de-gaze du Sud; German: Rotstirn-Bortenschwanz; Spanish: Ratona Emu Sureña.

PHYSICAL CHARACTERISTICS
6.2–7 in (15.7–17.8 cm); female 0.26–0.29 oz (7.4–8.3 g), male 0.19–0.32 oz (5.5–9 g). Males in breeding plumage have blue throat and breast that females, nonbreeding males, and immatures lack.

DISTRIBUTION
Disjunct populations along coast from western Australia to southern Queensland. *S. m. malachurus*: from Queensland to Victoria; *S. m. littleri*: confined to Tasmania and islands; *S. m. polionotum*: from Victoria and south Australia; *S. m. intermedius, S. m. halmaturnius, S. m. parimedia*: local distribution in south Australia; *S. m. westernensis*: in southwestern Western Australia; and *S. m. hartogi*: found on Dirk Hartog Island, Shark Bay, Western Australia.

HABITAT
Occurs in swamps, dunes, and coastal and high-altitude heathlands. Prefers low, dense vegetation.

BEHAVIOR
Usually found in small groups; secretive; weak flier and difficult to flush. Social organization poorly known.

FEEDING ECOLOGY AND DIET
Feeding ecology poorly known, but thought to be largely insectivorous, gleaning invertebrates from dense vegetation and ground. Will split open stems to get at insects; occasionally hawks flying insects.

REPRODUCTIVE BIOLOGY
Monogamous pairs hold breeding territory; female builds nest but fed by male. Clutch is 2–4 red-spotted eggs. Nest parasitized by several cuckoo species. Incubation mostly by female for 13–14 days; fledging in 11–15 days.

CONSERVATION STATUS
Not threatened as a species but adversely affected by drainage of swamps and clearing for agriculture. Altered fire regimes also a threat.

SIGNIFICANCE TO HUMANS
None known. ◆

Purple-crowned fairy-wren
Malurus coronatus

TAXONOMY
Malurus coronatus Gould, 1858, Victoria River, Northern Territory, Australia. Two subspecies..

OTHER COMMON NAMES
English: Lilac-crowned wren; French: Mérion couronné German: Purpurkopf-Staffelschwanz; Spanish: Ratona Australiana de Corona Morada.

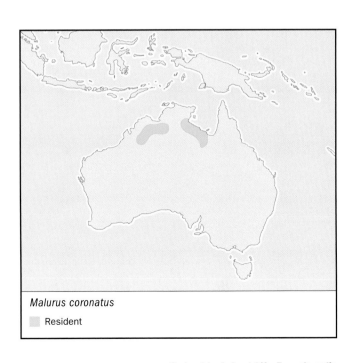

Stipiturus malachurus
■ Resident

Malurus coronatus
■ Resident

PHYSICAL CHARACTERISTICS

5.9 in (15 cm); female 0.31–0.44 oz (8.7–12.6 g), male 0.32–0.46 oz (9.2–13 g). Male in breeding plumage distinctive with purple crown bordered in black; female mostly buffy with chestnut face patch; nonbreeding males resemble females but have a dark face patch.

DISTRIBUTION

Restricted tropical range. *M. c. coronatus* found in northeastern Australia; *M. c. macgillivrayi* found in northwestern Australia.

HABITAT

Always close to rivers, streams, or permanent pools in thick vegetation, particularly pandanus leaves and litter.

BEHAVIOR

Territorial and monogamous; pairs advertise territory and reinforce pair-bond with dueting. Voice a high-pitched sequence of notes, as well as alarm and contact calls.

FEEDING ECOLOGY AND DIET

Glean mainly insects from dense vegetation along rivers or pond margins, and forage in leaf litter for worms and other invertebrates.

REPRODUCTIVE BIOLOGY

Monogamous and probably promiscuous. Maintain territory throughout year. Nest is domed and bulky. Clutch is two or three spotted, white eggs. Female incubates for about 14 days; fledging in about 10 days.

CONSERVATION STATUS

Not threatened but adversely affected by frequent fires and habitat loss.

SIGNIFICANCE TO HUMANS

None known. ◆

Red-backed fairy-wren
Malurus melanocephalus

TAXONOMY

Malurus melanocephalus Latham, 1801, central coast of New South Wales, Australia. Two subspecies.

OTHER COMMON NAMES

English: Red-backed wren; French: Mérion à dos rouge; German: Rotrücken-Staffelschwanz; Spanish: Ratona Australiana de Lomo Rojo.

PHYSICAL CHARACTERISTICS

5.7 in (14.5 cm); female 0.18–0.35 oz (5.0–10.0 g), male 0.21–0.35 oz (6.0–10.0 g). Breeding males are black with a bright red back. Nonbreeding males, females, and young drab brownish above, buff below.

DISTRIBUTION

Mostly tropical and along coastal belt. *M. m. cruentatus* across northern Australia; *M. m. melanocephalus* in northeastern Australia.

HABITAT

Tropical swamps, samphire flats, and woodlands with extensive grassy cover on coastal plains and adjacent mountains.

BEHAVIOR

Locally nomadic in nonbreeding season as preferred habitat of seasonal grass is subject to frequent fires; grazing and fires may inhibit formation of permanent territories.

Malurus melanocephalus

▨ Resident

FEEDING ECOLOGY AND DIET

Forage mainly by gleaning a broad spectrum of invertebrates from grass, leaves and twigs, but do take some fruit and seeds. Hop-search on open ground and hawk flying insect prey.

REPRODUCTIVE BIOLOGY

Territorial, cooperative breeders, but social system not well-understood. Female builds oval domed grass nest. Incubation is by the female for 13–14 days; fledging in 10–11 days.

CONSERVATION STATUS

Not threatened. Human habitat alteration may have actually increased available grassy habitat for this species.

SIGNIFICANCE TO HUMANS

None known. ◆

Splendid fairy-wren
Malurus splendens

TAXONOMY

Malurus splendens Quoy and Gaimard, 1830, King George Sound, Western Australia. Four subspecies.

OTHER COMMON NAMES

English: Splendid wren; French: Mérion splendide; German: Türkisstaffelschwanz; Spanish: Ratona Australiana Franjeada.

PHYSICAL CHARACTERISTICS

5.5 in (14 cm); female 0.27–0.36 oz (7.6–10.2 g), male 0.28–0.39 oz (7.9–11.1 g). Male in breeding plumage is brilliant blue with turquoise cheek patches and crown, black breast, face, and back markings. Nonbreeding males, females, and immatures drab olive above with blue tails and wings.

DISTRIBUTION

Populations scattered across Australia: *M. s. splendens* in the west, *M. s. musgravi* in the interior, *M. s. emmottorum* in Queensland, and *M. s. melanotus* in the east.

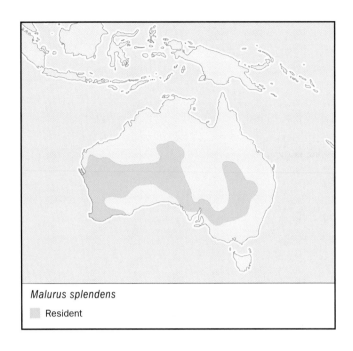

Malurus splendens

▨ Resident

Malurus cyaneus

▨ Resident

HABITAT
Mostly drier acacia woodlands and scrublands, including mulga, mallee, and saltbrush.

BEHAVIOR
Stronger fliers than most fairy-wrens, and more versatile foragers, often foraging in the canopy. A territorial, cooperative breeder, usually found in small groups that defend their territory. Voice a loud series of trills.

FEEDING ECOLOGY AND DIET
Insectivorous, mostly gleaning ants, grasshoppers, spiders, and insect larvae from the ground, litter, and foliage up to canopy height; also hop-search and pounce on prey, and may hawk flying insects.

REPRODUCTIVE BIOLOGY
Socially monogamous, but promiscuous, with males wandering into adjacent territories and often fathering less than half of the offspring from their own territory. Clutch is two to four red-spotted, white eggs. Female incubates for about two weeks, and fledging occurs in 10–13 days.

CONSERVATION STATUS
Not threatened but adversely affected by habitat destruction for agriculture and overgrazing.

SIGNIFICANCE TO HUMANS
None known. ◆

Superb fairy-wren
Malurus cyaneus

TAXONOMY
Malurus cyaneus Ellis, 1782, Adventure Bay, Tasmania. Six subspecies.

OTHER COMMON NAMES
English: Superb blue wren; French: Mérion superbe; German: Prachtstaffelschwanz; Spanish: Ratona Australiana Azul.

PHYSICAL CHARACTERISTICS
6.3 in (16 cm); female 0.28–0.35 oz (8–10 g), male 0.30–0.37 oz (8.5–10.5 g). Male in breeding plumage has brilliant turquoise crown, cheeks, and mantle, deep blue breast; other ages and sexes drab olive above, light gray below.

DISTRIBUTION
Restricted to southeastern Australia, usually to within 125 mi (200 km) of coast. The six subspecies recognized are *M. c. cyaneus*, *M. c. samueli*, *M. c. elizabethae*, *M. c. cyanochlamys*, *M. c. leggei*, *M. c. ashbyi*; many restricted to single islands.

HABITAT
Moist sclerophyll woodlands and forest with dense understory; also rainforest, swamps, coastal areas, river courses, and suburban gardens and parks.

BEHAVIOR
A cooperative breeding species, usually found in family groups of three to five birds, climbing through undergrowth and hopping across open ground, tail cocked. They are weak fliers. They maintain territories throughout the year. Calls consist of musical trills by both sexes.

FEEDING ECOLOGY AND DIET
Mainly insectivorous, eating ants, flies, weevils, grasshoppers, and insect larvae. Glean primarily ground, litter, and low foliage.

REPRODUCTIVE BIOLOGY
Socially monogamous and sexually promiscuous, with males wandering into adjacent territories, often carrying yellow flower petals to attract females; males may father less than half of the offspring produced in their territory. Clutch is three to four red-spotted, white eggs. Female incubates for two weeks; fledging in 10–14 days.

CONSERVATION STATUS
Not threatened but much native habitat has been converted to agriculture.

SIGNIFICANCE TO HUMANS
None known. ◆

Variegated fairy-wren
Malurus lamberti

TAXONOMY
Malurus lamberti Vigors and Horsefield, 1827, Sydney, New South Wales, Australia. Five subspecies.

OTHER COMMON NAMES
English: Variegated wren; French: Mérion de Lambert; German: Weissbauch-Staffelschwanz; Spanish: Ratona Australiana Variada.

PHYSICAL CHARACTERISTICS
5.5 in (14 cm); female 0.21–0.35 oz (5.9–10 g), male 0.21–0.41 oz (6.0–11.5 g). Breeding male with black throat and breast, russet back, and turquoise mantle, cap, and cheeks; white-tipped blue tail. Nonbreeding males, females, and immatures with blue heads and backs, white below.

DISTRIBUTION
Widely distributed across Australia. *M. l. assimilis* found throughout Australia except for southwest, southeast, and far north; *M. l. rogersi* in Kimberley Division, *M. l. dulcis* in Arnhem Land, *M. l. lamberti* in coastal southern Queensland and New South Wales, and *M. l. bernieri* is restricted to islands off Shark Bay in Western Australia.

HABITAT
A broad spectrum of habitats: in shrubby vegetation from coastal thickets through arid and semi-arid acacia woodlands and scrub; also, rocky escarpments and mallee.

BEHAVIOR
A weak flier that forages from ground to canopy by hopping. Group territorial defense throughout the year. Voice a fast, metallic trill.

FEEDING ECOLOGY AND DIET
Forages by gleaning on low shrubbery. Takes a broad spectrum of invertebrates including flies, caterpillars, grasshoppers, and spiders. May forage in canopy or on open ground.

REPRODUCTIVE BIOLOGY
Monogamous cooperative breeder, but probably promiscuous. Clutch is three or four red/brown-spotted, white eggs. Female incubates for 14–16 days; fledging in 10–12 days.

CONSERVATION STATUS
Widespread and not threatened, although adversely affected by clearing of habitat for agriculture, and by overgrazing.

SIGNIFICANCE TO HUMANS
None known. ◆

Black grasswren
Amytornis housei

TAXONOMY
Amytornis housei Milligan, 1902, central Kimberleys, Western Australia.

OTHER COMMON NAMES
French: Amytis noir; German: Schwarzkehl-Grasschlüpfer; Spanish: Ratona de la Hierba Negra.

PHYSICAL CHARACTERISTICS
8.3 in (21 cm); female 0.83–0.98 oz (23.5–27.9 g), male 1.0–1.1 oz (29.0–31 g). A large, dark grasswren, with rusty back and long, broad tail.

DISTRIBUTION
Rare and local in the Kimberley Division of northwestern Western Australia.

HABITAT
Found among tumbled sandstone outcrops and gorges, in spinifex and scrub.

BEHAVIOR
Poor fliers, and move about in groups by hopping among tussocks. Song is low-pitched and includes buzzing notes and trills.

Malurus lamberti
☐ Resident

Amytornis housei
☐ Resident

FEEDING ECOLOGY AND DIET
Forage mostly on ground for invertebrates and seeds of various grasses.

REPRODUCTIVE BIOLOGY
Breeding biology is poorly known.

CONSERVATION STATUS
Although not threatened by IUCN criteria, it is rare and local in distribution. May be threatened by frequent fires.

SIGNIFICANCE TO HUMANS
None known. ◆

Striated grasswren
Amytornis striatus

TAXONOMY
Amytornis striatus Gould, 1840, Liverpool Plains, New South Wales, Australia. Three subspecies.

OTHER COMMON NAMES
French: Amytis strié; German: Streifengrasschlüpfer; Spanish: Ratona de la Hierba Rayada.

PHYSICAL CHARACTERISTICS
5.7–6.9 in (14.5–17.5 cm); male 0.56–0.78 oz (16–22 g). Sexes similar but female has chestnut flanks.

DISTRIBUTION
Widely scattered populations across Australia. *A. s. rowleyi* is confined to a small area of central Queensland, *A. s. whitei* is found in Western Australia. *A. s. striatus* has at least four disjunct populations from New South Wales to Western Australia.

HABITAT
Found on spinifex-covered sandplains and rocky hills, sometimes with shrubby vegetation, of the arid interior.

BEHAVIOR
Poor fliers; hop about with tail cocked over open ground, or with tail horizontal when moving through dense vegetation. They are found singly or in small family groups. Melodious song of trills and whistles.

FEEDING ECOLOGY AND DIET
Forage mostly on the ground, taking insects, particularly ants and beetles, and seeds. They have been reported eating cactus flowers, and foraging by moonlight.

REPRODUCTIVE BIOLOGY
Breeding biology is virtually unknown for wild birds. Clutch is two or three red-spotted, white eggs. No helpers at the nest have been reported.

CONSERVATION STATUS
Not threatened. Adversely affected by clearing for agriculture, introduced herbivores, and overgrazing, as well as predation by introduced cats and foxes, and by extensive fires.

SIGNIFICANCE TO HUMANS
None known. ◆

Orange-crowned wren
Clytomyias insignis

TAXONOMY
Clytomyias insignis Sharp, 1879, Arfak Mountains, New Guinea. Two subspecies.

OTHER COMMON NAMES
English: Orange-crowned fairy wren; French: Mérion à tête rousse; German: Rotkopf-Staffelschwanz; Spanish: Ratona Australiana Rufa.

PHYSICAL CHARACTERISTICS
Female 0.42–0.49 oz (12–14 g), male 0.35–0.49 oz (10–14 g). Cock-tailed fairy-wren with orange crown, buffy orange below and orange/olive above.

Amytornis striatus
▨ Resident

Clytomyias insignis
▨ Resident

DISTRIBUTION
Found at 6,560–9,840 ft (2,000–3,000 m) along both flanks of the central cordillera of New Guinea. *C. i. insignis* occurs as an isolated population in far northwestern Irian Jaya.

HABITAT
Mountain rainforest, usually in thickets of vine and climbing bamboo, along tracks and in small clearings made by tree fall.

BEHAVIOR
Rarely flies, and moves in groups through dense foliage with tail half-cocked and partly spread. Does not join mixed-species foraging flocks. Groups remain in same area throughout year. Voice a high-pitched twitter.

FEEDING ECOLOGY AND DIET
Gleans underside of leaves for invertebrates.

REPRODUCTIVE BIOLOGY
Breeding biology is virtually unknown.

CONSERVATION STATUS
Not threatened but deforestation a potential threat.

SIGNIFICANCE TO HUMANS
None known. ◆

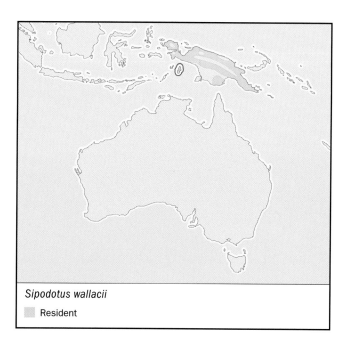

Sipodotus wallacii
▦ Resident

Wallace's wren
Sipodotus wallacii

TAXONOMY
Sipodotus wallacii G. R. Gray, 1862, Misool Island. Two subspecies.

OTHER COMMON NAMES
English: Wallace's fairy-wren; French: Mérion de Wallace; German: Rostnacken-Staffelschwanz; Spanish: Ratona Australiana de Wallace.

PHYSICAL CHARACTERISTICS
Female 0.23–0.28 oz (6.5–8 g), male 0.25–0.30 oz (7.0–8.5 g). Sexes similar in plumage. Long-billed, short-tailed fairy-wren, white below with white-streaked black cap and russet back.

DISTRIBUTION
Resident, lowland rainforests of New Guinea, sometimes to more than 3,280 ft (1,000 m) elevation. *S. w. wallacii* in north, *W. s. coronatus* in center and south.

HABITAT
Uses trees more than undergrowth in rainforest; in canopy to 130 ft (40 m); frequents tangles of vines and climbing bamboo at forest edge.

BEHAVIOR
Strong, undulating flight between trees. Found in family groups and in mixed species foraging flocks. Social organization poorly known. Does not cock its tail. Song is a series of high-pitched twittering notes.

FEEDING ECOLOGY AND DIET
Forages for insects by gleaning, probing, and hang-gleaning, mostly in foliage. Also hawks flying insects.

REPRODUCTIVE BIOLOGY
Domed nest suspended from vines 16.5–33 ft (5–10 m) from the ground. May have helpers, but little is known of reproductive biology.

CONSERVATION STATUS
Not threatened but deforestation a potential threat.

SIGNIFICANCE TO HUMANS
None known. ◆

Resources

Books
Blakers, M., S. J. J. F. Davies, and P. N. Reilly. *The Atlas of Australian Birds.* Melbourne: Melbourne University Press, 1984.

Higgins, P. J., J. M. Peters, and W. K. Steele. *Handbook of Australian, New Zealand and Antarctic Birds.* Vol. 5. Melbourne: Oxford University Press, 2001.

Rowley, I., and E. Russell. *Fairy-wrens and Grasswrens (Maluridae).* Oxford: Oxford University Press, 1997.

Sibley, C. G., and B. Monroe Jr. *Distribution and Taxonomy of Birds of the World.* New Haven: Yale University Press, 1990.

Schodde, R. *The Fairy-wrens.* Melbourne: Landsdowne, 1982.

William E. Davis, Jr

Australian warblers

(Acanthizidae)

Class Aves

Order Passeriformes

Suborder Passeri (Oscines)

Family Acanthizidae

Thumbnail description
Tiny to medium-sized, generally dull brown birds with fine bills, inhabiting canopy down to ground

Size
3.5–10 in (9–27 cm); 0.25–2.5 oz (7–70 g)

Number of genera, species
14–16 genera; 63–68 species

Habitat
Rainforests, eucalypt forests and woodlands, heathland and semi-arid woodland and scrub

Conservation status
Extinct: 1 species; Endangered: 2 species; Vulnerable: 6 species

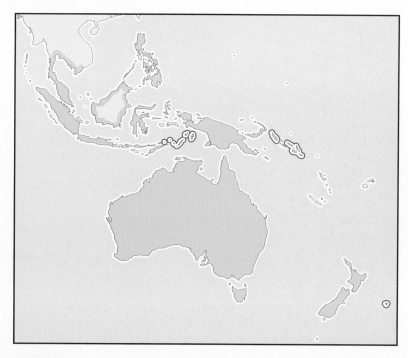

Distribution
New Zealand, Australia, New Guinea, Southeast Asia to Philippines, Vietnam, and Thailand

Evolution and systematics

The Australian warblers resemble in appearance and ecology the Old World warblers (Sylviidae). However, molecular studies show that they are more closely related to honeyeaters (Meliphagidae) and fairy-wrens (Maluridae) and are consequently part of the large Australian adaptive radiation of passerines. The pardalotes (*Pardalotus*) are closely related to the Acanthizidae and may be included in this family. The bristlebirds are sometimes placed in a separate subfamily (Dasyornithinae), but the pilotbird (*Pycnoptilus floccosus*) is intermediate between them and the other genera. The Mohuinae are a New Zealand group of uncertain affinities that are usually placed in the Acanthizidae. The most successful genera are the thornbills (*Acanthiza*) with about 12 species, and the gerygones (*Gerygone*) with about 20 species, though species limits are unclear due to many distinctive isolated forms. *Sericornis* also has many species, and at times has been expanded to include members of up to five other genera. Information from molecular biology and osteology, however, suggest the retention of a large number of genera with only one or two species.

Physical characteristics

Most species of acanthizids are tiny olive-green birds of the canopy or small brown birds of the understory. However, most have distinctive markings on the head or face, includ-

ing light eyebrows and streaks or spots, and some species have contrasting yellow or reddish rumps. Some of the thornbills and gerygones have yellow underparts while the pilotbird and rockwarbler (*Origma solitaria*) are reddish brown below. Most have slender bills, typical of small insectivores, while the weebill (*Smicrornis brevirostris*) has a short deep bill for prying off lerps (scales of psyllid insects). Whitefaces (*Aphelocephala*) also have deep bills, and the large-billed scrub-wren (*Sericornis magnirostris*) has a long pointed bill. All, except for the long-tailed bristlebirds (*Dasyornis*), have shortish tails. Bristlebirds have prominent rictal bristles.

Distribution

Australia has 43–45 species, with more than one species occurring at any point and five or more in many coastal regions. Thornbills tend to predominate in temperate regions and gerygones in the tropics, whereas scrubwrens occupy wetter areas. Twenty species are found in New Guinea, most of which are scrubwrens or gerygones. The gerygones alone have reached New Zealand, including the Chatham Islands, as well as Indonesia and South East Asia.

Habitat

Acanthizids occur in all habitats in Australia, including mangroves (the key habitat for three gerygones), rainforests

A large-billed scrubwren (*Sericornis magnirostris*) at its nest in Lamington National Park, Australia. (Photo by R. Brown/Animals Animals. Reproduced by permission.)

(where most of the scrubwrens are found), eucalypt forests and woodlands (weebill, several of the thornbills). The bristlebirds are found in dense coastal and upland heaths, whereas the rockwarbler, heathwrens (*Hylacola*), and fieldwrens (*Calamanthus*) also occur in habitats with heathy shrubs. The whitefaces and the redthroat (*Pyrrholaemus brunneus*) may be found in arid shrublands and deserts.

Behavior

The Australian warblers are almost always active, typically hopping on the ground or among the foliage of trees or bushes. Some species, especially the bristlebirds, can be quite cryptic. Breeding pairs are territorial in the breeding season, with many species joining flocks outside the breeding season. Thornbills, scrubwrens and gerygones are often nuclear species of mixed species feeding flocks in a wide range of habitats. Almost all species are sedentary or at most show very local movements, whereas the white-throated gerygone (*Gerygone olivacea*) migrates into southeastern Australia in spring. All species, except the gerygones, are rather weak fliers. A few species cock their tails. Many species have twittering calls and songs, though the wavering, downward warbles of the gerygones have earned them the name of bush canaries. The heathwrens and fieldwrens are also noted songsters and several species include mimicry in their calls.

Feeding ecology and diet

Most of the acanthizids are gleaners, taking small invertebrates from the foliage, from twigs, branches and trunks, or from the ground. Weebills and the gerygones include more active modes, such as snatching and hovering. Occasionally, insects are held under the feet. Within a habitat, different species often use different microhabitats. For instance, striated thornbills (*Acanthiza lineata*) glean from foliage in the canopy and subcanopy, while brown thornbills (*A. pusilla*) forage among shrubs and buff-rumped thornbills

(*A. reguloides*) feed from the ground or bark. Although the food of most species is predominantly invertebrates, seeds and fruits are occasionally eaten. Whitefaces consume large quantities of seeds.

Reproductive biology

Cooperative breeding is common in the family, and has been studied in several thornbills and scrubwrens. Mostly, though, acanthizids breed as pairs with some being assisted by a third or fourth bird. The breeding season is typically late winter to early summer and several attempts are made each season. Nests are domed, often with a clear hood above the entrance, and those of gerygones are pendant, attached to foliage. Although usually placed in trees or shrubs, crannies and hollows are also used and some species nest on the ground. The most common clutch is two eggs, though up to five eggs have been recorded. Egg color ranges from white in the rockwarbler, and white with sparse spotting in *Acanthiza* and *Gerygone*, to cream or buff with heavier spotting in many genera to plain chocolate in redthroat and speckled warblers (*Chthonicola sagittatus*). Eggs are laid at 48-hour intervals and are incubated by the female alone. Incubation periods and dependency of the fledglings are long. Both parents, and sometimes helpers, feed the young. Many nests are lost to predators, especially larger birds, and others are parasitized

A mountain thornbill (*Acanthiza katherina*) at its nest. (Photo by Hans and Judy Beste/Animals Animals. Reproduced by permission.)

by bronze-cuckoos (*Chrysococcyx*). Annual adult survival is 80% or more for some species; high for such small birds.

Conservation status

The species of gerygone on Lord Howe Island (*G. insularis*) is Extinct, that on Biak (*G. hypoxantha*) is Endangered, and the one on Norfolk Island (*G. modesta*) is Vulnerable. These island forms have suffered from major habitat loss as well as from introduced mammalian predators. The yellowhead (*Mohua ochrocephala*) of New Zealand is also Vulnerable. The 2000 Action Plan for Australian Birds lists a large number of acanthizids as threatened or nearly so. The bristlebirds have suffered from habitat loss, increased fire frequency, grazing by introduced mammals, and predation by mammalian carnivores. The western subspecies of rufous bristlebird (*Dasyornis broadbenti*) is Extinct, both subspecies of eastern bristlebird (*D. brachypterus*) are Endangered, one critically, whereas the western bristlebird (*D. longirostris*) is regarded as Vulnerable. Only the Coorong subspecies of rufous bristlebird is currently secure. Several isolated subspecies are threatened—with brown thornbills and scrubtits (*Acanthornis magnus*) on King Island being Critically Endangered. The chestnut-breasted whiteface (*Aphelocephala pectoralis*) and speckled warbler were considered Near Threatened throughout their range. Loss and degradation of habitat by agriculture and grazing are the main causes.

Significance to humans

As small, dull-colored and sometimes cryptic birds, acanthizids are generally not noticed much by people. Many are loosely called tits by country people, due to their superficial resemblance to members of the Paridae. The songs of bush canaries are well known, though the gerygones that produce the songs are less familiar to most people. Thornbills, scrubwrens and gerygones can provide a challenge to bird watchers, especially where there are local, indistinct species (e.g. Tasmanian thornbill [*Acanthiza ewingi*], Atherton scrubwren [*Sericornis keri*]), and the vocal but well-hidden bristlebirds can prove frustrating.

1. Bicolored mouse-warbler (*Crateroscelis nigrorufa*); 2. Yellow-rumped thornbill (*Acanthiza chrysorrhoa*); 3. Southern whiteface (*Aphelocephala leucopsis*); 4. Mangrove gerygone (*Gerygone levigaster*); 5. Rockwarbler (*Origma solitaria*); 6. White-browed scrubwren (*Sericornis frontalis*); 7. Eastern bristlebird (*Dasyornis brachypterus*); 8. Yellowhead (*Mohua ochrocephala*). (Illustration by Amanda Humphrey)

Species accounts

Eastern bristlebird
Dasyornis brachypterus

SUBFAMILY
Acanthizinae

TAXONOMY
Turdus brachypterus Latham, 1801, Sydney, New South Wales, Australia. Two subspecies.

OTHER COMMON NAMES
English: Bristle bird; French: Dasyorne brun; German: Braunkopf-Lackvogel; Spanish: Pájaro Cerdoso Común.

PHYSICAL CHARACTERISTICS
8.5 in (22 cm); 1.5 oz (c. 42 g). A gray-brown bird with small wing and sturdy legs and feet.

DISTRIBUTION
Eastern coastal Australia, with two isolated subspecies.

HABITAT
Dense, coastal and montane scrub, especially with grass tussocks.

BEHAVIOR
Solitary, shy and cryptic, mostly hidden in dense vegetation. May cock or fan tail. Sedentary, weak flier. Song is loud "it-wood-weet-sip" and harsh, abrupt call "zeip".

FEEDING ECOLOGY AND DIET
Gleans on ground, especially among leaf litter, taking insects and other arthropods, as well as seeds.

REPRODUCTIVE BIOLOGY
Breeds August to December. Domed nest is made of grass and plant tendrils and placed in a clump of grass. Two eggs, white to pale brown with gray and brown spots. Rarely more than one young raised.

CONSERVATION STATUS
Northern subspecies is Critically Endangered, with only a few dozen individuals. Southern subspecies is Endangered, with a populations of about 1,500 adults. Although most populations occur in national parks, frequent fires may kill the bird and render its habitat unsuitable for many years. Conversely, habitat that has not been burnt for a long time becomes unsuitable.

SIGNIFICANCE TO HUMANS
None known. ◆

White-browed scrubwren
Sericornis frontalis

SUBFAMILY
Acanthizinae

TAXONOMY
Saxicola longirostris Quoy and Gaimard, 1830, Western Port, Victoria, Australia. Up to 12 subspecies.

OTHER COMMON NAMES
English: Buff-breasted scrubwren, cartwheel bird; French: Séricorne à sourcils blancs; German: Weissbrauen-Sericornis; Spanish: Sericornis de Cejas Blancas.

PHYSICAL CHARACTERISTICS
4.5 in (120 cm); 0.4–0.5 oz (11–15 g). A relatively small bird with a distinctive white brow above the eyes, white chest, and brown back.

Dasyornis brachypterus
▨ Resident

Sericornis frontalis
▨ Resident

DISTRIBUTION
Coastal southwestern, southern, and southeastern Australia, including Kangaroo Island. Birds north and west of Adelaide have been regarded as a separate species (*S. maculatus*—spotted scrubwren). There is also physical variation on the east coast with northern birds being brighter and as many as 12 subspecies have been recognized.

HABITAT
Coastal heathlands and swamps, eucalypt forest with dense understory, rainforests. Occasionally in parks, gardens, and exotic pine plantations.

BEHAVIOR
An active ground and shrub dweller, which can sometimes be hidden but at other times quite visible and confiding. Typically in family groups, often with other scrubwrens and small birds. Quite sweet and complex, but rather erratic song. Harsh, scolding calls. Sedentary.

FEEDING ECOLOGY AND DIET
Mostly gleans from ground, including amongst leaf litter, and from shrubs. May forage in association with scrub-turkeys (*Alectura lathami*) who rake leaf litter into their mounds. Takes insects and other invertebrates, and some seeds.

REPRODUCTIVE BIOLOGY
Long breeding season from July to February. Rather large domed nest placed in low shrub or on the ground, and made of grass, bark, roots, and feather lining. Typically three, occasionally one, two or four eggs, which are pale blue, with brown freckles. May lay up to six clutches in a season. Some pairs have helpers. Incubation period is 17–22 days, fledging period is 12–18 days. Young are dependent on parents for a further six to seven weeks. Adults may live up to 17 years.

CONSERVATION STATUS
Not threatened.

SIGNIFICANCE TO HUMANS
None known. ◆

Gerygone levigaster

▨ Resident

Mangrove gerygone
Gerygone levigaster

SUBFAMILY
Acanthizinae

TAXONOMY
Gerygone levigaster Gould, 1843, Port Essington, Northern Territory, Australia.

OTHER COMMON NAMES
English: Mangrove warbler, buff-breasted warbler; French: Gérygone des mangroves; German: Mangrove-Gerygone; Spanish: Ratona Hada de los Mangles.

PHYSICAL CHARACTERISTICS
4 in (10–11 cm); 0.25 oz (7 g). A tiny bird with a brown back, white underparts, and a white brow.

DISTRIBUTION
Coasts of northern and eastern Australia south to Newcastle. Also small area in southeastern New Guinea.

HABITAT
Mangroves and neighboring vegetation, sometimes rainforests and gardens.

BEHAVIOR
Active and fairly tame. May be migratory at southern part of range. Attractive, whistling song and soft chattering notes.

FEEDING ECOLOGY AND DIET
Gleans, snatches, and hovers at outer foliage of mangroves and other trees for insects.

REPRODUCTIVE BIOLOGY
Breeds from September to April. Oval nest is suspended from foliage of mangroves and has a hooded side entrance. Two or three pinkish, speckled eggs.

CONSERVATION STATUS
Not threatened.

SIGNIFICANCE TO HUMANS
None known. ◆

Yellow-rumped thornbill
Acanthiza chrysorrhoa

SUBFAMILY
Acanthizinae

TAXONOMY
Acanthiza chrysorrhoa Quoy and Gaimard, 1830, New South Wales, Australia.

OTHER COMMON NAMES
English: Yellow-tailed thornbill, tom-tit; French: Acanthize à croupion jaune; German: Gelbbürzel-Dornschnabel; Spanish: Acanzisa de Cola Amarillo.

PHYSICAL CHARACTERISTICS
4 in (10 cm); 0.32 oz (9 g). Characterized by a bright yellow rump, black crown spotted white, and white brow.

DISTRIBUTION
Southern and central Australia, including Tasmania.

Acanthiza chrysorrhoa
Resident

Origma solitaria
Resident

HABITAT
Open woodland and edges, farmland and grassland with scattered trees or bushes, parks and gardens.

BEHAVIOR
Typically in family groups or small flocks, often with other thornbills. Active and noisy, with twittering songs and calls. Yellow rump very conspicuous in flight so that bird almost disappears when it lands. Sedentary or local movements only.

FEEDING ECOLOGY AND DIET
Primarily ground feeding, taking insects, other arthropods and sometimes seeds. Occasionally forages on shrubs and low trees.

REPRODUCTIVE BIOLOGY
Breeds from July to December, occasionally later. Domed nest is placed in a bush or sapling. Typically made of grass, lichen, and other plant fibers. The true entrance is at the side, and concealed by a hood, but there is a false cup-shaped nest placed on the top, possibly to confuse predators or cuckoos. Two to four pink eggs, lightly speckled. Both parents build nest, but only female incubates. Incubation period of 18–20 days, fledging period of 17–19 days. Parents are often assisted by helpers. Many nests fail due to predation or are parasitized by bronze-cuckoos.

CONSERVATION STATUS
Not threatened.

SIGNIFICANCE TO HUMANS
A well-known bird to many country dwellers. ◆

Rockwarbler
Origma solitaria

SUBFAMILY
Acanthizinae

TAXONOMY
Sylvia solitaria Lewin, 1808, Parramatta, New South Wales, Australia.

OTHER COMMON NAMES
English: Origma, rock robin, cave-bird; French: Origma des rochers; German: Steinhuscher; Spanish: Origma Piedra.

PHYSICAL CHARACTERISTICS
5.5 in (14 cm); c. 0.4 oz (12 g). Brown back with reddish brown underparts.

DISTRIBUTION
Very restricted range in sandstone and limestone country around Sydney, New South Wales.

HABITAT
Scrubby forest and heathland, especially near bare rock faces and cliffs.

BEHAVIOR
Solitary, but easily seen. Song is described as repeated "good-bye," also penetrating, rasping and twittering calls. Sedentary.

FEEDING ECOLOGY AND DIET
Typically feeds on rock surfaces or bare ground, including parking areas, etc. Gleans insects, and other arthropods, and occasionally seeds. Sometimes takes insects from bark or captures them in flight.

REPRODUCTIVE BIOLOGY
Breeds August to December. Nest is placed in a cave, or cleft in the rock, but also in man-made structures such as mineshafts. It is made of rootlets, bark, grass and moss and is spherical with a side entrance. Three or four white eggs.

CONSERVATION STATUS
Not threatened.

SIGNIFICANCE TO HUMANS
The only bird endemic to the state of New South Wales. ◆

Southern whiteface
Aphelocephala leucopsis

SUBFAMILY
Acanthizinae

TAXONOMY
Xerophila leucopsis Gould, 1841, Adelaide, South Australia.

OTHER COMMON NAMES
English: Chestnut-bellied whiteface, white-faced titmouse, eastern whiteface, tomtit; French: Gérygone blanchâtre; German: Fahlrücken-Weisstirnchen; Spanish: Ratona Blanca.

PHYSICAL CHARACTERISTICS
4.5 in (11 cm); 0.4 oz (11 g). Gray-brown upperparts and whitish underneath. Forehead is white with black edge extending to eye. Dark tail tipped white. Stubby, finch-like bill.

DISTRIBUTION
Southern and central Australia, though generally avoiding coastal areas.

HABITAT
Open eucalypt and acacia woodland, grassland and farmland with scattered trees.

BEHAVIOR
Typically in small to sometimes large groups, often with finches or small insectivores. Actively hops on ground, reminiscent of a sparrow. Song consists of musical, bell-like notes and twittering calls. Sedentary.

FEEDING ECOLOGY AND DIET
Mostly forages on the ground for small invertebrates and seeds, but also takes insects from shrubs and bark of trees.

REPRODUCTIVE BIOLOGY
Breeds from June to November. Nests are typically in shrubs or small trees but may be placed in hollows or even buildings. Two to five dull white eggs, with sparse brown or red speckles.

CONSERVATION STATUS
Not threatened.

SIGNIFICANCE TO HUMANS
Fairly familiar to many people in drier farming areas. ◆

Bicolored mouse-warbler
Crateroscelis nigrorufa

SUBFAMILY
Acanthizinae

TAXONOMY
Sericornis nigro-rufa Salvadori, 1894, Moroka, New Guinea.

OTHER COMMON NAMES
English: Black-backed mouse-warbler, mid-mountain mouse-babbler; French: Séricorne noir et roux; German: Schwarzrücken-Waldhuscher; Spanish: Ratona Semi Montañes.

PHYSICAL CHARACTERISTICS
5 in (12–13 cm); c. 0.35 oz (10 g). Rufous underparts from chin to lower breast. Abdomen rufous or black, depending on subspecies. Upperparts are black, flanks brown.

DISTRIBUTION
Scattered through central mountains of New Guinea.

HABITAT
Rainforest at mid-altitudes.

BEHAVIOR
Solitary, in pairs or small groups. Active, rapidly moving through an area, on the ground or in shrubs. When disturbed, bounces to and fro. Three-note whistling calls, and scolding alarms.

FEEDING ECOLOGY AND DIET
Forages on the ground, and in low shrubs.

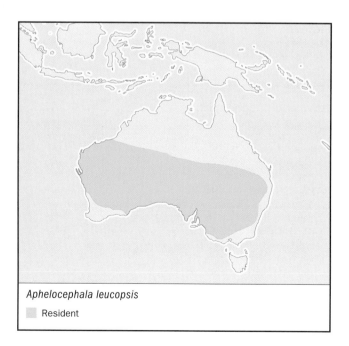

Aphelocephala leucopsis
◻ Resident

Crateroscelis nigrorufa
◻ Resident

REPRODUCTIVE BIOLOGY
Not well-known. Domed nest is placed in a bush, and is made of rootlets, moss, grass and feathers. Two white, lightly marked eggs.

CONSERVATION STATUS
Not threatened.

SIGNIFICANCE TO HUMANS
None known. ◆

Yellowhead
Mohua ochrocephala

SUBFAMILY
Mohuinae

TAXONOMY
Mohua ochrocephala Gmelin, 1798, Queen Charlotte Sound, New Zealand.

OTHER COMMON NAMES
English: Mohua, bush canary; French: Mohoua à tête jaune; German: Weissköpfchen; Spanish: Cabeza Amarilla.

PHYSICAL CHARACTERISTICS
6 in (15 cm); c. 0.7 oz (20 g). Brownish olive upperparts with bright yellow head and yellow breast.

DISTRIBUTION
South Island of New Zealand, including Marlborough, Nelson, Westland, western Otago, Southland and near Dunedin.

HABITAT
Forest, especially dominated by beech (*Nothofagus*).

BEHAVIOR
Pairs and trios occupy large home ranges in breeding season. Several families form larger flocks in nonbreeding season, which are joined by other bird species. Varied, mechanical call of six to eight notes rapidly repeated.

FEEDING ECOLOGY AND DIET
Spend most of the day foraging, in shaded canopy or upper subcanopy. Glean from foliage, branches and trunks and some-

Mohua ochrocephala

■ Resident

times rip into dead wood. Mostly insectivorous, especially taking larvae, but occasionally eat fruit, flowers and fungi.

REPRODUCTIVE BIOLOGY
Facultatively cooperative or possibly polygamous. Breeds October to February. Cup-shaped nests placed in holes. Three to four pinkish eggs, blotched with reddish brown. Incubated by female for 18–21 days, young fledge at 21 days and are fed by two or three adults up to 55 more days.

CONSERVATION STATUS
Declared Vulnerable, due to extensive decline as a result of forest loss. Also avoids edges, stunted and regrowth forests. Less vulnerable to nest predators than many New Zealand birds, due to hole nesting, but recently fledged young may be at risk.

SIGNIFICANCE TO HUMANS
None known. ◆

Resources

Books
Coates, B. J. *The Birds of Papua New Guinea, Including the Bismarck Archipelago and Bougainville.* Brisbane: Dove, 1990.

Garnett, S. T., and G. M. Crowley. *The Action Plan for Australian Birds 2000.* Canberra: Environment Australia, 2001.

Serventy, V. N., A. R. McGill, J. D. Pringle, and T. R. Lindsey. *The Wrens and Warblers of Australia.* Sydney: Angus and Robertson, 1982.

Schodde, R., and I. Mason. *The Directory of Australian Birds: Passerines.* Melbourne: CSIRO, 1999.

Periodicals
Magrath, R., et al. "Life in the slow lane: Reproductive life history of the white-browed scrubwren, an Australian endemic." *Auk* 117 (2000): 479–489.

Noske, R.A. "Nesting biology of the mangrove gerygone (*Gerygone levigaster*) in the monsoonal tropics." *Emu* 102 (in press).

Organizations
Birds Australia. 415 Riversdale Road, Hawthorn East, Victoria 3123 Australia. Phone: +61 3 9882 2622. Fax: +61 3 9882 2677. E-mail: mail@birdsaustralia.com.au Web site: <http://www.birdsaustralia.com.au>

Hugh Alastair Ford, PhD

Australian chats
(Epthianuridae)

Class Aves
Order Passeriformes
Suborder Passeri (Oscines)
Family Epthianuridae

Thumbnail description
Small, rotund, active birds with short tails, slender bills, and upright stance, sexually dimorphic with males often brightly colored

Size
4.3–5.5 in (11–14 cm); 0.3–0.6 oz (9–18 g)

Number of genera, species
2 genera; 5 species

Habitat
Desert, semi-arid, and coastal shrublands

Conservation status
Endangered: 1 subspecies; Critically Endangered: 1 subspecies of same species; Lower Risk (Least Concern): 1 subspecies

Distribution
Australia, except for northern and eastern coasts, and including Tasmania

Evolution and systematics

Australian chats resemble the muscicapid chats of Eurasia and Africa in appearance and behavior. Australian chats are related to the honeyeaters on the basis of their brush tongues. This relationship was confirmed in the 1980s and 1990s, and indeed they are now usually classified within the Meliphagidae. Honeyeaters and chats are related to a range of other largely Australian families such as the Petroicidae (Australian robins) and Maluridae (fairy-wrens).

Four species of chats are monotypic, probably because they are highly mobile around often-extensive ranges. The yellow chat (*Epthianura crocea*), in contrast, shows a series of well-differentiated and localized populations. It inhabits local wetlands, some of which are transient, in arid and semi-arid regions.

Physical characteristics

Australian chats are small birds with longish delicate legs. Bills are fine and, in some species, slightly decurved. Like the honeyeaters, they have brush tips to their tongues. The gibberbird (*Ashbyia lovensis*) is larger and more robust than the other species.

Males are brightly or conspicuously colored in the breeding season, with yellow, orange, or red underparts. The male white-fronted chat (*Epthianura albifrons*) is black, white, and gray. Juveniles, immatures, females, and, in some species, nonbreeding males show more subdued plumages.

Distribution

Chats occur throughout Australia, with the exception of the forested north and east coasts, and southwestern Tasmania. Orange (*Epthianura aurifrons*) and crimson (*E. tricolor*) chats are found throughout the center from the west coast to the western slopes of the Great Dividing Range and from the south coast to the tropics. However, in the moister parts of this range, they are irregular visitors. The white-fronted chat is found across southern Australia and is the only chat in Tasmania. The stony deserts of central Australia are where gibberbirds are found.

Yellow chats display one of the most scattered distributions of any bird species. One subspecies (*macgregori*) is restricted to a tiny range near the Fitzroy River on the central Queensland coast. A second subspecies (*tunneyi*) is found only in western Arnhem Land, Northern Territory. The dominate subspecies occurs in isolated populations scattered across the Kimberley region of western Australia, across Northern Territory, and into western Queensland. Reports in northeastern South Australia may refer to small local populations or vagrants.

Habitat

Chats are most strongly associated with chenopod shrubland, especially saltbush (*Atriplex*), bluebush (*Maireana*), and samphire (*Halosarcia*). They also occur in neighboring semi-arid woodland or shrubland, which is often dominated by acacias. Gibber plains, the home of the gibberbird, are stony deserts with a

A white-fronted chat (*Epthianura albifrons*) with its insect prey. (Photo by R. Brown/VIREO. Reproduced by permission.)

sparse cover of grass and saltbush. Yellow chats typically inhabit low vegetation close to swamps, floodplains, and bore drains.

Behavior

Chats typically occur in small, loose flocks, but pair up during the breeding season. They may defend breeding territories, though a detailed study of the white-fronted chat suggested that males defend the nest and their mate rather than a territory. They display in flight and from perches, where they dip their tails and raise the colorful feathers on their heads or back ends. Calls are mostly simple and metallic, with pretty twittering or piping songs, and, when threatened, harsh churring calls. Crimson and orange chats are highly mobile, displaying a north-south seasonal migration, as well as nomadic movements in response to local rainfall. Chats seem unable to drink saline water. During dry times, they may reach the coast. The other species may be more sedentary, but they also show poorly understood movements in response to local conditions.

Feeding ecology and diet

Insects and spiders are the principal food of the Australian chats, and are usually captured on the ground or from low shrubs. White-fronted chats also occasionally eat gastropods, crustaceans, and seeds. This species gleans its prey from the ground, dry and wet, or from shallow water, and may run after aerial prey. They rarely capture flying insects. The other chats show similar foraging methods, while seeds are a more

important component of the diet of gibberbirds. Crimson chats also take nectar.

Reproductive biology

Although there are detailed studies of the breeding biology of the white-fronted and the crimson chats, the breeding biology of the other species is less well known. This account is based mainly on the white-fronted chat. Chats have long breeding seasons, peaking in late winter and spring (August–November), and breeding again after the rainy season in late summer and fall (March–April). Up to five attempts may be made in a season. There is no evidence of polygamy or cooperative breeding among the chats. Nests are usually placed 1–4 ft (0.3–1.2 m) from the ground in small bushes, often saltbush or bluebush, and occasionally on the ground. Nests are cup-shaped, and made from grass, rushes, twigs, and plant fiber, and sometimes with mammal hair or fur and feathers. Eggs are fleshy or pinkish white with small reddish spots at the larger end. Clutches are of two to four eggs, maximum five (mean of 3.1 for white-fronted chats and 2.7 for crimson chats). Both males and females incubate the eggs, which hatch at 13–14 days. Both parents brood and feed the young, with a rate of seven visits per parent per hour. Young fledge at about 14 days in white-fronted chats, and a few days earlier in crimson and orange chats. Approximately 30% of nests succeed. Most failures are due to predation, and known predators include cats, foxes, snakes, and ravens. A small proportion of nests are parasitised by the Horsefield's bronze cuckoo (*Chrysococcyx basalis*).

Conservation status

Two subspecies of yellow chats are Endangered (one Critically Endangered) due to loss and degradation of their habitats. The third subspecies is secure as are all other species. Overgrazing and increased salinity in inland Australia may have benefited several species.

Significance to humans

Chats may have little significance to humans, although orange and crimson chats are conspicuous and colorful birds observed by many desert travelers.

Male crimson chat (*Epthianura tricolor*) with chicks in Western Australia. (Photo by Peter Slater. Photo Researchers, Inc. Reproduced by permission.)

Species accounts

Crimson chat

Epthianura tricolor

TAXONOMY

Epthianura tricolor Gould, 1841, Liverpool Plains, New South Wales, Australia.

OTHER COMMON NAMES

English: Tricolored chat, crimson tang, red canary; French: Epthianure tricolore; German: Scharlachtrugschmätzer; Spanish: Curruca Carmesí.

PHYSICAL CHARACTERISTICS

4.7 in (12 cm); 0.4 oz (11 g). Brownish upperparts with white throat and crimson crown and underparts.

Epthianura tricolor

DISTRIBUTION

Inland, western, and southern coasts of Australia, may break out into southeastern and eastern Australia.

HABITAT

Arid and semi-arid shrubland with saltbush, acacia, or other shrubs; occasionally, grassland or farmland.

BEHAVIOR

Small but highly mobile flocks in nonbreeding season. Metallic, whistling, and twittering calls.

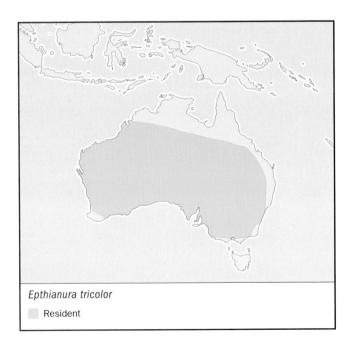

Epthianura tricolor

Resident

FEEDING ECOLOGY AND DIET

Takes insects and other invertebrates from the ground and low shrubs, occasionally from the air. Eats seeds and probes flowers for nectar.

REPRODUCTIVE BIOLOGY

Breeds as loosely associated pairs. Builds cup nest in low shrubs up to 3 ft (0.9 m). Clutches are of two to five eggs; incubation by both sexes for 10–14 days; fledges at 10 days. Both parents brood and feed young, and show distraction displays.

CONSERVATION STATUS

Not threatened. Common and widespread, but numbers vary greatly at any locality.

SIGNIFICANCE TO HUMANS

An attractive bird often observed by desert visitors. ◆

Gibberbird

Ashbyia lovensis

TAXONOMY

Epthianura lovensis Ashby, 1911, Leigh Creek, South Australia.

OTHER COMMON NAMES

English: Gibber chat, desert chat, desert bird; French: Epthianure d'Ashby; German: Wüstentrugschmätzer; Spanish: Curruca Desértica.

PHYSICAL CHARACTERISTICS

5 in (13 cm); 0.65 oz (18 g). Yellow cheek, throat, and underparts; crown and upperparts are sandy brown and used as camouflage on stony terrain.

Ashbyia lovensis

DISTRIBUTION

Borders of Queensland, New South Wales, South Australia, and Northern Territory in central Australia.

HABITAT

Gibber plains, which are sparsely vegetated, stony deserts, and occasionally claypans.

BEHAVIOR

Singly, in pairs, and small flocks. Run along ground and bob tail. Piping and twittering calls, attractive song.

FEEDING ECOLOGY AND DIET

Gleans invertebrates from ground and sometimes from the air. Eats seeds.

Ashbyia lovensis
▨ Resident

REPRODUCTIVE BIOLOGY
Breeds mostly in spring, but may at other times as well. Cup-shaped nests made in depressions in ground. Clutches are of two to four eggs; most aspects of breeding poorly known.

CONSERVATION STATUS
Not threatened. Has possibly benefited from grazing by livestock.

SIGNIFICANCE TO HUMANS
None known. ◆

Resources

Books

Higgins, P.J., J.M. Peter, and W.K. Steele, eds. *Handbook of Australian, New Zealand and Antarctic Birds,* Vol. 5, *Tyrant-flycatchers to Chats.* New York: Oxford University Press, 2000.

Sibley, C.G. and J.E. Ahlquist. *Phylogeny and Classification of Birds: A Study of Molecular Evolution.* New Haven: Yale University Press, 1990.

Periodicals

Major, R.E. "Breeding Biology of the White-fronted Chat *Ephthianura albifrons* in a Saltmarsh near Melbourne." *Emu* 91 (1991): 236–49.

Major, R.E. "Flocking and Feeding in the White-fronted Chat *Ephthianura albifrons*: The Relationship between Diet, Food Availability and Patch Selection." *Australian Journal of Ecology* 25 (1991): 395–407.

Parker, S.A. "The Tongues of *Ephthianura* and *Ashbyia*." *Emu* 73 (1973): 19–20.

Williams, C.K. "Ecology of Australian Chats (*Ephthianura* Gould): Reproduction in Aridity." *Australian Journal of Zoology* 27 (1979): 213–229.

Williams, C.K. and A.R. Main. "Ecology of Australian Chats (*Ephthianura* Gould): Aridity, Electrolytes and Water Economy." *Australian Journal of Zoology* 25 (1977): 673–691.

Hugh Alastair Ford, PhD

Logrunners and chowchillas
(Orthonychidae)

Class Aves
Order Passeriformes
Suborder Passeri (Oscines)
Family Orthonychidae

Thumbnail description
Small, stocky ground birds with powerful legs and shafts of tail feathers extending as spines

Size
7.3–11.8 in (18.5–30 cm); 0.10–0.47 lb (47–213 g)

Number of genera, species
1 genus; 3 species

Habitat
Rainforest and contiguous thick, low secondary growth

Conservation status
Not threatened

Distribution
Eastern Australia and central ranges of New Guinea

Evolution and systematics

The taxonomic placement of these unusual birds was long problematic. Traditionally they were included with other ground-frequenting passerines of the Australian and New Guinean region, such as quail-thrushes and whipbirds. The DNA-DNA hybridization studies by Charles Sibley and Jon Ahlquist in 1980–90s demonstrated that logrunners were quite distinct from these and other passerine families. Since then they have been separated at family level, an action supported by subsequent research, within the Australo-Papuan corvine radiation. They comprise one of the older groups, with fossil species known from Miocene and Quaternary deposits in Australia.

The distribution of the species has attracted much attention. The southern species, the southern logrunner (*Orthonyx temminckii*) occurs through central eastern Australia, then apparently leapfrogs the larger, darker chowchilla (*O. spaldingii*) of northeastern Australia, to reappear in the central ranges of New Guinea. This unusual pattern was illuminated in 2001 by Leo Joseph and his colleagues, who showed that the New Guinea populations were specifically distinct from those in southeastern Australia; they thus take the name New Guinea logrunner (*O. novaeguineae*). It is not certain, however, to which of the Australian species it is most closely related.

Earlier names for the Australian species were the southern and northern logrunners for *O. temminckii* and *O. spaldingii*, respectively. When the New Guinea populations were considered to be conspecific with *O. temminckii*, the common name of the latter was deemed inappropriate and shortened to logrunner; the name for *O. spaldingii* was changed to chowchilla, a rendering of its distinctive call. Now, with the logrunners split into two species, it seems appropriate to qualify the name of the Australian birds by reintroducing the traditional name, southern logrunner. The name logrunner is used collectively for the birds of this family.

The morphology of the hind limb and pelvis reflects the characteristic ground feeding method used by these birds. The femur has a distinctive hour-glass shape bestowed by the expansion of the distal end as a brace and the proximal end for muscle attachments. The pelvis is broad, with deep excavations for the powerful muscles that drive the legs.

Physical characteristics

Logrunners are stocky birds, with the chowchilla, the largest species, reaching a length of 12 in (30 cm); the other species are noticeably smaller (7.3–8.3 in; 18.5–21 cm). All have powerful legs and claws. The unspecialized bill is of

A female southern logrunner (*Orthonyx temminckii*) feeds in Lamington National Park, Australia. (Photo by R. Brown/Animals Animals. Reproduced by permission.)

moderate strength. Perhaps the most curious physical feature of these birds is the stiffened shafts of the 10 tail feathers, the tips of which protrude beyond the ends of the feathers as pliable spines. This led to the earlier name of "spine-tailed logrunner." While the chowchilla is simply colored with broad patches of unmarked black and white, and in one sex, orange, the other two species have much more complex patterns of brown, rufous, black, gray, and white. All three species have similar patterns of sexual dimorphism: males have white breasts, females orange ones.

Distribution

This family is found only in Australia and New Guinea. The southern logrunner is distributed along coastal areas and adjacent ranges in appropriate habitat from central eastern New South Wales north to southeastern Queensland. The chowchilla is restricted to the Atherton Tableland region in northeastern Australia, particularly above 1,475 ft (450 m). New Guinea birds occur along the central highlands in a zone 6,500–9,300 ft (1,980–2,840 m), in places above and below this. There are breaks in the distribution, some of which are real and others that may merely reflect our poor knowledge of this rare species.

Habitat

Logrunners are birds of the rainforest, although they will venture into contiguous habitats if the vegetation is sufficiently low and thick; this can include invasive, introduced plants.

Behavior

Logrunners live in permanent territories in pairs or small family parties. In 1999 Amy Jensen reported that these groups

of 2–5 birds occupied home ranges (the areas in which they foraged) of 1.7–9.8 acres (0.7–4.0 ha), substantially larger than the space actually defended. Territorial defense is strong, birds announcing themselves to neighboring families with loud calls. For the first hour after dawn they call constantly before settling down to feed. They also break into short bouts of calling throughout the day, particularly during encounters with neighbors along a border, and have a concluding burst at dusk. Mike McGuire investigated the vocalizations of several parties of chowchillas. His 1999 study recorded vocabularies of often complex calls with differing local dialects between adjoining groups.

These birds can be quite shy, quickly fleeing into cover, often with a loud shriek, when startled. At other times, they can be remarkably tame, ignoring quiet observers while they search for food—even walking across a person's feet while the bird is foraging.

Feeding ecology and diet

The purpose of the spine-tipped tail and oddly shaped femur become apparent when logrunners feed. A feeding bird spreads its tail and rests the spine-like quills against the ground. To expose the soil, the bird rakes a foot through the leaf litter and fallen sticks, throwing the leg to the side, perpendicular to the body. Once the ground is sufficiently cleared to feed, the legs are used alternately to scratch the soil front to back in search of insects, larvae and other invertebrates that form the diet, balancing on one leg while at the same time using the tail as a supporting prop. Pivoting around the tail, logrunners produce distinctive small cleared patches about 8 in (20 cm) in diameter on the forest floor. A foraging bird spends a brief period scratching at the one spot before moving on to another a short distance away. Smaller ground-feeding birds, such as the yellow-throated scrubwren (*Sericornis citreogularis*) and eastern whipbird (*Psophodes olivaceus*), may attend foraging logrunners, picking up small invertebrates that are unearthed by the scratching.

Reproductive biology

Unlike many Australian passerines, southern logrunners concentrate much of their breeding in the winter. Starting in about April or June their nesting season continues through to August or September, sometimes as late as October. The female is responsible for building the nest and incubating the eggs, and she provides most of the care for the young.

The nest is a globular dome, usually placed on the ground but on occasion may be situated in low vines or on a fallen log or a stump. The female builds a platform up to 2 in (5 cm) of short, thick sticks. She adds curving sides until these meet and join at the top, finally capping this with dry leaves and green moss. This roof hangs over the side entrance, partially concealing it and keeping the interior of the nest dry, even in heavy downpours. The clutch characteristically comprises two eggs, which are large and white. Usually these hatch after a 21–25 day incubation period. The female has most of the responsibility for feeding the nestlings, but she is assisted by the male; he passes food to her, which she in turn gives to the young birds. The young remain in nest for 16–18 days. After fledging they are fed by both parents. Juveniles have a mottled breast, which on the subsequent moult acquires the color appropriate to the bird's sex.

The breeding biology of the chowchilla, which was long unknown, was revealed in 1997 by Cliff Frith and his associates. Most aspects are similar to those of the southern logrunner, with some important differences. Nesting can occur in any month of the year, peaking in July through December. The nest is the same shape but considerably larger. Only a single egg is laid. Both incubation and nestling periods are longer: 25 days, with 75% of eggs hatching, and 22–27 days, with 67% of young successfully leaving the nest. More than one male may bring food to the female, but males do not themselves feed the chick.

What is known of the breeding in New Guinea logrunners is similar, with records from March and November of nests with single eggs.

Conservation status

The clearance of rainforest has eliminated southern logrunners from parts of the range; however, this species appears capable of persisting in small, isolated patches of rainforest. A population persists to the south of the metrop-olis of Sydney, New South Wales, well separated from the northern section of its range, and a pair of birds was found in 5 acres (2 ha) of rainforest that had been long-isolated within a surrounding countryside of dairy land. The possibility that proliferating exotic plants, such as lantana and blackberry, in and around the rainforest might force this species from its habitat did not eventuate. The southern logrunner seems to have adapted well to thick stands of these plants, using their cover to colonize along watercourses and even moving away from rainforests into areas where these vigorously invading plants have made inroads into cleared land.

While clearing has undoubtedly reduced the habitat available to chowchillas, this species remains common in northeastern Australia. The New Guinea logrunner appears to be uncommon to rare, but this may be due in part to its shy nature and the remoteness of much of its range. Only in 1987 was it discovered at Tari Gap, central New Guinea, where it had long been thought to be absent.

Significance to humans

The restricted habitat and cryptic behavior of these interesting birds make them unfamiliar to most people. The name "chowchilla" is the name used by the Dyirbal Aboriginal people, itself derived from this bird's call.

Species accounts

Southern logrunner
Orthonyx temminckii

TAXONOMY
Orthonyx temminckii Ranzani, 1822, Hat Hill, New South Wales, Australia.

OTHER COMMON NAMES
English: Spine-tailed logrunner; French: Orthonyx de Temminck; German: Stachelschwanzflöter; Spanish: Corretroncos Cola de Espinas.

PHYSICAL CHARACTERISTICS
7.3–8.3 in (18.5–21 cm); female 0.1–0.13 lb (46–58 g), male 0.13–0.15 lb (58–70 g). Gray and tan patterned plumage with black side-stripe. Males have white throats; orange throats in females.

Orthonyx temminckii

DISTRIBUTION
Central eastern Australia.

HABITAT
Rainforest, edges of contiguous wet sclerophyll forest, and dense fringing vegetation, including introduced species.

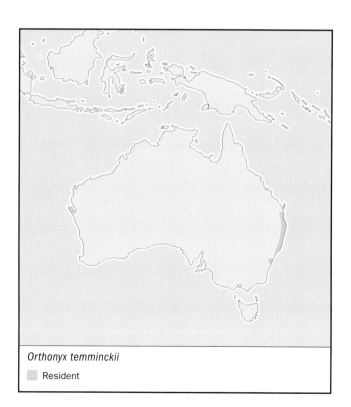

Orthonyx temminckii

▨ Resident

BEHAVIOR
Sedentary. Territorial throughout year, usually living in pairs or small family parties. Often shy, but generally ignores human observers when foraging. Generally unobtrusive except when giving loud, penetrating calls; most characteristic a lengthy rapid series of "weet" notes; also a piercing "kweek" when alarmed.

FEEDING ECOLOGY AND DIET
Eats insects and other small soil invertebrates uncovered by vigorous scratching; leaves characteristic shallow depressions in soil.

REPRODUCTIVE BIOLOGY
Breeds May to August, sometimes April to October; produces one or two broods per season. Female alone builds nest, incubates, and provides most of care for young. Nest is a dome of sticks and other vegetation with a side entrance overhung by moss; placed on or near ground against trunk or clump of vegetation. Two white eggs are laid. Incubation, 21–25 days; fledging period 16–18 days.

CONSERVATION STATUS
Not threatened. Common in northern part of range, decreasing southwards until rare at southern limits.

SIGNIFICANCE TO HUMANS
None known. ◆

New Guinea logrunner
Orthonyx novaeguineae

TAXONOMY
Orthonyx novae guineae Meyer, 1874, Arfak Mountains, New Guinea. Three subspecies.

OTHER COMMON NAMES
French: Orthonyx de Nouvelle Guinée; German: Neuguineaflöter; Spanish: Corretroncos de Nueva Guinea.

PHYSICAL CHARACTERISTICS
7.3 in (18.5 cm); female 0.10–0.13 lb (47–58 g), male 0.12–0.17 lb (53–75 g). Similar to southern longrenner.

Orthonyx novaeguineae

DISTRIBUTION
Scattered localities in mountains of New Guinea between 6,500–9,300 ft (1,980–2,840 m), probably up to 11,300 ft (3,450 m); occurs locally as low as 3,900 ft (1,200 m) in Irian Jaya. *O. n. novaeguineae*: northwestern New Guinea; *O. n. dorsalis*: western central New Guinea; *O. n. victoriana*: eastern New Guinea.

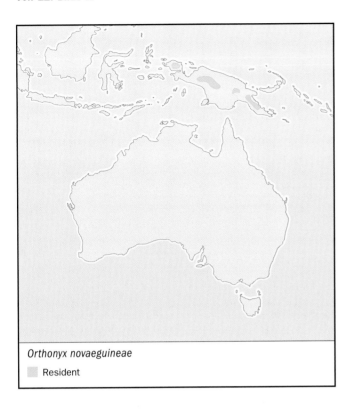

Orthonyx novaeguineae
▢ Resident

HABITAT
Mainly upper montane forest.

BEHAVIOR
Terrestrial. Quiet, cryptic, and easily overlooked. Usually seen in pairs or small parties of three or four birds. Song is series of four to six descending notes.

FEEDING ECOLOGY AND DIET
Scratches in ground litter for insects and other invertebrates.

REPRODUCTIVE BIOLOGY
The nest is a small dome of moss, root fibers, and plant stems, about 5 in (12.5 cm) wide placed on the ground. Nests with a single white egg found in March and November.

CONSERVATION STATUS
Generally scarce to rare, although not considered a threatened species.

SIGNIFICANCE TO HUMANS
None known. ◆

Chowchilla
Orthonyx spaldingii

TAXONOMY
Orthonix spaldingii Ramsay, 1868, Rockingham Bay, Queensland, Australia. Three subspecies.

OTHER COMMON NAMES
English: Northern logrunner; French: Orthonyx de Spalding; German: Schwarzkopfflöter; Spanish: Corretroncos de Spalding.

PHYSICAL CHARACTERISTICS
11–11.8 in (28–30 cm); female 0.25–0.32 lb (113–144 g), male 0.33–0.47 lb (150–213 g). Dark brown upperparts; breast is white in males and orange-brown in females. Thin white eye ring.

DISTRIBUTION
O. s. spaldingii: northeastern Australia; *O. s. melasmenus*: northeastern Australia, north of *O. s. spaldingii*. Generally above 1,470 ft (450 m), locally in lowlands with highest rainfall.

Orthonyx spaldingii

HABITAT
Rainforest.

BEHAVIOR
Territorial at all seasons, often living in family parties. Highly terrestrial. Often shy, but may be confiding when feeding. Generally unobtrusive except when calling. Loud, ringing "chow chowchilla chowry chook chook" or "chow chilla chow chow chilla"; also other growls, chucks.

FEEDING ECOLOGY AND DIET
Forages by vigorous scratching, excavating shallow depressions in ground. Eats insects and other terrestrial invertebrates; occasionally seeds.

REPRODUCTIVE BIOLOGY
Breeds almost year round, mainly April to August; one or two broods reared per season. Female lays a single white egg and is responsible for construction of nest, incubation of eggs, and

Orthonyx spaldingii
▢ Resident

care of nestlings. Male provisions female during activities. Nest is a dome of sticks, twigs, and moss, side entrance hooded by moss, platform of sticks leading to entrance; on or near ground in clump of vegetation. Incubation, 25 days; fledging 22–27 days.

CONSERVATION STATUS
Not threatened. Although some reduction of range has occurred through habitat loss, this species is still common.

SIGNIFICANCE TO HUMANS
None known. ◆

Resources

Books

Coates, Brian J. *The Birds of Papua New Guinea.* Vol. 2, *Passerines.* Alderley: Dove Publications, 1993.

Higgins, P. J., and J. M. Peter, eds. *Handbook of Australian, New Zealand and Antarctic Birds.* Vol. 6, *Pardalotes to Figbird.* Melbourne: Oxford University Press, forthcoming.

Schodde, R., and I. J. Mason. *The Directory of Australian Birds. Passerines.* Collingwood: CSIRO Publishing, 1999.

Periodicals

Frith, C. B., D. W. Frith, and A. Jensen. "The Nesting Biology of the Chowchilla *Orthonyx spaldingii* (Orthonychidae)." *Emu* 97 (1997): 18–30.

Hindwood, K. A. "The Spine-tailed Log-runner (*Orthonyx temminckii*)." *Emu* 33 (1934): 257–67.

Jensen, A. "Home Ranges and Group-territoriality in Chowchillas *Orthonyx spaldingii.*" *Emu* 99 (1999): 280–90.

Joseph, L., B. Slikas, D. Alpers, and R. Schodde. "Molecular Systematics and Phylogeography of New Guinean Logrunners (Orthonychidae)." *Emu* 101 (2001): 273–80.

McGuire, M. "Dialects of the Chowchilla *Orthonyx spaldingii* in Upland Rain Forest of North-eastern Australia." *Emu* 96 (1996): 174–80.

Organizations

Birds Australia. 415 Riversdale Road, Hawthorn East, Victoria 3123 Australia. Phone: +61 3 9882 2622. Fax: +61 3 9882 2677. E-mail: mail@birdsaustralia.com.au Web site: <http://www.birdsaustralia.com.au>

Walter E. Boles, PhD

▲
Quail thrushes and whipbirds
(Eupetidae)

Class Aves
Order Passeriformes
Suborder Passeri (Oscines)
Family Eupetidae

Thumbnail description
Medium small ground birds, either plump or slender, some with crests, all with powerful legs

Size
6.7–12.2 in (17–31 cm); 0.07–0.45 lb (30–205 g)

Number of genera, species
7 genera; 16–19 species

Habitat
Rainforest, forest, woodland, savanna, shrub steppe, heath, sandhills

Conservation status
Near Threatened: 2 species

Distribution
Australia, New Guinea, east and west Indonesia, Malaysia, south Thailand

Evolution and systematics

The birds in this heterogeneous family spend most of their time on the ground, but apart from their shared adaptations to this mode of life, they have few other obvious links. Once included with the Old World babblers (Timaliidae), they have been recognized as components of the autochthonous Australo-Papuan songbird assemblage; whether they constitute a natural group, however, is debatable. The affinities of the various genera within this family are puzzling, and it may be that this group is largely a convenient "catch-all" category. The logrunners were formerly included in this family, with the name Orthonychidae being applicable. With the segregation of the logrunners in their own family, the available name becomes Eupetidae, which has priority over the more commonly used Cinclosomatidae.

There are 16–19 species, divided among seven genera. There are five species of quail-thrushes (*Cinclosoma*); with several of the distinct populations sometimes treated as separate species. The three to four jewel-babblers (*Ptilorrhoa*) resemble quail-thrushes except for the color. Although they were once placed in the same genus as the rail-babbler (*Eupetes macrocerus*), the latter species is a distinctive bird of uncer-

tain affinities. The genus *Psophodes* comprises four to five species of Australian whipbirds and wedgebills. The two species of wedgebill are so similar in appearance and behavior that their distinctness, exhibited most notably in their songs, remained unappreciated for many years. The Papuan whipbird (*Androphobus viridis*) has not been studied but its relationships may lie with the Australian whipbirds. There are two species in the genus *Melampitta*; the lesser melampitta (*M. lugubris*) has been arguably placed with the birds of paradise (Paradisaeidae) by some authorities. The blue-capped ifrit (*Ifrita kowaldi*) looks and acts different from other members of the family. Some aspects of its skull morphology resemble those of the whistlers (Pachycephalidae).

Physical characteristics

Most species have long tarsi and powerful legs; plumage is thick and fluffy, the tail usually long and broad, and the bill relatively short. Quail-thrushes have bold patterns of black, white, brown, and orange, which are generally restricted to the underparts; the colors of the upper surface resemble the ground cover. Jewel-babblers somewhat resemble quail-thrushes, but have extensive areas of blue in the plumage. The rail-babbler is a long-necked, long-tailed bird with chestnut plumage, and a blue streak along the side of the neck. Whipbirds and wedgebills of Australia are slender, drab-colored, long-tailed birds with short but prominent pointed crests. The Papuan whipbird superficially resembles a small version of these birds, but lacks a crest. Melampittas are pitta-like birds with long legs, markedly short tails, and black plumage. The ifrit is small bird with a moderately short tail and rusty brown plumage, contrasting with its bright blue cap.

Distribution

Except for the rail-babbler, the family is restricted to Australia and New Guinea. Australia has eight to nine species (four quail-thrushes and four to five whipbirds). The genera have both mesic and arid representatives. Quail-thrushes occur across mainland Australia and Tasmania, with species generally replacing each other with minimal overlap. Whipbirds are distributed along the eastern and southern coasts, while wedgebills are birds of the interior. Among New Guinea's 8–9 species, the Papuan whipbird, melampittas, and ifrit are montane birds; the ifrit has been found at 12,145 ft (3,680 m). The painted quail-thrush (*C. ajax*) is a lowland species. The 3–4 jewel-babblers have representatives at all elevations, with species largely replacing each other at different altitudes. The rail-babbler inhabits lowland areas of peninsular Thailand and Malaysia, Sumatra, and Borneo.

Habitat

The rail-babbler and New Guinea species are birds of the rainforest, occupying a variety of closed forest types. The greater melampitta is restricted to forest on rugged karst environments. In Australia, only the eastern whipbird (*P. olivaceus*) is found in rainforest; it also occurs in wetter eucalypt forests and other low dense vegetation. The western whipbird (*P. nigrogularis*) occurs in dense heath and other thick, drier vegetation. Wedgebills are birds of dry woodlands, steppes, and heathlands. Other than the spotted quail-thrush (*C. punctatum*), which prefers dry sclerophyll forest with open understorey, the other Australian quail-thrushes live in a range of arid habitats, including dry woodlands, shrub steppe, stony plains, and sandhills.

Behavior

The ifrit is not particularly shy, but the other members of the family are usually secretive. Rainforest species often secret themselves in dense vegetation. When disturbed, quail-thrushes stand stationary or burst quail-like into flight. Upon alighting they either freeze or run swiftly on foot. Whipbirds and melampittas are curious and will often slowly approach a quiet observer for a better look.

These birds are much more often heard than seen. They have distinctive voices, which are often the first indication of their presence. The vocalizations vary from thin whistles to loud explosive notes. The male and female eastern whipbirds participate in antiphonal duets that give this species its name. The male makes a loud whistle like a whip passing through the air, to which the female immediately adds two loud cracks.

Feeding ecology and diet

Most of these birds feed on ground, walking slowly, shuffling or tossing ground litter aside, and picking at prey with bill; the legs are rarely, if ever, used to clear leaves. Insects and other invertebrates, and occasionally small vertebrates, are eaten; seeds may also be taken. The exception is the ifrit. It forages at any elevation in the forest, clambering about the trunks and branches, probing in the bark and moss for prey.

Reproductive biology

Most members of this family construct cup-shaped nests. These vary from that of quail-thrushes, which is made of dry vegetation placed in a small depression on the ground, to the bulky, thick-walled nest of the ifrit, situated about 10 ft (3 m) from the ground. In contrast, the lesser melampitta builds a domed nest with the entrance on the side and places it up the side of a tree fern trunk. Quail-thrushes, jewel-babblers, and the rail-babbler lay two eggs, the other species one. These are covered with dark spots and blotches against a pale background. The Australian whipbirds and wedgebills have light blue eggs strongly marked with black scribbles. The roles of the sexes and the lengths of incubation and nestling periods are poorly known for many members of this family.

Conservation status

In 2000, the IUCN listed the rail-babbler and western whipbird as Near Threatened through loss of habitat. The population of the western whipbird in the southwest Australia is endangered. Inappropriate fire regimes are blamed for the small distribution and small numbers. From a low of 17 pairs or less in the 1960s, this species has been slowly recovered through dedicated conservation efforts. Restriction of burning, captive breeding, and transfer of individuals have brought the population to over 500 individuals. The subspecies of the spotted quail-thrush from the Mount Lofty Ranges, South Australia, may have been extirpated through loss of habitat.

The other species in this family appear secure for the moment, but several populations are vulnerable, restricted to relatively small areas at higher elevations, or are sparsely distributed. The greater melampitta occurs throughout montane New Guinea, but is known from less than ten specimens. The Papuan whipbird is too poorly known to make a realistic assessment of its status.

Significance to humans

Most of these species are sufficiently cryptic or remote that they remain unknown to most people. The exception is the eastern whipbird, whose call is a well known and characteristic sound of the Australian bush. Even then, far more people have heard this bird than seen it.

The ifrit was shown in 2000 by Jack Dumbacher and colleagues to carry several types of toxins in its tissues, particularly the feathers. The purpose of this poison is unclear, but its acquisition appears to be related to the bird's diet. This is the second genus of New Guinea bird known to have such toxins, the other being the pitohuis *Pitohui* of the Pachycephalidae.

1. Blue-capped ifrit (*Ifrita kowaldi*); 2. Eastern whipbird (*Psophodes olivaceus*); 3. Blue jewel-babbler (*Ptilorrhoa caerulescens*); 4. Spotted quail-thrush (*Cinclosoma punctatum*); 5. Rail-babbler (*Eupetes macrocerus*). (Illustration by John Megahan)

Species accounts

Spotted quail-thrush
Cinclosoma punctatum

TAXONOMY
Turdus punctatus Shaw, 1794, New South Wales, Australia.

OTHER COMMON NAMES
English: Spotted ground-bird; French: Cinclosome pointillé;
German: Fleckenflöter; Spanish: Tordo Cordoniz Manchada.

PHYSICAL CHARACTERISTICS
10.2–11 in (26–28 cm); 2.4–3.1 oz (67–87 g). Mottled plumage
of white, buff, rufous, brown, and black. Light brown head
with white brow stripe. Throat black with white patch; pinkish
breast.

DISTRIBUTION
C. p. punctatum: S. E. Australia; *C. p. dovei*: Tasmania; *C. p.
anachoreta*: Mount Lofty Ranges, south central Australia.

HABITAT
Eucalypt forest with littered open floor, particularly on rocky
hillsides.

BEHAVIOR
Terrestrial, sedentary, and rather shy. When startled, takes off
from ground like a quail; upon landing it runs away or freezes,
relying on its coloration to conceal it. Vocalizations include a
repeated, double-note song and a high thin contact call, inaudible to many people.

FEEDING ECOLOGY AND DIET
Eats insects and other invertebrates, and occasionally small vertebrates and seeds. It searches for prey on the ground while
walking in a slow, meandering path.

REPRODUCTIVE BIOLOGY
Late July–August to December. The female builds the nest, a
cup of dry vegetation, which is placed in a depression in the
ground near the base of a tree, shrub, rock, or clump of grass.
She incubates the two spotted eggs. The male assists her in
feeding the chicks during and after the 19-day fledging period.
One to three broods may be raised in a season.

CONSERVATION STATUS
Generally sparse but locally common. The population in the
Mt. Lofty Ranges, South Australia, is Critically Endangered, if
not already extinct. This has been attributed to habitat clearance and fragmentation.

SIGNIFICANCE TO HUMANS
None known. ◆

Blue jewel-babbler
Ptilorrhoa caerulescens

TAXONOMY
Eupetes caerulescens Temminck, 1835, Lobo, Irian Jaya,
Indonesia.

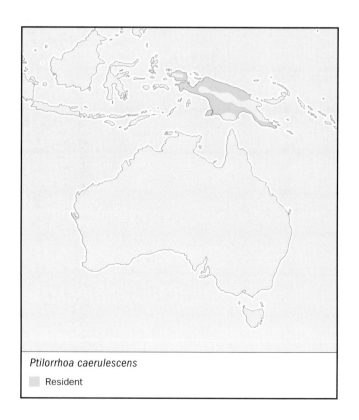

Cinclosoma punctatum
▨ Resident

Ptilorrhoa caerulescens
▨ Resident

OTHER COMMON NAMES
English: Lowland eupetes; French: Ptilorrhoa bleu; German: Blauflöter; Spanish: Hablantín del Valle.

PHYSICAL CHARACTERISTICS
8.7 in (22 cm); 1.7–2.2 oz (49–61 g). A smaller bird with mostly blue plumage. Black eye stripe from bill to breast, black primaries, and white throat and upper breast.

DISTRIBUTION
P. c. caerulescens: west New Guinea; *P. c. neumanni*: north New Guinea; *P. c. nigricrissa*: south New Guinea; *P. geislerorum*: northeast New Guinea (sometimes regarded as separate species). Lowlands and foothills up to about 990 ft (300 m).

HABITAT
Inhabits wetter forests, such as rainforest, monsoon forest, adjoining tall secondary growth, and gallery forest.

BEHAVIOR
Sedentary and territorial. Usually found in pairs or small groups. Terrestrial, walks daintily along the ground, with a bobbing head. Elusive, occupying areas of low vegetation cover. This species' song is among the more characteristic bird calls of the forest.

FEEDING ECOLOGY AND DIET
Feeds on insects and small invertebrates of the ground litter. When foraging, tosses leaves to aside using the bill.

REPRODUCTIVE BIOLOGY
The cup-shaped nest is placed in a depression in the ground at the base of a tree or between roots. The two eggs are spotted and streaked against a light brownish background. Breeding takes place from the mid dry to early wet season.

CONSERVATION STATUS
Generally common.

SIGNIFICANCE TO HUMANS
None known. ◆

Rail-babbler
Eupetes macrocerus

TAXONOMY
Eupetes macrocerus Temminck, 1831, Padang, Sumatra.

OTHER COMMON NAMES
English: Malaysian rail-babbler; French: Eupète à longue queue; German: Rallenläufer; Spanish: Hablantín Malayo.

PHYSICAL CHARACTERISTICS
11.4 in (29 cm). Brownish plumage with black legs and bill, white brow stripe and black eye stripe, sometimes with bluish stripe underneath.

DISTRIBUTION
E. m. macrocerus: Peninsular Thailand and Malaysia, Sumatra; *E. m. borneensis*: Borneo. Lives up to 3,300 ft (1,000 m).

HABITAT
Broad-leaved evergreen forest in lowland and hills.

BEHAVIOR
Very shy. Drawn-out monotone whistle heard much more than bird is seen. Walks on forest floor, jerking head like chicken; runs with great speed, holding the head up and the tail depressed.

Eupetes macrocerus
▨ Resident

FEEDING ECOLOGY AND DIET
Eats insects and other small ground invertebrates.

REPRODUCTIVE BIOLOGY
Breeds January–July, laying two eggs.

CONSERVATION STATUS
Scarce to locally fairly common resident. Considered Near Threatened by the IUCN and Birdlife International owing to extensive logging in parts of its range.

SIGNIFICANCE TO HUMANS
None known. ◆

Eastern whipbird
Psophodes olivaceus

TAXONOMY
Corvus olivaceus Latham, 1801, New South Wales, Australia.

OTHER COMMON NAMES
English: Stockwhip bird, green-backed whipbird; French: Psophode à tête noire; German: Schwarzschopf-Wippflöter; Spanish: Pájaro Látigo Verde.

PHYSICAL CHARACTERISTICS
10.4–12 in (26.5–30.5 cm); 1.6–2.5 oz (47–72 g). Dark olive-green body, black head with small black crest, white cheeks, and long tail.

DISTRIBUTION
P. o. lateralis: northeast Australia; *P. o. olivaceus*: central east and southeast Australia.

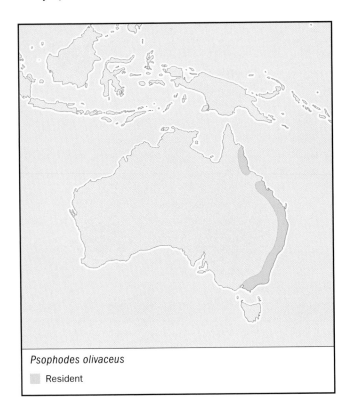

Psophodes olivaceus
▨ Resident

HABITAT
Rainforest, eucalypt forest, riverine vegetation, other low dense vegetation.

BEHAVIOR
Sedentary and territorial at all seasons. Usually stays on the ground or in low vegetation but may sing from high perch. Furtive and usually difficult to observe, but curious, approaching a quiet observer. Song an antiphonal duet, male producing soft swish then drawn out, explosive whip-crack to which female adds two quick, low notes.

FEEDING ECOLOGY AND DIET
Eats insects and other small invertebrates obtained by rummaging through leaves and other litter on forest floor with the bill.

REPRODUCTIVE BIOLOGY
One to two broods are reared in the season, which runs July–December. Two or three eggs, bluish white with black and gray spots or scribbles, are laid in the cup-shaped nest placed in dense vegetation near the ground. The female incubates and cares for the young, with the male assisting in their feeding. Incubation 18 days; young leave nest prematurely at 11–12 days.

CONSERVATION STATUS
Common but affected by clearing.

SIGNIFICANCE TO HUMANS
The whip-crack song is one of Australia's most familiar bird sounds, known to many people who have never seen the bird, and the source of many colloquial names for this species. ◆

Blue-capped ifrit
Ifrita kowaldi

TAXONOMY
Todopsis kowaldi De Vis, 1890, Owen Stanley Mountains, New Guinea. Two subspecies.

OTHER COMMON NAMES
English: Blue-capped babbler; French: Ifrita de kowald; German: Blaukappenflöter; Spanish: Hablantín de Gorra Azul.

PHYSICAL CHARACTERISTICS
6.5 in (16.5 cm); 1.1–1.3 oz (30–36 g). A small bird with olive upperparts and light brownish golden cheeks and underparts. White eye stripe, black bill, and blue cap from forehead to nape.

DISTRIBUTION
T. k. kowaldi: central and east New Guinea; *T. k. brunnea*: New Guinea: central and west New Guinea. From 4,818 to 12,144 ft (1,460 to 3,680 m), mainly 6,600–9,570 ft (2,000–2,900 m).

HABITAT
Montane forest, particularly mossy forest.

BEHAVIOR
Active; unlike most species in this family, it is not shy and can be easily observed. Usually seen in pairs or small groups, sometimes encountered in mixed-species foraging parties.

FEEDING ECOLOGY AND DIET
Feeds mainly on insects, occasionally soft fruit. Forages from near ground level on fallen logs to branches in the upper canopy. Creeps up trunks and along branches like a nuthatch-like fashion, probing in the moss for food and bending around branches to inspect the undersides.

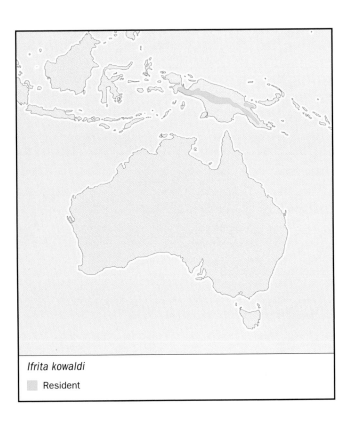

Ifrita kowaldi
▨ Resident

REPRODUCTIVE BIOLOGY

The nest is placed about 12 ft (3.6 m) from the ground. It consists of moss made into a bulky, deep cup. A single egg is laid; this is white with dark spots most densely placed at the large end. Records of active nests indicate that the breeding season is the mid-late dry season to early and late wet season.

CONSERVATION STATUS

Generally fairly common, though thinly distributed in many places.

SIGNIFICANCE TO HUMANS

Recognized by the local people as unsuitable for eating. It was recently found that this bird has a toxin in its feathers and soft tissues. ◆

Resources

Books

Coates, Brian J. "Passerines." In *The Birds of Papua New Guinea*. Vol. 2. Alderley: Dove Publications, 1993.

Higgins, P. J. and J. M. Peter, eds. *Handbook of Australian, New Zealand and Antarctic Birds*. Vol. 6, *Pardalotes to Figbird*. Melbourne: Oxford University Press (in preparation).

Schodde, R. and I. J. Mason. "Passerines." In *The Directory of Australian Birds*. Collingwood: CSIRO Publishing, 1999.

Periodicals

Dumbacher, J. P., T. F. Spande, and J. W. Daly. "Batrachotoxin Alkaloids from Passerine Birds: A Second Toxic Bird Genus (*Ifrita kowaldi*) from New Guinea."

Proceedings of the National Academy of Sciences USA. 97 (2000): 12970–12975.

Smith, G. T. "Ecology of the Western Whipbird *Psophodes nigrogularis* in Western Australia." *Emu* 91 (1991): 145–157.

Organizations

Birds Australia. 415 Riversdale Road, Hawthorn East, Victoria 3123 Australia. Phone: +61 3 9882 2622. Fax: +61 3 9882 2677. E-mail: mail@birdsaustralia.com.au Web site: <http://www.birdsaustralia.com.au>

Walter E. Boles, PhD

Fantails
(Rhipiduridae)

Class Aves

Order Passeriformes

Suborder Passeri (Oscines)

Family Rhipiduridae

Thumbnail description
Small to medium-small slim birds with flat, triangular bills, double rows of long rictal bristles, and prominent long, rounded, fanned tails.

Size
5.5–8.5 in (14–21.5 cm); 0.2–0.9 oz (6–25 g)

Number of genera, species
2 genera; 40 species

Habitat
Rainforest, forest, mangroves, woodland, savanna, shrub steppe, heath, grasslands, areas near human habitation

Conservation status
Vulnerable: 2 species; Near Threatened: 5 species

Distribution
Australia, New Zealand, New Guinea, southwest Pacific islands, Micronesia, Philippines, Indonesia, South and Southeast Asia

Evolution and systematics

Fantails, also called wagtail flycatchers, are a component of the large Australo-Papuan songbird group, the so-called "corvine radiation." They are closest to the drongos of the Dicruridae family and monarch flycatchers of the Monarchinae family. The fantails are either classified in their own family or as a subfamily of the monarch flycatchers, or in a broader drongo-monarch grouping. There is no doubt that they represent a distinctive group of insectivorous birds. Generally the 40 or more species are referred to the single genus *Rhipidura*, although the yellow-breasted fantail (*R. hypoxantha*) is sometimes placed in the monotypic genus *Chelidorhynx*. Within the genus *Rhipidura*, there are several distinct subgroups.

The largest of the fantails, the willie wagtail (*R. leucophrys*), is perhaps the most atypical member of the genus. It spends much more time on the ground than other species. The northern fantail (*R. rufiventris*) is one of several more sedate species; although capturing prey in flight, these do so without the aerial flourishes of most fantails. The thicket-fantails, like the sooty thicket-fantail (*R. threnothorax*), are shy and secretive inhabitants of dense rainforest understory. The remaining species of fantail, however, are active and conspicuous. These are distributed across the family's range. The gray fantail (*R. albiscapa*), streaked fantail (*R. spilodera*), and rufous fantail (*R.*

rufifrons) species-groups each include a number of species spread across substantial parts of Indonesia, Australasia and the southwestern Pacific. Other species have less obvious connections with other fantails. The tiny yellow-bellied fantail has a unique color pattern and other minor differences, and the blue-headed fantail (*R. cyaniceps*) and blue fantail (*R. superciliosa*) are also unusual in their color; otherwise these are fairly typical members of this family.

A species may be distributed on mainlands and many islands across a wide area. This makes it difficult to differentiate between species when they are so closely related. The species limits among the 13 or so members of the rufous fantail group are sometimes rather arbitrary. In his detailed 1987 review of Australian populations of the gray fantail, Julian Ford recognized the mangrove-inhabiting population of the northern coasts as a distinct species, the mangrove fantail (*R. phasiana*). Subsequently, studies by Richard Schodde and Ian Mason of the gray fantail both in and beyond Australia recommended that the New Zealand populations be specifically classified as New Zealand fantails (*R. fuliginosa*), and the Australian birds as *R. albiscapa*. Richard Holdaway and his colleagues went even further, dividing the fantails of New Zealand's North, South, and Chatham Islands into three different species. Multi-island distributions have given rise to

Fantail display. (Illustration by Barbara Duperron)

numerous subspecies in widespread species; the northern fantail, for example, has 24 named forms scattered across the Bismarck Archipelago, northern Australia, New Guinea, and islands of the Moluccas and Lesser Sundas.

On occasion, the two small flycatchers in the genus *Culicapa* have been associated with the fantails. Although the proper taxonomic placement of these south and southeast Asian species is uncertain, it does not appear to belong to the genus *Rhipidura*.

Physical characteristics

Fantails are small birds, ranging from the yellow-bellied fantail at 5.5 in (15 cm) and 0.2 oz (6 g), to the larger willie wagtail at 8.5 in (21.5 cm) and 0.9 oz (25 g). The obvious feature of this family, from which the group name is derived, is the distinctive long, rounded, fan-shaped tail, which may represent 50% or more of a bird's total length. Fantails also have the flat, triangular bill characteristic of many aerial insectivores. The bill is surrounded by rictal bristles, in an unusual double row arrangement. With the exception of the more terrestrial species, fantails have short tarsi and small feet. The wings are rather rounded, sacrificing speed but allowing a highly maneuverable, usually very buoyant flight.

Fantails are generally not brightly plumaged, with the major colors being brown, rufous, white, gray, or black, or a combination of these. This generalization breaks down along the northwestern and western periphery of the family's distribution, where two species have large amounts of blue in the plumage and another one, yellow. The overall color of many species is muted or drab, but a few species, such as the black-and-cinnamon fantail (*R. nigrocinnamomea*), have boldly contrasting colors. Males and females rarely exhibit dimorphism in their plumage; an exception is the black fantail (*R. atra*) of New Guinea, with black males and rufous females. The New Guinea dimorphic fantail (*R. brachyrhyncha*) also has two color phases: a dark one in which the tail is black and rufous, and a light phase with a light gray tail. Young fantails look like adults with washed-out colors and rusty edges to some feathers, particularly the wing coverts. Widespread species, particularly those that are found over a number of islands, show marked plumage variation among the populations.

Distribution

Although primarily an Australasian family, the fantails distribute well beyond this region. Species are found in eastern Pakistan, India, Sri Lanka, the Himalayas, southern China,

A willie wagtail (*Rhipidura leucophrys*) in flight. (Photo by C.H. Greenewalt/ VIREO. Reproduced by permission.)

southeast Asia, the Philippines, Indonesia, New Guinea, Australia, New Zealand, and the islands of the southwestern Pacific, east of Samoa and north to Micronesia. Several species may coexist, particularly in New Guinea, with up to seven found in the same locality. Some species are quite widespread, such as the white-throated fantail (*R. albicollis*), which is found from eastern Pakistan through south and southeast Asia to Borneo. Another example is the rufous fantail, which ranges from the Moluccas to the Santa Cruz islands in the east and to Micronesia in the north. Other species are restricted to only a single small island; the Ponapé fantail (*R. kubaryi*) and Matthias fantail (*R. matthiae*) are endemic to Ponapé, Micronesia, and to Mussau, in the Bismark Archipelago, respectively.

Habitat

The fantail family has representatives in many types of habitats, but most species are found in rainforest, either exclusively or as part of their broader range of habitat preferences. The mangrove fantail, as the name suggests, is restricted to mangroves. In Australia, the rufous fantail is primarily a bird of rainforest and wet sclerophyll forest in the breeding and non-breeding seasons; during migration, individuals may wander into a range of more open habitats, including city centers. The greatest diversity of habitat preferences is that of the willie wagtail. Although it favors open situations, this species occurs in all but the densest rainforest, and seems equally at home in deserts and city parks. Where several species of fantails occupy the same habitat, they favor different elevations in the forest or different parts of the tree. Some, like the sooty thicket-fantail, choose low, dense thickets. The willie wagtail is the most terrestrial species, spending much of its time foraging on the ground.

Behavior

The tail is usually held cocked, alternately fanned and closed, and often swung from side to side while a bird is perched or moving about the foliage. This tail posture is also used when species engage in highly aerobatic, looping flights to capture flying insects. So active are these flights at times that they have been described as "hysterical." These species are restless, rarely perching still for long. This behavior gave the gray fantail the colloquial nickname of "mad fan". Some species, such as the northern fantail, are more sedate, spending more time perched and making more direct sallies for insects. The willie wagtail rarely fans the tail, but swings it energetically from side to side.

Many fantail species are tame and confiding towards humans. They may come close to capture insects flushed by a moving observer. Other large, harmless animals are used in a similar fashion; willie wagtails frequently use domestic cattle both as a perch and to flush insects. This behavior contrasts with the response to animals perceived as predators or territorial intruders. Fantails can be boldly aggressive towards larger birds, fearlessly attacking birds of prey, even landing on their backs. Aggressive behavior in a willie wagtail is signaled by its rasping, scolding calls and a greatly expanded white eyebrow. If the conflict involves a territorial dispute, the losing bird will shrink the eyebrow until it is no longer visible. The thicket-fantails, unlike most species, are shy and secretive in the presence of humans and can be difficult to see as they skulk in dense undergrowth.

As a general rule, populations near the equator do not migrate. Thus, most tropical species are sedentary, remaining in the same area throughout the year. In more southern temperate regions, and at higher elevations, birds may exhibit marked seasonal movements. In Australia, the rufous fantail regularly moves north and south along the east coast. The gray fantail has more pronounced shifts, with southeastern populations moving extensive distances north and northwest

A willie wagtail (*Rhipidura leucophrys*) broods its chicks during a rainstorm. (Photo by Peter Slater. Photo Researchers, Inc. Reproduced by permission.)

in winter. Species like the white-throated and yellow-bellied fantails, which spend the summer in the Himalayas, move to lower altitudes at the end of the season.

Fantails are not particularly noted songsters, although their voices are by no means unpleasant. They are not strong and the calls are simple. Songs, which can be uttered frequently, are rapid and enthusiastic. That of the gray fantail has been likened to the notes of a violin. The exception to possessing a soft song is the willie wagtail. Both its scolding call and song are robust compared to those of other fantails and can be heard for some distance. The song, transliterated as "sweet pretty creature", may be repeated for quite extended periods, particularly on moonlit nights.

Feeding ecology and diet

Other than that of a few larger species, such as the willie wagtail, the bill of fantails is not strong enough to handle large, robust prey. The major food is small insects and other invertebrates. Larger prey, such as moths, may be hammered on a branch to subdue and render it suitable to be eaten. The willie wagtail has captured and eaten small skinks. Most prey are caught in the air, gleaned from foliage, or, less often, pursued on the ground. The flight of a gray fantail for a flying insect can be an impressively dizzying aerial pursuit. Rapid loops and sudden changes of direction appear to threaten to break the bird apart. The tail is held cocked and spread as a

A gray fantail (*Rhipidura albiscapa*) at its nest. (Photo by R. Brown/ VIREO. Reproduced by permission.)

bird moves through the foliage, and it has been suggested that this may assist in flushing insects.

In a study of three coexisting Australian fantails, Elizabeth Cameron found that they partitioned the environment, each selecting different heights and sections of the substrate, and using different foraging techniques, thus reducing overlap and competition. During the summer, gray fantails fed across a broad range of elevations, occasionally in the canopy at heights of about 130 ft (40 m). The predominant foraging method was to search for prey from a lookout perch, then pursue it in the air, before returning to a new vantage point. Rufous fantails generally fed lower, in the understory and shrub strata, and spent more time actively moving through the foliage flushing insects. Willie wagtails remained within 10 ft (3 m) of the ground and often on it, either capturing prey in the air with short flights or pursuing it on foot on the ground. In winter, when the rufous fantails migrated away from the area, gray fantails became more aerial and willie wagtails more terrestrial.

Reproductive biology

Some or all aspects of the breeding biology are unknown for a number of the rarer or more remote species. Conversely, several of the common, widespread species have been particularly well studied. Most fantails have quite similar breeding patterns. Both sexes construct the nest, a small, tidy cup of fine grass stems bound together by a thick external coating of cobwebs. This is placed in a horizontal fork or sometimes in a human-made structure or other suitable site, from 3–50 ft (1–15 m) from the ground, but usually within less than 10 ft (3 m). Most species for which the nest is known attach a dangling "tail" of nesting material to its underside. The clutch includes 2–4 eggs. These are pale or cream, marked with brown or gray blotches and spots, which form a wreath at the larger end or around the midline. Eggs of the yellow-bellied fantail differ somewhat by being cream or pinkish cream with a cap on the larger end consisting of pinkish brown stippling. Both parents incubate the eggs for 12–14 days. After hatching, chicks remain in the nest for 13–15 days, attended by both parents.

Because there is little attempt at concealment, fantail nests may be subject to considerable predation by larger birds. Richard Major and his team studied predation rates and the major perpetrators by building artificial nests with eggs made from modeling clay. Through direct observation and by identifying the bite marks left in the false eggs, he determined that more than ten avian species and several small mammals attempted to steal the eggs. The major predator was the pied currawong (*Strepera graculina*), which was responsible in more than half of the directly observed raids. Fantails also suffer from nest parasitism by cuckoos. A rufous fantail can be the host of a vigorously begging young pallid cuckoo (*Cuculus pallidus*), a bird that is up to eight times its weight.

Conservation status

Many species of fantails are common, and those found on larger landmasses appear not to be under threat. Several island populations, however, have been identified as in threatened or approaching this unfortunate status. The IUCN and Birdlife International have listed two species as Vulnerable. The Malaita fantail (*R. malaitae*), on Malaita, in the Solomon Islands, have a small estimated population, with recent records from only two locations. For this species, the threat factors affecting it are uncertain at present. The Manus fantail (*R. semirubra*), from the Admiralty Islands (Papua New Guinea), also has a small range. Once common on Manus Island, there have been no records since 1934. Although it is still found on neighboring islands, the reasons for this decline are not known.

Five species are considered Near Threatened: cinnamon-tailed fantail (*R. fuscorufa*) and long-tailed fantail (*R. opistherythra*), both of the Tanimbar Islands; Cockerell's fantail (*R. cockerelli*), Solomon Islands; dusky fantail (*R. tenebrosa*), San Christobal, Solomon Islands; and Matthias fantail of the Mussau, Bismarck Archipelago. Significant amounts of logging across critical parts of these species' ranges has been implicated as a threatening process in all cases.

Because distribution of some species is restricted to remote islands or high elevations, it is difficult to assess the status of some populations. These may be not be under current threat but, because of small population sizes, they are potentially susceptible to introduced species or habitat alteration.

Significance to humans

In places, fantails are well known because they are common and active. Combined with their tame and confiding nature around humans, this has made them particular favorites among birdwatchers and the general public. In parts of New Guinea, the willie wagtail is thought to be the ghost of a paternal relative. A singing bird around a newly planted garden means that crops will flourish. Elsewhere, it has the reputation as a gossip; it is not persecuted, but important business is not discussed when the bird is present.

1. Northern fantail (*Rhipidura rufiventris*); 2. Yellow-bellied fantail (*Rhipidura hypoxantha*); 3. Dimorphic fantail (*Rhipidura brachyrhyncha*); 4. Gray fantail (*Rhipidura albiscapa*); 5. Rufous fantail (*Rhipidura rufifrons*); 6. White-throated fantail (*Rhipidura albicollis*); 7. Blue-headed fantail (*Rhipidura cyaniceps*); 8. Streaked fantail (*Rhipidura spilodera*); 9. Willie wagtail (*Rhipidura leucophrys*); 10. Sooty thicket-fantail (*Rhipidura threnothorax*). (Illustration by Jacqueline Mahannah)

Species accounts

Yellow-bellied fantail
Rhipidura hypoxantha

TAXONOMY
Rhipidura hypoxantha Blyth, 1843, Darjeeling, India.

OTHER COMMON NAMES
English: Yellow-bellied fantail flycatcher; French: Rhipidure à ventre jaune; German: Goldbauch-Fächerschwanz; Spanish: Cola de Abanico de Vientre Amarillo.

PHYSICAL CHARACTERISTICS
4.5–4.9 in (11.5–12.5 cm); 0.2 oz (5–6 g). Yellow forehead, brow, and underparts; upperparts are olive-gray. Black mask and long, white-tipped gray tail. Female's mask is dark olive in color.

DISTRIBUTION
Northern India, southeastern Tibet, southwestern China, Nepal, Sikkim, Bhutan, northern Bangladesh, northern Myanmar, northern Thailand, and northern Vietnam. Occurs at 4,950–12,210 ft (1,500–3,700 m), locally down to 594 ft (180 m) in winter.

HABITAT
Can be found in various kinds of forest and secondary jungle but prefers moist evergreen forest. In the breeding season, occurs mostly in mixed coniferous and birch or rhododendron forests.

BEHAVIOR
This is a very active and restless bird. It is constantly in motion, flicking its wings and fanning its tail. While foraging, it continually utters a high, thin "sip sip."

FEEDING ECOLOGY AND DIET
Feeds on small flying insects caught in the air. Often found in mixed-species feeding flocks.

REPRODUCTIVE BIOLOGY
Breeds in Apr.–Jul. The nest is a compact deep cup, attached to the upperside of a branch, 10–20 ft (3–6 m) above ground. The clutch consists of three cream to pinkish-cream eggs with tiny dark reddish speckles, which usually form a ring around the larger end.

CONSERVATION STATUS
Not threatened.

SIGNIFICANCE TO HUMANS
None known. ◆

Blue-headed fantail
Rhipidura cyaniceps

TAXONOMY
Muscipeta cyaniceps Cassin, 1855, Mt Makiling, Luzon, Philippine Islands. Four subspecies.

OTHER COMMON NAMES
French: Rhipidure à tête bleue; German: Blaukopf-Fächerschwanz; Spanish: Cola de Abanico Azul.

PHYSICAL CHARACTERISTICS
6.5–7.5 in (16.5–19 cm). Head, throat, and breast grayish blue. White brow and under eye stripe. Lower back, rump, and belly to undertail is rufous. Black central tail feathers.

Rhipidura hypoxantha
▢ Resident

Rhipidura cyaniceps
▢ Resident

DISTRIBUTION
R. c. pinicola: northwest Philippine Islands; *R. c. cyaniceps*: northeast Philippine Islands; *R. c. sauli*: western central Philippine Islands; *R. c. albiventris*: eastern central Philippine Islands.

HABITAT
This species can be found in all forest types, up to 6,600 ft (2,000 m).

BEHAVIOR
Usually found in family parties or mixed feeding flocks, this fantail is a noisy and conspicuous bird in the understory of the forest.

FEEDING ECOLOGY AND DIET
This species captures much of its insect food by hawking in the upper canopy. It is a frequent member of mixed-species foraging flocks.

REPRODUCTIVE BIOLOGY
Birds in breeding condition have been recorded in February and March. Two eggs are laid in a cup-shaped nest built on a branch.

CONSERVATION STATUS
Not threatened.

SIGNIFICANCE TO HUMANS
None known. ◆

White-throated fantail
Rhipidura albicollis

TAXONOMY
Platyrhynchos albicollis Vieillot, 1818, Bengal, India. Eleven subspecies.

OTHER COMMON NAMES
English: White-throated fantail flycatcher; French: Rhipidure à

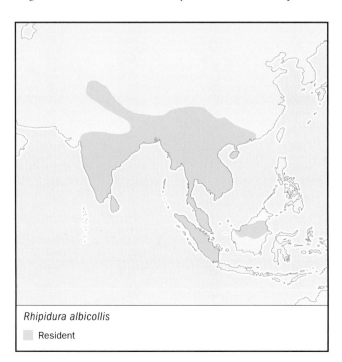

Rhipidura albicollis
☐ Resident

gorge blanche; German: Weisskehl-fächerschwanz; Spanish: Cola de Abanico de Garganta Blanca.

PHYSICAL CHARACTERISTICS
6.9–8.1 in (17.5–20.5 cm); 0.3–0.45 oz (9–13 g). Mostly gray with white throat, brow, and tip of tail.

DISTRIBUTION
Northeastern Pakistan, India, southeastern Tibet, southern China, Myanmar, Thailand, Asia, Peninsular Malaysia, Sumatra, Borneo. Foothills and adjacent plains up to 10,000 ft (3,000 m).

HABITAT
Inhabits broad-leaved evergreen forest and can be found locally in human-modified areas such as bamboo, parks, secondary regrowth, and wooded gardens.

BEHAVIOR
This is a typical fantail in its behavior—restless, constantly fanning its tail, and conspicuous. It is a bird of the understory and middle growth. In winter, it exhibits marked altitudinal migration, moving from higher elevations to foothills and plains. The song consists of thin, high-pitched notes.

FEEDING ECOLOGY AND DIET
Works along branches, as well as outside of foliage, feeding on small flying insects it disturbs. It may be encountered singly, in pairs, or as part of mixed hunting parties.

REPRODUCTIVE BIOLOGY
The breeding season varies throughout the range, from Feb.–May to Mar.–Aug; two broods may be raised. Both sexes build the nest and incubate the three spotted eggs. The small cup-shaped nest is made of fine grass stems held together by an external coating of cobwebs; it has a dangling "tail" of grasses below. Incubation period 12–13 days; fledging 13–15 days.

CONSERVATION STATUS
Not threatened.

SIGNIFICANCE TO HUMANS
None known. ◆

Rufous fantail
Rhipidura rufifrons

TAXONOMY
Muscicapa rufifrons Latham, 1801, Sydney, New South Wales, Australia. Twenty subspecies.

OTHER COMMON NAMES
English: Rufous-fronted fantail; French: Rhipidure roux; German: Fuchsfächerschwanz; Spanish: Cola de Abanico Rufo.

PHYSICAL CHARACTERISTICS
5.5–6.7 in (14–17.5 cm); 0.25–0.35 oz (7.2–10 g). Rufous-brown plumage with mottled throat. Tail color is rufous at base, to brown middle and whitish tips. White patch from chin to cheek and upper throat.

DISTRIBUTION
Moluccas and Lesser Sunda Islands, coastal eastern Australia, New Guinea, Bismarck Archipelago, Solomon Islands, Santa Cruz Islands, and Micronesia. Located from sea level to 6,600 ft (2,000 m).

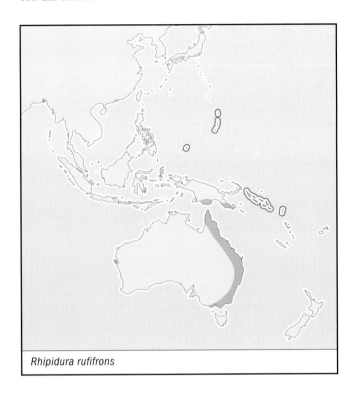

Rhipidura rufifrons

HABITAT
Rainforest, monsoon forest, mid-montane forest, riparian vegetation, swamp woodland, wet eucalypt forest, mangroves; on migration may occur in open or urban situations.

BEHAVIOR
Usually seen singly, sometimes in pairs. Frequents lower substage to mid-tree levels. Confiding. Restless; moves with rapid jerky movements. Strongly migratory in southeast Australia; tropical birds apparently not sedentary, but nature of movements uncertain.

FEEDING ECOLOGY AND DIET
Insectivorous. Favors lower and middle levels of foliage. Forages mainly by gleaning, occasionally by flycatching. Works quietly along slender outer branches, from the inner section to the extreme tip. May join mixed feeding parties.

REPRODUCTIVE BIOLOGY
Breeding season varies across range: Oct.–Feb. (Australia), Aug.–Jan. (New Guinea region), Feb.–Mar. (Micronesia). Sexes share nest construction, incubation, and care of young, rearing 1–2 broods per season. The nest is a compact cup of fine grass bound with spider web, with a pendant "tail" about 3 in (7.5 cm) long. The nest is placed in thin fork up to 16.5 ft (5 m) above the ground. The clutch includes 2–3 eggs; these are cream with small warm brown and lavender dots, mostly at large end. Incubation takes about 14 days.

CONSERVATION STATUS
In Australia, this species is common in the north, decreasing in abundance southwards; it is not considered under threat. Several of the Micronesian populations exist in low numbers and are potentially susceptible to loss of habitat and introduced predators.

SIGNIFICANCE TO HUMANS
None known. ◆

Dimorphic fantail
Rhipidura brachyrhyncha

TAXONOMY
Rhipidura brachyrhyncha Schlegel, 1871, Arfak Mountains, New Guinea. Two subspecies.

OTHER COMMON NAMES
English: Dimorphic rufous fantail; French: Rhipidure dimorphe; German: Zweiphasen-Fächerschwanz; Spanish: Cola de Abanico Dimorfo.

PHYSICAL CHARACTERISTICS
5.9–6.5 in (15–16.5 cm); 0.3–0.4 oz (8.3–10.7 g). Crown to upper back dark brown, fading to rufous lower back, rump, and wing edges. Underparts buffy gray with either pale or dark undertail, depending on color phase. Bill brownish black, feet pale brown.

DISTRIBUTION
R. b. brachyrhyncha: Northwestern New Guinea; *R. b. devisi*: western, central, and eastern New Guinea. Occurs mainly at 6,600–12,144 ft (2,000–3,680 m), as low as 3,828 (1,160 m) and as high as 12,870 ft (3,900 m).

HABITAT
Lives in mountain forest, particularly stunted moss forest.

BEHAVIOR
A tame and inquisitive bird, usually encountered in pairs, sometimes singly. Active.

FEEDING ECOLOGY AND DIET
Obtains most of its food by catching flying insects, although it will also glean items from foliage while hovering. One foraging

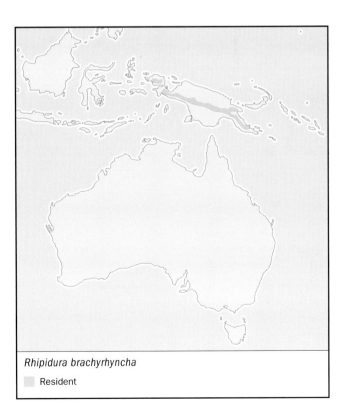

Rhipidura brachyrhyncha

☐ Resident

technique is to crash through the foliage, catching any flushed insects.

REPRODUCTIVE BIOLOGY
Nesting habits undescribed.

CONSERVATION STATUS
Not threatened.

SIGNIFICANCE TO HUMANS
None known. ◆

Streaked fantail
Rhipidura spilodera

TAXONOMY
Rhipidura spilodera G.R. Gray, 1870, Vanua Lava, New Hebrides. Five subspecies.

OTHER COMMON NAMES
French: Rhipidure tacheté; German: Fleckenfächerschwanz; Spanish: Cola de Abanico Moteado.

PHYSICAL CHARACTERISTICS
6.3–7.1 in (16–18 cm); female 0.3–0.35 oz (9.2-10.2 g), male 0.4 oz (11-12 g). Olive-brown upperparts with gray-brown tail feathers and two white wing bars. White brow stripe and throat, with scaly brown pattern on white breast. Abdomen is whitish buff.

DISTRIBUTION
R. s. verreauxi: New Caledonia and Loyalty Islands; *R. s. spilodera*: Central and northern Vanuatu; *R. s. layardi*: western Fiji; *R. s. erythronota*: northern Fiji; *R. s. rufilateralis*: northeastern Fiji.

HABITAT
True forest and forest edge, occasionally secondary growth and gardens. Occurs up to 3,960 ft (1,200 m).

BEHAVIOR
An active, restless, bold, and inquisitive bird. It frequents lower levels in the forest. The song is a pleasant series of 4–5 notes.

FEEDING ECOLOGY AND DIET
This fantail is an active feeder in the understory, capturing insects by gleaning and hawking. It frequently joins mixed-species feeding flocks.

REPRODUCTIVE BIOLOGY
The breeding season on New Caledonia and Vanuatu is Oct.–Jan. Two spotted eggs are laid in the small cup nest, which is situated in the lower strata, 6.6–10 ft (2–3 m) from the ground. Both parents share the nesting duties.

CONSERVATION STATUS
Common.

SIGNIFICANCE TO HUMANS
None known. ◆

Gray fantail
Rhipidura albiscapa

TAXONOMY
Rhipidura albiscapa Gould, 1840, Tasmania. Eight subspecies.

OTHER COMMON NAMES
English: White-shafted fantail; French: Rhipidure à collier; German: Graufächerschwanz; Spanish: Cola de Abanico Gris.

PHYSICAL CHARACTERISTICS
5.5–6.6 in (14.0–16.8 cm); 0.2–0.3 oz (6–9 g). Plumage color ranges from light gray to dark brown-gray. White from chin to top of throat; also white brow and tail tips.

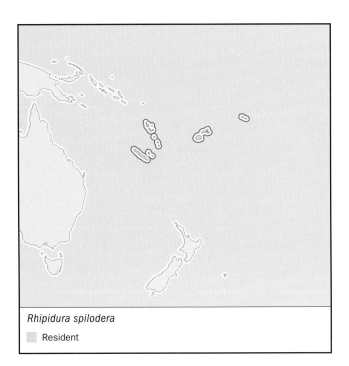

Rhipidura spilodera

    Resident

Rhipidura albiscapa

    Breeding   Nonbreeding

DISTRIBUTION

R. a. preissi: southwestern Australia, reaching northwestern Australia on migration; *R. a. alisteri*: south central, southeastern, and east central Australia, on migration to northern Australia and possibly New Guinea; *R. a. albiscapa*: Tasmania and Bass Strait islands, southeastern Australia on migration; *R. a. keasti*: northeastern Australia; *R. a. albicauda*: C. Australia; *R. a. pelzelni*: Norfolk Island; *R. a. bulgeri*: New Caledonia and Lifu, Loyalty Islands; *R. a. brenchleyi*: Vanuatu, Banks Islands; San Cristobal, Solomon Islands.

HABITAT

Can be found in almost any wooded habitat, including urban situations, rainforest, eucalypt forest and woodland, semi-arid scrublands, mangroves, and riverine vegetation.

BEHAVIOR

Territorial; usually solitary or in pairs; often joins mixed feeding parties. Extremely active and conspicuous; inquisitive and confiding. Frequents all levels of foliage. Populations of southern Australia and Tasmania are migratory, those of central Australia are nomadic and those in northeastern Australia and on Pacific islands are sedentary. Song is a series of high-pitched and thin but attractive notes, given in a "see-saw" cadence and sometimes likened to a violin.

FEEDING ECOLOGY AND DIET

Eats small insects. Forages mainly by hawking on the wing in involved intricate acrobatic chases; sometimes gleans from foliage.

REPRODUCTIVE BIOLOGY

In Australia, breeds Jul.–Jan.; produces one, two, or often three broods in a season. The sexes share the breeding duties. The nest is the typical fantail structure: a small cup of thin grass bound with spider web, with pendent tail up to 6 in (15 cm) long, placed in a fork. This is usually 6.5–16.5 ft (2–5 m), sometimes up to 33 ft (10 m), above the ground. The eggs, 2–4 in a clutch, are cream with small light brown and underlying pale gray spots form a wreath at the larger end. Incubation takes 14 days, age at fledging 10–12 days.

CONSERVATION STATUS

Common in most of its range.

SIGNIFICANCE TO HUMANS

A favorite with Australian birdwatchers. ◆

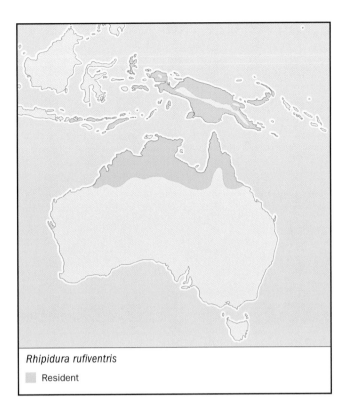

Rhipidura rufiventris

▨ Resident

Northern fantail

Rhipidura rufiventris

TAXONOMY

Platyrhynchos rufiventris Vieillot, 1818, Timor. Twenty-four subspecies.

OTHER COMMON NAMES

English: White-throated fantail; French: Rhipidure à ventre chamois; German: Witwen-Fächerschwanz; Spanish: Cola de Abanico Norteño.

PHYSICAL CHARACTERISTICS

6.5–7.1 in (16.5–18.0 cm); 0.35–0.6 oz (10–17 g). Gray-brown plumage with buff belly and white lateral tail feathers. White streaks on breast; white brow and chin to throat.

DISTRIBUTION

Northern Australia, New Guinea, Bismarck Archipelago, Moluccas, Lesser Sunda Islands. Found mainly in the lowlands and hills, locally up to 5,400 ft (1,640 m) on New Britain.

HABITAT

This species can be found in a range of habitats, including open eucalypt woodland, rainforest fringes, mangrove forest, monsoon forest, riverine vegetation, wooded swamps, tall secondary growth, forest edges, and garden areas.

BEHAVIOR

Territorial, usually solitary or in pairs. Often conspicuous when it chooses exposed perches. Often joins mixed-species feeding flocks. Less active than other fantails; spends more time perching and engages in more sedate aerial pursuits. Generally quiet, unobtrusive, undemonstrative. Typical stance upright with the tail held vertically.

FEEDING ECOLOGY AND DIET

Eats a variety of insects obtained almost entirely by hawking, although sometimes gleaned from branches and leaves. Commonly joins mixed-species foraging flocks of other small insectivorous songbirds.

REPRODUCTIVE BIOLOGY

A pair may produce 1–2 broods per season, which runs from Aug.–Jan. in Australia and from the mid-dry to the mid-wet season in New Guinea. Both parents build the nest, incubate, and care for the young. The female lays two spotted eggs in the small cup nest. As with most fantails, this has a tail about 2.8 in (7 cm) long hanging from the underside.

CONSERVATION STATUS

Generally common to fairly common; not threatened.

SIGNIFICANCE TO HUMANS

None known. ◆

Sooty thicket-fantail
Rhipidura threnothorax

TAXONOMY
Rhipidura threnothorax S. Müller, 1843, Lobo, Triton Bay, New Guinea. Two subspecies.

OTHER COMMON NAMES
French: Rhipidure fuligineux; German: Rosenberg-Fächerschwanz; Spanish: Cola de Abanico Sombrío.

PHYSICAL CHARACTERISTICS
16.5–18 cm. Crown and back dull olive-brown with blackish rump. Breast is black with distinctive white spots, with remaining underparts dark gray to black. White chin and brow stripe.

DISTRIBUTION
R. t. threnothorax: New Guinea and satellite islands; *R. t. fumosa*: Japen Island, New Guinea. Occurs in lowlands and hills up to about 3,630 ft (1,100 m).

HABITAT
Inhabits wet forests with particularly dense undergrowth, where it is largely restricted to the understory, usually no more than 10 ft (3 m) from the ground.

BEHAVIOR
Sedentary. Unlike most fantails, this shy species hides itself in dense undergrowth.

FEEDING ECOLOGY AND DIET
It feeds on small insects, most of which are obtained by gleaning from leaf litter and foliage and by hawking. It may accompany larger terrestrially-feeding birds, catching insects that they disturb.

REPRODUCTIVE BIOLOGY
Nesting habits and eggs undescribed.

CONSERVATION STATUS
Generally common to very common, but secretive and infrequently seen.

SIGNIFICANCE TO HUMANS
None known. ◆

Willie wagtail
Rhipidura leucophrys

TAXONOMY
Turdus leucophrys Latham, 1801, Sydney, Australia. Three subspecies.

OTHER COMMON NAMES
English: Black-and-white fantail; French: Rhipidure hochequeue; German: Gartenfächerschwanz; Spanish: Cola de Abanico Blanco y Negro.

PHYSICAL CHARACTERISTICS
7.1–8.7 in (18–22 cm); 0.6–0.8 oz (17–24 g). A large bird with black plumage and white brow and breast.

DISTRIBUTION
R. l. melanoleuca: Moluccas, New Guinea, and surrounding islands, Bismarck Archipelago, Solomon Islands; *R. l. picata*: northern Australia; *R. l. leucophrys*: southwest, southern, central, and southeast Australia.

HABITAT
This species can be found in almost any habitat except the densest rainforest or eucalypt forest; it prefers relatively open areas, from sea level to 9,240 ft (2,800 m).

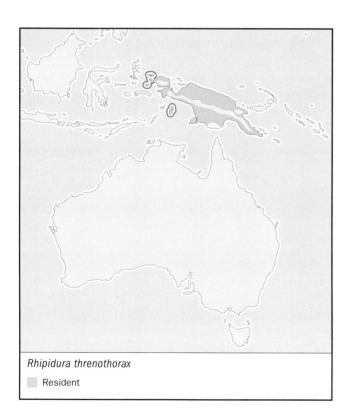

Rhipidura threnothorax
■ Resident

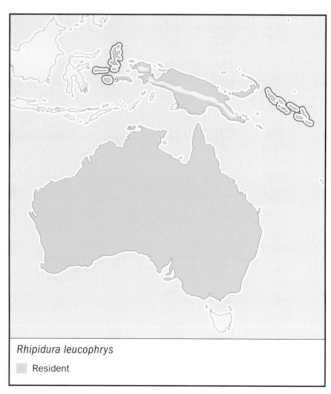

Rhipidura leucophrys
■ Resident

BEHAVIOR

Terrestrial for much of its time, running, walking, or hopping on the ground. As it does so, the tail is usually held elevated but not often fanned. When pausing, the tail is constantly waved from side to side and up and down. The willie wagtail is usually seen singly, although mates often are nearby. A conspicuous, active, and bold bird that often draws attention to itself by harassing or attacking larger animals that are considered as predators or enter territory during breeding. An aggressive individual, it conspicuously expands its white eyebrows. In Australia, this species is mainly sedentary or locally nomadic, while in New Guinea it may leave some areas in the dry season, returning to breed during the rains. The song, rendered as "sweet pretty creature", may be heard incessantly during breeding, often throughout a moonlit night. There is also a harsh scolding note given when a bird is agitated.

FEEDING ECOLOGY AND DIET

Feeds on insects, larvae, and occasionally small lizards. Much food is obtained by hawking from perches for insects on the wing, or snatching them from ground after short runs.

REPRODUCTIVE BIOLOGY

Although breeding occurs mainly in Jul.–Feb. (Australia), this species can nest in any month, conditions allowing. This may yield up to four or more broods to be reared in a season. Both parents share nest building, incubation, and care of the young. The nest is made of grass and fine bark strips, covered with spider web, but it lacks the tail of most fantails; this is placed on horizontal fork or in man-made structure or other suitable site, usually less than 16.5 ft (5 m) above the ground. The eggs are cream with brown and gray speckles forming a wreath at larger end. Incubation, 14–15 days; age at fledging 14 days.

CONSERVATION STATUS

Common throughout its range.

SIGNIFICANCE TO H UMANS

A well-known and popular bird in Australia. In parts of New Guinea it is considered to be a gossip or the ghost of a paternal kinsperson bringing good luck. ◆

Resources

Books

Coates, Brian J. *"The Birds of Papua New Guinea.* Vol. 2. *Passerines."* Alderley: Dove Publications, 1993.

Higgins, P.J. and J.M. Peter, eds. *Handbook of Australian, New Zealand and Antarctic Birds.* Vol. 6. *Pardalotes to Figbird.* Melbourne: Oxford University Press (in preparation).

Schodde, R. and I.J. Mason. *The Directory of Australian Birds. Passerines.* Collingwood: CSIRO Publishing, 1999.

Periodicals

Beck, J. and K. Chan. "Habitat Preference of Grey Fantails *Rhipidura fuliginosa* Wintering in Central Queensland." *Sunbird* 29 (1999): 41–51.

Major, R.E., G. Gowing, G. and C.E. Kendal. "Nest Predation in Australian Urban Environments and the Role of the Pied Currawong, *Strepera graculina." Australian Journal of Ecology* 21 (1996): 399–409.

Reis, K.R and R.S. Kennedy. "Review of the Montane Bird Species from Mindanao, Philippines: Part 1—Black-and-Cinnamon Fantail, *Rhipidura nigrocinnamomea." Bulletin of the British Ornithologist's Club* 119 (1999): 103–109.

Webb-Pullman, B.Z., and M.A. Elgar. "The Influence of Time of Day and Environmental Conditions on the Foraging Behaviours of Willie Wagtails, *Rhipidura leucophrys." Australian Journal of Zoology* 46 (1998): 137–144.

Organizations

Birds Australia. 415 Riversdale Road, Hawthorn East, Victoria 3123 Australia. Phone: +61 3 9882 2622. Fax: +61 3 9882 2677. E-mail: mail@birdsaustralia.com.au Web site: <http://www.birdsaustralia.com.au>

Walter E. Boles, PhD

Monarch flycatchers

(Monarchidae)

Class Aves

Order Passeriformes

Suborder Passeri (Oscines)

Family Monarchidae

Thumbnail description
Small to medium size insectivorous songbirds

Size
5–21 in (13–53 cm).

Number of genera, species
17 genera; 96 species

Habitat
Forest and open woodland

Conservation status
Critically Endangered: 5 species; Endangered: 6; Vulnerable: 9 species; Near Threatened: 14 species; Extinct: 2 species

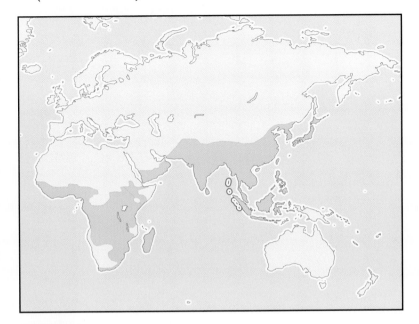

Distribution
Tropical Old World from sub-Saharan Africa to Indonesia

Evolution and systematics

The paradise-flycatchers and monarchs are the most impressive and colorful of the flycatcher-like birds. Formerly considered to be part of the 'true' flycatchers (Muscicapidae), they have now been separated into a family of their own, though DNA-DNA hybridization studies suggest that they are closely related to the crows (Corvidae). The Monarchidae have differently shaped jaw muscles from the true flycatchers and no 'turdine thumb' on the syrinx. They are probably of Asian origin, but have spread several thousand miles southwest and southeast and are closely related to the fantails of Asia, Australasia, and Polynesia. Within the order, the *Elminia* and *Erythrocercus* genera have different anatomical characteristics, and future studies may suggest that further reclassification is necessary.

Physical characteristics

Paradise-flycatchers and monarchs are small to medium in size, from 5–21 in (13–53 cm), although the tail can account for 6 in (15 cm) of the largest species. The bill is wide, typically blue-gray, with a fringe of well-developed bristles around the gape, and many species have a keeled ridge on the upper mandible. The Monarchidae have short legs, small and sharp, curved claws, and relatively long wings, but the tail length varies from short to incredibly long. The plumage is often brightly colored, and many species have impressive crests. Most show no plumage differences between males and females, although in a few species the sexual dimorphism is dramatically exaggerated.

Distribution

The family is found throughout the tropical Old World, from Senegal to Indonesia. However, while representatives of the order are widely distributed, 30 species are endemic to single small islands or archipelagoes, even tiny islets, such as Sangihe and Talaud in Indonesia (caerulean paradise-flycatcher, *Hypothymis coelestis*) and Annobón, in the Atlantic Ocean, several hundred miles off the coast of Gabon (Annobón paradise-flycatcher, *Terpsiphone smithii*).

Habitat

Monarchidae forest-dwellers many frequenting open areas, clearings, and the forest edge, with some species regularly feeding in parks, gardens and fruit plantations.

Behavior

Most monarch and paradise-flycatcher species live in pairs throughout the year, although a few live in loosely organized groups, with nonbreeding 'helpers' assisting pairs to rear their young. When excited or alarmed, many species fan their tail.

Feeding ecology and diet

Many Monarchidae are gleaners, busily searching for insects, particularly arthropods, in the foliage of trees and shrubs, although most species can feed aerially, catching flying insects on the wing in the style of the 'true' flycatchers.

Spectacled monarch flycatcher (*Monarcha trivirgatus*) nesting in Australian rainforest. (Photo by Frithfoto/Olympus. Bruce Coleman Inc. Reproduced by permission.)

Reproductive biology

Monarchidae are not shy birds, the males being noisy and showy in their efforts to attract a mate, but some species are found deep in forests where few people ever witness the display. The nest is a tiny, deep cup, woven of fibers, moss, lichens and spider webs, often fastened in the fork of a branch or twig. In most species, males and females incubate the eggs, which hatch after about 14 days. Juveniles are plain, often brown, only gaining their colorful plumage after the completion of the first molt.

Conservation status

Thrity-six species of Monarchidae are listed in *The 2000 Red List of Threatened Species*. Many are endemic to single islands in Southeast Asia. Five species are Critically Endangered, the rarest being the Tahiti monarch (*Pomarea nigra*), with only a handful of individuals remaining by the late 1990s. Loss of wooded habitat is the principal cause of population decline among the family. Two species are already extinct, in-

cluding the Guam flycatcher (*Myiagra freycineti*), which was common on the Pacific island of Guam until the 1970s but crashed to extinction in 1983 as a result of predation by the introduced brown tree snake (*Boiga irregularis*).

Significance to humans

None known.

The long tailfeathers of the African paradise-flycatcher (*Terpsiphone viridis*) are visible when it sits on its nest. (Photo by W. Tarboton/ VIREO. Reproduced by permission.)

1. Male African paradise-flycatcher (*Terpsiphone viridis*) in breeding plumage; 2. White-tipped monarch (*Monarcha everetti*); 3. Black-naped monarch (*Hypothymis azurea*); 4. African blue-flycatcher (*Elminia longicauda*); 5. Biak monarch (*Monarcha brehmii*). (Illustration by Emily Damstra)

Species accounts

African blue flycatcher
Elminia longicauda

TAXONOMY
Myiagra longicauda Swainson, 1838, Senegal. Two subspecies. Sometimes considered a superspecies with the white-tailed blue flycatcher (*E. albicauda*).

OTHER COMMON NAMES
English: Blue flycatcher, northern fairy flycatcher; French: Gobemouche bleu; German: Türkis Elminie; Spanish: Eliminia Azul.

PHYSICAL CHARACTERISTICS
Male is cobalt blue above and on the throat, paler below, with a long tail. Females and immatures are grayer, with a narrow band of blue on edges of wing and tail-feathers.

DISTRIBUTION
The widest distribution of the *Elminia* flycatchers, from the Atlantic coast of West Africa to Lake Victoria in the east. There are apparent gaps in its distribution, but this probably reflects a lack of human knowledge, even of such a colorful bird.

HABITAT
Resident in woodlands, from sea-level plains to 7,200 ft (2,400 m), inhabits clearings and edges of open secondary forest, swampy or riverine forests, and wooded savannas. Also found in cleared and abandoned fruit and cocoa plantations.

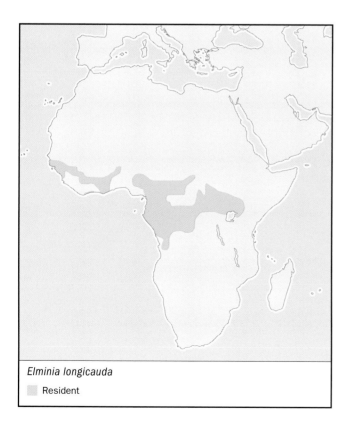

Elminia longicauda

▨ Resident

Live at lower altitudes than the similar white-tailed blue flycatcher in most places.

BEHAVIOR
Live in groups of up to 12, but usually 3–7 birds. Largest groups comprise several pairs or trios of adults with immatures or dependent young. A 'leader' defends the territory, singing frequently, with its crest erect, wings flicking and tail fanned.

FEEDING ECOLOGY AND DIET
Always on the move, picking small insects from the foliage in the top of trees and shrubs. Sometimes fly from a perch, catch a flying insect, and return to the branch, in true flycatcher fashion.

REPRODUCTIVE BIOLOGY
Appear to be monogamous, though some may be polygamous. Female lays and incubates one or two white eggs in a compact, cup-shaped nest. Both birds in the pair feed the young, with assistance from nonbreeding adults or immatures in the group.

CONSERVATION STATUS
Relatively common in good quality habitat, although endemic and breeding at a relatively low density.

SIGNIFICANCE TO HUMANS
None known. ◆

African paradise-flycatcher
Terpsiphone viridis

TAXONOMY
Muscicapa cristata Gmelin, 1789, based on Brisson, 1760, Senegal. Ten subspecies.

OTHER COMMON NAMES
French: Moucherolle de paradise; German: Graubrust-Paradeisschnäpper; Spanish: Monarca Paraíso Africano.

PHYSICAL CHARACTERISTICS
Largest paradise-flycatcher in Africa, with tail that constitutes two-thirds of its length. Large head and crest glossy blue-black, eye surrounded by bright blue ring. Upperparts and graduated, outer tail feathers russet-brown, underparts gray. Most obvious markings are broad bar in each wing (white in some races, black in others) and two white, central tail streamers, which can be 3.5 in (9 cm) long in largest males. Females similar, but smaller, and colors are less bold and glossy.

DISTRIBUTION
Widespread south of the Sahara, the only flycatcher found in most of eastern and southern Africa.

HABITAT
Commonest flycatcher in Africa, found in almost all habitats except arid zones. Equally at home in savanna woodland, open forest, and plantations, prefers edges and avoids dense forest. Birds also nest and feed in orchards, parks and well-established gardens, up to 8,202 ft (2,500 m) in east Africa.

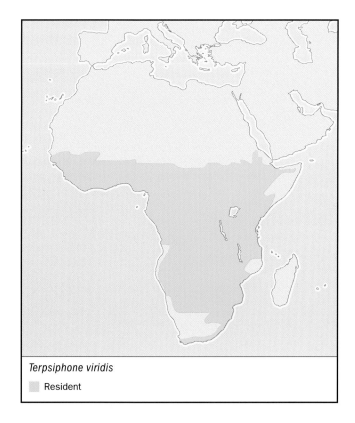

Terpsiphone viridis

Resident

Monarcha brehmii

Resident

BEHAVIOR
Usually found in ones or twos, often tame and vocal but usually unobtrusive. Long tail makes for a distinctive and graceful flight, often slow and undulating, with tail streamers waving. Some populations migrate within Africa, usually between habitats during the dry season.

FEEDING ECOLOGY AND DIET
Glean insects from the foliage, flit among the leaves to catch prey, or perch in wait for a passing insect, catching it in mid air. Feed on caterpillars, beetles, moths, butterflies, flying ants and termites, up to 1.2 in (3 cm).

REPRODUCTIVE BIOLOGY
Males patrol and defend territory with loud calls and songs, especially at dawn and dusk. Tail streamers and crest play important role in courtship display, combined with shivering wing tips, calls, and sometimes 'dancing' on a perch. Neat, tight cup nest is fixed with cobwebs to fork of a branch, into which two or three creamy white, oval eggs are laid. Both parents incubate eggs for about 15 days. Young stay in the nest for 11–15 days and remain dependent on the parents for a further week.

CONSERVATION STATUS
Common and widespread in a range of habitats.

SIGNIFICANCE TO HUMANS
None known. ◆

Biak monarch
Monarcha brehmii

TAXONOMY
Monarcha brehmii Schlegel, 1871, Biak Island, New Guinea.

OTHER COMMON NAMES
French: Monarche de Brehm; German: Falbschwanzmonarch; Spanish: Monarca Biak.

PHYSICAL CHARACTERISTICS
5.2 in (13 cm). One of the smallest monarchs. Males have black upperparts, throat, and breast, with pale yellow underparts, tail, and wing bars. Along with a short crest, its most significant feature is pale yellow crescent on each side of the head. Females similar, but have a whitish patch on throat and breast.

DISTRIBUTION
Endemic to the islands of Biak and Supiori in Cenderawasih (Geelvink) Bay, Papua (formerly Irian Jaya) in Indonesia.

HABITAT
The few birds found were in lowland forest, but it is suspected that they favor thick, elevated forests. The rainforests of Biak are very important for wildlife, with 13 endemic or near-endemic birds, 18 endemic butterflies, and five endemic mammals.

BEHAVIOR
Few individuals have been seen, so little is known. Its grating call is characteristic of the Monarchidae.

FEEDING ECOLOGY AND DIET
Appear to gain most of their insect food by gleaning from foliage in the subcanopy of the forest, and by hawking aerial insects from a perch.

REPRODUCTIVE BIOLOGY
Not known.

CONSERVATION STATUS
Listed as Endangered by the IUCN and BirdLife International, with a declining and small population. Recorded only four times between 1980 and 2000, though there have been no recent expeditions to the interior forests, where it may be more common. Logging for timber and clearing the land for farming has already destroyed the primary forest in southern Biak and threatens the northern areas. Fortunately, Supioro's forests are steeper, with fewer people, and so at lower risk.

New expeditions to find and protect the most important areas of forest are a priority for their conservation.

SIGNIFICANCE TO HUMANS
None known. ◆

Black-naped monarch
Hypothymis azurea

TAXONOMY
Muscicapa coeruleocephala Sykes, 1832, Decan, India. 19 subspecies recognized.

OTHER COMMON NAMES
English: Pacific monarch; French: Tchitrec azuré; German: Schwarzgenickschnäpper; Spanish: Monarca Azul de Nuca Negra.

PHYSICAL CHARACTERISTICS
6 in (16 cm). Long-tailed, slender. Similar in shape to fantails, but coloration is bolder. Males bright blue, with black patch on nape and black band on throat. Females grayer, lack black markings.

DISTRIBUTION
From India in the west, across southern Asia to southern China and the island groups of Hainan, Taiwan, Sundas, Philippines, Andamans, and Nicobars.

HABITAT
Common in scrub, forest and overgrown plantations, mostly in the lowlands, but up to 4,265 ft (1,300 m) in some areas. Birds tend to be found in the medium or lower canopy, but nests can be close to the ground.

BEHAVIOR
Vocal and showy, with harsh, chirping call and rattling trill or whistle, which is easily attracted to an imitation. Outside

breeding season, pairs rove widely, often joining with other Black-naped monarchs or other species to form small flocks.

FEEDING ECOLOGY AND DIET
Feed on variety of small insects, including small crickets, butterflies, and moths, mostly by gleaning from foliage, but also by snatching from mid-air.

REPRODUCTIVE BIOLOGY
Nest is deep cup woven of thin strips of bark, plant fibers, moss, and spider webs, wedged in the upright fork of a tree, sometimes just a few feet above the ground. Females lay two or three buff- or cream-colored eggs, with red-brown spots. During breeding season, usually in the first half of the year, males defend territory from intrusion by conspecifics.

CONSERVATION STATUS
Common and widespread in a range of habitats.

SIGNIFICANCE TO HUMANS
None known. ◆

White-tipped monarch
Monarcha everetti

TAXONOMY
Monarcha everetti Hartert, 1896, Tanahjampea, Flores Sea. Considered an allospecies with the nearby Flores monarch (*M. sacerdotum*). There is scant knowledge about its relationship with the other species in the genus.

OTHER COMMON NAMES
English: Buru Island monarch, Everett's monarch, white-tailed monarch; French: Monarque d'Everett; German: Weisschwanzmonarch; Spanish: Monarca de Puntas Blancas.

PHYSICAL CHARACTERISTICS
5.5 in (14 cm); smaller than Flores monarch. Adult head, chest, upperparts, and central tail feathers are black, all other feathers are white. Immature birds grayish or brown above and white below, with rusty-colored wash and often a buff rump.

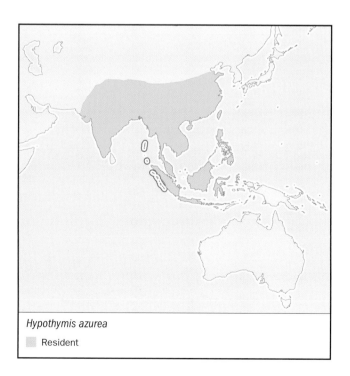

Hypothymis azurea
▨ Resident

Monarcha everetti
▨ Resident

DISTRIBUTION
Endemic to Tanahjampea, a small island in the Flores Sea, south of Sulawesi.

HABITAT
Common in forested areas, also found in scrub and mangroves with scattered, large trees.

BEHAVIOR
Found in pairs throughout the year, but outside breeding season they join small flocks with same and other species. Noisy, especially when foraging, and like many other Monarchidae, cock or fan the tail when alarmed, accompanied by harsh, scolding call.

FEEDING ECOLOGY AND DIET
Not known.

REPRODUCTIVE BIOLOGY
A tremulous, plaintive whistle may be associated with the male's courtship or territorial display.

CONSERVATION STATUS
Not scarce on Tanahjampea, but tiny global range puts the species at risk from habitat loss and degradation; fewer than 2,500 pairs probably remain. Logging is the common threat to all forest birds in Indonesia, and has been underway on this island since the 1920s, initially to create grazing areas for cattle. Although white-tipped monarchs appear to survive in logged or nonforest habitats, their densities are much lower, putting them at greater risk to other threats. As of 2001, no specific measures had been taken to identify or to protect their core breeding areas.

SIGNIFICANCE TO HUMANS
None known.◆

Resources

Books
Beehler, B. M., T. K. Pratt, and D. A. Zimmerman. *Birds of New Guinea.* Princeton, NJ: Princeton University Press, 1996.

BirdLife International. *Threatened Birds of Asia: The BirdLife International Red Data Book.* Cambridge: BirdLife International, 2001.

BirdLife International. *Threatened Birds of the World.* Cambridge: BirdLife International, 2000.

Coates, B. J., and K. D. Bishop. *A Guide to the Birds of Wallacea.* Queensland, Australia: Dove Publications, 1997.

King, B., M. Woodcock, and E. C. Dickinson. *Birds of South-East Asia.* London: HarperCollins, 1975.

MacKinnon, J. *Field Guide to the Birds of Java and Bali.* Yogyakarta, Indonesia: Gadjah Mada University Press, 1988.

Sibley, C. G., and B. L. Monroe, Jr. *Distribution and Taxonomy of Birds of the World.* New Haven: Yale University Press, 1990.

Urban, E. K., H. C. Fry, and S. Keith, eds. *The Birds of Africa,* Vol. 5. London: Academic Press, 1997.

Wikramanayake, E., E. Dinerstein, and C. J. Loucks. *Terrestrial Ecoregions of the Indo-Pacific: A Conservation Assessment.* Washington, DC: Island Press, 2001.

Organizations
BirdLife International. Wellbrook Court, Girton Road, Cambridge, Cambridgeshire CB3 0NA United Kingdom. Phone: +44 1 223 277 318. Fax: +44-1-223-277-200. E-mail: birdlife@birdlife.org.uk Web site: <http://www.birdlife.net>

Julian Hughes

Australian robins

(Petroicidae)

Class Aves
Order Passeriformes
Suborder Passeri (Oscines)
Family Petroicidae

Thumbnail description
Small, generally plump birds, with big heads and short tails, upright stance, long legs, and delicate feet, that prey on insects

Size
4–10 in (10–25 cm); 0.4–1.4 oz (12–38 g)

Number of genera, species
15 genera; 35 species

Habitat
Forests, woodlands, mangroves, and semiarid scrub

Conservation status
Endangered: 1 species; Vulnerable: 2 subspecies; Near Threatened: 1 subspecies

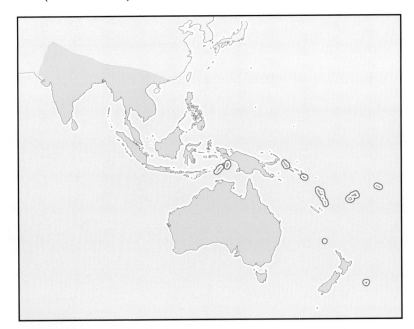

Distribution
New Guinea, Australia, New Zealand, and other Pacific Islands; south and Southeast Asia

Evolution and systematics

Molecular data indicate Australian robins are not closely related to the robins and flycatchers of Eurasia and Africa (Muscicapidae). Rather, their similar appearance and behavior are adaptations to similar ecological niches. The Petroicidae belong within the major adaptive radiation of the endemic Australian passerines, which include lyrebirds and honeyeaters. Petroicidae's nearest relatives are not clear, and the family is probably quite ancient. The number of species and genera is undecided and separation into subfamilies poorly resolved, though scrub robins are clearly separate. The family name Eopsaltriidae also has been used but Petroicidae has precedence.

Physical characteristics

Most species of Australian robins are similar in structure, although distinct in plumage. The genus *Petroica* has males that are black or gray and white with pink to red breasts, whereas females are grayish brown. Related New Zealand forms lack the red coloring. Many other robins are black and white, and several species are gray and yellow. Almost all are plump with upright stances and resemble the Eurasian robins, flycatchers, and chats. They have small bills, prominent rictal (on the bill) bristles, and short tails. The scrub robins differ from the family in being larger with long tails. They forage and nest entirely on the ground.

Distribution

New Guinea has 25 species of Petroicidae and Australia 20 species. There are three species of Australian robins in New Zealand, and the scarlet robin (*Petroica multicolor*) is spread widely across the western Pacific. One species of *Microeca* is found only on the Tanimbar Islands of Indonesia. The flycatchers of *Culicicapa* occur as far as the Himalayas, Sri Lanka, China, and the Philippines.

Habitat

Most Australian robins are found in forests and woodlands, with the red-capped robin (*P. goodenovii*) occurring in semiarid scrub. A few species also occur in mangroves, whereas others may be found in agricultural regions, especially during winter.

Behavior

Pairs or family groups defend territories ranging from about 1 to 10 acres (0.5–4 ha) but expand their ranges outside the breeding season. Many species engage in tail and wing flicking, especially when agitated, but the behavior also could be used to aid foraging. The flame robin (*P. phoenicea*) of upland Australia and the gray-headed flycatcher (*Culicicapa ceylonensis*) show short distance and altitudinal migration. Songs are usually attractive whistling or piping notes with some harsh alarm calls.

A pale-yellow robin (*Tregellasia capito*) adult and chick at their nest. (Photo by R. Brown/VIREO. Reproduced by permission.)

Feeding ecology and diet

Australian robins employ the perch-and-pounce method of foraging, typically sitting on a low branch or sideways across a tree trunk before flying onto the ground to capture a beetle, larva, or other insect. Some species, such as the hooded robin (*Melanodryas cucullata*) often use artificial perches, such as fence posts or overhead wires. The *Microeca* flycatchers and the smaller *Petroica* species also sally from perches after flying insects, especially during warmer months. Scrub robins remain on the ground, from which they glean their insect prey. Spiders, earthworms, and, more unusually, mollusks, crabs, and leeches also are eaten.

Reproductive biology

Australian robins are probably all socially monogamous, with helpers in cooperative breeding occurring in a few species. These birds build neat, cup-shaped nests in a fork of a tree or on a horizontal branch. Lichen and strips of bark are often added to conceal them. Females build nests and incubate the eggs, although the females often are fed by the males. Clutches in most species are comprised of two or three eggs, although clutches containing only a single egg occur in some New Guinea species and in the southern scrub robin (*Drymodes brunneopygia*). Breeding biology is poorly known

for many species. Many nests of the Australian robins suffer predation, especially from large birds, so that breeding success is often low.

Conservation status

The Chatham Islands black robin (*P. traversi*) was rescued from the brink of extinction by imaginative management. Many of the robins in Australia have suffered declines due to clearing and degradation of woodlands and forests for agriculture; they may experience difficulty moving between vegetation remnants and suffer increased nest predation due to their simplified habitat. Several New Guinea species are poorly known and may be threatened.

Significance to humans

The tame and trusting nature of Australian robins, as well as the bright colors and attractive songs of some species, make them favorites among bird watchers.

Yellow robin (*Eopsaltria australis*) eggs and nest. (Photo by R. Brown/VIREO. Reproduced by permission.)

1. Chatham Island black robin (*Petroica traversi*); 2. Scarlet robin (*Petroica multicolor*); 3. Gray-headed flycatcher (*Culicicapa ceylonensis*); 4. Jacky winter (*Microeca fascinans*); 5. White-winged robin (*Peneothello sigillatus*); 6. Eastern yellow robin (*Eopsaltria australis*); 7. Southern scrub-robin (*Drymodes brunneopygia*); 8. Gray-headed robin (*Heteromyias albispecularis*). (Illustration by Emily Damstra)

Species accounts

Jacky winter
Microeca fascinans

SUBFAMILY
Petroicinae

TAXONOMY
Sylvia leucophaea Latham, 1801, Sydney, New South Wales, Australia. Four subspecies.

OTHER COMMON NAMES
English: Brown flycatcher; French: Miro enchanteur; German: Weissschwanzschnäpper; Spanish: Tordo Australiano de Cola Blanca.

PHYSICAL CHARACTERISTICS
5–5.5 in (12.5–14 cm); 0.5–0.65 oz (14–18 g). Sandy crown and back, white throat and belly, brownish gray and white wings and tail, white around eye and black stripe through eye.

DISTRIBUTION
Most of Australia, except for central and western deserts, and northern Cape York; absent from Tasmania and Kangaroo Island. Also around Port Moresby, New Guinea.

HABITAT
Wide variety of woodlands and open scrub, lightly timbered farmland and occasionally in gardens.

BEHAVIOR
Generally quiet but active and tame. Often wags tail side to side or spreads feathers. Sedentary or showing local movements. Song is a loud, repeated "peter-peter"; also makes whistling calls.

FEEDING ECOLOGY AND DIET
Mostly sallies for flying insects from a perch on a branch, fence post, or overhead wires but also pounces onto ground after larvae, beetles, and worms, occasionally hovering just above the ground.

REPRODUCTIVE BIOLOGY
Breeds from July to December and occasionally at other times. Nest is made of grass and roots and placed in a horizontal fork on a living or dead tree branch. Clutch of two to three pale blue eggs, blotched with brown and lavender. Eggs are incubated for 16–17 days. Young are fed by both parents and fledge at 14–17 days.

CONSERVATION STATUS
Common in many parts of range but has declined in agricultural regions where most of the native vegetation has been lost or degraded.

SIGNIFICANCE TO HUMANS
Despite dull colors, the jacky winter's trusting and lively habits and distinctive song make it a popular bird. ◆

Scarlet robin
Petroica multicolor

SUBFAMILY
Petroicinae

TAXONOMY
Muscicapa multicolor Gmelin, 1789, Norfolk Island. Eighteen subspecies.

OTHER COMMON NAMES
French: Miro écarlate; German: Scharlachschnäpper; Spanish: Tordo Australiano Carmín.

Microeca fascinans
░ Resident

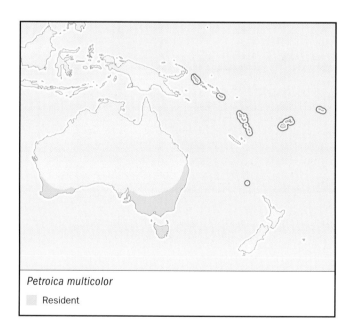

Petroica multicolor
░ Resident

PHYSICAL CHARACTERISTICS
5 in (13 cm); 0.4–0.5 oz (12–14 g). Some subspecies are smaller. Black throat, bill, and upperparts; white forehead, wing coverts, and under tail coverts.

DISTRIBUTION
Southwestern and southeastern Australia, including Tasmania, Kangaroo Island, and Norfolk Island. Widespread in the Pacific, including Fiji, Samoa, Vanuatu, Solomon Islands, and Bougainville.

HABITAT
Dry eucalypt forests and woodlands, with some understory. Forests, edges, clearings, and gardens on Pacific islands.

BEHAVIOR
Perches in a conspicuous location, although usually quiet, and may flick wings and raise and lower tail, perhaps when agitated. Territorial in breeding season and wandering more widely at other times. Song is pretty, trilling "wee-cheedalee-dalee"; also makes ticking calls.

FEEDING ECOLOGY AND DIET
Mostly pounces onto ground from a low branch for insects and spiders. Also sallies for flying insects in warm weather and gleans from branches and occasionally foliage.

REPRODUCTIVE BIOLOGY
Breeding season from August to January, with repeated attempts. Nest is made from bark and lichen and is placed in a tree fork or sometimes a shallow cavity. Usually three eggs in a clutch. Female is fed by male on nest, and both parents feed young. Incubation and fledging periods last about 15 days. Only 10% of nests succeeded in one New South Wales study.

CONSERVATION STATUS
Common in many areas but has declined in agricultural regions due to habitat loss. Norfolk Island subspecies is classified as Vulnerable.

SIGNIFICANCE TO HUMANS
A popular bird with bird watchers. ◆

Eopsaltria australis

▨ Resident

Eastern yellow robin
Eopsaltria australis

SUBFAMILY
Petroicinae

TAXONOMY
Motacilla australis White, 1790, New South Wales, Australia.

OTHER COMMON NAMES
English: Yellow robin, gray-breasted robin, western yellow robin; French: Miro à poitrine jaune; German: Goldbauch-schnäpper, Graumantelschnäpper; Spanish: Tordo Australiano Amarillo.

PHYSICAL CHARACTERISTICS
6–7 in (15–17 cm); 0.6–0.8 oz (18–23 g). Yellow underparts with gray throat and grayish brown crown, head, and wings.

DISTRIBUTION
Eastern Australia from southern Cape York to southeastern Australia. Two well-defined forms in north and south, with clinal variation.

HABITAT
Rainforest, eucalypt forest and woodland, mallee (low, scrubby evergreen *Eucalyptus* trees found in western Australia) and acacia woodland.

BEHAVIOR
Territorial in breeding season but expands home range in non-breeding season. Fairly quiet and tame. Often perches sideways on trunks. Sedentary. Song comprised of piping notes and a "chop...chop" call, especially in early morning; harsher calls when threatened.

FEEDING ECOLOGY AND DIET
Pounces on ground from low perches to capture larvae, beetles, other insects, and spiders. Occasionally gleans from bark or sallies for flying insects.

REPRODUCTIVE BIOLOGY
Breeds from July to January, with repeated attempts. Cup-shaped nest placed in fork and made from bark, decorated with lichens, and lined with grass and dead leaves. Clutch contains two to three eggs, which are incubated for about 15 days. Female is fed by male while incubating, and young fed by both parents and sometimes helpers. Young fledge at 10–14 days. About a quarter to a third of nests are successful.

CONSERVATION STATUS
Not threatened. Common in wetter forests but less so in drier woodland; has declined in agricultural areas.

SIGNIFICANCE TO HUMANS
A familiar bush bird. ◆

Gray-headed robin
Heteromyias albispecularis

SUBFAMILY
Petroicinae

TAXONOMY
Pachycephala albispecularis Salvadori, 1876, New Guinea.

OTHER COMMON NAMES
English: Ashy robin, black-cheeked robin; French: Miro cendré German: Farnschnäpper; Spanish: Tordo Australiano Terrestre.

Heteromyias albispecularis
Resident

PHYSICAL CHARACTERISTICS
6–8 in (15–20 cm); 1.1–1.4 oz (31–40 g). Gray crown with medium-brown upperparts, dark brown wings with white-tipped secondary coverts, and grayish white underparts.

DISTRIBUTION
Northeastern Queensland, Australia, and mountains of New Guinea. Three subspecies in New Guinea and a well-defined one in Australia.

HABITAT
Rainforest, 2,500–8,000 ft (850–2,600 m) elevation in New Guinea, 800 ft (250 m) and up in Queensland.

BEHAVIOR
Territorial and sedentary. Perches low on trunks and vertical saplings; often jerks tail up and down. Breeding females have a wing-flutter display. Song is a series of piping whistles, as well as harsh alarm calls.

FEEDING ECOLOGY AND DIET
Pounces onto insects and their larvae, centipedes, earthworms, and leeches.

REPRODUCTIVE BIOLOGY
Breeding season is from August to January. Nest is an untidy cup, made of rootlets, tendrils, and twigs and covered with moss on the outside; nest is placed in climbing palms or sapling branches. Female builds nest, incubates eggs, and broods and feeds young, but she is fed by the male. Clutch comprised of one or two eggs, incubated for 17–19 days. Young fledge at 12–13 days. Young fledged from 53% of nests in one Queensland study.

CONSERVATION STATUS
Not threatened.

SIGNIFICANCE TO HUMANS
None known. ◆

Chatham Island black robin
Petroica traversi

SUBFAMILY
Petroicinae

TAXONOMY
Miro traversi Buller, 1872, Chatham Islands, New Zealand.

OTHER COMMON NAMES
English: Chatham Islands robin, Chatham robin; French: Miro des Chatham; German: Chathamschnäpper; Spanish: Tordo de Chatham.

PHYSICAL CHARACTERISTICS
5.5–6 in (13–15 cm); 0.8 oz (23 g). All black with stocky build.

DISTRIBUTION
Chatham Islands, New Zealand.

HABITAT
Scrubby forest on islands.

BEHAVIOR
Tame and tolerant of human intrusion, but defends small territories while breeding.

FEEDING ECOLOGY AND DIET
Hops on ground or low branches, taking invertebrates, such as cockroaches and wetas, grubs, and worms.

REPRODUCTIVE BIOLOGY
Breeds from October to January, with several attempts. Nests placed in tangles of low vines or in cavities in tree trunks and limbs, occasionally in old blackbird (*Turdus merula*) nests. Clutch is comprised of two, occasionally three, eggs. Incubated for 18 days, young fledge after 23 days and depend on parents for six more weeks.

CONSERVATION STATUS
Endangered. Declined on Chatham Islands after European settlement due to loss of habitat and introduced mammals, such as rats and cats. By 1980, there were only seven birds, with just

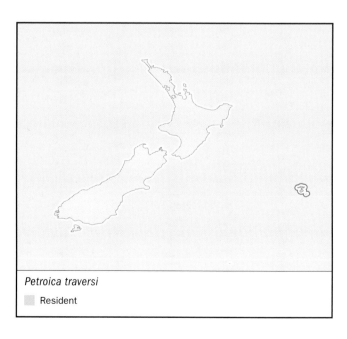

Petroica traversi
Resident

one successfully breeding pair. All birds were moved to another island, and egg-manipulation and cross-fostering of eggs and young to warblers (*Gerygone albofrontata*) and tomtits (*Petroica macrocephala chathamensis*) allowed the population to recover to about 200 birds.

SIGNIFICANCE TO HUMANS
The rescue of the Chatham Island black robin from extinction attracted widespread interest in New Zealand, so much so that the death of the old breeding female ("Old Blue"), who was used to save the species, was announced in a press release from the Minister of Internal Affairs. A detailed account of the struggle to save what was then the rarest bird species in the world is told by David Butler and Don Merton (1992). ◆

Gray-headed flycatcher
Culicicapa ceylonensis

SUBFAMILY
Petroicinae

TAXONOMY
Platyrhynchus ceylonensis Swainson, 1820, Ceylon (Sri Lanka).

OTHER COMMON NAMES
English: Gray-headed canary flycatcher; French: Gobemouche à tête grise; German: Graukopf-Kanarienschnäpper; Spanish: Papamoscas Canario de Cabeza Gris.

PHYSICAL CHARACTERISTICS
4–5.25 in (11–13 cm); c. 0.4 oz (11 g). Gray head, throat, and breast. Yellow back and belly; and dark grayish brown wings and tail.

DISTRIBUTION
Breeds patchily in India and along Himalayas from northern Pakistan across to Bangladesh, Myanmar, southern China, Thailand, Laos, Cambodia, Vietnam, peninsular Malaysia, Sumatra, Java, Borneo, Bali, Lombok, Sumba, and Flores. Leaves higher altitudes in winter and spreads into plains in India and into southern Thailand and Vietnam. Four subspecies.

HABITAT
Primary and tall secondary lowland and hill forests; riverine forests; remnant forests; and cultivated trees, including coffee. Up to 9,840 ft (3,000 m) elevation in the Himalayas. Moves onto plains in India in winter and out of higher altitudes.

BEHAVIOR
Singly or in pairs. Usually in mid-story, perched on bare branches or vines. May join flocks of mixed species. Calls are comprised of a rising series of two or three short whistles, as well as a trilling song.

FEEDING ECOLOGY AND DIET
Sallies from branches for flying insects. Flutters and hovers among foliage.

REPRODUCTIVE BIOLOGY
Breeds March to June. Cup-shaped nest of moss and lichen, often placed on epiphytes on trees or rocks. Lays two to four pale buff eggs, with gray blotches.

CONSERVATION STATUS
Not threatened. Some populations are common, but subspecies on Lombok and Flores are apparently uncommon or rare.

SIGNIFICANCE TO HUMANS
None known. ◆

White-winged robin
Peneothello sigillatus

SUBFAMILY
Petroicinae

TAXONOMY
Poecilodryas sigillata De Vis, 1890, southeastern New Guinea.

OTHER COMMON NAMES
English: White-winged thicket-flycatcher; French: Miro á ailes blanches; German: Spiegeldickichtschnäpper; Spanish: Tordo Australiano de Alas Blancas.

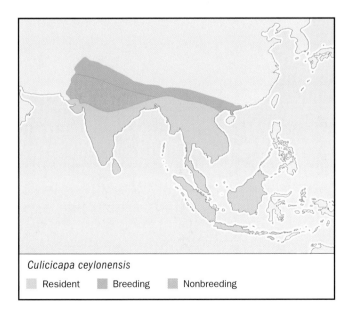

Culicicapa ceylonensis

▨ Resident ▨ Breeding ▨ Nonbreeding

Peneothello sigillatus

▨ Resident

PHYSICAL CHARACTERISTICS
5.7 in (14.5 cm); c. 0.6 oz (16 g). All black with white on secondaries.

DISTRIBUTION
Central ranges of New Guinea and the mountains of Huon Peninsula. Four subspecies.

HABITAT
Mid-montane and subalpine forests and adjacent shrubs, between 6,500 and 12,000 ft (2,150–3,900 m) elevation.

BEHAVIOR
Tame; found singly, in pairs, or small family groups. Perches on mossy branches. Song is comprised of trilling and piping notes that rise and fall. Metallic notes and sharp alarm call.

FEEDING ECOLOGY AND DIET
Gleans and snatches from branches, trunks, and the ground for insects. Also takes some fruit.

REPRODUCTIVE BIOLOGY
Breeds from September to January. Bulky nest is made of green moss, dried fern, and rootlets, placed in a tree fork. Clutch is comprised of a single, light-olive-colored egg, sparsely marked with brown.

CONSERVATION STATUS
Not threatened. Common in its habitat.

SIGNIFICANCE TO HUMANS
None known. ◆

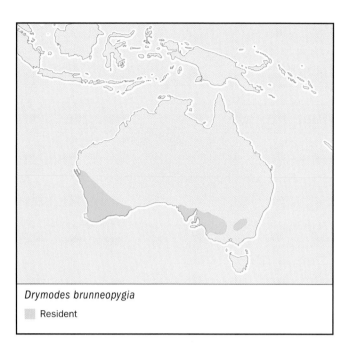

Drymodes brunneopygia
▨ Resident

Southern scrub robin
Drymodes brunneopygia

SUBFAMILY
Drymodinae

TAXONOMY
Drymodes brunneopygia Gould, 1841, Murray River, South Australia.

OTHER COMMON NAMES
French: Drymode à croupion brun; German: Mallee-Scheindrossel; Spanish: Tordo de Lomo Castaño.

PHYSICAL CHARACTERISTICS
8–9 in (21–23 cm); 1.25–1.35 oz (36–38 g). A large bird with white to buff underparts, dark brown wing and light gray-brown upperparts with white-tipped tail.

DISTRIBUTION
Southwestern Australia, south-central Australia into western Victoria and southwestern New South Wales.

HABITAT
Mallee, especially with broombush (*Melaleuca*) or heathy understory; other semiarid scrub; coastal tea tree thickets.

BEHAVIOR
Shy and often hidden, heard more often than seen. Territorial year-round. Flicks or gently raises and lowers tail. Call is a thin "pee...pee" or a musical "chip...chip...par-ee."

FEEDING ECOLOGY AND DIET
Forages entirely on the ground, gleaning insects, especially ants, termites, and beetles; occasionally picks fruit from low shrubs.

REPRODUCTIVE BIOLOGY
Breeds from July to December, building a cup-shaped nest that is placed on or near ground. Clutch is comprised of one pale green egg, blotched with brown and black. Female builds nest and incubates the egg for 16 days. Both sexes feed the young bird, which fledges at nine to 12 days. Nest success was 64% in Western Australia study. Male whistles and draws predators away from the nest.

CONSERVATION STATUS
Not threatened. Common in suitable habitat but has declined due to extensive clearing of mallee for agriculture.

SIGNIFICANCE TO HUMANS
None known. ◆

Resources

Books

Boles, W.E. *The Robins and Flycatchers of Australia.* Sydney: Angus and Robertson, 1988.

Butler, D., and D. Murton. *The Black Robin: Saving the World's Most Endangered Bird.* Oxford: Oxford University Press, 1992.

Marchant, S. "Breeding of the Eastern Yellow Robin *Eopsaltria australis.*" In *Birds of Eucalypt Forests and Woodlands: Ecology, Conservation, Management,* edited by A. Keast, H.F. Recher, H.A. Ford, and D.A. Saunders. Sydney: Surrey Beatty, 1985.

Resources

Periodicals

Brooker, B. "Biology of the Southern Scrub-robin (*Drymodes brunneopygia*) at Peron Peninsula, Western Australia." *Emu* 101 (2001): 181–190.

Frith, D.W., and C.B. Frith. "The Nesting Biology of the Gray-headed Robin *Heteromyias albispecularis* (Petroicidae) in Australian Upland Tropical Rainforest." *Emu* 100 (2000): 81–94.

Robinson, D. "The Nesting Ecology of Sympatric Scarlet Robin *Petroica multicolor* and Flame Robin *P. phoenicea* in Open Eucalypt Forest." *Emu* 90 (1990): 40–52.

Other

Black Robin. New Zealand Department of Conservation. 26 March 2002. <http://www.doc.govt.nz/Conservation/001Plants-and-Animals/001Native-Animals/Black-Robin.asp>

Hugh Alastair Ford, PhD

Whistlers
(Pachycephalidae)

Class Aves
Order Passeriformes
Suborder Passeri (Oscines)
Family Pachycephalidae

Thumbnail description
Small to medium-small birds with large, rounded heads, stout bills, and loud, cheerful songs

Size
5–11 in (12.5–28 cm); 0.03–.24 lb (12.5–110 g)

Number of genera, species
13 genera; 52–59 species

Habitat
Rainforest, forest, mangroves, woodland, savanna, shrub steppe, heath, grasslands, human habitation

Conservation status
Extinct: 1 species; Critically Endangered: 1 species; Vulnerable: 1 species; Near Threatened: 3 species

Distribution
Australia, New Zealand, New Guinea, Southwest Pacific islands, Micronesia, Philippines, Indonesia, South and Southeast Asia

Evolution and systematics

Early ornithologists focused on the hooked, rather robust bills of these birds to associate them with various groups of flycatchers or shrikes from other parts of the world. Whistlers and their kin are now believed to be part of the largely endemic Australo-Papuan songbird assemblage. Generic relationships within the family are poorly understood. As currently delimited, the Pachycephalidae comprises a central cluster of three genera (whistlers *Pachycephala*, shrike-thrushes *Colluricincla*, and pitohuis *Pitohui*) and several smaller ones, some whose affinities may lie elsewhere. With 26 to 33 recognized species, whistlers constitute the largest and most widespread genus. Four species are sometimes merged with *Pachycephala* or, alternatively, each is segregated into its own monotypic genus, the rufous-naped whistler (*Aleadryas rufinucha*) and mottled whistler (*Rhagologus leucostigma*) of New Guinea and the yellow-flanked whistler (*Hylocitrea bonensis*) and maroon-backed whistler (*Coracornis raveni*) of Sulawesi.

The six species of shrike-thrushes resemble oversized whistlers and the pitohuis seem larger versions yet. The crested bellbird (*Oreoica gutturalis*) suggests a large shrike-thrush and is probably closely related. The three isolated populations of shrike-tits (*Falcunculus*) have been considered separate species or a single species. They share with the ploughbill (*Eulacestoma nigropectus*) a peculiarly shaped bill. It has not been determined if this similarity in bill structure is convergent or indicative of a close relationship among genera. The goldenface (*Pachycare flavogrisea*) has traditionally been placed with whistlers, but this proposed relationship has received little scrutiny.

Two problematic New Zealand genera have also been allied to this family. *Mohoua*, with which *Finschia* has been united, has three species that were long uncomfortably placed with various warbler groups. Biochemical studies point to *Mohoua* being part of the whistler assemblage. The extinct piopio (*Turnagra capensis*) is of uncertain affinities. There is competing evidence that it is closest to the bowerbirds (Ptilonorhynchidae) or to this family. Sometimes it is maintained in its own family.

A feature of the group that has attracted attention is the considerable geographic variation exhibited by some species.

The rufous-naped whistler (*Aleadryas rufinucha*) makes its nest of moss, ferns, and rootlets. (Photo by W. Peckover/VIREO. Reproduced by permission.)

This phenomenon reaches its apex in the golden whistler (*Pachycephala pectoralis*). About 70 subspecies are accepted. The complex pattern of variation disguises connections among populations, and some subspecies differ more from each other in plumage than from related species.

Physical characteristics

Birds in this family range in length from the 5 in (12.5 cm) goldenface to the 11 in (28.0 cm) rusty pitohui (*Pitohui ferrugineus*). They are characterized by a robust body and large, rounded head, the latter the reason for the earlier name "thickhead" bestowed to whistlers and as a group name for the family. Legs and feet are strong, wings broad and rounded, and tail unadorned. The bills are robust, although often of moderate length and, in some of the larger species, rather imposing, especially when attached to the finger of an unwary handler. There is a shrike-like hook at the tip, part of the origin of the name "shrike-thrush" for the thrush-sized species in the genus *Colluricincla*. Together with a well-developed notch, this makes the bill efficient at grasping prey. The most specialized bills are those of the shrike-tits and ploughbill. These are strongly laterally compressed, making them much deeper than wide. The shrike-tits, in particular, are endowed with powerful jaw muscles. Rictal bristles are generally not strongly developed in this family and may be absent in some forms. Members of this family rarely have adornments of the plumage or other parts of the body. The crested bellbird, crested pitohui (*Pitohui cristatus*), and the shrike-tits have distinct crests. The adult male ploughbill has two round, pink wattles at the base of its bill, while the bare-throated whistler

(*Pachycephala nudigula*) has a naked patch of colored skin on the throat.

The predominating plumage colors are rather somber—various browns, rufous, gray, and olive—with white and black less frequent, and in a few species, bright yellow. Brightly colored species often have boldly contrasting patterns. The variable pitohui (*Pitohui kirhocephalus*) has large areas of black and dark rufous, and a number of whistlers have a white throat, bordered by a black breast band, which separates it from differently colored underparts. Species with bright males are often sexually dimorphic, with the female being plain brownish or grayish, occasionally with streaked underparts. More drably plumaged species usually show few differences between the sexes. In the unusually patterned mottled whistler, females are much more heavily spotted than males. Chicks and juveniles of most species have an unmarked rufous plumage. Parts of this are retained in immatures, particularly on secondaries and secondary coverts; these feathers are very noticeable in some young whistlers.

The geographic variation in plumage in the golden whistler is remarkable. Across this species' distribution, the head may be gray, black, or olive; nape collar yellow or absent; throat white, yellow, or black; breast band black or absent; and back gray, olive, or black. Each population combines a different selection of these to produce marked variation in color patterns. In birds of Norfolk Island, the bright plumage of the male has been lost and both sexes resemble the brown female. Females exhibit more limited geographical variation. They can range from dull grays or browns to bright yellows approaching colors of the males.

Distribution

Although the Pachycephalidae reaches its greatest diversity in New Guinea and Australia, some species have a much wider distribution and the family ranges from southeast India through Malaysia, Borneo, and Sulawesi, north to the Philippines and Palau in Micronesia, and east to Fiji, Tonga, Samoa, and other Pacific islands. The endemic New Zealand genera *Mohoua* and *Turnagra* may also belong in this group. On the periphery of the family distribution, there is usually only a single species at any locality. In the core area, several can coexist. New Guinea has 25 species and Australia, 14–16.

Habitat

This family is represented in a variety of habitats. Most species in the tropics inhabit rainforest; those in more temperate regions are found in forests and woodlands; some, like the crested bellbird, occur throughout the arid zones of Australia. Wide-ranging species may be found in a number of habitat types; the gray shrike-thrush, for example, can be encountered in almost any wooded habitat except dense rainforest. Other species have quite specific habitat requirements. The mangrove whistler (*Pachycephala grisola*) and white-breasted whistler (*Pachycephala lanioides*) are restricted to stands of mangrove, and the sandstone shrike-thrush (*Colluricincla woodwardi*) is found only on sandstone escarpments dissected by gorges.

Behavior

During the breeding season, these birds are strongly territorial, and both sexes help defend the territory. In the non-breeding season, individuals are generally solitary, other than in a few more social species, such as some pitohuis, which are always encountered in small parties.

Most species appear to be sedentary or, at best, locally nomadic. The notable exception is the rufous whistler (*Pachycephala rufiventris*). Populations in the southeast of mainland Australia are strongly migratory. On her study site in the tablelands of northeastern New South Wales, Lynda Bridges found that birds arrived in early September and remained through summer, departing in April.

The name "whistler" is well earned: they and the shrike-thrushes are among the most outstanding avian songsters in this part of the world. Each species has a song that differs in its phrases yet is usually sufficiently characteristic to be immediately recognizable as belonging to a member of this family. The enthusiastic songs of whistlers are pitched higher than those of shrike-thrushes and are longer, but those of shrike-thrushes are the stronger and richer. Duetting has been recorded for at least two species of pitohui and one whistler. One of the most arresting sounds of inland Australia is the song of the crested bellbird. Its odd bell-like notes and ventriloquial quality are distinctive. Shrike-tits songs lack the power and quality of these other birds.

During the breeding season, birds advertise territories with frequent, loud bursts of song. Birds are vocal during the rest of the year, but less regularly. Males, females, and even the young of many species are enthusiastic singers. Some whistlers have an interesting reaction to a loud, sudden noise, such as a rifle shot or roll of thunder: they burst into a short, loud outpouring of song.

Feeding ecology and diet

Whistlers and their relatives are, for the most part, rather sedate feeders. They search foliage and limbs in a methodical fashion, gleaning prey from leaves or bark, and some pick items from the ground by pouncing. Because these species do not pursue flying insects, most lack rictal bristles of more aerial insectivorous birds. Most species feed in the top to middle of the canopy, but some like the rufous-naped whistler and olive whistler (*Pachycephala olivacea*) forage in low dense understory. The larger shrike-thrushes and, more frequently, the crested bellbird feed on the ground, hopping in a thrush-like manner. The ploughbill and shrike-tits use their strong bills to strip bark from branches, feeding on insects they expose.

The main prey are insects and other small invertebrates. The mangrove-inhabiting white-breasted whistler frequently eats small crabs and small mollusks. The larger species of shrike-thrushes opportunistically take eggs, baby birds, and small vertebrates. Many species include fruit, usually berries, in their diets. Some species of pitohuis include considerable amount of fruit, and the mottled whistler and yellow-flanked whistler are predominantly frugivorous.

A male golden whistler (*Pachycephala pectoralis*) feeds chicks in the nest. (Photo by Eric Lindgren. Photo Researchers, Inc. Reproduced by permission.)

Members of this family generally feed solitarily. But shrike-tits forage gregariously, several individuals maintaining a small group as they hunt for insects. Pitohuis commonly join mixed feeding parties with similarly colored species.

Reproductive biology

For many New Guinea species, breeding begins late in the dry season, extending into the first half of the wet season. In temperate regions, breeding is from late winter-early spring through mid to late summer. Arid zone species are more opportunistic, breeding as conditions permit. Most have a single brood per season; some attempt two or three clutches.

In the golden whistler and gray shrike-thrush, both male and female contribute to nest construction, incubation of eggs, and care of young. The rufous whistler is similar, except that the female builds the nest. In shrike-tits, the female does most nest construction and incubation, and both adults care for young; additional birds serve as helpers at the nest. Helpers are also known for the whitehead (*Mohoua albicilla*), in which the female builds the nest. Incubation may be by both parents at some nests or by the female at others. The latter tend to be nests with helpers. After hatching, the chicks are fed by the parents and the helpers.

In the whistlers, the nest can range from the substantial bowl built by the red-lored whistler (*Pachycephala rufogularis*) to the thin, flimsy cup of the mangrove whistler. Twigs and bark comprise much of the coarsely constructed nest of many species. In habitats with taller trees, nests may be higher, up to 33 ft (10 m) in the case of the rufous whistler. Species from more arid areas, where trees are shorter, and those favoring low, dense shrubs, often place the nest within 3 ft (100 cm) of the ground. Nests may be situated in a tree fork, shrub, or dense vegetation. The nests of shrike-thrushes, pitohuis, and the crested bellbird resemble those of whistlers, although they are generally larger. The sandstone shrike-thrush, which lives on rugged sandstone escarpments with few trees, places its nest of porcupine grass rootlets on a cliff edge or in a crevice. Compared to the nests of most members of this family, those of shrike-tits are made of finer material. The deep cup or goblet is constructed of finely shredded bark bound with spider web and lined with bark and grass. Nests of the New Zealand species of *Mohoua* are also made of finer material. Moss, lichens, bark strips, and leaf skeletons are used to make a compact cup, bound with spider web. This is usually placed in a fork, but may be suspended from small branches. The crested bellbird has the unusual habit of placing paralyzed caterpillars around the rim of the nest during incubation; the purpose of this behavior is unknown. Clutch size, where known, is two to three, sometimes four, eggs. These are covered with spots and blotches, but the color of these, and that of the background, vary considerably within the family. The background may be white, light to dark pink, cream, buff, olive, or salmon, with markings of black, browns, brick, or lavender. Incubation and fledgling periods are unrecorded for many, perhaps most, species. Where these are known, incubation varies from 14 in some whistlers to 21 days in the brown creeper (*Mohoua novaeseelandiae*). The nesting period also lasts 14–21 days.

Conservation status

The Sangihe shrike-thrush (*Colluricincla sanghirensis*) is rated as Critically Endangered by the IUCN. Known only from a single 19th century specimen until rediscovered in 1995, this species may have fewer than 100 individuals left. There has been almost total loss of forest on its small Indonesian island of Sangihe.

The yellowhead (*Mohoua ochrocephala*) occurs on the North Island of New Zealand. Periodic irruptions of the introduced stoat result in massive losses (50–100%) of eggs, chicks, and adult females. The range is now fragmented through extirpation of local populations. IUCN lists this species as Vulnerable.

The piopio was common on both New Zealand islands in the 1870s, but started a rapid decline in the 1880s. The last sightings were in 1950–60s. Its extinction has been attributed to predation by introduced rats and loss of habitat.

Three species are considered Near Threatened. The red-lored whistler of Australia, Tongan whistler (*Pachycephala jacquinoti*) of Tonga, and white-bellied pitohui (*Pitohui incertus*) of New Guinea almost meet the criteria for listing as Vulnerable. Contributing factors include loss of habitat, introduced animals, and small known ranges.

Significance to humans

A few species are shy but most are curious and tame. In parks and inhabited areas, this confiding and vocal nature draws attention. Many species readily respond to human whistles and squeaks, and so can be readily attracted. The gray shrike-thrush will nest in potted plants around houses. For the most part, however, these birds remain unfamiliar to most of the public.

New Guinea inhabitants reported that pitohuis were bad eating birds. In 1992, Jack Dumbacher and his colleagues discovered that these birds had a strong toxin in their feathers and soft tissue, which confer this unpalatibility.

1. Golden whistler (*Pachycephala pectoralis*); 2. Whitehead (*Mohoua albicilla*); 3. Regent whistler (*Pachycephala schlegelii*); 4. Crested bellbird (*Oreoica gutturalis*); 5. Little shrike-thrush (*Colluricincla megarhyncha*); 6. Eastern shrike-tit (*Falcunculus frontatus*); 7. Rufous-naped whistler (*Aleadryas rufinucha*); 8. Gray shrike-thrush (*Colluricincla harmonica*); 9. Rufous whistler (*Pachycephala rufiventris*); 10. Variable pitohui (*Pitohui kirhocephalus*). (Illustration by Emily Damstra)

Species accounts

Eastern shrike-tit
Falcunculus frontatus

SUBFAMILY
Pachycephalinae

TAXONOMY
Lanius frontatus Latham, 1801, Sydney, New South Wales, Australia.

OTHER COMMON NAMES
English: Crested shrike-tit; French: Falconelle à casque; German: Meisendickkopf; Spanish: Carbonero Verdugo Crestado.

PHYSICAL CHARACTERISTICS
6.3–7.5 in (16–19 cm); 0.05–0.07 lb (24–32 g). A short-billed bird with distinctive black crest, white head, and black stripe from eye to shoulder.

DISTRIBUTION
East and southeast Australia.

HABITAT
Eucalypt forest and woodland, occasionally other vegetation types.

BEHAVIOR
Sedentary, territorial, usually seen in small family groups. Generally quiet, unobtrusive, but often produces a great deal of noise when foraging.

FEEDING ECOLOGY AND DIET
Strictly arboreal. Forages for insects and spiders among outer foliage by gleaning and under loose bark by prising it away with bill.

REPRODUCTIVE BIOLOGY
Breeds August–January; one brood per year. Female does most nest construction and incubation; both adults care for young. Helpers at the nest have been reported. Nest is a deep cup or goblet of finely shredded bark bound with spider web and placed in an upright, usually three-pronged fork, 33–100 ft (10–30 m) above ground. The two to three spotted eggs take 16–19 days to hatch; nesting period 15–17 days.

CONSERVATION STATUS
Not threatened, but potentially vulnerable to habitat loss.

SIGNIFICANCE TO HUMANS
None known. ◆

Rufous-naped whistler
Aleadryas rufinucha

SUBFAMILY
Pachycephalinae

TAXONOMY
Pachycephala rufinucha Sclater, 1874, New Guinea, mountains of Vogelkop. Five subspecies.

OTHER COMMON NAMES
French: Siffleur à nuque rousse; German: Rotnacken-Dickkopf; Spanish: Chiflador de Nuca Rufa.

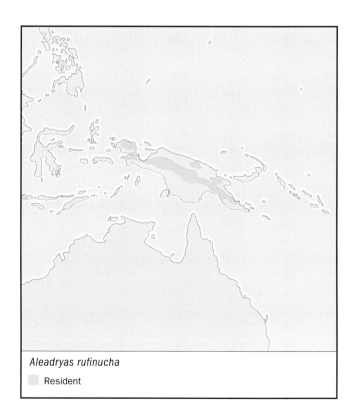

Falcunculus frontatus
▨ Resident

Aleadryas rufinucha
▨ Resident

PHYSICAL CHARACTERISTICS
6.5–7 in (16.5–18 cm); 0.03–0.09 lb (37.5–42.5 g). Gray crown with rufous nape, yellow throat, and olive-brown upperparts.

DISTRIBUTION
New Guinea from 3,960 to 11,880 ft (1,200 to 3,600 m), mainly 4,600–8,500 ft (1,400–2,600 m); *A. r. rufinucha*: northwest New Guinea; *A. r. niveifrons*: west and central New Guinea; *A. r. gamblei*: southeast New Guinea; *A. r. prasinonota*: Herzog Mountains, northeast New Guinea; *A. r. lochmia*: Huon Peninsula, northeast New Guinea.

HABITAT
Mountain forest and secondary growth.

BEHAVIOR
Sedentary. Spends most of its time on the ground or in low vegetation, sometimes moving to higher elevations in the forest.

FEEDING ECOLOGY AND DIET
Eats mainly insects and small invertebrates, also fruit.

REPRODUCTIVE BIOLOGY
Two white eggs with dark markings laid in a deep cup nest of moss, ferns, and rootlets, usually placed low in a sapling, sometimes in a higher site.

CONSERVATION STATUS
Not threatened. Moderately common but relatively secretive.

SIGNIFICANCE TO HUMANS
None known. ◆

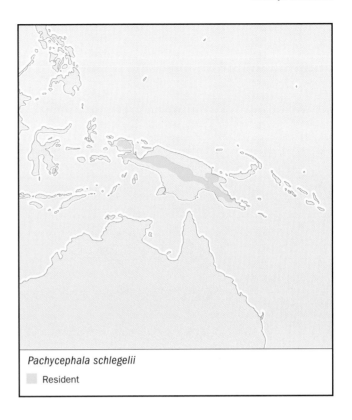

Pachycephala schlegelii
▨ Resident

FEEDING ECOLOGY AND DIET
Gleans insects, spiders, and seeds.

REPRODUCTIVE BIOLOGY
Nesting habits undescribed. Lays two white eggs spotted with black and lavender.

CONSERVATION STATUS
Not threatened. Common to abundant in middle elevations and above, but scarce lower down.

SIGNIFICANCE TO HUMANS
None known. ◆

Regent whistler
Pachycephala schlegelii

SUBFAMILY
Pachycephalinae

TAXONOMY
Pachycephala schlegelii Schlegel, 1871, Arfak Mountains, New Guinea. Three subspecies.

OTHER COMMON NAMES
English: Schlegel's whistler; French: Siffleur de Schlegel; German: Schlegeldickkopf; Spanish: Chiflador de Schlegel.

PHYSICAL CHARACTERISTICS
5.9–6.5 in (15–16.5 cm); 0.04–0.05 lb (19.5–24.8 g). Black head, throat, and wing and tail coverts; orange belly; and brownish mantle with yellow stripe.

DISTRIBUTION
New Guinea from about 4,290 to 12,000 ft (1,300 to 3,500 m), mainly above 6,100 ft (1,850 m). *P. s. schlegelii*: northwest New Guinea; *P. s. cyclopum*: north central New Guinea; *P. s. obscurior*: central, east, and northeast New Guinea.

HABITAT
Inhabits forest and forest edges.

BEHAVIOR
Usually seen singly or in pairs. During display the crown feathers are raised and the nape feathers erected across back of the head.

Golden whistler
Pachycephala pectoralis

SUBFAMILY
Pachycephalinae

TAXONOMY
Muscicapa pectoralis Latham, 1801, Port Jackson, Sydney, New South Wales, Australia. Up to 70 subspecies.

OTHER COMMON NAMES
English: Golden-breasted whistler; French: Siffleur doré; German: Gelbbauch-Dickkopf; Spanish: Chiflador Dorado.

PHYSICAL CHARACTERISTICS
5.9–7.5 in (15–19 cm); 0.05–0.06 lb (21–28 g). Black head and bill with golden yellow underparts and white patch on throat.

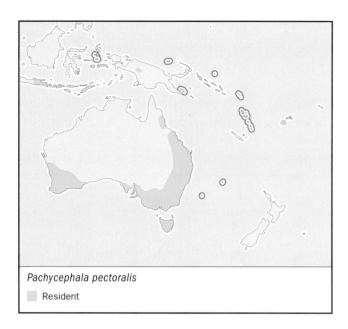

Pachycephala pectoralis

Resident

DISTRIBUTION
Widely distributed from east Indonesia through New Guinea and Australia to southwest Pacific islands as far as Fiji. From sea level to above 6,900 ft (2,100 m) on some islands.

HABITAT
A broad range of habitat types, from rainforest, secondary growth, eucalypt forest, and drier woodlands, occasionally mangroves and urban areas.

BEHAVIOR
Territorial; usually solitary, sometimes in pairs. Easily observed and can be attracted by whistles. Mainly found in lower and middle storys, less often in lower canopy. Some winter movements of birds in southeast Australia to areas north and west, including dispersal of young and altitudinal movement from higher altitudes; mainly sedentary elsewhere in range. Behavior generally quiet, unobtrusive except when breeding, when very vocal.

FEEDING ECOLOGY AND DIET
Forages primarily in thickets and low and middle storys. Obtains food mainly by gleaning from branches, twigs, and foliage; occasionally hover-gleaning or hawking. Food includes insects, spiders, and berries.

REPRODUCTIVE BIOLOGY
In Australia, breeding takes place from September to January, with one brood being reared. The male and female build the cup-like nest of sticks, twigs, grass, and bark, bound with spider web, placing it in fork or thick shrub up to 20 ft (6 m) above ground. The clutch is one to three eggs. These can be quite variable in color, with a background of cream, cream-buff, pale olive or salmon, marked with olive brown or reddish brown and gray, and underlying lavender blotches, which often form a cap at the larger end. Both sexes share incubation (14–17 days) and care of young (fledging period 10–13 days).

CONSERVATION STATUS
Not threatened. Common to moderately common in most parts of range.

SIGNIFICANCE TO HUMANS
None known. ◆

Rufous whistler
Pachycephala rufiventris

SUBFAMILY
Pachycephalinae

TAXONOMY
Sylvia rufiventris Latham, 1801, Sydney, New South Wales, Australia. Five subspecies.

OTHER COMMON NAMES
English: Rufous-breasted whistler; French: Siffleur itchong; German: Schlichtmantel-Dickkopf; Spanish: Chiflador Rufo.

PHYSICAL CHARACTERISTICS
6.5–7.1 in (16.3–18.1 cm); 0.7–1.0 oz (20–27 g). Gray with rufous underparts and white patch on throat.

DISTRIBUTION
P. r. rufiventris: east, central, west and south Australia; *P. r. pallida*: northeast Australia; *P. r. minor*: Melville and Bathurst Islands; *P. r. falcata*: northwest and north central Australia; *P. r. xanthetraea*: New Caledonia.

HABITAT
Open eucalypt forest and woodland, drier woodland and scrubs.

BEHAVIOR
Territorial. Very vocal during breeding and can be stimulated into bouts of singing by sudden loud noises. Tame, easily attracted by whistling. Migratory in southeast Australia; elsewhere partly migratory, nomadic, or sedentary.

FEEDING ECOLOGY AND DIET
Forages sedately in middle to upper vegetation layers for insects, spiders, and berries.

Pachycephala rufiventris

Resident

REPRODUCTIVE BIOLOGY
Breeds Sep.–Feb., one or two broods per season. Female builds nest, but both sexes incubate and care for young. Nest is a bowl of twigs and grasses, bound with spider web, and lined with grass and rootlets; it is placed up to 33 ft (10 m) above ground. Eggs, usually two or three in a clutch, are light olive-green with dark brown and lavender-gray blotches. Incubation 15 days.

CONSERVATION STATUS
Not threatened. Common and widespread.

SIGNIFICANCE TO HUMANS
None known. ◆

Little shrike-thrush
Colluricincla megarhyncha

SUBFAMILY
Pachycephalinae

TAXONOMY
Muscicapa megarhyncha Quoy and Gaimard, 1830, Dorey, Vogelkop, New Guinea. Thirty subspecies.

OTHER COMMON NAMES
English: Rufous shrike-thrush; French: Pitohui châtain; German: Waldgudilang; Spanish: Charlatán Verdugo Rufo.

PHYSICAL CHARACTERISTICS
6.7–7.5 in (17–19 cm); 1.1–1.5 oz (30–43 g). Gray-brown upperparts and bill; whitish throat and rufous-beige underparts.

Colluricincla megarhyncha
  Resident

DISTRIBUTION
New Guinea and surrounding islands; north and east Australia. Mainly in the lowlands, hills, and lower mountains up to about 6,100 ft (1,850 m), locally up to 7,590 ft (2,300 m).

HABITAT
Wide range of humid timbered habitats, including rainforest, tall secondary growth, mangroves, swamp and riverine vegetation, and coastal woodland.

BEHAVIOR
Sedentary, territorial at all seasons. Behavior generally quiet, unobtrusive, heard far more often than seen. Becomes quite vocal when breeding.

FEEDING ECOLOGY AND DIET
Food is mainly insects, spiders, small snails, and occasionally fruit, obtained mostly by gleaning.

REPRODUCTIVE BIOLOGY
In Australia, breeding season is Sep.–Feb. with one brood per season; in New Guinea, there are two breeding periods, one in the late dry to early wet season, and a second one in the late wet to early dry season. The nest, a deep cup of bark and dry leaves bound with spider web, is placed in upright fork or dense tangle of vegetation. The female lays two to three white to pale pinkish cream eggs, adorned with brown spots and lilac blotches. Incubation at about 19 days; fledging at about 12 days.

CONSERVATION STATUS
Not threatened. Common to abundant in New Guinea; common in northeast Australia, becoming scarcer and local southwards, uncommon in north central Australia.

SIGNIFICANCE TO HUMANS
None known. ◆

Gray shrike-thrush
Colluricincla harmonica

SUBFAMILY
Pachycephalinae

TAXONOMY
Turdus harmonicus Latham, 1801, Sydney, New South Wales, Australia. Six subspecies.

OTHER COMMON NAMES
Brown or western shrike-thrush; French: Pitohui gris; German: Graubrust-Gudilang; Spanish: Charlatán Verdugo Gris.

PHYSICAL CHARACTERISTICS
8.9–9.8 in (22.5–25.0 cm); 0.13–0.16 lb (58–74 g). Gray head, tail coverts and underparts with brownish back and wings.

DISTRIBUTION
C. h. brunnea: northwest and north Australia; *C. h. superciliosa*: northeast Australia; *C. h. tachycrypta*: southeast New Guinea; *C. h. harmonica*: southeast and east central Australia; *C. h. strigata*: Tasmania; *C. h. rufiventris*: southwest Australia.

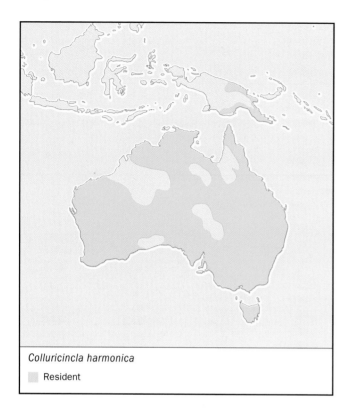

Colluricincla harmonica

◼ Resident

HABITAT
May be found in many habitat types, including rainforest, eucalypt forest and woodland, mangroves, dry open woodlands, riverine vegetation, urban parks, and gardens, from sea level up to about 5,610 ft (1,700 m).

BEHAVIOR
Often tame in east, shy in west. Sedentary, although young of year may disperse some distance. Territorial in all seasons. Generally unobtrusive except when singing; most vocal when breeding. Loud musical song varies geographically and among individuals, who have a wide repertoire.

FEEDING ECOLOGY AND DIET
Forages deliberately on the ground, logs, limbs and trunks of trees, and in foliage. Eats insects, spiders and other invertebrates, small vertebrates including lizards, nestling birds, and small mammals. Prey obtained mainly by gleaning from bark and leaves, sometimes from the ground.

REPRODUCTIVE BIOLOGY
Breeds July–February, opportunistically in drier areas, producing one to two broods per season. Sexes share nest building, incubation, and care of young. Nest is a cup of bark, grass, other dry vegetation, lined with rootlets. It is placed in fork or tangle of vegetation, in crevice, or on stump, ledge or ground, usually within 20 ft (6 m) of ground. The two to four white eggs are blotched and spotted with gray and olive-brown. Incubation 17–18 days.

CONSERVATION STATUS
Not threatened. Common in Australia, although in west decreasing round human habitation; fairly common to scarce in New Guinea, but possibly expanding range.

SIGNIFICANCE TO HUMANS
None known. ◆

Crested bellbird
Oreoica gutturalis

SUBFAMILY
Pachycephalinae

TAXONOMY
Falcunculus gutturalis Vigors and Horsfield, 1827, Bass Strait. Two subspecies.

OTHER COMMON NAMES
French: Carillonneur huppé; German: Haubengudilang; Spanish: Campanero Crestado.

PHYSICAL CHARACTERISTICS
8.3–9.1 in (21–23 cm); 0.13–0.15 lb (57–67 g). Brownish upperparts with white around bill, black crest tip and breast, and buff belly.

DISTRIBUTION
O. g. gutturalis: South, southwest and inland southeast Australia; *O. g. pallescens*: northern, west central, and central Australia.

HABITAT
Arid inland and coastal woodlands and scrubs.

BEHAVIOR
Sedentary or locally nomadic in more arid areas. Territorial through year, rather solitary except when breeding. Unobtrusive, keeping to cover, except when male sings from elevated perch. Unusual and distinctive song consists of two slow, then three fast ringing notes, reminiscent of a cowbell; has a ventriloquial effect, making the singing bird difficult to locate.

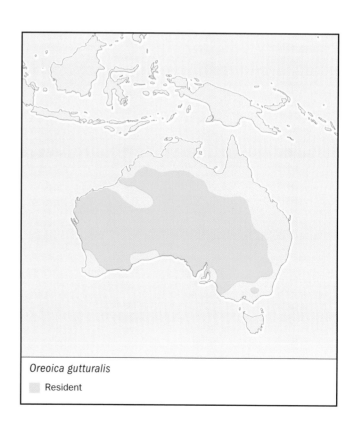

Oreoica gutturalis

◼ Resident

FEEDING ECOLOGY AND DIET
Insects and seeds. Forages on the ground, moving by hops.

REPRODUCTIVE BIOLOGY
Breeding season August–January, usually earlier in south, later in north, and often opportunistically in more arid regions. Depending on conditions, a second brood may be raised. The nest, a deep cup of bark strips, short twigs, leaves, and grass, is placed in fork, on top of broken stump, or in dense shrub, 3.3–10 ft (1–3 m) from ground. An odd and unexplained habit is the placement of paralyzed caterpillars around rim of the nest during incubation. The two to four eggs are pale bluish white with black, brown, and lavender-gray blotches. Both sexes incubate (14–17 days). Fledging period 11–12 days.

CONSERVATION STATUS
Common in interior, decreasing towards coastal districts and more humid regions, populations may be contracting in some regions.

SIGNIFICANCE TO HUMANS
None known. ◆

Variable pitohui
Pitohui kirhocephalus

SUBFAMILY
Pachycephalinae

TAXONOMY
Lanius kirhocephalus Lesson and Garnot, 1827, Dorey, Vogelkop, New Guinea. Twenty-one subspecies.

Pitohui kirhocephalus
◻ Resident

OTHER COMMON NAMES
French: Pitohui variable; German: Okkerpitohui; Spanish: Pitojuí Variable.

PHYSICAL CHARACTERISTICS
9–10 in (23–25.5 cm); 0.19–0.22 lb (85–100 g). Black head and primary feathers with orange underparts, rufous back and secondaries, and black legs.

DISTRIBUTION
New Guinea and satellite islands.

HABITAT
Thick vegetation along forest edges and in secondary growth and disturbed forest up to 3,630 ft (1,100 m), in places to 4,950 ft (1,500 m).

BEHAVIOR
Gregarious, usually seen in pairs or small parties. Secretive, hides in dense vegetation. Has a loud musical voice; duetting between birds has been recorded.

FEEDING ECOLOGY AND DIET
Insects and fruit. Often joins other species in mixed-species foraging flocks.

REPRODUCTIVE BIOLOGY
Nesting habits and eggs are undescribed.

CONSERVATION STATUS
Not threatened. Abundance variable across range; very common in places, rare or absent in others.

SIGNIFICANCE TO HUMANS
Avoided as food because of unpleasant taste, known to be due to toxin in tissues. ◆

Whitehead
Mohoua albicilla

SUBFAMILY
Mohouinae

TAXONOMY
Fringilla albicilla Lesson, 1830, Bay of Islands, North Island, New Zealand.

OTHER COMMON NAMES
French: Mohoua à tête blanche; German: Weissköpfchen; Spanish: Cabeza Blanca.

PHYSICAL CHARACTERISTICS
5.9 in (15 cm); female 0.4–0.6 lb (12.3–16.4 g), male 0.6–0.7 oz (16.6–19.3 g). Light reddish brown nape and upperparts with white head to creamy underparts.

DISTRIBUTION
North Island, New Zealand.

HABITAT
Native forest and scrub, occasionally woodland and exotic pines, up to 4,620 ft (1,400 m).

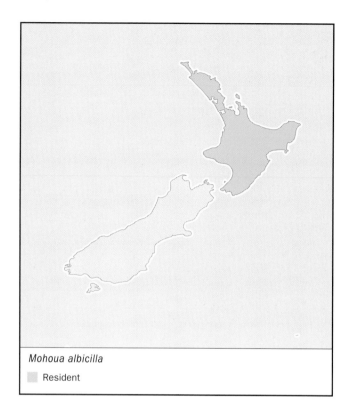

Mohoua albicilla

Resident

BEHAVIOR
Very social in nonbreeding season, with flocks up to 70 birds. In breeding season, form family parties and flocks of up to 10 birds.

FEEDING ECOLOGY AND DIET
Eat insects, as well as seeds and fruit, foraging from ground to canopy. Most prey gleaned from foliage and branches. Bills used to flake off bark fragments.

REPRODUCTIVE BIOLOGY
The nest is a compact cup in which two to four eggs are laid. Eggs variable in color from white to deep pink with yellowish to reddish brown markings. Male and female, or mainly female, incubate for about 17 days with fledging at 16–17 days. Nest commonly tended by three to four birds, including helpers.

CONSERVATION STATUS
Not threatened. Once abundant across North Island, now in moderate numbers only in forested areas. Plentiful on some offshore islands.

SIGNIFICANCE TO HUMANS
None known. ◆

Resources

Books

Coates, B. J. *The Birds of Papua New Guinea.* Vol. 2, *Passerines.* Aderley: Dove Publications, 1993.

Higgins, P. J. and J. M. Peter, eds. *Handbook of Australian, New Zealand and Antarctic Birds.* Vol. 6, *Pardalotes to Figbird.* Melbourne: Oxford University Press, in progress.

Schodde, R. and I. J. Mason. *The Directory of Australian Birds. Passerines.* Collingwood: CSIRO Publishing, 1999.

Periodicals

Bridges, L. "Breeding Biology of a Migratory Population of the Rufous Whistler *Pachycephala rufiventris.*" *Emu* 94 (1994): 106–115.

Dilks, P. "Recovery of a Mohoua (*Mohoua ochrocephala*) Population following Predator Control in the Eglinton Valley, Fiordland, New Zealand." *Notornis* 46 (1999): 323–332.

Dumbacher, J. P.; "Evolution of Toxicity in Pitohuis: 1. Effects of Homobatrachotoxin on Chewing Lice (Order Phthiraptera)." *Auk* 116 (1999): 957–963.

Galbraith, I. C. J. "Variation, Relationships and Evolution in the *Pachycephala pectoralis* Superspecies (Aves, Muscicapidae)." *Bulletin of the British Museum of Natural History* 4 (1956): 133–225.

McDonald, P. G. "The Function of Vocalisations and Aggressive Behaviour Used by Male Rufous Whistlers, *Pachycephala rufiventris.*" *Emu* 101 (2001): 65–72.

Rozendaal, F. G. and F. R. Lambert. "The Taxonomic and Conservation Status of *Pinarolestes sanghirensis* Oustalet 1881." *Forktail* 15 (1999): 1–13.

Organizations

Birds Australia. 415 Riversdale Road, Hawthorn East, Victoria 3123 Australia. Phone: +61 3 9882 2622. Fax: +61 3 9882 2677. E-mail: mail@birdsaustralia.com.au Web site: <http://www.birdsaustralia.com.au>

Walter E. Boles, PhD

Pseudo babblers

(Pomatostomidae)

Class Aves

Order Passeriformes

Suborder Passeri (Oscines)

Family Pomatostomidae

Thumbnail description
Large babbler-like songbirds with long tail and scimitar-like bill; plumage is plain russet or patterned with white brows, throat, and tail tip

Size
7–10.5 in (18–27 cm); 1.6–3.2 oz (45–90 g)

Number of genera, species
2 genera; 5 species

Habitat
Shrubberies of woodland, open forest, and tropical rainforest

Conservation status
Not threatened

Distribution
Mainland Australia and lowland New Guinea

Evolution and systematics

So similar in appearance are Australasian pseudo babblers and Asian scimitar babblers that both were formerly grouped together in the same tribe of the babbler family, Timaliidae. However, molecular evidence and examination of the skeleton have shown that the similarities are superficial. Pseudo babblers are divergent members of a quite different assemblage of crow-like songbirds that radiated massively within Australia at least 30 million years ago. Like other crow-like birds, pseudo babblers have a single pneumatised depression (*fossa*) in the head of the humerus. The sternum (breast bone) is shallowly keeled and processes on the pelvic girdle are much attenuated, both signs that pseudo babblers live more on leg than wing. They have other unique skeletal traits in the palate and skull that interact to support the bill when probing and digging.

The five living pomatostomid species belong to two genera. *Garritornis* contains a single species (*G. isidorei*, rufous babbler) endemic to the lowland rainforests of New Guinea. *Pomatostomus* comprises four species. Gray-crowned babblers (*Pomatostomus temporalis*) and Hall's babblers (*P. halli*) are centered in the woodlands of Torresian north Australia, and white-browed babblers (*P. superciliosus*) and chestnut-crowned babblers (*P. ruficeps*) are centered in similar habitat across southern Australia. The history of their evolution is unclear.

Physical characteristics

Pseudo babblers are medium-sized songbirds that measure 7–10.5 in (18–27 cm) long and weigh 1.6–3.2 oz (45–90 g). Their wings are short and rounded and equipped with ten well-developed primary feathers and 9–12 uniquely varying secondaries. The tail is long, fan-shaped, and 12-feathered. The bill, vestigially bristled, is long, slender, and downcurved. It lacks terminal notching and is better fitted for probing and digging than for grasping and grabbing. Legs are long and powerful with stoutly muscled calves. Pseudo babblers have scutellate, bilamini-plantar tarsi ("feet") that are adapted for prolonged hopping. Plumage is either plain russet (rufous babblers) or brown and boldly patterned with broad white brows, throat, and tail tip (*Pomatostomus*). Sexes are alike, and immature birds resemble adults except for their shorter bill. Eyes, which are brown in most species, become light cream in adults of rufous babblers and gray-crowned babblers. Feet are dusky in all species. The bill is yellowish in rufous babblers and mainly dull brown in *Pomatostomus*.

Distribution

Pseudo babblers live in lowland New Guinea up to 1,500 ft (500 m) altitude, Misool in the western Papuan Islands, and all of mainland Australia except the extreme southeast and southwest coasts and in central and northwest sand deserts. Northern and southern Australian groups of *Pomatostomus* overlap widely in the central arid zone of Australia.

A white-browed babbler (*Pomatostomus superciliosus*) at its nest in Australia. (Photo by Ken Stepnel. Bruce Coleman Inc. Reproduced by permission.)

Habitat

Pseudo babblers inhabit tall open shrubbery under eucalyptus woodland and open forest and, in New Guinea, lowland rainforest. In inland Australia, several species extend widely through mulga (*Acacia*) woodland; Hall's babblers are confined to it in the Great Artesian basin. Where species overlap there is little obvious partitioning of habitat that might avert competition. Habitat that is little disturbed and has a fairly intact layer of ground litter is critical.

Behavior

Noisy, gregarious, and sedentary, pseudo babblers quarter their foraging grounds energetically in coordinated groups of up to a dozen or more. The group does everything together, including feeding, preening, dust bathing, resting, and roosting. Groups keep within cover as they travel, bounding in powerful hops over the ground and among shrubbery. To keep contact, they call constantly in growls, mews, whistles, and chatterings. Flight, a fluttering of wings broken by long direct glides with short wings fully stretched, is limited to dashes between cover in follow-the-leader fashion. Throughout their territories, pseudo babblers also maintain one or more dormitory nests that are similar in structure and positioning to breeding nests but bulkier; each night, the whole group crams into a single nest to sleep.

Feeding ecology and diet

Pseudo babblers are scansorial insectivores. Each group holds to permanent feeding territories of about 124 acres (5–50 ha) or more year round, depending on productivity, and spends as much as 75% of the day foraging. They use the long bill to toss about in litter, dig and hammer in the ground, and probe the trunks and branches of shrubs and small trees. They eat a range of small to medium-sized arthropods, including beetles, katydids, spiders, crickets, centipedes, termites, grubs, and caterpillars. Seeds, buds, and small reptiles are eaten occasionally. Large prey may be shared among members of a group.

Reproductive biology

Territorial groups usually consist of a senior pair and their offspring and siblings. Breeding may occur erratically year-round in New Guinea but is more limited to spring and early summer in Australian species. The senior pair carries out most of the nest construction. The female incubates and broods alone and is fed on and off the nest by others of her group. All members of the group help in feeding the nestlings, which fledge in about 20–21 days. Nests are rough bulky domes of twigs and fiber that are lined with fine vegetable fiber and animal wool. They are wedged in the upper branches of shrubs and small trees at 6.6–26.2 ft (2–8 m) above the ground. Rufous babblers in New Guinea usually suspend their nest at the ends of palm fronds. Eggs, in clutches of two to five, are pale gray and scribbled all over with fine dusky lines; they hatch in 16–23 days depending on species.

Conservation status

Although no species is threatened, species occurring in southern Australian states are declining due to loss and alienation of habitat (e.g., gray-crowned babblers in South Australia and Victoria and Hall's babblers in New South Wales).

Significance to humans

None known.

A pair of white-browed babblers (*Pomatostomus superciliosus*) share a branch in South Australia. (Photo by R. Drummond/VIREO. Reproduced by permission.)

1. Gray-crowned babbler (*Pomatostomus temporalis*); 2. Rufous babbler (*Garritornis isidorei*). (Illustration by Marguette Dongvillo)

Species accounts

Rufous babbler
Garritornis isidorei

TAXONOMY
Pomatorhinus isidorei Lesson, 1827, Dorei Harbor (Manokwari, Cendrawasi). Two subspecies.

OTHER COMMON NAMES
English: New Guinea babbler, Isidore's babbler; French: Pomatostome Isidore; German: Beutelsäbler; Spanish: Hablantín de Isidore.

PHYSICAL CHARACTERISTICS
Slender, medium-sized pseudo babbler, 9–10 in (23–25 cm); 2.2–2.6 oz (65–75 g). Adults and immature birds are uniformly rich russet-brown all over, with yellowish bill and dusky feet; eyes are pale cream in adults, brown in immature birds.

DISTRIBUTION
All lowland New Guinea and Misool Island up to about 1,500 ft (500 m) altitude.

HABITAT
Interior lower stages and floor of primary and tall secondary rainforest, usually within 33–49 ft (10–15 m) of ground.

BEHAVIOR
In permanent territorial groups of usually 5–10 birds, mixing with other species in foraging parties in under-shrubbery and low trees, traveling quickly by powerful hopping. Groups tight and call continually with soft and loud whistles, rasps, and yodels. They apparently roost communally at night in one nest that is used for a season.

FEEDING ECOLOGY AND DIET
Forages mainly by probing bark and crannies on trunks and branchlets of forest substage but also digs in litter of jungle floor. Diet includes a range of arthropods; small reptiles also taken.

REPRODUCTIVE BIOLOGY
Poorly documented. Nests are pensile, massively elongate, and slung from the ends of fronds (usually rattan palms) at 10–26 ft (3–8 m) above the forest floor. Nests are built by the senior pair and helpers. The clutch, probably incubated by the female alone, is usually of two eggs, about 1.1 by 0.7 in (28 by 18 mm), and scribbled all over as in other pseudo babblers. Both parents, at least, feed the young.

CONSERVATION STATUS
Not threatened.

SIGNIFICANCE TO HUMANS
Some totemic significance for some lowland tribal groups in New Guinea. ◆

Gray-crowned babbler
Pomatostomus temporalis

TAXONOMY
Pomatorhinus temporalis Vigors & Horsfield, 1827, Shoalwater Bay, Queensland, Australia. Two subspecies.

OTHER COMMON NAMES
English: Red-breasted babbler, cackler, chatterer, happy jack, pinebird, temporal babbler; French: Pomatostome à calotte gris; German: Grauscheitelsäbler; Spanish: Hablantín de Corona Gris.

Garritornis isidorei
▢ Resident

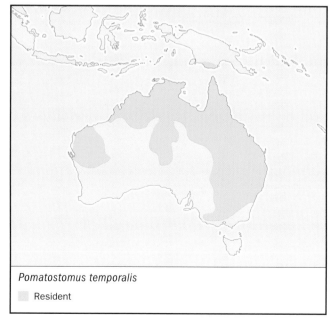

Pomatostomus temporalis
▢ Resident

PHYSICAL CHARACTERISTICS

Largest pseudo babbler, 9.5–10.5 in (24–27 cm); 2.2–3.2 oz (65–90 g). Adults and immature birds dull brown with diffuse white brows, broadly white-tipped tail, and a diagnostic pale rufous patch in the outer wings that shows only in flight. Varying according to subspecies, throat is white grading to dull brown over the belly or to deep russet-brown over the breast; upper backs and center crowns are respectively gray or dusky. Bill is brown with a bone-colored ridge on upper mandible, feet are dusky black, and eyes are pale cream in adults and brown in immature birds.

DISTRIBUTION

Drier coastal and subcoastal northern and eastern Australia, northwest to Kimberley Division, and southeast to central west Victoria, with outliers in the Trans-Fly of south New Guinea and in central-west and central Australia.

HABITAT

Open to dense woodlands with trees of moderate height and under-shrubberies that are sparse to only moderately dense. Dominant trees are species of eucalyptus, paperbarks (*Melaleuca*), mulga, and cypress (*Callitris*).

BEHAVIOR

In territorial groups of about 12, gray-crowned babblers hold foraging territory of 25–37 acres (10–15 ha), or larger in the arid zone, keeping in contact and warning of predators with loud chatterings and whistles. In song, they chorus antiphonally, the lead female braying "ya" and the male responding "ahoo", one following the other in rapid succession. At nightfall, all members of a group cluster to sleep in one of several dormitory nests in their territory.

FEEDING ECOLOGY AND DIET

Foraging extends well up in shrubs and trees, to as much as 66 ft (20 m) above the ground, because of the nature of the habitat. Arthropods of all kinds are picked from crannies in bark and under branches and from tossing litter and digging on the ground; the babbler even laboriously flies up in attempts to catch winged termites.

REPRODUCTIVE BIOLOGY

Breeds in spring and early summer in most areas, but also in autumn in Western Australia. Nests are coarse round domes of twigs wedged in forks in upper branches of tall shrubs and small trees at 9.8–32.8 ft (3–10 m) above the ground. Eggs, incubated by senior females, are about 1.1 x 0.7 in (28 x 20 mm) and covered with the fine scribblings characteristic of the family. Two to four eggs form a clutch. Young hatch in 18–23 days and fledge in about 20–22 days. When occasional groups contain several breeding pairs, the pairs use the same nest together and females share incubation of clutches of up to ten or more eggs. Recruitment nevertheless is low, which in turn is offset by longevity; once they have reached adulthood, individual birds may live for 15 years or more.

CONSERVATION STATUS

Although most populations of both subspecies are not threatened, those of *P. t. temporalis* have withdrawn from much of their range in southeast Australia; they are extinct in South Australia and almost extinct in Victoria.

SIGNIFICANCE TO HUMANS

None known. ◆

Resources

Books

Beehler, B.M., T.K. Pratt, and D.A. Zimmerman. *Birds of New Guinea*. Princeton: Princeton University Press, 1986.

Deignan, H.G. "Timaliinae, Babblers." In *Check-List of Birds of the World*. Vol. 10. Cambridge, MA: Museum of Comparative Zoology, 1964.

Pizzey, G., and F. Knight. *The Field Guide to the Birds of Australia*. Sydney: Angus and Robertson, 1997.

Schodde, R., and I.J. Mason. *The Directory of Australian Birds: Passerines*. Melbourne: CSIRO Publishing, 1999.

Richard Schodde, PhD

Australian creepers

(Climacteridae)

Class Aves

Order Passeriformes

Suborder Passeri (Oscines)

Family Climacteridae

Thumbnail description
Small to medium-sized, brown to blackish streaked birds with long toes and claws and curved bills that climb up tree trunks and branches

Size
5.7–6.9 in (14.5–17.5 cm); 0.75–1.15 oz (21–32 g)

Number of genera, species
2 genera; 7 species

Habitat
Forests, including rainforests, woodlands, and tall shrublands

Conservation status
Near Threatened: 1 subspecies of one species, and two subspecies of another species; remaining subspecies of both are of Least Concern

Distribution
Australia (except treeless regions), mountains of New Guinea, absent from Tasmania

Evolution and systematics

Australian treecreepers resemble, in appearance and habits, the Northern Hemisphere creepers (Certhiidae) and a few other ecologically similar species elsewhere. However, this resemblance is the result of convergence due to occupying a similar niche. Details of their anatomy and behavior, and especially molecular studies, indicate that they are related to the large group of other Australian passerines. They do not seem to be closely related to any other family, but their nearest relatives are probably the lyrebirds and scrub-birds.

One genus, identified by Sibley and Monroe, consists of a superspecies including the polytypic white-throated treecreeper (*Cormobates leucophaea*) and Papuan treecreeper (*C. placens*), the only species that occurs outside Australia. The other genus, *Climacteris*, is remarkably homogeneous.

Physical characteristics

Australian treecreepers are sparrow-sized or slightly larger, fairly robust, with a short neck and longish, decurved bills. Although their legs are short, all toes are long with very long and curved claws, which are presumably adaptations to tree climbing.

Their coloration ranges from rufous through brown to almost black. All species are streaked, sometimes boldly un-

derneath, with white, black, and brown. Several have pale or white throats, and all have a pale stripe through the wing, which is prominent in flight. Sexual dimorphism is subdued, with females typically having orange patches on neck, throat, or breast.

A rufous treecreeper (*Climacteris rufa*) peers into an opening in a dead tree. (Photo by G. Reynard/VIREO. Reproduced by permission.)

A white-throated treecreeper (*Climacteris leucophaea*) at its nest. (Photo by Marshall Sklaß. Photo Researchers, Inc. Reproduced by permission.)

Distribution

Treecreepers are found across Australia, but are absent from sandy and stony deserts and grasslands. These treeless areas often separate subspecies or populations of one species. For instance, the Great Sandy Desert in northwestern Australia lies between two subspecies of the black-tailed treecreeper (*Climacteris melanura*). The Nullarbor Plain lies between the southwestern and Eyre Peninsula populations of the rufous treecreeper (*Climacteris rufa*), though they maintain contact along a thin strip of wooded country on the edge of the Great Victoria Desert. Tasmania has no treecreepers despite having extensive rainforests, eucalypt forests, and woodlands, possibly because treecreepers are poor fliers and Tasmania had less forest when it became isolated from mainland Australia.

Despite its extensive forests, New Guinea only has treecreepers in some of its mountains. The Papuan treecreeper has an inexplicable gap in distribution of about 250 mi (400 km) in central New Guinea.

Habitat

Only one species of treecreeper occurs in rainforest in Australia, with most of the other species in eucalypt forest and woodland. Brown (*Climacteris picumnus*) and rufous treecreepers are also found in mallee, which are low woodland with multi-stemmed eucalypts. However, they generally avoid areas with dense understory. Black-tailed and white-browed treecreepers (*Climacteris affinis*) are found in acacia woodland

in semi-arid regions, with the latter also having a strong affinity with sheoke trees (*Casuarina*).

Behavior

Treecreepers are solitary or occur as pairs and family groups. Territorial defense may be strong at pair or group boundaries, involving much calling and chasing. They are strongly sedentary with no indication of migration, though young birds may disperse several miles (kilometers).

Most species have whistling or clicking calls. Displays include tail clicking and flicking.

Feeding ecology and diet

Most treecreepers forage by climbing up or along trunks and branches, typically of rough barked trees. Their strong claws allow them to cling upside-down, however, they rarely move down a trunk or branch. They capture insects by gleaning from the surface or probing into fissures in the bark or in peeling bark. Some species display a significant amount of ground foraging. Ants feature prominently in the diet, but

A brown treecreeper (*Climacteris picumnus*) clings to a tree in Australia. (Photo by R. Drummond/VIREO. Reproduced by permission.)

they take other insects, especially beetles, larvae, spiders, and nectar and seeds on rare occasions.

Reproductive biology

Although some species breed as pairs, most are cooperative breeders. Helpers consist of young males, rarely females, from earlier years. There is often a close relationship among neighboring groups, with males often only dispersing one or two territories from their homes, and helping may occur among groups. The breeding season ranges from August to January with repeated attempts, and sometimes two broods reared in a year. All species nest in tree hollows, at which females perform an unusual behavior of sweeping snakeskin, in-

sect wings, or even plastic around the entrance. Clutches are usually two or three white to pinkish eggs with brown markings. Incubation takes from 14–24 days, and fledging from 25–27 days.

Conservation status

Clearance and degradation of woodland has led to the decline of several species, at least locally, with three subspecies being regarded as Near Threatened.

Significance to humans

None known.

1. Red-browed treecreeper (*Climacteris erythrops*); 2. White-throated treecreeper (*Cormobates leucophaeus*); 3. Rufous treecreeper (*Climacteris rufa*). (Illustration by Gillian Harris)

Species accounts

White-throated treecreeper
Cormobates leucophaeus

TAXONOMY
Certhia leucophaea Latham, 1802, Port Jackson, New South Wales, Australia.

OTHER COMMON NAMES
English: Papuan treecreeper, little treecreeper; French: Échelet Leucophée; German: Weisskehl-Baumrutscher; Spanish: Sube Palo de Garganta Blanca.

PHYSICAL CHARACTERISTICS
6 in (15 cm); 0.8 oz (22 g). Brownish gray upperparts with white throat and rufous patch at lower cheek; breast is buff streaked with white.

DISTRIBUTION
Eastern Australia from north Queensland to Adelaide region. Distinct subspecies in northeast Queensland, Clarke Range of central coastal Queensland, and Mount Lofty Ranges of South Australia.

HABITAT
Tropical, subtropical, and temperate rainforests, eucalypt forests, and woodlands.

BEHAVIOR
Solitary or in pairs in strongly defended territories. Strident whistling and piping, and more subdued chattering calls, display by rapidly flicking tail open and closed with audible clicking sound.

FEEDING ECOLOGY AND DIET
Climbs up trunks and along branches of rough-barked eucalyptus and other trees, gleaning from the surface and probing into bark for ants and other arthropods.

REPRODUCTIVE BIOLOGY
Breeds August–January, but mostly September/October, in hollows in the trunks or branches of living or dead trees, often eucalypts. Two or three eggs, incubated by female for 22–24 days. Nestlings fed by both parents, but only females brood chicks; fledging at 26 days.

CONSERVATION STATUS
Secure, common and copes well with habitat fragmentation.

SIGNIFICANCE TO HUMANS
None known. ◆

Red-browed treecreeper
Climacteris erythrops

TAXONOMY
Climacteris erythrops Gould, 1841, Liverpool Plains, New South Wales, Australia.

OTHER COMMON NAMES
English: Red-eyebrowed treecreeper; French: Échelet à sourcils roux; German: Rostbrauen-Baumrutscher; Spanish: Sube Palo de Cejas Rojas.

PHYSICAL CHARACTERISTICS
6.3 in (16 cm); 0.8 oz (23 g). Underparts brown streaked with white; gray head with rufous around eye, white throat, brown mantle, and grayish tail.

DISTRIBUTION
Southeastern Australia from Brisbane to Melbourne.

Cormobates leucophaeus
▨ Resident

Climacteris erythrops
▨ Resident

HABITAT

Eucalypt forests of the Great Dividing Range, less commonly in woodlands to the west of the range. Occasionally, temperate and subtropical rainforest. Especially associated with eucalypts that have peeling bark or that shed bark on their lower branches and trunks.

BEHAVIOR

Lives in family groups in a large home range, but highly sedentary. Aggression and conflict between neighbors much less frequent than in the white-throated treecreeper. Rather soft, chattering calls.

FEEDING ECOLOGY AND DIET

Typically probe into bark of rough-barked trees, and especially into accumulations of peeling bark on gums and boxes (subgenus *Symphyomyrtus*) for insects such as spiders and especially ants.

REPRODUCTIVE BIOLOGY

Breeds September–January in cooperative groups of three to four birds. Nests are placed in tree hollows, typically in a spout of a living tree. Clutch size is strictly two eggs, incubated by the breeding female for about 18 days. Young are fed by parents and helpers, and fledge at 25 days, with a high success rate (74% in New South Wales).

CONSERVATION STATUS

Secure, but cope poorly with habitat fragmentation, and have contracted from the west of their range where woodland has been extensively cleared.

SIGNIFICANCE TO HUMANS

None known. ◆

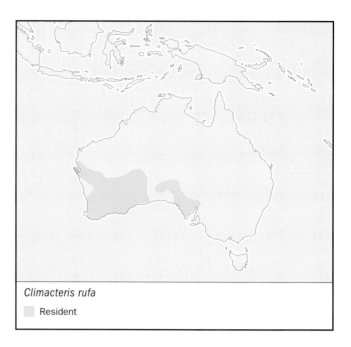

Climacteris rufa

■ Resident

Rufous treecreeper
Climacteris rufa

TAXONOMY

Climacteris rufa Gould, 1841, Swan River, Western Australia.

OTHER COMMON NAMES

English: Allied rufous treecreeper, wheelbarrow; French: Échelet roux; German: Rostbauch-Baumrutscher; Spanish: Sube Palo Rufo.

PHYSICAL CHARACTERISTICS

6.7 in (17 cm); 1.1–1.2 oz (30–33 g). Rufous brow and cheek with black eye strip; underparts rufous streaked with white; gray-brown upperparts with rufous-brown tail.

DISTRIBUTION

Southwestern Australia, Eyre Peninsula of South Australia, with loose links between these populations.

HABITAT

Eucalypt woodland and forest; mallee.

BEHAVIOR

Lives in family groups, consisting of a breeding pair and offspring from previous breeding seasons. Sedentary. Peeping calls, churring calls at predators.

FEEDING ECOLOGY AND DIET

Forages on trunks and lower branches of eucalypts and casuarinas, and also commonly on the ground, especially among fallen timber. Eats insects, especially ants, as well as centipedes and snails, small reptiles, and seeds.

REPRODUCTIVE BIOLOGY

Breeds August–January in hollows in branches, stumps, and fallen logs. Female incubates one to three eggs for 17 days. Young are fed by parents and helpers, and fledge at 26 days, with high success (78% in one Western Australia study).

CONSERVATION STATUS

Not threatened. Secure, but populations have declined or gone locally extinct in parts of the heavily cleared Wheatbelt of western Australia.

SIGNIFICANCE TO HUMANS

None known. ◆

Resources

Books

Garnett, S.T., and G.M. Crowley. *The Action Plan for Australian Birds 2000.* Canberra: Environment Australia, 2000.

Higgins, P.J., J.M. Peter, and W.K. Steele, eds. *Handbook of Australian, New Zealand and Antarctic Birds.* Vol. 5, *Tyrant-flycatchers to Chats.* Melbourne: Oxford University Press, 2001.

Sibley, C.G. and J.E. Ahlquist. *Phylogeny and Classification of Birds: A Study of Molecular Evolution.* New Haven: Yale University Press, 1990.

Periodicals

Luck, G., A. Charmantier, and P. Ezanno. "Seasonal and Landscape Differences in the Foraging Behaviour of the Rufous Treecreeper *Climacteris rufa.*" *Pacific Conservation Biology* 7 (2001): 9–20.

Resources

Noske, R. "Co-Existence of Three Species of Treecreepers in North-Eastern New South Wales." *Emu* 79 (1979): 120–128.

Noske, R.A. "Intersexual Niche Segregation among Three Bark-Foraging Birds of Eucalypt Forests." *Australian Journal of Ecology* 11 (1986): 255–267.

Noske, R.A. "A Demographic Comparison of Cooperatively Breeding and Non-Cooperative Treecreepers." *Emu* 93 (1993): 73–86.

Walters, J.R., H.A. Ford, and C.B. Cooper. "The Ecological Basis of Sensitivity of Brown Treecreepers to Habitat Fragmentation: A Preliminary Assessment." *Biological Conservation* 90 (1999): 13–20.

Hugh Alastair Ford, PhD

Long-tailed titmice

(Aegithalidae)

Class Aves

Order Passeriformes

Suborder Passeri (Oscines)

Family Aegithalidae

Thumbnail description

Small tits with relatively long tails and loose feathering that gives a fluffy appearance. They are generally dark above, either gray or brown and lighter, and often white below. Many species have a black mask, and some show hints of pink in their feathering

Size

3.5–6.3 in (8.9–16 cm); 0.14–0.32 oz (4–9 g)

Number of genera, species

3 genera, 7 species

Habitat

Woodland and forest

Conservation status

Near Threatened: 2 species

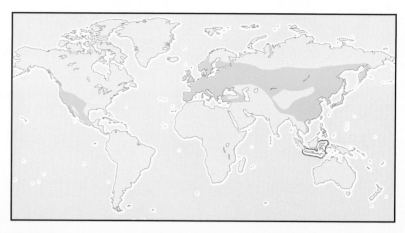

Distribution

Western Europe to the Himalayas and the Far East; western North America and Mexico

Evolution and systematics

The long-tailed tit family (Aegithalidae) consists of three genera and seven species. Sibley and Ahlquist (1990) place the family in the superfamily Sylvioidea: this includes true tits, penduline tits, treecreepers, wrens, nuthatches, and others.

Physical characteristics

Family members range from 6.3 in (16 cm) to a mere 3.5 in (8.9 cm) in the aptly named pygmy tit (*Psaltria exilis*). All have long tails, particularly the long-tailed tit (*Aegithalos caudatus*), whose tail can make up half its total body length. Adult males and females have a similar plumage. They are generally dark above (gray or brown and lighter) and often white below. Many species have black mask and some show hints of pink in their feathering. A loose arrangement of body feathers makes them appear fluffy and endears them to many observers.

Distribution

Of the seven species, five are found in the Himalayas or mountainous parts of western China. The most widespread species is the long-tailed tit, with a range through western Europe and Asia, as far eastward as China and Japan. The most restricted species is the pygmy tit, which is endemic to Java. The only New World representative of the group is the bushtit (*Psaltriparus minimus*), found in western North America and Mexico.

A long-tailed titmouse (*Aegithalos caudatus*) clings to a branch. (Photo by B. Gadsby/VIREO. Reproduced by permission.)

A long-tailed titmouse nest. (Illustration by Jonathan Higgins)

Habitat

Long-tailed tits are primarily birds of edges and shrub layers of woodland and forest. In the Himalayas and mountains of China, they are found between 4,000 and 8,860 ft (1,200 and 2,700 m), or, in the case of the white-throated tit (*Aegithalos niveogularis*), up to the tree line at 13,100 ft (4,000 m).

Behavior

Birds in the long-tailed tit family spend most of their time in single-species flocks. For individual long-tailed tits, these flocks may be composed largely of related birds. Observers often first notice their presence on hearing constant chattering, the contact calls described as *tsee-tsee-tsee* (long-tailed tits) or *pit-pit-pit* (bushtit). Following this, a procession of single birds may typically be seen flying from one bush to another. In the evening, birds roost communally, with small groups lining up together on a suitable branch. If it is cold, they huddle shoulder to shoulder, with the flock's most dominant birds toward the middle of the row where most heat is retained. Long-tailed tits also have been observed roosting in holes in the ground.

Feeding ecology and diet

Birds of the long-tailed tit family spend much of their time in feeding flocks, searching for invertebrates and occasionally fruit and seed. Like other near relatives, they are extremely dextrous birds, comfortable hanging acrobatically from the thinnest of branches, holding an item of food in one claw while picking at it with the fine stubby bill.

Reproductive biology

Breeding season is from January (bushtit) to July. The diminutive pygmy tit on Java has a further season, from August to November.

During breeding, larger feeding and roosting flocks break down as individual birds pair together. In early parts of the breeding season, birds often still roost together; during a cold spell, feeding flocks may reform. Once the nest has been constructed, its warmth and security provide adequate roosting space for the pair alone.

Nests are enclosed oval or more elongated structures woven from moss, lichen, spider silk, and plant material. Once complete, they are quite light in color, possibly an attempt by the builders to camouflage them against light background breaks in the woodland canopy. Toward the top, each nest

Long-tailed titmouse (*Aegithalos caudatus*) feeding young at its nest. (Photo by Stephen Dalton. Photo Researchers, Inc. Reproduced by permission.)

has an entrance hole and is furnished with a soft lining that can include more than 2,000 feathers. They are commonly located low in the woodland shrub layer, suspended among or in the forks of suitable branches.

Clutch size is 2–12 eggs. The birds incubate for 12–18 days. Once hatched, youngsters are cared for by the parents and, in some cases, other members of the flock, often individuals whose own breeding attempts have failed. The young fledge within three weeks of hatching and remain with the parents' flock over the first winter.

Conservation status

Birds in the long-tailed tit family are common over much of their range and are not threatened. However, harsh winters can decimate the population by up to 80%. Himalayan and Chinese mountain species are common to locally common across their range with two exceptions: the sooty tit (*Aegithalos fuliginosus*) and white-throated tit (*Aegithalos niveogularis*) are scarce and listed as Near Threatened. Javanese endemic pygmy tits are locally common, but the ever-present threat of deforestation is of concern for this forest species.

Significance to humans

Bushtits visit garden feeders; long-tailed tits are rarely seen at feeders but their appearance in garden trees and parks is always popular.

1. Bushtit (*Psaltriparus minimus*); 2. Black-throated tit (*Aegithalos concinnus*); 3. Long-tailed tit (*Aegithalos caudatus*). (Illustration by Gillian Harris)

Species accounts

Long-tailed tit
Aegithalos caudatus

TAXONOMY
Aegithalos caudatus Linnaeus, 1758. Nineteen subspecies are recognized.

OTHER COMMON NAMES
French: Mésange à longue queue; German: Schwanzmeise; Spanish: Satrecito de Cola Larga.

PHYSICAL CHARACTERISTICS
5–6.3 in (13–16 cm). Small tit with extremely long tail, plumage variable across range but generally a mix of black, white, and pink.

DISTRIBUTION
The most widespread of the long-tailed tit family with a range from western Europe through Asia, and into China and Japan.

HABITAT
Woodland, deciduous, and mixed with plenty of scrub in which to forage and nest.

BEHAVIOR
A gregarious and acrobatic species, often first picked up on call (*tsee-tsee-tsee*). Flock frequently observed flying in single file, one bird at a time, from bush to bush. Roost communally on branches, huddled together in cold weather.

FEEDING ECOLOGY AND DIET
Largely invertebrates, especially insects and spiders.

REPRODUCTIVE BIOLOGY
Nests March to June. Oval ball-shaped nest of moss and lichen, lined with feathers, located low in bushes and shrubs. Clutch 8–12 eggs, incubation 12–18 days, fledging 14–18 days. Nonbreeding birds may assist parents with feeding of young.

CONSERVATION STATUS
Not threatened. Common across range, suffers after harsh winters, and takes a few years to recover population.

SIGNIFICANCE TO HUMANS
None known. ◆

Black-throated tit
Aegithalos concinnus

TAXONOMY
Aegithalos concinnus Gould, 1855. Six subspecies are recognized.

OTHER COMMON NAMES
English: Red-headed tit; French: Mésange à tête rousse; German: Rostkappen-schwanzmeise; Spanish: Satrecito de Cabeza Roja.

PHYSICAL CHARACTERISTICS
4 in (10 cm); 0.14–0.32 oz (4–9 g). Small and very attractive tit with a rufous crown, black mask, white moustache, and black bib. Gray upper and buff-to-white underparts.

DISTRIBUTION
Himalayas, upland Myanmar, and Indochina.

HABITAT
Broadleaf forest.

BEHAVIOR
A gregarious species foraging in the shrub layers of forests, often in mixed species flocks. Roosts communally. Contact call *psip*, *psip* and *si-si-si*.

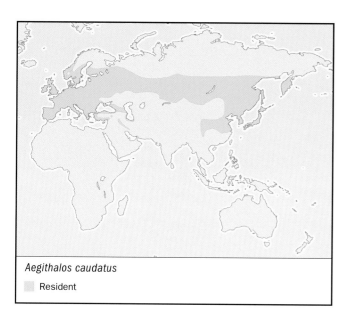

Aegithalos caudatus
▨ Resident

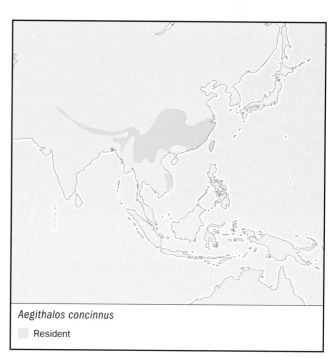

Aegithalos concinnus
▨ Resident

FEEDING ECOLOGY AND DIET
Feeds on insects, seed, and fruit.

REPRODUCTIVE BIOLOGY
Nests February to May. Nest a typical ball of moss and lichen positioned low in a bush. Clutch 3–9, incubation 15–16 days.

CONSERVATION STATUS
Not threatened.

SIGNIFICANCE TO HUMANS
None known. ◆

Bushtit
Psaltriparus minimus

TAXONOMY
Psaltriparus minimus Townsend, 1837. Eleven subspecies recognized.

OTHER COMMON NAMES
English: Common bushtit, black-eared bushtit; French: Mésange masquée; German: Buschmeise; Spanish: Satrecito Común.

PHYSICAL CHARACTERISTICS
4–4.5 in (10–11.4 cm); 0.18–0.21 oz (5–6 g). Tiny birds with a variable plumage range. Generally gray above with paler gray underparts. Coastal birds have brown caps and black-eared forms have black masks extending back to ear coverts.

DISTRIBUTION
Western United States (extending a little northward into Canada) and Mexico.

HABITAT
Deciduous and mixed woodlands, parks, and gardens.

BEHAVIOR
A gregarious and active species that forages in large flocks. Roosts communally, as with other bushtits.

FEEDING ECOLOGY AND DIET
Feeds on insects, spiders, seed, and fruit.

Psaltriparus minimus

▨ Resident ▨ Nonbreeding

REPRODUCTIVE BIOLOGY
Nests January to June. Nest cucumber-shaped construction of twigs, moss, and lichen hung from the end of a branch. Clutch 5–7, incubation 12 days, fledging 14–15 days. Occasionally parents will be helped by other birds, as in long-tailed tits.

CONSERVATION STATUS
Not threatened. Common, and increasing in some parts of its range.

SIGNIFICANCE TO HUMANS
None known. ◆

Resources

Books
Harrap, S. and D. Quinn. *Chickadees, Tits, Nuthatches and Treecreepers.* Princeton: Princeton University Press, 1996.

Periodicals
Hansell, M. H. "The demand for feathers as building material by woodland nesting birds." *Bird Study* 42 (1995): 240–245.

Hansell, M. H. "The function of lichen flakes and white spider cocoons on the outer surface of bird's nests." *Journal of Natural History* 30 (1996): 303–311.

Hatchwell, B. J., C. Anderson, D. J. Ross, M. K. Fowlie, and P. G. Blackwell. "Social organization of cooperatively breeding long-tailed tits: kinship and spatial dynamics." *Journal of Animal Ecology* 70 (2001).

Tony Whitehead, BSc

▲
Penduline titmice
(Remizidae)

Class Aves
Order Passeriformes
Suborder Passeri (Oscines)
Family Remizidae

Thumbnail description
A variable group of small passerines with short wings and tails, and delicate heads. Plumage is uniform and dull-colored in both males and females. Bill is not curved but straight

Size
3–4.3 in (7.5–11 cm); 0.16–0.44 oz (4.6–12.5 g)

Number of genera, species
4 genera; 10 species

Habitat
Varied: deserts, wetlands, scrub and forest

Conservation status
Not threatened

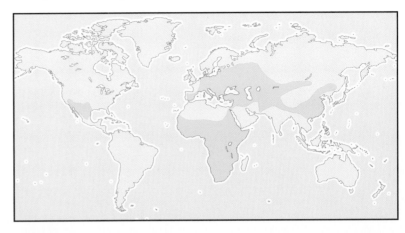

Distribution
Wide distribution from Africa through Europe and into Asia; one species in North America

Evolution and systematics

There are 10 species of penduline tit in four genera, according to Peters (1934–1986). However, classification is problematic. Traditionally they are placed in their own family, the Remizidae, but more recent classification based on DNA analysis suggests they may be better placed as a subfamily (Remizinae) in a larger Paridae family, which also includes the true tits, i.e. chickadees (Sibley and Ahlquist 1990).

Physical characteristics

The penduline tits are small passerines ranging in length from the 3 in (75 mm) tit-hylia (*Pholidornis rushiae*) to the larger 4.3 in (110 mm) European penduline tit (*Remiz pendulinus*). Their color and form is quite variable, reflecting the group's diversity. Upper parts range from gray to chestnut, to olive green, and underparts from white to yellow. Some, such as those in the *Remiz* genus, show distinctive black masks. The African penduline tits (genus *Anthroscopus*) are similar to leaf warblers (Sylviidae). Tail length is variable; some of the African penduline tits and fire-capped tits (*Cephalopyrus flammiceps*) have quite short tails, whereas those of the European penduline tits are relatively long. Perhaps the most constant family feature is the bill, which in most species is conical and sharply pointed.

Distribution

With the exception of the North American verdins (*Auriparus flaviceps*), the penduline tits are primarily an Old World family with a wide distribution from Africa through Europe and into Asia.

Habitat

Penduline tits are found in a range of habitats from the deserts of Arizona to the expansive reedbeds of the black-headed penduline tits (*Remiz macronyx*) and the sub-Saharan scrub and forests of the African penduline tits.

Behavior

Penduline tits are extremely active and agile birds. Typically, they are found in pairs or in small groups. Forest-dwelling species spend much time in the tree canopy, using their agility to move through branches, sometimes nimbly making their way along the undersides of twigs and boughs. Like long-tailed tits (Aegithalidae), they may roost in groups at night. Verdins are of particular interest in their construction of roosting nests. These are similar to breeding nests but lack soft linings and are generally less "finished."

Penduline tits from north temperate areas migrate outside the breeding season; other species are largely sedentary.

Calls are high pitched and songs range from the rich vocalizations of the *Remiz* to the rather more repetitious *Anthoscopus*. They are territorial in the breeding season, but the territory is largely confined to the immediate environs of the nest. This small, defended area leaves room for possible semi-colonial nesting in species such as the European penduline tit.

A Eurasian penduline titmouse (*Remiz pendulinus*) at its nest. (Photo by J. Peltomaki/VIREO. Reproduced by permission.)

Feeding ecology and diet

Penduline tits feed on a variety of invertebrates, fruit, and seeds. Like true tits, they are able to grasp food in one foot and peck at the item with their bill. Often they search for food, such as spider's nests, in crevices and holes in trees.

Reproductive biology

The mating system, where studied, has proven highly complex. European penduline tits can be monogamous, polygamous, and polyandrous. With the African penduline tit, the presence of unusually large clutches in single nests may be evidence of the attentions of more than one female.

In northern temperate species, breeding takes place from April through July. In Africa, breeding depends on local climatic conditions, with some species nesting in the rainy season and others in the dry season.

Penduline tits derive their family name from their free-hanging, pendulous nests. These are found in a variety of locations in the branches of trees and shrubs or, in the case of European penduline and black-headed tits, from groups of reeds.

They are teardrop or pear-like in shape with a convenient entrance hole towards the top. In some genera, such as the *Anthoscopus*, this entrance has a ledge sometimes visited by other birds. The entrance is fastened together when not in use as a defense against predators. In addition, when the entrance is closed in this way the ledge gives the appearance of a confusing false entrance.

Nests are constructed from plant matter in most cases, compressed to produce a durable exterior and lined with softer grasses, mosses, and lichens inside for a snug home. The conspicuous nest of the verdin differs a little in construction in that the exterior is woven from hundreds to as many as 2,000 thorny twigs. The fire-capped tit is the only one without a pendulous nest, preferring an altogether more conservative cup-shaped structure hidden a tree hole.

Most penduline tit eggs are white. Clutch size varies from two to nine. Where observed, incubation takes between 13 and 17 days. Once the chicks hatch, care of the young is often shared between parents. In some species, care may be cooperative within a larger group (e.g. tit-hylia).

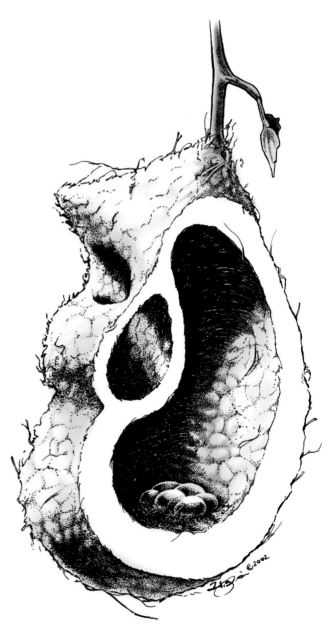

Penduline titmouse nest. (Illustration by Jonathan Higgins)

Conservation status

The European penduline tit considerably expanded its range westwards between the 1930s and the 1980s. Likewise, the Chinese penduline tit (*Remiz consobrinus*) has also increased in observed numbers on migration and on its wintering sites. Others, such as the white-crowned (*Remiz coronatus*) and black-headed tits, are not faring as well in face of the relentless intensification of agriculture and development of land. Some species though, especially the *Anthosco-pus*, are described as uncommon but may simply be overlooked by observers.

Significance to humans

Admired as architects, they are otherwise of little significance to humans. However, the nests of European penduline tits have been used in Eastern Europe as slippers for children.

1. Fire-capped tit (*Cephalopyrus flammiceps*); 2. European penduline tit (*Remiz pendulinus*); 3. Verdin (*Auriparus flaviceps*); 4. African penduline tit (*Anthoscopus caroli*). (Illustration by Gillian Harris)

Species accounts

European penduline tit
Remiz pendulinus

TAXONOMY
Remiz pendulinus Linnaeus, 1758. Four subspecies.

OTHER COMMON NAMES
English: Eurasian penduline tit, penduline tit; French: Rémiz penduline; German: Beutelmeise; Spanish: Baloncito Común.

PHYSICAL CHARACTERISTICS
4.3 in (110 mm); 0.28–0.44 oz (8–12.5 g). A small tit with, for the group, a relatively long tail. Nominate race: sexes similar, though female paler; prominent black face mask contrasting with otherwise gray head; chestnut brown back, pale buff beneath; juvenile lacks mask.

DISTRIBUTION
A Eurasian bird with a breeding range from western Europe to western Siberia and south to Turkey. Northern populations migratory, moving to southern Europe and the Middle East in winter.

HABITAT
Largely found in wetlands with a mix of reed and scrub, including willow and alder.

BEHAVIOR
Gregarious in flocks of up to 60 birds, especially out of season and on migration. Males aggressively defend immediate territory around nest in breeding season, but this small area allows for possible semi-colonial nesting.

FEEDING ECOLOGY AND DIET
Feeds on invertebrates, including insects and spiders; reed seeds important in winter.

REPRODUCTIVE BIOLOGY
Constructs pendulous (free hanging) pear-shaped nest from compressed plant material and lined with softer items. Suspended from branches or a number of reed stems. Clutch size is two to seven eggs; incubation 13–14 days; fledging 18–26 days.

CONSERVATION STATUS
Not threatened. Locally common in suitable habitat. Considerable westward expansion between 1930s and 1980s.

SIGNIFICANCE TO HUMANS
None known. ◆

African penduline tit
Anthoscopus caroli

TAXONOMY
Anthoscopus caroli Sharpe, 1871. Eleven subspecies.

OTHER COMMON NAMES
English: Gray penduline tit; French: Rémiz de Carol; German: Weisstirn-Beutelmeise; Spanish: Baloncito Africano.

PHYSICAL CHARACTERISTICS
3.5 in (90 mm); 0.21–0.24 oz (6–6.9 g). A variable but bland warbler-like species with a typical conical pointed bill and short tail. Upperparts range between species from olive-green to gray; underparts from pale yellow to cream.

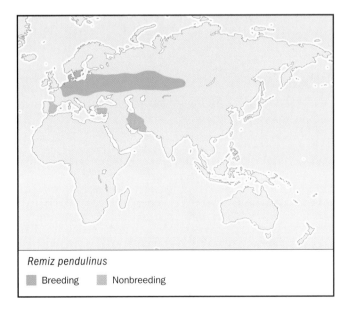

Remiz pendulinus
■ Breeding ■ Nonbreeding

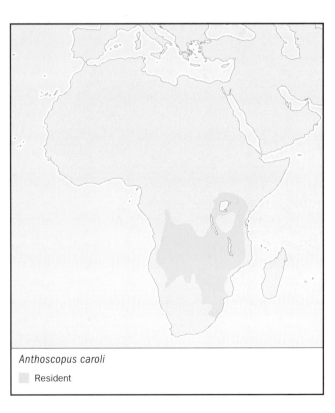

Anthoscopus caroli
■ Resident

DISTRIBUTION
Most widespread of the *Anthoscopus* spp. with a range from Kenya to South Africa.

HABITAT
Woodland.

BEHAVIOR
Found in pairs or small groups, sometimes foraging in the tree canopy with other species. Very active.

FEEDING ECOLOGY AND DIET
Feeds on invertebrates and fruit.

REPRODUCTIVE BIOLOGY
Breeds throughout the year across its wide range. Nest typical of the subfamily; pear shaped pendulous sack of compressed plant material provided with a entrance hole and ledge. The hole is fastened together when not in use. Clutch 4–6 white eggs, occasionally more possibly due to two females laying in the same nest. Incubation and fledging periods not known.

CONSERVATION STATUS
Not threatened, but some contraction along coastal areas. May be overlooked.

SIGNIFICANCE TO HUMANS
None known. ◆

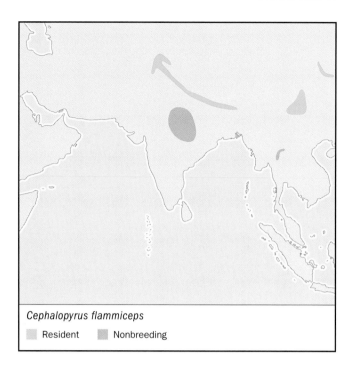

Cephalopyrus flammiceps
☐ Resident ☐ Nonbreeding

Fire-capped tit
Cephalopyrus flammiceps

TAXONOMY
Cephalopyrus flammiceps Burton, 1836. Two subspecies.

OTHER COMMON NAMES
French: Rémiz tête-de-feu; German: Flammenstirnchen; Spanish: Baloncito de Capa en Llamas.

PHYSICAL CHARACTERISTICS
4 in (100 mm); 0.25 oz (7 g). A small short-tailed tit. Sexes similar out of breeding season, olive-green above and olive-yellow below. Breeding male has distinctive orange red forecrown (the "fire cap"), chin, and throat. Juveniles dull, but similar to adults.

DISTRIBUTION
Himalayas and parts of Western China. Northern India and northern parts of southeast Asia outside breeding season.

HABITAT
Moist, temperate woodland and forest.

BEHAVIOR
Active and agile bird that forages in single species or mixed flocks in the tree canopy.

FEEDING ECOLOGY AND DIET
Invertebrates plus flower and leaf buds.

REPRODUCTIVE BIOLOGY
Nests April to June in holes in trees. Clutch of three to five blue-green eggs; incubation and fledging periods not known.

CONSERVATION STATUS
Not threatened. Scarce over much of its breeding and wintering range.

SIGNIFICANCE TO HUMANS
None known ◆

Verdin
Auriparus flaviceps

TAXONOMY
Auriparus flaviceps Sundevall, 1850. Six subspecies.

OTHER COMMON NAMES
French: Auripare verdin; German: Goldköpfchen; Spanish: Baloncito Verdín.

PHYSICAL CHARACTERISTICS
4–4.5 in (100–110 mm); 0.21–0.29 oz (6–8.2 g). Small with sharper bill than its relatives. Dull yellow head and throat, dark grey upperparts, lighter grey beneath.

DISTRIBUTION
Southwestern United States and Mexico.

HABITAT
Open desert with scattered bushes and cacti.

BEHAVIOR
Sprightly birds normally found in pairs and family groups. More solitary than other penduline tits. Interesting in its construction of roosting nests.

FEEDING ECOLOGY AND DIET
Feeds on invertebrates, seeds, and fruit.

REPRODUCTIVE BIOLOGY
Breeds from March to June. Nest a spherical construction up to 7.9 in (200 mm) diameter of layered thorny and thornless

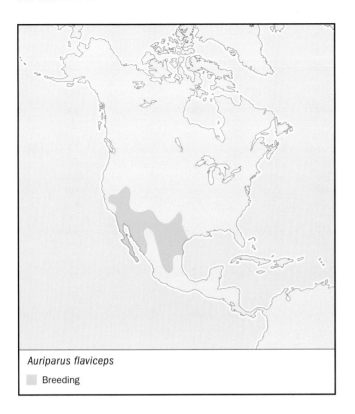

Auriparus flaviceps

Breeding

twigs lined with softer material and located in scrub towards the end of branches. Clutch of two to four blue-green eggs; incubation 14–17 days; fledging 17–19 days.

CONSERVATION STATUS
Not threatened. Common and increasing.

SIGNIFICANCE TO HUMANS
None known. ◆

Resources

Books

Campbell, B. and E. Lack. *A Dictionary of Birds.* San Diego: T and A D Poyser, 1985.

Harrap, S. and D. Quinn. *Chickadees, Tits, Nuthatches and Treecreepers.* Princeton: Princeton University Press, 1996.

Snow, D.W. and C.M. Perrins. *Birds of the Western Palearctic.* Concise Edition. Vol. 2, *Passerines.* New York: Oxford University Press, 1998.

Tony Whitehead, BSc

Titmice and chickadees
(Paridae)

Class Aves
Order Passeriformes
Suborder Passeri (Oscines)
Family Paridae

Thumbnail description
Small, compact, and agile birds with short, stout bills; sexes usually similar in size and plumage, typically brightly colored with contrasting plumage features

Size
3.9–8 in (10–20.5 cm); 0.2–1.7 oz (5–49 g)

Number of genera, species
5 genera; 55–58 species

Habitat
Forest, woodlands, parks, orchards, gardens, and scrub

Conservation status
Vulnerable: 1 species; Near Threatened: 3 species

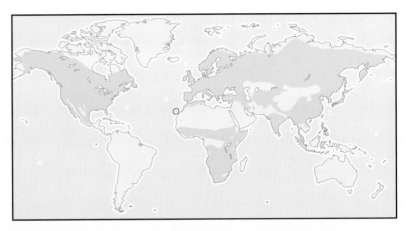

Distribution
Europe, Asia, Africa, North America, and Mexico

Evolution and systematics

The titmice and chickadees are oscine passerines (perching birds). The oscines are divided into two parvorders: Corvida and Passerida. The Passerida are further divided into three superfamilies, including the Sylvioidea, to which tits belong, along with swallows, bulbuls, warblers, and babblers. The titmice and chickadees are usually considered to be a subfamily (Parinae) of the Paridae, the latter also containing the Remizinae, or penduline tits. Genetic evidence suggests that the two groups of tits are closely related, but behavioral differences mean that some authors still treat them as separate families in their own right: the Paridae and Remizidae. The Aegithalidae, or long-tailed tits, are also sometimes included within the Paridae, although genetic comparisons have shown that they are best treated as a distinct family.

The Parinae is divided into five genera: *Parus*, the largest group, including most of the tits; *Poecile*, including the 12–15 species of black- and brown-capped tits and chickadees (sometimes treated as a subgenus of *Parus*); *Baeolophus*, the five species of American tufted titmice; and two monospecific genera, *Sylviparus* and *Melanochlora*. (Although Peters only recognizes four genera: *Parus*, *Baeolophus*, *Sylviparus*, and *Melanochlora*).

Physical characteristics

Fast-moving, agile and small—this is the classic description of the titmice. With short wings and tail, and short legs with very strong feet, they are well-adapted to their arboreal life-style. There is little variation in size within the family and sexes of the same species are typically similar, although fe-

males can be slightly smaller. The smallest of the family is the yellow-browed tit (*Sylviparus modestus*), while the largest is the sultan tit (*Melanochlora sultanea*). Throughout the family, the bill is typically short and stout. This appearance is accentuated in some species that also have relatively large heads (e.g., the willow tit [*Parus montanus*]), which either specialize in feeding on hard seeds or excavate their own nest holes. Within a single species, the bill shape can vary according to the habitats in which individuals live and the types of seeds and nuts available to them. Bill and leg colors, and iris color in most species, are dark and drab, although two of the African species have distinctive pale yellow irises.

Generally, members of the different subgenera share plumage characteristics. The *Poecile* are typically dark-capped, brown-backed and cream-breasted with white cheeks and varying amounts of rufous on the flanks. The spot-winged tits have characteristic white tips on the wing coverts, which give the spot-winged appearance. The crested tits are represented by two species that both possess head-crests but vary in color patterns. The six species of African black tits are typically all black with striking white wing-coverts and white edges to the flight feathers. The rufous-bellied tits are three closely related species all characterized by having rufous underparts and dark upperparts. The African gray tits are gray in overall coloration ranging from the paler Miombo tit, *Parus griseiventris*, to the duskier southern gray tit, *Parus afer*. Like the rufous-bellied tits, the gray tits have well-patterned wings formed from white edging to all of the wing feathers. The great tit, *Parus major*, and its allies are mostly brightly colored tits. All have black heads with white cheeks and a black vertical breast stripe, though back and breast coloration is highly variable, includ-

Blue titmouse (*Parus caeruleus*) feeding young. (Photo by IFA Bilderteam. Bruce Coleman Inc. Reproduced by permission.)

ing species with gray, green, blue, black, or yellow colors. The three species making up the blue tit group are similar in general shape, but vary dramatically in color from the brightly colored blue tit, *Parus caeruleus*, to the very pale azure tit, *Parus cyanus*. The American tufted titmice are mostly drab birds: gray or brown with pale underparts and darker upperparts. The yellow-browed tit is a small, all-green bird with pale-yellowish underparts and a faint yellow eye-ring, lacking the high-contrast plumage features typical of nearly all other tits.

The majority of species, especially those with wide geographical distributions, have many subspecies described and these often differ in plumage coloration to some degree.

Distribution

True tits are absent only from the treeless Arctic zone, South America, the desert regions of Africa and Asia, and Australasia.

Twelve species occur in North America and Mexico; seven *Poecile* species and all of the *Baeolophus* species. The black-capped (*Poecile atricapilla*) and boreal (*P. hudsonica*) chickadees are the most widespread, occurring from coast to coast across North America through its boreal forests and temperate woodlands. All of the other North American and Mexican species are more restricted in range. For example, the tufted titmice of the genus *Baeolophus* have largely south and southwestern distributions, many overlapping into Mexico, as does the Mexican chickadee, *Poecile sclateri*.

Within Europe and temperate Asia, many species have wide distributions across the two continents throughout a variety of forest habitats. However, some species are more specialized in habitat use, such as the Siberian tit, *Parus cinctus*, which is restricted to the boreal forests (and also ranges into arctic North America) and the somber tit, *Parus lugubris*, which occurs in the European eastern Mediterranean and across into Iran. Many of the temperate Asiatic species occur in the highlands of central Asia and have relatively restricted distributions. One species with a particularly restricted range is Père David's tit, *Parus davidi*, which occurs only in mixed forests at 7,000–11,000 ft (2,135–3350 m) in three mountain ranges in central China. In the Far East, some species are restricted to island groups, such as the yellow-tit in Taiwan, and the three Philippine species. Some of the Asian species have ranges that take them into tropical Asia, such as the sultan tit, which occurs from the Himalayas to Indochina, and the black-lored tit, which occurs from the Himalayas to southern India.

While some of the European species also occur in the far northwest of Africa, the truly African species all have sub-Saharan distributions and are endemic. The majority have relatively restricted distributions with only the white-shouldered tit (*Parus guineensis*) having a cross-continent range from coastal Senegal east to Ethiopia.

Habitat

Those species with wide distributions tend to be more generalist in habitat use, utilizing a wide variety of different wood-

A mountain chickadee (*Poecile gambeli*) at its nest. (Photo by B. Randall/VIREO. Reproduced by permission.)

land types from conifers to evergreen broad-leaved woodland to deciduous broad-leaved woodland. Open woodlands, parks, hedgerows, gardens, vineyards, orchards, and scrub are also widely used. Some species are conifer-forest specialists, such as the boreal tit and Siberian tit, while others are deciduous-woodland species, like many of the African tits. Some occur in arid forests, such as the somber tit, which is found in dry, open woodlands in southern Europe. Tits are present from lowland woods at sea level up to high-altitude forests and sub-alpine scrub at 14,764 ft (4,500 m) in mountainous areas.

Behavior

All of the tits are extremely agile and acrobatic in the way they move through the canopy, hopping quickly from branch to branch and frequently hanging upside down to pry under bark on the underside of branches. They typically make only short, fast flights, which are either direct or gently undulating depending on the distance covered.

All species are largely resident and sedentary, but a few make regular seasonal altitudinal movements, some are irruptive under certain conditions, and some may be truly partially migratory. Most species live in pairs or small groups, being territorial while breeding, but typically joining mixed-species flocks outside of the breeding period. Some flocks can

contain up to 100 individuals of one species, and often include other tit species. Some species are highly aggressive in flocks and chases are common, especially when competing for food provided by man. A variety of threat displays have developed, such as the head-up posture in species with black bibs or crest-raising in the crested species.

Song development is generally poor throughout the family and song displays are rare, but most species have a wide variety of calls that are often loud and given frequently, such as the *chick-a-dee* call given by many of the black- and brown-capped species.

Feeding ecology and diet

Tits are arboreal species, foraging predominantly in the canopies of trees and scrub, and in the shrub layer. However, for some species, ground-foraging is also important. All tits eat a wide variety of invertebrates. Many also eat seeds, nuts, and fruit, especially outside of the breeding season, and some will take nectar. Food storage is a very important behavior for many of the northern species of black- and brown-capped, spot-winged, crested, and American tufted tits; both insects and seeds are stored. Some species may remember precise locations of stores while others store food in a scattered way and relocate it by chance. Some species are able to enter a state of regulated hypothermia overnight to conserve energy.

Reproductive biology

All of the tits nest in cavities, mostly in trees, but also between rocks, in walls, on raised ground, and even in pipes. Many species excavate their own nests in rotting wood, while others use natural holes or those made by other species such as woodpeckers. Many will take readily to nesting in nest-boxes specially created for them by man, although some require saw-dust filling so that they can still perform excavation behavior. Cavities are lined with various soft nesting materials. Most species nest between March and July, but Southern Hemisphere African tits may nest year-round or

Black-capped chickadees (*Poecile atricapilla*) are common throughout North America. (Photo by T. Vezo/VIREO. Reproduced by permission.)

Bridled titmouse (*Baeolophus wollweberi*) in flight. (Photo by Joe Mc-Donald. Bruce Coleman Inc. Reproduced by permission.)

in most species, dispersing after just two to three weeks. Some species, like the coal, great, and blue tits, regularly have two broods per year. Breeding success is highly variable, but has been good in many studies. However, all species are vulnerable to nest loss from specialist predators like squirrels, woodpeckers, and snakes.

Conservation status

The majority of tits are common with wide continental distributions. However, the white-naped tit is listed as Vulnerable under IUCN criteria. Endemic to India, it occurs in small, fragmented populations in tropical, dry, thorn-scrub habitat and dry deciduous forests. Three other species, all restricted to small island groups, are listed as Near Threatened: the Palawan tit, *Parus amabilis*, which is restricted to primary forests on three islands in the Philippines; the white-fronted tit, which occurs on two islands in the Philippines; and the yellow tit, which is found in broad-leaved forest in the central mountain regions of Taiwan.

seasonally. Clutch size varies between the different subgenera, but most have four to 10 eggs. The African species have relatively small clutches of three to six eggs, while the blue and great tits have the largest clutches with up to 13 eggs. Eggs are typically white, or blushed pink, with some red-brown spotting at the larger end. Incubation periods are generally around 14 days, while the brood period is between 14 and 24 days. The young become independent very quickly

Significance to humans

With the possible exception of the yellow tit, the tits are of no known economic significance, but are of contemporary cultural importance in some areas due to the close relationship with humans shown by many species. Also, many species readily visit gardens to take foods provided for them, and will also readily use nest-boxes.

1. Black-capped chickadee (*Poecile atricapilla*); 2. Yellow-browed tit (*Sylviparus modestus*); 3. Siberian tit (*Parus cinctus*); 4. Male white-naped tit (*Parus nuchalis*); 5. Male sombre tit (*Parus lugubris*); 6. Male yellow-bellied tit (*Parus venustulus*); 7. Rufous-bellied tit (*Parus rufiventris*); 8. Bridled titmouse (*Baeolophus wollweberi*); 9. Male sultan tit (*Melanochlora sultanea*); 10. Male great tit (*Parus major*). (Illustration by Emily Damstra)

Species accounts

Sombre tit
Parus lugubris

SUBFAMILY
Parinae

TAXONOMY
Parus lugubris Temminck, 1820. Five subspecies.

OTHER COMMON NAMES
French: Mésange lugubre; German: Trauermeise; Spanish: Carbonero Lúgubre.

PHYSICAL CHARACTERISTICS
5.5 in (14 cm); 0.5–0.7 oz (15–19 g); general plumage color typical of the 'black-capped' tits; sexes similar.

DISTRIBUTION
P. l. lugubris: Balkans and Greece; *P. l. anatoliae*: Greek island of Lesbos, Asia Minor, southern Transcaucasia, Levant and northern Iraq. *P. l. hyrcanus*: southeast Transcaucasia and northern Iran; *P. l. dubius*: southwest Iran and northeast Iraq; *P. l. kirmanensis*: southern Iran.

HABITAT
Open broadleaf and conifer woodlands, parkland, orchards, gardens, vineyards, and scrub habitats at higher elevations including olive (*Olea*) groves.

BEHAVIOR
Resident and largely sedentary. Seasonally territorial with members of pair remaining together throughout year, but may join mixed-species flocks. Song bouts short and infrequent.

FEEDING ECOLOGY AND DIET
Forages throughout crowns of trees and shrubs, and on ground. Diet is large range of invertebrates and some seeds. Unlike many other tits, does not store food.

REPRODUCTIVE BIOLOGY
Nests in cavities in trees or occasionally rocks; uses nest-boxes. Most subspecies use existing cavities, but *P. l. hyrcanus* excavates own nest cavity (both sexes taking part). Eggs laid March through April, with second clutches May through June. Clutch size five to 10 eggs (average is seven). Incubation: 12–14 days, by female alone. Brood period to fledging: 21–23 days.

CONSERVATION STATUS
Not threatened. Has a relatively restricted range and occurs at low breeding densities, but population size in European part of range estimated to be 130,000–640,000 pairs, about 75% of world population.

SIGNIFICANCE TO HUMANS
None known. ◆

Black-capped chickadee
Poecile atricapilla

SUBFAMILY
Parinae

TAXONOMY
Parus atricapillus Linnaeus, 1766. Nine subspecies.

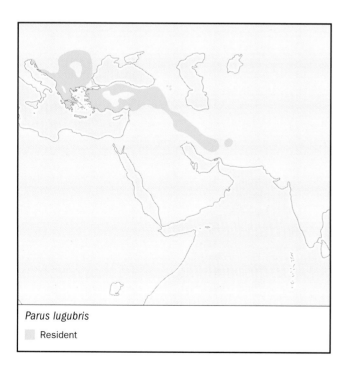

Parus lugubris
░ Resident

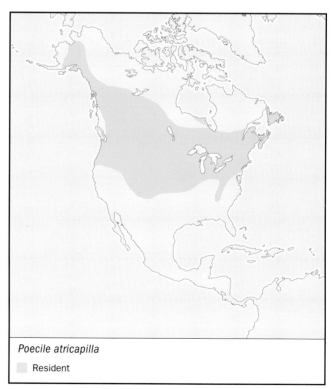

Poecile atricapilla
░ Resident

OTHER COMMON NAMES
English: Black-capped tit; French: Mésange à tête noire; German: Schwarzkopfmeise; Spanish: Carbonero de Capucha Negra.

PHYSICAL CHARACTERISTICS
4.8–5.7 in (12.3–14.6 cm); 0.3–0.5 oz (10–14 g); plumage characteristic of the 'black-capped' tits; sexes similar.

DISTRIBUTION
P. a. atricapillus: northeastern United States and eastern Canada; *P. a. practicus*: Appalachian mountains in eastern United States; *P. a. bartletti*: Newfoundland; *P. a. turneri*: Alaska and northwestern Canada; *P. a. septentrionalis*: mid-continental North America; *P. a. occidentalis*: western coast of United States; *P. a. fortuitus*: western United States to east of Cascade Mountains; *P. a. nevadensis*: Great Basin area of United States; *P. a. garrinus*: Rocky Mountains of United States.

HABITAT
Prefers deciduous or mixed woodland, including open habitats such as parks and gardens, and willow and cottonwood thickets. Frequently associated with birch (*Betula*) and alder (*Alnus*).

BEHAVIOR
Typically resident and territorial, but in mountain areas may show seasonal movements to lower elevations. Forms mixed-species flocks outside of breeding periods with distinct dominance-hierarchies. Has a variety of calls, including the familiar loud *chick-a-dee*.

FEEDING ECOLOGY AND DIET
Forages throughout tree canopy and especially on bark in winter. Rarely forages on the ground. Diet comprises a wide range of invertebrates as well as fruits and seeds. Like many other tits, stores food in autumn for use in the winter. Can save energy overnight by regulated hypothermia, lowering body temperature by up to 53.6°F (12°C).

REPRODUCTIVE BIOLOGY
Nests in cavities in trees, excavating own hole (both sexes) or using natural holes, and will use nest-boxes that are partly filled with sawdust. Lays eggs mid-April to late May, usually a single clutch of six to eight eggs (maximum 13). Female incubates for 12–13 days and broods for 12–16 days. Young birds disperse from parents' territory after three to four weeks and form flocks with unrelated adults or become floaters.

CONSERVATION STATUS
Not threatened. A common and very widespread species, with a typical density of 0.6 pairs/acre (0.25 pairs/ha).

SIGNIFICANCE TO HUMANS
None known. ◆

Siberian tit

Parus cinctus

SUBFAMILY
Parinae

TAXONOMY
Parus cinctus Boddaert, 1783. Four subspecies.

OTHER COMMON NAMES
English: Gray-headed chickadee, Taiga tit; French: Mésange lapone; German: Lapplandmeise; Spanish: Carbonero Lapón.

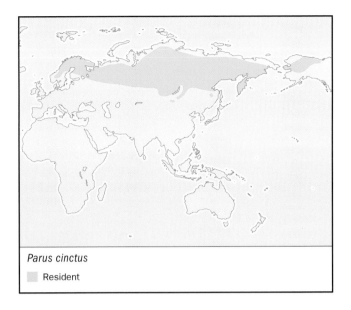

Parus cinctus

Resident

PHYSICAL CHARACTERISTICS
5.3 in (13.5 cm); 0.4–0.6 oz (11–16 g), males larger than females; general plumage color typical of the 'brown-capped' tits; sexes similar.

DISTRIBUTION
Unique within the Paridae as it occurs in both the Old and New Worlds. *P. c. lapponicus*: Fenno-Scandia east to northern European Russia; *P. c. cinctus*: Siberia from Urals east to Bering Sea, and south to Lake Baykal, northeast Mongolia and northeast China; *P. c. sayanus*: central Asia; *P. c. lathami*: Alaska and northwest Canada.

HABITAT
Boreal conifer forests and broadleaf woodland within these forests, the latter more so in winter.

BEHAVIOR
Largely sedentary, although some populations nomadic during winter months. Adults territorial, but readily forms mixed-species flocks outside of breeding season, often associated with willow tit, *Parus montanus*. Song simple and infrequent, not given in display.

FEEDING ECOLOGY AND DIET
Forages predominantly amongst leaves of trees, but explores all parts of tree and frequently forages on the ground. Diet is a variety of invertebrates and seeds; food stored throughout the year at ends of twigs, in cracks in bark and among lichens. Estimated that each bird may store up to 7 lb (15 kg) of food per year, but needs only 15% of this to survive over the winter period.

REPRODUCTIVE BIOLOGY
Nests in cavities in trees. Single clutch of six to 10 eggs laid May through June. Incubation: 15–18 days by female only. Brood period to fledging is 19–20 days.

CONSERVATION STATUS
A common, widespread species. The total European population estimated to be 170,000–1,200,000 pairs, but this represents only a small proportion of total world population. Range in northwest Europe has receded northward in twentieth

century due to changes in forestry practice, loss of habitat, and possibly climate warming.

SIGNIFICANCE TO HUMANS
None known. ◆

Bridled titmouse
Baeolophus wollweberi

SUBFAMILY
Parinae

TAXONOMY
Parus wollweberi Bonaparte, 1850. Four subspecies.

OTHER COMMON NAMES
French: Mésange arlequin; German: Zügelmeise; Spanish: Herrerillo Enmascarado.

PHYSICAL CHARACTERISTICS
4 in (10 cm); 0.3–0.4 oz (8–12 g). Head with striking black and white pattern, and with gray crown edged black, forming crest; sexes similar.

DISTRIBUTION
B. w. vandevenderi: central Arizona to southwest New Mexico, United States; *B. w. phillipsi*: southeastern Arizona to Sonora, Mexico; *B. w. wollweberi*: Mexico; *B. w. caliginosus*: southwestern Mexico.

HABITAT
Mid- to high-elevation woodlands of oak and pine, often mixed with juniper. At lower elevations, savanna-like oak woodlands are occupied, as are open grassland areas with scattered deciduous trees and juniper.

BEHAVIOR
Territorial in the breeding season, but in winter joins mixed-species flocks occupying a home range. Birds at higher elevations may move to lower areas in winter. Song is simple, but is given frequently during breeding period.

FEEDING ECOLOGY AND DIET
Forages throughout tree canopy at all times, and also frequently on ground in winter. Diet is mainly insects, but will eat pulp of acorns and in some habitats spiders become an important food. Not known to store food.

REPRODUCTIVE BIOLOGY
A hole-nesting species, but does not excavate own cavity. Eggs are laid April to June and second clutches are laid only if first is lost. Clutch size is typically five to seven eggs. Females incubate 13–14 days followed by 18–20 days till fledging. In North America, the bridled titmouse is the only tit species to have a helper breeding system.

CONSERVATION STATUS
A species with a restricted range, but common within this range, reaching densities of 7.7–10.3 birds/25 acres (10 ha) in summer. Loss and fragmentation of oak woodlands due to conversion to arable land known to have caused decline in central Mexico.

SIGNIFICANCE TO HUMANS
None known. ◆

Yellow-bellied tit
Parus venustulus

SUBFAMILY
Parinae

TAXONOMY
Parus venustulus Swinhoe, 1870. Monotypic.

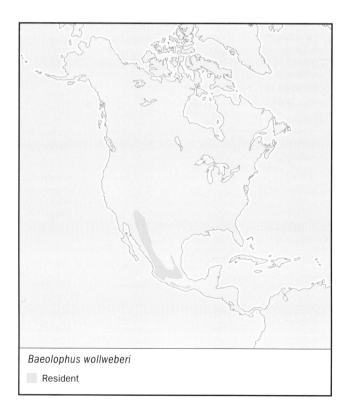

Baeolophus wollweberi
■ Resident

Parus venustulus
■ Resident

OTHER COMMON NAMES
French: Mésange gracieuse; German: Schmuckmeise; Spanish: Carbonero de Vientre Amarillo.

PHYSICAL CHARACTERISTICS
4 in (10 cm); 0.3–0.45 oz (9–12.5 g); plumage like that of the 'spot-winged' tits, with yellow breast and gray-blue mantle; sexually dichromatic.

DISTRIBUTION
Endemic to central and eastern China.

HABITAT
A wide range of forest and woodland types, predominantly in mountainous areas, and around human habitation and cultivated land where trees are present.

BEHAVIOR
Typically resident, but known to be migratory in parts of range. Little known about territoriality, but does join mixed-species flocks outside of the breeding season. Song is simple.

FEEDING ECOLOGY AND DIET
Little known about foraging behavior, but diet largely invertebrates with some fruits taken. Makes stores of food.

REPRODUCTIVE BIOLOGY
Nests in natural cavities, but does not excavate own hole. Eggs laid between May and June and clutch size is five to seven eggs. The incubation period is 12 days and the young fledge 16–17 days after hatching.

CONSERVATION STATUS
Not threatened. With the exception of the great tit, this is probably the most widespread of all the southern Asiatic species.

SIGNIFICANCE TO HUMANS
None known. ◆

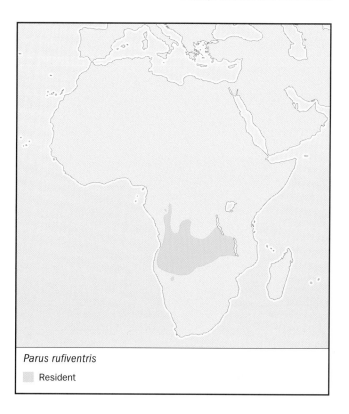

Parus rufiventris
▨ Resident

Rufous-bellied tit
Parus rufiventris

SUBFAMILY
Parinae

TAXONOMY
Parus rufiventris Barboza du Bocage, 1877. Three subspecies.

OTHER COMMON NAMES
English: Cinnamon-breasted tit; rufous tit; French: Mésange à ventre cannelle; German: Rostbauchmeise; Spanish: Herrerillo de Vientre Rufo.

PHYSICAL CHARACTERISTICS
6 in (15 cm); an attractive species with a black head, dark gray upperparts and breast, black tail, and rufous belly. The iris is a distinctive yellow in the adults; sexes similar.

DISTRIBUTION
P. r. rufiventris: western and central Africa (Angola, Zaïre, Zambia); *P. r. maskukuensis*: central Africa (Zambia and Malawi); *P. r. diligens*: western Africa (Namibia and Angola).

HABITAT
Main habitat is miombo woodland, which is moist savanna-woodland of *Brachystegia*. Also found in other woodlands similar to miombo woodland, dry evergreen forests and drier

savanna-woodlands. Typically found from 1,969–6,562 ft (600–2,000 m).

BEHAVIOR
Resident. May hold territories during breeding season, but recorded in mixed-species flocks in winter. As with most tits, the song is simple, but a variety of calls are used.

FEEDING ECOLOGY AND DIET
Forages mainly among twigs and leaves on the outer-most parts of trees in the mid and upper canopy. Diet is invertebrates, especially moth larvae.

REPRODUCTIVE BIOLOGY
Poorly known. Nests in holes in trees or stumps, but does not excavate own hole. Breeding season September to November. Clutch size is three or four eggs. Nothing is known of incubation or brood period. May have a cooperative breeding system.

CONSERVATION STATUS
Not threatened. A widespread species considered common throughout most of range, although probably rarer in southern areas of range.

SIGNIFICANCE TO HUMANS
None known. ◆

Great tit
Parus major

SUBFAMILY
Parinae

TAXONOMY
Parus major Linnaeus, 1758. Up to 31 subspecies recognized.

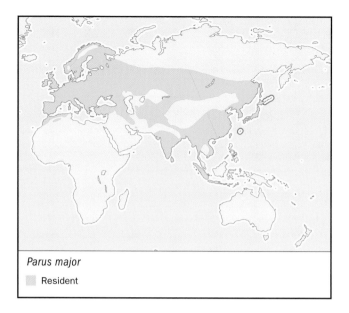

Parus major
■ Resident

and as late as May in the north. Frequently lays two clutches per year, rarely three. Clutch size highly variable: three to 18 eggs laid. Incubation is by female alone and takes 12–15 days, fledging in 16–22 days.

CONSERVATION STATUS
A very common species, but some subspecies may have relatively small populations. In Europe alone, the estimated population is between 41 and 180 million pairs.

SIGNIFICANCE TO HUMANS
No commercial significance, but of contemporary cultural significance in some areas, especially Europe, where closely associated with humans. ◆

OTHER COMMON NAMES
French: Mésange charbonnière; German: Kohlmeise; Spanish: Carbonero Común.

PHYSICAL CHARACTERISTICS
5.5 in (14 cm); 0.5–0.8 oz (14–22 g), males typically slightly larger than female; very variable species with plumage characterized by black crown, throat, and vertical breast stripe; white cheeks; green back; blue wings, rump, and tail; and yellow breast.

DISTRIBUTION
The great tit is possibly the most widespread of all the parids; up to 31 subspecies have been identified across Eurasia and into Southeast Asia and northern China.

HABITAT
Occurs in a very wide range of different woodland types, but generally favors lowland broad-leaved woodlands, especially those with good shrub growth. Dislikes pure conifer forests. Also found in open woodland habitats, including parks, gardens, cemeteries, and hedges.

BEHAVIOR
Resident throughout much of its range, but shows seasonal altitudinal migration in some mountain areas and local migration in others. Irruptive movements can occur in years with good breeding success and/or poor beech *Fagus* mast crop. Occasionally territorial throughout year, but may join hierarchically-organized, mixed-species flocks outside of breeding season. Song frequent, simple, and variable, with each male having several repertoires.

FEEDING ECOLOGY AND DIET
Forages on all parts of trees and shrubs, but shows preference for foraging among leaves. However, foraging behavior highly adaptable and shows remarkable learning ability, including the use of tools (pine needles used to extract insects from holes). Does not store food. Diet comprises an extremely wide variety of invertebrates, seeds, nuts, and fruits.

REPRODUCTIVE BIOLOGY
Nests in pre-existing tree-cavities, walls, burrows, and nest-boxes. Egg-laying starts in February in southern populations

White-naped tit
Parus nuchalis

SUBFAMILY
Parinae

TAXONOMY
Parus nuchalis Jerdon, 1845. Monotypic.

OTHER COMMON NAMES
English: White-winged tit, white-naped black tit, collared tit; French: Mésange à ailes blanches; German: Weissflügelmeise; Spanish: Carbonero de Nuca Blanca.

PHYSICAL CHARACTERISTICS
4.75 in (12 cm); 0.45–0.5 oz (13–14.5 g); contrasting patterns of black and white throughout except for pale yellow underparts.

DISTRIBUTION
Endemic to India with small, fragmented populations in the northwest and south of the country.

Parus nuchalis
■ Resident

HABITAT

Occurs in lowland dry thorn-scrub forests in northwest India up to 1,969 ft (600 m), as well as dry to moist deciduous woodlands in south India. Also found in gardens, along stream-beds and around irrigated croplands.

BEHAVIOR

Resident, but makes local movements. Little known about territoriality, but will join mixed-species flocks. Song is a repeated monosyllabic call, but a variety of other calls are made.

FEEDING ECOLOGY AND DIET

Forages throughout the canopy and shrub layer, with diet comprising insects and spiders.

REPRODUCTIVE BIOLOGY

Poorly known. Nests in tree cavities, typically those made by woodpeckers. Clutch size, incubation, and brooding behavior not known. Timing of breeding is during the monsoons: May-August.

CONSERVATION STATUS

Vulnerable. The population is highly fragmented and declining, and is estimated to be 2,500–10,000 pairs. Most serious threat is continued degradation, loss, and fragmentation of thorn forest habitat.

SIGNIFICANCE TO HUMANS

None known. ◆

Sultan tit
Melanochlora sultanea

SUBFAMILY
Parinae

TAXONOMY
Melanochlora sultanea Hodgson, 1837. Four subspecies.

Melanochlora sultanea
▨ Resident

OTHER COMMON NAMES

French: Mésange sultane; German: Sultanmeise; Spanish: Carbonero Sultán.

PHYSICAL CHARACTERISTICS

8 in (20.5 cm); 1.2–1.7 oz (34–49 g); the largest of the tits, with striking yellow and glossy black plumage in males; duller in females.

DISTRIBUTION

M. s. sultanea: eastern Himalayas, northeastern India, Myanmar, northern Thailand; *M. s. flavocristata*: southern Myanmar, Malaysia, Sumatra; *M. s. seorsa*: southern China, Hainan, northern Indochina; *M. s. gayeti*: central Vietnam, southern Laos.

HABITAT

Wide variety of forest types, preferring light evergreen forests and forest edges. Found mainly below 3,200 ft (1,000 m), but recorded in mountainous areas up to 6,600 ft (2,000 m).

BEHAVIOR

Resident. During breeding, found in pairs and probably territorial. Outside of breeding season, forms small flocks, often mixing with other species. Song is series of loud, clear whistles, and a variety of other calls are also given.

FEEDING ECOLOGY AND DIET

Forages mostly in upper canopy, but in some habitats (e.g., bamboo) will forage close to ground. Foods include invertebrates, fruits and seeds.

REPRODUCTIVE BIOLOGY

Poorly known. Nests in holes and other openings in trees. Timing of nesting April through July and clutch size five to seven eggs. No information about incubation, brooding behavior, or breeding success.

CONSERVATION STATUS

Not threatened. A relatively widespread species. Has declined in western part of range due to habitat loss, and is now rare.

SIGNIFICANCE TO HUMANS

None known. ◆

Yellow-browed tit
Sylviparus modestus

SUBFAMILY
Parinae

TAXONOMY
Sylviparus modestus Burton, 1836. Three subspecies.

OTHER COMMON NAMES

French: Mésange modeste; German: Laubmeise; Spanish: Carbonero de Cejas Amarillas.

PHYSICAL CHARACTERISTICS

4 in (10 cm); 0.2–0.3 oz (5–9 g); small greenish bird with a short bill, pale yellow eye ring and short crest; sexes similar.

DISTRIBUTION

S. m. simlaensis: northwestern Himalayas; *S. m. modestus*: central and eastern Himalayas, Myanmar, southwestern China, northern Thailand, northern Laos, northern Vietnam; *S. m. klossi*: southern Vietnam.

HABITAT

Inhabits alpine scrub and forest, occurring principally from 3,900–11,150 ft (1,200–3,400 m), but down to 1,475 ft (450 m)

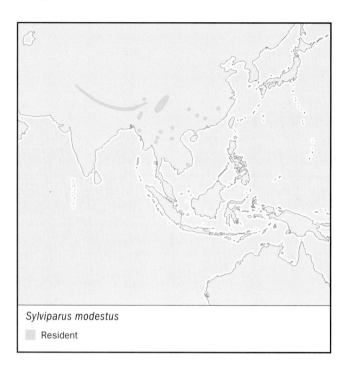

Sylviparus modestus

☐ Resident

in parts of range. Very localized distribution outside of the Himalayas, including coniferous forests in China.

BEHAVIOR

Resident, but undertakes seasonal altitudinal movements, generally moving to lower elevations outside of the breeding season. Territoriality not understood, but joins mixed-species flocks outside of the breeding season like most other tits. Has a wide variety of calls.

FEEDING ECOLOGY AND DIET

Forages throughout the canopy and occasionally in scrub beneath the tree canopy. Described as both acrobatic and restless while foraging. Diet mainly invertebrates, but also takes some seeds.

REPRODUCTIVE BIOLOGY

Poorly understood. Nests in natural cavities in trees. Breeds April through May. Clutch size four to six eggs, but no information on incubation and nestling periods.

CONSERVATION STATUS

Not threatened. A common species, but with fragmented and highly localized distribution outside of the Himalayas.

SIGNIFICANCE TO HUMANS

None known. ◆

Resources

Books

Cicero, C. "Oak Titmouse (*Baeolophus inornatus*) and Juniper Titmouse (*Baeolophus ridgwayi*)." In *The Birds of North America*. No. 485, edited by A. Poole and F. Gill. Philadelphia: The Birds of North America, Inc., 2000.

Cramp, S., and C. M. Perrins, eds. *The Birds of the Western Palearctic.* Vol. VII. Oxford, U.K.: Oxford University Press, 1993.

Ficken, M., and J. Nocedal. "Mexican Chickadee." In *The Birds of North America*, No. 8, edited by A. Poole, P. Stettenheim, and F. Gill. Philadelphia: The Academy of Natural Sciences; Washington, DC: The American Ornithologists' Union, 1992.

Ficken, M., M. A. McLaren, and J. P. Hailman. "Boreal Chickadee (*Parus hudsonicus*)." In *The Birds of North America*, No. 254, edited by A. Poole and F. Gill. Philadelphia: The Academy of Natural Sciences; Washington, DC: The American Ornithologists' Union, 1996.

Hailman, J. P., and S. Haftorn. "Siberian Tit (*Parus cinctus*)." In *The Birds of North America*, No. 196, edited by A. Poole and F. Gill. Philadelphia: The Academy of Natural Sciences; Washington, DC: The American Ornithologists' Union, 1995.

Harrap, S., and D. Quinn. *Tits, Nuthatches and Treecreepers.* London, U.K.: Christopher Helm, 1996.

McCallum, D. A., R. Grundel, and D. L. Dahlsten. "Mountain Chickadee (*Poecile gambeli*)." In *The Birds of North America*, No. 453, edited by A. Poole and F. Gill. Philadelphia: The Birds of North America, Inc., 1999.

Nocedal, J., and M. S. Ficken. "Bridled Titmouse (*Baeolophus wollweberi*)." In *The Birds of North America*, No. 375, edited by A. Poole and F. Gill. Philadelphia: The Academy of Natural Sciences, 1998.

Sibley, C. G., and B. L. Monroe, Jr. *Distribution and Taxonomy of Birds of the World.* New Haven & London: Yale University Press, 1990.

Smith, S. M. "Black-capped Chickadee." In *The Birds of North America*, No. 39, edited by A. Poole, P. Stettenheim, and F. Gill. Philadelphia: The Academy of Natural Sciences; Washington, DC: The American Ornithologists' Union, 1993.

Tucker, G. M., and M. F. Heath. *Birds in Europe: Their Conservation Status.* Cambridge, U.K.: BirdLife International (BirdLife Conservation Series no. 3), 1994.

Helen Baker, PhD

Nuthatches and wall creepers
(*Sittidae*)

Class Aves
Order Passeriformes
Suborder Passeri (Oscines)
Family Sittidae

Thumbnail description
Small, large-headed, short-tailed perching birds that clamber upwards or downwards on the surface of tree trunks or rocks to obtain their food of invertebrates

Size
Nuthatches: 3.5–7.5 in (8.5–19 cm), sittellas: 4.3–4.8 in (11–12 cm), and wall creepers: about 6.3 in (16 cm) in body length

Number of genera, species
3 genera; about 27 species

Habitat
Mostly forest and woodlands

Conservation status
Endangered: 2 species; Vulnerable: 2 species; Near Threatened: 2 species

Distribution
North America, Eurasia, Africa, Southeast Asia, and Australasia

Evolution and systematics

The family Sittidae is within the Passeriformes, the most diverse order of birds. As considered here, the Sittidae includes three subfamilies: the typical nuthatches (subfamily Sittinae), the sittellas (Neosittinae), and the wall creeper (Tichodrominae). There are about 23 species of nuthatches (genus *Sitta*), three of sittellas (genus *Daphoenositta*), and one wall creeper (*Tichodroma muraria*).

Physical characteristics

The nuthatches (Sittinae) range in body length from 3.5 to 7.5 in (8.5–19 cm). They have a compact body, a large head, a short squared tail, a short neck, and a thin, chisel-shaped, slightly upturned bill. They are generally colored blue-gray above, white or brownish below, with a dark top of the head, and a white stripe over the eye. The subfamily Neosittinae consists of one genus and two species of sittellas. They have a body length of 4.3–4.8 in (11–12 cm), a compact body, a large head, a short tail, and a thin, chisel-shaped bill. The body coloration is brown-streaked or black with a red face. The subfamily Tichodrominae consists of only one species, the wall creeper. It has a body length of about 6.5 in (16 cm). It has a compact body, a short tail, and a strong, slightly down-turned bill. It is colored brownish above, white below with brown streaks, has a white throat, and an ochre patch around the eyes. The claw of its hind toe is rather long, and its feet can grasp even slight projections on a rock face. In most species the sexes are similar in morphology and coloration.

A pygmy nuthatch (*Sitta pygmaea*) at its nest hole in Eastern Sierras, California. (Photo by Richard R. Hansen. Photo Researchers, Inc. Reproduced by permission.)

A red-breasted nuthatch (*Sitta canadensis*) feeds on a pine cone. (Photo by H.P. Smith Jr./VIREO. Reproduced by permission.)

Distribution

Species occur in North America, Eurasia, Africa, Southeast Asia, and Australasia.

Habitat

Most species occur in forest or woodland habitat, and several in rocky scrubland.

Behavior

The nuthatches forage on the bark of trees. They are the only birds that regularly clamber head-downwards instead of only upwards while seeking food on this kind of substrate (they will also climb upwards, and sideways). They move downwards by reaching down with one foot while holding themselves with the other one, and continuing this with alternate feet. Their climbing movements are usually on a slant, resulting in a spiral or zigzag pathway being followed.

Feeding ecology and diet

Nuthatches use their bill to glean insects, spiders, snails, and other invertebrates from the surface and crevices of tree trunks, sometimes chiseling loose bark away to expose prey beneath. They may also forage on rocks and in epiphytic mosses and lichens. They also glean arthropods from foliage, from the ground, and may even catch them in flight. During the spring and summer they mostly feed on invertebrates, but in the winter they also eat small fruits and oil-rich seeds of various kinds of plants. To open an enclosed seed, such as that of the sunflower, they wedge it into a cleft and hammer the top with their bill. Nuthatches break open snail shells in a similar manner. At times when food is abundant, they store it in clefts for use later in times of scarcity.

Reproductive biology

Established pairs of nuthatches are monogamous and, unless disturbed, occupy a permanent territory. They nest in

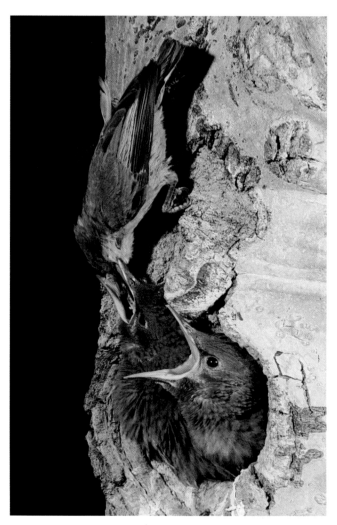

A pygmy nuthatch (*Sitta pygmaea*) feeds young starlings in California. (Photo by Bob & Clara Calhoun. Bruce Coleman Inc. Reproduced by permission.)

A brown-headed nuthatch (*Sitta pusilla*) brings food to its nest hole. (Photo by F. Truslow/VIREO. Reproduced by permission.)

A white-breasted nuthatch (*Sitta carolinensis*) perched on a branch in Michigan. (Photo by L. West. Bruce Coleman Inc. Reproduced by permission.)

natural cavities in trees or in ones excavated and later abandoned by woodpeckers. They typically narrow the diameter of the entrance hole by plastering it with mud, which hardens as it dries. This is done to keep predators and competitors for the nest site at bay. Some species nest in rock cavities. Within the cavity, the nest itself is usually a loose structure of bark flakes and leaves. They lay four to 10 eggs, which are white and flecked with brown or red. The eggs are incubated by the female, but both sexes feed the young birds.

Conservation status

The IUCN considers two species Endangered, two species Vulnerable, and two species Near Threatened. The Algerian nuthatch (*Sitta ledanti*) of Algeria is Endangered, as is the white-browed nuthatch (*S. victoriae*) of Myanmar. Vulnerable species are the giant nuthatch (*S. magna*) of Myanmar, Thailand, and adjacent China; and the beautiful nuthatch (*S. formosa*) of India, Bhutan, Myanmar, southern China, and Laos. Near Threatened species include the Yunnan nuthatch (*S. yunnanensis*) of southwestern China; the yellow-billed nuthatch (*S. solangiae*) of Vietnam and southeastern China.

Significance to humans

Nuthatches are not of direct importance to humans, other than the indirect economic benefits of ecotourism and birdwatching focused on seeing birds in natural habitats.

1. Red-breasted nuthatch (*Sitta canadensis*); 2. White-breasted nuthatch (*Sitta carolinensis*); 3. Nuthatch (*Sitta europaea*); 4. Giant nuthatch (*Sitta magna*); 5. Brown-headed nuthatch (*Sitta pusilla*); 6. Wall-creeper (*Tichodroma muraria*); 7. Black sittella (*Daphoenositta miranda*). (Illustration by John Megahan)

Species accounts

Red-breasted nuthatch
Sitta canadensis

SUBFAMILY
Sittinae

TAXONOMY
Sitta canadensis Linnaeus, 1776.

OTHER COMMON NAMES
French: Sittelle à poitrine rousse; German: Kanadakleiber;
Spanish: Saltapalo Canadiense.

PHYSICAL CHARACTERISTICS
4 in (10 cm), with a short tail. The back is colored blue-gray,
the undersides red-brown, crown black, throat white, and with
a white line over the eye and a black one through it.

DISTRIBUTION
Occurs widely in southern and northwestern Canada and most
of the United States.

HABITAT
Breeds in mature, boreal or montane, conifer-dominated forest.
Winters in a wider range of conifer dominated forest.

BEHAVIOR
Occurs as pairs that defend a breeding territory. It is an irregular migrant that spends some winters in the breeding range.
Often flocks with chickadees in winter. The song is a series of
nasal notes.

FEEDING ECOLOGY AND DIET
Gleans invertebrates from tree bark and foliage, and also eats
fruits and seeds in winter.

REPRODUCTIVE BIOLOGY
Pairs nest in a tree cavity. The female incubates the eggs but
both sexes feed the young.

CONSERVATION STATUS
Not threatened. A widespread and abundant species.

SIGNIFICANCE TO HUMANS
None known. ◆

White-breasted nuthatch
Sitta carolinensis

SUBFAMILY
Sittinae

TAXONOMY
Sitta carolinensis Latham, 1790.

OTHER COMMON NAMES
French: Sittelle à poitrine blanche; German: Carolinakleiber;
Spanish: Saltapalo Blanco.

PHYSICAL CHARACTERISTICS
5 in (12.7 cm), with a short tail. The back is colored blue-gray,
the crown black, and undersides and throat white.

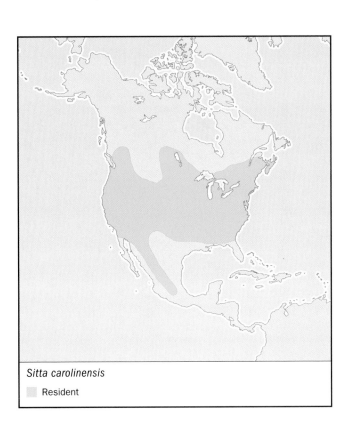

Sitta canadensis
  □ Resident   ■ Breeding   ■ Nonbreeding

Sitta carolinensis
  □ Resident

DISTRIBUTION
Occurs widely in extreme southern Canada, through most of the United States, and into western Mexico. Does not migrate, except for northernmost populations that may move somewhat south.

HABITAT
Occurs in a wide range of mature, deciduous-dominated forest types.

BEHAVIOR
Occurs as pairs that defend a breeding territory. Typically does not migrate. Often flocks with chickadees in winter. The song is a series of nasal notes.

FEEDING ECOLOGY AND DIET
Gleans invertebrates from tree bark and foliage, and also eats fruits and seeds in winter.

REPRODUCTIVE BIOLOGY
Pairs nest in a tree cavity. The female incubates the eggs but both sexes feed the young.

CONSERVATION STATUS
Not threatened. A widespread and abundant species.

SIGNIFICANCE TO HUMANS
None known. ◆

Sitta pusilla
▨ Resident

Brown-breasted nuthatch
Sitta pusilla

SUBFAMILY
Sittinae

TAXONOMY
Sitta pusilla Latham, 1790.

OTHER COMMON NAMES
French: Sittelle à tête brune; German: Braunkopfkleiber; Spanish: Sita de Cabeza Castaña.

PHYSICAL CHARACTERISTICS
3.5 in (9 cm), with a short tail. The back is colored blue-gray, the crown brown, and the undersides and throat are white washed with brown.

DISTRIBUTION
Occurs widely in the southeastern United States.

HABITAT
Occurs in mature, pine-dominated forest types.

BEHAVIOR
Occurs as pairs that defend a breeding territory. Does not migrate. Occurs in mixed-species flocks with chickadees and warblers during the nonbreeding season. The song is a series of soft twitterings.

FEEDING ECOLOGY AND DIET
Gleans invertebrates from tree bark and foliage, especially on branches. Also eats fruits and seeds in winter.

REPRODUCTIVE BIOLOGY
Pairs nest in a tree cavity. The female incubates the eggs but both sexes feed the young.

CONSERVATION STATUS
Not threatened. A widespread and abundant species.

SIGNIFICANCE TO HUMANS
None known. ◆

Nuthatch
Sitta europaea

SUBFAMILY
Sittinae

TAXONOMY
Sitta europaea Linnaeus, 1758. Twenty-five subspecies.

OTHER COMMON NAMES
English: Eurasian nuthatch, wood nuthatch; French: Sittelle torchepot; German: Kleiber; Spanish: Trepador Azul.

PHYSICAL CHARACTERISTICS
5.5 in (14 cm), with a short tail. The back is colored blue-gray, the crown blue-gray, undersides brown to white, throat white, and with a black line through the eye. However, coloration varies considerably among the approximately 25 geographic subspecies of this wide-ranging species.

DISTRIBUTION
Occurs widely in temperate Eurasia, from the Atlantic to Pacific coasts.

HABITAT
Occurs in a wide range of mature temperate forests, ranging from deciduous- to conifer-dominated types.

Sitta europaea

Resident

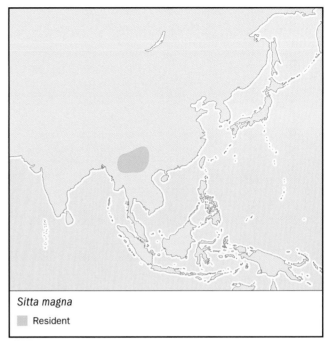

Sitta magna

Resident

BEHAVIOR
Occurs as pairs that defend a breeding territory. Does not migrate. Occurs in mixed-species flocks with tits (or chickadees) in the nonbreeding season. The song is a varied series of loud calls.

FEEDING ECOLOGY AND DIET
Gleans invertebrates from tree bark and foliage, especially on branches. Also eats fruits and seeds in winter.

REPRODUCTIVE BIOLOGY
Pairs nest in a tree cavity. The female incubates the eggs but both sexes feed the young.

CONSERVATION STATUS
Not threatened. A widespread and abundant species.

SIGNIFICANCE TO HUMANS
None known. ◆

Giant nuthatch
Sitta magna

SUBFAMILY
Sittinae

TAXONOMY
Sitta magna Ramsay, 1876.

OTHER COMMON NAMES
French: Sitelle géante; German: Riesenkleiber; Spanish: Sita Gigante.

PHYSICAL CHARACTERISTICS
8 in (20 cm), with a short tail. The back is colored blue-gray, the crown blue-gray, undersides light gray with chestnut beneath tail, throat white, and with a black line through the eye.

DISTRIBUTION
Occurs in southwestern China, east-central Myanmar, and northwestern China.

HABITAT
Occurs in open montane forest with pines present or dominant, at altitudes of 3,900–8,200 ft (1,200–2,500 m) or higher.

BEHAVIOR
Occurs as pairs that defend a breeding territory. Does not migrate. Occurs in mixed-species flocks with tits (or chickadees) and other nuthatch species in the nonbreeding season. The song is a series of loud trisyllabic calls.

FEEDING ECOLOGY AND DIET
Gleans invertebrates from tree bark and foliage and also eats fruits and seeds in winter.

REPRODUCTIVE BIOLOGY
Pairs nest in a tree cavity. The female incubates the eggs but both sexes feed the young.

CONSERVATION STATUS
A rare species that is listed as Vulnerable. This species is much reduced in abundance and range. It has been greatly affected by habitat loss caused by agricultural disturbance and conversion and by fuelwood harvesting. Its remaining critical habitats must be protected.

SIGNIFICANCE TO HUMANS
None known. ◆

Black sittella
Daphoenositta miranda

SUBFAMILY
Neosittinae

TAXONOMY
Daphoenositta miranda De Vis, 1897.

Daphoenositta miranda
☐ Resident

OTHER COMMON NAMES
English: Pink-faced sittella; French: Néositte noire; German: Prachtkleiber; Spanish: Sita de Cara Rosada.

PHYSICAL CHARACTERISTICS
4.3 in (11 cm), with a short tail, a black body, and red-pink face and tip of tail.

DISTRIBUTION
Occurs in New Guinea.

HABITAT
Occurs in mossy cloud forest at altitudes of 6,600–11,800 ft (2,000–3,600 m) or higher.

BEHAVIOR
Occurs as pairs or in small flocks. Does not migrate. Like nuthatches, it will clamber head-downwards on tree trunks. The song is a faint series of squeaky notes.

FEEDING ECOLOGY AND DIET
Gleans invertebrates from tree branches and foliage and also eats fruits.

REPRODUCTIVE BIOLOGY
Builds a deep cup-shaped nest on a lichen-covered branch (rather than nesting in tree or rock cavities like other members of the Sittidae). The nest consists of spider and cocoon silk, with pieces of bark to camouflage the exterior. There are usually three eggs, which are similar in color to the lichens near the nest. Cooperative breeding occurs, in which not only the parents take part in nest building and the feeding of the young, but also other individuals. The "helpers" are likely close relatives, such as non-breeding young or siblings of the breeding pair.

CONSERVATION STATUS
An uncommon species but not listed as threatened.

SIGNIFICANCE TO HUMANS
None known. ◆

Wall creeper
Tichodroma muraria

SUBFAMILY
Tichodrominae

TAXONOMY
Tichodroma muraria Linaeus, 1766.

OTHER COMMON NAMES
French: Tichodrome échelette; German: Mauerlä; Spanish: Treparriscos.

PHYSICAL CHARACTERISTICS
6.5 in (16.5 cm), with a short tail, and a long, black, down-curved bill. The back crown of the head is gray, the chin and tail are black, and the wings are white-spotted gray with red patches.

DISTRIBUTION
Occurs widely in Eurasia, from the Pyrenees, Alps, northern Apennines and Carinthian mountains, over the Balkans to Syria, the Himalayas and adjacent mountainous China.

HABITAT
Occurs in mountainous regions in the vicinity of cliffs and rocky gorges as high as the snow line. There must, however, be vegetation in the vicinity of the rocky places where they breed. Migrates to somewhat lower habitats in the winter.

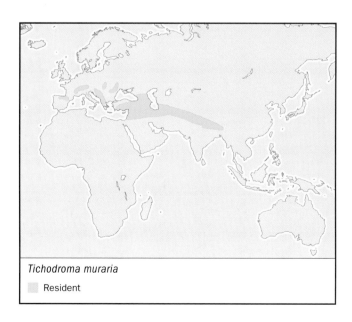

Tichodroma muraria
☐ Resident

BEHAVIOR

Wall creepers have a solitary lifestyle outside of the breeding season. They frequently flit their wings and have a light, butterfly-like flight pattern. They have an altitudinal migration. The song is a series of rising, high-pitched notes.

FEEDING ECOLOGY AND DIET

Forages for invertebrates in crevices of steep rockfaces and walls. Also eats small fruits and seeds.

REPRODUCTIVE BIOLOGY

Nests in a rocky crevice. Lays three or four white eggs.

CONSERVATION STATUS

Not threatened. A widespread species within its habitat, but not particularly abundant.

SIGNIFICANCE TO HUMANS

None known. ◆

Resources

Books

BirdLife International. *Threatened Birds of the World.* Barcelona and Cambridge: Lynx Edicions and BirdLife International, 2000.

Harrap, S., and D. Quinn. *Tits, Nuthatches and Treecreepers of the World.* London: Christopher Helm, 1996.

Matthysen, E. *The Nuthatches.* San Diego: Academic Press: 1998.

Periodicals

Norris, R. A. "Comparative Biosystematics and Life History of the Nuthatches *Sitta pygmaea* and *Sitta pusilla.*" *University of California Publications in Zoology* 56 (1958): 119–300.

Organizations

BirdLife International. Wellbrook Court, Girton Road, Cambridge, Cambridgeshire CB3 0NA United Kingdom. Phone: +44 1 223 277 318. Fax: +44-1-223-277-200. E-mail: birdlife@birdlife.org.uk Web site: <http://www .birdlife.net>

IUCN–The World Conservation Union. Rue Mauverney 28, Gland, 1196 Switzerland. Phone: +41-22-999-0001. Fax: +41-22-999-0025. E-mail: mail@hq.iucn.org Web site: <http://www.iucn.org>

Bill Freedman, PhD

▲
Treecreepers
(Certhiidae)

Class Aves

Order Passeriformes

Suborder Passeri (Oscines)

Family Certhiidae

Thumbnail description
Small birds with long, thin curved bills, long tails, and highly cryptic coloration

Size
5–6 in (12–15 cm)

Number of genera, species
2 genera; 7 species

Habitat
Woodlands and forests

Conservation status
Not threatened; slight local declines for some species

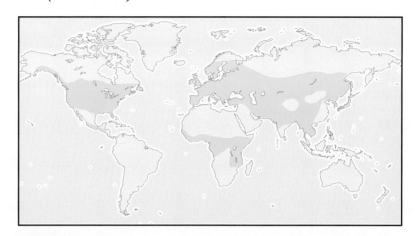

Distribution
Widespread across the Northern Hemisphere, with occurrences in central and southwest Africa

Evolution and systematics

The taxonomy of the Certhiidae is somewhat problematic, and on the basis of DNA-DNA analysis comparisons some workers suggest placing 22 genera and about 100 species in the family. Sibley and Ahlquist have further divided the family into three subfamilies: wrens (Troglodytinae); gnatcatchers, gnatwrens, and verdins (Poliotilinae); and treecreepers (Certhiinae). Most workers, including the American Ornithological Union, still consider the family to consist of "typical" treecreepers of the Northern Hemisphere (*Certhia*) and spotted creepers (*Salpornis spilonotus*) of India and Africa.

Physical characteristics

Size, coloration, and markings allow creepers to blend almost flawlessly with their preferred habitat, the rough brown bark of trees. Treecreepers are small, about 5–6 inches (12–15 cm) in length, and have a long, thin, curved bill, a slender tail, and streamlined, teardrop-shaped bodies. The twelve stiff, pointed tail feathers have shafts that project beyond the vanes. In all but the bar-tailed treecreeper (*Certhia himalayana*), this results in giving the tail a bristly appearance at the terminus. Treecreepers use their highly specialized tails to help them climb, but spotted creepers, which hold their tails away from the trunk, do not.

The legs of treecreepers appear disproportionately short, but this very functional design, along with long toes and claws, enables them to cling closely to the side of trees as they search for food.

Plumage variation among species can be subtle and confusing. The cryptic upperparts are mostly shades of brown

with streaks or streaky spots of black, gray, buff, and white. All species have a noticeable buff or white eyebrow, though it is less pronounced in some. Underparts are buff or white, with shades of cinnamon, rufous, or gray less common. Males and females are similar in appearance in both size and color. Juvenal plumage is somewhat duller and more streaky on the upperparts, but first-year birds are otherwise identical to adults. Spotted creepers are also highly cryptic but distinctively barred and more prominently spotted than treecreepers.

Distribution

Treecreepers are found throughout the forests and woodlands of Central America and temperate regions of the Northern Hemisphere. Short-toed (*C. brachydactyla*), Eurasian (*C. familiaris*), rusty-flanked (*C. nipalensis*), brown-throated (*C. discolor*), and bar-tailed treecreepers are found in Europe, central Asia, and Asia. Brown creepers (*C. americana*) are found in North and Central America. The spotted creeper is found in the woodlands of India and Africa in sparse and scattered populations.

Habitat

Depending on the species, creepers can be found anywhere from sea-level to mountain regions and from temperate to tropical climates. Creepers generally require mature forests and woodlands, which provide the type of bark that houses the small invertebrates on which they forage and the structures for securing their nests. Lack of suitable habitat for nesting may be the limiting factor in the expansion of breeding range in some species.

Behavior

Creepers forage singly, in pairs, or in mixed-species flocks. Foraging behavior is distinctive and consists of flying to the base of a tree, then gleaning and probing rough bark for insects while climbing the trunk. Creepers also forage for prey on limbs by clinging to the underside and creeping outward from the trunk almost to the tip of the main branch. Treecreepers typically climb the trunk in a jerky, spiral motion.

Treecreepers roost singly, in pairs, or sometimes in small groups during the nonbreeding season, probably for warmth. Some species roost arranged in a small circle with individuals facing one another.

Feeding ecology and diet

The Certhiidae use their thin, thorn-like bill to probe beneath bark for food, which primarily consists of small insects, spiders, and other small invertebrates. Treecreepers also eat seeds and nuts, usually in wintertime when invertebrates are scarce. There is only one known record of treecreepers storing food, which consisted of sunflower seeds taken from feeders one winter.

Reproductive biology

Male creepers establish and defend the breeding territory through vocalization. Songs are generally weak and quiet trills, and calls are high-pitched and thin. Though they may often go unnoticed, songs and calls may be the best way of locating these highly cryptic birds.

Breeding biology of the rusty-flanked treecreeper and the brown-throated treecreeper is not very well known. In all other treecreeper species, nests are usually constructed beneath a loose piece of bark on dead or dying trees. Occasionally creepers will construct nests on buildings, walls, crevices in trees, or in heavy vegetation such as ivy. Some species also infrequently use nest boxes. Height of nests from the ground varies from species to species and can range from 1.6 to 52 ft (0.5 m to 16 m).

There are four to six eggs in the treecreeper clutch, and they are white and faintly spotted red or reddish brown; Spotted creepers have fewer eggs in each clutch (one to three) and the eggs are usually gray, greenish gray, bluish, or greenish blue with black or brown spots. Only the female incubates the eggs, but both sexes feed the young. Duration of brooding varies from species to species, but lasts from 13 to 17 days. Treecreeper pairs frequently double brood. Treecreepers usually form family parties for two to three weeks after the young fledge.

Conservation status

Brown creepers might be in slight decline in some regions of North America. Competition with wood ants for prey may be causing a decline in some populations of Eurasian treecreepers.

Significance to humans

None known.

1. Brown creeper (*Certhia americana*); 2. Short-toed treecreeper (*Certhia brachydactyla*); 3. Eurasian treecreeper (*Certhia familiaris*). (Illustration by Michelle Meneghini)

Species accounts

Eurasian treecreeper
Certhia familiaris

TAXONOMY
Certhia familiaris Linnaeus, 1758.

OTHER COMMON NAMES
English: Northern, common treecreeper; French: Grimpereau des bois; German: Waldbaumläufer; Spanish: Agateador Norteño.

PHYSICAL CHARACTERISTICS
Length 5 in (12.5 cm). Upperparts brown, spotted and streaked with white or buff, rufous rump; underparts white to buff.

DISTRIBUTION
Widespread in Europe, Central Asia, and Asia.

HABITAT
Forest and woodland.

BEHAVIOR
Roost singly, in pairs, or sometimes in small groups during the nonbreeding season.

FEEDING ECOLOGY AND DIET
Spiders, insects, and other small invertebrates.

REPRODUCTIVE BIOLOGY
Nest behind loose bark or in crevice; will also use dead leaves and heavy vegetation to hide nest. Three to five eggs, eggs white with pink or reddish spots.

CONSERVATION STATUS
Not threatened. Common and widespread.

SIGNIFICANCE TO HUMANS
None known. ◆

Brown creeper
Certhia americana

TAXONOMY
Certhia americana Bonaparte, 1838, North America.

OTHER COMMON NAMES
English: American treecreeper; French: Grimpereau brun; German: Andenbaumläufer; Spanish: Trepador Americano.

PHYSICAL CHARACTERISTICS
Length 5.25 in (13 cm). Populations highly variable, but generally upperparts dark brown, spotted and streaked with white, buff, or pale gray; cinnamon-colored rump and undertail coverts, underparts white to buff; western and Mexican populations darker and smaller.

Certhia familiaris
◻ Resident

Certhia americana
◻ Resident

DISTRIBUTION
North and Central America.

HABITAT
Forest and woodland.

BEHAVIOR
Fly to base of tree and search for insects while climbing upwards.

FEEDING ECOLOGY AND DIET
Spiders, insects, and other small invertebrates; sometimes seeds from feeders in winter.

REPRODUCTIVE BIOLOGY
Nest behind loose bark or crevice within 197 ft (60 m) of water. Nest cup shaped with "horns." Five to six eggs, eggs white with reddish brown spots.

CONSERVATION STATUS
Possibly declining, but not threatened at this time.

SIGNIFICANCE TO HUMANS
None known. ◆

Short-toed treecreeper
Certhia brachydactyla

TAXONOMY
Certhia brachydactyla Brehm, 1820.

OTHER COMMON NAMES
French: Grimpereau des jardins; German: Gartenbaumläufer; Spanish: Agateador Común.

PHYSICAL CHARACTERISTICS
Length 5 in (12.5 cm). Resembles Eurasian treecreeper and is hard to distinguish on the basis of physical appearance.

DISTRIBUTION
Western Europe.

HABITAT
Mixed forest and woodland.

Certhia brachydactyla
▪ Resident

BEHAVIOR
Forage for prey by clinging to the underside of tree branches.

FEEDING ECOLOGY AND DIET
Spiders, insects, and other small invertebrates.

REPRODUCTIVE BIOLOGY
Nest behind loose bark or in crevice. Will use nest box. Five to seven eggs, white with reddish purple or reddish brown spots.

CONSERVATION STATUS
Not threatened.

SIGNIFICANCE TO HUMANS
None known. ◆

Resources

Books
Harrap, Simon, and David Quinn. *Chickadees, Tits, Nuthatches, and Treecreepers.* Princeton: Princeton University Press, 1995.

Kaufmann, Kenn. *The Lives of North American Birds.* New York: Houghton Mifflin, 1996.

Sibley, Charles G., and Burt L. Monroe, Jr. *Distribution and Taxonomy of Birds of the World.* New Haven: Yale University Press, 1990.

Periodicals
Aho, Teija, Markku Kuitunen, Jukka Suhonen, Ari Jäntti, and Tomi Hakkari. "Reproductive Success of Eurasian

Treecreepers, *Certhia familiaris*, Lower in Territories with Wood Ants." *Ecology* 80 (1999): 998–1007.

Aho, Teija, Markku Kuitunen, Jukka Suhonen, Ari Jäntti, and Tomi Hakkari. "Behavioral Responses of Eurasian Treecreepers, *Certhia familiaris*, to Competition with Ants." *Animal Behavior* 54 (1997): 1283–1290.

Baptista, Luis F., and Robin Krebs. "Vocalizations and Relationships of Brown Creepers *Certhia americana*: A Taxonomic Mystery." *Ibis* 142 (2000) :457–465.

Davis, Cheyleen. "A Nesting Study of the Brown Creeper." *The Living Bird* 17 (1978): 237–263.

Susan L. Tomlinson, PhD

Philippine creepers
(Rhabdornithidae)

Class Aves
Order Passeriformes
Suborder Passeri (Oscines)
Family Rhabdornithidae

Thumbnail description
Medium-sized birds with long, slender bills and brush-tipped tongues; shaded and marked with brown, black, gray, and white

Size
6–7 in (15–17 cm), 3–4 oz (80–95 g)

Number of genera, species
1 genus, 3 species

Habitat
Forest

Conservation status
Not threatened

Distribution
Endemic to several major and minor Philippine Islands

Evolution and systematics

The rhabdornises are an obscure family, still only partially known. They have been difficult to classify, and no significant DNA-DNA hybrid, mitichondrial DNA, or other molecular comparison studies have been made.

Some ornithologists once placed the rhabdornises as a subfamily within the treecreepers (family Certhiidae), but the idea has been discarded since the birds show no real resemblance to the true treecreepers and behave quite differently. When foraging, treecreepers run along the tops of tree branches and crawl about over tree bark on the trunks and main limbs, but rhabdornises hop and jump between branches, and show no significant "creeping" behavior. Rhabdornises also have brush-tipped tongues, a feature unknown among the Certhiidae. As of 2002, ornithologists generally concur in placing rhabdornises in a discrete family, Rhabdornithidae. Their closest relations are probably the babblers (family Timaliidae).

As of 2002, the life sciences recognize three species of rhabdornises within a single genus: The greater rhabdornis (*Rhabdornis grandis*) which is not recognized by Peters, the

stripe-breasted rhabdornis (*Rhabdornis inornatus*), and the stripe-headed rhabdornis (*Rhabdornis mysticalis*).

Physical characteristics

Rhabdornises are medium-sized birds, very similar in size, coloring, and behavior, so much so that they may be indistinguishable to untrained eyes. The coloring and patterns are somewhat cryptic, favoring browns, red-browns, blacks, and whites, with much dark streaking on the upperparts and lighter streaking elsewhere.

Distribution

The major Philippine Islands of Luzon, Samar, Leyte, Mindanao, Negros, and Panay, and several smaller islands: Catanduanes, Masbate, Calicoan, Dinagat, Basilan, and Bohol.

Habitat

Dense, tropical, primary and secondary lowland and montane forest, in the upper levels and crowns of trees.

Behavior

Rhabdornises are diurnal and social, spending their days foraging in flocks. These flocks may be mixed species, either of rhabdornis or non-rhabdornis species. At dusk, rhabdornis flocks settle down in groups among the upper branches of forest trees for the night.

Feeding ecology and diet

The main rhabdornis fare is insects, which the birds search for on the bark of tree trunks and branches and among flowers, but all add nectar, fruit, and seeds to the diet.

Reproductive biology

Little is known about rhabdornis reproductive biology. They nest in tree holes, but there is no information on what they use for nest materials or on the appearance of the eggs. Breeding probably begins in March.

Conservation status

The stripe-breasted rhabdornis and stripe-headed rhabdornis are common throughout their ranges, while the greater rhabdornis is relatively rare and confined to mountainous regions in Luzon. None are listed on the IUCN 2000 Red List of Threatened Species. In the longer run, all may be endangered, since the native forests of the Philippines are being destroyed at an accelerating rate.

Significance to humans

None known.

1. Greater rhabdornis (*Rhabdornis grandis*); 2. Stripe-breasted rhabdornis (*Rhabdornis inornatus*); 3. Stripe-headed rhabdornis (*Rhabdornis mysticalis*). (Illustration by John Megahan)

Species accounts

Greater rhabdornis
Rhabdornis grandis

TAXONOMY
Rhabdornis grandis Salomonsen, 1953.

OTHER COMMON NAMES
English: Long-billed rhabdornis; French: Rhabdornis á long bec; German: Langschnabel-Rhabdornis; Spanish: Trepador Filipino Grande.

PHYSICAL CHARACTERISTICS
This species has an average length of 6.7 in (17 cm) and weighs about 3.3 oz (93 g). The male is larger than the female, but the sexes are similar in coloring, with black or dark brown bills and brown eyes, and olive-gray legs. Upperparts are patterned with gray, brown, black, and white, in characteristic streak patterns. The face has a black mask with white lines above and below. The lower breast and belly are white.

DISTRIBUTION
The Cordillera and Sierra Madre on Luzon Island.

HABITAT
Middle-elevation (330–3,300 ft; 100–1,000 m), tropical forest.

BEHAVIOR
The greater rhabdornis forages in groups or mixed-species flocks in the upper levels of forests. Greater rhabdornises will often flock with stripe-headed rhabdornises in flowering or fruiting trees.

FEEDING ECOLOGY AND DIET
Forages like other rhabdornis species, hopping and jumping along branches, searching for and eating insects among leaves, bark, and flowers. It will vary its diet with nectar, seeds, and fruit.

REPRODUCTIVE BIOLOGY
Nests in tree holes. Enlarged gonads in May.

CONSERVATION STATUS
Uncommon, but not threatened.

SIGNIFICANCE TO HUMANS
None known. ◆

Stripe-breasted rhabdornis
Rhabdornis inornatus

TAXONOMY
Rhabdornis inornatus Ogilvie-Grant, 1896.

OTHER COMMON NAMES
English: Plain-headed rhabdornis, plain-headed creeper; French: Rhabdornis á tête brune; German: Braunkopf-Rhabdornis; Spanish: Trepador Filipino de Cabeza Lisa.

PHYSICAL CHARACTERISTICS
These birds are 6.3 in (16 cm) long, and weigh about 3 oz (86 g). As with its bigger relative, the sexes show little in the way of

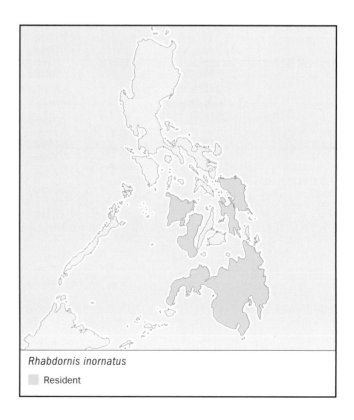

Rhabdornis grandis
■ Resident

Rhabdornis inornatus
■ Resident

color differences. Both sexes have a black bill, dark brown eyes, and legs ranging from grayish yellow to horn. The upperparts are medium-brown, while the crown, tail, and primaries are darker brown. The mask is dark brown, bordered above and below by white. The throat is grayish white and the center of the belly is white, while the breast, flanks, and undertail feathers are streaked white and brown. The female has lighter brown upperparts and mask.

DISTRIBUTION
Endemic to the islands of Samar, Basilan, Leyte, Mindanao, Negros, and Panay.

HABITAT
Prefers tropical forest over 2,600 ft (800 m) above sea level, but will sometimes range into lower levels.

BEHAVIOR
Behavior is similar to that of the stripe-headed rhabdornis, but stripe-breasteds prefer smaller foraging flocks of 4–5 individuals. The birds will sometimes roost in large numbers in the isolated canopy of a group of trees in a clearing. The call is a high-pitched note of "tzit" repeated often, sometimes in a rapid staccato.

FEEDING ECOLOGY AND DIET
Forages in middle to higher altitude forests, but may range lower. Its diet covers a wide selection of insects and fruits. On Mindanao, field researchers saw one individual snag and eat a tree frog. In eastern Mindanao, residents witnessed over 100 *Rhabdornis inornatus* individuals gathering in a great flock to flycatch among seasonal swarms of flying insects.

REPRODUCTIVE BIOLOGY
Nests in April; enlarged gonads in March and April. Eggs not described.

CONSERVATION STATUS
Not threatened.

SIGNIFICANCE TO HUMANS
None known. ◆

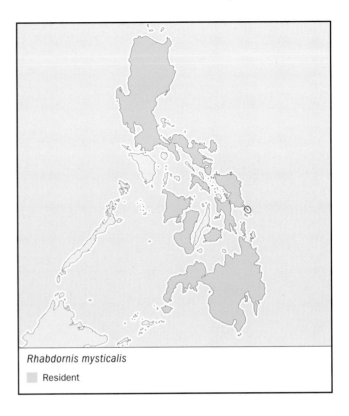

Rhabdornis mysticalis

▨ Resident

Stripe-headed rhabdornis
Rhabdornis mysticalis

TAXONOMY
Rhabdornis mysticalis Temminck, 1825. Two subspecies.

OTHER COMMON NAMES
English: Striped-headed creeper; French: Rhabdornis á tête striée; German: Streifenkopf-Rhabdornis; Spanish: Trepador Filipino de Cabeza Rayada.

PHYSICAL CHARACTERISTICS
The two subspecies of the stripe-headed rhabdornis vary somewhat in size. *Rhabdornis mysticalis mysticalis* is 6.2 in (15.8 cm) long and weighs about 3 oz (85 g). *Rhabdornis mysticalis minor* is 5.7 in (14.5 cm) long, at 2.75 oz (78 g).

The sexes are similarly colored, and males are larger than females. Both subspecies have black bills, dark brown eyes, and dark horn legs. Males have a blackish brown crown and nape with numerous white streaks, a broad stripe through the

eye, while the face and the rest of the neck are blackish brown. The female differs in having a lighter brown crown and face.

DISTRIBUTION
Philippine islands of Luzon, Negros, Panay, Masbate, Contanduenes, Leyte, Mindanao, Samar, Basilan, Bohol, Calicoan, and Dinagat.

HABITAT
Tropical forest from sea level up to 3,900 ft (1,200 m) above sea level.

BEHAVIOR
Diurnal and quite active, in canopy and middle story of forests, forest edges, and secondary growth in groups or in mixed flocks of up to 25 individuals. At dusk, they may form large roosting groups of several hundred individuals. The call is an unmelodious, high-pitched "tsee tsee WICK tsee," the "tsee" notes very soft and the "WICK" sharp and loud.

FEEDING ECOLOGY AND DIET
Foraging and diet are like that of the other rhabdornis species, with insects as main fare varied with nectar, fruit, and seeds.

REPRODUCTIVE BIOLOGY
Breeding biology is poorly known. The species is thought to nest in tree cavities as does the stripe-breasted rhabdornis. Enlarged gonads in March.

CONSERVATION STATUS
Not threatened.

SIGNIFICANCE TO HUMANS
None known. ◆

Resources

Books

Fisher, Tim, and Nigel Hicks. *A Photographic Guide to the Birds of the Philippines.* Sanibel Island, FL: Ralph Curtis Publishing Inc., 2000.

Kennedy, Robert S., Pedro C. Gonzales, Edward C. Dickinson, Hector C. Miranda Jr., and Timothy H. Fisher. *A Guide to the Birds of the Philippines.* Oxford: Oxford University Press, 2000.

Organizations

BirdLife International. Wellbrook Court, Girton Road, Cambridge, Cambridgeshire CB3 0NA United Kingdom. Phone: +44 1 223 277 318. Fax: +44-1-223-277-200. E-mail: birdlife@birdlife.org.uk Web site: <http://www.birdlife.net>

Haribon Foundation for the Conservation of Natural Resources. 9A Malingap Cot, Malumanay Streets, Teachers' Village, 1101 Diliman, Quezon City, Philippines. Phone: +63 2 9253332. E-mail: info@haribon.org.ph Web site: <http://www.haribon.org.ph>

IUCN–World Conservation Union, USA Multilateral Office. 1630 Connecticut Avenue, Washington, DC 20009 USA. Phone: (202) 387-4826. E-mail: postmaster@iucnus.org Web site: <http://www.iucn.org/places/usa/inter.html>

Oriental Bird Club, American Office. 4 Vestal Street, Nantucket, MA 02554 USA. Phone: (508) 228-1782. E-mail: rkennedy@mmo.org Web site: <http://www.orientalbirdclub.org>

Kevin F. Fitzgerald

Flowerpeckers

(Dicaeidae)

Class Aves

Order Passeriformes

Suborder Passeri (Oscines)

Family Dicaeidae

Thumbnail description

Flowerpeckers (genera *Prionochilus* and *Dicaeum*) are very small, often brightly colored birds with short, usually straight, bills and short stubby tails; keeping to higher levels of trees, they are noisy, singing chattering songs and making piping calls; berrypeckers are similar but larger with longer and more slender bills

Size

2.2–8.3 in (5.6–21 cm); 0.14–2.8 oz (4–80 g)

Number of genera, species

6 genera; 52 species

Habitat

Tall forest; all habitats from sea-level up to more than 12,000 ft (3,700 m)

Conservation status

Critically Endangered: 1 species; Vulnerable: 2 species; Near Threatened: 5 species

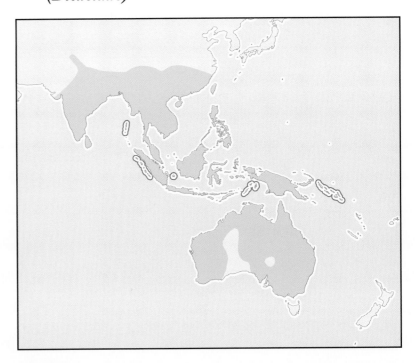

Distribution

Indian sub-continent, Sri Lanka, Myanmar, Thailand, Vietnam, Cambodia, Laos, south China, Hainan Island, Taiwan, the Malay peninsular, Indonesia, and the Philippines, Sulawesi, the Moluccas, New Guinea and its surrounding islands, and Australia

Evolution and systematics

Apart from true flowerpeckers (genera *Prionochilus* and *Dicaeum*), the composition and affinities of the Dicaeidae are disputed. On the basis of DNA hybridization studies, no other genera in the family were admitted. However, berrypeckers (genera *Melanocharis*, *Rhamphocharis*, *Oreocharis* and *Paramythia*) of New Guinea and the eight species of pardalotes, or diamondbirds (*Pardalotus* spp.), found in Australia and Tasmania, are often included within the flowerpecker family. Berrypeckers, included here with flowerpeckers, have simply constructed tongues. The pardalotes differ from the other groups as they lack serrations on the bills and have simple tongues. Other close relatives of flowerpeckers include sunbirds (Nectariniidae) and the white-eyes (Zosteropidae).

Physical characteristics

Mostly small, these passerines are 2.2–8.3 in (5.6–21 cm) in length. Flowerpeckers in the genera *Prionochilus* and *Dicaeum* are small with short bills and short stubby tails and have the distal third of the upper mandibles serrated. Tongues of *Prionochilus* are split at the end and each prong is further subdivided by a cleft. Tongues of the *Dicaeum* species are similar but longer and more variable, some having their edges

curled up to form two tubes to facilitate uptake of nectar. Species in both genera have frilly outer edges, termed fimbriations, to their tongues. Some species are dull in plumage but others are brightly colored with patterns of red or yellow contrasting with black or dark blue feathering. In most cases, plumages of females are duller than those of males.

Berrypeckers vary from the small tit-like Arfak berrypecker (*Oreocharis arfaki*) to the biggest member of the family, the thrush-like crested berrypecker (*Paramythia montium*). They have simple tongues, elongated straight bills, and lack specializations of the gut that those flowerpeckers that deal with mistletoe berries have. *Melanocharis* spp. and *Rhamphocharis crassirostris* have pectoral tufts.

Distribution

Flowerpeckers of the genus *Prionochilus* are found in Thailand, Vietnam, the Malay Peninsula, Indonesia, and the Philippines. The genus *Dicaeum* occurs in the same areas but also extends to the Indian sub-continent, Sri Lanka, Myanmar, Cambodia, Laos, south China, Hainan Island, Taiwan, Sulawesi, the Moluccas, New Guinea and its surrounding islands, and Australia. Berrypeckers (genera *Melanocharis*, *Rhamphocharis*, *Oreocharis*, and *Paramythia*) are restricted to New Guinea.

The fan-tailed berrypecker (*Melanocharis versteri*) shows its distinctive tail. (Photo by W. Peckover/VIREO. Reproduced by permission.)

Habitat

Flowerpeckers and berrypeckers are mostly birds of forests, but some are found at sea-level and others occur very high up in mountains where vegetation is sparse. Many species are partial to secondary growth or even cultivations and the scarlet-backed flowerpecker (*Dicaeum cruentatum*) frequents towns and villages. The mistletoebird (*D. hirundinaceum*) is the only Australian representative, and it can be found in habitats ranging from rainforest to arid woodland.

Behavior

Flowerpeckers are active and agile birds, twisting and turning among foliage, flicking their wings, twittering, and calling sharply. They are usually single, or in pairs or small groups, but they will join mixed-species bird parties. They sometimes sit motionless on perches for long periods. Songs are mostly simple high-pitched chirps and clicks, but the musical ability of the mistletoebird is remarkable and mimics songs of many different Australian birds.

In contrast, berrypeckers such as the crested berrypecker (*Paramythia montium*) are highly social birds, often occurring in groups of up to 12 birds; a large flock of 75 has been observed. They sometimes stand with tail cocked and raise their crests if excited or frightened. They fly in a jerky manner and are noisy in flight making low, short, calls.

Feeding ecology and diet

Flowerpeckers are fond of treetops where they seek out mistletoe berries, fruits, nectar, pollen, small insects, and spi-

ders. The fleshy part of large mistletoe berries is stripped off and eaten with the seeds discarded. There is an elaborate and amusing dance-like behavior as the birds try to deposit the seed on a branch and separate themselves from it and its sticky threads. Smaller fruits are eaten whole and pass very rapidly through the specialized gut that permits the berries to by-pass the stomach and enter the intestine directly; insects and spiders on the other hand are digested via the stomach. Insects may be caught by hawking, flycatcher-fashion. Berrypeckers are mostly frugivorous, but also take insects and spiders, occasionally hover-gleaning to catch them.

Reproductive biology

The reproductive behavior of flowerpeckers is little known and the eggs of some species have yet to be described. The mistletoebird is territorial, with males chasing intruders in weaving flights over their boundaries. For courtship, they flit around a female, calling and fanning their tails. Both sexes of most flowerpeckers are involved in building nests, incubation, and feeding of the young. Nests are neat purse-shaped bags with slit entrances near the tops and are suspended from bushes or trees. They are made of vegetable material, lichen, dried flowers, feathers, small roots, or grass, held together with cobwebs, and lined with vegetable down. Some nests are decorated with insect excreta or other debris. Most eggs are white, but those of a few species are spotted. The usual clutch is two but may be up to four. Nests of berrypeckers are cup-shaped and placed

A pair of Arfak berrypeckers (*Oreocharis arfaki*) share a perch. (Photo by W. Peckover/VIREO. Reproduced by permission.)

in thick shrubbery. Only one egg is laid in the nests of *Paramythia montium*.

Conservation status

The Cebu flowerpecker (*Dicaeum quadricolor*) is Critically Endangered with only a tiny population of less than 50 birds surviving in three forest fragments on the island of Cebu in the Philippines. Long considered to be Extinct, the species was re-discovered in 1992. There are two other globally threatened species of flowerpecker in the Philippines: the black-belted or Visayan flowerpecker (*D. haematostictum*) and the scarlet-collared flowerpecker (*D. retrocinctum*). The black-belted flowerpecker is threatened as its small range in the western Visayas Islands is becoming increasingly fragmented by deforestation. The scarlet-collared flowerpecker also exists only in forest fragments, mostly in Mindoro, and is threatened by dynamite blasting for marble and encroachment by slash-and-burn agriculture. Also within the Philippines are two other flowerpeckers that are Near Threatened: the whiskered flowerpecker (*D. proprium*) is endemic to Mindanao

and the flame-crowned flowerpecker (*D. anthonyi*) occurs only on Mindanao and Luzon. Other Near Threatened flowerpeckers include the white-throated flowerpecker (*D. vincens*) that is confined to Sri Lanka, while the scarlet-breasted flowerpecker (*Prionochilus thoracicus*) and the brown-backed flowerpecker (*D. everetti*) occur in Malaysia, Indonesia, and Brunei.

The obscure berrypecker (*Melanocharis arfakiana*) was thought to be Endangered for many years, but field-work in Papua New Guinea has shown it to be quite common in some areas, including at one site near Port Moresby. The species is apparently able to survive in degraded forest, so it may be adaptable in the face of logging and agricultural clearances of its native forest.

Significance to humans

Some species are accorded pest status as they spread parasitic mistletoes on trees of economic importance. The crested berrypecker is prized as food in the highlands of New Guinea.

1. Fire-breasted flowerpecker (*Dicaeum ignipectus*); 2. Gray-sided flowerpecker (*Dicaeum celebicum*); 3. Scarlet-breasted flowerpecker (*Prionochilus thoracicus*); 4. Red-capped flowerpecker (*Dicaeum geelvinkianum*); 5. Fan-tailed berrypecker (*Melanocharis versteri*); 6. Yellow-vented flowerpecker (*Dicaeum chrysorrheum*); 7. Mistletoebird (*Dicaeum hirundinaceum*); 8. Plain flowerpecker (*Dicaeum concolor*); 9. Midget flowerpecker (*Dicaeum aeneum*); 10. Thick-billed flowerpecker (*Dicaeum agile*); 11. Scarlet-backed flowerpecker (*Dicaeum cruentatum*). (Illustration by Bruce Worden)

Species accounts

Scarlet-breasted flowerpecker
Prionochilus thoracicus

TAXONOMY
Pardalotus thoracicus Temminck and Laugier, 1836, Borneo.

OTHER COMMON NAMES
French: Dicée à poitrine écarlate; German: Rubinkehl-Mistelfresser; Spanish: Pica Flor de Pecho Carmín.

PHYSICAL CHARACTERISTICS
4 in (10.2 cm); 0.32 oz (9 g). Dark brown head, wings, and tail with red patch on throat and crown. Back and rump yellow.

DISTRIBUTION
South Vietnam, Thailand, Malaysia, Borneo, and Sumatra.

HABITAT
Beaches at sea level up to 4,200 ft (1,280 m) in secondary forest, at forest edge, in rubber plantations, heath forest, and swamp forest.

BEHAVIOR
Varied calls include clicks and high-pitched sequence of notes sounding like an insect. Feeds at all heights, sometimes on tree trunks.

FEEDING ECOLOGY AND DIET
Takes spiders from their webs, insects, and berries.

REPRODUCTIVE BIOLOGY
Builds nest in low bush, breeds January–March and July–October.

CONSERVATION STATUS
Uncommon and listed as Near Threatened. Continuing loss of primary forest is the main threat but species can find refuges in montane and secondary forests.

SIGNIFICANCE TO HUMANS
None known. ◆

Thick-billed flowerpecker
Dicaeum agile

TAXONOMY
Fringilla agilis Tickell, 1833, Bengal, India. Eleven subspecies.

OTHER COMMON NAMES
English: Striped flowerpecker; French: Dicée à bec épais; German: Dickschnabel-Mistelfresser; Spanish: Pica Flor de Pico Ancho.

PHYSICAL CHARACTERISTICS
4 in (10.2 cm); 0.32 oz (9 g). Olive-gray upperparts with whitish, lores, breast, belly, and undertail coverts.

DISTRIBUTION
Pakistan, India, Sri Lanka, Bangladesh, Myanmar, Thailand, Laos, north Vietnam, Malaysia, Sumatra, Java, and the Philippines.

HABITAT
Forests, woods, plantations, and gardens from sea-level up to 9,800 ft (3,000 m).

BEHAVIOR
Acrobatic feeder that joins mixed-species groups. Characteristically wags its tail when perched.

FEEDING ECOLOGY AND DIET
Feeds in bushes and trees up to 100 ft (30 m) on wild fruits and berries of figs and mistletoes, spiders, and insects.

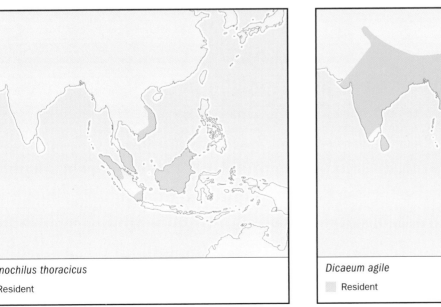

Prionochilus thoracicus
Resident

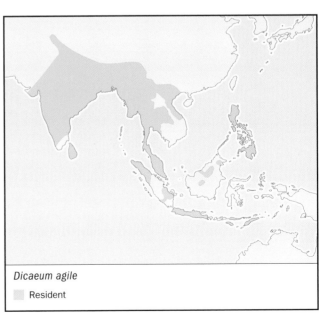

Dicaeum agile
Resident

REPRODUCTIVE BIOLOGY

Nest looks like a dead leaf suspended in a tree and resembles a small pouch made of fibers, buds, vegetable down, and spiders' webs with a side entrance. Lays clutch of two to four pale pink eggs with speckles of red during January–August.

CONSERVATION STATUS

Not threatened. Common in most of its range but rare in Myanmar and Indonesia.

SIGNIFICANCE TO HUMANS

None known. ◆

Yellow-vented flowerpecker

Dicaeum chrysorrheum

TAXONOMY

Dicaeum chrysorrheum Temminck and Laugier, 1829, Java. Two subspecies.

OTHER COMMON NAMES

French: Dicée cul-d'or; German: Gelbsteiss-Mistelfresser; Spanish: Pica Flor de Rabo Amarillo.

PHYSICAL CHARACTERISTICS

4 in (10.2 cm); 0.32 oz (9 g). Bright green upperparts with white streaked underparts and thin, dark eye stripe.

DISTRIBUTION

D. c. chrysochlore: Nepal to Bhutan, Myanmar, Indochina, and Thailand; *D. c. chrysorrheum*: southern Thailand to Sumatra, Bali, Java, and Borneo.

HABITAT

Lowland and hilly forest up to 6,600 ft (2,000 m) in Sikkim, woods, orchards, and gardens.

BEHAVIOR

Feeds at all heights. Aggressive.

FEEDING ECOLOGY AND DIET

Food consists of berries of mistletoes, figs, nectar, and insects.

REPRODUCTIVE BIOLOGY

Nest is well-hidden below 26 ft (8 m). Male and female assist in nest construction and incubation of two or three white eggs laid April–August.

CONSERVATION STATUS

Not threatened. Common in India, rare elsewhere.

SIGNIFICANCE TO HUMANS

None known. ◆

Plain flowerpecker

Dicaeum concolor

TAXONOMY

Dicaeum concolor Jerdon, 1840, Malabar coast, India. Seven subspecies.

OTHER COMMON NAMES

French: Dicée concolore; German: Einfarbmistelfresser; Spanish: Pica Flor Descolorido.

PHYSICAL CHARACTERISTICS

3.3 in (8.4 cm); 0.14–0.28 oz (4–8 g). Drab in appearance with buff breast and belly and darker upperparts.

DISTRIBUTION

D. c. borneanum: Malaysia, Borneo, and Sumatra; *D. c. concolor*: southwest India; *D. c. minullum*: Hainan Island; *D. c. olivaceum*: India, Myanmar, China, Thailand, northern Malaysia; *D. c. sollicitans*: Bali and Java; *D. c. uchidai*: Taiwan; *D. c. virescens*: Middle and southern Andaman Islands.

HABITAT

Forests, secondary growth, open areas, and plantations, from sea-level to 11,800 ft (3,600 m).

BEHAVIOR

Restless, usually seen singly in treetops, but also forages down low.

Dicaeum chrysorrheum
  Resident

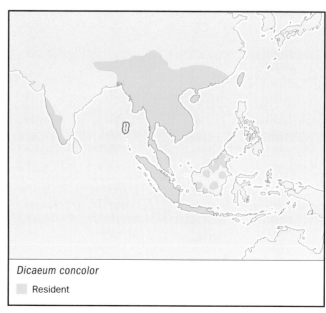

Dicaeum concolor
  Resident

FEEDING ECOLOGY AND DIET
Heavily dependent on mistletoes for berries and nectar, but also eats insects and spiders.

REPRODUCTIVE BIOLOGY
Nest is a small purse of down and fibers suspended 3.3–39 ft (1–12 m) up in a tree, bush, or low growth, in which are laid two or three white eggs from January–September. Male and female are involved in nest-building, incubation, and rearing of young.

CONSERVATION STATUS
Not threatened. Common in most of its range.

SIGNIFICANCE TO HUMANS
Spreads mistletoes. ◆

Red-capped flowerpecker
Dicaeum geelvinkianum

TAXONOMY
Dicaeum geelvinkianum Meyer, 1874, Japan. Eleven subspecies.

OTHER COMMON NAMES
English: Red-crowned flowerpecker; French: Dicée de Geelvink; German: Papuamistelfresser; Spanish: Pica Flor de Gorra Roja.

PHYSICAL CHARACTERISTICS
2.4 in (6 cm). Brownish with scarlet crown, nape, and breast; white throat patch and vents.

DISTRIBUTION
New Guinea, Geelvink Bay Islands, and the D'Entrecasteaux Archipelago.

HABITAT
Forests, forest edges, thick savanna, gardens, and plantations. Recorded as high as 7,700 ft (2,350 m).

Dicaeum geelvinkianum

Resident

BEHAVIOR
Restless. Frequents treetops.

FEEDING ECOLOGY AND DIET
Fruits, sometimes eaten whole, and spiders.

REPRODUCTIVE BIOLOGY
Two or three white eggs are laid November–December, in pear-shaped nests made of strips of ferns and animal hair suspended from low tree.

CONSERVATION STATUS
Common, not threatened.

SIGNIFICANCE TO HUMANS
None known. ◆

Midget flowerpecker
Dicaeum aeneum

TAXONOMY
Dicaeum aeneum Pucheran, 1853, San Jorge, Solomon Islands. Three subspecies.

OTHER COMMON NAMES
English: Solomons flowerpecker; French: Dicée des Salomon; German: Bronzemistelfresser; Spanish: Pica Flor de la Isla Salomon.

PHYSICAL CHARACTERISTICS
2.2 in (5.6 cm); 0.25–0.31 oz (7.1–8.8 g). Grayish upperparts with white throat patch, scarlet breast patch and yellow-green flanks.

DISTRIBUTION
D. a. aeneum: Bougainville, Choiseul, and Ysabel in northern Solomon Islands; *D. a. becki*: Florida and Guadacanal; *D. a. malaitae*: Malaita Island.

HABITAT
All habitats in the Solomon Islands, up to 1,640 ft (500 m).

Dicaeum aeneum

Resident

BEHAVIOR
Bobs head up and down when lands on perch, after rapid flight. Usually alone or in pairs.

FEEDING ECOLOGY AND DIET
Occasionally hovers to feed. Eats fruits and insects.

REPRODUCTIVE BIOLOGY
Nest is pear-shaped with a rounded base and side entrance, made of vegetable matter including grass, suspended low down in a bush. Male and female are involved in feeding young and taking away fecal sacs.

CONSERVATION STATUS
Common, not threatened.

SIGNIFICANCE TO HUMANS
None known. ◆

Fire-breasted flowerpecker
Dicaeum ignipectus

TAXONOMY
Myzante ignipectus Blyth, 1843, Nepal and Bhutan. Seven subspecies.

OTHER COMMON NAMES
English: Buff-bellied flowerpecker; French: Dicée à gorge feu; German: Feuerbrust-Mistelfresser; Spanish: Pica Flor de Lomo Verde.

PHYSICAL CHARACTERISTICS
3.5 in (8.9 cm); 0.14–0.28 oz (4–8 g). Black crown and upperparts with dark brown cheek, scarlet breast, and buff throat and belly.

DISTRIBUTION
D. i. apo: Mindanao and Negros; *D. i. beccarii*: Sumatra; *D. i. bonga*: Samar in the Philippines; *D. i. cambodianum*: Cambodia, northeast and southeast Thailand; *D. i. dolichorhynchum*: peninsular Malaysia; *D. i. formosum*: Taiwan; *D. i. ignipectum*: Kash-

mir, northeast India, Nepal, Bhutan, Sikkim, northern Myanmar, northern Indochina, southern China, southeast Tibet.

HABITAT
Montane forests, oak woodlands, rhododendrons, and cultivations up to 12,950 ft (3,950 m).

BEHAVIOR
Active at tops of trees. Joins parties and mixed-species flocks in nonbreeding season.

FEEDING ECOLOGY AND DIET
Nectar, fruits and berries of mistletoes, insects, and spiders.

REPRODUCTIVE BIOLOGY
Two or three white eggs are laid in a purse-shaped nest made of vegetable material including rootlets, grass, and moss kept together with cobwebs and suspended in a tree 10–29 ft (3–9 m) up.

CONSERVATION STATUS
Not threatened.

SIGNIFICANCE TO HUMANS
None known. ◆

Gray-sided flowerpecker
Dicaeum celebicum

TAXONOMY
Dicaeum celebicum S. Müller, 1843, Celebes. Five subspecies.

OTHER COMMON NAMES
English: Black-sided flowerpecker; French: Dicée des Célèbes; German: Schwarzwangen-Mistelfresser; Spanish: Pica Flor de Dorso Negro.

PHYSICAL CHARACTERISTICS
2.4 in (6 cm). Blue-black crown and upperparts with scarlet throat and breast, grayish sides and pale vents.

Dicaeum ignipectus
■ Resident

Dicaeum celebicum
■ Resident

DISTRIBUTION
D. c. celebicum: Bangka, Butung, Lembeh, Manadotua, Muna, Sulawesi, and Togian; *D. c. kuehni*: the Archipelago of Tukangbesi; *D. c. sanghirense*: Sangihe; *D. c. sulaense*: Banggai and Sula; *D. c. talautense*: Talaud.

HABITAT
Varied environments including forests, forest edges, scrub, cultivations, gardens, and villages from sea level to 3,300 ft (1,000 m).

BEHAVIOR
Holds itself upright and shakes wings before swallowing fruit. Calls are high-pitched repetitions of unmusical notes and a series of chirps in flight.

FEEDING ECOLOGY AND DIET
Swallows small fruits of mistletoes and cherries whole, but larger ones are pierced and the contents squeezed out. Also takes insects, spiders, nectar, and pollen.

REPRODUCTIVE BIOLOGY
Three white eggs are laid in a nest shaped like a purse with a slit entrance near the top, sometimes with an overhanging porch. Made of vegetable down, grass, twigs, leaves, and cobwebs, the nest is suspended low in a tree.

CONSERVATION STATUS
Not threatened.

SIGNIFICANCE TO HUMANS
None known. ◆

Mistletoebird
Dicaeum hirundinaceum

TAXONOMY
Motacilla hirundinaceum Shaw and Nodder, 1792, New Holland (Australia). Four subspecies.

OTHER COMMON NAMES
English: Australian flowerpecker, Australian flower swallow, mistletoe flowerpecker; French: Dicée hirondelle; German: Rotsteiss-Mistelfresser; Spanish: Pica Flor del Muérdago.

PHYSICAL CHARACTERISTICS
3.7–4.3 in (9.5–11 cm); 0.28–0.35 oz (8–10 g). Blue-black upperparts with red throat, breast, and vents. White belly with black patch.

DISTRIBUTION
D. h. fulgidum: Tanimbar Islands; *D. h. hirundinaceum*: Australia (except Tasmania); *D. h. ignicolle*: Aru Island; *D. h. keiense*: Kai, Tayundu, and Watubele.

HABITAT
Forests, woodlands, savanna, scrub, and mangroves.

BEHAVIOR
Utters characteristic two- or three-note calls, various flight notes, and song; is also a remarkable vocal mimic of many other birds. Keeps upright on perch. Restless, fast flier that is nomadic in search of fruiting mistletoes. When searching for food, tends to flick its wings, moving rapidly among upper branches of trees; also hawks for insects.

FEEDING ECOLOGY AND DIET
Heavily dependent on mistletoes; also feeds on insects, spiders, fruits, nectar, and pollen.

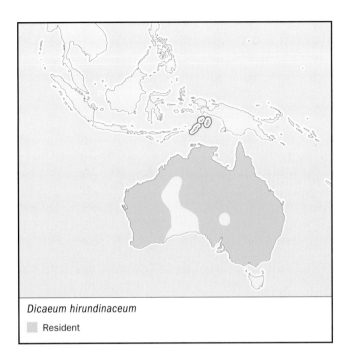

Dicaeum hirundinaceum

▢ Resident

REPRODUCTIVE BIOLOGY
Maintains territories by chasing intruders and singing from high perches. Courtship involves male chasing female in flight, landing next to her, and fanning tail. Three or four white eggs are laid in purse-shaped nest with slit entrance at the side. Female alone incubates for 12 days; both sexes then feed young in nest for two weeks. Breeding season tied to fruiting period of mistletoes, September–April.

CONSERVATION STATUS
Not threatened.

SIGNIFICANCE TO HUMANS
Spreads mistletoes. ◆

Scarlet-backed flowerpecker
Dicaeum cruentatum

TAXONOMY
Certhia cruentata Linnaeus, 1758, Bengal (India). Seven subspecies.

OTHER COMMON NAMES
English: Pryer's flowerpecker; French: Dicée à dos rouge; German: Scharlachmistelfresser; Spanish: Pica Flor de Lomo Carmín.

PHYSICAL CHARACTERISTICS
3.5 in (8.9 cm); 0.19–0.28 oz (5.5–8 g). Red stripe from forehead to base of tail; blue-black wings and tail; and whitish belly.

DISTRIBUTION
D. c. batuense: Mentawai Islands; *D. c. cruentatum*: northeast India to southern China, eastern Thailand, Indochina, Sumatra, and Riau Archipelago; *D. c. ignitum*: peninsular Malaysia; *D. c. niasense*: Nias Island; *D. c. nigrimentum*: Borneo; *D. c. simalurense*: Simeulue Island; *D. c. sumatranum*: Sumatra and surrounding islands.

HABITAT
Forests, secondary growth, orchards, plantations, and gardens up to 3,300 ft (1,000 m) mostly, but attains more than 6,900 ft (2,100 m) in Nepal.

Dicaeum cruentatum

Resident

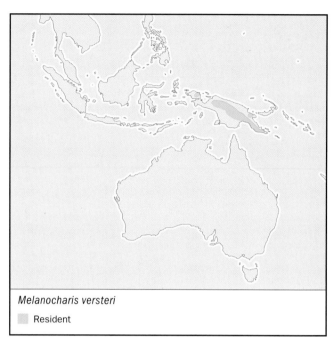

Melanocharis versteri

Resident

BEHAVIOR
Strong flier, aggressive and restless in treetops but occasionally feeds low in bushes; hawks for insects. Moves up and down mountains with the seasons.

FEEDING ECOLOGY AND DIET
Fruits, especially of mistletoes, figs, berries, seeds, nectar, spiders, and insects.

REPRODUCTIVE BIOLOGY
Two to four gray eggs laid in tiny oval pouch, made by both sexes, of grass and suspended 6.6–49 ft (2–15 m) up in tree; entrance near top sometimes has porch.

CONSERVATION STATUS
Not threatened. Common in most of range but rare in Bhutan and Nepal.

SIGNIFICANCE TO HUMANS
None known. ◆

Fan-tailed berrypecker
Melanocharis versteri

TAXONOMY
Pristorhamphus versteri Finsch, 1876, New Guinea. Four subspecies.

OTHER COMMON NAMES
French: Piquebaie éventail; German: Fächerschwanz-Beerenpicker; Spanish: Cerezero Cola de Abanico.

PHYSICAL CHARACTERISTICS
5.5–6 in (14–19 cm). Female larger (wing length (2.6–2.8 in [6.6–7.1 cm]) and heavier (0.56–0.7 oz [16–20 g]) than male (2.32–2.52 in [5.9–6.4 cm]; 0.44–0.53 oz [12.5–15 g]). Whitish

underparts and lateral feathers on distinctively long tail. Brownish black upperparts.

DISTRIBUTION
From 4,500 to 10,800 ft (1,400–3,300 m) up in mountains in New Guinea in both Indonesia (Irian Jaya) and Papua New Guinea. *M. v. maculiceps*: southeast New Guinea; *M. v. meeki*: northwest New Guinea, Weyland, and Snow Mountains; *M. v. versteri*: northwest New Guinea, Vogelkop; *M. v. virago*: northern and northeast New Guinea.

HABITAT
Montane forest, tree-fern heaths, and alpine thickets. Sometimes occurs in middle strata of forest but usually keeps to undergrowth.

BEHAVIOR
Shy, usually solitary or in pairs, and an active feeder. Shows white in tail when flying. Acrobatic. Harsh song; it also utters squeaks and nasal scolding calls.

FEEDING ECOLOGY AND DIET
Small berries, eaten whole, and insects. Sometimes hover-gleans to feed.

REPRODUCTIVE BIOLOGY
Nest large in relation to the bird's size. A solid deep cup, 3 in (8 cm) across and 4 in (10 cm) high, placed in the fork of a branch, and well-camouflaged.

CONSERVATION STATUS
Not threatened.

SIGNIFICANCE TO HUMANS
None known. ◆

Resources

Books

Beehler, B. M., T. K. Pratt, and D. A. Zimmerman. *Birds of New Guinea*. Princeton: Princeton University Press, 1986.

BirdLife International. *Threatened Birds of the World*. Barcelona and Cambridge: Lynx Edicions and BirdLife International, 2000.

Cheke, R. A., C. F. Mann, and R. Allen. *Sunbirds: A Guide to the Sunbirds, Flowerpeckers, Spiderhunters, and Sugarbirds of the World*. New Haven: Yale University Press, 2001.

Rand, A. L., and E. T. Gilliard. *Handbook of New Guinea Birds*. London: Weidenfeld and Nicholson, 1967.

Sibley, C. G., and B. L. Monroe. *Distribution and Taxonomy of Birds of the World*. New Haven: Yale University Press, 1990.

Robert Alexander Cheke, PhD

Pardalotes

(Pardalotidae)

Class Aves
Order Passeriformes
Suborder Passeri (Oscines)
Family Pardalotidae

Thumbnail description
Small, short-tailed, dumpy birds with short, scoop-shaped bills

Size
3.2–4.7 in (8–12 cm)

Number of genera, species
1 genus; 8 species

Habitat
Eucalypt forest and woodland

Conservation status
Endangered: 1 species

Distribution
Australia and Tasmania

Evolution and systematics

Recent DNA studies suggest an affinity between the pardalotes and acanthizids that has resulted in placing both in the same family in most taxonomic treatments. However, genomic distances are large, and the two groups have substantial differences in morphology and behavior, including wing structure, nesting behavior, osteological features, and plumage patterns. Hence, each group as been accorded family status in the most recent taxonomic treatment (1999).

Physical characteristics

Pardalotes are noisy, generally bright colored, conspicuous, beautiful, dumpy little birds, with short bills and tails. They have 12 rectrices and short pointed wings with a vestigial 10th primary. All have some combination of yellow, brown, and black with white spots or streaking. The sparkling color combinations give them the name "diamond bird". Most species are monomorphic in plumage, with juvenile plumage muted.

Distribution

The pardalotes are endemic to Australia, and are found throughout the continent where suitable habitat occurs; they are only missing from a few patches of southern desert. One

species is endemic to Tasmania and has a very restricted distribution; other species are found across the continent.

Habitat

A broad spectrum of woodlands and forest, primarily eucalypt and acacia, from wet coastal to arid conditions.

Behavior

Solitary during nesting, but gregarious during winter and migratory or nomadic post-breeding dispersal. They often forage in mixed species flocks in winter. They actively forage, searching foliage and gleaning scale insects and other invertebrates. Their calls consist of generally loud two- to five-note whistles, often repeated endlessly. One species is monotypic, the others polytypic.

Feeding ecology and diet

Scurry about foliage, gleaning a wide variety of invertebrates from leaves and twigs with their scoop-shaped bills. They prefer soft-bodied invertebrates, including small wasps, spiders, weevils, and termites. They also eat lerps (sugary secretions of psillid insects).

A striated pardalote (*Pardalotus substriatus*) holds an insect in its bill. (Photo by R. Brown/VIREO. Reproduced by permission.)

Reproductive biology

Pairs defend nesting territories that they advertise with repetitive two- to five-note note whistles. They are monogamous, some maintaining pair bonds throughout the year, and some may have helpers at the nest. Nests are cup-shaped, sometimes domed, of plant fibers in tunnels burrowed into banks or horizontal ground; sometimes in tree hollows.

Conservation status

The forty-spotted pardalote (*Pardalotus quadragintus*) has six small disjunct populations that are confined to southeastern Tasmania. It is Endangered and has been the subject of an intensive recovery plan. With protection of suitable habitat, its prospects are encouraging. The other pardalote species are geographically widespread and are not considered threatened.

Significance to humans

None known.

1. Red-browed pardalote (*Pardalotus rubricatus*); 2. Striated pardalote (*Pardalotus striatus*); 3. Forty-spotted pardalote (*Pardalotus quadragintus*); 4. Spotted pardalote (*Pardalotus punctatus*). (Illustration by Wendy Baker)

Species accounts

Forty-spotted pardalote
Pardalotus quadragintus

TAXONOMY
Pardalotus quadragintus Gould, 1838, Tasmania.

OTHER COMMON NAMES
English: Golden-rumped diamondbird; French: Pardalote de Tasmanie; German: Tasmanpanthervogel: Spanish: Pardalote de Cuarenta Manchas.

PHYSICAL CHARACTERISTICS
3.5–3.9 in (9–10 cm); 0.38 oz (10.7 g). Sexes similar in plumage.

DISTRIBUTION
Restricted to southeastern Tasmanian coastal forests and woodlands.

HABITAT
Dry sclerophyll forest and woodlands; prefers white gum (*Eucalyptus viminalis*).

BEHAVIOR
Nests in loose colonies but forms flocks in winter, often with other species. Soft two-note call.

FEEDING ECOLOGY AND DIET
Gleans eucalypt foliage and twigs. Forages on a broad spectrum on invertebrates, but apparently does not take many scale insects as other pardalotes.

REPRODUCTIVE BIOLOGY
Often nest within 33 ft (10 m) of other nests in loose associations. The nest is a dome or cup of plant fibers, lined with grass or feathers, placed in a tree hollow up to 66 ft (20 m) above the ground. Clutch is three to five white eggs; incubation period 16 days, with fledging about a month later.

CONSERVATION STATUS
Endangered. Population declined in twentieth century, probably due to competition with spotted pardalotes (*Pardalotus punctatus*). Current population is probably about 3,000 birds in about six disjunct populations (several on islands). Has been the focus of a recovery team project. With habitat protection, population will probably stabilize.

SIGNIFICANCE TO HUMANS
None known. ◆

Red-browed pardalote
Pardalotus rubricatus

TAXONOMY
Pardalotus rubricatus Gould, 1838, Australia. Two subspecies.

OTHER COMMON NAMES
English: Fawn-eyed diamond bird; French: Pardalote à sourcils rouge; German: Rotbrauen-Panthervogel; Spanish: Pardalote de Cejas Rojas.

PHYSICAL CHARACTERISTICS
3.9 in (10 cm) 0.32–0.39 oz (9–11 g). Sexes similar in plumage; immature birds are drab pale-olive above, light gray below.

DISTRIBUTION
Arid areas across most of northern and central Australia. *P. r. yorki* confined to Cape York Peninsula.

HABITAT
Eucalypts and acacias in drier woodlands, forest, and scrub. Frequent eucalypts along river beds.

Pardalotus quadragintus
Resident

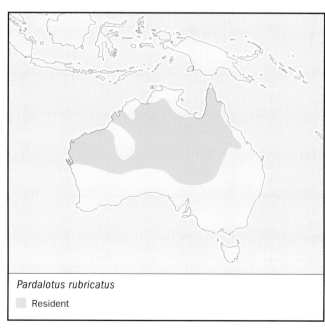

Pardalotus rubricatus
Resident

BEHAVIOR
Often found with other pardalote species; often feeds in trees with sparse vegetation. Distinctive five-note call.

FEEDING ECOLOGY AND DIET
Gleans a broad spectrum of invertebrates from foliage and twigs of primarily eucalypts.

REPRODUCTIVE BIOLOGY
Cup-shaped nest of plant fibers lined with grass at end of 1.6–2.3 ft (50-70 cm) burrow. Clutch two to four white eggs.

CONSERVATION STATUS
Populations widespread and currently not threatened. Overgrazing and habitat alteration pose potential threats.

SIGNIFICANCE TO HUMANS
None known. ◆

Spotted pardalote
Pardalotus punctatus

TAXONOMY
Pardalotus punctatus Shaw, 1792, New Holland (Australia). Three subspecies.

OTHER COMMON NAMES
English: Diamond bird; bank diamond; French: Pardalote pointillé German: Fleckenpanthervogel; Spanish: Pardalote Moetado.

PHYSICAL CHARACTERISTICS
3.5 in (9 cm); 0.32 oz (9g). Sexually dimorphic in plumage: males have yellow throats and breast.

DISTRIBUTION
Australia. *P. p. punctatus*: coastal belt from southern Queensland to eastern South Australia and southwestern West Australia and in Tasmania; *P. p. millitaris*: coastal belt of north central

Queensland; *P. p. xanthopygae*: mallee and mulga areas of Victoria to Western Australia.

HABITAT
Eucalypt forests and woodlands, residential parks and gardens. *P. p. xanthopygae* mostly in mallee and mulga woodlands.

BEHAVIOR
Territorial during breeding season, but forms flocks of a dozen birds or more in winter, and may join mixed species foraging flocks. Two- or three-note call.

FEEDING ECOLOGY AND DIET
Gleans twigs and foliage for scale insects, and a wide variety of other invertebrates.

REPRODUCTIVE BIOLOGY
Nest is of plant fibers about 19.7 in (50 cm) in tunnel burrowed into earthen bank or flat ground, often near a creek. The usual clutch is three to five white eggs incubated by both parents; hatching is in 14-16 days, fledging in about a month.

CONSERVATION STATUS
Not threatened, but habitat alteration is a potential threat.

SIGNIFICANCE TO HUMANS
None known. ◆

Striated pardalote
Pardalotus striatus

TAXONOMY
Pardalotus striatus Gmelin, 1789, Tasmania. Six subspecies.

OTHER COMMON NAMES
English: Black-headed pardalote; stripe-crowned pardalote; red-tipped pardalote; yellow-tipped pardalote; French: Pardalote à point jaune; German: Streifenpanthervogel; Spanish: Pardalote Estriado.

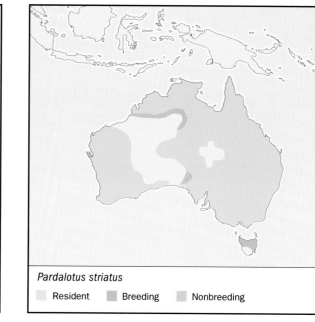

Pardalotus punctatus
   Resident   Nonbreeding

Pardalotus striatus
   Resident   Breeding   Nonbreeding

PHYSICAL CHARACTERISTICS
3.5–4.5 in (9–11.5 cm); 0.42 oz (12 g). Sexes similar in plumage, immatures with muted head color.

DISTRIBUTION
Australia except for desert sections of interior. *P. s. uropygialis*: northern Australia; *P. s. melvillensis*: Melville and Bathurst Islands; *P. s. melanocephalus*: coastal belt of Queensland; *P. s. ornatus*: coastal New South Wales; *P. s. striatus*: Tasmania; *P. s. substriatus*: Australia from New South Wales to Western Australia.

HABITAT
Widely distributed through eucalypt woodlands and forest, but also in rainforest and mangroves.

BEHAVIOR
Form flocks during winter. The races *striatus*, *substriatus*, and *ornatus* are nomadic or migratory; other races tend to be sedentary. Loud and repetitive two- to three-note call.

FEEDING ECOLOGY AND DIET
Gleans foliage and twigs, primarily in eucalypts and acacias, for a broad spectrum of invertebrates.

REPRODUCTIVE BIOLOGY
Nest is cup-shaped, partly or completely domed, of plant fibers, at the end of an earthen burrow or in a tree hollow. Both parents contribute to burrow excavation and nest construction. Typical clutch is three to five white eggs, incubated by both parents.

CONSERVATION STATUS
Not threatened; found in a broad spectrum of environments and across a wide geographic range.

SIGNIFICANCE TO HUMANS
None known. ◆

Resources

Books

Christidis, L., and W. E. Boles. *The Taxonomy and Species of Birds of Australia and its Territories.* Hawthorn East: Royal Australasian Ornithologists' Union, Monograph 2, 1994.

Longmore, W. *Honeyeaters and Their Allies.* North Ryde, NSW: HarperCollins, 1991.

Pizzy, G., and F. Knight. *The Graham Pizzy & Frank Knight Field Guide to the Birds of Australia.* Sydney: HarperCollins Publishers, 1997.

Schodde, R., and I. J. Mason. *The Directory of Australian Birds: Passerines.* Canberra: CSIRO Wildlife and Ecology, 1999.

Simpson, K., and N. Day. *Birds of Australia.* Princeton: Princeton University Press, 1999.

William E. Davis, Jr

Sunbirds
(Nectariniidae)

Class Aves
Order Passeriformes
Suborder Passeri (Oscines)
Family Nectariniidae

Thumbnail description
Small passerines, often brightly colored with iridescent plumage, and short, almost straight, to long and markedly decurved, bills; vocal, singing chattering songs, and often calling between feeds

Size
3.5–11 in (9–27 cm); 0.14–0.92 oz (4–26 g)

Number of genera, species
16 genera; 130 species

Habitat
Forest, woodlands, savanna, mountains, scrubland, coastal zones, and gardens

Conservation status
Endangered: 2 species; Vulnerable: 4 species; Near Threatened: 8 species

Distribution
Sub-Saharan Africa, Nile Valley north to coast of Mediterranean Sea in Egypt, Madagascar, Israel, Arabian Peninsula, Socotra Island, Comoro Islands, Seychelles, South and Southeast Asia, Pacific Islands, and Northeast Australia

Evolution and systematics

Although ecological equivalents of the nonpasserine hummingbirds, sunbirds as passerines are quite unrelated to them. On morphological grounds, sunbirds were considered close relatives of the honeycreepers (Meliphagidae) and the white-eyes (Zosteropidae), but on the basis of analyses of their DNA, Sibley and Monroe (1990) placed them with the flowerpeckers and sugarbirds. Irwin (1999) revised the sunbirds and concluded that they are of African origin with the short-billed, mostly insectivorous, genera *Deleornis*, and *Anthreptes* being the most primitive.

Evolution of the long, curved bills associated with nectar-feeding members of *Nectarinia* and other genera probably came about as a consequence of seeking insects in flowers. In addition to *Deleornis* and *Anthreptes*, Irwin accepted the validity of the genera *Chalcomitra*, *Cyanomitra*, *Cinnyris*, and *Leptocoma*, which Delacour (1944) had grouped in *Nectarinia*. Other genera within the Nectariniidae currently recognized include *Chalcoparia*, *Hedydipna*, *Hypogramma*, *Anabathmis*, *Dreptes*, *Anthobaphes*, *Drepanorhynchus*, *Aethopyga*, and *Arachnothera*.

Some sources recognize 130 species in 16 genera, but Peters recognizes 5 genera and 117 species. The sunbirds have radiated into most habitats throughout sub-Saharan Africa and tropical Asia. They have also penetrated to extremely high altitudes on both continents: Gould's sunbird (*Aethopyga gouldiae*), for instance, breeds up to 14,100 ft (4,300 m). In order to cope with the freezing conditions they encounter on high mountains, some sunbirds have the ability to lower their body temperatures while roosting. There are close associations between some sunbirds and particular groups of plants. For example, there has been coevolution between genera of mistletoes and the long-billed sunbirds that pollinate them, and the orange-breasted sunbird (*Anthobaphes violacea*) is dependent on proteas and heaths in its fynbos habitat.

Physical characteristics

Most male sunbirds and many of the females are brightly colored, with iridescent plumage covering varying proportions of their bodies. The color of the iridescence changes with the angle of incident light such that a blue may suddenly appear green or black. Many have marked contrasts in their colors, especially the double-collared group amongst the genus *Cinnyris*, who have broad red bands across their chests. Brightly colored pectoral tufts, usually yellow or red, are a feature of many species, particularly among males that use them in courtship and aggressive displays. The predominantly black bills of sunbirds are nearly all decurved, but the extent of the curvature varies from very slight in the genus *Deleornis* to the sickle-shaped bill of the golden-winged sunbird (*Drepanorhynchus reichenowi*). The birds' tongues are long and may be extruded far beyond the tip of the bill. The tongues vary in size and shape, with tubular structures and serrations

Female malachite sunbird (*Nectarinia famosa*) in its nest. (Photo by Kenneth W. Fink. Bruce Coleman Inc. Reproduced by permission.)

islands such as Mauritius, sunbirds are found in Madagascar and the Comoros. There is also an endemic species in the Seychelles, the Seychelles sunbird (*Cinnyris dussumieri*) and another on Socotra Island, the Socotra sunbird (*Chalcomitra balfouri*).

The olive-backed sunbird (*Cinnyris jugularis*), the species that occurs in northeast Australia, is the most widespread of the Asiatic species. The collared sunbird (*Hedydipna collaris*), scarlet-chested sunbird (*Chalcomitra senegalensis*), variable sunbird (*Nectarinia venusta*), and copper sunbird (*Cinnyris cupreus*) are the most widespread species in Africa, all having populations in central, eastern, southern, and West Africa.

Habitat

Most sunbirds are birds of forest, woodlands, and savanna regions, where there is an ample supply of flowering plants and insects. However, some species such as the dusky sunbird (*Cinnyris fuscus*) of southern Africa and the Nile Valley sunbird (*Hedydipna metallica*) of northeast Africa and Arabia are found in semidesert habitats. The altitudinal range of

at the tips being most common. Tails may be short and square-ended, or graduated and elongated, with males of the genus *Nectarinia*, *Drepanorhynchus* and *Aethopyga* having extended central tail feathers. No sunbirds have truly forked tails, even the fork-tailed sunbird (*Aethopyga christinae*) gets its name from central tail feathers that are elongated into a forked shape. The legs are long and thin and usually black, with feet having curved claws.

The smallest sunbird is the crimson-backed sunbird (*Leptocoma minima*), which may be only 3.5 in (9 cm) long and weigh as little as 0.14 oz (4 g). The largest sunbird is the São Tomé sunbird (*Dreptes thomensis*), males of which may be 9 in (23 cm) long and weigh 0.9 oz (26 g). The 10 species of spiderhunters in the genus *Arachnothera* are larger than almost all of the other sunbirds and are restricted to Asia. Their sexes are similar and lack any iridescent plumage. Their decurved bills are very long, being at least twice the lengths of their heads.

Distribution

Sunbirds are found throughout sub-Saharan Africa, along the Nile valley to the coast of the Mediterranean Sea in Egypt, and eastwards through the Middle East to most of south and Southeast Asia, including many of the Pacific islands, and one species reaches Australia. Although absent from Indian Ocean

A golden-winged sunbird (*Drepanorhynchus reichenowi*) in Kenya. (Photo by A.J. Deane. Bruce Coleman Inc. Reproduced by permission.)

sunbirds is extensive, ranging from sea level to at least 14,700 ft (4,500 m) in Afro-alpine moorlands, where the scarlet-tufted malachite sunbird (*Nectarinia johnstoni*) occurs. Sunbirds require food sources of nectar from flowers and insects in their habitats, and plant material, usually grass, with which to make their nests. Lack of a year-round supply of flowers in one place can be circumvented by local movements, including altitudinal shifts or long-distance migrations, but most species are opportunistic and will exploit a source of nectar in whatever habitat they find it. Forest-dwelling species are often found in the canopy of the tallest trees, taking nectar and insects.

Behavior

Sunbirds are diurnal and active from dawn to dusk. Their high metabolic rate and small size necessitate almost constant searches for food, but they sometimes rest on exposed perches to preen, wipe their bills, or sing. Characteristically, they draw attention to themselves by their high-pitched calls as they flit from one flower to another, but they also catch insects on the wing and may hover in front of flowers as they probe them for nectar. Sunbirds are aggressive to competitors of their own species, but they also attack other species of sunbirds, especially at feeding areas.

Sunbirds are mostly found on their own, in pairs, or in family groups, but they are also social, coming together to feed at abundant sources of nectar or joining in mixed-species groups in forest canopies.

Apart form localized movements in response to shifts in food availability and seasonal dispersal up and down mountains, some sunbirds make regular migrations. The African subspecies of the Palestine sunbird (*Cinnyris oseus decorsei*) migrates with the movement of the rains during the year, as do some populations of the beautiful sunbird (*Cinnyris pulchellus*), the pygmy sunbird (*Hedydipna platura*), and the scarlet-chested sunbird. A female of the latter species has been recovered 220 mi (360 km) away from where it was banded, and scarlet-chested sunbirds were among the species in a migratory flock of sunbirds seen in Botswana moving northeastward at a rate of 500 birds per hour.

Although not renowned for their singing, the vocalizations of sunbirds include quite complex and occasionally very melodious songs. These are chattering warbles, interspersed with whistles and wheezes, uttered from prominent perches such as dead trees. There is regional variation in the songs of a given species, and local dialects have been identified in the voices of the splendid sunbird (*Cinnyris coccinigaster*) in West Africa. Sunbird calls are also distinct, mostly bisyllabic chirps or whistles uttered as contact notes or alarm signals. The songs of sunbirds are used to advertise their territories, which they defend vigorously around nests and feeding zones. Courtship displays involve exposure of pectoral tufts by males of those species that possess them, and elaborate bowing rituals, with wings partly open and quivering, and tails cocked or fanned.

Variable sunbird (*Nectarinia venusta*) feeding on *Leonotis* flowers at Lake Nakuru National Park, Kenya. (Photo by John Shaw. Bruce Coleman Inc. Reproduced by permission.)

Feeding ecology and diet

The most primitive genus *Deleornis* feeds almost exclusively on insects and spiders, but takes a few small fruits as well. All other genera of sunbirds take these items as food, but also probe flowers for nectar. Energy-rich nectar is a very important part of the diet of species such as the bronze sunbird (*Nectarinia kilimensis*) and the golden-winged sunbird. The proportions of nectar or invertebrates in the diet of sunbirds varies from species to species and with the seasons. Spiders are regularly taken by most species, and small fruits, seeds, and pollen are also common constituents of sunbird meals. There are close associations between groups of flowering plants, such as mistletoes, proteas, and aloes, that rely on the long-billed sunbirds to pollinate them. Short-billed sunbirds may also feed on nectar from flowers that should be beyond their reach by nectar-robbing activities such as probing into the base of flowers rather than through the more usual open route. Despite their name, spiderhunters have diets similar to those of other sunbirds, taking nectar, pollen, small fruits, and insects, in addition to spiders, which may be snatched from their webs.

Reproductive biology

Some forest species such as the western olive sunbird (*Cyanomitra obscura*) gather in leks, with up to eight males displaying to each other in the absence of females. Although usually monogamous, there is evidence that females copulate with more than one male, and 36% of the broods of the Palestine sunbird (*Cinnyris oseus*) are sired by males other than the female's main partner.

Sunbirds defend breeding territories by singing and active pursuit or by attacks on intruders. Although male spiderhunters help with incubation, females of the other sunbirds are responsible for most of the nest-building and all of the incubation, although males become involved with feeding young. This may continue for a few days after fledging, which occurs about two weeks after hatching. There are many records of cloaca-pecking by males, suggestive of much extra-pair copulation, but this has been documented only in the Palestine sunbird and the purple-rumped sunbird (*Leptocoma zeylonica*). The Seychelles sunbird is sometimes polygamous.

Some spiderhunters construct cup-shaped nests, but all the other sunbirds enclose their eggs and young in pouch- or pear-shaped nests with side entrance holes. For the most part, these are made of grass, plant fibers, twigs, lichen, and moss, all held together with threads from spider webs, and are suspended from a bush or tree, although the orange-breasted sunbird places its nest directly in a bush. Many of the suspended nests have porches built above their entrances, trailing vegetation hanging below their bases, decorations of dead leaves and other debris as camouflage, and are lined with wool or feather down. Sunbirds may often be double- or triple-brooded, laying one to three eggs in each clutch. The eggs take about two weeks to hatch. Cuckoos regularly parasitize nests of sunbirds.

Conservation status

In Africa, the Endangered Amani sunbird (*Hedydipna pallidigaster*) has a very restricted range in isolated forest pockets in the Arabuko-Sokoke forest of Kenya and the Usambaras and Udzungwa Mountains in Tanzania, where it is threatened by tree-felling. Another Endangered species is the elegant sunbird (*Aethopyga duyvenbodei*), which is restricted to Sangihe Island, north of Sulawesi, in Indonesia. Its forest home on this tiny island is also threatened by tree-felling, but the bird has adapted somewhat to feeding in plantations.

Habitat loss is another major threat facing sunbirds of conservation concern such as the Vulnerable banded sunbird (*Anthreptes rubritorques*), São Tomé sunbird, Rockefeller's sunbird (*Cinnyris rockefelleri*), and rufous-winged sunbird (*Cinnyris rufipennis*), all of which survive in small forest remnants. Other sunbirds treated as Near Threatened are the plain-backed sunbird (*Anthreptes reichenowi*), the red-throated sunbird (*Anthreptes rhodolaema*), Ursula's sunbird (*Cinnyris ursulae*), Neergaard's sunbird (*Cinnyris neergaardi*), Moreau's sunbird (*Cinnyris moreaui*), the gray-hooded sunbird (*Aethopyga primigenius*), the Apo sunbird (*Aethopyga boltoni*), and Lina's sunbird (*Aethopyga linaraborae*). Among the spiderhunters, only Whitehead's spiderhunter (*Arachnothera juliae*), restricted to the uplands of Borneo, is of any conservation concern.

Significance to humans

Sunbirds are important to humans for aesthetic, biological, and economic reasons. The beauty and charm of the brilliantly colored species bring pleasure to many in wilderness areas, cultivations, and gardens throughout the Old World tropics. Sunbirds are important biologically as pollinators of many plants, including some of economic value, and as predators of nuisance insects. The pollination habits of some species leads them to be given pest status. For instance, the scarlet-chested sunbird helps to propagate the mistletoe (*Tapinanthus bangwensis*), a scourge of cocoa plantations in West Africa, and the purple sunbird (*Cinnyris asiaticus*) plays a similar role with mistletoes affecting timber trees in Asia and is also implicated in attacks on cultivated grapes.

1. Reichenbach's sunbird (*Anabathmis reichenbachii*); 2. Scarlet-tufted sunbird (*Deleornis fraseri*); 3. Collared sunbird (*Hedydipna collaris*); 4. Purple-naped sunbird (*Hypogramma hypogrammicum*); 5. São Tomé sunbird (*Dreptes thomensis*); 6. Ruby-cheeked sunbird (*Chalcoparia singalensis*); 7. Green sunbird (*Anthreptes rectirostris*); 8. Orange-breasted sunbird (*Anthobaphes violacea*); 9. Green-headed sunbird (*Cyanomitra verticalis*); 10. Plain-throated sunbird (*Anthreptes malacensis*). (Illustration by Barbara Duperron)

1. Olive-bellied sunbird (*Cinnyris chloropygius*); 2. Greater double-collared sunbird (*Cinnyris afra*); 3. Yellow-eared spiderhunter (*Arachnothera chrysogenys*); 4. Golden-winged sunbird (*Drepanorhynchus reichenowi*); 5. Gould's sunbird (*Aethopyga gouldiae*); 6. Malachite sunbird (*Nectarinia famosa*); 7. Scarlet-chested sunbird (*Chalcomitra senegalensis*); 8. Olive-backed sunbird (*Cinnyris jugularis*); 9. Crimson-backed sunbird (*Leptocoma minima*); 10. Purple sunbird (*Cinnyris asiaticus*). (Illustration by Barbara Duperron)

Species accounts

Ruby-cheeked sunbird
Chalcoparia singalensis

SUBFAMILY
Nectariniinae

TAXONOMY
Motacilla singalensis Gmelin, 1788, Malacca. Eleven subspecies.

OTHER COMMON NAMES
English: Rubycheek; French: Souimanga à joues rubis; German: Rubinwangen-Nektarvogel; Spanish: Nectarina de Mejillas Rojas.

PHYSICAL CHARACTERISTICS
4.5 in (11.5 cm); 0.29–0.32 oz (8.2–9.1 g). Differs from all other sunbirds by unique structure of its tongue, which is covered in a horny plate with deeply-notched tip. Green upperparts with orange throat and yellow breast and belly.

DISTRIBUTION
C. s. assamensis: East Nepal to Bangladesh, northern Myanmar, northern Thailand, and western and southern Yunnan; *C. s. bantenensis*: western Java; *C. s. borneana*: Borneo; *C. s. internota*: southern Myanmar and southern Thailand; *C. s. interposita*: southern Thailand; *C. s. koratensis*: eastern Thailand, Laos, and Vietnam; *C. s. pallida*: north Natuna Islands; *C. s. panopsia*: islands off west coast of Sumatra; *C. s. phoenicotis*: eastern and central Java; *C. s. singalensis*: Malay Peninsula; *C. s. sumatrana*: Sumatra and Belitung.

HABITAT
Forests, scrub, clearings, mangroves, coasts, and well-vegetated riverbanks and gardens.

BEHAVIOR
Active, tit-like behavior. Sometimes forages in small flocks.

FEEDING ECOLOGY AND DIET
Feeds on nectar, fruits, pollen, and insects, which it may take from spider webs.

REPRODUCTIVE BIOLOGY
Male sings shrill song from perch on tall tree or low bush to defend territory, jerking tail as it does so, and feeds female during courtship. Lays two eggs in pear-shaped nest, January through August.

CONSERVATION STATUS
Not threatened. Common in most of range but rare in Nepal and Bhutan.

SIGNIFICANCE TO HUMANS
None known. ◆

Scarlet-tufted sunbird
Deleornis fraseri

SUBFAMILY
Nectariniinae

TAXONOMY
Anthreptes fraseri Jardine and Selby, 1843, Bioko, Equatorial Guinea. Three subspecies.

Chalcoparia singalensis
☐ Resident

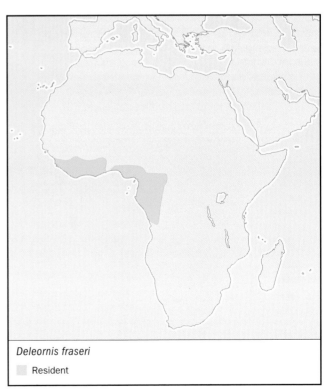

Deleornis fraseri
☐ Resident

OTHER COMMON NAMES
English: Fraser's sunbird; French: Souimanga de Fraser; German: Laubnektarvogel; Spanish: Nectarina Roja.

PHYSICAL CHARACTERISTICS
4.5–5.0 in (11.5–12.7 cm); 0.35–0.54 oz (10–15.3g). Plumage non-metallic, uniform bright green; sexes alike except for orange-yellow pectoral tufts on male only. Immature birds like adults but olive-green above and paler below.

DISTRIBUTION
Central and West Africa from Sierra Leone to Angola. *D. f. cameroonensis*: southern Nigeria to northwestern Angola; *D. f. fraseri*: Bioko, Equatorial Guinea; *D. f. idius*: Sierra Leone to Togo.

HABITAT
Forests, forest edges, and cocoa plantations.

BEHAVIOR
Forages like a warbler, seeking insects among leaves, rarely seen at flowers.

FEEDING ECOLOGY AND DIET
Feeds on small insects and spiders.

REPRODUCTIVE BIOLOGY
Males defend territories aggressively, displaying with shrill calls while jerking head and tail forward and exposing scarlet pectoral tufts. Young fed by both sexes, but nest and eggs unknown.

CONSERVATION STATUS
Not threatened. Common in Liberia, elsewhere locally common.

SIGNIFICANCE TO HUMANS
None known. ◆

Plain-throated sunbird
Anthreptes malacensis

SUBFAMILY
Nectariniinae

TAXONOMY
Certhia malacensis Scopoli, 1786, Malacca. Seventeen subspecies.

OTHER COMMON NAMES
English: Brown-throated sunbird, gray-throated sunbird; French: Souimanga á gorge brune; German: Braunkehl-Nektarvogel; Spanish: Nectarina de Garganta Descolorida.

PHYSICAL CHARACTERISTICS
5.5 in (14 cm); 0.26–0.48 oz (7.4–13.5 g). Metallic green from head to back with dark brown cheek, throat, and wings; purplish tail and yellow underparts.

DISTRIBUTION
Two main groups of subspecies. The *malacensis* group includes 15 subspecies occurring in various islands of the west and southwest Philippines, Borneo, the Celebes, Myanmar, Indochina, the Malay Peninsula, Thailand, and Indonesia. The *griseigularis* group includes two subspecies found in the eastern and northern Philippines.

HABITAT
Forest, woodland, mangroves, bamboos, coastal vegetation, swamps, coconut groves, and gardens from sea level up to 3,900 ft (1,200 m).

Anthreptes malacensis

  ▢ Resident

BEHAVIOR
Aggressive toward other small birds. Often in low bushes but feeds at all heights, sometimes in large groups or in mixed-species parties.

FEEDING ECOLOGY AND DIET
Feeds on nectar of citrus, mistletoes, and other flowers. Also takes fruits, insects, and spiders.

REPRODUCTIVE BIOLOGY
Males sing from prominent perches. Two eggs laid in pear-shaped nest suspended 3.3–42.6 ft (1–13 m) up, made of grass, plant fibers, and cobwebs. Slivers of bark on outside, lined with tree-cotton or moss. Young fledge two weeks after hatching.

CONSERVATION STATUS
Not threatened.

SIGNIFICANCE TO HUMANS
Pollinates *Nicolaia elatior*, the buds of which are used as spices. ◆

Green sunbird
Anthreptes rectirostris

SUBFAMILY
Nectariniinae

TAXONOMY
Certhia rectirostris Shaw, 1811–12, Ashanti, Ghana. Two subspecies.

OTHER COMMON NAMES
English: Yellow-chinned sunbird, gray-chinned sunbird, banded sunbird, banded green sunbird; French: Souimanga á bec droit; German: Goldband-Nektarvogel; Spanish: Nectarina Verde.

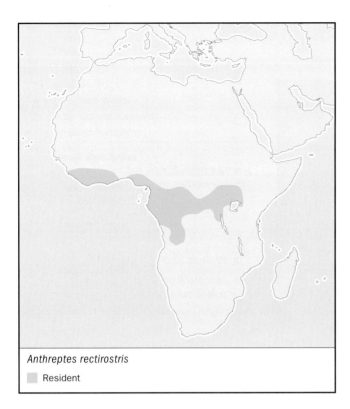

Anthreptes rectirostris
□ Resident

PHYSICAL CHARACTERISTICS
3.54–3.94 in (9–10 cm); 0.25–0.42 oz (7–12 g). Tiny, short-billed bird with lime-green head to back and darker wings. Yellow throat, orange band across chest, and light underparts.

DISTRIBUTION
A. r. rectirostris: Sierra Leone to Ghana; *A. r. tephrolaema*: southern Nigeria to Democratic Republic of Congo, southern Sudan, and western Kenya.

HABITAT
Upper stories of primary and secondary forest, gallery forest, forest plantations, and mountains.

BEHAVIOR
Occurs singly or in family groups of up to seven birds, usually more than 66 ft (20 m) up trees. Joins mixed-species flocks. Searches for insects on or below leaves, and along tree trunks and branches like a warbler. Sometimes catches flying insects like a flycatcher.

FEEDING ECOLOGY AND DIET
Feeds on small fruits, berries, seeds, insects, and spiders.

REPRODUCTIVE BIOLOGY
Males are territorial, singing from high perches. Species may be cooperative breeder as more than two adults feed young, but only female incubates. Two ovate eggs, gray with violet or gray-green markings, laid in nest suspended low from vine or 131 ft (40 m) up in a tree. Nest globular, made of fibers, lichens, or moss and lined with vegetable silk.

CONSERVATION STATUS
Uncommon, but not threatened.

SIGNIFICANCE TO HUMANS
None known. ◆

Collared sunbird
Hedydipna collaris

SUBFAMILY
Nectariniinae

TAXONOMY
Cinnyris collaris Vieillot, 1819, Gamtoos River, Cape Province, South Africa. Nine subspecies.

OTHER COMMON NAMES
French: Soiumanga à collier; German: Waldnektarvogel; Spanish: Nectarina Acollarada.

PHYSICAL CHARACTERISTICS
3.9–4.1 in (10–10.5 cm); male 0.19–0.39 oz (5.3–11.0 g); female 0.19–0.34 oz (5.4–9.7 g). Small and short-billed with green head to back, yellow belly with purplish dark stripe across breast.

DISTRIBUTION
H. c. collaris: eastern Cape Province to southern Kwazulu-Natal, South Africa; *H. c. djamdjamensis*: southwestern Ethiopia; *H. c. elachior*: coastal and inland Kenya, coastal Tanzania, Sudan and Somalia and Zanzibar; *H. c. garguensis*: western Kenya, southern Sudan, Uganda, Rwanda, Burundi, western Tanzania, Zambia, Angola; *H. c. hypodila*: Bioko, Equatorial Guinea; *H. c. somereni*: from southeastern Nigeria to north-western Angola, northern Democratic Republic of the Congo to southwestern Sudan, west of the River Nile; *H. c. subcollaris*: Senegal to the delta of the River Niger, Nigeria; *H. c. zambesiana*: Angola to southern Democratic Republic of the Congo, Zambia, Zimbabwe, southwestern Tanzania, Zanzibar, Malawi, Mozambique, and Botswana; *H. c. zuluensis*: northeastern Kwazulu-Natal, eastern Swaziland, southern Mozambique, and Zimbabwe.

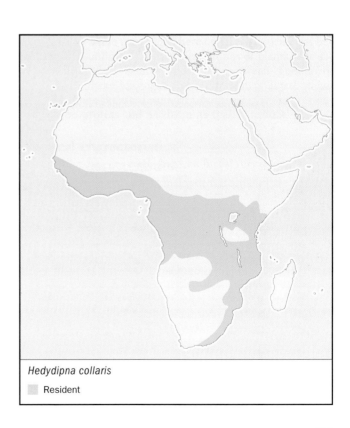

Hedydipna collaris
□ Resident

HABITAT
Varied. Occurs in forests, plantations, and swamps, but most common in open habitats such as clearings, savanna, thickets, and gardens.

BEHAVIOR
Commonly a member of mixed-species parties with other sunbirds, white-eyes, and warblers. Feeds acrobatically, and seen "anting."

FEEDING ECOLOGY AND DIET
Forages in low bushes but also up to 82 ft (25 m) high in forest canopy. Takes insects like a warbler does, and by aerial captures. Feeds mostly on insects, but also eats small spiders, snails, seeds, and fruits, and probes wide range of flowers for nectar.

REPRODUCTIVE BIOLOGY
Males defend territories with short whistling song and chase other males while making sounds with wing-flicks. May be polyandrous. Regular host of Klaas's cuckoo (*Chrysococcyx klaas*). Nest pear-shaped, made of grass, dead leaves, and cobwebs, sometimes decorated with lichen, bark, or flowers. Clutch one to four eggs with white background marked with various shades of grays and browns.

CONSERVATION STATUS
Not threatened; common and widespread.

SIGNIFICANCE TO HUMANS
None known. ◆

Purple-naped sunbird
Hypogramma hypogrammicum

SUBFAMILY
Nectariniinae

TAXONOMY
Hypogramma hypogrammicum S. Muller, 1843, Sumatra and Borneo. Five subspecies.

OTHER COMMON NAMES
English: Blue-naped sunbird; French: Souimanga strié German: Streifennektarvogel; Spanish: Nectarina de Nuca Azul.

PHYSICAL CHARACTERISTICS
6 in (15 cm); male 0.27–0.54 oz (7.8–15.2 g); female 0.34–0.48 oz (9.7–13.5 g). Medium-length bill with red eye and streaked yellowish underparts. Purple nape and rump with olive head, wings, and tail.

DISTRIBUTION
H. h. hypogrammicum: Sumatra and Borneo; *H. h. lisettae*: northern Myanmar, northern Thailand, northern and central Indochina, and western Yunnan; *H. h. mariae*: Cambodia and southern Indochina; *H. h. natunense*: northern Natuna Islands; *H. h. nuchale*: southern Myanmar, southern Thailand, and peninsular Malaysia.

HABITAT
Forests, plantations, swamps, and gardens.

BEHAVIOR
Usually keeps under cover, feeding up to 16 ft (5 m) high in trees. Fans and flicks tail.

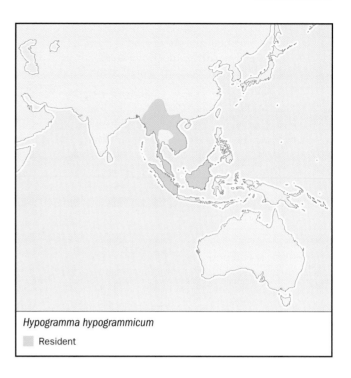

Hypogramma hypogrammicum
▨ Resident

FEEDING ECOLOGY AND DIET
Takes nectar from gingers and other plants. Also feeds on small fruits, seeds, small insects, and spiders.

REPRODUCTIVE BIOLOGY
Lays two or three whitish eggs with lilac wash and gray and black lines and blotches in ball- or pear-shaped shaped nest made of grass or of bark, lichen, moss, and leaves, held together with cobwebs.

CONSERVATION STATUS
Not threatened. Common in parts of Laos, Sumatra, and Borneo, but uncommon elsewhere.

SIGNIFICANCE TO HUMANS
None known. ◆

Reichenbach's sunbird
Anabathmis reichenbachii

SUBFAMILY
Nectariniinae

TAXONOMY
Nectarinia reichenbachii Hartlaub, 1857.

OTHER COMMON NAMES
French: Souimanga de Reichenbach; German: Reichenbach-nektarvogel; Spanish: Nectarina de Reichenbach.

PHYSICAL CHARACTERISTICS
4.7–5.5 in (12–14 cm); 0.35–0.46 oz (9.8–13 g). Metallic blue from crown to breast, with grayish belly and yellow vents. Upperparts olive-brown with gray band at tail tip.

DISTRIBUTION
Liberia to Angola.

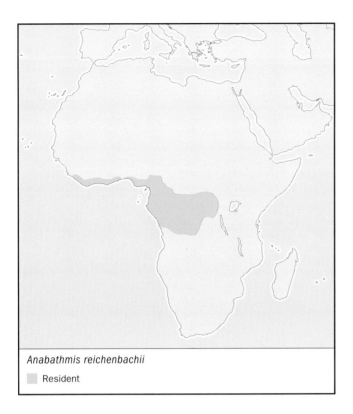

Anabathmis reichenbachii
▨ Resident

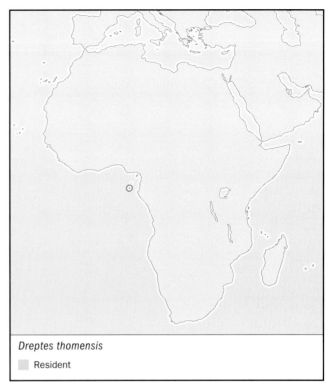

Dreptes thomensis
▨ Resident

HABITAT
Coastal zones in West Africa, but penetrating far inland in central Africa range, especially in Democratic Republic of the Congo where found along sides of rivers.

BEHAVIOR
Forages in vegetation overhanging water. Also catches insects in mid-air, returning to same perch where may dally for long periods.

FEEDING ECOLOGY AND DIET
Takes insects and probes flowers, including mistletoes, for nectar.

REPRODUCTIVE BIOLOGY
Defends territories all year, singing complex, high-pitched, jangling song from prominent perch. Lays single light-brown, speckled egg in small nest made of grass, leaves, and fibers, held together by cobwebs and suspended above water.

CONSERVATION STATUS
Not threatened. Common in Cameroon and Republic of the Congo but scarce elsewhere.

SIGNIFICANCE TO HUMANS
None known. ◆

São Tomé sunbird
Dreptes thomensis

SUBFAMILY
Nectariniinae

TAXONOMY
Nectarinia thomensis Barbosa du Bocage, 1889, St. Miguel, São Tomé.

OTHER COMMON NAMES
English: Giant sunbird, São tomé giant sunbird, dusky são tomé sunbird; French: Souimanga de são tomé; German: Riesennektarvogel; Spanish: Nectarina de Santo Tomé.

PHYSICAL CHARACTERISTICS
Male 7.9–9 in (20–23 cm), female 7.1–7.5 in (18–19 cm); male 0.92 oz (26 g), female 0.63–0.67 oz (18.0–18.9 g). Largest of the sunbirds, with long bill and tail feathers. Dark purplish upperparts with brown belly and breast. White-tipped tail.

DISTRIBUTION
São Tomé.

HABITAT
Montane forest, secondary forest, scrub along streams, and cultivations.

BEHAVIOR
Usually occurs singly or in pairs, but up to seven birds may congregate at flowers. Constantly moves from plant to plant, and sometimes feeds on bark of trees like a treecreeper (*Certhia* sp.).

FEEDING ECOLOGY AND DIET
Gleans leaves for insects, hovers beneath leaves to feed, probes bark and flowers. Food consists of insects, nectar, and fruit pulp.

REPRODUCTIVE BIOLOGY
Territorial. Possibly polygynous. Two long white eggs with red spots, laid September through January in nest suspended from end of branch. Nest made of moss and plant fibers with small porch and long trailing "beard" of plant fibers below base.

CONSERVATION STATUS
Vulnerable. Restricted to forests of São Tomé and threatened by forest clearances.

SIGNIFICANCE TO HUMANS
None known. ◆

Orange-breasted sunbird
Anthobaphes violacea

SUBFAMILY
Nectariniinae

TAXONOMY
Certhia violacea Linnaeus, 1766, Cape of Good Hope, South Africa.

OTHER COMMON NAMES
English: Violet-headed sunbird, wedge-tailed sunbird; French: Souimanga orangé; German: Goldbrust-Nektarvogel: Spanish: Nectarina de Pecho Anaranjado.

PHYSICAL CHARACTERISTICS
Male 5.7–6.5 in (14.5–16.5 cm), female 4.9–5.3 in (12.5–13.5 cm); male 0.32–0.4 oz (9–11.3 g), female 0.3–0.34 oz (8.6–9.7 g). Head metallic green with brown upperparts, orange breast, and blue band across base of throat.

DISTRIBUTION
West Cape Province to Cape Town, South Africa.

HABITAT
Restricted to the fynbos of South Africa, where found in heathlands and protea stands, but also occurs in parks and gardens.

BEHAVIOR
Found singly or in pairs during breeding season, but congregates in flocks of up to 100 birds in the nonbreeding period. Migrates from lower to higher altitudes during the southern summer in search of flowering plants. Tame.

FEEDING ECOLOGY AND DIET
Closely associated with *Erica* heaths, taking their nectar by probing into florets while clinging to stems. Also feeds on insects, often taking them in the air during spectacular forays from perches, and on spiders.

REPRODUCTIVE BIOLOGY
Timing of breeding linked to flowering by *Erica* heaths with peak activity in May through August. Males defend territories aggressively, attacking and chasing intruders. Nest unusual for a sunbird as is ball-shaped and placed in bush and not suspended. Only female builds, using rootlets, leaves from heaths, twigs and grass, with cobwebs as adhesive. One or two eggs, mostly white with brown markings, hatch two weeks after being laid. Both sexes feed young but female does two-thirds of the work. After fledging, parents tend young for three weeks.

CONSERVATION STATUS
Not threatened; common in appropriate habitat. Threatened by urbanization, agricultural developments, and fires.

SIGNIFICANCE TO HUMANS
None known apart from role in pollinating proteas, some of which are sold commercially, and heaths. ◆

Green-headed sunbird
Cyanomitra verticalis

SUBFAMILY
Nectariniinae

TAXONOMY
Certhia verticalis Linnaeus, 1790, Senegal. Four subspecies.

OTHER COMMON NAMES
English: Green-headed olive sunbird, olive-backed sunbird; French: Souimanga à tête verte; German: Grünkopf-Nektarvogel; Spanish: Nectarina de Cabeza Verde.

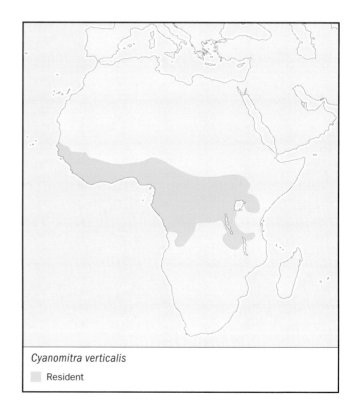

Anthobaphes violacea
▢ Resident

Cyanomitra verticalis
▢ Resident

PHYSICAL CHARACTERISTICS
5.1–5.7 in (13–14.5 cm); male 0.34–0.55 oz (9.7–15.5 g), female 0.38–0.55 oz (10.7–15.5 g). Long bill with shorter tail; head is actually metallic blue. Olive upperparts with gray breast to undertail.

DISTRIBUTION
C. v. boehndorffi: southern Cameroon through inland areas to Angola; *C. v. cyanocephala*: coasts from mainland Equatorial Guinea to northwest Democratic Republic of the Congo; *C. v. verticalis*: Senegal to Nigeria; *C. v. viridisplendens*: southern Sudan and northeast Democratic Republic of the Congo, east to Uganda and northwest Kenya, western Tanzania, northern Malawi, and northeastern Zambia.

HABITAT
Forests and well-wooded savanna, coastal habitats, plantations, and gardens.

BEHAVIOR
Usually forages high in canopy, sometimes in mixed-species flocks. Males may congregate in fruiting trees and defend feeding territories with aggressive displays, including showing pectoral tufts.

FEEDING ECOLOGY AND DIET
Searches leaves and tree bark for insects and also catches insects in flight. Eats small fruits and seeds and oil-palm sap, as well as nectar from wide variety of flowers.

REPRODUCTIVE BIOLOGY
Territorial. Nest globular, made of grass, bark, and fibers woven with cobwebs. Suspended, often over water, and some with long (1.6 ft; 0.5 m) streamers below base. Clutch of two white or pink eggs with dark dots. Both parents feed young.

CONSERVATION STATUS
Not threatened.

SIGNIFICANCE TO HUMANS
None known. ◆

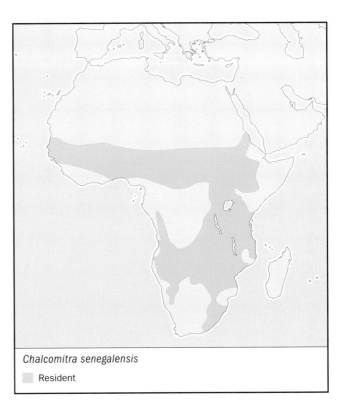

Chalcomitra senegalensis

▨ Resident

Scarlet-chested sunbird
Chalcomitra senegalensis

SUBFAMILY
Nectariniinae

TAXONOMY
Certhia senegalensis Linnaeus, 1766, Senegal. Five subspecies.

OTHER COMMON NAMES
English: Scarlet-breasted sunbird, scarlet-throated sunbird; French: Souimanga à poitrine rouge; German: Rotbrust-Glanzköpfchen; Spanish: Nectarina de Pecho Carmín.

PHYSICAL CHARACTERISTICS
5.1–5.9 in (13–15 cm); male 0.26–0.61 oz (7.5–17.2 g), female 0.24–0.54 oz (6.8–15.3 g). Mostly black, fading to brownish black. Bright red throat and chest with iridescent blue speckles. Glossy green forehead, crown, and chin.

DISTRIBUTION
C. s. acik: northern Cameroon to western and southern Sudan, Central African Republic, northeastern Democratic Republic of the Congo and northeastern Uganda; *C. s. cruentata*: southeast-

ern Sudan, Ethiopia, Eritrea, northern Kenya; *C. s. gutturalis*: northern Democratic Republic of the Congo, Angola, northern Namibia, northern Botswana, northern and eastern South Africa, Swaziland, Mozambique, Zimbabwe, Zambia, Malawi and Zanzibar; *C. s. lamperti* : Democratic Republic of the Congo, Burundi, Rwanda, Uganda, Kenya, southern Sudan, and western Tanzania; *C. s. senegalensis*: Senegal to Nigeria.

HABITAT
Wooded savanna, thorn scrub, gallery forests, inselbergs, coastal habitats, farmland, plantations, parks, and gardens.

BEHAVIOR
Partial migrant, with some birds moving with rain fronts; a banded bird traveled 224 mi (360 km) in Zimbabwe. Noisy and conspicuous, forming groups of up to 20 birds at flowering trees. Aggressive, apparently defends feeding territories as well as breeding domains. Territorial; male advertises presence from tall perch while singing agitated twittering song.

FEEDING ECOLOGY AND DIET
Feeds on nectar and insects such as ants, beetles, moths, and termites, as well as spiders. Hovers in front of leaves for insects and flowers for nectar.

REPRODUCTIVE BIOLOGY
Clutch of one to three whitish eggs with brown markings laid in domed nest with prominent porch of dried grasses above entrance hole, suspended from tree or human-made support. May be double- or triple-brooded, sometimes reuses same nest for each brood. Both parents feed young. Nests parasitized by Klaas's cuckoo, emerald cuckoo (*Chrysococcyx cupreus*), and by honeyguides, including greater honeyguide (*Indicator indicator*).

CONSERVATION STATUS
Not threatened.

SIGNIFICANCE TO HUMANS
Minor pest in Zimbabwe where it damages commercial crops of proteas. Also pollinates mistletoe pests of cocoa in West Africa. ◆

Crimson-backed sunbird
Leptocoma minima

SUBFAMILY
Nectariniinae

TAXONOMY
Cinnyris minima Sykes, 1832, Ghauts, Dukhun, India.

OTHER COMMON NAMES
French: Souimanga menu; German: Däumlingsnektarvogel; Spanish: Nectarina Chica.

PHYSICAL CHARACTERISTICS
3.5–3.9 in (8.9–9.9 cm); 0.14–0.21 oz (4–6 g). Small bird with brown wings and tail, lime-green forehead to nape, metallic purple throat and rump, reddish back and breast, and black around eye.

DISTRIBUTION
Western India from north of Bombay to hills of southern Kerala. Possibly also occurs in Sri Lanka.

HABITAT
Forest, plantations, and gardens.

BEHAVIOR
While singing squeaky song 5 to 10 seconds long, male twists from side to side. Active, acrobatic when feeding, hovers in front of flowers and feeds from upside-down poses. Defends feeding territories against conspecifics and against flowerpeckers. Makes seasonal short-distance movements.

FEEDING ECOLOGY AND DIET
Takes nectar from mistletoes, *Erythrina* trees, and other flowers. Also feeds on insects and spiders.

REPRODUCTIVE BIOLOGY
Clutch of two white eggs with reddish spots laid in neat pouch nest made of fibers, moss, and cobwebs suspended from bush.

CONSERVATION STATUS
Not threatened.

SIGNIFICANCE TO HUMANS
None apart from pollination roles. ◆

Malachite sunbird
Nectarinia famosa

SUBFAMILY
Nectariniinae

TAXONOMY
Certhia famosa Linnaeus, 1766, Cape of Good Hope, South Africa. Two subspecies.

OTHER COMMON NAMES
English: Yellow-tufted malachite sunbird, green sugarbird, long-tailed emerald sunbird; French: Souimanga malachite; German: Malachitnektarvogel; Spanish: Nectarina de Copete Amarillo.

PHYSICAL CHARACTERISTICS
Male 9.4–10.6 in (24–27 cm), female 5.1–5.9 in (13–15 cm); male 0.42–0.79 oz (12.0–22.5 g), female 0.32–0.62 oz (9.1–17.5 g). Mostly dark green with long bill and short tail with two elongated tail feathers. Blackish wings with small yellow patch.

Leptocoma minima
▨ Resident

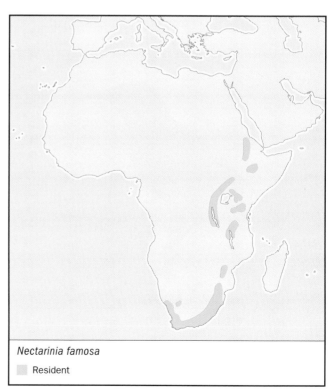

Nectarinia famosa
▨ Resident

DISTRIBUTION
N. f. cupreonitens: highlands of Eritrea, Ethiopia, southern Sudan, eastern Democratic Republic of the Congo, Tanzania, Uganda, Kenya, Zimbabwe, northern Malawi, and northern Mozambique; *N. f. famosa*: Namibia, South Africa, Lesotho, western Swaziland, and Zimbabwe.

HABITAT
In South Africa from coast to 9,200 ft (2,800 m) high in fynbos, karoo vegetation, alpine moorland, and gardens, but not in forest. Elsewhere found in open areas, moorland, bamboo zone, and at forest edges.

BEHAVIOR
Often seen singly but may congregate in flocks of more than 1,000 birds in patches of favorite food such as *Leonotis leonurus*. Aggressive, defending feeding areas against conspecifics involving physical duels in mid-air, other species of sunbirds, and wide variety of other birds. Can lower body temperature during cold nights.

Territorial. Males perform elaborate display flights, involving dive-bombing rivals from high up or twisting flights with wings stretched out. Song sometimes accompanied by pointing head upward and displaying pectoral tufts with wings half open. Courtship display by males involves drooping wings and whistling, followed by fast warbling and flapping of wings and showing of pectoral tufts, before vertical flight and landing on female to copulate.

FEEDING ECOLOGY AND DIET
Feeds on flowers to take nectar, especially from proteas, red-hot pokers, and giant lobelias. Also takes wide variety of insects, sometimes catching them in mid-air like a flycatcher.

REPRODUCTIVE BIOLOGY
Up to three greenish eggs with dark mottles laid in oval nest, often with porch of grass above entrance hole. Nest may be suspended or placed in a bush. Female incubates for two weeks. After two-week nestling period, both parents feed fledglings, who return to nest for roosting. May be double- or triple-brooded, sometimes reusing same nest. Parasitized by Klaas's cuckoo and by red-chested cuckoo (*Cuculus solitarius*).

CONSERVATION STATUS
Not threatened. Locally common in highland areas.

SIGNIFICANCE TO HUMANS
Pollinator of proteas and other flowers. ◆

Golden-winged sunbird
Drepanorhynchus reichenowi

SUBFAMILY
Nectariniinae

TAXONOMY
Drepanorhynchus reichenowi Fischer, 1884, Lake Naivasha, Kenya. Three subspecies.

OTHER COMMON NAMES
English: Golden-wing sunbird; French: Souimanga à ailes dorées; German: Gelbschwanz-Sichelhopf; Spanish: Nectarina de Alas Doradas.

PHYSICAL CHARACTERISTICS
Male 6.3–9.4 in (16–24 cm), female 5.5–5.9 in (14–15 cm); male 0.45–0.62 oz (12.8–17.5 g), female 0.39–0.56 oz (11–15.9 g).

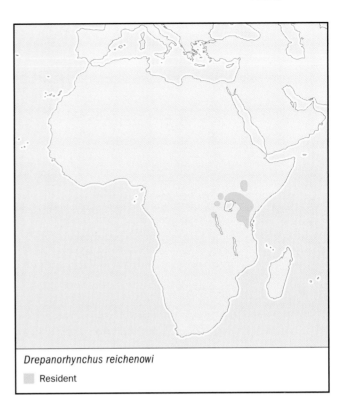

Drepanorhynchus reichenowi

■ Resident

Mostly chestnut-brown with decurved bill and yellow-golden wing and lateral tail feathers. Two elongated feathers extended from tail.

DISTRIBUTION
D. r. lathburyi: northern Kenya on isolated mountain ranges; *D. r. reichenowi*: southern and western Uganda, western and central Kenya, mountains in northeastern Tanzania; *D. r. shelleyi*: highlands of Democratic Republic of the Congo northwest of Lake Tanganyika.

HABITAT
Grasslands, clearings, forest edges, and gardens in highlands above 3,900 ft (1,200 m).

BEHAVIOR
Altitudinal migrant, following flowering seasons. Sometimes forms large feeding flocks at nectar sources. Has groove in crown that collects pollen. Territorial; can obtain three-quarters of energy needs from within territory. Song consists of short twittering and high-pitched "chi-chi-chi."

FEEDING ECOLOGY AND DIET
Associated in particular with lion's claw flower (*Crotalaria agatiflora*) and *Leonotis* spp. Feeds from aloes and jacarandas, also on insects that are sometimes caught in flight.

REPRODUCTIVE BIOLOGY
Single white egg, mottled with gray-brown, laid in globular nest made of grass and other plant material with porch over entrance hole.

CONSERVATION STATUS
Not threatened.

SIGNIFICANCE TO HUMANS
None known. ◆

Olive-bellied sunbird

Cinnyris chloropygius

SUBFAMILY
Nectariniinae

TAXONOMY
Nectarinia chloropygia Jardine, 1842, Aboh, River Niger, Nigeria.
Four subspecies.

OTHER COMMON NAMES
French: Souimanga à ventre olive; German: Olivbauch-Nektar-
vogel; Spanish: Nectarina de Vientre Olivo.

PHYSICAL CHARACTERISTICS
4.1–4.3 in (10.5–11.0 cm); male 0.17–0.28 oz (4.7–8.0 g), female 0.18–0.26 oz (5–7.5 g). Glossy dark green forehead to
back and throat; scarlet breast, brown wings and tail, and olive
belly.

DISTRIBUTION
C. c. bineschensis: southwestern Ethiopia; *C. c. chloropygius*:
southeastern Nigeria to Angola; *C. c. kempi*: Senegal to southwestern Nigeria; *C. c. orphogaster*: Congo River basin and
northeastern Angola through Democratic Republic of the
Congo to Burundi, southern Sudan, Uganda, Kenya, and
northwestern Tanzania.

HABITAT
Lower levels of trees and bushes at edges of forests and in
clearings, plantations, mangroves, farmland, well-wooded savanna, parks, and gardens.

BEHAVIOR
Noisy, active, tame birds with a rapid flight. Territorial and
aggressive to other sunbirds.

FEEDING ECOLOGY AND DIET
Commonly feeds on flowers of *Hibiscus* spp., bougainvillea, and
other garden flowers, but also attracted to banana, cassava, and
coffee flowers; takes small insects, spiders, and seeds.

REPRODUCTIVE BIOLOGY
One to three white or gray eggs laid in untidy oval nest made
from grass, dead leaves, and bark, and suspended from bush or
palm. Parasitized by emerald cuckoo and by Cassin's honeyguide (*Prodotiscus regulus*).

CONSERVATION STATUS
Not threatened.

SIGNIFICANCE TO HUMANS
None known. ◆

Greater double-collared sunbird

Cinnyris afra

SUBFAMILY
Nectariniinae

TAXONOMY
Certhia afra Linnaeus, 1766, Cape of Good Hope, South
Africa. Two subspecies.

OTHER COMMON NAMES
English: Larger double-collared sunbird, red-breasted sunbird;
French: Souimanga à plastron rouge; German: Grosser Hals-
band-Nektarvogel; Spanish: Nectarina de Dos Collares
Grande.

PHYSICAL CHARACTERISTICS
4.5–4.9 in (11.5–12.5 cm); male 0.32–0.64 oz (9.0–18.0 g), female 0.29–0.49 oz (8.1–14.0 g). Similar in coloring to olive-

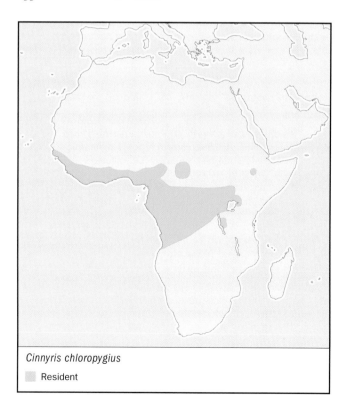

Cinnyris chloropygius
▨ Resident

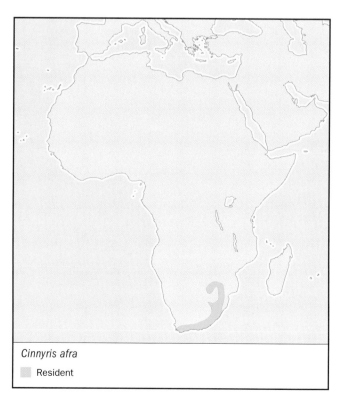

Cinnyris afra
▨ Resident

bellied sunbird, with a longer bill, larger size, and purplish bands cross throat and rump.

DISTRIBUTION
C. a. afra: Cape and Western Cape Provinces of South Africa; *C. a. saliens*: Eastern Cape, Free State, Northern, and Kwazulu-Natal Provinces of South Africa, Lesotho, and western Swaziland.

HABITAT
Open scrubland, plains, protea moorland, fynbos, parks, gardens, and forest edges.

BEHAVIOR
Aggressive with both sexes chasing each other. Males may fan tail and display pectoral tufts, and they sing to one another when perched close and swing bodies from side to side, while pointing head skyward. Displaying birds sometimes swing upside-down on perch. Bathe in birdbaths and attack reflections in windows. Males perform elaborate courtship behaviors, bobbing heads up and down and swaying sideways; also indulge in display flights.

FEEDING ECOLOGY AND DIET
Takes nectar from flowers such as aloes, proteas, and figs, but also feeds on juices from figs and grapes, insects, and spiders. May take latter from their webs in hovering flight. Catches insects on the wing.

REPRODUCTIVE BIOLOGY
Monogamous, but extra-pair copulations probably common, judging by frequency of cloaca-pecking. Female builds nest using spider webs to hold together grasses, bark twigs, rags, feathers, wool, and other debris into oval shape. Nest lined with feathers and wool, decorated with large leaves, lichen, and even cloth, and either placed in bush or suspended. Two heavily marked whitish eggs are laid at any time of year and incubated by female only for two weeks. Nestlings cared for by both parents for two weeks. Fledglings return to nest to roost for first few nights. May be triple-brooded. Parasitized by Klaas's cuckoo.

CONSERVATION STATUS
Not threatened.

SIGNIFICANCE TO HUMANS
None known. ◆

Purple sunbird
Cinnyris asiaticus

SUBFAMILY
Nectariniinae

TAXONOMY
Certhia asiatica Latham, 1790, Gurgaon, India. Three subspecies.

OTHER COMMON NAMES
French: Souimanga Asiatique; German: Purpurnektarvogel; Spanish: Nectarina Asiática.

PHYSICAL CHARACTERISTICS
4–4.5 in (10.1–11.4 cm); male 0.24–0.39 oz (6.9–11.0 g), female 0.17–0.35 oz (5–10 g). Medium-sized decurved bill. Male all dark, iridescent during breeding. Female brown and yellow.

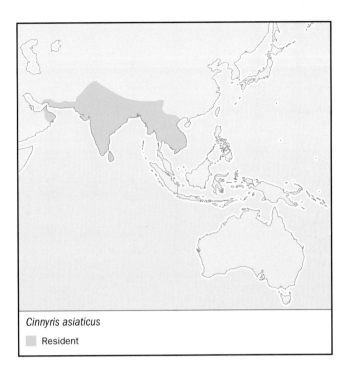

Cinnyris asiaticus
■ Resident

DISTRIBUTION
C. a. asiaticus: India south of Himalayas, except northwest and north, and Sri Lanka; *C. a. brevirostris*: southeastern Oman, southern Iran, Baluchistan, and Pakistan, to western India at Rajasthan and western Gujarat; *C. a. intermedius*: India in northern Andhra Pradesh, Orissa, Assam, and Bangladesh to Indochina and southern Yunnan.

HABITAT
Deciduous forest, thorn-scrub, farmland and gardens up to 7,875 ft (2,400 m) in hills.

BEHAVIOR
Aggressive, active, and noisy. Probes flowers including mistletoes, and catches insects like a flycatcher. In India migrates northwards March through April, returning August through September. Also altitudinal migrant, traveling up after breeding. Migratory in Oman.

Male displays to female with slightly open wings to expose pectoral tufts, raises head and flutters while singing excited "cheewit-cheewit" song.

FEEDING ECOLOGY AND DIET
Mistletoe fruits, nectar, grapes, and small insects.

REPRODUCTIVE BIOLOGY
Clutch of one to three grayish-white, streaked chocolate, eggs laid in oblong, purse-shaped nest made of grass, fibers, leaves, and cobwebs. Nest sometimes decorated with caterpillar droppings, bark, and other debris and is usually suspended. Only female incubates, but both parents feed young. Parasitized by plaintive cuckoo (*Cacomantis merulinus*).

CONSERVATION STATUS
Not threatened.

SIGNIFICANCE TO HUMANS
Has pest status in grape-growing areas of India as it pierces the fruit and sucks out juices. ◆

Olive-backed sunbird
Cinnyris jugularis

SUBFAMILY
Nectariniinae

TAXONOMY
Certhia jugularis Linnaeus, 1766, Philippines. Twenty-one sub-species.

OTHER COMMON NAMES
English: Yellow-bellied sunbird, yellow-breasted sunbird, black-breasted sunbird, black-throated sunbird; French: Souimanga à dos vert; German: Grünrücken-Nektarvogel; Spanish: Nectarina de Lomo Olivo.

PHYSICAL CHARACTERISTICS
4.5 in (11.4 cm); male 0.24–0.37 oz (6.7–10.5 g), female 0.21–0.32 oz (6.0–9.1 g). Dull olive-brown upperparts with contrasting yellow underparts. Metallic forehead, throat, and upper breast. All underparts bright yellow in females.

DISTRIBUTION
Myanmar, Thailand, Indochina, Malaysia, southeastern China, Philippines, New Guinea, and northeastern Australia.

HABITAT
Scrub, mangroves, forest, woodland, farmland, plantations, and gardens.

BEHAVIOR
Tame but restless. Often in mixed-species groups. Aggressive. Male displays underneath female, exposing black breast and pectoral tufts while moving head from side to side and calling.

FEEDING ECOLOGY AND DIET
Often feeds low down. Sometimes hovers in front of flowers, leaves, and cobwebs to take nectar, insects, and spiders respectively. Also eats small fruits and pollen.

REPRODUCTIVE BIOLOGY
Oval, purse-shaped nest with hooded side entrance built by female of grass, moss, lichens, and other vegetable matter. One to three grayish eggs incubated for two weeks. Young fledge after further two weeks. In Australia, parasitized by Gould's bronze cuckoo (*Chrysococcyx russatus*).

CONSERVATION STATUS
Not threatened.

SIGNIFICANCE TO HUMANS
Sometimes nests near or on houses, otherwise none known. ◆

Gould's sunbird
Aethopyga gouldiae

SUBFAMILY
Nectariniinae

TAXONOMY
Cinnyris gouldiae Vigors, 1831, Simla-Almora District, Himalayas. Four subspecies.

OTHER COMMON NAMES
English: Mrs. Gould's sunbird; French: Souimanga de Gould; German: Gouldnektarvogel; Spanish: Nectarina de la Gould.

PHYSICAL CHARACTERISTICS
4.3 in (11 cm), but male's tail may be 1.75 in (4.5 cm) longer; male 0.23–0.28 oz (6.5–8.0 g), female 0.14–0.21 oz (4.0–6.1 g). A glossy purple head and tail; red back with two stripes to the bill on each side. Wings dull brown with yellow underparts and rump.

DISTRIBUTION
A. g. annamensis: southern Vietnam, southern Laos, and Thailand; *A. g. dabryii*: eastern Nagaland, west central and southern

Cinnyris jugularis
▨ Resident

Aethopyga gouldiae
▨ Resident

China, southeastern Tibet, Manipur, Myanmar; *A. g. gouldiae*: Himalayas from Sutlej Valley to Aruchanel Pradesh and southeastern Tibet; *A. g. isolata*: south of River Brahmaputra in northern Assam, Ngaland, Manipur, and south to Chittagong Hills and northwestern Myanmar.

HABITAT
Highlands. Coniferous forest, oaks, scrub jungle, and rhododendrons.

BEHAVIOR
Energetic but shy.

FEEDING ECOLOGY AND DIET
Takes nectar from mistletoes and rhododendrons, and also eats insects and spiders. Drinks readily from pools.

REPRODUCTIVE BIOLOGY
Breeds as high as 14,000 ft (4,270 m). Clutch of two or three white eggs with small reddish brown marks laid mid-March to August. Nest oval and composed of grass, cobwebs, moss, fibers, and other vegetable matter, lined with down, and suspended from fern or low bush. In India parasitized by Asian emerald cuckoo (*Chrysococcyx maculatus*).

CONSERVATION STATUS
Not threatened; but uncommon.

SIGNIFICANCE TO HUMANS
None known. ◆

Arachnothera chrysogenys

▨ Resident

DISTRIBUTION
A. c. chrysogenys: southern Myanmar, southern Thailand, peninsular Malaysia, western Borneo, Java, and Sumatra; *A. c. harrissoni*: eastern Borneo.

HABITAT
Tops of trees in primary forest, secondary forests, plantations, villages, and gardens. Found up to 5,900 ft (1,800 m).

BEHAVIOR
Forages singly or in pairs. Sometimes hovers and also hangs upside-down when feeding.

FEEDING ECOLOGY AND DIET
Feeds on nectar, pollen, seeds, small fruits, insects, and spiders.

REPRODUCTIVE BIOLOGY
Nest is a neatly woven structure of vegetable matter sewn under banana leaf or palm frond, sometimes with long entrance tunnel. Inner cup, where eggs laid, made of finer material than rest of nest. Clutch two white eggs with black lines and speckles laid February to September.

CONSERVATION STATUS
Scarce, but not threatened.

SIGNIFICANCE TO HUMANS
None known. ◆

Yellow-eared spiderhunter
Arachnothera chrysogenys

SUBFAMILY
Arachnotherinae

TAXONOMY
Nectarinia chrysogenys Temminck, 1826, Bantam District, Java. Two subspecies.

OTHER COMMON NAMES
English: Lesser yellow-eared spiderhunter; French: Arachnothère à joues jaunes; German: Gelbwangen-Spinnenjäger: Spanish: Arañera de Orejas Amarillas Menor.

PHYSICAL CHARACTERISTICS
7 in (17.8 cm). Long-billed bird with greenish head fading to brown upperparts and yellow underparts. Yellow ear patch and ring around eye.

Resources

Books

Cheke, R. A., C. F. Mann, and R. Allen. *Sunbirds: A Guide to the Sunbirds, Flowerpeckers, Spiderhunters, and Sugarbirds of the World.* New Haven: Yale University Press, 2001.

Fry, C. H., S. Keith, and E. K. Urban, eds. *The Birds of Africa.* Vol. 6. London: Academic Press, 2000.

Sibley, C. G., and B. L. Monroe. *Distribution and Taxonomy of Birds of the World.* New Haven: Yale University Press, 1990.

Resources

Skead, C. J. *The Sunbirds of Southern Africa; Also the Sugarbirds, the White-Eyes, and the Spotted Creeper.* Cape Town: A. A. Balkema, 1967.

Stattersfield, Alison J., and David R. Capper, eds. *Threatened Birds of the World: The Official Source for Birds on the IUCN Red List.* Barcelona: Lynx Ediciones, 2001.

Periodicals

Delacour, J. A. "Revision of the Family Nectariniidae (Sunbirds)." *Zoologica* 29 (1944): 17–38.

Irwin, M. P. S. "The Genus *Nectarinia* and the Evolution and Diversification of Sunbirds: An Afrotropical Perspective." *Honeyguide* 45 (1999): 45–58.

Robert Alexander Cheke, PhD

White-eyes

(Zosteropidae)

Class Aves
Order Passeriformes
Suborder Passeri (Oscines)
Family Zosteropidae

Thumbnail description
A fairly uniform group of small perching birds, mostly with a white eye-ring and a brush-tipped tongue, distributed widely in the Old World except in Europe and in arctic and arid regions

Size
4–6 in (10–15 cm); 0.3–1.1 oz (8–31 g)

Number of genera, species
13 genera; 86 species

Habitat
Woodlands, forest edges, and gardens

Conservation status
Recently Extinct: 2 species; Critically Endangered: 6 species; Endangered: 1 species; Vulnerable: 14 species

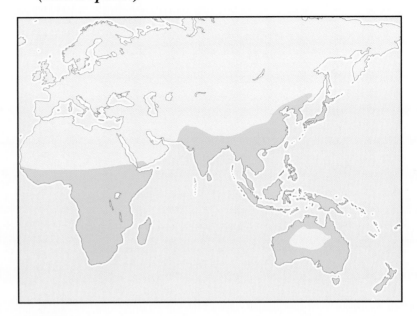

Distribution
Sub-Saharan Africa, Asia, New Guinea, Australia, and Oceania, introduced to Hawaii and Tahiti

Evolution and systematics

In 1766, Linnaeus named two species of white-eye, but in early years they were variously placed within wagtails (*Motacilla*), northern treecreepers (*Certhia*), flycatchers (*Muscicapa*), warblers (*Sylvia*), and flowerpeckers (*Dicaeum*). In 1826 Vigors and Horsfield created a new genus, *Zosterops*, for those species with a white eye-ring; and Gadow, who figured the tongue of an Australian species (*Z. lateralis*), placed the genus in Meliphagidae (honeyeaters) because of the brush-tipped tongue. In 1891 Sharpe elevated it to Zosteropidae, on the mistaken grounds that the tongue of *Zosterops* resembled that of a tit (*Parus*) and that it had no similarity to the brush tongue of Meliphagidae. In 1888 Newton also separated them as a single family. Recent molecular work places Zosteropidae under a passerine superfamily Sylvioidea and includes in this family the Endangered golden white-eye, *Cleptornis marchei*, of Mariana Islands and the Bonin white-eye, *Apalopteron familiare*, of Ogasawara Islands, which had been placed under Meliphagidae (honeyeaters) previously.

Those with a white eye-ring, from which the English name white-eye, the German name *Brillenvogel* (spectacle-bird), and the genus name *Zosterops* are derived, consist of four African continental, two Gulf of Guinea island, seven Indian Ocean island, and 49 Asia-Pacific species. This forms an extraordinarily uniform genus. For example, the Madagascar white-eye, *Zosterops maderaspatanus*, is almost indistinguishable from the New Guinea white-eye, *Z. novaeguineae*; the Annobon white-eye of Gulf of Guinea, *Z. griseovirescens*, is very similar to the Christmas Island white-eye from south of Java, *Z. na-*

talis; and the Australian yellow white-eye, *Z. luteus*, has the same plumage color as the East African subspecies of the white-breasted white-eye, *Z. abyssinicus flavilateralis*. Indeed, the relationships between the African and Asian species of *Zosterops* are yet to be clarified.

Physical characteristics

White-eyes have somewhat rounded wings with only nine functional primaries (the outermost primary is much reduced), and a brush-tipped tongue, quadrifid and fimbriated both at the sides and at the tip, showing a high degree of specialization for nectar feeding. In nearly all species, a white eye-ring appears soon after fledging, formed of minute silky white feathers. The ring is interrupted usually by the blackish lore. Iris color ranges from gray to brown and does not relate to age, sex, or race. The apposition of melanins and yellow carotenoids produces various shades of yellow-green, depending on the amount and distribution of these pigments and the structure of the feathers. The upperparts are green to greenish yellow, with gray upperback in some species, and the underparts are yellowish from throat to undertail coverts in some species or grayish to white in others. Flank color varies from light gray to dark brown. There are no seasonal changes in plumage color and the sexes are similar, though males tend to be larger and more brightly colored or darker than females in colder parts of Australia and New Zealand (*Z. lateralis*). Their short, thin bill is blackish, slightly decurved, and sharply pointed. The legs are grayish to brownish, the

Silvereye (*Zosterops lateralis*) at its nest in an orange tree. (Photo by Peter Slater. Photo Researchers, Inc. Reproduced by permission.)

first-year birds having darker colors and young fledglings and old birds having a pinkish color. Older birds (older than 5 years) molt into longer wing and tail feathers.

Continental species typically have short wings (2.2–2.6 in [55–65 mm]) and weigh 0.3–0.5 oz (9–15 g), with a cline following Bergmann's rule (larger in higher latitudes), but the derived island species and races tend to increase bill, leg, and body sizes. For example, in the eastern races of silvereyes (*Z. lateralis*) in Australia and New Zealand, a comparison between local populations indicates that the South Island population of New Zealand is the largest and the northern Australian population the smallest. At northern sites the winter populations are diluted by large southern migrants, except on wooded islands of southern Great Barrier Reef, where the Capricorn race (*Z. lateralis chlorocephalus*) is resident and much larger (body weight 0.49–0.53 oz [14–15 g]) than the mainland races (0.3 oz [10 g]) that migrate. Another genus of this family, *Lophozosterops*, with six species, mostly mountain birds of Indonesian and Philippine islands (Java, Bali, Sumbawa, Flores, Sulawesi, Seram, and Mindanao), contains not only very large species (wing greater than 2.8 in [70 mm]) but also atypical forms. They have a gray or brown crown, which is often striped or has small white spots. One species, *Lophozosterops dohertyi*, has a crest. In the giant white-eye, *Megazosterops palauensis*, of Palau island, the wing length reaches 3.3 in (85 mm) and weighs up to 1.1 oz (31 g). In the mountain blackeye, *Chlorocharis emiliae*, of northern Borneo and Bonin white-eye, the feathers around the eye are black.

Distribution

White-eyes are distributed widely in sub-Saharan Africa and the Asia-Pacific region. In Africa four polytypic species occur on the continent, with one species extending its range to Yemen. The Gulf of Guinea islands support an endemic genus and several species. In the Indian Ocean white-eyes are found in Mauritius, Réunion, Comoros, Aldabra, Madagascar, and Seychelles. On the Asian continent they occur from

Pakistan and Afghanistan in the west to the Amur River in the north. In the Pacific they occur on Japanese islands in the north to New Zealand in the south. Practically all wooded islands of the Asia-Pacific region support white-eye populations, which extend eastward to Samoa. They have been introduced to Hawaii (*Zosterops japonica*), Tahiti (*Z. lateralis*), and Pulo Luar in the Cocos-Keeling Group (*Z. natalis*).

Habitat

White-eyes live in forest edges and canopy, and they frequent bushes in gardens. They occur from sea level (*Z. luteus* is associated with mangroves and coastal vegetation) up to 9,800 ft (3,000 m) of high mountains in Java (*Lophozosterops javanicus*). *Z. citrinellus* and *Z. lateralis chlorocephalus* live on many wooded coral cays that are too small to support a breeding population of other passerine birds. Other offshore islands are also colonized by white-eyes. In New Zealand, explorers and settlers noticed silvereyes (*Z. lateralis*) in the South Island in 1832 (Milford Sound) and 1851 (Otago), and eventually large numbers appeared in the North Island in the winter of 1856, when the species, not previously known to the Maori, was given the name *tauhou* (meaning stranger). This was indeed the Tasmanian race (*Z. lateralis lateralis*). Half a century later, Dr. Metcalfe of Norfolk Island, who regularly corresponded with the Australian Museum, wrote in 1904 that a new bird had just arrived on Norfolk Island and that he was sending the first two specimens to the museum. It turned out to be the brown-sided Tasmanian race, considered to have colonized there from New Zealand at that time. Today, we know that the Tasmanian race contains both migrants and residents and that migrants do not migrate every year. Occasionally they must also attempt dispersal across the ocean.

Behavior

White-eyes are highly social birds, often seen huddling together. Foraging, bathing (including bathing in dew on leaves), resting, and roosting activities are governed by social factors. Sunning is done individually. Allopreening (mutual preening) occurs between sexual pairs, parent-offspring, young siblings, and prospective partners in pair formation, which starts as early as one month of age. Silvereyes in Australia normally mate for life and remain in pairs in winter flocks. On Heron Island in the Great Barrier Reef, older birds tend to stay near their summer territories while younger birds range widely in flocks. They continue to improve their foraging skills until their third year. They exhibit ritualized forms of aggression, such as wing fluttering and bill clattering, leading to dominance hierarchies within flocks. Dominance contributes significantly to differential survival and reproduction among individuals. Courtship involves horizontal wing quivering and sometimes motions of nest building without nest material. Males sing for up to 20 minutes at dawn throughout the breeding season. Some also sing for a shorter period at dusk and sporadically through the day. Their melodious and rich warbles are similar across the different species. Males also have a soft courtship warble and both sexes have a long-drawn plaintive call and short calls. Incessant exchanges of high-pitched short calls that characterize migratory flocking and take-off may also be heard in the predawn sky during the migration. In addition, they have a distinctive alarm call, a roosting call, a soft huddling call, a begging call, variable aggressive calls, and a distress call.

Continental species in the temperate region migrate to warmer areas in winter. *Zosterops erythropleura*, a chestnut-flanked species, breeds along the lower Amur and Ussuriland in the high latitudes that no other species of the family has reached. It migrates a distance of 2,200 mi (3,500 km) to winter in mountain forests of Myanmar, Thailand, and southern China. In Australia, for a long time Sydney ornithologists were familiar with yellow-throated and fawn-flanked silvereyes in summer and gray-throated and chestnut-sided silvereyes in winter, and thought that silvereyes had different plumage in summer and winter, which was accepted as fact since this species had two molts a year, spring and autumn. Allen Keast found that only some members of the Sydney population had brown flanks in winter and that in the aviary individual birds did not change plumage color over an entire year. He also noted that the brown-flanked birds did not appear until late April, long after the molt of the local birds, and thus concluded that the gray-throated, brown-sided birds were migrants from Tasmania and did not breed locally in the Sydney area. Large-scale banding of silvereyes started in 1958, and soon one bird banded in Sydney became the first of many to be recovered across Bass Strait to prove the migration theory. Silvereyes are easy to trap and have become the most popular species among the Australian bird banders. Between 1953 and 1997, 285,345 silvereyes were banded, yielding long-distance recoveries of up to 1,000 mi (1,600 km) and longevity of up to 11 years.

Feeding ecology and diet

Silvereyes in Australia are highly flexible foragers. Foliage gleaning is the most common mode of foraging, but they also hawk, snap prey from a substrate (even small insects caught in spiders' webs), probe small clefts in clumps of leaves, bark, buds, flowers, and nests of other birds by forcefully opening the bill to widen the clefts in search of arthropod prey, and scavenge on the ground in tourist resort areas. They are the most common bird at bird tables (bird feeders) in New Zealand. Flocking in winter helps to locate sources of food in woodlands as well as to detect predators. They collect nectar with a brush-tipped tongue, peck succulent fruit, and swallow berries. They are known to disperse figs and other seeds of trees and shrubs.

Reproductive biology

In Africa and Australia white-eyes have a long breeding season, normally starting in September or October (triggered by summer rain) and lasting up to six months. The territory density of monogamous silvereyes varies from 3.4–8.7 per 25 acres (10 ha) in Canberra suburbs to 75 per 25 acres (10 ha) at the height of season on Heron Island. Nests and eggs are known from only about half the species. The nest is cup-

Yellow white-eye (*Zosterops senegalensis*) feeding a chick at its nest. (Photo by Peter Davey. Bruce Coleman Inc. Reproduced by permission.)

shaped and mostly made of plant fibers. It is usually slung in a slender fork under cover of vegetation at any height. Eggs are pale blue to bluish green and somewhat glossy (with spots in four species), measuring 0.55 by 0.43 in (14 by 11 mm) to 0.79 by 0.59 in (20 by 15 mm). Incubation starts when two eggs are laid out of a normal clutch of three (varies 1–5) and normally lasts 10 to 12 days. The nestling period is 11 to 15 days depending on food supply. Both parents take part in nest building, incubation, and feeding of young. Young are nidicolous (reared for a time in the nest) and weigh about 0.07 oz (2 g) when born. They are fed with insects when small, but fruit usually becomes a significant portion of their diet by fledging. Fledglings are fed for three weeks or more, and the second clutch usually starts while the first clutch fledglings are still fed. A new nest is constructed for each clutch and older pairs may produce up to five successful clutches in a season while some first-year birds may not breed at all.

Conservation status

Small isolated populations, particularly on islands, are under threat of extinction from the disappearance of habitats. Two recently Extinct species are both insular species: the robust white-eye, *Zosterops strenuus*, of Lord Howe Island became extinct between 1919 and 1938 as a result of predation by ship rats, which became abundant on the island and caused extinction of 14 endemic forms of birds on the island, and the white-chested white-eye, *Z. albogularis*, of Norfolk Island, which has not been seen since 1980 and is now considered extinct as a result of habitat destruction. Of six other species, which are Critically Endangered, four are on the islands of Comoro, Mauritius, Seychelles, and Mariana, and two are known from the mountains of Kenya. One species Endangered is an island species (Truk Islands), and 14 considered

Vulnerable are also mostly island birds. Population viability analysis of island birds shows that they are at risk of extinction in one hundred years even without destruction of habitats, if mortality increases greatly as a result of introduction of new predators or frequent severe storms. For example, the golden white-eye, *Cleptornis marchei*, of Northern Mariana Islands was very common until the brown tree snake, *Boiga irregularis*, was introduced to Saipan in early 1990s. This white-eye species is now nearly extinct.

Significance to humans

White-eyes are valued as cage birds for their songs in some Asian countries. Because they are difficult to breed in aviaries they are trapped legally or illegally in the wild each winter. The trappers and dealers must know how to tell the sex of the birds, as only the males sing. They are also considered pests in vineyards and orchards in Southern Africa and Australia, though they consume large quantities of aphids and other pest insects as well as soft fruit.

1. Japanese white-eye (*Zosterops japonicus*); 2. Seychelles gray white-eye (*Zosterops modestus*); 3. Cape white-eye (*Zosterops pallidus*). (Illustration by Wendy Baker)

Species accounts

Cape white-eye
Zosterops pallidus

TAXONOMY
Zosterops pallidus Swainson, 1838, Rustenburg, Transvaal, South Africa. Seven subspecies.

OTHER COMMON NAMES
English: African pale white-eye, pale white-eye; French: Zostérops du Cap; German: Kapbrillenvogel; Spanish: Ojiblanco Pálido.

PHYSICAL CHARACTERISTICS
Length, 3.9–5.1 in (10–13 cm); weight, 0.28–0.53 oz (8–15 g); wing, 2.0–2.7 in (52–68 mm); tail, 1.6–2.2 in (40–56 mm); tarsus, 0.6–0.8 in (15–20 mm); culmen, 0.4–0.6 in (9–15 mm). Underparts gray, upperparts pale green.

DISTRIBUTION
Southern Africa to Ethiopia.

HABITAT
Forests, woodlands, savanna, exotic plantations, and suburban gardens.

BEHAVIOR
Some migrate while others remain sedentary. Readily come to bird feeders.

FEEDING ECOLOGY AND DIET
Feed on insects, spiders, soft fruit, berries, nectar.

REPRODUCTIVE BIOLOGY
Nests at the start and end of the monsoon season, with a peak between September and December. Two to three eggs per clutch, incubated for 11–12 days, and nestlings fed for 12–13 days.

CONSERVATION STATUS
Not threatened. Abundant in woods and suburban areas.

SIGNIFICANCE TO HUMANS
A popular species among the bird banders of southern Africa. Band recoveries have begun to demonstrate the complex nature of movements, molt patterns, and longevity. ◆

Seychelles gray white-eye
Zosterops modestus

TAXONOMY
Zosterops modestus Newton, 1867, Mahé, Seychelles. Monotypic.

OTHER COMMON NAMES
English: Seychelles white-eye, Seychelles brown white-eye; French: Zostérops des Seychelles, German: Mahébrillenvogel; Spanish: Ojiblanco de Seychelles.

PHYSICAL CHARACTERISTICS
Length 3.9 in (10 cm). A dull olive-gray bird with obscure eye-ring.

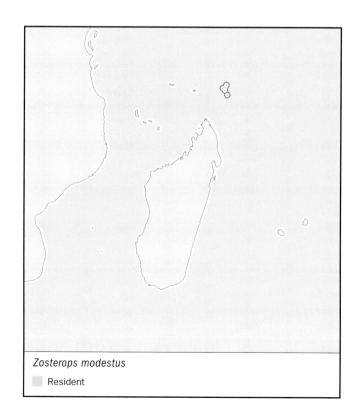

Zosterops pallidus
▢ Resident

Zosterops modestus
▢ Resident

DISTRIBUTION
Mahé and Conception Islands of Seychelles.

HABITAT
Mixed secondary forest between 1,000 and 2,000 ft (300–600 m), confined to about 2 mi² (5 km²) on Mahé.

BEHAVIOR
Very active in small flocks, giving contact calls frequently.

FEEDING ECOLOGY AND DIET
Feed on small insects and berries among shrubs, and search bark crevices of trees for arthropod prey.

REPRODUCTIVE BIOLOGY
Nests in Oct–Nov and Feb–Mar at start and end of monsoon season. Clutch size is two.

CONSERVATION STATUS
Critically Endangered. Fewer than 100 left on Mahé but common on the small island of Conception.

SIGNIFICANCE TO HUMANS
Has the conservation value of being the only species of white-eye left on Seychelles, as the Seychelles yellow white-eye, the Seychelles race of the chestnut-sided white-eye, *Z. mayottensis semiflavus*, on Marianne is extinct since about 1890. ◆

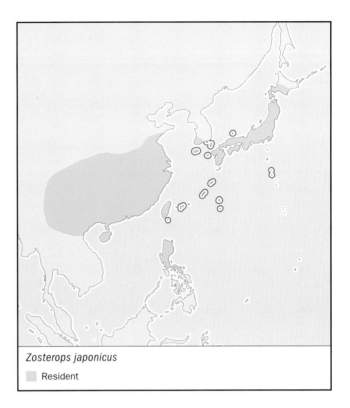

Zosterops japonicus

☐ Resident

Japanese white-eye
Zosterops japonicus

TAXONOMY
Zosterops japonicus Temminck & Schlegel, 1847, Nagasaki, Japan. 11 subspecies on different islands.

OTHER COMMON NAMES
French: Zosterops des Japon; German: Japanbrillenvogel; Spanish: Ojiblanco Japonés.

PHYSICAL CHARACTERISTICS
Length, 4.7 in (12 cm); weight, 0.4 oz (11 g); wing, 20.5–25.6 in (52–65 cm); tail, 13.4–18.1 in (34–46 cm); tarsus, 5.5–7.5 in (14–19 cm); culmen, 3.5–5.1 in (9–13 cm). Olive-green back and pale gray under, with lemon-yellow throat and undertail coverts.

DISTRIBUTION
Japanese islands, China, Taiwan, Hainan Island, and the Philippines. Introduced into Hawaii and Bonin Island.

HABITAT
Broadleaf evergreen forests and deciduous forests on lowlands and foothills of mountains.

BEHAVIOR
Form small flocks after breeding, sometimes move in mixed flocks, to hunt arthropod prey, and soft fruit and berries. Partial migrant, appearing in villages and suburban gardens in winter.

FEEDING ECOLOGY AND DIET
Apart from arthropod prey, they feed on fruit and nectar. Famous examples include nectar from camellia on warm temperate islands and fruit of ripe persimmon on main islands of Japan.

REPRODUCTIVE BIOLOGY
Breeds in spring, each pair holding a small nesting territory. The cup-shaped nest is hung from a fork in shrubs and 3–4 eggs are incubated for 11 days.

CONSERVATION STATUS
Not threatened. Common in most parts, but some isolated populations on islands are vulnerable.

SIGNIFICANCE TO HUMANS
Males were kept in cages for songs. One of the familiar birds of the countryside in Japan, often featured in various forms of art and literature. ◆

Resources

Books

Collar, N. J., M. J. Crosby, and A. J. Stattersfield. *Birds to Watch 2, The World List of Threatened Birds.* Cambridge, United Kingdom: BirdLife International, 1994.

Kikkawa, J. "Social Relations and Fitness in Silvereyes." In *Animal Societies: Theories and Facts,* edited by Y. Ito, J. L. Brown, and J. Kikkawa, 253–266. Tokyo: Japan Scientific Societies Press, 1987.

Periodicals

Brook, B. W., and J. Kikkawa. "Examining Threats Faced by Island Birds: A Population Viability Analysis on the Capricorn Silvereye using Long-Term Data." *Journal of Applied Ecology* 35 (1998): 491–503.

Mees, G. F. "A Systematic Review of the Indo-Australian Zosteropidae Parts I–III." *Zoologische Verhandelingen* 35 (1957): 1–204; 50 (1961): 1–168; 102 (1969): 1–390.

Resources

Moreau, R. E. "Variation in the Western Zosteropidae." *Bulletin of the British Museum (Natural History), Zoology* 4, no. 7 (1957): 311–433.

Prys-Jones, R. P. "Movements, Mortality and The Annual Cycle of White-eyes in Southern Africa." *Safring News* 14 (1985): 25–35.

Robertson, B. C., S. M. Degnan, J. Kikkawa, and C. C. Moritz. "Genetic Monogamy in the Absence of Paternity Guards: The Capricorn Silvereye, *Zosterops lateralis chlorocephalus*, on Heron Island." *Behavioral Ecology* 12 (2001): 666–673.

Jiro Kikkawa, DSc

Australian honeyeaters
(Meliphagidae)

Class Aves
Order Passeriformes
Suborder Passeri (Oscines)
Family Meliphagidae

Thumbnail description
Mostly small, but some tiny and others jay-sized; typically dull greenish, olive, or brown; often seen probing flowers for nectar with decurved bill; typically active, sometimes noisy and aggressive

Size
3–20 in (7–50 cm); 0.25–7 oz (7–200 g)

Number of genera, species
42 genera; 182 species

Habitat
Tropical, subtropical, and temperate rainforests, eucalyptus forest, monsoonal forest, woodlands with eucalyptus, casuarina, native pine, and acacia, semi-arid woodland and scrub, desert shrub-steppe, and coastal and upland heaths; generally absent from grassland; several species are common in parks and gardens

Conservation status
Near Threatened: 1 species, 4 subspecies; Endangered: 2 species; Critically Endangered: 2 species, 1 subspecies; Vulnerable: 5 species; Data Deficient: Many Indonesian species

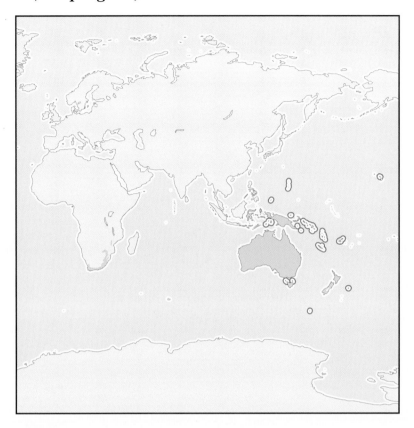

Distribution
Throughout Australia, New Guinea, Melanesia, Moluccas, and Lesser Sundas; west to Bali, Micronesia, New Caledonia, and New Zealand; through Polynesia to Hawaiian Islands; sugarbirds occur in southern Africa

Evolution and systematics

Honeyeaters belong to the Australo-Papuan lineage of passerines. Their nearest relatives are probably fairy wrens (Maluridae) and Australian warblers (Acanthizidae). The sugarbirds (*Promerops*) of South Africa have been classified as honeyeaters but are probably starlings. The dwarf and long-bill honeyeaters of New Guinea have been shown by molecular analysis to be more closely related to flowerpeckers (Dicaeidae) than to honeyeaters, whereas the Bonin Island honeyeater (*Apalopteron familiare*) is a white-eye (Zosteropidae). However, MacGregor's bird of paradise (*Macgregoria pulchra*) appears to be a honeyeater on the basis of mitochondrial DNA.

Relationships within the honeyeaters have not yet been fully unraveled. A few genera, such as *Myzomela*, *Meliphaga*, and *Lichenostomus*, contain numerous very similar species, whereas others are artificial and contain unrelated species. Honeyeaters of New Zealand and Hawaii do not have obvious close relatives in Australia or New Guinea, and perhaps some are not honeyeaters. Most genera with more than one

or two species are centered in Australia. However, *Lichmera* and *Philemon* have many species in Wallacea (Moluccas and Lesser Sundas). *Myzomela* is the most widespread genus, occurring from Australia to Sulawesi, Micronesia, and Fiji.

Physical characteristics

Many smaller honeyeaters are olive, green, or brown and often have yellow on the underparts, as an ear patch, or as neck plumes. Several smaller species are black and white. Almost half of the species are notably sexually dichromatic. Some of the larger species are gray, black, dark green, or streaked brown. Some honeyeaters have distinct juvenal plumages, although in many species the differences from adults are subtle.

Most honeyeaters have colored bare skin in the form of a discrete eyepatch, a somewhat swollen gape, elaborate wattles, knobs on the bill, or a bald head. Many of these features change in color or conspicuousness with age, and sometimes there are seasonal changes in relation to breed-

A brown-backed honeyeater (*Ramsayornis modestus*) at its nest. (Photo by Frithfoto. Bruce Coleman Inc. Reproduced by permission.)

ing. Bill and legs may also be distinctively colored. The bill ranges from short and straight to slightly decurved to quite long and markedly decurved. All honeyeaters have a brush tongue, which is quadrifid, with numerous bristles at the tip; this morphology is an adaptation to feeding on nectar and other sugary solutions. Digestion of sugary solutions is very rapid, and most sugar is absorbed before watery feces are expelled. Honeyeaters have strong legs and feet and often have sharp claws.

Distribution

One or more species occurs everywhere in Australia, except in extensive grassland with no trees and shrubs. The greatest diversity of species occurs in eastern Australia, where many habitats occur close together, and in the rainforests of New Guinea. Tasmania has four endemic species, all with fairly clearly identifiable relatives in mainland Australia. A few species that are widespread in New Guinea have a toehold in extreme north Queensland. Within New Guinea there is a changeover of species with altitude. Sulawesi has three hon-

eyeaters and Timor has six, with from one to four species elsewhere on each Lesser Sunda or Moluccan islands.

Two of the honeyeaters in New Zealand are quite widespread in native vegetation and less commonly in modified vegetation. The other species is restricted to offshore islands. New Caledonia, Bougainville and Kadavu Island in the Fiji group all have endemic honeyeaters. All of the main Hawaiian islands once had honeyeaters, but most are extinct.

The two species of sugarbirds are found in a limited area of southern Africa.

Habitat

Honeyeaters occur in subalpine scrub in New Guinea, through all forest types to the mangroves. In Australia three species are largely restricted to mangroves. It is not uncommon to find ten or more species at a location in forests and coastal heathlands. Woodlands, mallee, and other semi-arid scrubs can also be rich in species. Within forests and woodlands, most species occupy the canopy, with some of the more nectarivorous species feeding more in the shrub layer. Most of the Wallacean and Pacific island honeyeaters are found in rainforest or monsoonal woodland. Honeyeaters in many parts of their range have successfully colonized suburban areas, especially where native vegetation remains or has been replanted. African sugarbirds occupy fynbos (equivalent to coastal Australian heaths), which are dominated by the Proteaceae.

Behavior

Honeyeaters are rarely solitary and often occur in family groups or loose flocks. Miners occupy colonies that have a complex structure. Migratory species may occur in large flocks. Large aggregations gather at rich nectar sources such as ironbarks (e. g., *Eucalyptus sideroxylon*), banksias, and bottlebrushes (*Callistemon*). These groups are frequently noisy, with much chasing and displacement. Larger species such as wattlebirds and friarbirds often dominate such gatherings, with smaller species being displaced to less rich areas. Aggression is often displayed to other nectar feeders. While breeding and also when molting (which follows breeding), even aggressive species can become quiet and inconspicuous.

Whereas some honeyeaters are sedentary, most species show some seasonal movements; these may consist of local wandering outside the breeding season to visit flowering or fruiting trees and shrubs. Quite likely there are fairly sedentary and more mobile components to the populations of many species. A few species show clear migratory behavior with large flocks flying north in autumn and south in spring. Many honeyeaters, especially those of arid and semi-arid habitat, have been described as nomadic. Whereas their movements may be complex, they are probably not random but may follow fairly predictable routes. Where the birds settle, though, probably depends on the availability of nectar or other food sources.

The calls and songs of honeyeaters range from beautiful to harsh and grating. Smaller species have whistling calls and twittering songs. Medium-sized spiny-cheeked (*Acanthagenys rufogularis*) and striped (*Plectorhyncha lanceolata*) honeyeaters have attractive repertoires, whereas large wattlebirds emit harsh cackling and coughing calls. Tuis (*Prosthemadera novaeseelandiae*) of New Zealand are regarded as one of the finest singers in the world, while the friarbirds and forest honeyeaters of the Pacific have loud bugling and even comical calls.

Feeding ecology and diet

All species of honeyeaters consume varying amounts of nectar and invertebrates, especially insects. Honeyeaters visit a wide range of native and exotic flowers, with eucalyptus, banksias, bottlebrushes, grevilleas, and mistletoes being among the most popular. Sugarbirds favor proteas. Other sweet food sources include honeydew from bugs (Hemiptera), manna (sugary exudate from damaged foliage), and lerp (the sugary coating on scale insects of the family Psyllidae). Honeyeaters also consume sap exuding from scars on branches caused by gliding possums. Smaller honeyeaters may consume tiny insects that they capture in flight, and they also glean caterpillars and beetles from foliage. Strong-billed (*Melithreptus validirostris*) and white-eared honeyeaters (*Lichenostomus leucops*) frequently probe bark for insects and honeydew. Spiders are also taken, and more unusual foods include crustaceans and small lizards. A few larger species prey on eggs and nestling birds. Miners (*Manorina*) and tawny-crowned honeyeaters (*Phylidonyris melanops*) often forage on the ground and walk well. Fruit is a major food for honeyeaters in the wetter forests, especially in New Zealand and New Guinea.

Reproductive biology

Many, possibly most, honeyeaters are monogamous, although polygamy or a mixed mating system seems likely in species with great sexual dimorphism and in colonial species. Perhaps a third of species are cooperative breeders, ranging from occasional helpers to complex colonies.

Honeyeaters generally have long breeding seasons. Most Australian species have breeding seasons for six or more months. Breeding tends to peak in late winter to late spring (August to October) for many species. However, in areas where nectar is most abundant in winter, breeding may peak in July and August. Several of the friarbirds have a short breeding season, as do the arid-inhabiting black (*Certhionyx niger*) and pied (*C. variegatus*) honeyeaters. Some species show differences in breeding season in different parts of their range (typically earlier in the north than in the south) and from year to year.

The placement of honeyeater nests ranges from low bushes, sometimes almost on the ground, to the tops of tall trees. Most nests are carefully woven and placed in forks or suspended from foliage. Spider webs, plant down, animal hair, wool, feathers, artificial materials, and even human hair may be added to nests. The stitchbird (*Notiomystis cincta*) of New Zealand and at least one of the Hawaiian honeyeaters are hollow nesters. Eggs range in color from white to pale pink, with

A New Holland honeyeater (*Phylidonyris novaehollandiae*) feeding. (Photo by Michael Morcombe. Bruce Coleman Inc. Reproduced by permission.)

purple, red, brown, or black spots and blotches. The typical clutch size for most species is two eggs, but miners and friarbirds usually lay three or four eggs and some species lay only one. Eggs are probably laid at intervals of about 24 hours and are incubated, mostly by the female, for 12 to 17 days.

Both parents and often helpers feed the young; insects compose most of the diet, but the young of some species are fed nectar. The fledging period ranges from 11 to 20 days, but it may be as long as 32 days in the hollow-nesting stitchbird. Success rate of nests varies widely among species, years, and locations. Most failures are due to predation by corvids, currawongs (*Strepera*), butcherbirds (*Cracticus*), and a range of other birds, as well as snakes and introduced mammals. Honeyeaters themselves occasionally destroy eggs. Ants also attack nestlings, as do the parasitic maggots of bot flies (*Passeromyia*). Nests are sometimes blown down in storms or drenched by heavy rain. A range of cuckoos are brood parasites.

Conservation status

Many species of honeyeaters have declined due to the clearing of forests and woodlands or to the degradation of their habitat in other ways. Avian malaria, carried by introduced mosquitoes, contributed to the decline and extinction of the Hawaiian honeyeaters, whereas introduced mammals have affected the stitchbird and to a lesser extent the two other honeyeaters in New Zealand. The brown tree snake *Boiga irregularis*, from Australia has affected the Micronesian honeyeater (*Myzomela cardinalis*) on Guam.

Regent honeyeaters (*Xanthomyza phrygia*) have become rare in southeastern Australia, but habitat is being protected and replanted and a captive population has been established. Black-eared miners (*Manorina melanotis*) have hybridized with yellow-throated miners (*M. flavigula*), and few pure colonies of the former remain. However, release of captured birds into new areas of good habitat has proved successful.

Some of the larger honeyeaters have adapted well to fragmented habitat and to urban and suburban areas. Noisy (*M. melanocephala*) and bell miners (*M. melanophrys*) have become locally abundant and pose a threat to other honeyeaters and insectivores, which they drive out of their colonies.

Significance to humans

Honeyeaters are a conspicuous part of the avifauna, especially in Australia where many species are common in gardens and parks. Several species are regarded as minor pests of grapes and other fruit crops. A few of the larger honeyeaters have been hunted by humans. Feathers from the stitchbird and the Hawaiian honeyeaters were used in Polynesian cloaks. Bell and noisy miners are sometimes culled where they are perceived to be a threat to other birds. Honeyeaters are major pollinators of many native plants. They also disperse seeds, including some exotic weeds. They consequently play an important role in the ecosystems that they occupy.

1. Greater Sulawesi honeyeater (*Myza sarasinorum*); 2. Western spinebill (*Acanthorhynchus superciliosus*); 3. Bishop's oo (*Moho bishopi*); 4. Cape sugarbird (*Promerops cafer*); 5. Rufous-banded honeyeater (*Conopophila albogularis*); 6. Common smoky honeyeater (*Melipotes fumigatus*); 7. Tui (*Prosthemadera novaeseelandiae*); 8. New Holland honeyeater (*Phylidonyris novaehollandiae*); 9. Belford's melidectes (*Melidectes belfordi*) 10. Puff-backed meliphaga (*Meliphaga aruensis*). (Illustration by Emily Damstra)

1. Red myzomela (*Myzomela cruentata*); 2. Brown honeyeater (*Lichmera indistincta*); 3. Strong-billed honeyeater (*Melithreptus validirostris*); 4. Stitchbird (*Notiomystis cincta*); 5. Regent honeyeater (*Xanthomyza phrygia*); 6. Yellow-tufted honeyeater (*Lichenostomus melanops*); 7. Noisy friarbird (*Philemon corniculatus*); 8. Bell miner (*Manorina melanophrys*); 9. Striped honeyeater (*Plectorhyncha lanceolata*); 10. Red wattlebird (*Anthochaera carunculata*). (Illustration by Emily Damstra)

Species accounts

Red wattlebird
Anthochaera carunculata

TAXONOMY
Merops carunculata Shaw, 1790, Port Jackson, New South Wales, Australia.

OTHER COMMON NAMES
English: Wattled honeyeater, gillbird; French: Méliphage barbe-rouge; German: Rotlappen-Honigfresser; Spanish: Filemón Rojo.

PHYSICAL CHARACTERISTICS
14 in (35 cm); 4 oz (120 g). Female noticeably smaller than male. Buff underparts with white streaks and yellowish patch on belly. Upper feathers darker and white-tipped. White patch under eye with thin red band across chin.

DISTRIBUTION
Southern Australia, with possible gap across Nullarbor Plain; separate subspecies in southwest and Mount Lofty Ranges of South Australia, Kangaroo Island, and eastern Australia. Vagrant to New Zealand.

HABITAT
Woodland and open forest, typically with eucalyptus, mallee, heathland, parks, and gardens.

BEHAVIOR
Occur in pairs and family groups and sometimes loose flocks. Can be noisy and aggressive but are quiet while breeding. Male makes a harsh, raucous cough, to which female replies with a more musical *plew...plew...plew* call. Make other harsh rasping calls when nest or young threatened. Short distance migrant and nomad.

FEEDING ECOLOGY AND DIET
Take nectar from flowers, especially eucalyptus, but also mistletoes and a wide variety of shrubs. Also eat insects from foliage, bark, or the ground and by aerial capture. Eat fruit less commonly, and rarely take small reptiles and young birds.

REPRODUCTIVE BIOLOGY
Long breeding season, mostly August to December. Lay two eggs in a stick nest in a tree. Both adults feed young, which hatch at about 16 days and fledge after a further 16 days. Parasitized by pallid cuckoo (*Cuculus pallidus*).

CONSERVATION STATUS
Not threatened.

SIGNIFICANCE TO HUMANS
One of the few Australian passerines that was hunted for food, gillbird pie was once a favorite. They are sometimes shot when feeding on cultivated fruit. They are successful suburban birds. ◆

Striped honeyeater
Plectorhyncha lanceolata

TAXONOMY
Plectrorhyncha lanceolata Gould, 1838, New South Wales, Australia.

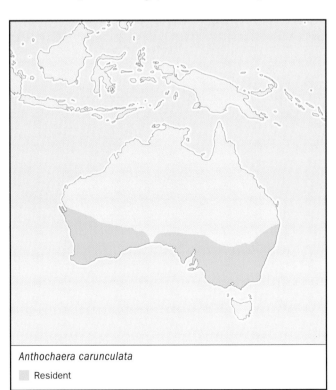

Anthochaera carunculata
☐ Resident

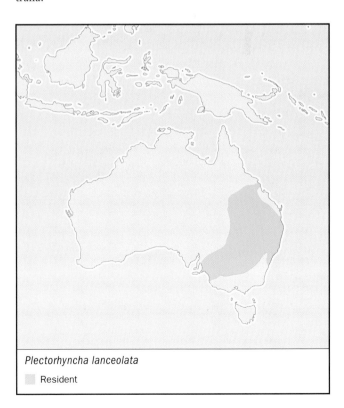

Plectorhyncha lanceolata
☐ Resident

OTHER COMMON NAMES
English: Lanceolated honeyeater; French: Méliphage lancéolé;
German: Strichelhonigfresser; Spanish: Pájaro Azúcar Gris.

PHYSICAL CHARACTERISTICS
8.5 in (22 cm); 1.4 oz (40 g). Cheek and forehead to nape is
dark with white stripes. Underparts a pinkish buff with grayish
upperparts and tail.

DISTRIBUTION
Eastern Australia, from mid-north Queensland to northern
Victoria and west to Yorke Peninsula, especially inland from
Great Dividing Range.

HABITAT
Riparian woodland with *Casuarina* and mallee and other semi-
arid woodlands with eucalyptus, acacia, and native pine.

BEHAVIOR
In pairs or small groups, emit an attractive whistling song.
Generally sedentary, but exhibit some, probably local, move-
ments.

FEEDING ECOLOGY AND DIET
Take nectar from eucalyptus, mistletoes, and other plants, and
occasionally eat fruits and seeds. Insects and spiders are
gleaned from foliage and bark or captured in the air.

REPRODUCTIVE BIOLOGY
Breed from August to January in nest suspended from drooping foliage, often near nests of gray butcherbird (*Cracticus
torquatus*). Clutch from two to five eggs (usually three). Both
parents apparently incubate and occasionally have helpers feeding young, which hatch at 16–17 days and fledge at 16–17
days. Parasitized by pallid cuckoo.

CONSERVATION STATUS
Not threatened.

SIGNIFICANCE TO HUMANS
Occasionally regarded as a pest at orchards. ◆

Noisy friarbird
Philemon corniculatus

TAXONOMY
Merops corniculatus Latham, 1790, Endeavour River, Queensland, Australia.

OTHER COMMON NAMES
English: Leatherhead, four-o'clock; French: Polochion criard;
German: Lärmlederkopf; Spanish: Filemón Gritón.

PHYSICAL CHARACTERISTICS
13 in (33 cm); 3.4–4.4 oz (96–122 g). Bare black head and upright knob on bill. White throat, breast, and tail, with darker
upperparts.

DISTRIBUTION
Eastern Australia from coast to edge of plains and along Murray River. Trans-Fly region of New Guinea. Separate subspecies in north Queensland and New Guinea.

HABITAT
Woodlands and open forests, coastal heathland, farmland,
parks and gardens.

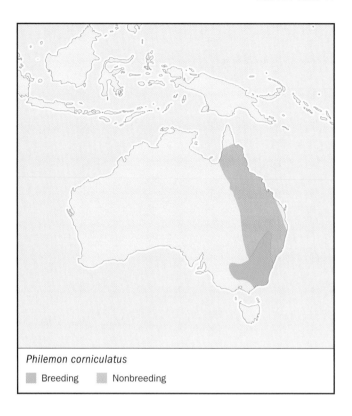

Philemon corniculatus
■ Breeding   ■ Nonbreeding

BEHAVIOR
Can be noisy and quarrelsome at flowering trees, but also quiet
and unobtrusive when breeding. Comical calls, like *four o'clock,
tobacco, chocka-lock*, etc. by males. Calls become more complex
when pairs chase around during breeding season. Pairs may
duet. Also emit softer calls and harsh alarm calls. May strike
predators, including humans, near nest.

FEEDING ECOLOGY AND DIET
Nectar of eucalyptus, mistletoes, or other plants; sometimes eat
native or cultivated fruit. Insects and spiders taken by gleaning
from foliage or capturing in flight, including large scarab beetles and cicadas. Exceptionally take eggs, young birds, and fish.

REPRODUCTIVE BIOLOGY
May breed from August to March, but season can be short in
any locality. Nests are large and conspicuous and globular,
made of bark, grass, and often wool and placed on outer
branches of eucalyptus and other trees or in saplings. Probably
only female builds nest and incubates. Usually lay three eggs
(occasionally two or four), which hatch after about 16 days.
Young fledge at 16–17 days.

CONSERVATION STATUS
Not threatened.

SIGNIFICANCE TO HUMANS
Sometimes shot at wineries and orchards when feeding on
fruit. ◆

Regent honeyeater
Xanthomyza phrygia

TAXONOMY
Merops phrygius Shaw, 1794, Sydney, New South Wales, Australia.

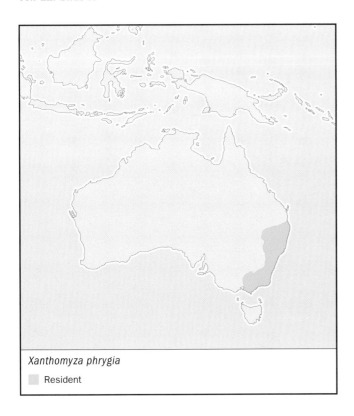

Xanthomyza phrygia

☐ Resident

OTHER COMMON NAMES
English: Warty-faced honeyeater; French: Mélephage régent; German: Warzenhonigfresser; Spanish: Pájaro Azúcar Real.

PHYSICAL CHARACTERISTICS
9 in (22.5 cm); 1.4–1.6 oz (39–45 g). Dark head with orange eye patch; dark wings and tail with yellowish tips; white underparts with black scaly pattern.

DISTRIBUTION
Southeastern Australia from near Brisbane to central Victoria. Vagrant into Adelaide region and in many parts of range. Main breeding populations are on northwest slopes of New South Wales, west of the Blue Mountains, and in northeast Victoria.

HABITAT
Open forests and woodlands, especially with ironbark, riparian woodland, coastal heathland, and tall eucalyptus forest.

BEHAVIOR
Often occur in small groups, formally in large flocks, and may roost communally. Active and sometimes aggressive at flowering trees, but also unobtrusive at times. Vocalizations include bubbling, tinkling, soft song, bill snaps, trilling, mewing, and sometimes mimicry of other honeyeaters. Display complex movements that involve some regularity, particularly to breeding areas, but also more nomadic outside breeding season.

FEEDING ECOLOGY AND DIET
Nectar of eucalyptus, especially ironbarks, mistletoes, banksia, grevillea, and other shrubs. Sometimes eat fruit as well as lerp and manna. Insects are taken from foliage and by sallying.

REPRODUCTIVE BIOLOGY
Breeding season July to January, but mostly September to November, occasionally at other times. May be loosely colonial. Nest of sticks and bark in high branches and forks of tall trees or in mistletoes. Clutch of two to three eggs are incubated for

14 days. Young are fed insects, nectar, and lerp and fledge at about 16 days.

CONSERVATION STATUS
Endangered. Populations have declined greatly in abundance during the twentieth century, with contraction in breeding range out of South Australia and western and central Victoria. Some critical habitats are protected or are being reestablished, but habitats and foods used in nonbreeding season are poorly known, which makes conservation difficult. Breed well in captivity, and releases of captive-bred birds into the wild have exhibited short-term success.

SIGNIFICANCE TO HUMANS
Formerly shot as a pest in orchards, but their striking coloration and endangered status make them high profile birds. The New South Wales town of Barraba has adopted the regent honeyeater as its emblem. ◆

Bell miner
Manorina melanophrys

TAXONOMY
Turdus melanophrys Latham, 1802, Port Jackson, New South Wales, Australia.

OTHER COMMON NAMES
English: Bellbird; French: Mélephage à sourcils noirs; German: Glockenhonigfresser; Spanish: Manorina Campanera.

PHYSICAL CHARACTERISTICS
7.3 in (18.5 cm); 0.9–1.25 oz (25–35 g). Mostly olive-green with darker wings and yellow-orange eye patch.

DISTRIBUTION
Southeastern Australia from north of Brisbane to Melbourne.

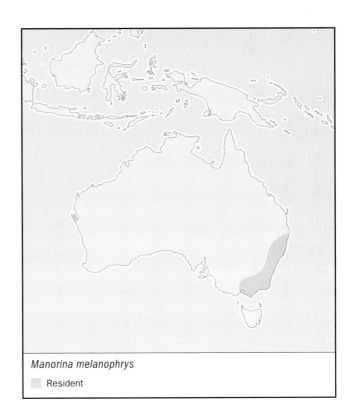

Manorina melanophrys

☐ Resident

HABITAT

Eucalyptus forests in the ranges and coastal plains, typically in wet gullies with a good shrub layer. Less commonly in eucalyptus woodlands.

BEHAVIOR

Highly colonial, with intense aggression keeping out most other small and medium-sized birds. Larger birds and mammals are mobbed. Colonies are easy to locate with the constant bell-like calls and chipping and chucking calls. Highly sedentary, with only occasional movements out of the colony.

FEEDING ECOLOGY AND DIET

Glean insects, especially psyllids and their lerp, from foliage. Less commonly forage on bark. Feed on nectar when available in their colony.

REPRODUCTIVE BIOLOGY

Coteries in colonies are groups of breeding pairs, among which there is strong cooperation. May breed at any time of year but mostly June to November in the north and August to January in the south. Nests are typically placed in the understory at 3–10 feet (1–3 m) high and are made of grass and twigs and are built only by the female. Usually two eggs (occasionally one or three), which are laid and incubated by the female for about 14 days. About half of feeding visits to the young are by the parents; the remainder are by numerous helpers, mostly male, some of which are breeding themselves. Young fledge at 12 days.

CONSERVATION STATUS

Not threatened.

SIGNIFICANCE TO HUMANS

Bellbirds' aggressive behavior means that they can deter threatened species from remaining in an area. ◆

Lichenostomus melanops

▨ Resident

Yellow-tufted honeyeater

Lichenostomus melanops

TAXONOMY

Turdus melanops Latham, 1801, Port Jackson, New South Wales, Australia.

OTHER COMMON NAMES

English: Black-faced honeyeater, golden-tufted honeyeater; French: Méliphage cornu; German: Gelbstirn-Honigfresser; Spanish: Melífago de Moño Amarillo.

PHYSICAL CHARACTERISTICS

6.5–8.3 in (16.5–21 cm); 0.8–1.3 oz (22–36 g). Black head with yellow crest to nape and chin to throat. Grayish underparts and dark secondary wing feathers fading to yellowish primaries.

DISTRIBUTION

Southeastern Australia from southeastern Queensland to extreme southeast South Australia. One well-defined subspecies (*L. m. cassidix*; helmeted honeyeater) near Melbourne has been regarded as a separate species, and several other subspecies exist.

HABITAT

Woodland and open forest dominated by eucalyptus, typically with well-developed understory and often in gullies. Sometimes in heathland, mallee, or woodland with native pine.

BEHAVIOR

Often found in loose colonies in which birds may cooperate to drive out other honeyeaters and small birds and mob predators.

Mostly sedentary although some movements occur. The single note calls *tsup* or *see* emitted frequently may act as a territorial call. They also emit a trilling flight call. Noisy calls when birds gather into corroborees.

FEEDING ECOLOGY AND DIET

Take insects, lerp, and manna from eucalyptus foliage and sometimes exudates from trunks and branches. May sally and hover. Also take nectar from eucalyptus and shrubs and occasionally eat fruit.

REPRODUCTIVE BIOLOGY

May breed throughout the year but mostly September to November. Apparently monogamous, though with helpers at the nest. Nests are in shrubs a few feet above the ground. Usually lay two eggs (occasionally one or three). Females do most incubation, for about 15 days. Young are fed by both parents and helpers on nectar and insects and fledge at about 16 days.

CONSERVATION STATUS

Critically Endangered helmeted honeyeaters are confined to one population of about 100 adults east of Melbourne. Other subspecies are not threatened.

SIGNIFICANCE TO HUMANS

Helmeted honeyeater is Victoria's state bird. ◆

Strong-billed honeyeater

Melithreptus validirostris

TAXONOMY

Haematops validirostris Gould, 1837, Tasmania.

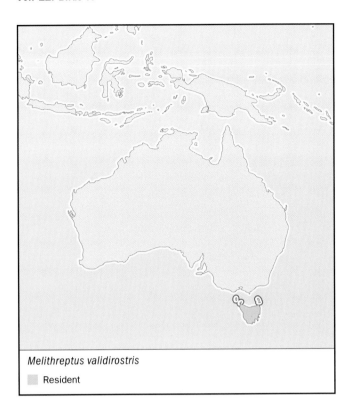

Melithreptus validirostris
☐ Resident

OTHER COMMON NAMES
English: Black-capped honeyeater; French: Méliphage à bec fort; German: Starkschnabel-Honigfresser; Spanish: Pájaro Miel Picudo.

PHYSICAL CHARACTERISTICS
6 in (15 cm); 0.9 oz (25 g). Black head and chin with white band from eye to nape. White throat and buff belly with dull olive-brown upperparts.

DISTRIBUTION
Endemic to Tasmania, including King, Flinders, and Cape Barren Islands.

HABITAT
Eucalyptus forests, especially wet gullies, but also in dry forest, sometimes in cool temperate rainforest, and occasionally coastal heathland, parks, and gardens.

BEHAVIOR
In pairs, family groups, or small flocks. Sometimes noisy and aggressive. Emit cheeping and churring calls. Sedentary, although exhibit local movements.

FEEDING ECOLOGY AND DIET
Forage on bark more than most honeyeaters do, probing into rough or peeling bark on trunks and branches of eucalyptus. Use long, sturdy bill to flake, lever, or tear off pieces or strips of bark to catch insects and spiders. Less commonly feed on nectar.

REPRODUCTIVE BIOLOGY
Poorly known, but nests found from July to January. They suspend a nest in eucalyptus or tea-tree foliage. Usually lay three eggs that are incubated by both parents. Incubation and nestling periods not known. Probably cooperative breeders.

CONSERVATION STATUS
Not threatened.

SIGNIFICANCE TO HUMANS
None known. ◆

Stitchbird
Notiomystis cincta

TAXONOMY
Meliphaga cincta Du Bus, 1839, North Island, New Zealand.

OTHER COMMON NAMES
English: Hihi; French: Méliphage hihi; German: Hihi; Spanish: Pájaro Puntado.

PHYSICAL CHARACTERISTICS
7 in (18 cm); female 1.1 oz (30 g), male 1.3 oz (36 g). Dark head, throat, and upperparts with white patch behind eye. Yellow breast with white belly; wings black with yellow and white tips.

DISTRIBUTION
Once widespread on North Island, but the only natural population is on Little Barrier Island. They have been introduced to several other islands, including Hen, Cuvier, Kapiti, and Tiritiri Matingi off North Island and Mokoia Island in Lake Rotorua.

HABITAT
Warm temperate rainforest, especially in gullies.

BEHAVIOR
In pairs in breeding season and small groups at other times. Male has an aggressive display with tail and ear-tufts raised and showing off golden shoulders. Although pairs usually mate in the conventional manner, males sometimes force the female to copulate by holding her on her back on the ground. Such males are typically not the partner, and this method of copulating is apparently unique among birds. High-pitched contact calls, males give a loud whistling song. Roost in cavities.

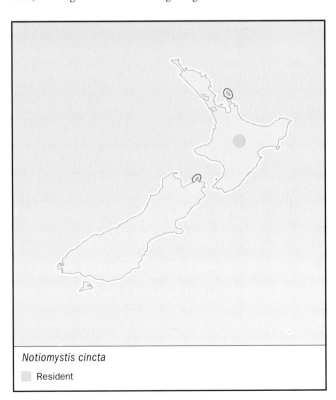

Notiomystis cincta
☐ Resident

FEEDING ECOLOGY AND DIET
Insects are taken by gleaning and sallying. Forage extensively on nectar and fruit of a wide variety of trees and shrubs, with proportions varying seasonally.

REPRODUCTIVE BIOLOGY
Breed as monogamous pairs but with frequent extra-pair copulations. Exceptional among honeyeaters in nesting in hollows of dead or living trees. Will use nesting boxes on islands with insufficient natural hollows. Complex nest built of sticks, fern rhizomes, and rootlets. The nest is mostly built by female (occasionally more than one), but male may also collect sticks. Two to five eggs are incubated solely by female for about 16 days. Young are fed by both parents and fledge at 26–32 days (an exceptionally long period for a honeyeater). Young join a creche one week after fledging and are not fed by their parents.

CONSERVATION STATUS
Vulnerable, with population of about 5,000 on Little Barrier Island. Populations reestablished on other islands remain small and precarious. Previous threats included over-collection, loss of nesting hollows, and introduced predators such as black rats (*Rattus rattus*).

SIGNIFICANCE TO HUMANS
Feathers of a large number of these birds were incorporated into Maori ceremonial cloaks. ◆

Red myzomela
Myzomela cruentata

TAXONOMY
Myzomela cruentata Meyer, 1875, Arfak Mountains, New Guinea.

Myzomela cruentata
▨ Resident

OTHER COMMON NAMES
English: Red honeyeater; French: Myzomèle vermillion; German: Bluthonigfresser; Spanish: Meloncillo Rojo.

PHYSICAL CHARACTERISTICS
4 in (11 cm), 0.25–0.30 oz (7–9 g). Small bird with entirely scarlet plumage, with exception of dark undertail and wing coverts.

DISTRIBUTION
Mountains of New Guinea from 2,000–4,800 feet (600–1,500 m), Yapen Island, Bismarck Archipelago. One subspecies in New Guinea and five more on different islands of the Bismarcks.

HABITAT
Rainforest.

BEHAVIOR
Generally quiet and inconspicuous. Emit a high-pitched *tseet*.

FEEDING ECOLOGY AND DIET
Feed on flowers in tall forest trees as well as in sago. Insects captured in the air and by probing into moss on high tree branches. Typically forage in the forest canopy.

REPRODUCTIVE BIOLOGY
Poorly known.

CONSERVATION STATUS
Not threatened; locally common.

SIGNIFICANCE TO HUMANS
None known. ◆

Brown honeyeater
Lichmera indistincta

TAXONOMY
Meliphaga indistincta Vigors and Horsfield, 1827, King George Sound, Western Australia.

OTHER COMMON NAMES
English: Least honeyeater, warbling honeyeater; French: Méliphage brunâtre; German: Braunhonigfresser; Spanish: Pájaro Miel Castaño.

PHYSICAL CHARACTERISTICS
4.3–6.3 in (11–16 cm); 0.3–0.4 oz (9–11 g). Drab olive-brown upperparts with a whitish patch behind eye. Underparts fade from light brown throat to whitish belly.

DISTRIBUTION
Most of Australia, except for the southeast, southern coast, and parts of the center. Trans-Fly region of southern New Guinea, Aru Islands, and Lesser Sundas from Tanimbar Islands west to Bali. The only honeyeater to cross Wallace's Line. The form found in the Lesser Sundas is sometimes regarded as a distinct species (*L. limbata*). Three subspecies recognized in Australia.

HABITAT
Mangroves, paperbark, and eucalyptus forests and woodlands, heathlands, semi-arid shrublands in Australia; also in parks and gardens. Mangroves, monsoon woodlands, scrub, secondary growth, and cultivated areas in Lesser Sundas, from sea level to 8,000 feet (2,600 m).

Lichmera indistincta

■ Resident

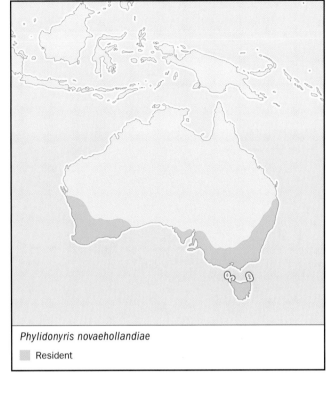

Phylidonyris novaehollandiae

■ Resident

BEHAVIOR

Usually singly or in pairs, sometimes in small groups. Present year-round in many areas but also show local movements in response to flowering, and occasionally found outside normal range. A renowned singer with loud song described as rich, cheerful, pleasing, and musical.

FEEDING ECOLOGY AND DIET

Feed on nectar from flowers of trees and shrubs, including mangroves, eucalyptus, tea trees, and grevilleas. Insects are taken from foliage and by aerial capture.

REPRODUCTIVE BIOLOGY

Long breeding season, although birds in northern Australia breed in the dry season from April to September and those in the south breed in spring and summer (July to January). Tightly woven cup-shaped nest is usually within 6 ft (2 m) of the ground. Clutch of two (occasionally one or three) eggs. Probably only the female incubates, but both sexes feed young. Incubation period lasts 12–14 days, and fledging occurs at about 14 days.

CONSERVATION STATUS

Not threatened; common in many parts of range.

SIGNIFICANCE TO HUMANS

Successful in towns, where their song is appreciated by many. ◆

New Holland honeyeater

Phylidonyris novaehollandiae

TAXONOMY

Certhia novaehollandiae Latham, 1790, Port Jackson, New South Wales, Australia. Five subspecies.

OTHER COMMON NAMES

English: Yellow-winged honeyeater, white-bearded honeyeater, white-eyed honeyeater; French: Mélephage de Nouvelle-Hollande; German: Weissaugen-Honigfresser; Spanish: Pájaro Azúcar de Alas Amarillas.

PHYSICAL CHARACTERISTICS

6.3–7.9 in (16–20 cm); 0.7 oz (20 g). Black head and upperparts, white spots above and below eye. Black wings have yellow at tips. Underparts black under bill fading to white under belly.

DISTRIBUTION

Southwestern Australia and southeastern Australia north to about Brisbane, Tasmania.

HABITAT

Open forest and woodland with dense understory, heathland, mallee heath.

BEHAVIOR

One of the best-studied honeyeaters. They are conspicuous, active, and aggressive. Breeding males defend areas near the nest by perching in conspicuous locations, calling, and chasing intruders. May perform corroborees where up to about a dozen birds gather closely together with much calling and wing fluttering. This may involve intraspecific interaction or be a response to potential predators. Other displays involve spreading the white ear-plumes, holding the bill open, and tail flicking. Show complex but probably mostly local movements in response to pattern of flowering. Calls mostly simple whistles, with stronger calls during corroboree and in response to predators, and a warbling song flight.

FEEDING ECOLOGY AND DIET

Take nectar from a wide range of native plants, including eucalyptus, banksias, heaths, and mistletoes. May defend feeding territories. Insects mostly taken by sallying, although sometimes taken from foliage or bark. Also take manna, lerp, and honeydew.

REPRODUCTIVE BIOLOGY
May breed at almost any time of year but is mostly concentrated in late winter to spring (July to September) with a secondary peak in autumn in some years and places. Several attempts are made per year. The cup-shaped nest is placed in a low shrub. Clutch size is typically two, sometimes three, and occasionally one egg. Incubation and fledging periods last about 14 days. Parents are occasionally helped by other birds.

CONSERVATION STATUS
Not threatened; abundant in many areas.

SIGNIFICANCE TO HUMANS
A well-known honeyeater in parks and gardens in southern Australia. ◆

Rufous-banded honeyeater
Conopophila albogularis

TAXONOMY
Entomophila albogularis Gould, 1843, Port Essington, Northern Territory, Australia.

OTHER COMMON NAMES
English: Rufous-breasted honeyeater; French: Méliphage à gorge blanc; German: Rostband-Honigfresser; Spanish: Pájaro Azúcar de Bandas.

PHYSICAL CHARACTERISTICS
5.1 in (13 cm); 0.4 oz (12 g). Black head, wings, and upperparts with some yellow on wings. Underparts white, with wide rufous band at breast.

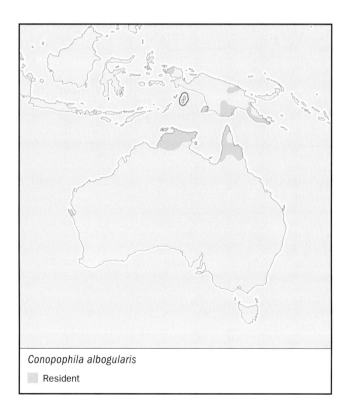

Conopophila albogularis
▨ Resident

DISTRIBUTION
North of Northern Territory and Queensland. Patchily distributed in lowland New Guinea and on Aru Islands.

HABITAT
Riparian forests of paperbark (*Melaleuca*), eucalyptus woodlands, mangroves, vine thickets, town parks, and gardens.

BEHAVIOR
Poorly known. Territorial in breeding season, unobtrusive at other times. Resident, but some evidence of local nomadism. Emit an attractive, melodious, but squeaky song. Other calls include *zzheep*, twittering, and chipping.

FEEDING ECOLOGY AND DIET
Highly insectivorous, foraging from outer foliage, especially of wattles. Also take nectar from eucalyptus and paperbark flowers and eat the arils that attach wattle seeds to the pod.

REPRODUCTIVE BIOLOGY
May breed at any time but main peaks are in late dry (September–November) and late wet (January to March) seasons. The purse-shaped nest is suspended from outer twigs of wattle or paperbark, often over water. Usually two to three eggs (occasionally one or four) that are incubated for about 14 days. Both adults feed young that fledge at 14 days.

CONSERVATION STATUS
Not threatened. Very common in suburban Darwin.

SIGNIFICANCE TO HUMANS
None known. ◆

Western spinebill
Acanthorhynchus superciliosus

TAXONOMY
Acanthorhynchus superciliosus Gould, 1837, Perth, Western Australia.

OTHER COMMON NAMES
English: Western spinebilled honeyeater; French: Méliphage festonné; German: Buntkopf-Honigfresser; Spanish: Pico de Espina Occidental.

PHYSICAL CHARACTERISTICS
5.5 in (14 cm); 0.35 oz (10 g). Head black, back and wings grayish. Rufous band behind neck, and from throat to breast. Underparts light gray with black and white bands below breast. White bands behind bill and eye.

DISTRIBUTION
Southwestern Australia.

HABITAT
Heathland, woodland, and open forest with heathy understory. Sometimes in mallee, rarely in gardens.

BEHAVIOR
Active. White outer tail feathers conspicuous in flight. Produce audible wing beats. Exhibit flight displays, frequent rapid and erratic chases, and male displays to female by fanning tail. Emit twittering and whistling calls and song. Poorly understood movements, perhaps local and in response to flowering patterns of plants.

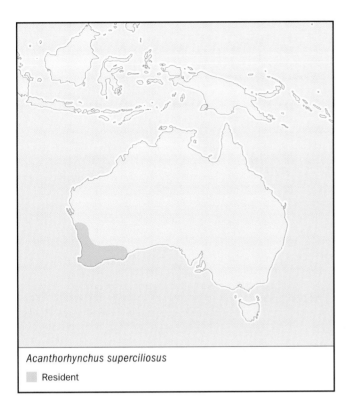

Acanthorhynchus superciliosus

 Resident

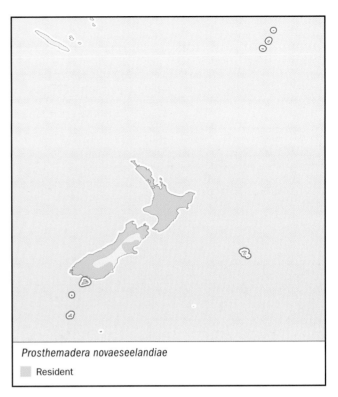

Prosthemadera novaeseelandiae

Resident

FEEDING ECOLOGY AND DIET

Probe flowers of banksias, eucalyptus, and numerous shrubs, including *Dryandra*, *Grevillea*, *Adenanthos*, and *Calothamnus*, as well as kangaroo paws (*Anigozanthos*). Also take insects, mostly captured in the air but also gleaned from plants.

REPRODUCTIVE BIOLOGY

Breed July to December, occasionally later. The rounded cup-shaped nest is placed in a shrub or small tree. Female mostly incubates the clutch of one or two eggs, but both parents feed young. Incubation and fledging periods not known.

CONSERVATION STATUS

Not threatened, but has declined in northeastern part of range due to extensive clearing of habitat; also adversely affected by fire.

SIGNIFICANCE TO HUMANS

None known. ◆

Tui

Prosthemadera novaeseelandiae

TAXONOMY

Merops novaeseelandiae Gmelin, 1788, Queen Charlotte Sound, New Zealand.

OTHER COMMON NAMES

English: Parson bird; French: Tui cravate-frisée; German: Tui; Spanish: Pájaro Sacerdote.

PHYSICAL CHARACTERISTICS

10.5–12.5 in (27–32 cm); male 4.3 oz (120 g), female 3 oz (85 g). Dark, iridescent plumage with two white throat tufts. Back and flanks dark reddish brown. White collar and wing bars.

DISTRIBUTION

New Zealand, including North Island, western and southeastern South Island (patchy on the rest of South Island), Stewart Island, other offshore islands, Chatham Islands, Raoul Island in Kermadecs, and possibly vagrant on Snares and Auckland Islands.

HABITAT

Podocarp, broadleaf, and beech forest, including remnants and regrowth. Also in dense exotic vegetation and in parks and gardens.

BEHAVIOR

Hold breeding territories and occur in loose groups outside breeding season, when males dominate females and tuis dominate other honeyeaters. May perform corroborees in small groups near the ground. Resident, but with local movements; more common in winter in urban areas. Males display song flights. Song is highly complex, and tuis have been deemed among the best singers in the world. The song is rich, melodious, and includes soft liquid warbling notes, bell-like calls, and chimes interspersed with sighs, sobs, coughs, laughs, sneezes, etc. Noisy flight is due to wing slots.

FEEDING ECOLOGY AND DIET

Feed on nectar from a wide range of plants, including *Metrosideros*, *Fuchsia*, and New Zealand flax (*Phormium tenax*). Glean insects, especially from foliage, and even sandhoppers (Amphipoda). Fruit is also an important component of the diet from late summer to winter. Honeydew is consumed from scale insects on beech trees.

REPRODUCTIVE BIOLOGY

Breed mostly October to January. The large, untidy nest is placed in fork or shrub or tree. The clutch of two to four (occasionally five) eggs is incubated by the female. Incubation and fledging periods are about 14 days. Young are fed by both parents, but more by female. Nests preyed on by introduced mammals and birds.

CONSERVATION STATUS
Not threatened, but has declined in many areas during nineteenth and twentieth centuries due to habitat loss, hunting, and predation by introduced mammals and birds.

SIGNIFICANCE TO HUMANS
Hunted by Maoris, especially consumed at feasts, and Europeans, who made pasties (pies) from them. Skins have been used in ladies' hats. Occasionally kept as pets. ◆

Bishop's oo
Moho bishopi

TAXONOMY
Acrulocercus bishopi Rothschild, 1893, Molokai Island, Hawaii.

OTHER COMMON NAMES
English: Molokai oo; French: Moho de Bishop; German: Ohrbüschelmoho; Spanish: Oo Obispo.

PHYSICAL CHARACTERISTICS
12 in (31 cm). Smoky black neck, back, and underparts with narrow white shaft lines on feathers. Wings and tail black. Tufts of golden feathers at ear coverts, undertail, and axillary.

DISTRIBUTION
Maui Island, formerly Molokai Island, Hawaiian Islands.

HABITAT
Dense rainforest in mountains.

BEHAVIOR
Inquisitive but timid and alert. Very loud *owow, owow-ow* call. The long graduated tail of male oos may have been used, along

with the yellow feathers on wing, neck, and tail coverts, to display to the female.

FEEDING ECOLOGY AND DIET
Feed on nectar from lobelia flowers. Also take insects from upper canopy.

REPRODUCTIVE BIOLOGY
Not known, but possibly a hollow nester like the Kauai oo.

CONSERVATION STATUS
Critically Endangered, last seen on Molokai in 1904. Rediscovered on Maui in 1981. Probably wiped out on Molokai due to habitat loss and introduced malaria. Perhaps restricted to highest parts of Maui due to introduced malaria in the lowlands.

SIGNIFICANCE TO HUMANS
Snared by native Hawaiians for its yellow plumes, which were used for ceremonial cloaks. ◆

Greater Sulawesi honeyeater
Myza sarasinorum

TAXONOMY
Myza sarasinorum Meyer and Wiglesworth, 1895, Matinan Mountains, north Celebes (Sulawesi).

OTHER COMMON NAMES
English: White-eared myza, greater streaked honeyeater, spot-headed honeyeater; French: Méliphage à points; German: Sarasinhonigfresser; Spanish: Pájaro Azúcar de Cabeza Moteada.

PHYSICAL CHARACTERISTICS
8 in (20 cm); weight about 1 oz (30 g). Long body and long, decurved bill. Head black with pinkish white patch behind eye.

Moho bishopi
◼ Resident

Myza sarasinorum
◼ Resident

Rufous from chin to vent, with breast and back mottled rufous brown. Brown rump and tail.

DISTRIBUTION
Sulawesi, with separate subspecies in northern, central, and southeastern parts of the island.

HABITAT
Montane and elfin moss forest, at 5,000–8,500 ft (1,700–2,800 m).

BEHAVIOR
Occur singly or in pairs, active and pugnacious. Described as squirrel-like in the way that they scurry about the branches of moss-draped trees. Wide variety of calls, including *zunk*, *kep*, *kik*, *kuik*, *zip*, and *tuck*. Also emit a high-pitched squeaky song.

FEEDING ECOLOGY AND DIET
Forage from the understory to the treetops, taking arthropods from epiphytes and probing flowers, especially flowers of gingers.

REPRODUCTIVE BIOLOGY
Not known.

CONSERVATION STATUS
Not threatened. Common in suitable habitat.

SIGNIFICANCE TO HUMANS
None known. ◆

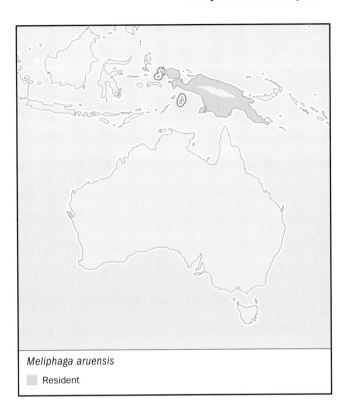

Meliphaga aruensis
▨ Resident

Puff-backed meliphaga
Meliphaga aruensis

TAXONOMY
Meliphaga auriculata Salvadori, 1881, Utanata River, New Guinea.

OTHER COMMON NAMES
English: Puff-backed honeyeater; French: Méliphage bouffant; German: Aruhonigfresser; Spanish: Melífago de Lomo Inflado.

PHYSICAL CHARACTERISTICS
6.7 in (17 cm); 1.1 oz (30 g). Olive-green upperparts with narrow yellowish rictal spot on sides of head; blackish lores; and gray underparts washed yellow, darker on flanks. Feet dull gray, bill blackish brown.

DISTRIBUTION
New Guinea from sea level to 4,000 ft (1,200 m), Aru, Yapen, West Papuan, D'Entrecasteaux, and Trobriand Islands.

HABITAT
Lowland rainforest, especially in dense regrowth after disturbance in both secondary and primary forest.

BEHAVIOR
Shy, usually hidden and very hard to identify from several similar meliphagas with which they may occur. Voice is a rapid series of mellow staccato notes.

FEEDING ECOLOGY AND DIET
Take fruit, nectar, and insects by gleaning in the understory and lower canopy.

REPRODUCTIVE BIOLOGY
Breeding recorded March through April and August through December. The cup-shaped nest of dead leaves and bark is suspended in shrub or sapling. Female incubates one or two eggs, and both adults feed young.

CONSERVATION STATUS
Not threatened.

SIGNIFICANCE TO HUMANS
None known. ◆

Belford's melidectes
Melidectes belfordi

TAXONOMY
Melirrhophetes belfordi De Vis, 1890, Mount Knutsford, New Guinea. Seven subspecies.

OTHER COMMON NAMES
English: Belford's honeyeater; French: Méliphage de Belford; German: Belfordhonigfresser; Spanish: Pájaro Miel de Belford.

PHYSICAL CHARACTERISTICS
10.7 in (27 cm); female 2.1 oz (60 g), male 2.6 oz (74 g). Light blue bare face skin and whitish wattles. Crown, lores, cheeks, and throat black, with white brow stripe and broad white streaks at side of throat. Gray underparts with brownish belly and undertail.

DISTRIBUTION
Central ranges of New Guinea from 5,250 to 12,500 ft (1,600 to 3,800 m). They also hybridize with yellow-browed melidectes (*M. rufocrissalis*) in the eastern part of range.

HABITAT
Mountain rainforests, extending into pockets of shrubbery in subalpine grasslands.

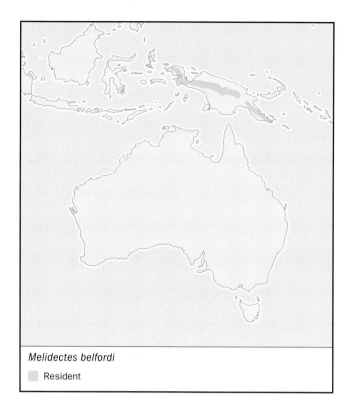

Melidectes belfordi

☐ Resident

Melipotes fumigatus

☐ Resident

BEHAVIOR
Boisterous and aggressive toward other species at flowering trees. Sail across openings with wings held widely spread, with infrequent flaps. Noisy, loud, and repeated calls, with gurgling, coughing, cackling, cawing, and piping notes.

FEEDING ECOLOGY AND DIET
Feed in canopy on nectar, arthropods, and some fruit. Glean from vines and foliage and probe into moss and epiphytes.

REPRODUCTIVE BIOLOGY
Breeding season April to July or later. The deep cup-shaped nest is made of moss, twigs, and hairs from the fronds of tree ferns. Egg apparently undescribed.

CONSERVATION STATUS
Not threatened.

SIGNIFICANCE TO HUMANS
None known. ◆

Common smoky honeyeater
Melipotes fumigatus

TAXONOMY
Melipotes fumigatus Meyer, 1886, southeast New Guinea.

OTHER COMMON NAMES
English: Common melipotes, smoky honeyeater; French: Méliphage enfumé; German: Aschbrust-Honigfresser; Spanish: Pájaro Azúcar Común.

PHYSICAL CHARACTERISTICS
8.6 in (22 cm); female 1.9 oz (52 g), male 2.1 oz (58 g). Orange-yellow face skin and ear wattle; blackish brown upperparts. Un-

derparts are dark gray-brown, paler on throat and abdomen. Bill blackish; legs sooty gray.

DISTRIBUTION
Central ranges of New Guinea, and Kumawa, Fakfak, Foya, and Cyclops ranges, at 3,000–13,000 ft (1,000–4,200 m).

HABITAT
Montane rainforest, forest edge, secondary growth, and gardens.

BEHAVIOR
Quiet and sluggish for a honeyeater, although sometimes aggressive. Call is a weak, monotonous *swit...swit...swit....*

FEEDING ECOLOGY AND DIET
Feed mostly on small fruits in mid-story and canopy. Also eat small insects by gleaning and sallying.

REPRODUCTIVE BIOLOGY
Nest August to October. The nest is loose and cup-shaped and made of leaves and moss and is held together with black fibers from a fungus. Females lay one egg. Young are fed on fruit.

CONSERVATION STATUS
Not threatened.

SIGNIFICANCE TO HUMANS
None known. ◆

Cape sugarbird
Promerops cafer

TAXONOMY
Merops cafer Linnaeus, 1758, Cape of Good Hope, South Africa.

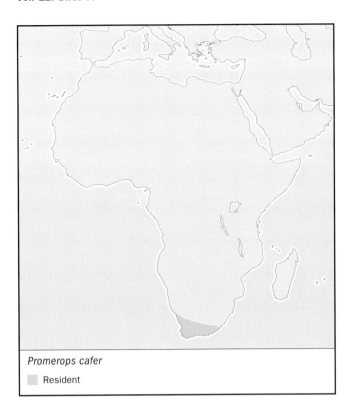

Promerops cafer

Resident

PHYSICAL CHARACTERISTICS
Females 9.5–11.5 in (24–29 cm), males 14.5–17 in (37–44 cm), including long tail; 1.5 oz (42 g). Rufous head and breast. Distinctive long bill and long, brownish tail feathers. Chin is white with a moustachial dark streak. Abdomen is whitish, vent yellow.

DISTRIBUTION
South Cape Province, South Africa.

HABITAT
Fynbos (coastal heathland).

BEHAVIOR
Single or paired, sometimes in small flocks. Males in breeding season perform aerial displays in which wings are clapped together and tail is held high. Both sexes defend flowering bushes from other sugarbirds and sunbirds. Song is a sustained jumble of discordant notes. Emit harsh alarm call of chicks and clatters. Show local and altitudinal movements in response to flowering of proteas.

FEEDING ECOLOGY AND DIET
Forage on nectar, especially of Proteaceae, and on insects captured in flight or gleaned from plants.

REPRODUCTIVE BIOLOGY
Breed from February to August, but varies with local timing of flowering. The deep cup-shaped nest is placed in a bush or low tree and is made from grass and twigs and lined with plant down. Lay two buff to reddish brown eggs with brown spots, streaks, and blotches.

CONSERVATION STATUS
Not threatened.

SIGNIFICANCE TO HUMANS
None known. ◆

OTHER COMMON NAMES
English: Sugarbird; French: Promérops du cap; German: Kaphonigfresser; Spanish: Pájaro Azúcar de el Cabo.

Resources

Books

Beehler, B.M., T.K. Pratt, and D.A. Zimmerman. *Birds of New Guinea*. Princeton: Princeton University Press, 1986.

Ford, H.A., and D.C. Paton. *The Dynamic Partnership: Birds and Plants in Southern Australia*. Adelaide: Government Printer, 1986.

Higgins, P.J., J.M. Peter, and W.K. Steele. *Handbook of Australian, New Zealand and Antarctic Birds*. Vol. 5. *Tyrant-flycatchers to Chats*. Melbourne: Oxford University Press, 2001.

Longmore, W. *Honeyeaters and their Allies of Australia*. Angus and Robertson: Sydney, 1991.

Periodicals

Ford, H.A., and S. Tremont. "Life History Characteristics of Two Australian Honeyeaters (Meliphagidae)." *Australian Journal of Zoology* 48 (2000): 21–32.

Franklin, D.C., and R.A. Noske. "Local Movements of Honeyeaters in a Sub-coastal Vegetation Mosaic in the Northern Territory." *Corella* 22 (1998): 97–103.

Franklin, D.C., and R.A. Noske. "The Nesting Biology of the Brown Honeyeater *Lichmera indistincta* in the Darwin Region of Northern Australia, with Notes on Tidal Flooding of Nests." *Corella* 24 (2000): 38–44.

Noske, R.A. "Breeding Biology, Demography and Success of the Rufous-banded Honeyeater, *Conopophila albogularis*, in Darwin, a Monsoonal Tropical City." *Wildlife Research* 25 (1998): 339–256.

Noske, R.A., and D.C. Franklin. "Breeding Seasons of Land Birds in the Australian Monsoonal Tropics: Diverse Responses to a Highly Seasonal Environment." *Australian Biologist* 12 (1999): 72–90.

Wilson, K., and H.F. Recher. "Foraging Ecology and Habitat Selection of the Yellow-plumed Honeyeater, *Lichenostomus ornatus*, in a Western Australian Woodland: Implications for Conservation." *Emu* 101 (2001): 89–94.

Hugh Alastair Ford, PhD

Vireos and peppershrikes

(Vireonidae)

Class Aves
Order Passeriformes
Suborder Passeri (Oscines)
Family Vireonidae

Thumbnail description
A group of small, plain-colored songbirds of forests and woodlands; they feed by gleaning invertebrates from plants and may also eat small berries

Size
4–7 in (10–18 cm)

Number of genera, species
4 genera; 43 species

Habitat
A wide range of types of forests, woodlands, and shrublands, from boreal to humid tropical

Conservation status
Critically Endangered: 1 species; Endangered: 1 species; Vulnerable: 1 species; Near Threatened: 2 species

Distribution
Boreal regions of North America, tropics of Central and South America, and temperate woodlands of southern South America

Evolution and systematics

Vireonidae is comprised of 43 species of small arboreal songbirds, divided into four genera. The phylogenetic relationships of the Vireonidae with other families are not totally clear, but they appear to be closely related to the New World warblers (family Parulidae).

Vireonidae is divided into three subfamilies: the Vireoninae, consisting of the true vireos (genus *Vireo*) and greenlets (*Hylophilus*); the Cyclarhidinae or peppershrikes (*Cyclarhis*); and the Vireolaniinae or shrike-vireos (*Vireolanius*).

Physical characteristics

The vireos are small, plain-colored songbirds; the name vireo is derived from the Latin word for "greenish." The bird has a short, somewhat heavy, pointed beak, with a small hook at the end formed by the overhanging tip of the upper mandible. The wings may be rounded at the end, or may be more pointed. The legs are short but strong. Vireos are typically colored olive gray, olive brown, greenish, or yellowish above; they have a white, light gray, yellow, or yellow-washed

breast and belly. Most species have a black line through the eye and a white stripe above it, but some have a light-colored eye-ring. A pale wing bar is generally present. The sexes are colored similarly.

Distribution

Vireos and their allies range widely over the Americas, from the boreal and temperate regions of North America, through the tropics of Central and South America, to the temperate woodlands of southern South America. They occur in northern Canada; throughout most of the United States, Mexico, and Central America; and in most of South America, as far south as Uruguay, northern Argentina, and northern Chile. Northern species of vireos are migratory, while those breeding in the tropics and subtropics are resident.

Habitat

These birds inhabit a wide range of forests and woodlands, including boreal, temperate, and tropical types.

A black-capped vireo (*Vireo atricapillus*) feeds its young. (Photo by S. & S. Rucker/VIREO. Reproduced by permission.)

Behavior

Vireos are usually solitary, or they appear as a breeding pair or family group. They sometimes participate in mixed-species foraging flocks during the nonbreeding season. They are deliberate but active birds. Their song is generally loud and melodic, and consists of several repeated phrases. Songs are varied and rather complex, and species may have from about 10 to more than 100 song types. The males sing frequently and persistently, often while foraging; they may even sing while on the nest.

Feeding ecology and diet

Vireos and their allies feed by gleaning insects, spiders, and other invertebrates from foliage, flowers, bark, and other plant surfaces. They also may eat small berries and other fruits.

Reproductive biology

Migratory species pair up soon after their arrival on the breeding grounds in the spring, with the male delimiting and defending the territory by song. The open, cup-shaped nest is made of spider and silkworm webbing, fine grass stems, other plant fibers, lichens, mosses, and feathers. The nest is usually located at the fork of a branch, hanging below the place of attachment, either close to the ground or high in the canopy.

The clutch size is two to five, and the eggs vary in color from whitish to speckled. Both sexes share in incubating the eggs and caring for the young. The incubation period is typically 12–14 days, and the nestlings fledge at nine to 11 days. The fledglings cannot fly well at first but are good at scrambling on branches and in shrubs. They are fed by the parents for about three weeks after leaving the nest.

Migratory species try to nest two to three times each season. Nesting vireos are highly vulnerable to parasitism by species of cowbirds, and to predation by small mammals, snakes, and predatory birds.

Conservation status

The IUCN lists five species in the Vireonidae as being at risk due to loss of habitat as a result of agricultural conversion of forest, logging, and other human activities. The Saint Andrew vireo (*Vireo caribaeus*) of the Caribbean island of Saint Andrew is listed as Critically Endangered. The Choco vireo (*Vireo masteri*) is known only from a tiny range in Colombia; it is Endangered because of the loss of most of its montane tropical rainforest habitat. The black-capped vireo (*Vireo atricapillus*) of Mexico and the southern United States is Vulnerable. The Noronha vireo (*Vireo gracilirostris*) of Brazil and the Blue Mountain vireo (*Vireo osburni*) of Jamaica are Near Threatened, but they too have suffered large population declines due to the loss and fragmentation of their habitats.

Significance to humans

Vireos and their allies are not of direct importance to humans. It is important that research be undertaken to better understand the biology and habitat needs of the rare and endangered species of vireonids.

Red-eyed vireo (*Vireo olivaceus*) nest with a cowbird's egg. (Photo by Mary M. Thacher. Photo Researchers, Inc. Reproduced by permission.)

1. Red-eyed vireo (*Vireo olivaceus*); 2. Bell's vireo (*Vireo bellii*); 3. Warbling vireo (*Vireo gilvus*); 4. Rufous-browed peppershrike (*Cyclarhis guja-nensis*); 5. Lemon-chested greenlet (*Hylophilus thoracicus*); 6. Slaty-capped shrike-vireo (*Vireolanius leucotis*); 7. Black-capped vireo (*Vireo atri-capillus*). (Illustration by Michelle Meneghini)

Species accounts

Red-eyed vireo
Vireo olivaceus

TAXONOMY
Vireo olivaceus Linneaus, 1766.

OTHER COMMON NAMES
English: Noronha vireo, yellow-green vireo; French: Viréo aux yeux rouges; German: Rotaugenvireo; Spanish: Vireo de Ojos Rojos.

PHYSICAL CHARACTERISTICS
6 in (15 cm). Olive green upperparts; white breast and belly. Dark tips on wings, but no wing-bars. Dark red iris; a black line through the eye and a white one above. Head has a gray crown.

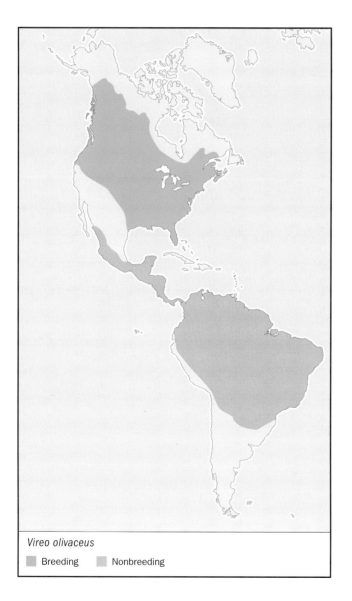

Vireo olivaceus
■ Breeding ■ Nonbreeding

DISTRIBUTION
Breeds throughout much of the United States and Canada; winters in Central America and South America as far south as Argentina.

HABITAT
Deciduous-dominated forest and woodlands.

BEHAVIOR
A migratory species that defends a breeding territory. The robinlike song is a series of loud, high-pitched, melodious, whistled phrases.

FEEDING ECOLOGY AND DIET
Eats small invertebrates. Usually forages on foliage, flowers, and limb surfaces, but will also capture flying insects; feeds on small fruits when invertebrates are not abundant.

REPRODUCTIVE BIOLOGY
Builds a small, cup-shaped nest that hangs from a fork in a tree branch. Lays two to four eggs, incubated by both parents for 11–14 days. Often raises two broods per season.

CONSERVATION STATUS
Not threatened, but its numbers are declining in some parts of its range.

SIGNIFICANCE TO HUMANS
Birdwatchers may look for this species. ◆

Bell's vireo
Vireo bellii

TAXONOMY
Vireo bellii Audubon, 1844.

OTHER COMMON NAMES
French: Viréo de Bell; German: Braunaugenvireo; Spanish: Vireo de Bell.

PHYSICAL CHARACTERISTICS
4 in (11 cm). A relatively small and dull-colored vireo, with olive-gray upperparts, white underneath, a broken (or incomplete) white eye-ring, and faint whitish wing-bars.

DISTRIBUTION
Breeds in the southwestern and central United States and south to parts of northern Mexico; winters from Baja California to Honduras.

HABITAT
Shrubby vegetation, such as thickets of willows and mesquite, especially in riparian habitat near streams and rivers.

BEHAVIOR
A migratory species that defends a breeding territory. The song is a loud, high-pitched, melodious series of simple phrases.

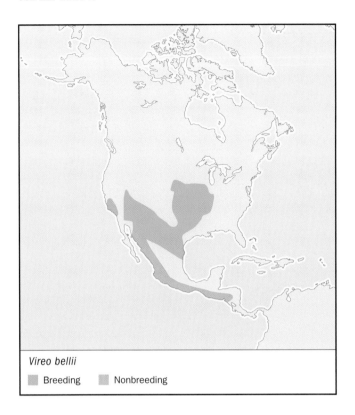

Vireo bellii

■ Breeding ■ Nonbreeding

Vireo atricapillus

■ Breeding ■ Nonbreeding

FEEDING ECOLOGY AND DIET
Feeds actively on invertebrates in foliage, flowers, and limb surfaces. Also eats small berries when invertebrates are not abundant.

REPRODUCTIVE BIOLOGY
Builds a small, cup-shaped nest that hangs from a fork in a tree branch. Lays three or four eggs, incubated by both parents for about 14 days.

CONSERVATION STATUS
Not threatened by IUCN criteria; relatively widespread and abundant. Some populations are threatened by habitat loss due to agriculture, mining, flood-control projects, and reservoir construction. In 1986, the U.S. Fish and Wildlife Service placed a subspecies, the least Bell's vireo (*Vireo bellii pusillus*), on the U.S. Endangered Species List. The Fish and Wildlife Service regulates human activities in riparian habitat used by the least Bell's vireo in southern California and elsewhere in its range.

SIGNIFICANCE TO HUMANS
None known. ◆

Black-capped vireo
Vireo atricapillus

TAXONOMY
Vireo atricapillus Woodhouse, 1852.

OTHER COMMON NAMES
French: Viréo à tête noire; German: Schwarzkopfvireo; Spanish: Vireo de Capa Negra.

PHYSICAL CHARACTERISTICS
About 4.5 in (12 cm); a small vireo with olive-colored upperparts, white underneath, a yellow wash beneath the wings, yellow wing-bars, and a red iris. Males have a glossy black head and white eye-rings that resemble eyeglasses. Females are similar, but the head is slate gray.

DISTRIBUTION
Breeds in parts of Texas, New Mexico, Oklahoma, Missouri, and north-central Mexico; winters along the west coast of Mexico.

HABITAT
Open, grassy woodlands with clumps of shrubs and trees. Prefers the low-lying vegetation that grows back after a fire in juniper and oak woodlands.

BEHAVIOR
A short-distance migrant. Defends a breeding territory. The song is a series of slowly repeated, husky, complex phrases.

FEEDING ECOLOGY AND DIET
Primarily feeds on invertebrates gleaned from leaves of trees and shrubs, but also eats small fruit.

REPRODUCTIVE BIOLOGY
Builds a cup-shaped, hanging nest at a branch fork of a shrub or low tree. Lays three or four eggs, which are incubated for 14–19 days.

CONSERVATION STATUS
Not threatened by IUCN criteria, though some populations are threatened by habitat loss caused by agriculture, mining, flood-control projects, and reservoir construction. Subject to severe nest parasitism by cowbirds. In 1986, the U.S. Fish and Wildlife Service placed it on the U.S. Endangered Species List in recognition of extensive habitat loss and damage. Efforts are underway in Texas and Oklahoma to trap cowbirds and remove

them from nesting areas of the black-capped vireo. The Fish and Wildlife Service regulates human activities in critical riparian habitat used by the black-capped vireo.

SIGNIFICANCE TO HUMANS
None known. ◆

Warbling vireo
Vireo gilvus

TAXONOMY
Vireo gilva Vieillot, 1808.

OTHER COMMON NAMES
English: Eastern warbling-vireo, brown-capped vireo, Western warbling-vireo; French: Viréo mélodieux; German: Sägervireo, Braunkappenvireo; Spanish: Vireo Chipe.

PHYSICAL CHARACTERISTICS
About 5.5 in (14 cm). The upper body is uniformly gray, with white underparts, light-yellow flanks, and a white line over the eye.

DISTRIBUTION
Breeds widely in southern and western Canada, through most of the United States, and in part of north-central Mexico; winters in Mexico and Guatemala.

HABITAT
Hardwood-dominated forests.

BEHAVIOR
A migratory species that defends a breeding territory. The song is a slow warble.

FEEDING ECOLOGY AND DIET
Feeds on invertebrates gleaned from leaves, flowers, and branches, and also eats small fruits.

REPRODUCTIVE BIOLOGY
Builds a small, cup-shaped nest that hangs from a forked tree branch. Lays three or four eggs, incubated by both parents for about 14 days.

CONSERVATION STATUS
Not threatened. A widespread and abundant songbird, but its numbers are declining in some parts of its breeding range.

SIGNIFICANCE TO HUMANS
None known, except indirect economic benefits of birding. ◆

Lemon-chested greenlet
Hylophilus thoracicus

TAXONOMY
Hylophilus thoracicus Temminck, 1822.

OTHER COMMON NAMES
English: Rio de Janeiro greenlet; French: Viréon à plastron; German: Gelbbrustvireo; Spanish: Vireillo de Pecho Limón.

PHYSICAL CHARACTERISTICS
About 5 in (13 cm). A slim-bodied bird with a short, pointed beak. The upperparts are uniformly olive-green, the chest yellow, the belly and chin light gray, and the iris white.

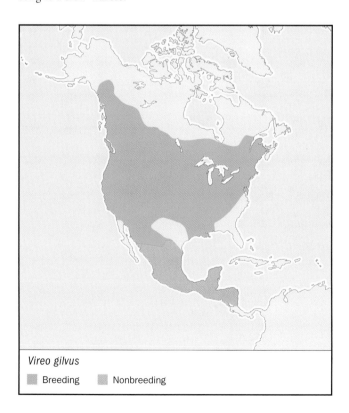

Vireo gilvus
■ Breeding   ■ Nonbreeding

Hylophilus thoracicus
■ Resident

DISTRIBUTION
Northern South America, southeastern Venezuela and Colombia, eastern Peru, northern Bolivia, and southeastern and Amazonian Brazil.

HABITAT
Humid tropical forest and forest edges, as high as about 2,000 ft (600 m).

BEHAVIOR
A nonmigratory species that defends a breeding territory. The song is a series of rapid, rolling "peeer" phrases.

FEEDING ECOLOGY AND DIET
An extremely active forager for insects and other invertebrates, gleaned from foliage, flowers, and limbs; also eats small fruits. Occurs in mixed-species foraging flocks.

REPRODUCTIVE BIOLOGY
Builds a cup-shaped nest that hangs from a forked tree branch. Both parents incubate the eggs and care for the young.

CONSERVATION STATUS
Not threatened.

SIGNIFICANCE TO HUMANS
None known. ◆

Rufous-browed peppershrike
Cyclarhis gujanensis

TAXONOMY
Cyclarhis gujanensis Gmelin J.F., 1789.

OTHER COMMON NAMES
French: Sourciroux mélodieux; German: Rostbrauenvireo; Spanish: Alegrín de Cejas Rojizas.

PHYSICAL CHARACTERISTICS
5.5–6 in (14–15 cm). The body is relatively heavy, the head large, and the beak stout. The back is dark olive-green, the chest and flanks are yellow, the belly white, the top of the head gray, and a broad rufous stripe over the eye.

DISTRIBUTION
Widespread from southeastern Mexico to central Argentina, but missing from most of Amazonia.

HABITAT
Dry and moist forest borders, scrub, and clearings with trees present, as high as about 6,600 ft (2,000 m).

BEHAVIOR
A nonmigratory species; defends a breeding territory. The song—a repeated, musical phrase—is given throughout the year. Individual birds have several song types, and there are regional dialects.

FEEDING ECOLOGY AND DIET
Feeds on insects and other invertebrates gleaned from foliage, flowers, and limbs; also eats small fruits. Occurs in mixed-species foraging flocks.

Cyclarhis gujanensis
▨ Resident

REPRODUCTIVE BIOLOGY
Pairs stay together throughout the year. Builds a cup-shaped nest that hangs from a forked tree branch. Both parents incubate the eggs and care for the young.

CONSERVATION STATUS
Not threatened.

SIGNIFICANCE TO HUMANS
None known. ◆

Slaty-capped shrike-vireo
Vireolanius leucotis

TAXONOMY
Vireolanius leucotis Swainson, 1838.

OTHER COMMON NAMES
French: Smaragdan oreillard; German: Scieferkopfvireo; Spanish: Follajero de Capa Oscura.

PHYSICAL CHARACTERISTICS
About 5.5 in (14 cm). The body is relatively heavy, the head large, and the beak stout. The back is olive green, the chest and belly are yellow, the top of the head and cheeks gray, with a broad yellow stripe over the eye.

Vireolanius leucotis

☐ Resident

DISTRIBUTION
Widespread in northern South America, including parts of Colombia, Ecuador, southern Venezuela, the Guianas, much of Amazonian Brazil, northern Bolivia, and eastern Peru.

HABITAT
Humid tropical forest, as high as about 6,250 ft (1,800 m).

BEHAVIOR
A nonmigratory species that defends a breeding territory. The song is a loud, repeated, simple phrase.

FEEDING ECOLOGY AND DIET
Feeds on insects and other invertebrates gleaned from foliage, flowers, and limbs.

REPRODUCTIVE BIOLOGY
Builds a cup-shaped nest that hangs from a forked tree branch. Both parents incubate the eggs and care for the young.

CONSERVATION STATUS
Not threatened.

SIGNIFICANCE TO HUMANS
None known. ◆

Resources

Books

Bent, A.C. *Life Histories of North American Wagtails, Shrikes, Vireos, and Their Allies.* New York: Dover Publications, 1965.

BirdLife International. *Threatened Birds of the World.* Barcelona and Cambridge, UK: Lynx Edicions and BirdLife International, 2000.

Ridgely, R. S., and G. Tudor. *The Birds of South America.* Vol. 1. *The Oscine Passerines.* Austin: University of Texas Press, 1994.

Periodicals

Greenberg, R., D. K. Niven, S. L. Hopp, and C. A. Boone. "Frugivory and Coexistence in a Resident and Migratory Vireo on the Yucatan Peninsula." *Condor* 95, no. 4 (1993): 990–999.

Hopp, S. L., C. A. Boone, and A. Kirby. "Banding Returns, Arrival Pattern and Site Fidelity of White-Eyed Vireos." *Wilson Bulletin* 111 (1999): 46–55.

Organizations

BirdLife International. Wellbrook Court, Girton Road, Cambridge, Cambridgeshire CB3 0NA United Kingdom. Phone: +44 1 223 277 318. Fax: +44-1-223-277-200. E-mail: birdlife@birdlife.org.uk Web site: <http://www.birdlife.net>

IUCN–The World Conservation Union. Rue Mauverney 28, Gland, 1196 Switzerland. Phone: +41-22-999-0001. Fax: +41-22-999-0025. E-mail: mail@hq.iucn.org Web site: <http://www.iucn.org>

Other

The Vireo Homepage. 2002. <http://eebweb.arizona.edu/faculty/hopp/vireo.html>

Bill Freedman, PhD
Melissa Knopper, MS

New World finches
(Emberizidae)

Class Aves
Order Passeriformes
Suborder Passeri (Oscines)
Family Emberizidae

Thumbnail description
Small to medium-sized, highly vocal songbirds with short, conical bill and short to medium-length tail; most species have brown, olive, or gray plumage, but many are brightly colored

Size
4–9.5 in (10–24 cm); 0.3–2.6 oz (8–75 g)

Number of genera, species
72 genera; 291 species

Habitat
Grasslands, marshes, arctic tundra, alpine meadows, open woodlands, park land, and hedgerows

Conservation status
Critically Endangered: 6 species; Endangered: 7 species; Vulnerable: 9 species; Near Threatened: 2 species

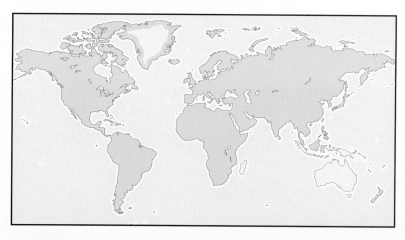

Distribution
Practically worldwide; absent from extreme Southeast Asia, Australasia (introduced into New Zealand), and Madagascar

Evolution and systematics

Buntings and New World sparrows are variously treated as a family (Emberizidae; New World finches) or as a subfamily (Emberizinae) of the Fringillidae. Most commonly, however, these are recognized as distinct families. No direct information exists about the origin of these closely related families, but scientists believe that they must have evolved in the New World. Currently all but the Emberizidae are restricted to the New World, and even the buntings are most diverse in the Americas (only 42 species in five genera are found elsewhere). All of the known fossils are from the New World and date from the Lower Pliocene to the Holocene.

Physical characteristics

Emberizids range in size from petite blue-black grassquits (*Volatinia jacarina*) that weigh as little as 0.28 oz (8 g) to California towhees (*Pipilo crissalis*) that weigh up to 2 oz (60 g). Many species have subdued brown, beige, or gray colors, commonly with complex facial patterns of black, white, buff, and sometime yellow stripes. Males generally are somewhat larger than females. In many species the sexes are alike in plumage pattern, but in others males and females look strikingly different. Many buntings feed on the ground and have medium-sized legs with rather large feet; some hop on the ground whereas others run. Relatively short, conical bills are characteristic.

Distribution

Emberizids are found throughout the New World, from Greenland to the islands in Arctic Canada and south to Cape Horn. Several species live in the Galápagos Islands. In the Old World, buntings are found across Eurasia south to India, Malaysia, the Philippines, and Borneo (in winter), and throughout Africa (but not Madagascar). They are not found in extreme southeastern Asia or in New Guinea or Australia; two species have been introduced into New Zealand.

Habitat

Buntings and New World sparrows live in a variety of habitats, but most inhabit fairly open, brushy, or grassland areas. Of the many species of buntings that breed and winter in grasslands, some inhabit short, dry grasslands, whereas others live in tall grass in wet meadows. Some emberizids inhabit arctic or alpine tundra; others breed in marshes or woodlands.

Behavior

Although sparrows commonly sing at night during breeding season, they are diurnal. Most sparrows and buntings are territorial, and territorial males use song, chasing, and fights to defend their territories, which generally serve as a place to build a nest and to forage. When singing, males often sit in a conspicuous place and throw back their head to sing. Commonly, they sing with the feathers of their crown or rump

1. Lapland longspur female (*Calcarius lapponicus*) solicitation display; 2. Snow bunting (*Plectrophenax nivalis*) song-flight; 3. Reed bunting (*Emberiza schoeniclus*) singing with rump feathers ruffed; 4. Corn bunting (*Emberiza calandra*) dangling-legs display. (Illustration by Bruce Worden)

ruffed. Many species, especially those that live in tundra or prairies, sing their songs while in flight as a part of an elaborate flight display. When soliciting food or copulation, sparrows generally point their head forward, more-or-less parallel with the ground, elevate their tail, and shiver their wings.

During migration and winter, sparrows may be seen in small, loose, often mixed-species flocks. Some species, however, form large flocks.

Feeding ecology and diet

Most sparrows feed on or near the ground as they pick up insects or fallen grass seeds. Some scratch away leaf litter to find food. Towhees (*Pipilo*) use a distinctive double scratch when feeding: they remain stationary while scratching backward simultaneously with both feet. The conical bill efficiently handles and shells seeds; large-billed species can crack hard seeds that many other birds could not eat. Some species, especially seedeaters (*Sporophila*) and grassquits (*Volitinia* and *Tiaris*), feed almost exclusively on seeds; they cling to tall grass and pick seeds from seed heads. Although seeds are an important part of the winter diet of almost all species, during nesting seasons most feed insects to their nestlings. Fruit is eaten when available. Woodpecker finches (*Cactospiza pallida*) of the Galápagos Islands hold a cactus spine or stick in their bill and use it to pry insects and their larvae from dead branches.

Reproductive biology

Most buntings and New World sparrows are socially monogamous (during any single breeding season a single male is associated with a single female), but there are exceptions. In some species, and in some populations within a species,

A painted bunting (*Passerina ciris*) bathes in shallow water. (Photo by B. Schorre/VIREO. Reproduced by permission.)

A black-throated sparrow (*Amphispiza bilineata*) drinks from an outside faucet. (Photo by Sam Fried/VIREO. Reproduced by permission.)

A woodpecker finch (*Cactospiza pallida*) uses a small twig to retrieve insects from a hole in a branch. One of "Darwin's finches," the woodpecker finch is found in the Galápagos Islands. (Photo by Alan Root/Okapia. Photo Researchers, Inc. Reproduced by permission.)

A large cactus ground-finch (*Geospiza conirostris*) feeds on a cactus flower. (Photo by K. Schafer/VIREO. Reproduced by permission.)

Le Conte's sparrow (*Ammodramus leconteii*) singing in Montana. (Photo by Anthony Mercieca. Photo Researchers, Inc. Reproduced by permission.)

A small ground-finch (*Geospiza fuliginosa*) picks ticks from a marine iguana on Fernandina Island in the Galápagos Islands. (Photo by Tui De Roy. Bruce Coleman Inc. Reproduced by permission.)

males are polygynous (one male mated with two or more females). Polygyny in sparrows often seems to occur when birds nest in high density. In a few sparrow species no pair bond is formed. Rather, during any breeding season individuals of both sexes mate promiscuously. In all sparrows the female alone incubates the eggs, but in general, both members of a pair help to feed and care for the young.

Conservation status

Seven emberizid species are listed as Endangered, six as Critically Endangered, nine as Vulnerable, and two as Lower Risk/Near Threatened. For example, pale-headed brush-finches (*Atlapetes pallidiceps*), a species found only in southwestern Ecuador, is threatened by the near total removal of vegetation in its tiny range, and it may be extinct. Sierra Madre sparrows (*Xenospiza baileyi*) of south central Mexico were never widespread but now are limited to a small area of bunch-grass habitat just south of Mexico City. Cuban sparrows (*Torreornis inexpectata*) are scrub-dwelling birds found in three different areas of Cuba; each of these populations is small,

and they are threatened by habitat destruction. In eastern North America, many sparrows are declining in numbers or have disappeared as marginal farms have been abandoned and reverted to woodlands or as urban development has replaced meadows. Similarly, in northwestern Europe, some species that inhabit open country have declined in numbers, probably because of changes in farming practices. In central and western North America, many grassland species are declining as a consequence of habitat degradation. Many of the Central and South American seedeaters are popular cage birds, and intense commercial trapping, particularly in northern Argentina, has led to substantial declines in many populations.

Significance to humans

Many species of buntings have pretty songs and are popular as cage birds in many parts of the world. Although many species eat substantial numbers of insects that may be agricultural pests, they probably do not destroy insects in sufficient quantities to be of economic significance.

1. Chipping sparrow (*Spizella passerina*); 2. Savannah sparrow (*Passerculus sandwichensis*); 3. Blue-black grassquit (*Volatinia jacarina*); 4. Lark bunting (*Calamospiza melanocorys*); 5. Chestnut-capped brush-finch (*Atlapetes brunneinucha*); 6. Variable seedeater (*Sporophila americana*); 7. Plumbeous sierra-finch (*Phrygilus unicolor*); 8. Bachman's sparrow (*Aimophila aestivalis*); 9. Eastern towhee (*Pipilo erythrophthalmus*); 10. Woodpecker finch (*Cactospiza pallida*). (Illustration by Bruce Worden)

1. Reed bunting (*Emberiza schoeniclus*); 2. White-throated sparrow (*Zonotrichia albicollis*); 3. Snow bunting (*Plectrophenax nivalis*); 4. Rock bunting (*Emberiza cia*); 5. Corn bunting (*Miliaria calandra*); 6. Yellowhammer (*Emberiza citrinella*); 7. Lapland longspur (*Calcarius lapponicus*); 8. Dark-eyed junco (*Junco hyemalis*); 9. Nelson's sharp-tailed sparrow (*Ammodramus nelsoni*); 10. Song sparrow (*Melospiza melodia*); 11. Crested bunting (*Melophus lathami*). (Illustration by Bruce Worden)

Species accounts

Nelson's sharp-tailed sparrow
Ammodramus nelsoni

TAXONOMY
Ammodramus caudacutus var. *nelsoni* Allen, 1875, Cook Co., Illinois. Three subspecies.

OTHER COMMON NAMES
English: Nelson's sparrow, Acadian sparrow, sharp-tailed sparrow; French: Bruant des Nelson; German: Nelsonammer; Spanish: Gorrión Coliagudo.

PHYSICAL CHARACTERISTICS
4.5–5 in (11–13 cm); 0.14–0.74 oz (14–21 g). Small, sharp-tailed sparrows with broad median crown stripe outlined by brown lateral crown stripe, a yellow-ochre stripe above the eye, a brown stripe through the eye, and gray ear coverts and nape. The back is dark brown, usually with distinct grayish or white stripes. Underparts are pale yellow with faint brown streaks on side of throat and flanks. Birds breeding in the maritimes of Canada and coastal Maine are less brightly colored. Sexes are alike. Juveniles lack the grayish color.

DISTRIBUTION
A. n. nelsoni breeds from north Alberta, central Saskatchewan, and south Manitoba to northeast South Dakota and winters along the Gulf Coast and southern coast of California. *A. n. alterus* breeds on the south coast of Hudson Bay and James Bay and winters along the coast of the United States from New York to southern Texas. *A. n. subvirgatus* breeds along the coast of the Gulf of St. Lawrence, the coast of Quebec, Nova Scotia, and New Brunswick to central Maine and winters from coastal Massachusetts to northern Florida.

HABITAT
In the prairies, breed in freshwater marshes where cordgrass and whitetop are common. Along the shores of James and Hudson Bays, they are found in dense sedge bogs, generally where there are a few dwarf birch trees. In the maritime provinces and Quebec they are found in rank, wet grassland and salt marshes. During migration, they may be found in wet fields and marshes, often in cattail. In winter they are found along the coast in freshwater and salt marsh habitats, but most commonly in brackish marshes.

BEHAVIOR
Males sing persistently during breeding season and commonly sing at night. They may sing from a bush, from the ground, or during an elaborate aerial display. On the ground, they commonly run.

FEEDING ECOLOGY AND DIET
During breeding season they principally feed on insects and other arthropods but also on small mollusks. In fall and winter seeds become important. They feed while walking through dense grasses or while clambering in vegetation.

REPRODUCTIVE BIOLOGY
Nonterritorial and promiscuous. Only the female provides parental care. The nest is usually a simple cup of coarse grass placed on the ground or just above water in marshes. Nesting takes place in June and July. Incubation of four to five eggs lasts about 12 days, and the young fledge after about 10 days.

CONSERVATION STATUS
Not threatened, though marsh degradation and loss have caused serious problems in some populations.

SIGNIFICANCE TO HUMANS
None known. ◆

Ammodramus nelsoni

☐ Resident ▨ Breeding ▨ Nonbreeding

Song sparrow
Melospiza melodia

TAXONOMY
Fringilla melodia Wilson, 1810, Philadelphia, Pennsylvania. Thirty-eight subspecies.

OTHER COMMON NAMES
French: Bruant chanteur; German: Singammer; Spanish: Gorrión Cantor.

PHYSICAL CHARACTERISTICS
5–7 in (12–17 cm); 0.67–1.5 oz (19–42 g). Song sparrows are medium to large sized sparrows with a long, round tail. The head is brown to light rusty with paler median crown stripe, grayish stripe above the eye, conspicuous brown malar stripes, a brown mottled back, and heavily streaked breast with a dark central breast spot. Sexes are alike. Juveniles have brown crowns, are heavily streaked below, and are generally more buff in color than adults. This species is highly variable geographically.

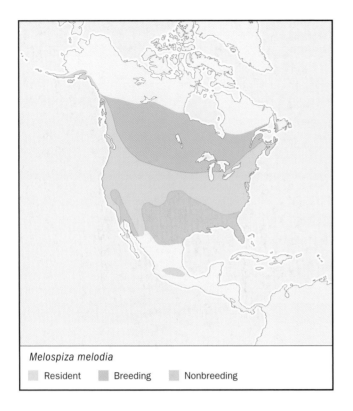

Meloospiza melodia

■ Resident   ■ Breeding   ■ Nonbreeding

CONSERVATION STATUS
Not threatened.

SIGNIFICANCE TO HUMANS
None known. ◆

White-throated sparrow
Zonotrichia albicollis

TAXONOMY
Fringilla albicollis Gmelin, 1789, Philadelphia, Pennsylvania.

OTHER COMMON NAMES
English: White throat, Canada bird; French: Bruant à gorge blanche; German: Weisskehlammer; Spanish: Gorrión Gorjiblanco.

PHYSICAL CHARACTERISTICS
6–7 in (15–17 cm); 0.9 oz (26 g). Adults have a pale or white stripe above the eye that is yellow in front of the eye, a pale or white median crown stripe, a white throat, brown or rusty brown back, and a pale grayish brown breast that often is slightly streaked. Sexes similar. Juveniles have an indistinct median crown stripe and their breast and flanks are heavily streaked with brown.

DISTRIBUTION
Breed from central Yukon to northern Manitoba and across northern Ontario through central Quebec and Newfoundland; also south to New York, Michigan, Manitoba, and British Columbia. Winter along the Pacific Coast and in the east from Ontario, Michigan, and Colorado to Texas, the Gulf Coast, and Florida.

DISTRIBUTION
Breed from the Aleutian Islands, along the southern coast of Alaska, east across southern Nunavut, northern Ontario, and central Quebec to southwest Newfoundland, and south to Georgia, Missouri, Nebraska, New Mexico, Arizona, and California. Locally resident in Baja California and central Mexico. Resident in Alaska and along the Pacific coast, but most northern-breeding birds migrate in winter to southern Florida, the Gulf Coast, northern Mexico, and southern Baja California.

HABITAT
Generally found in open brushy habitats, often near ponds, streams, or marshes. In winter, they are found in brush and woodland edge.

BEHAVIOR
Generally stay low in vegetation, but they often perch conspicuously in a tree, bush, or on top of a weed when singing. In flight they appear to pump the tail, and they hop or run on the ground. They defend territories with chases and fights. In winter, they can be found in loose flocks that often contain other species of sparrows.

FEEDING ECOLOGY AND DIET
In summer, they eat primarily insects and other invertebrates, but in winter they eat mostly seeds. Song sparrows feed on the ground or by picking food from vegetation.

REPRODUCTIVE BIOLOGY
Socially monogamous. The nest is a bulky cup of leaves, strips of bark, grass, and other plants, commonly placed on the ground among grasses, low in a bush, or rarely in a cavity. They usually lay three to six eggs. Nesting takes place from late February (in the southern parts of their range) into August. Incubation takes 10–14 days, and young fledge in 7–14 days. Both parents feed the young.

Zonotrichia albicollis

■ Resident   ■ Breeding   ■ Nonbreeding

HABITAT
Inhabit brush during all seasons. They most often breed in fairly open mixed woodlands, commonly where spruce, balsam fir, birch, and aspens predominate. In winter they are found in dense deciduous thickets or brush piles, often along the edge of woodlands or in woodland clearings.

BEHAVIOR
During breeding season, males sing persistently, usually from an inconspicuous perch. In winter, they are often found in small, loose flocks and sometimes associate with other species of sparrow.

FEEDING ECOLOGY AND DIET
In breeding season, they principally eat insects, but during migration and in winter they eat mostly seeds.

REPRODUCTIVE BIOLOGY
Monogamous. The nest is placed on or near the ground in areas where there are small trees interspersed with low vegetation. Nesting occurs from late May through early July. Three to seven (usually four) eggs are incubated for 11–13 days, and young fledge after 8–9 days. Both parents feed the young.

CONSERVATION STATUS
Not threatened.

SIGNIFICANCE TO HUMANS
None known. ◆

Dark-eyed junco
Junco hyemalis

TAXONOMY
Fringilla hyemalis Linnaeus, 1758, South Carolina. Sixteen subspecies.

OTHER COMMON NAMES
English: Gray-headed junco, pink-sided junco, red-backed junco, Schufeldt's junco, slate-colored junco, Thurber's junco, Townsend's junco; French: Junco ardoisé German: Junko; Spanish: Junco Ojioscuro.

PHYSICAL CHARACTERISTICS
5–6.5 in (13–17 cm); 0.5–0.88 oz (15–25 g). Medium-sized sparrows that lack breast streaking and have white outer tail feathers. They are geographically variable, and the 16 subspecies can be divided into five subspecies groups. Slate-colored juncos (*J. h. hyemalis*) are found in eastern North America west to Alaska and the mountains of British Columbia. Adults have pink bills and are uniformly gray above, with a white belly; females are similar to males, but are paler gray, often washed with brownish. White-winged juncos (*J. h. aikeni,*) breed in southeast Montana, western South Dakota, and northwest Nebraska. They are grayish above, with a white belly and two white wing bars. Pink-sided juncos (*J. h. mearnsi*) breed in southeast Alberta, southwest Saskatachewan, and south to southeast Idaho. They have a dull brown back and pink flanks. Gray-headed juncos (*J. h. caniceps*) breed in the Rocky Mountains. They have a gray head with dark gray around the eye and a rusty-red mantle. The *J. h. oreganus* group have a dark gray head, cinnamon brown upperparts, and pinkish washed flanks. Juvenile juncos are dusky, and heavily streaked both on the back and breast, with whitish bellies.

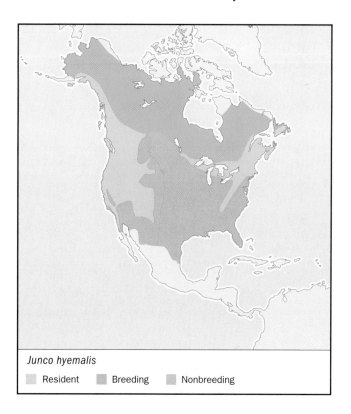

Junco hyemalis

⬜ Resident ⬛ Breeding ⬛ Nonbreeding

DISTRIBUTION
Breeds north to the limit of trees in Alaska and Canada and south to northern Georgia, northern Ohio, northern Minnesota, central Saskatchewan, and in the mountains to central New Mexico and Arizona, and northern Baja California, Mexico. Winters along the Pacific Coast of southern Alaska, southern Yukon, and northeast British Columbia, east through central British Columbia, southern Manitoba, southern Quebec, and southern Newfoundland, south to south Florida, the Gulf Coast, and northern Mexico.

HABITAT
Breed in a variety of habitats, but especially in open coniferous or mixed woodlands. In winter, they are found in brush, woodland edge, and hedgerows.

BEHAVIOR
Territorial and found in pairs or family groups during the breeding season. Males sing from an exposed perch in a tree, often from near the top of a small conifer. In winter, they often occur in loose flocks and frequently associate with other species of sparrows. They hop or run on the ground.

FEEDING ECOLOGY AND DIET
Juncos feed on the ground. In summer their diet is mostly insects; seeds are the principal winter food.

REPRODUCTIVE BIOLOGY
Monogamous. The nest, which is a cup of woven grasses and rootlets, usually is placed on the ground but occasionally is found low in a bush. Nesting takes place from May through July. They lay three to six eggs. Incubation lasts 11–13 days, and the young fledge after 9–13 days. Both parents feed the young.

CONSERVATION STATUS
Not threatened.

SIGNIFICANCE TO HUMANS
None known. ◆

Lapland longspur
Calcarius lapponicus

TAXONOMY
Fringilla lapponica Linnaeus 1758, Lapland. Two subspecies.

OTHER COMMON NAMES
English: Alaska longspur, lapland bunting; French: Bruant lapon; German: Spornammer; Spanish: Escribano Lapón.

PHYSICAL CHARACTERISTICS
5.5–7 in (23.5–17 cm); 0.95 oz (27 g). Lapland longspurs are large, strong-flying sparrows. The sexes differ in color. Breeding males have a black face outlined with buff white stripes above the eye and behind the ear, the back of the head and nape are bright rusty, and the bill is bright yellow with a black tip; they also have a black bib. Females have a buff stripe over the eye and buff ear coverts that are outlined in black, and the throat is blackish. Juveniles have streaking on the crown, in the buff line over the eye, and on their underparts.

DISTRIBUTION
Circumpolar. In Eurasia, they breed from Finland west across northern Russia and Siberia to the Kamchatka Peninsula and winter from northern Europe and northern Asia south to the British Isles, France, southern Russia, Mongolia, and northern China. In North America, they breed from the Aleutian Islands, Alaska east across arctic Alaska and Canada (including the arctic islands) to the coast of eastern Greenland and winter from the central Great Plains and southern Wisconsin, southern Ontario, and central Nova Scotia south to the Gulf Coast and northern Florida. *C. l. alascensis* breeds in north and west Alaska including the Aleutian Islands and islands in the Bering Sea. *C. l. coloratus* breeds in eastern Siberia, the Kamchatka Peninsula, and Commander Islands, and occasionally east to Attu Island, Alaska.

HABITAT
They are generally the most common birds of the high arctic where they can be found in a variety of tundra habitats. During migration and winter, they can be found in fallow fields, short pastures, and along beaches.

BEHAVIOR
Males arrive on the breeding ground before females and start defending and advertising a territory by giving a flight song and chasing intruders from their territory. They also sing from a rock, the top of a sedge tussock, or phone wires. During courtship, the pair engages in reciprocal chasing. During migration and winter, they are often found in large flocks, sometimes of more than a million individuals. They may also be found with horned larks (*Eremophila alpestris*), pipits (*Anthus*), other longspurs, or snow buntings (*Plectrophenax nivalis*).

FEEDING ECOLOGY AND DIET
They feed on the ground and, in summer, they eat insects and other invertebrates. In winter they eat primarily seeds and grain.

REPRODUCTIVE BIOLOGY
Monogamous. The nest is a cup of rather coarse sedges placed in a depression in the ground. They lay one to six (usually five) eggs; incubation requires 10–14 days, and the young fledge after 8–10 days. Both parents feed the young. Nesting takes place from late May through early July.

CONSERVATION STATUS
Not threatened.

SIGNIFICANCE TO HUMANS
None known. ◆

Snow bunting
Plectrophenax nivalis

TAXONOMY
Emberiza nivalis Linnaeus 1758, Lapland. Two subspecies.

OTHER COMMON NAMES
English: Snowflake, snowbird; French: Bruant blanc; German: Schneeammer; Spanish: Escribano Nival.

PHYSICAL CHARACTERISTICS
6–7.5 in (15–19 cm); 1.5 oz (42 g). Sexes differ in color. Males in summer have a white head, a black back sometimes mottled with brown, a black rump mottled with white, white outer tail feathers partially tipped with black, and white underparts. In winter, the white areas are washed with pale rusty brown.

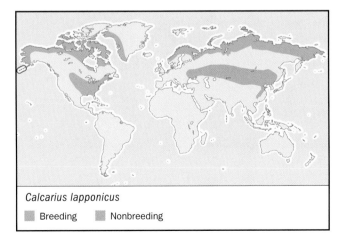

Calcarius lapponicus
  ■ Breeding   ■ Nonbreeding

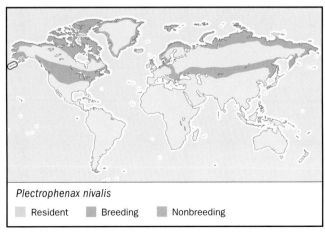

Plectrophenax nivalis
  ■ Resident   ■ Breeding   ■ Nonbreeding

Females in summer resemble breeding males, but the crown is dusky and black areas are paler, often brownish. In winter they resemble winter males. Juveniles are grayish with pale bellies.

DISTRIBUTION
Circumpolar. Breeds from Iceland, northern Scotland, the mountains of Norway and Sweden, Spitzbergen, Franz Joseph Land, north Kola Peninsula, Novaya Zemlya, northern Russia and northern Siberia east to Wrangel Island, the Bering Strait, and south to east Kamchatka, northern Alaska and mountains of Alaska, northern Canada north to Labrador, and the coast of Greenland. Winters south to British Isles, coast of northern France, Denmark, Germany, Poland, southern Russia, Manchuria, Korea, Kuril Islands, and Hokkaidō, and in North America to western and southern Alaska and from central and southern Canada south along the Pacific coast to northern California, the central Plains, and coastal North Carolina. *P. n. insulae* breeds in Iceland, and *P. n. vlasowae* breeds in northeast Russia east through Siberia and to the Bering Strait.

HABITAT
Breed in the high Arctic in sparse, dry, rocky areas such as shores, mountain slopes, and rocky outcrops. During migration and winter they are characteristically found in field, pastures, roadsides, and along the shore.

BEHAVIOR
Males arrive on the breeding grounds well before females. When the weather begins to warm, they establish territories, and chasing, flight-singing, and fights are common. When on the ground, they run rather than hop. In winter they are often found in fairly large flocks. As they move through a field, they appear to roll along like blowing snow as the birds at the back of the flock leap-frog over those toward the front. Although they generally stay on the ground, they sometimes will fly up into a tree. They are sometimes associated with horned larks (*Eremophila alpestris*) and Lapland longspurs (*Calcarius lapponicus*).

FEEDING ECOLOGY AND DIET
They feed on the ground. In summer they take insects and other invertebrates, but in winter they eat principally seeds and grain.

REPRODUCTIVE BIOLOGY
Most are monogamous, but individuals of either sex may have two mates. Nesting takes place from late May through July. The nest, which is a large thick-walled bulky cup of dried sedges, grasses, and lichens, is placed on the ground, often in a crevice in rocks. They lay three to nine (usually four to seven) eggs. Incubation lasts 10–15 days, and the young fledge after 10–17 days. Both parents feed the young.

CONSERVATION STATUS
Not threatened.

SIGNIFICANCE TO HUMANS
None known. ◆

Yellowhammer
Emberiza citrinella

TAXONOMY
Emberiza citrinella Linnaeus, 1758, Sweden. Three subspecies.

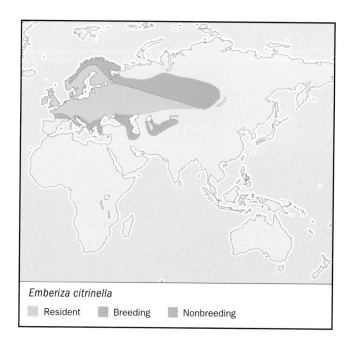

Emberiza citrinella

Resident ▮ Breeding ▮ Nonbreeding

OTHER COMMON NAMES
French: Bruant jaune; German: Goldammer; Spanish: Escribano Cerillo.

PHYSICAL CHARACTERISTICS
6.3–6.5 in (16–16.5 cm); 0.88–1.1 oz (25–31 g). Sexes differ in color. Males have a streaked yellow head, a cinnamon-washed black and breast, and a chestnut rump. Females and juveniles have much less yellow and are more heavily streaked.

DISTRIBUTION
E. c. citrinella breeds from southeast England, northern and western Europe east to Russia and south to northern Portugal and Spain, central Italy, northern Greece, Bulgaria, Rumania, Poland, and south in Russia to Moscow. In eastern and south-eastern Europe they intergrade with *E. c. erythrogenys*, which is found west to central Siberia. They winter from central Europe south to northern Africa, Iraq, and northern Mongolia. *E. c. caliginosa* is resident in Ireland, Scotland, Wales, and northern and western England.

HABITAT
Open country in hedgerows, the edge of woods, or in bushes. In winter they are found in cultivated fields or woodland edge.

BEHAVIOR
During the breeding season they are territorial, and males defend the territory by singing from an exposed perch, such as a wire. In fall and winter they form loose flocks, sometimes of up to 1,000 individuals.

FEEDING ECOLOGY AND DIET
Feed almost entirely on the ground, in pastures, cultivated ground, stubble, or waste ground. In summer they eat seeds, insects, and other invertebrates, but in winter they chiefly eat seeds.

REPRODUCTIVE BIOLOGY
Socially Monogamous. Eggs are laid from April through early September. The nest is nearly always placed on the ground and is well hidden among the vegetation; two to six (usually four to

five) eggs are laid. Incubation lasts 12–14 days, and the young fledge at 11–13 days.

CONSERVATION STATUS
Not threatened. Their numbers have declined in many European countries but have increased in Finland, all apparently as a consequence of changing farming practices.

SIGNIFICANCE TO HUMANS
None known. ◆

Rock bunting
Emberiza cia

TAXONOMY
Emberiza cia Linnaeus, 1766, Austria. Ten subspecies.

OTHER COMMON NAMES
French: Bruant fou; German: Zippammer; Spanish: Escribano Montesino.

PHYSICAL CHARACTERISTICS
6.3 in (16 cm); 0.7–0.9 oz (20–25 g). Sexes differ in color. Males have a gray, black, and white patterned head with a gray throat, a chestnut-streaked back, a rufous breast, belly, and rump, and a white-edged tail. Females and juveniles are much duller.

DISTRIBUTION
Breeds from southern France and southern Germany south to Spain, Portugal, North Africa, Italy, and Greece, Turkey, north and central Iran, southern Turkistan, northwest India, western Himalayas, and western Nepal. Some birds move south or to lower elevation in winter.

HABITAT
Inhabit dry, rocky hillsides, and, in winter, weedy or grassy areas with hedges and trees. They often roost in trees.

BEHAVIOR
Territorial during breeding season. They defend their territory with song, which is emitted from a perch on the top of a bush or rock. Outside of the breeding season they occur singly or in small groups. Wintering flocks form in late summer and may contain other *Emberiza* species.

FEEDING ECOLOGY AND DIET
Primarily feed on the ground among rocks and shrubs. In summer they take insects and other invertebrates (including small snails), as well as seeds. Winter diet is principally seeds, which they find in pastures, cultivated land, and gardens.

REPRODUCTIVE BIOLOGY
Monogamous. Eggs are laid from May through July. The nest is placed on or close to the ground and is well concealed in vegetation. They lay three to six (usually four to five) eggs; incubation lasts 12–14 days, and young fledge at 10–13 days of age. Both parents feed the young.

CONSERVATION STATUS
Not threatened.

SIGNIFICANCE TO HUMANS
None known. ◆

Reed bunting
Emberiza schoeniclus

TAXONOMY
Emperiza schoeniclus Linnaeus, 1758, Sweden. Fourteen subspecies.

OTHER COMMON NAMES
French: Bruant des roseaux; German: Rohrammer; Spanish: Escribano Palustre.

PHYSICAL CHARACTERISTICS
6–6.5 in (15–16.5 cm); 0.6–0.8 oz (17.5–22 g). Sexes differ in

Emberiza cia
  ▢ Resident   ▨ Breeding   ▨ Nonbreeding

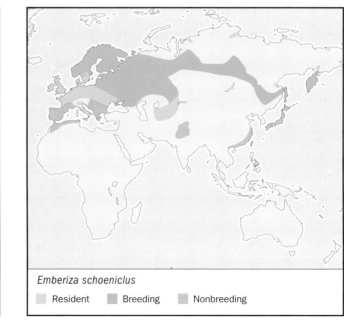

Emberiza schoeniclus
  ▢ Resident   ▨ Breeding   ▨ Nonbreeding

color. Males have a black head and throat and a white collar. The back and rump are brown and streaked, and the belly is whitish with varying amounts of black streaking, especially on the flanks. Females have a brown head with a buff stripe above the eye and around the ear coverts, a streaked brown back, and a light buff breast with brown streaks. Young birds resemble females.

DISTRIBUTION
Breeds throughout Europe (except for southern Greece) east into Russia (northern Kazakhstan) and southwest Siberia, southwest Mongolia, and locally in China (Xinjiang (*E. s. pyrrhuloides*), Heilongjiang (*E. s. minor*), and in northeast Siberia, Kamchatka, Kurile Island, and Hokkaidō (*E. s. pyrrhulina*). Winters in western and southern Europe, Turkey, southeast China, and Japan.

HABITAT
Breed in tall reeds, swamps, sewage farms, and marshy areas adjacent to fresh and brackish marshes. In winter, they are found in cultivated fields, farm lands, and open woodlands, often not near water.

BEHAVIOR
Territorial in summer, and males advertise their territory by singing. They regularly form flocks outside the breeding season, often flocking with other *Emberiza* species. These flocks form in September.

FEEDING ECOLOGY AND DIET
Feed on the ground or in vegetation. They commonly catch flying insects in sally flights. They sometimes make holes in bullrush stems to extract insect larvae. In the breeding season, they principally eat invertebrates (especially insects) but in winter take many seeds.

REPRODUCTIVE BIOLOGY
Monogamous, but extrapair fertilizations are common and bigamy occurs in some populations (probably where densities are high). The nesting season is geographically variable. The well-hidden nest is usually placed on the ground but may be placed in a bush or small tree. They lay two to six (usually five) eggs. Incubation lasts 12–15 (commonly 13) days, and the young fledge after 9–13 (usually 10–12) days. Both parents feed the young.

CONSERVATION STATUS
Not threatened. There are no consistent population trends, and the species is common in many parts of its range.

SIGNIFICANCE TO HUMANS
None known. ◆

Corn bunting
Miliaria calandra

TAXONOMY
Emberiza calandra Linnaeus, 1758, Sweden. Several subspecies have been described, but geographic variation is slight.

OTHER COMMON NAMES
French: Bruant proyer; German: Grauammer; Spanish: Triguero.

PHYSICAL CHARACTERISTICS
7 in (18 cm); 1.2–1.75 oz (35–50 g). A large bunting with a stout bill. Sexes are alike in color, but males are larger than females. They are uniformly brownish and streaked with

Miliaria calandra
☐ Resident ☐ Breeding ☐ Nonbreeding

brown, with underparts paler than upperparts; some have a blackish spot on the breast.

DISTRIBUTION
Breeds from Britain, southern Sweden, and Lithuania southeast across Russia to the Caspian Sea and south through all of eastern Europe and the Mediterranean islands to the Canary Islands, North Africa, Syria, northern Iraq, Iran, northern Afghanistan and Tajikistan, and western Tianshan, China. Winters in the breeding range south to Israel and southern Iraq and Iran.

HABITAT
Live in open country with few bushes, especially in farmlands.

BEHAVIOR
In the breeding season, males advertise by singing. They often give a flight song with their legs dangling in a distinctive way and their wings uplifted. They flock outside of the breeding season, often with other seed-eating birds; flocks start to form after the end of the breeding season.

FEEDING ECOLOGY AND DIET
Feed on the ground in fields, damp meadows, and in short grass. In the breeding season they take animal material (arthropods, snails, earthworms) and seeds, and in winter they eat seeds (especially cereals).

REPRODUCTIVE BIOLOGY
Mostly monogamous, but in some populations about 20% of the males are polygynous. The nest is placed on the ground in thick tangled grass or in a shrub or depression. They lay two to seven (usually four to five) eggs. Incubation takes 12–14 days, and young fledge in 9–13 days. The female does most of the feeding, but males with more than one mate tend to feed more than males in a monogamous pair.

CONSERVATION STATUS
Not threatened, though populations of corn buntings have declined in most places in Europe, due mainly to changes in agricultural practices.

SIGNIFICANCE TO HUMANS
None known. ◆

Crested bunting
Melophus lathami

TAXONOMY
Emberiza lathami J. E. Gray, Canton, Kwangtung, China. Two subspecies are little different and not recognized by all authors.

OTHER COMMON NAMES
French: Bruant huppé; German: Haubenammer; Spanish: Pinzón Capitón.

PHYSICAL CHARACTERISTICS
6.5–7 in (16.5–17 cm); 0.63–0.88 oz (18–25 g). Sexes differ in color. Males are blackish with a prominent crest and rusty wings and tail, which is black-tipped; females are brown, crested, and have rusty edges to the wing and tail feathers. Juveniles are paler than females with thin streaks on the breast.

DISTRIBUTION
M. l. lathami is resident in central, southeast, and southwest China south to North Vietnam. *M. l. subcristatus* is resident from western Pakistan east through northern India, Nepal, Sikkim, Bhutan, and Myanmar. There is some altitudinal migration.

HABITAT
Inhabit rocky, grassy hillsides with sparse shrubs and dry rice paddies in China. In southeast China they often nest in tea fields.

BEHAVIOR
Gregarious outside of the breeding season, when they form loose flocks.

FEEDING ECOLOGY AND DIET
Feed on the ground on seeds and invertebrates.

REPRODUCTIVE BIOLOGY
Monogamous. Nesting takes place between April and August during the local wet season. The nest is a neat cup of woven grass placed on the ground under vegetation or a rock or in a crevice. They lay three to five eggs; no data on incubation and fledging. Both parents feed the young.

CONSERVATION STATUS
Not threatened.

SIGNIFICANCE TO HUMANS
None known. ◆

Blue-black grassquit
Volatinia jacarina

TAXONOMY
Tanagra jacarina Linnaeus, 1766, northeastern Brazil. Two subspecies.

OTHER COMMON NAMES
French: Jacarini noir; German: Jacariniammer; Spanish: Semillerito Negriazulado.

PHYSICAL CHARACTERISTICS
4–4.3 in (10–4,3 cm); 0.34 oz (9.7 g). Sexes differ in color. Males are uniformly blue-black. Females are brown, paler below, with dark-streaked chest. Juveniles resemble females. Males obtain breeding plumage at about one year of age.

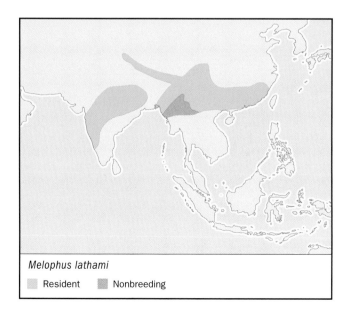

Melophus lathami
  ▢ Resident ▨ Nonbreeding

Volatinia jacarina
  ▢ Resident

DISTRIBUTION
V. j. jacarina resident from central and eastern Brazil south to central Argentina and northern Chile; *V. j. splendens* resident from central Mexico south to Venezuela and Colombia; also on Grenada in the Lesser Antilles.

HABITAT
Low, seasonally wet grassland, arid lowland scrub, and weedy fields.

BEHAVIOR
Males sing conspicuously from fences and display by jumping upward with a flick of their wings. In winter, they are found in flocks, some containing a few hundred birds; they sometimes flock with other small seed-eating birds.

FEEDING ECOLOGY AND DIET
The diet is almost exclusively grass seeds, although they eat some insects and berries. They feed by picking seeds from grass seed heads and will pick grit and seeds from roads.

REPRODUCTIVE BIOLOGY
Monogamous. The nest is placed low to the ground to 10 ft (3 m) high. Commonly three, less often two, eggs are laid from May through October. Data on incubation and fledging not available.

CONSERVATION STATUS
Not threatened; locally abundant.

SIGNIFICANCE TO HUMANS
None known. ◆

Cactospiza pallida

▨ Resident

Woodpecker finch
Cactospiza pallida

TAXONOMY
Cactornis pallida Slavin and Slavin, 1870, Galápagos Islands.

OTHER COMMON NAMES
French: Géospize pique-bois; German: Spechtfink; Spanish: Chimbito Pálido.

PHYSICAL CHARACTERISTICS
6 in (15 cm). Sexes similar in color. Upperparts brown or olive-brown, with little streaking. Underparts whitish with slight streaking on breast.

DISTRIBUTION
Galápagos Islands.

HABITAT
Mixed cactus and trees with dense understory; less common in arid areas.

BEHAVIOR
Solitary or occurring in small groups, probably family groups.

FEEDING ECOLOGY AND DIET
Eat largish insects that they find behind bark. They probe into dead wood, in bark, or by flaking off bark using a cactus spine or twig,

REPRODUCTIVE BIOLOGY
Not available.

CONSERVATION STATUS
Not threatened. Fairly common on several islands.

SIGNIFICANCE TO HUMANS
None known. ◆

Variable seedeater
Sporophila americana

TAXONOMY
Loxia americana Gmelin, 1789, Cayenne. Three subspecies.

OTHER COMMON NAMES
English: Wing-barred seedeater, black seedeater; French: Sporophile variable; German: Wechselpfäffchen; Spanish: Semillero Variable.

PHYSICAL CHARACTERISTICS
4.2–4.5 in (11–11.5; cm); 3.5–0.4 oz (10–11 g). Sexes differ in color. Males from Mexico to Costa Rica are black with white at the base of their primaries; males from Costa Rica to northeast Peru and Amazonian Brazil are highly variable, with white on the throat and side of the neck, a black chest band, and a white belly and gray rump (and in some places in South America two thin white wing bars). Females uniformly olive-brown. Juveniles resemble females.

DISTRIBUTION
S. a. americana resident from northeast Venezuela south to northeast Brazil; *A. a. aurita* resident from western Costa Rica

Sporophila americana

■ Resident

east to northeast Venezuela; *A. a. corvina* resident from eastern Mexico south to western Panama.

HABITAT
Second-growth scrub, weedy fields, woodland edge, and secondary forest.

BEHAVIOR
Males sing from a bush or a tree, usually from a low perch. When not breeding they occur in small flocks in which young birds and females usually are more common than males.

FEEDING ECOLOGY AND DIET
Feed almost exclusively on grass seeds. They sit on grasses and pick seeds from grass heads; sometimes they fly to a grass head and bend it to the ground, making feeding easier. They also feed on grass growing in water well away from land.

REPRODUCTIVE BIOLOGY
Monogamous. The nest cup is low to mid-level in a bush. Commonly two, but occasionally three, eggs are laid during the time of year when their food is most abundant, which varies seasonally and geographically. The incubation period is about 12 days; young fledge after 11–13 days.

CONSERVATION STATUS
Not threatened; locally abundant.

SIGNIFICANCE TO HUMANS
None known. ◆

Chestnut-capped brush-finch
Atlapetes brunneinucha

TAXONOMY
Embernagra brunnei-nucha Lafresnaye, 1839, Veracruz, Mexico. Nine subspecies.

OTHER COMMON NAMES
French: Tohi à nuque brune; German: Braunkopf-Buschammer; Spanish: Saltón Gorricastiaño.

PHYSICAL CHARACTERISTICS
7–7.5 in (17.19.5 cm); 1.2 oz (33 g). Sexes similar in color. A medium-sized sparrow with a chestnut cap, black forehead and side of face, a black band across the throat (lacking in some populations), a white throat, gray flanks, and olive-green back, wings, and tail. Juveniles have a blackish brown head with brownish olive upperparts, white mottling on the throat, and a yellow-streaked belly.

DISTRIBUTION
Resident from southern Veracruz and Guerrero south to Panama, and in South America from northern Venezuela and Colombia to southern Peru.

HABITAT
Montane evergreen forest and secondary forest.

Atlapetes brunneinucha

■ Resident

BEHAVIOR
Do not form flocks, but individuals of a pair stay together, and when young are fledged the family group feeds together for some time. They live in dense undergrowth and rarely sing from an exposed perch. They commonly puff out their white throats.

FEEDING ECOLOGY AND DIET
Forage on the ground primarily for insects. They lurk on the outskirts of a column of army ants where they presumably capture insects that are flushed by the ants.

REPRODUCTIVE BIOLOGY
Monogamous, and the individuals of the pair are always found close together. They seem to produce one brood per year, and the clutch contains one to two eggs. Nests are placed near the ground. Incubation and fledging periods are unknown.

CONSERVATION STATUS
Not threatened.

SIGNIFICANCE TO HUMANS
None known. ◆

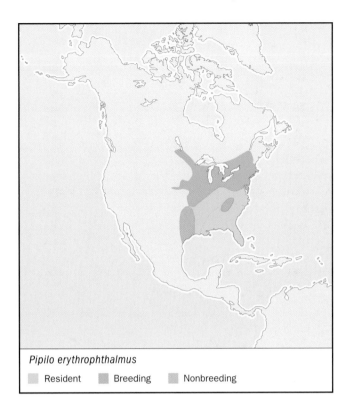

Pipilo erythrophthalmus

■ Resident   ■ Breeding   ■ Nonbreeding

Eastern towhee
Pipilo erythrophthalmus

TAXONOMY
Fringilla erythrophthalma Linnaeus, 1758, South Carolina. Four subspecies.

OTHER COMMON NAMES
English: Red-eyed towhee, rufous-sided towhee; French: Tohià flancs roux; German: Rötelgrundammer; Spanish: Rascador Ojirrojo.

PHYSICAL CHARACTERISTICS
7–8 in (17–20.5 cm); 1.2–1.8 oz (35–50 g). Sexually dimorphic, large, long-tailed sparrows. Males have a black head, throat back, and tail, white at the base of the primary wing feathers, rusty flanks, and a white belly. Females are patterned as males but are a warm brown where the males are black. Eye color varies from bright red in the north to yellowish or whitish in the southeast United States. Juveniles are brownish and are streaked below.

DISTRIBUTION
Breed from southern Manitoba, east across southern Canada, and south to Florida and the Gulf Coast (west to Texas). In winter, they migrate from the northern part of the breeding range southward and to eastern Texas and central Oklahoma. The four subspecies differ slightly in size and in eye color.

HABITAT
Inhabit dense deciduous thickets or edges of woodlands. In the South they may be found in scrub palmetto.

BEHAVIOR
Solitary or occur in pairs or family groups. Males often sing from a conspicuous perch in the top of a tall bush or from a tree. When not singing they can be difficult to see.

FEEDING ECOLOGY AND DIET
Forage on the ground by scratching in leaf litter for insects or seeds.

REPRODUCTIVE BIOLOGY
Monogamous. The nest is placed in a depression in the ground under a bush or occasionally low in a bush or vine. Nesting occurs between April and mid-August. Clutch size is two to six (usually three to four) eggs. Incubation takes 10–12 days and young leave the nest after 8–10 days. Both parents feed the young.

CONSERVATION STATUS
Not threatened.

SIGNIFICANCE TO HUMANS
None known. ◆

Plumbeous sierra-finch
Phrygilus unicolor

TAXONOMY
Emberiza unicolor Lafresnaye and d'Orbigny, 1837, Tacora, Tacna, Peru. Six subspecies.

OTHER COMMON NAMES
French: Phrygile gris-de-plomb; German: Bleiämmerling; Spanish: Frigilo Plomizo.

PHYSICAL CHARACTERISTICS
6 in (15 cm); 0.75 oz (21 g). Sexually dimorphic, rather large sparrows of the high mountains. Males are uniformly leaden gray; females are streaked throughout with a coarsely streaked breast and grayish rump. Juveniles resemble females.

DISTRIBUTION
Resident in the Andes from western Venezuela south to Argentine Tierra del Fuego.

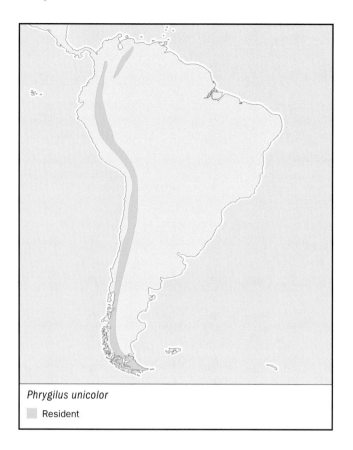

Phrygilus unicolor
Resident

Aimophila aestivalis
Resident

HABITAT
Found in pastures, meadows, shrubby edges, especially along rivulets in paramo, and rarely up to the snowline.

BEHAVIOR
Not shy, they allow close approach; they crouch low to the ground before flushing. After flying a short distance they drop back into the paramo vegetation where they can be difficult to see. Found in pairs or in small groups.

FEEDING ECOLOGY AND DIET
Feed on the ground, probably on seeds and invertebrates

REPRODUCTIVE BIOLOGY
Probably monogamous. Breeding season varies geographically.

CONSERVATION STATUS
Not threatened.

SIGNIFICANCE TO HUMANS
None known. ◆

Bachman's sparrow
Aimophila aestivalis

TAXONOMY
Fringilla aestivalis Lichtenstein, 1823, Georgia, U.S.A. Three subspecies.

OTHER COMMON NAMES
English: Pine-wood sparrow; French: Bruant des pinèdes; German: Bachmanammer; Spanish: Zacatonero de Bachman.

PHYSICAL CHARACTERISTICS
5–6 in (12–16 cm); 0.7 oz (20 g). A fairly large, large-billed, round-tailed sparrow with reddish brown lateral crown stripes, streaked scapulars and back, gray chin and throat, and unstreaked breast. Sexes are alike in color. Birds from the southeastern United States are more rufous in coloration than are western birds.

DISTRIBUTION
A. a. aestivalis and *A. a. bachmani*, which are very similar in appearance, breed from Virginia to Florida and west to Louisiana. *A. a. illinoensis* breed from southern Missouri to central Louisiana and eastern Texas. They migrate from the northern part of their range in winter.

HABITAT
Inhabit open pine woods with fairly rank understory of wiregrass, palmettos, and weeds. They also occur in oak-palmetto scrub and in grasslands away from pine woods and in degraded pastures.

BEHAVIOR
In the breeding season, males sing persistently from an exposed perch, commonly in a pine tree. At other times they are secretive and hard to see and may run rather than fly when pursued.

FEEDING ECOLOGY AND DIET
Feed on the ground, eating insects, other invertebrates, and seeds. In winter they principally eat seeds.

REPRODUCTIVE BIOLOGY
Socially monogamous. Nests are placed on the ground, usually in dense cover, and are well concealed by vegetation. Two to five (usually four) eggs are laid from mid-April through July (commonly in June). Incubation lasts 12–14 days, and the young fledge after about 10 days.

CONSERVATION STATUS

Near Threatened. Has declined in numbers since the 1930s and is extirpated in the northern parts of its range.

SIGNIFICANCE TO HUMANS

None known. ◆

Chipping sparrow

Spizella passerina

TAXONOMY

Fringilla passerina Bechstein, 1798, Quebec. Six subspecies.

OTHER COMMON NAMES

English: Chippy; French: Bruant familier; German: Schwirrammer; Spanish: Gorrión Cejiblanco.

PHYSICAL CHARACTERISTICS

4.5–5.5 in (12–14 cm); 0.4 oz (12 g). A small, slim sparrow with a long, notched tail. Sexes are similar in color. Adults have a rufous cap with a white stripe over the eye, a black eyeline stripe, a gray nape and rump, and pale gray, unstreaked underparts. Juveniles are like adults but buff, with a streaked, brown cap.

DISTRIBUTION

Breeds from southeast Alaska east across Canada to southwest Newfoundland and south to Florida, the Gulf Coast west to northern Baja California, and south in the highlands of Mexico to Guatemala. Winters in southern United States and Mexico.

HABITAT

Breed in dry, open woodlands and woodland edge with fairly open understory and in urban parks and golf courses. They are found in deciduous, coniferous, or mixed woods.

BEHAVIOR

During breeding season, males sing persistently from a tree, usually not from an exposed perch. During migration they often occur in large, loose flocks. Their flight is strong, fast, and direct.

FEEDING ECOLOGY AND DIET

Feed both in trees and on the ground. During migration, they often feed on the ground in loose flocks. In summer, their diet consists principally of insects; in winter, they eat mainly seeds.

REPRODUCTIVE BIOLOGY

Monogamous. The cup-like nest is placed in a tree, commonly a conifer, from 3 to 56 ft (1–19 m) high; nest rarely found on the ground. Three to five (usually four) eggs are laid from March through July. Incubation takes 11–14 days, and the young fledge after 9–12 days. Both parents feed the young.

CONSERVATION STATUS

Not threatened.

SIGNIFICANCE TO HUMANS

None known. ◆

Lark bunting

Calamospiza melanocorys

TAXONOMY

Fringilla bicolor J. K. Townsend, 1837, Nebraska. *Calamospiza melanocorys*, Stejneger, 1885; *Fringilla bicolor* preoccupied.

OTHER COMMON NAMES

English: White-winged blackbird, prairie bobolink; French: Bruant noir et blanc; German: Prärieammer; Spanish: Gorrión Alipálido.

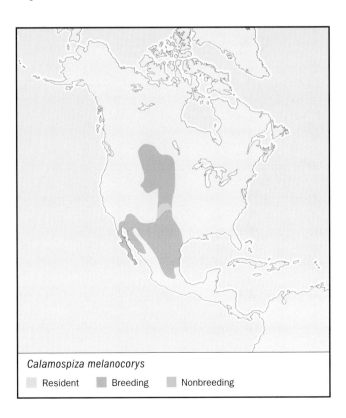

Spizella passerina

■ Resident ■ Breeding ■ Nonbreeding

Calamospiza melanocorys

■ Resident ■ Breeding ■ Nonbreeding

PHYSICAL CHARACTERISTICS
5.5–7 in (14–18 cm); 1.4 oz (40 g). A large, chunky sparrow with a large bill. Sexes differ in color. Males in breeding plumage are black with conspicuous white patches in the wing and white corners to the tail. Females are heavily streaked with chocolate-brown, with whitish buff in the wings and white or light buff corners to the tail. Juveniles are similar in color to females but with a yellowish cast to their plumage. Males in winter resemble females but usually have some black feathers.

DISTRIBUTION
Breed from the southern Canadian prairies south to eastern New Mexico and northwest Texas. Winters from southern Utah, Arizona, New Mexico, and central Texas south through Baja California and northern Mexico.

HABITAT
Breed in shortgrass prairie interspersed with sage or other shrubs. In winter then are found in weedy, dry grasslands or open farmland.

BEHAVIOR
On the breeding ground, they are conspicuous birds, with males frequently giving an elaborate stiff-winged flight display. They run or hop on the ground. In winter they are found in flocks.

FEEDING ECOLOGY AND DIET
Forage on the ground, eating mostly insects in the summer and seeds in the winter.

REPRODUCTIVE BIOLOGY
Usually monogamous, although some males, especially where density is high, have two or more mates. The cup-shaped nest is placed on the ground, under a bush or in taller vegetation where it is protected from the sun. Three to seven (usually four to five) eggs are laid from mid-May through mid-July. Incubation lasts 12 days, and the young leave the nest after 8–9 days.

CONSERVATION STATUS
Not threatened.

SIGNIFICANCE TO HUMANS
None known. ◆

Savannah sparrow
Passerculus sandwichensis

TAXONOMY
Emberiza sandwichensis Gmelin, 1789, Unalaska, Alaska. Twenty-one subspecies.

OTHER COMMON NAMES
English: Belding's sparrow, Bryant's sparrow, Ipswich sparrow, large-billed sparrow, marsh sparrow, San Benito sparrow, seashore sparrow; French: Bruant des prés; German: Grasammer; Spanish: Sabanero Zanjero.

PHYSICAL CHARACTERISTICS
4.6–6 in (11–15 cm); 0.6–1.1 oz (17–30 g). Typically brown or dark brown streaked on the back and breast, with pink legs, a

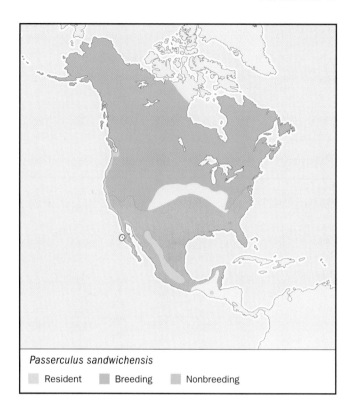

Passerculus sandwichensis
▨ Resident ▨ Breeding ▨ Nonbreeding

yellow stripe above the eye, a pale median crown stripe, and a rather short, notched tail. Sexes are alike in color, but they are geographically variable.

DISTRIBUTION
Breeds from northern Alaska, northern Canada (absent on the arctic islands), south to northern Georgia, the central Great Plains, and south in the mountains to Guatemala. They winter along the east coast of the United States, west through the central Plains, and south to northern Central America. They are resident along the west coast from southern British Columbia south to southern Baja California, along the west coast of Mexico, south to central Sinaloa, and in the highlands of central Mexico.

HABITAT
Inhabit open country, such as grassy meadows, cultivated fields, pastures, roadsides, sedge bogs, the edge of salt marshes, and tundra.

BEHAVIOR
Found in pairs and family groups in summer. Territorial males typically sing from an exposed perch. On the ground they run or hop. During migration and winter they can be found in small, loose flocks. Flight is strong and direct.

FEEDING ECOLOGY AND DIET
In summer, they eat a variety of invertebrates, especially insects, but also some seeds and fruit. In winter, they principally eat seeds. They forage on the ground, low in bushes and weeds, or along the tide line on beaches and in beach wrack.

REPRODUCTIVE BIOLOGY
Generally monogamous, but in some populations a male may have two mates (bigamy). The nest is a woven cup of grasses

and other vegetation that is placed on the ground or in a slight depression and is partly covered by grasses and other vegetation. Two to six (usually four to five) eggs are laid from February to August. Incubation usually lasts about 12 days, and young usually fledge after 10–12 days. Both parents feed the young.

CONSERVATION STATUS
Common throughout its range, but declining in eastern North America as marginal pastures are reverting to woodland.

SIGNIFICANCE TO HUMANS
None known. ◆

Resources

Books

Byers, C., U. Olsson, and J. Curson. *Buntings and Sparrows.* Sussex: Pica Press, 1995.

Collar, N.J., LP. Gonzaga, N. Krabbe, A. Madroño Nieto, L.G. Naranjo, T.A. Parker III, and D.C. Wrege. *Threatened Birds of the Americas.* 3rd ed. Washington, DC: Smithsonian Institution Press, 1993.

Cramp, S., and C.M. Perrins, eds. *Handbook of the Birds of Europe, the Middle East and North Africa.* Vol. 9. Oxford: Oxford University Press, 1994.

Ridgely, R.S., and G. Tudor. *The Birds of South America.* Vol. 1. Austin: University of Texas Press, 1989.

Rising, J.D., and D.D. Beadle. *A Guide to the Identification and Natural History of the Sparrows of the United States and Canada.* London: Academic Press, 1996.

James David Rising, PhD

New World warblers

(Parulidae)

Class Aves
Order Passeriformes
Suborder Passeri (Oscines)
Family Parulidae

Thumbnail description
Small- to medium-sized songbirds with nine primaries (outer flight feathers); very fine, thin bills; and often with colorful plumage

Size
4–7.5 in (10–19 cm)

Number of genera, species
About 28 genera; 126 species

Habitat
Varies among species from heavily wooded deciduous and coniferous forests, marshes, and swamps to semi-open shrubby areas

Conservation status
Critically Endangered: 3 species; Endangered: 5 species; Vulnerable: 6 species; Near Threatened: 7 species

Distribution
North, Central, and South America, and the West Indies

Evolution and systematics

The wood warbler family (Parulidae) was once, and occasionally still is, split into two subfamilies. The "wood warblers proper" (Parulinae) includes about 26 genera and 116 species, and are most numerous in North America. The "bananaquits" (Coerebinae) has one species, which is common in the Caribbean. Many classifications, including the one currently in use by the American Ornithologists' Union (AOU), now place that species in its own family, the Coerebidae. In addition, some classifications also place the nine species of conebills (*Conirostrum* spp.) in the Parulidae, which would change the family totals to 27 genera and 125 species. Various additional changes have occurred over the years, and several will be addressed below.

As one of the largest bird families, the Parulidae presents a number of challenges to taxonomists, and the most vexing of these center around whether a group of birds warrants the title of species, subspecies, or simple variant. Perhaps the most well-known example involves a group of birds that have fallen under four common names: the blue-winged, golden-winged, Brewster's, and Lawrence's warblers. The two prominent groups—the blue- and golden-wings—look quite

different. A blue-wing (*Vermivora pinus*) has a golden head and underside, with a darker back, a black eye stripe, and a blue-gray tail and wings, which feature two conspicuous white wing bars. A golden-wing (*Vermivora chrysoptera*) has a yellow crown and two yellow wing bars, and the rest of the bird resembles the white, black, and gray pattern of a chickadee, with a gray back and wings, white underside, and black eye mask and throat. Historically, the two birds were separated geographically, with the blue-wings tending toward the central United States and the golden-wings remaining in the East. Both preferred the low bushes of semi-open, shrubby woodlands often at the edges of forests, and remained fairly well separated as long as large expanses of open field existed between them. The open fields persisted into the early and mid-1800s, but woodlands slowly began to take over the fields, and the blue-wings expanded their range to overlap that of the golden-wings. Although the two looked dissimilar and had distinct songs, they interbred and produced fertile offspring. The confusion escalated when birders identified the offspring as two additional species, the Brewster's and Lawrence's warblers. A Brewster's looks like a bluewing, except that its wing bars are yellow instead of white, and it has a whitish instead of yellow underside. The

Sexually dimorphic plumage in black-throated blue warblers (*Dendroica caerulescens*). The female is on the nest, while the male perches on a nearby branch. (Illustration by Gillian Harris)

Lawrence's, on the other hand, resembles the golden-wing, but with a yellow rather than white underside.

For years, taxonomists were left scratching their heads. The first to consider the possibility that blue- and golden-wings were the same species, and were producing hybrid offspring, was probably John James Audubon, who apparently examined one of the earliest collected specimens of a Brewster's warbler. Through his astute observational skills, he noted the similarities to the blue-winged and golden-winged warblers, and asserted in a letter dated 1835 that the three were likely the same species. His later writings made no reference to this contention, however, and the puzzle continued for several decades until the early 1900s when birder Walter Faxon found a mating pair made up of a female blue-winged and male golden-winged, and discovered the progeny all to be Brewster's warblers. Lawrence's warblers were later similarly discovered to be hybrids. Although the findings settled the classification of Lawrence's and Brewster's, the status of the blue-winged and golden-winged warblers was much less certain. Currently, taxonomists regard them as two separate species, as they do with some other occasional interbreeders, like the hermit (*Dendroica occidentalis*) and Townsend's warblers (*D. townsendi*). On the other hand, taxonomists have determined that two formerly separate species, known as the myrtle and Audubon's warblers, are actually one species with two plumage patterns. The birds, which still retain their common names (Audubon's warbler in western North America and myrtle warbler in the East), are now both listed as the yellow-rumped warbler (*D. coronata*).

The phylogenetic relationship of Parulidae to other bird families is also less than clear-cut. Some taxonomists have placed them closest to the tanager family, Thraupidae (sometimes treated as a subfamily, Thraupinae, of Emberizidae), while others feel they are nearest to the New World finches

family, Emberizidae. In some classifications, the wood warblers are actually listed as a subfamily, Parulinae, within Emberizidae.

Without a substantial fossil trail to follow, taxonomists in the past relied mainly on anatomical, morphological, and behavioral characteristics to deduce the family's evolutionary history. DNA-comparison technology, however, has now allowed scientists to obtain a different view of phylogenies. One group compared the genetic code of different birds, and determined that the wood warblers are genetically so similar to other groups of birds, including the blackbirds, buntings, cardinals, and tangers that they together should make up just one of three subfamilies within the family Fringillidae. While the work is intriguing, most birders, including the AOU, still use the more traditional arrangement with the wood warblers in their own family, the Parulidae. The AOU has, however, determined that one of the warblers is different enough from the other wood warblers to justify its own family. That bird is the olive warbler (*Peucedramus taeniatus*), which may reside alone in the Peucedramidae family.

Other taxonomists have used DNA to determine how long ago the wood warblers split from their ancestral rootstock into the many species that occur today. For example, one group compared the timing of the speciation of New World wood warblers vs. Old World warblers (Sylviidae) by studying differences in one specific gene, the mitochondrial cytochrome *b* gene. From the 13 wood warblers and 8 Old World warblers they studied, they determined that the wood warblers experienced an explosive radiation in the late Pliocene or early

A prairie warbler (*Dendroica discolor*) adult watches its chick eating. (Photo by B. Henry/VIREO. Reproduced by permission.)

Different foraging methods in warblers: American redstart (*Setophaga ruticilla*) (top) sallying into the air to hawk insects; northern parula (*Parula americana*) (middle left) hangs from a branch tip to glean foliage; bay-breasted warbler (*Dendroica castanea*) (middle right) cleans branches and foliage; northern waterthrush (*Seiurus noveboracensis*) (bottom) hunts through leaf debris and at the water's edge. (Illustration by Gillian Harris)

Pleistocene, much later than the other families within the Passeriformes order, including the Old World warblers. The research team suggested that changing environmental conditions in the late Pliocene opened new habitats to the ancestral wood warblers, and speciation followed. Another study conducted by a separate research group also used analyses of mitochondrial DNA to take a closer look at wood warbler phylogeny. In this study, the scientists investigated the *Dendroica* genus, which includes more than two dozen wood warbler species, and concluded that the *Dendroica* speciation dated back as far as the late Miocene or early Pliocene, making the extant species the current end-points of ancient lineages.

Although DNA studies may ultimately produce a definitive phylogeny for the wood warblers, much work remains as scientists examine additional species, struggle to sift through the mounting, and sometimes conflicting, genetic data, and determine how to apply the new knowledge appropriately.

Physical characteristics

The Parulidae is a vast group, yet the wood warblers share a number of traits. Most have slender or flat beaks that are pointy, but short. Some wood warblers, like the yellow-breasted chat (*Icteria virens*) and yellow-rumped warbler, have

A Connecticut warbler (*Oporornis agilis*) in flight. (Photo by Doug Wechsler/VIREO. Reproduced by permission.)

more robust bills, while the American redstart (*Setophaga ruticilla*) sports distinctive rictal bristles, which are stiff, modified feathers at the base of its flat bill. Most wood warblers are small birds, tending toward the smaller side of the family's 4–7.5 in (10.2–19.1 cm) range. They are characterized by legs that look no more sturdy than a toothpick, and have the typical three-toes-forward foot structure of other birds in the Passeriformes order.

Wood warbler plumage ranges widely in color. These "jewels of the forest" generally have yellow, red, black, gray, or green areas of plumage, with yellow and olive the predominant colors within the family. Males of the temperate species are usually much brighter in color and have sharper patterns than do the females, but males of some species become duller and resemble the females in fall and winter. Juvenal plumage is frequently similar to the female's, but duller still. Among the more tropical species, the males and females generally look alike. In some species of wood warbler, like the yellow warbler (*Dendroica petechia*), the male is slightly larger than the female. A unifying trait among all wood warblers is the presence of nine functional primary feathers. Other songbirds typically have 10.

Another identifying feature of many wood warblers is their seemingly constant movement. They flit from branch to branch, usually giving observers only a quick glimpse of color before they fly off to another spot. A birder can spend hours in the field listening to a common yellowthroat (*Geothlypis trichas*) or yellow warbler singing nearby, but may be unable to focus the binoculars on anything more than a branch left swaying by the bird that just left.

Finally, one of the most distinguishing characteristics of the wood warblers is their songs, which add greatly to na-

ture's symphony. Prime examples include the Louisiana waterthrush (*Seiurus motacilla*), an insignificant-looking ground dweller that often utters its loud, pleasant tune to a background of rushing water; and the yellow-breasted chat, which has a loud, flutelike and gurgling song. The ovenbird (*Seiurus aurocapillus*) is much less melodious, with a blunt cry that sounds like "teacher, teacher." In contrast, the feeble "heeebsss" song of the blue-winged warbler sounds like an insect and is often identified as such by inexperienced human observers.

Distribution

The wood warblers are distributed over almost all of the New World from Alaska and northern Canada to northern Chile and Argentina. More than 50 species breed within the central and eastern portions of North America from about the middle half of the United States to the southern half of Canada.

Many wood warblers are wide-ranging. The yellow warbler and common yellowthroat, for example, breed from coast to coast in North America, throughout the United States and well into Canada and Mexico. On the opposite end of the spectrum, the Kirtland's warbler (*Dendroica kirtlandii*) occupies a tiny breeding range in north-central Michigan.

Habitat

These songbirds occupy a great number of habitats from thick to semi-open woodlands, to marshes and swamps, to forest edges. They generally prefer areas that have dense shrubs or thickets, where they often spend much of their time. While most favor woodlands or woodland edges, others tend toward more exotic habitats. The hermit warbler (*Dendroica occidentalis*), for instance, lives in the fairly open coniferous forests along the west coast of the United States, whereas the cerulean warbler (*Dendroica cerulea*) inhabits the dense deciduous forests in the eastern half of the United States, and the Kirtland's warbler breeds in stands of young jack pines in Michigan. The northern waterthrush (*Seiurus noveboracensis*) lives in the swamps of Canada, Alaska, and the northernmost United States, while the worm-eating warbler (*Helmitheros vermivorus*) exists in dry woodlots in the central-eastern United States, and Lucy's warbler (*Vermivora luciae*) makes a living in the mesquite deserts of the southwestern United States and Mexico.

This family has infiltrated many niches, much as the Old World warblers have done in the Eastern Hemisphere. While individual species of wood warblers are not gregarious during the breeding season, several different species frequently coexist. From a good vantage point along the edge of a forest in the northeastern United States, a birder can expect to hear upwards of eight to 10 different species of wood warblers singing in early summer.

It is important to point out that most of what is known of the wood warblers is based on information collected during the North American species' breeding seasons. Although the species spend about eight months on average in their winter homes and only four in their breeding areas, North America has many more birders, as well as organized research projects, in place to study them.

Behavior

Most North American wood warblers are long-distance migrants. Their regular passage is one of the most striking features of North American bird life. During the height of the spring migration in May, it is possible to see up to 30 species in a day in eastern North America. The North American species winter mainly in Central America and northern South America, but a few resistant species winter in the United States, while others move as far south as Argentina.

Although the birds are fairly small, they travel very long distances between their summer breeding grounds and winter homes, with some species traveling 3,000 mi (4,800 km) and more each way—quite a feat for such a small bird. The birds use almost their entire fat stores during the flight. An American redstart with summer territory in New Hampshire, for example, loses about 50% of its premigration weight during its 1,800 mi (2,900 km) autumn flight to its wintering site in Jamaica. Yellow warblers travel even farther. With residences in Central and South America, but summer breeding grounds reaching into northern Canada and Alaska, the birds can face treks upwards of 3,000 mi (4,800 km).

The birds generally migrate from dusk to dawn during their spring and fall migrations. With top speeds of 30 mph (48 kph), the birds can cover considerable ground—often more than 100 mi (160 km) a night—when wind and weather conditions are right. Upon landing, these energy-starved birds spend little time resting, instead filling the day by filling their stomachs. During these flights, different species will often flock together, presenting a spectacular viewing opportunity for human observers. At night, the warblers are visible as they fly in front of a bright moon. During the day, the multispecies flocks land en masse, making quite a racket while searching for insects. Males and females will chip and chirp in flight and on their feeding forays, and males add to the daytime chatter in the spring by performing a few prebreeding songs.

Once the birds reach the breeding grounds, most wood warblers become territorial toward other birds of the same species. While a 20-acre (8-hectare) New England forest may contain 10 or more different warbler species, males of the same species separate themselves from one another. Territory sizes depend on the availability of food and nesting sites, but usually range from about a third of an acre (0.13 ha) to five acres (2 ha) or more. Males typically rely on their songs and visual displays to ward off intruding males. American redstarts have a particularly elaborate display, involving a succession of circling movements that serve to confirm their territorial boundaries.

In addition to visual cues, wood warblers communicate through their songs, which are used only by the males in all but a few species, and through their calls. Bird calls are the nonmelodic "chips" that both males and females utter year-round. Wood warblers, like all other songbirds in the Oscines

A black-and-white warbler (*Mniotilta varia*) perches on a small branch. (Photo by G. Bailey/VIREO. Reproduced by permission.)

suborder of the Passeriformes, learn their songs. The learning begins within the first week after hatching, with the males learning to sing songs, and the females learning to recognize them. An exception to the rule is the female painted redstart (*Myioborus pictus*), whose song rivals that of the males of the species.

Feeding ecology and diet

Wood warblers feed mainly on insects, which they seek busily everywhere, much like the Old World leaf warblers. While searching for food, they move on the ground amid grass, on bushes, or in the foliage—in some cases also on tree bark. Several species are able to live and feed together in overlapping habitats because of the availability of their primary food source, as well as the slightly different feeding habits many exhibit.

Although small arthropods are the main food of wood warblers, some species occasionally take berries, seeds, or even the juice inside fruits. Regardless of their fare, wood warblers have bills suited to the job at hand. Their typically thin and pointy bills allow them to extricate insects and spiders from tiny splits in bark or spaces between grass blades. The sharper bill of many other warblers expands their feeding range into seeds and berries. Species like the myrtle warbler/yellow-rumped warbler can only survive the winter in eastern North America by feeding on berries and sometimes seeds. Others, like the blue-winged warbler, have longer bills that provide access to insects buried within flowers. The American redstart is unusual in its bill structure and its feeding habits. Its relatively short and broad bill combine with its quick reflexes and excellent flying ability to give it the tools required for in-flight insect capture, much like the flycatchers (family Tyrannidae).

A pair of chestnut-sided warblers (*Dendroica pensylvanica*) at their nest with chicks. (Photo by B. Henry/VIREO. Reproduced by permission.)

While wood warblers retain their general feeding habits all year, they typically are territorial among their own species only when they are on summer breeding grounds. During the migratory flights and at their winter residences, most warblers are communal, if not necessarily gregarious. In some species, the males and females separate and tend toward different habitats in the winter, so an observer might spot a group of males in the woods and a small flock of females in a field.

Reproductive biology

The reproductive biology of wood warblers, like most other birds, is highly cyclical. In general, wood warblers follow this annual pattern: following a spring migration, males arrive in their breeding grounds and use their voice and visual displays to set up territories and attract mates. Males and females form pairs that are monogamous at least for the season. Although yearlings engage in the mating activities, they are typically less successful than older birds.

Of all of the warblers' activities, their songs draw the most attention. According to several studies, warblers appear to use one variation of their song to establish and defend territory boundaries, and a second variation to attract females. Several studies have investigated the importance of this song in the female's choice of a mate, as well as the long-term effects of such sexual selection. One study has suggested that a superior singing voice in a male songbird is related to the level of development of specific learning areas in the brain, and that development hinges on how well the male overcame nutritional stresses as a hatchling and young fledgling when those learning areas were forming. The birds that fared best, the author argues, have the most well-developed learning areas

with which to build singing and perhaps other skills. By selecting the best singers, then, the females are actually choosing the highest-quality mates.

After a pair forms, the majority of the 53 North American wood warblers remain monogamous for the breeding season, while Central and South American pairs may persist for longer periods. Among some species, females may pair socially with one male, but mate with another. As a result some males help rear hatchlings that are not their own. Among yellow warblers, one of the most common of the North American warblers, females that engage in this so-called "extra-pair mating" most frequently do so with males that are large in size and have extra brown streaking on their yellow breasts, thus perpetuating these male traits through sexual selection.

Nesting is species-specific in wood warblers, and experienced ornithologists can often identify the species just by looking at its nest. Wood warbler nests are frequently cup-shaped, and may be on the ground, among grass, in bushes, or on trees, often more than 50 ft (15 m) high. They are usually firm, densely woven structures, generally with an inner lining layer of rootlets or moss. Nonetheless, nests among species in this family exhibit considerable variation. The prothonotary warbler (*Protonotaria citrea*) makes its nest in tree cavities, while the Louisiana waterthrush prefers a hole dug into a stream bank. Grace's warbler (*Dendroica graciae*) builds its small cup high on pine branches, but the ovenbird makes a roof of leaves to hide its ground-level nest.

The clutch size varies a great deal within the family, but four eggs per nest is typical of North American warblers and three is common among the tropical genera. Food supply often regulates clutch size, with sparse years producing fewer eggs per pair. The eggs are usually white with pale reddish brown or black markings, sometimes arranged as in a wreath. If the birds breed early enough, they sometimes have time to lay another clutch if the first fails. Through the incubation period, the female stays with the eggs, and the male brings her food. Eggs, on average, hatch in 12 days, at which time both the female and male take on feeding duty. The young fledge in about 10 days, and although they have already reached 90% of their adult weight, they lack adequate flight muscles or coordination. The parents continue feeding them for several more weeks, with the male generally responsible for half the fledglings and the female accountable for the other half. After that, the young birds are on their own.

Conservation status

In the wood warbler family, the IUCN lists seven as Near Threatened, seven as Vulnerable (including one conebill), five as Endangered, and three as Critically Endangered species. Of the Critically Endangered, two are feared to already be Extinct. They are the Semper's warbler (*Leucopeza semperi*) of the mountains of St. Lucia in the West Indies, and Bachman's warbler (*Vermivora bachmanii*), a once abundant North American breeding bird that wintered in West Indies' forests. As might be expected, the remaining at-risk species are typified by small ranges, where even small habitat destruction can be disastrous. Of the at-risk species, many breed in North America, including one Near Threatened, five Vulnerable, three Endangered, and two Critically Endangered species.

Vulnerable species include: Pirre warbler (*Basileuterus ignotus*), elfin-wood warbler (*Dendroica angelae*), Kirtland's warbler, pink-headed warbler (*Ergaticus versicolor*), Altamira yellowthroat (*Geothlypis flavovelata*), white-winged (ground) warbler (*Xenoligea montana*), and Tamarugo conebill (*Conirostrum tamarugense*). The Endangered species include: gray-headed warbler (*Basileuterus griseiceps*), whistling warbler (*Catharopeza bishopi*), golden-cheeked warbler (*Dendroica chrysoparia*), black-polled yellowthroat (*Geothlypis speciosa*), and Paria redstart (*Myioborus pariae*), which also goes by the common names of Paria whitestart and yellow-faced redstart. Besides the Semper's and Bachman's warblers, Belding's yellowthroat (*Geothlypis beldingi*) is Critically Endangered and faces extinction from the handful of small sites it occupies in the Baja California peninsula of Mexico.

In addition to the at-risk species, scientists are also concerned about other songbirds, even those with more vast ranges. In the 1970s and 1980s, birders began to notice and report precipitous declines in populations of many common wood warblers, such as the cerulean warbler and the Kentucky warbler (*Oporornis formosus*), both of which breed in the central and eastern United States. The reports prompted research, which has since found the major culprit to be habitat destruction in the wintering grounds combined with habitat fragmentation in the breeding grounds. Many North American wood warblers share wintering grounds in Costa Rica, Guatemala, and other Neotropical and tropical countries that have undergone extensive deforestation, so scientists have taken a particularly close look at the relationship between habitat loss and songbird declines. A study of American redstarts, a black-and-orange bird that breeds in a swath across North America from western Canada south and east to the Atlantic coast of Canada and the United States, showed that the birds' success in the breeding season relied on the quantity and quality of their wintering habitat in Jamaica. With less habitat to go around, some redstarts wintered in lush, rich forests, while others were relegated to dry scrublands. Those in the rich forests found more food, faced less overall stress, and were in prime condition to leave on their spring migrations up to four weeks earlier than the less fortunate birds eking out a living in the scrublands. The earlier arrival in North America of the healthier warblers meant less competition for choice nesting sites and mates.

Other research has placed the blame for songbird declines on changes in breeding habitat, particularly fragmentation that involves breaking up large expanses of open land, usually with developments. The fragmentation of habitat has obvious impacts on species that require large territories, but the primary detriment to songbirds is that fragmentation paves the way for the many predatory animals that thrive along habitat edges. One particularly devastating animal is the brown-headed cowbird (*Molothrus ater*), a brood parasite that lays its eggs in the nests of other birds, including warblers. The cowbird hatchlings are much larger and can outcompete with their nest mates for food from the warbler parents, which typically

Prairie warbler (*Dendroica discolor*) female pants to keep her eggs cool in Perry County, Pennsylvania. (Photo by Jeff Lepore. Photo Researchers, Inc. Reproduced by permission.)

feed whichever hatchling makes its mouth most available. The cowbird hatchlings become stronger and bigger, and those young warblers that avoid being crushed or simply pushed out of the nest by the cowbird hatchlings usually starve to death. Besides cowbirds, egg-eating raccoons and opossums also flourish in edge areas, and can devastate the warblers that nest on the ground.

Many efforts are now under way to combat warbler declines, and several have been quite successful. The Kirtland's warbler is one example. Due to habitat restoration efforts, populations have risen in recent years, and the bird has been upgraded from its 1994 ranking as Endangered to a 1996 placement at Vulnerable, where it has remained.

Significance to humans

Too tiny to be much of a source of food or feathers, warblers still play an important role in human existence. Perhaps their greatest significance lies in their aesthetic beauty. Nothing lifts the spirit quite like a common yellowthroat alighting just for a moment on a sun-streaked branch, or the bright, tinkling song of a yellow warbler on a warm, late spring morning in the northern woods. These birds fill the skies during migration, breaking the spell of winter in the north, or dance back home to Central and South America in autumn. No other group of animals could take the place of these bundles of energy in inviting people to come outside and revel in nature's splendor.

1. Kirtland's warbler (*Dendroica kirtlandii*); 2. Cerulean warbler (*Dendroica cerulea*); 3. Yellow warbler (*Dendroica petechia*); 4. Hooded warbler (*Wilsonia citrina*); 5. Yellow-breasted chat (*Icteria virens*); 6. Blue-winged warbler (*Vermivora pinus*); 7. Golden-winged warbler (*Vermivora chrysoptera*); 8. Black-and-white warbler (*Mniotilta varia*); 9. Ovenbird (*Seiurus aurocapillus*); 10. Prothonotary warbler (*Protonotaria citrea*). (Illustration by Gillian Harris)

Species accounts

Yellow-breasted chat
Icteria virens

SUBFAMILY
Parulinae

TAXONOMY
Icteria virens Linnaeus, 1758, South Carolina.

OTHER COMMON NAMES
French: Ictérie à poitrine jaune; German: Gelbbrust-Wald-sänger; Spanish: Reinita Grande.

PHYSICAL CHARACTERISTICS
With a length of up to 7.5 in (19 cm), it is the largest wood warbler. Upperparts are olive, throat and breast are bright yellow, and the rest of the underside is white. Face is blackish with a white eye stripe and eye circle, and it sports a plump, conical bill unlike the typical warbler bill of the family. The female and male are similar in appearance.

DISTRIBUTION
Breeds from east to west in the United States, north into southern Canada and south into Mexico. Winters from Mexico south to Panama.

HABITAT
Prefers to be near water, spending most of its time hidden among tangles and thickets.

BEHAVIOR
Unlike other warblers that flit among branches, the yellow-breasted chat takes more of a bull-in-a-china-shop approach—flopping, falling, and tail-swooning through the thickets. Its songs, some of which it performs at night, combine a few melodious bursts interspersed with gobbles, caws, and other less delicate sounds. It also has been described as a ventriloquist in that it can throw its voice and fool even experienced birders.

FEEDING ECOLOGY AND DIET
Mainly insects, but also eats berries.

REPRODUCTIVE BIOLOGY
Cup-like nests rest in the tangles and brambles, and typically hold up to six brown-and-purple-speckled eggs. More than one pair may share a nesting site.

CONSERVATION STATUS
Not threatened. Widespread and fairly common.

SIGNIFICANCE TO HUMANS
Song and behavior provide enjoyment for birders. ◆

Golden-winged warbler
Vermivora chrysoptera

SUBFAMILY
Parulinae

TAXONOMY
Vermivora chrysoptera Linnaeus, 1766, Pennsylvania.

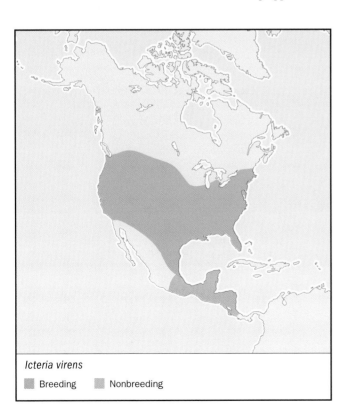

Icteria virens
▮ Breeding ▮ Nonbreeding

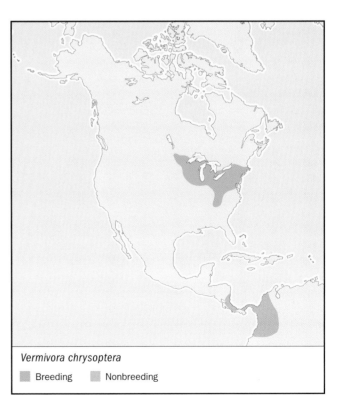

Vermivora chrysoptera
▮ Breeding ▮ Nonbreeding

OTHER COMMON NAMES
French: Paruline à ailes dorées; German: Goldflügel-Wald-sänger; Spanish: Reinita Gusanera.

PHYSICAL CHARACTERISTICS
4.5–5 in (11.4–12.7 cm). The very thin-billed golden-winged warbler has a yellow crown, and two yellow wing bars. The remaining plumage has a white, black, and gray pattern reminiscent of a chickadee, with a gray back and wings, white underside, and black eye mask and throat. The female is similar, but duller, with more olive coloration than gray and black.

DISTRIBUTION
Breed in northern United States, winter in Central America.

HABITAT
Forest edges and swamps.

BEHAVIOR
Has a buzzing song. Often seen hanging upside-down on branches, much as chickadees do.

FEEDING ECOLOGY AND DIET
Mainly insects.

REPRODUCTIVE BIOLOGY
Breeds in low bushes, building cup-shaped nests on or close to the ground. Average clutch is five to seven speckled eggs that hatch in less than two weeks. Sometimes occurs side by side with blue-winged warblers, and the two produce viable hybrid offspring, known by the common names of Brewster's and Lawrence's warblers.

CONSERVATION STATUS
Not threatened, but appears to be on the decline where its habitat overlaps with that of blue-winged warblers.

SIGNIFICANCE TO HUMANS
None known. ◆

Vermivora pinus
■ Breeding ■ Nonbreeding

Blue-winged warbler

Vermivora pinus

SUBFAMILY
Parulinae

TAXONOMY
Vermivora pinus Linnaeus, 1766, Pennsylvania.

OTHER COMMON NAMES
English: Blue-winged yellow warbler; French: Paruline à ailes bleues; German: Blauflügel-Waldsänger; Spanish: Reinita Aliazul.

PHYSICAL CHARACTERISTICS
4.5–5 in (11.4–12.7 cm). Golden head with a black eye stripe, golden underside, darker back, blue-gray tail and wings, and two white wing bars. Males and females have similar appearance.

DISTRIBUTION
Breeds in northern United States, winters from Mexico to Panama in Central America.

HABITAT
Forest edges and swamps.

BEHAVIOR
Moves rather deliberately through trees, sometimes hanging upside-down from branches. Often pauses for lengths of time on favorite perches.

FEEDING ECOLOGY AND DIET
Mainly insects.

REPRODUCTIVE BIOLOGY
Breeds in low bushes, building cup-shaped nests on or close to the ground. Average clutch is four to seven finely speckled eggs that hatch in less than two weeks. Sometimes occurs side by side with golden-winged warblers, and the two produce viable hybrid offspring, known by the common names of Brewster's and Lawrence's warblers.

CONSERVATION STATUS
Not threatened. Appears to be replacing golden-winged warblers in areas where they overlap.

SIGNIFICANCE TO HUMANS
None known. ◆

Prothonotary warbler

Protonotaria citrea

SUBFAMILY
Parulinae

TAXONOMY
Protonotaria citrea Boddaert, 1783, Louisiana.

OTHER COMMON NAMES
French: Paruline protonotaire; German: Orangefleck-Waldsänger; Spanish: Reinita Cabecidorada.

PHYSICAL CHARACTERISTICS
5–5.5 in (12.7—14 cm). Long, thin bill, bright golden yellow and gray plumage, and a fairly short tail. Its name comes from

Protonotaria citrea

Breeding   Nonbreeding

Mniotilta varia

Breeding   Nonbreeding

a resemblance of its plumage to the garb of court clerks, known as prothonotaries. The female is similar but less brightly colored.

DISTRIBUTION
Breed from Florida west to Texas and north to Minnesota, southern Ontario and Pennsylvania. Winters in pristine rain-forest habitats from southern Mexico to Venezuela.

HABITAT
Wooded swamps.

BEHAVIOR
Song is a repeated series of "tweets." Often seen poking into the saturated logs of the swamps, where it resides.

FEEDING ECOLOGY AND DIET
Mainly insects.

REPRODUCTIVE BIOLOGY
Unusual for warblers, it builds its mainly moss nests in tree holes such as old woodpecker holes, or in nest boxes. Average clutch size is three to eight, with speckled light pink eggs that hatch in about two weeks.

CONSERVATION STATUS
Not threatened.

SIGNIFICANCE TO HUMANS
None known. ◆

Black-and-white warbler
Mniotilta varia

SUBFAMILY
Parulinae

TAXONOMY
Mniotilta varia Linnaeus, 1766, Hispaniola.

OTHER COMMON NAMES
English: Black-and-white nuthatch; French: Paruline noir et blanc; German: Kletterwaldsänger; Spanish: Reinita Trepadora.

PHYSICAL CHARACTERISTICS
4.5–5.5 in (11.4–14 cm). The only bird in its genus, it is an abundantly striped black-and-white bird. Males have a black bib; females do not. The rear toe and claw are unusually long, and ideal for its creeper-like habits.

DISTRIBUTION
Breeds from the southern and central United States and far north into western Canada. Winters from the far southern Gulf states to the northern reaches of South America.

HABITAT
Coniferous or deciduous forests, particularly common in more northern areas.

BEHAVIOR
It has a soft, often-repeated "wee-ee" song. It pecks its food out of clefts in the bark of trees, which it searches carefully, usually by working over one tree trunk after another from near the ground by creeping upwards, much as nuthatches (*Sitta* spp.) do.

FEEDING ECOLOGY AND DIET
Mainly arthropods.

REPRODUCTIVE BIOLOGY
It is one of the earliest wood warblers to return to North America in the spring. It builds its nest of strips of bark, moss, grass, and other materials, behind a piece of bark on the lower part of a tree trunk or more frequently on the ground beneath a tree where it covers it partly with leaves. Produces four to five speckled eggs, which hatch in a week and a half.

CONSERVATION STATUS
Not threatened. Widespread and common in northern North America.

SIGNIFICANCE TO HUMANS
None known. ◆

Yellow warbler
Dendroica petechia

SUBFAMILY
Parulinae

TAXONOMY
Dendroica petechia Linnaeus, 1766, Barbados. More than six subspecies.

OTHER COMMON NAMES
English: Summer yellow bird, wild canary; French: Paruline jaune; German: Goldwaldsänger; Spanish: Reinita Amarilla.

PHYSICAL CHARACTERISTICS
4.5–5.5 in (11.4–14 cm). Primarily bright yellow bird with olive hint on upperparts. Breast of male is streaked with rusty color, female has no or only light streaking. Two yellow wing bars are present on its yellow-and-olive-colored wings.

DISTRIBUTION
Breeds in most of North America from Alaska and far northern Canada south to northwest Mexico. Winters from Mexico to Peru.

HABITAT
Often seen in gardens, its wild habitat includes woods and brushy areas near water.

BEHAVIOR
Songs are bright and repeated during the breeding season and beyond, often beginning before sunrise. The songs draw the

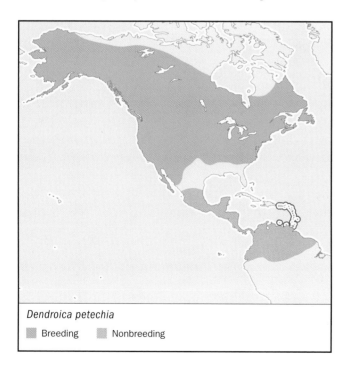

Dendroica petechia
■ Breeding   ■ Nonbreeding

attention of birders, but the birds seldom sit still long enough for a thorough viewing.

FEEDING ECOLOGY AND DIET
Mainly arthropods.

REPRODUCTIVE BIOLOGY
Begin mating shortly after arriving on the breeding grounds in early spring. Occasionally females will socially pair with one male, but mate with another. Eggs can be sired by either male. Nests are typically constructed of grass and plant materials in the crook of a branch about 6–8 ft (1.8–2.4 m) above the ground in small trees or shrubs. Eggs number four or five, often have a grayish or greenish white hue, and hatch in about a week and a half.

CONSERVATION STATUS
Not threatened. Their numbers, however, have been affected by cowbird brood parasitism. Yellow warblers often readily rear cowbird young, but occasionally will reject the parasitized brood altogether and built a new nest, often on top of the old one.

SIGNIFICANCE TO HUMANS
None known. ◆

Ovenbird
Seiurus aurocapillus

SUBFAMILY
Parulinae

TAXONOMY
Seiurus aurocapillus Linnaeus, 1766, at sea off Haiti. Two subspecies.

OTHER COMMON NAMES
English: Teacherbird; French: Paruline couronnée; German: Pieperwaldsänger; Spanish: Reinita Hornera.

PHYSICAL CHARACTERISTICS
5.5–6.5 in (14–16.5 cm). Brownish upperparts, two black stripes on crown of head with a rusty orange patch between. White throat with brown stripes. Underparts are white with mottled-brown breast, somewhat similar in appearance to a wood thrush.

DISTRIBUTION
Breeds in the northern half of the United States from the eastern coast through the Midwest and into Montana, and well into Canada from the Atlantic Ocean to Alberta.

HABITAT
Mixed and deciduous woods lacking dense bushy undergrowth.

BEHAVIOR
Instead of flitting from tree to tree in characteristic warbler fashion, it runs about the fallen leaves and tosses them aside in its search for hidden arthropods. It responds to the presence of an intruder, particularly when the latter is near the nest, by flying up, uttering "tick" calls or even by erecting the orange-colored feathers of its crown patch. It also sometimes mimics an injured bird to draw away the intruder. Its song is one of the best-known North American bird songs, and consists of a sequence of "teach-er, teach-er teach-er," which gets progressively louder. The popular name "teacherbird" derives from

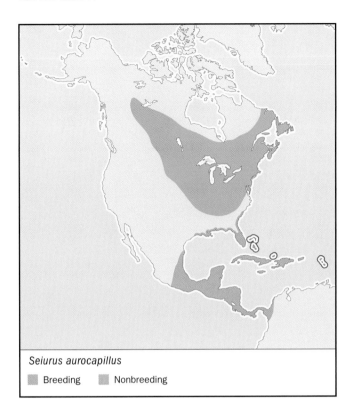

Seiurus aurocapillus

■ Breeding ■ Nonbreeding

Wilsonia citrina

■ Breeding ■ Nonbreeding

the song, which is delivered from the ground or from low to fairly high-placed branches.

FEEDING ECOLOGY AND DIET
Mainly arthropods, earthworms, and snails.

REPRODUCTIVE BIOLOGY
Courtship involves in-flight singing, posturing, and pursuit by the males. The ovenbird makes its nests on the ground, using grass and stems topped with leaves for the roof, and grass and hair for the interior lining. The nest resembles a Dutch oven, and is responsible for the bird's common name. Typically lay three to six speckled eggs, which hatch about two weeks later.

CONSERVATION STATUS
Not threatened. Their nests, however, are becoming more vulnerable to predators that prefer fragmented habitat, which is becoming more plentiful in their breeding grounds.

SIGNIFICANCE TO HUMANS
Particularly valued for their songs—they are seldom seen but frequently heard. ◆

Hooded warbler
Wilsonia citrina

SUBFAMILY
Parulinae

TAXONOMY
Wilsonia citrina Boddaert, 1783, Louisiana.

OTHER COMMON NAMES
French: Paruline à capuchon; German: Kapuzenwaldsänger; Spanish: Reinita Encapuchada.

PHYSICAL CHARACTERISTICS
5–6 in (12.7–15.3 cm). Dark olive above, yellow below, and a yellow face. Conspicuous bristles at the angles of its fairly wide bill. Male also has a black cowl that extends from his throat to the top of his head.

DISTRIBUTION
Breeds in the eastern half of the United States from the Gulf states north to central Iowa, Ohio, and parts of New York. Winters in Central America.

HABITAT
Undergrowth of mature deciduous woods.

BEHAVIOR
Like a flycatcher, the male often catches flying arthropods on the wing. The females, however, generally forage on foliage and branches nearer ground level. The male's loud ringing song, "wee te wee tee o," proclaims its presence in thickets, often near water. Unlike most other warblers, the sexes frequently segregate in the wintering grounds.

FEEDING ECOLOGY AND DIET
Mainly arthropods.

REPRODUCTIVE BIOLOGY
It builds its nest among bushes and climbers. The nest itself is firmly built of leaves and grasses and lined with rootlets or fine grass. The female lays three or four speckled eggs, similar to those of many other wood warblers. The young hatch a little more than a week later. This species typically produces two broods a year.

CONSERVATION STATUS
Not threatened.

SIGNIFICANCE TO HUMANS
None known. ◆

Kirtland's warbler
Dendroica kirtlandii

SUBFAMILY
Parulinae

TAXONOMY
Dendroica kirtlandii Baird, 1852, Ohio.

OTHER COMMON NAMES
French: Paruline de Kirtland; German: Michiganwaldsänger;
Spanish: Reinita Kirtland.

PHYSICAL CHARACTERISTICS
A dark gray and yellow bird that reaches 6 in (15.3) in length.
The body is essentially split by color, with gray above and yellow below. The underparts are mostly yellow with a bit of gray
streaking along the sides. Males are more brightly colored than
the females and sport a dark mask, which the females lack.

DISTRIBUTION
Breeds in numbers only in central Michigan. Winters primarily
in the Bahamas.

HABITAT
Jack pine stands, preferring those with younger trees.

BEHAVIOR
Frequently spotted among jack pine trees, searching for insects
and pecking at sap, which they also consume. The males have a
distinctive, loud, and somewhat low-pitched song.

FEEDING ECOLOGY AND DIET
Mainly insects, but also some plant materials and berries.

REPRODUCTIVE BIOLOGY
Breeding begins shortly after the birds' arrival in the spring.
They build their cup-shaped nests of grass and other plant materials, and conceal them on the ground beneath the branches
of young jack pines. Clutch size is generally four or five speckled eggs that are either white or tinged with pink. Hatching
occurs about two weeks later.

CONSERVATION STATUS
The IUCN Red List for 2000 lists the birds as Vulnerable, an
improvement from their 1990 and 1994 rankings of Endangered. The upgraded status results, at least in part, from efforts
to improve their habitat. These efforts have included periodic
burns of jack pine stands to trigger the stubborn cones to
open, release their seeds, and start new growth. Despite the increase in breeding pairs, the species' range is still small enough
to warrant the Vulnerable status. It is also one of the birds that
has fallen victim to the cowbird's brood parasitism.

SIGNIFICANCE TO HUMANS
They have become the source of some economic benefit in
Michigan, as the birds attract visitors, who wish to catch a
glimpse of this "poster child" of United States at-risk
species. ◆

Cerulean warbler
Dendroica cerulea

SUBFAMILY
Parulinae

TAXONOMY
Dendroica cerulea Wilson, 1810, Hispaniola.

OTHER COMMON NAMES
French: Paruline d'azur; German: Pappelwaldsänger; Spanish:
Reinita Cerúlea.

PHYSICAL CHARACTERISTICS
A blue, black, and white bird that reaches 4–5 in (10.2–12.7
cm) long. Its upperparts are blue to blue-gray with a few black

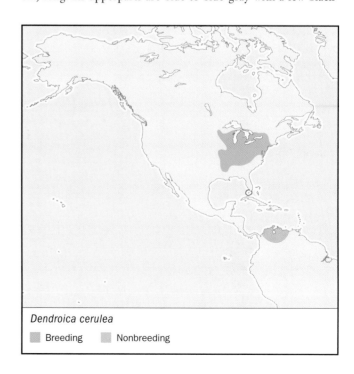

Dendroica kirtlandii
■ Breeding   ■ Nonbreeding

Dendroica cerulea
■ Breeding   ■ Nonbreeding

streaks, and the underparts are mostly white. The wings feature two white bars. The female looks similar, but substitutes a soft olive for the blue plumage, and lacks the black neck ring the males have.

DISTRIBUTION
Breeds mostly in the eastern half of the United States, north into the southernmost points in Ontario, south to North Carolina, and southwest to Louisiana. Winters primarily in rainforests of northern South America.

HABITAT
Deciduous forests, particularly among maples, elms, and black ash.

BEHAVIOR
These birds remain among the treetops most of the time, much to the chagrin of birders. They are always on the move, seldom staying still for more than a few minutes before moving to another perch. The birds sing throughout the day.

FEEDING ECOLOGY AND DIET
Insects.

REPRODUCTIVE BIOLOGY
The cup-shaped, neat nest, which the bird builds high up in the trees, consists mostly of moss and lichens. The white to greenish white, speckled eggs usually number four, and they hatch in about two weeks.

CONSERVATION STATUS
Although not threatened under IUCN criteria, cerulean warbler populations have declined by as much as 70% over the last three decades. The primary cause appears to be habitat destruction in both their breeding and wintering grounds. Efforts are currently under way to protect their northern and southern habitats.

SIGNIFICANCE TO HUMANS
They have some economic benefit, as they bring to the northern woods birders who are seeking a challenge in bird observation. ◆

Resources

Books

Bent, Arthur C. *Life Histories of North American Wood Warblers.* New York: Dover Publications, Inc., 1963.

Cassidy, James, ed. *Book of North American Birds.* New York: The Reader's Digest Association, Inc., 1990.

Dock Jr., George. "Yellow-Breasted Chat." In *Audubon's Birds of America.* New York: Harry N. Abrams, Inc., 1979.

Ehrlich, Paul R., David S. Dobkin, and Darryl Wheye. *The Birder's Handbook.* New York: Simon and Schuster, Inc. (Fireside Books), 1988.

Garrett, Kimball L., and John B. Dunning Jr. "Wood-Warblers." In *The Sibley Guide to Bird Life and Behavior,* edited by Chris Elphick, John B. Dunning Jr., and David Allen Sibley. New York: Alfred A. Knopf, 2001.

Peterson, Roger Tory. *A Field Guide to the Birds of Eastern and Central North America.* Boston: Houghton Mifflin Co., 1980.

Periodicals

Berger, Cynthia. "Exposed: Secret Lives of Warblers." *National Wildlife* 23 (2000): 46–52.

Dunaief, Daniel. "Taking Back the Nest." *Discover* 16 (1995): 34.

Heist, Annette. "Singing in the Brain." *Natural History* 109, no. 8 (2000): 14–16.

Lichtenstein, G. and S. G. Sealy. "Nestling Competition, Rather Than Supernormal Stimulus, Explains the Success of Parasitic Brown-Headed Cowbird Chicks in Yellow Warbler Nests." *Proceedings of the Royal Society of London B* 265, no. 1392 (2000): 249–254.

Line, Les. "Tale of Two Warblers." *National Wildlife* 32 (1994): 16–19.

Lovette, I. J., and E. Bermingham. "Explosive Speciation in the New World Dendroica Warblers." *Proceedings of the Royal Society of London B* 266, no. 1429 (1999): 1629.

Nowicki, Stephen, Susan Peters, and Jeffrey Podos. "Song Learning, Early Nutrition, and Sexual Selection in Songbirds." *American Zoologist* 38, no. 1 (1998): 179–190.

Price, T., H. L. Gibbs, L. de Sousa, and A. D. Richman. "Different Timing of the Adaptive Radiations of North American and Asian Warblers." *Proceedings of the Royal Society of London B* 265 (1998): 1969–1975.

Raikow, Robert J. "Phylogeny and Evolution of the Passerine Birds." *BioScience* 50, no. 6 (2000): 487–499.

Sibley, C. G., and J. E. Ahlquist. "Reconstructing Bird Phylogeny by Comparing DNAs." *Scientific American* (Feb. 1986): 82–92

Tangley, Laura. "A Good Place in the Sun: Tropical Winter Habitat Proves Critical to North America's Migratory Songbirds." *U.S. News & World Report* 125, no. 23 (1998): 63.

Van Buskirk, J. "Independent Evolution of Song Structure and Note Structure in American Wood Warblers." *Proceedings of the Royal Society of London B* 264, no. 1382 (1997): 755–761.

Weidensaul, Scott. "Jewels in the Treetops." *Country Journal* 23 (1996): 58–61.

Yezerinac, M., and P. J. Weatherhead. "Extra-Pair Mating, Male Plumage Coloration, and Sexual Selection in Yellow Warblers *(Dendroica petechia).*" *Proceedings of the Royal Society of London B* 264, no. 1381 (1997): 527–532.

Organizations

National Audubon Society Population & Habitat Program. 1901 Pennsylvania Ave. NW, Suite 1100, Washington, DC 20006 USA. Phone: (202) 861-2242. E-mail: population@audubon.org Web site: <http://www.audubonpopulation.org>

Smithsonian Migratory Bird Center, Smithsonian National Zoological Park. 3001 Connecticut Avenue, NW, Washington, DC 20008 USA. Phone: (202) 673-4800. E-mail: nationalzoo@nzp.si.edu Web site: <http://www.natzoo.si.edu>

Resources

The Songbird Foundation. 2367 Eastlake Ave. East, Seattle, WA 98102 USA. Phone: (206) 374-3674. Fax: (206) 374-3674. E-mail: kim@songbird.org Web site: <http://www.songbird.org>

Other

"Threatened Animals of the World." Listing of at-risk species prepared by the UNEP World Conservation Monitoring Centre. (19 January 2002). <http://www.wcmc.org.uk/data/database/rl_anml_combo.html>

Leslie Ann Mertz, PhD

New World blackbirds and orioles

(Icteridae)

Class Aves

Order Passeriformes

Suborder Passeri (Oscines)

Family Icteridae

Thumbnail description

Medium to large-sized songbirds with short and massive to rather long and slender conical bills; many are colorful and sexually dimorphic in both size and coloration

Size

6.7–21.5 in (17–55 cm); 0.6 oz–1.2 lb (16–528 g)

Number of genera, species

25 genera; 103 species

Habitat

Forest, woodlands, marshes, and grassland

Conservation status

Critically Endangered: 3 species; Endangered: 4 species; Vulnerable: 4 species; Near Threatened: 1 species; Extinct: 1 species

Distribution

North, Central, and South America (Alaska to Cape Horn)

Evolution and systematics

The New World blackbirds, troupials, and meadowlarks (Icteridae) include the beautiful orioles (*Icterus*) and caciques (*Cacicus*), as well as many other colorful songbirds. The icterids are variously treated as a family, subfamily, or tribe in a more inclusive family (Emberizidae or Fringillidae) that also contains a number of other songbirds having only nine primary wing feathers. Most sources separate these at the familial level. Other groups of birds that are closely related to the blackbirds are the olive warbler (*Peucedramus taeniatus*), bananaquit (*Coereba flaveola*), and other New World warblers (Parulidae), and the tanagers, buntings, and other New World finches (Emberizidae). These birds no doubt evolved in the New World, and today all, with the exception of the Emberizidae, are confined in their distribution to the New World. The Icteridae almost certainly evolved in South America where today the largest number of species occur. One large genus of blackbirds, the orioles, may be of Central American origin. These families probably differentiated rather recently in geological time, perhaps during the Pliocene. All of the fossils of icterids, however, are from the Pleistocene or Holocene.

Molecular studies delineate five distinct lineages within the Icteridae. These are: (1) the grackles and allies (including *Agelaius*, *Euphagus*, *Quiscalus*, *Molothrus*, and 11 other genera), (2) the caciques and oropendolas (*Cacicus* and *Psarocolius*), (3) the orioles (*Icterus*), (4) the meadowlarks and allies (*Dolichonyx*, *Xanthocephalus*, *Sturnella*, and *Leistes*), and (5) the cup-nesting caciques (*Amblycercus holosericeus*).

Physical characteristics

Icterids are remarkably diverse in size. They range from the slim orchard oriole (*Icterus spurius*), which weighs as little as 0.6 oz (16 g), to the robust Montezuma's oropendola (*Psarocolius montezuma*), which can weigh as much as 1.2 lb (528 g), making it one of the largest songbirds. Most species are brightly colored, commonly with black (often with a bright metallic gloss) and yellow, but many species have prominent red coloration, and some of the oropendolas are greenish. Some species have a neck ruff; others, a sparse crest.

The family name comes from the Greek *ikteros* which was a yellow bird, perhaps the golden oriole (*Oriolus*) which was

A Scott's oriole (*Icterus parisorum*) sings from its perch. (Photo by R. & N. Bowers/VIREO. Reproduced by permission.)

thought to cure jaundice. The English vernacular name for the New World orioles (*Icterus*) was given to them by early naturalists who thought them to be related to the *Oriolus* orioles of the Old World, but today it is realized that these birds are not closely related.

There is a great deal of variation in bill size and shape in the Icteridae. Many blackbirds have rather long, stout, and pointed bills; the ridge of the upper mandible is often straight in profile, and may extend onto the forehead, forming a frontal shield. Some of the orioles have slim, gracile bills that are slightly downturned, and the bobolink (*Dolichonyx oryzivorus*) and some of the cowbirds (*Molothrus*) have a finch-like conical bill. Blackbirds—particularly those species that forage extensively on the ground—have rather large legs and long claws.

Many species of blackbirds are strikingly sexually dimorphic in size. This is especially true in species where males are commonly polygynous (a male will be mated to two or more females in a single breeding season), e.g., *Psarocolius*, *Agelaius*, *Quiscalus*, and *Xanthocephalus*. As well, many blackbirds are sexually dimorphic in color. In dimorphic species, the males tend to have a bright display plumage whereas females are dull and relatively inconspicuous; the young tend to resemble females, and males may not develop their full display plumage until they are two years of age. Plumage dimorphism is characteristic of icterids that breed in north temperate areas. Within the orioles, the species that breed in North America, including some of the Mexican species, are dimorphic in

color, whereas the tropical species (including the three South American species) are monomorphic (although the female is sometimes somewhat less brightly colored).

Distribution

New World blackbirds are widespread in the New World, and, with the exception of a few records of vagrant individuals, they are confined to the New World. The species with the northern-most range, the rusty blackbird (*Euphagus carolinus*), breeds from the limit of trees in northern Alaska, northern Yukon, southwestern Nunavut, and Labrador southward into central Alberta and into northern-most parts of the northeastern United States. The species with the southern-most range, the austral blackbird (*Curaeus curaeus*), is found south to Cape Horn, and the long-tailed meadowlark (*Sturnella loyca*) breeds from Chile and Argentina south to the Beagle Channel Islands in Tierra del Fuego and in the Falkland Islands. Between these extremes, there is no place in the Americas where there is not at least one breeding icterid. The areas of the greatest richness of species are southern Mexico, northwestern South America, and southern Brazil, Paraguay, Uruguay, and northeastern Argentina. The genus with the largest number of species, *Icterus*, reaches its greatest richness in Central America where nine species breed, and an additional three species winter. The oropendolas (*Psarocolius*), which are forest dwelling, are confined to the Neotropics; two species are found north to the humid forests of southern Mexico. Nine of the 10 species of caciques likewise are limited to the Neotropics, although one species does range north to the Pacific coastal lowlands of central Sinaloa. The genus *Euphagus*, with two species, is the only genus restricted in range to North America (one winters south to central Mexico). The grackles (*Quiscalus*), which are closely related to the *Euphagus* blackbirds, are also found in North America, but some species are also found in Central and northern South America and in the Caribbean. Several studies have indicated that the blackbirds that are presently placed in the genus *Agelaius* should be divided into two genera, one found in North and Central America and the Caribbean, and the other confined to South America.

Habitat

New World blackbirds are found in a variety of different habitats, such as forests, savanna, grasslands, deserts, marshes, and bogs. The caciques and oropendolas are all found in forests, commonly humid lowland tropical forests. Many species, however, are associated with forest edge or clearings and are not generally found in the forest interior, and some are found in montane cloud forests. The yellow-winged cacique (*Cacicus melanicterus*), which occurs north to at least central Sinaloa, Mexico, may be found in thorn scrub, plantations (coconuts, mangos) and other settled areas, and in mangroves.

Most of the orioles (*Icterus*), the melodious, Brewer's, and yellow-shouldered blackbirds (*Dives dives*, *Euphagus cyanocephalus*, and *Agelaius xanthomus*), and the common and greater Antillean grackles (*Quiscalus quiscula* and *Q. niger*) are found in open woods, early successional habitats, riparian woods, savannas, or urban areas. The bobolink, meadowlarks (*Sturnella*), and the brown-headed cowbird (*Molothrus ater*) are found in grasslands, the yellow oriole (*I. nigrogularis*) and Bolivian blackbird (*Oreopsar bolivianus*) are found deserts, and the rusty blackbird occurs in boreal bogs. Lastly, a large number of blackbirds nest in marshes. Included in this group are the widespread and familiar North American red-winged blackbird (*Agelaius phoeniceus*) and other species in this genus, the yellow-headed blackbird (*Xanthocephalus xanthocephalus*), the marshbirds (*Pseudoleistes*), and the scarlet-headed blackbird (*Amblyramphus holosericeus*) of South America, and several of the grackles. Many of these marsh-nesting blackbirds are colonial or breed in grouped territories.

Behavior

Although they may occasionally sing during the night, blackbirds are diurnal. During migration and winter, many species form flocks—sometimes huge flocks—when they are both foraging and roosting. In the mid-1970s, there were 723 major roosting flocks in the United States containing an estimated 438 million blackbirds. An estimated 200 million icterids winter in Arkansas, Louisiana, Tennessee, and Mississippi. In the west, particularly in rice growing areas of east Texas and California, another 139 million icterids winter, and in North Carolina about 76 million winter. There are advantages to the birds in these large roosting flocks. There is strength in numbers: there are more birds to sight potential predators, and it can be dangerous for flying predators to enter a large flock of birds. There is also protection from rain, wind, and heat loss in a large flock. A disadvantage is that there is not sufficient food close to a large roost to feed the large numbers of birds present, and some individuals must travel up to 60 mi (100 km) each direction every day to reach feeding areas where the food has not already been depleted. Migrating and wintering blackbirds often forage in flocks as well, but foraging flocks are generally smaller than roosting ones. New World blackbirds walk rather than hop on the ground.

Blackbirds perform a wide variety of displays, many of which are quite interesting to watch. One common and widespread icterid display is the "song-spread" (or "rough-out") display. In this display, the bird (most commonly a male, but females of some species also do this) spreads its wings somewhat, and raises the feathers on its shoulders, back, and neck; on the red-winged blackbird this displays the bright red epaulets prominently. Some birds, such as some of the cowbirds (*Molothrus*), have a neck ruff that is exaggerated during this display. The "song-spread" display is accompanied by song. The "song-spread" appears to function both in territorial defense and for mate attraction. The "bill-tilt" display is another display that many blackbirds use. In this display, the feathers are sleeked, and the bill, head, and body are pointed upward. "Bill-tilting" is aggressive, and is used during encounters both within and between sexes. Many blackbirds have "flight-song" displays, and in some species these displays can be spectacular. Male white-browed blackbirds (*Sturnella superciliaris*) fly to a height of about 60 ft (20 m) and para-

Baltimore oriole (*Icterus galbula*) female builds her nest. (Photo by Gregory K. Scott. Photo Researchers, Inc. Reproduced by permission.)

chute to the ground while giving their song. The "flight-song" of the bobolink is also spectacular. The male, with a fluttery flight and wings bowed—not rising above the horizontal, and tail pointed downward, fly up to perhaps 30 ft (10 m) from the ground while uttering their bubbling song. Male red-winged blackbirds commonly give a rapid series of notes, "tseee tch-tch-tch-tch chee-chee-chee" while in flight. Female blackbirds characteristically arch their backs, and with their heads down raise their tails and shiver their wings in a pre-copulatory display. Injury feigning distractions are given by some blackbirds, but are not common in the family. Male oropendolas perform a "bow-display." In this display, the male sits on a branch, partially opens his wings, thrusts his head downward and tail upward, until he is completely upside down, with his tail at right a right angle to his body, then rights himself as he finishes his song, and shakes his wings.

Feeding ecology and diet

The largely arboreal oropendolas, caciques, and orioles commonly forage by gleaning food from vegetation in trees. These birds, like many other blackbirds, often feed by

An eastern meadowlark (*Sturnella magna*) sings in eastern Michigan. (Photo by John Shaw. Bruce Coleman Inc. Reproduced by permission.)

gaping—inserting their closed bills into a flower, fruit, or stem, then opening the bill to expose food or facilitate nectar feeding. Twigs and branches often harbor small insects, especially ants. By gaping into these branches, birds capture these insects. Bark can be pried loose by gaping, and yellow-backed orioles (*Icterus chrysater*) gape under the scales of pine cones, as well as flake away bark, in the search for insects. Black caciques, such as the Ecuadorian cacique (*Cacicus sclateri*), split the slender stems of bamboo to find insects, but like most blackbirds are generalists when foraging, eating fruit and nectar in addition to insects. An oriole often gapes when eating fruit: it thrusts its closed beak into the fruit, then opens it and uses its tongue to drink the laked juice. They also forage for nectar (or sugar in hummingbird or oriole feeders) and many commonly forage on the ground.

The species of blackbirds that live in grasslands feed mostly on the ground, and the marsh-dwelling species also commonly feed on the ground, especially in the winter. Meadowlarks commonly feed by gaping. They insert their long bills into the ground, then open the mandibles to expose insect prey or grain; the muscles used for opening the bill are particularly well-developed in these birds. Particularly during the winter, many blackbirds forage extensively on the ground for grain and other seeds, and some species eat small vertebrate animals. Red-winged blackbirds have been observed to eat Nelson's sharp-tailed sparrows (*Ammodramus nelsoni*), and rusty blackbirds and grackles have been observed eating birds as large as white-throated sparrows (*Zonotrichia albicollis*). Grackles are also adept at catching small fish in shallow waters, and they can crush acorns with their powerful bills. Black-backed orioles (*I. abeillei*) are one of the few birds that will eat the of-

ten toxic monarch butterflies (*Danaus plexippus*). The toxicity of these butterflies is derived from some of the plants that they eat, and the majority of the butterflies are only weakly toxic. The orioles taste each butterfly that they capture, and release the more toxic ones. Blackbirds are not adept at flycatching, but they do take some low-flying insects. In winter, many species of blackbirds forage in flocks on the ground. Commonly these flocks are composed of mixed species. In North America, flocks often contain a combination of cowbirds, grackles, red-winged blackbirds and, in the west, yellow-headed blackbirds and Brewer's blackbirds. When the birds are foraging, these flocks appear to roll along as the birds toward the back of the flock fly to the front of the flock as it moves along.

Reproductive biology

Most icterids build an open cup-shaped nest, woven into supporting vegetation or in a crotch in a shrub. Hole nesting is extremely rare in blackbirds, but occasionally nests are placed cavities. The only species that regularly nests in holes is the Bolivian blackbird, which nests in arid areas on the eastern slope of the Andes, where there are few trees or shrubs. Thus, it is not surprising that they place their nests in crevices in cliffs. Baywings (formerly called the bay-winged cowbird) (*Agelaioides badius*) usually lay their eggs in nests built by other species of birds, and they seem to prefer the domed nests of ovenbirds (Furnariidae); they also will nest in nest boxes. Orioles, caciques, and oropendolas weave distinctive pendant bag nests, which are suspended from the branches of trees, although in Louisiana, where they often nest in marshes; orchard orioles (*Icterus spurius*) often weave their nests into supporting reeds. Nests of oropendolas may be over 50 in (125 cm) long. The nests of grassland icterids usually are placed on the ground, often a nest woven of fine grass, and placed at the base of a tuft of grass that may overhang the nest. The nests of some of the meadowlarks are domed, and there may be runways in the grass leading to the nest. The cowbirds (*Molothrus*) are brood parasites, that is, they lay their eggs in the nests of other species of bird. Some of these, such as the brown-headed and shiny cowbirds (*M. ater* and *M. bonariensis*) parasitize a large number of host species, but the screaming cowbirds (*M. rufoaxillaris*) lay their eggs only in the nests of baywings, and a few other species of blackbirds. Many species of blackbirds are colonial, or at least have grouped territories. Nest building is done exclusively or principally by the female. The number of eggs and their shape and coloration varies among species. Some oropendolas may lay only a single egg, whereas the parasitic brown-headed cowbird may lay a great many eggs in any season (perhaps 30; one captive female laid 77), but these tend to be laid daily, in sequences of one to seven eggs. They vary in shape from rather round to elongated; they are glossy and tend to be marked with spots and blotches, although some cowbirds may lay unmarked eggs.

Most blackbirds are at least socially monogamous, but a great many species are polygynous. Polygyny seems to be particularly frequent in colonial marsh-nesting blackbirds and many of the colonial caciques and oropendolas. In the polyg-

Red-winged blackbirds (*Agelaius phoeniceus*) in flight. (Photo by A. Morris/VIREO. Reproduced by permission.)

ynous crested oropendola (*Psarocolius decumanus*), females greatly outnumber males, perhaps by as much as 10 to one in some populations. The much-studied red-winged blackbird commonly is polygynous, but some individuals are monogamous; in this species—and probably many others—extra-pair copulations are common. Females appear to solicit such copulations. A male red-winged blackbird may have as many as 15 mates. Not surprisingly, males give more assistance to females in monogamous pairs than in polygynous ones. The marsh-nesting scarlet-headed blackbird of South America, and some other marsh-nesting species, however, are monogamous.

Conservation status

Many species in the Icteridae are among the most numerous and successful of American songbirds. They are a conspicuous component of the fauna of many grasslands and marshes, and some, such as many of the grackles and cowbirds, have adjusted well to human-induced changes to the environment. Nonetheless, some 7% of the species are classified as being under some level of threat, ranging from endangered to further protection desirable.

The Martinique oriole (*Icterus bonana*) is Vulnerable, and Forbes's blackbird (*Curaeus forbesi*) is listed as Critically En-

dangered. The oriole is endemic on the West Indian island of Martinique where it was originally distributed throughout the forested areas of the island at lower elevations. The principal threat to this species is brood-parasitism from the shiny cowbird, which colonized the island during the late 1940s and is increasing in abundance. Today, about 75% of the nests are parasitized by cowbirds. Forbe's blackbird is known with certainty from only two regions near the Atlantic coast of Brazil. In the 1880s, when this species was first described, it was said to be local in occurrence, but apparently common where found; however, they are very similar in appearance to the chopi blackbird (*Gnorimopsar chopi*), so there is some uncertainty about their status because many so-called records of Forbe's blackbirds may have been of chopi blackbirds. Forbe's blackbird is threatened by deforestation and by pressure from parasitism by shiny cowbirds.

The pampas meadowlark (*Sturnella defilippii*) is Vulnerable, and the Baudó oropendola (*Psarocolius cassini*) and red-bellied grackle (*Hypopyrrhus pyrohypogaster*) are listed as Endangered. The meadowlark was apparently always rare in Uruguay and Brazil, but was formerly common in central-eastern Argentina. Today, however, cultivation and overgrazing has resulted in a serious decline in their numbers. The red-bellied grackle is endemic to the Andes of Colombia. Although in the past these were not uncommon birds in humid

A hooded oriole (*Icterus cucullatus*) feeds on a flower. (Photo by H.P. Smith Jr./VIREO. Reproduced by permission.)

tropical forests, habitat destruction has led to declines in numbers, and today they are local in distribution, although common in a few sites. The saffron-cowled blackbird (*Xanthopsar flavus*) is a Vulnerable bird of moist grasslands, marshes, agricultural fields, and dry bushy areas of southern Brazil, Paraguay, eastern Argentina, and Uruguay. For reasons that are not well known, but include habitat destruction, their numbers have declined sharply in the last century, especially since the 1970s. The selva cacique (*Cacicus koepckeae*) is Vulnerable as well. This cacique is known only from the type-locality (the site where it was first collected), in the lowlands of Peru, although there are other sight records from Manú National Park in Peru. Little is known of their biology, but they are probably extremely rare. Although only seven of the approximately 103 species of blackbirds are listed, three of these represent monotypic genera (that is, they are the only species in that genus). In other words, if those three species were to go extinct, three of 25 genera, or approximately 12%, would be lost, leading to a fairly substantial loss of the diversity that has evolved in the Icteridae.

Significance to humans

Many North American blackbirds are considered agricultural pests in many areas where flocks of them descend to feed in grain fields (rice, corn) or feed lots, and thus compete with humans and their livestock for food. As well, they often congregate in immense flocks in migration and winter, and when these flocks gather in densely inhabited areas—as they frequently do—concern is often expressed that they could constitute a threat to human health. Additionally, many people find the noise associated with a blackbird flock to be unpleasant and the excrement unaesthetic, particularly when they are roosting in a city park. Consequently, there is often pressure on government and game officials for mass extermination, and millions have been destroyed.

New World blackbirds have also played an important part in biological research, in part because many are common, conspicuous, and easily observed, and in part because of the great diversity of lifestyles and behaviors represented in the family. Therefore, they have been important subjects of many major studies of behavior, breeding biology, mating systems, and ecology. In North America, studies of the yellow-headed and red-winged blackbirds have been important in the development and testing of theories about territoriality and mating systems of birds, and the paternity of young. These studies have helped to answer many general questions in biology.

1. Rusty blackbird (*Euphagus carolinus*), nonbreeding plumage; 2. Bobolink (*Dolichonyx oryzivorus*); 3. Chopi blackbird (*Gnorimopsar chopi*); 4. Great-tailed grackle (*Quiscalus mexicanus*); 5. Western meadowlark (*Sturnella neglecta*); 6. Brown-headed cowbird (*Molothrus ater*); 7. Baywing (*Agelaioides badius*); 8. Brown-and-yellow marshbird (*Pseudoleistes virescens*); 9. Melodious blackbird (*Dives dives*); 10. Long-tailed meadowlark (*Sturnella loyca*). (Illustration by Gillian Harris)

1. Jamaican blackbird (*Nesopsar nigerrimus*); 2. Yellow-billed cacique (*Amblycercus holosericeus*); 3. Yellow-headed blackbird (*Xanthocephalus xanthocephalus*); 4. Baltimore oriole (*Icterus galbula*); 5. Yellow-hooded blackbird (*Agelaius icterocephalus*); 6. White-browed blackbird (*Sturnella superciliaris*); 7. Red-winged blackbird (*Agelaius phoeniceus*); 8. Yellow-rumped cacique (*Cacicus cela*); 9. Female Montezuma oropendola (*Psarocolius montezuma*); 10. Oriole blackbird (*Gymnomystax mexicanus*). (Illustration by Emily Damstra)

Species accounts

Montezuma's oropendola
Psarocolius montezuma

TAXONOMY
Cacicus montezuma Lesson, 1830, Mexico.

OTHER COMMON NAMES
French: Cassique de Montezuma; German: Montezumastirnvogel; Spanish: Chacarero de Montezuma, Oropéndola de Moctezuma.

PHYSICAL CHARACTERISTICS
15–20 in (38–50.5 cm); female 7–9 oz (198–254 g), male 12.5 oz–1.2 lb (353–528 g). Sexes similar in color. A large chestnut oropendola with a black head, a bare blue cheek patch, a small pink patch at the base of the lower bill, orange-tipped black bill, and yellow lateral tail feathers.

DISTRIBUTION
Resident from southeastern Mexico south to central Panama.

HABITAT
Tropical lowland forests and secondary forests, to 3,300 ft (1,000 m).

BEHAVIOR
During the breeding season they are highly colonial. Females forage away from the colony in small groups. Males often feed solitarily. Males have an elaborate courtship display: they bow forward, pointing their bill downward, and fan their tail, while uttering a characteristic vocalization.

FEEDING ECOLOGY AND DIET
Forage in trees; their diet is almost exclusively fruit.

REPRODUCTIVE BIOLOGY
Highly polygamous; successful males mate with several different females. Females weave a long basket-like nest of grasses and other fibers, which is suspended from the tips of branches of a tree; nests are up to 47 in (120 cm) in length. Oropendolas are highly colonial with over 60 nests in some colonies. Nesting trees are characteristically tall trees that are away from other trees (so that it is difficult for monkeys and other arboreal predators to get to them). Generally two eggs are laid January–May. Incubation about 15 days; young fledge about 30 days.

CONSERVATION STATUS
Not threatened. Montezuma oropendolas are locally common on the Caribbean coast of Central America, but generally local on the Pacific coast of Nicaragua, northwestern Costa Rica, and Panama.

SIGNIFICANCE TO HUMANS
None known. ◆

Yellow-rumped cacique
Cacicus cela

TAXONOMY
Parus cela Linnaeus, 1758, Surinam.

OTHER COMMON NAMES
French: Cassique cul-jaune; German: Gelbbürzelkassike; Spanish: Charro de Rabadilla Amarilla.

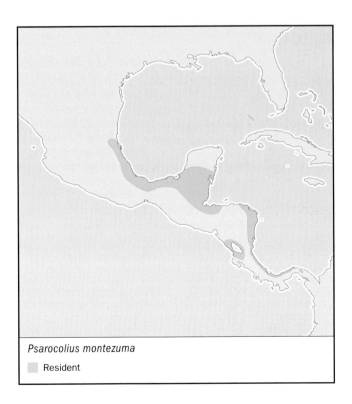

Psarocolius montezuma
　■ Resident

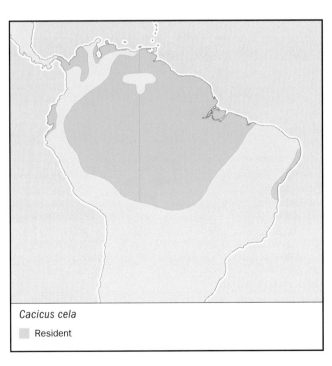

Cacicus cela
　■ Resident

PHYSICAL CHARACTERISTICS

8–11 in (22–29 cm); female 2.4–3.9 oz (67–110 g), male 2.9–4.3 oz (81–121 g). Sexes similar in color. A large black cacique with a yellow rump, undertail coverts, and wingbar. The bill is pale yellow.

DISTRIBUTION

Resident from southern Panama to northern and central South America, east of the Andes south to southern Bolivia and central and eastern Brazil, and west of the Andes in western Ecuador and northwestern Peru.

HABITAT

Tropical lowland forest edge and river-edge forest.

BEHAVIOR

Yellow-rumped caciques often nest colonially, and at colonies females outnumber males. Males display by roughing out their feathers, especially the yellow rump feathers, and with their body in a horizontal position, they flutter their wings and thrust their head downward while vocalizing.

FEEDING ECOLOGY AND DIET

They feed in trees, primarily in the outer foliage or in the canopy. Their food is primarily insects, but they also eat fruit. They often feed in pairs or small groups, but males often feed singly.

REPRODUCTIVE BIOLOGY

Successful males mate with several different females in a season. Within both sexes, dominance hierarchies are established, with the largest individuals at the top of the hierarchy. Dominant males obtain the most mates, and dominant females can occupy prime nesting sites, near wasp nests; they often nest on islands. The nest is a hanging basket, averaging about 17 in (43 cm), woven of palm strips, and suspended from a tree branch. Generally two eggs are laid; eggs can be laid at any season, but most nesting takes place in the driest times of the year. Incubation 13–14 days; fledging takes place after 24–30 days.

CONSERVATION STATUS

Not threatened. Widespread and common in suitable habitat.

SIGNIFICANCE TO HUMANS

None known. ◆

Yellow-billed cacique

Amblycercus holosericeus

TAXONOMY

Sturnus holosericeus Deppe, 1830, Veracruz, Mexico.

OTHER COMMON NAMES

French: Cassique à bec jaune; German: Gelbschnabelkassike; Spanish: Charro Piquihueso; Cacique Piquiclaro; Pico de Plata.

PHYSICAL CHARACTERISTICS

8.5–10 in (22–25 cm); female 2 oz (56 g), male 2.5 oz (71 g). Sexes similar in color. Entirely black, with a whitish yellow bill; eyes yellow.

DISTRIBUTION

Resident from coastal eastern Mexico south to northern Colombia; in South America in the mountains from northern Venezuela south to northern Bolivia.

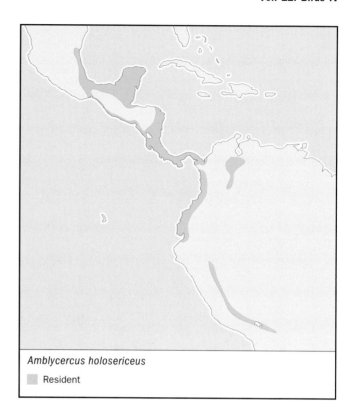

Amblycercus holosericeus

Resident

HABITAT

Lowland and montane evergreen forests; secondary forests to 9,800 ft (3,000 m). In the highlands, it is characteristic of bamboo thickets.

BEHAVIOR

The yellow-billed cacique is a skulking bird of dense undergrowth. They typically wander in the undergrowth in pairs or family groups.

FEEDING ECOLOGY AND DIET

Forage mostly in low thickets where they poke into rolled leaves, bamboo shoots, or hammer at branches to get insects. They have been observed following army ants to pick up insect prey.

REPRODUCTIVE BIOLOGY

Monogamous. Nests are a cup of leaves and vines, placed within 3 ft (1 m) of the ground. Generally 1–2 eggs are laid. Breeding season varies geographically; in Costa Rica nesting takes place in February–June; in South America, in November–April. Incubation and fledging times not reported.

CONSERVATION STATUS

Not threatened. They are locally common in thickets, clearings and second growth, habitats that are created by clearing of primary forests.

SIGNIFICANCE TO HUMANS

None known. ◆

Baltimore oriole

Icterus galbula

TAXONOMY

Coracias galbula Linnaeus, 1758, Virginia.

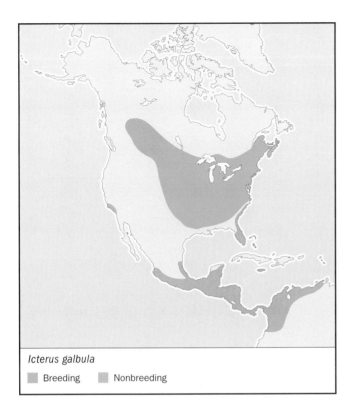

Icterus galbula

■ Breeding   ■ Nonbreeding

OTHER COMMON NAMES
English: Northern oriole, black-backed oriole, Bullock's oriole;
French: Oriole de Baltimore; German: Baltimoretrupial; Span-
ish: Bolsero de Baltimore.

PHYSICAL CHARACTERISTICS
7–8 in (18–20 cm); female 1–1.4 oz (28–41 g), male 1.1–1.4 oz
(31–40 g). Sexually dimorphic in color. Males with a black
head, wings, and middle tail feathers, and yellow-orange on the
breast, belly, shoulder, and the tips of the tail, with white
markings in their wings. Males in their second year resemble
females. Females, which are variable, are yellowish or orangish
green, usually with some black on the head, and greenish gray
wings. Juveniles resemble females, but lack black, and are usually duller in coloration.

DISTRIBUTION
Breeds in eastern North America, from central Alberta and
southern Quebec south to northern Louisiana and central
Georgia. Winters from central Mexico south to northern
South America, Florida, Jamaica, and along the coast of south-
ern California.

HABITAT
Woodland edge and open woodlands.

BEHAVIOR
Territorial during the breeding season. Males defend a territory with songs and chasing. In winter, solitary or found in
small groups.

FEEDING ECOLOGY AND DIET
Forage mostly in trees, gleaning insects, or eating fruit and insects. Also forage on ground or low in vegetation.

REPRODUCTIVE BIOLOGY
Monogamous. Females (sometimes with some assistance
from males) weave a bag-like nest of grasses, which is sus-

pended from the branches of a tree. Generally 4–5 eggs are
laid from May to mid-June. Incubation 11–14 days; fledging
11–14 days. Single brooded, but replacement clutches may be
produced.

CONSERVATION STATUS
Not threatened. Widespread and common, but numbers declining in many areas. Destruction of suitable habitat for them
on the wintering grounds may be affecting numbers.

SIGNIFICANCE TO HUMANS
Can be an important predator on defoliating insects; one of the
few birds that eat significant numbers of tent caterpillars
(*Malacosoma*). ◆

Jamaican blackbird
Nesopsar nigerrimus

TAXONOMY
Icterus nigerrimus Osburn, 1859, Jamaica.

OTHER COMMON NAMES
French: Carouge de la Jamaïque; German: Bromelienstärling;
Spanish: Pradero Jamaicano.

PHYSICAL CHARACTERISTICS
7 in (18 cm); 1.4 oz (39 g). Sexes similar in coloration. Uniformly black.

DISTRIBUTION
Resident in Jamaica.

HABITAT
Wet montane forests.

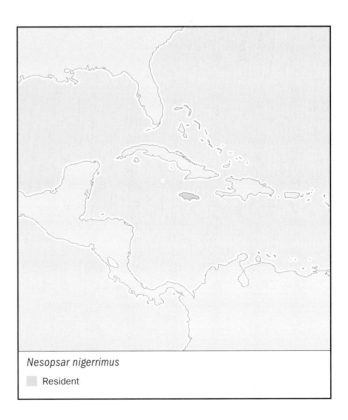

Nesopsar nigerrimus

■ Resident

BEHAVIOR
Territorial. Jamaican blackbirds spend most of their time foraging in the forest canopy. They vocalize frequently.

FEEDING ECOLOGY AND DIET
Forage in trees, searching epiphytes for invertebrate food.

REPRODUCTIVE BIOLOGY
Monogamous. The bulky nest is constructed of rootlets and epiphytic orchids, and is placed against the trunk of a tree in the lower canopy. Two eggs are laid in May–July. Incubation is about 14 days. Single brooded.

CONSERVATION STATUS
Endangered. Although locally common, they are found only in places where there is mature rainforest, habitat that is being destroyed for coffee plantations.

SIGNIFICANCE TO HUMANS
None known. ◆

Oriole blackbird

Gymnomystax mexicanus

TAXONOMY
Oriolus mexicanus Linnaeus, 1766, Cayenne.

OTHER COMMON NAMES
French: Carouge loriot; German: Nacktaugentrupial; Spanish: Tordo Maicero.

PHYSICAL CHARACTERISTICS
10.5–12 in (27–30 cm); 3.3 oz (93 g). Sexes similar in coloration, but males somewhat larger. Head and underparts, and epaulets (shoulders) bright yellow, and black otherwise; blackish patches of skin around the eyes and on the malar area. Young are like adults, but have a black cap.

Gymnomystax mexicanus
■ Resident

DISTRIBUTION
Resident in much of Venezuela, central Colombia, and along the Amazon River from eastern Ecuador east to the Atlantic Coast. Sparse in the Guianas.

HABITAT
Marshy area and open areas with scattered trees and palms.

BEHAVIOR
Territorial during the breeding season. Oriole blackbirds are usually observed in pairs or small flocks, but sometimes roost in large flocks. They typically sit on low perches, such as fence posts or low branches in trees, and walk on the ground. When singing, males rough out the feathers of the neck, fan their tails, and droop their wings.

FEEDING ECOLOGY AND DIET
They mostly forage on the ground, slowly walking and picking up grain, or probing in the ground. They pick fruit from trees.

REPRODUCTIVE BIOLOGY
Monogamous. The nest is a bulky cup of grasses and other vegetation, built by the female with some help from the male, and placed 20–25 ft (6–8 m) up in a tree. Generally three eggs are laid in August–December in Guyana, and March in Peru. Incubation 18–20 days; fledging time not reported.

CONSERVATION STATUS
Not threatened. Oriole blackbirds are common in appropriate habitat. Clearing of forests probably creates habitat that is suitable for them.

SIGNIFICANCE TO HUMANS
They frequently raid corn fields, and locally oriole blackbirds may be pests. ◆

Yellow-headed blackbird

Xanthocephalus xanthocephalus

TAXONOMY
Icterus icterocephalus Bonaparte, 1826, Nance Co., Nebraska.

OTHER COMMON NAMES
French: Carouges à tête jaune; German: Brillenstärling; Spanish: Tordo Cabeciamarillo, Tordo Cabecidorado.

PHYSICAL CHARACTERISTICS
9–10 in (23–25 cm); female 1.5–2 oz (42.5–56 g), male 2.6–3 oz (72.5–85.5 g). Sexually dimorphic in color. Males black, with a bright yellow head, throat, and breast, black between the eye and bill and on the chin, and a large white patch on the primary wing coverts. Females are brown above, with a golden yellow throat and stripe above the eye, and heavy brown streaks on their breast; in worn (summer) individuals, the yellow is paler and somewhat brighter. Young birds resemble females, and males in their first summer resemble summer females.

DISTRIBUTION
Breeds from eastern British Columbia and the Canadian prairies, east locally to southern Ontario, and south through the American west to eastern Washington and California, south to northern New Mexico. Resident in the lower Colorado River valley, and locally in south-central California. Winters from central Arizona, New Mexico, and western Texas south to central Mexico.

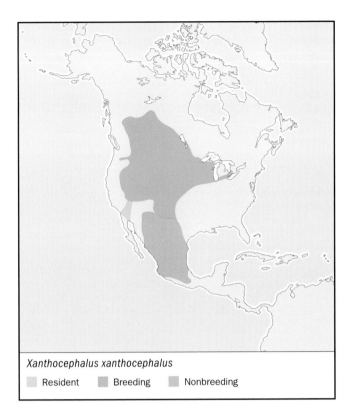

Xanthocephalus xanthocephalus

☐ Resident ■ Breeding ■ Nonbreeding

HABITAT
Freshwater marshes, with cattails, tules, or bullrushes. In migration and winter, found in agricultural fields.

BEHAVIOR
Territorial during the breeding season. Males defend a territory with songs and chasing. Males perform a "song spread display," in which the wings are held up in a V as they vocalize. Both sexes perform a display in which the wings and tail are partially spread, the head turned, and the bill pointed upward. In winter, found in flocks, often enormous ones that contain several different species of blackbirds as well as European starlings (*Sturnus vulgaris*).

FEEDING ECOLOGY AND DIET
Forage low in vegetation in a marsh or on the ground. Their food consists of invertebrate animals (especially insects), grain, and other seeds.

REPRODUCTIVE BIOLOGY
Males commonly are simultaneously paired to up to six females; females are not monogamous, and will solicit copulation from males other than their principal mate. Females build the nest, which is a bulky cup of woven vegetation, commonly woven to several stalks of emergent vegetation in marshes. Three to five eggs are laid in May–July. Incubation 12–13 days; fledging at about 12 days. Usually single brooded, but replacement clutches may be produced.

CONSERVATION STATUS
Not threatened. Yellow-headed blackbirds are common, and data indicate that their numbers are stable.

SIGNIFICANCE TO HUMANS
Considered an agricultural pest in some areas. They commonly gather in large roosts in urban areas in winter, where their droppings may be a concern for public health. ◆

Yellow-hooded blackbird
Agelaius icterocephalus

TAXONOMY
Oriolus icterocephalus Linnaeus, 1766, Cayenne. Two geographically distinct forms are recognized.

OTHER COMMON NAMES
French: Carouge à capuchon; German: Gelbkopfstärling; Spanish: Turpial de Agua.

PHYSICAL CHARACTERISTICS
6.5–7 in (17–18 cm); female 0.8–1.1 oz (24–31 g), male 1.1–1.4 oz (31.5–40 g). Sexually dimorphic in color. Males are black with a yellow hood and black around the bill. Females are grayish olive above, and have a brownish belly, flecked with black, and a dusky yellow hood, with the yellow on the throat and the stripe over the eye brighter.

DISTRIBUTION
Resident of northern South America and along the Amazon River, from northern Colombia to central Brazil, east to northeastern Peru.

HABITAT
Freshwater marshes and tall, wet grasslands. Although characteristically a bird of the lowlands, they are found to about 8,500 ft (2,600 m) in the Andes of Colombia.

BEHAVIOR
Territorial during the breeding season. Males display to other males using a "song-spread" display, much like that of the North American red-winged blackbird. Males approach females with a distinctive fluttering flight; receptive females follow males to nests constructed by the males. During all seasons they are commonly seen in small loose flocks; large numbers may congregate in roosts. During the breeding season, males form colonies in marshes and start building nests.

FEEDING ECOLOGY AND DIET
Yellow-hooded blackbirds feed in marshes or in pastures, where they eat seeds and capture invertebrates.

Agelaius icterocephalus

☐ Resident

REPRODUCTIVE BIOLOGY
Successful males mate with up to five different females in a single season. Males build a nest in emergent aquatic vegetation; the female adds the lining to the nest after the pair is formed. Mated males stay with their mate until incubation begins, then they build another nest and seek an additional mate. Generally 2–3 eggs are laid in May–October in Trinidad and October–November in Venezuela. Incubation 10–11 days; young fledge at about 11 days.

CONSERVATION STATUS
Not threatened. They are locally common.

SIGNIFICANCE TO HUMANS
They commonly forage in rice paddies and other agricultural lands, and cause some crop damage. ◆

Red-winged blackbird
Agelaius phoeniceus

TAXONOMY
Oriolus phoeniceus Linnaeus, 1766, Charleston, South Carolina. Presently, 20 geographically discrete forms are recognized.

OTHER COMMON NAMES
English: Cuban red-winged blackbird; French: Carouges à épaulettes; German: Rotschulterstärling; Spanish: Tordo Alirrojo, Tordo Capitán, Mayito de la Diénaga, Sargento.

PHYSICAL CHARACTERISTICS
7–9.5 in (18–24 cm); female 1–1.9 oz (29–55 g), male 1.9–2.9 oz (53–81 g). Sexually dimorphic in color. Males black with red epaulets edged with a yellow bar. Young males are dark brown

and heavily streaked below with some red in the wing. Males in their second year resemble older males, but many of their feathers are edged with brown. Females are streaked with dark brown. Juveniles resemble females.

DISTRIBUTION
Breeds from eastern Alaska and Great Bear Lake, east to James Bay, southern Quebec, and southwestern Newfoundland, south to Costa Rica; local in Central America, and absent as a breeder from western Mexico and central and southern Baja California. Winters from southern Canada south to Costa Rica, including western Mexico and Baja California.

HABITAT
Marshes and rank, moist thickets.

BEHAVIOR
Territorial during the breeding season. Males defend a territory with songs and chasing. In winter, found in flocks, often enormous ones that contain several different species of blackbirds as well as European starlings (*Sturnus vulgaris*).

FEEDING ECOLOGY AND DIET
Forage low in vegetation or on the ground.

REPRODUCTIVE BIOLOGY
Males are commonly simultaneously paired to several females. Females build the nest, which is a bulky cup of woven vegetation, commonly woven to several stalks of emergent vegetation in marshes. Clutch size is smaller in the south than in the north and ranges in size from 4 to 5 eggs, which are laid from May to mid-June. Incubation 11–13 days; fledging at about 11–12 days. Usually single brooded, but replacement clutches may be produced.

CONSERVATION STATUS
Not threatened. One of the most abundant North American birds.

SIGNIFICANCE TO HUMANS
Considered an agricultural pest in many areas. Much effort and money have been expended in attempts to control blackbird roosts, and consequently humans are one of the major causes of adult mortality. ◆

Agelaius phoeniceus

Resident   Breeding   Nonbreeding

White-browed blackbird
Sturnella superciliaris

TAXONOMY
Trupialis superciliaris Bonaparte, 1851, Matto Grosso, Brazil.

OTHER COMMON NAMES
English: Red-breasted blackbird; French: Sturnelle à sourcils blancs; German: Rotbruststärling, Weissbrauenstärling; Spanish: Pecho Colorado, Pechicolorado Chico.

PHYSICAL CHARACTERISTICS
6.5–7 in (17–18 cm); female 1.4 oz (39.5 g), male 1.9 oz (53 g). Sexually dimorphic in color. Males are black with a bright red throat, breast, and shoulders (visible in flight), and a prominent white stripe behind the eye. Females have a pale buff stripe above the eye, dark brown crown stripes and a dark brown stripe behind the eye, with a brown, mottled back, and buff undersides, with streaks on the throat and belly, but not on the breast, which is variously tinged with pink.

Sturnella superciliaris

☐ Resident   ■ Breeding

Long-tailed meadowlark
Sturnella loyca

TAXONOMY
Trupialis loyca Bonaparte, 1851, Peru. At present, four geographically discrete forms are recognized.

OTHER COMMON NAMES
French: Sturnelle australe; German: Langschwanz-Soldatenstärling; Spanish: Loica Común.

PHYSICAL CHARACTERISTICS
9.5–10 in (24–25.5 cm); 4 oz (113 g). The sexes are similar in color, but the females are somewhat duller. The throat, breast, and belly are red, and there is a dash of red from above the eye to the bill. There is a white line running from the top of the eye back to the nape.

DISTRIBUTION
Resident in western South America from central Chile and in foothills and intermontane valleys in western Argentina, and in the steppes of central Argentina south to Tierra del Fuego and the Falkland Islands. Some Chilean and Patagonian birds move northward in winter.

HABITAT
Arid grassy areas with scattered shrubs; pastures and cultivated fields.

BEHAVIOR
Territorial during the breeding season. Males defend a territory with songs and chasing; they generally sing from bushes or trees. Long-tailed meadowlarks typically perch in trees in flocks and behave more like many other blackbirds rather than other meadowlarks, which characteristically spend most of their time on the ground. In winter, solitary or found in small groups.

FEEDING ECOLOGY AND DIET
Forage on ground where they take seeds and insects.

DISTRIBUTION
Resident from southeastern Peru and west-central Brazil south to northern Argentina, and also in the lowlands of eastern Brazil. Withdraw from the southernmost part of the range in Argentina in winter. Found from sea level to over 8,200 ft (2,500 m).

HABITAT
Grasslands and damp pastures.

BEHAVIOR
Gregarious, especially in the nonbreeding season, but they often nest in loose colonies. Territorial during the breeding season. Males have a spectacular butterfly-like aerial display: they fly to a height of over 66 ft (20 m) over their territory, then parachute downward while singing. They also sing from the ground. Females are difficult to see.

FEEDING ECOLOGY AND DIET
Forage on ground where they take seeds and insects. They often feed among livestock and agricultural fields.

REPRODUCTIVE BIOLOGY
Monogamous. Nest is placed on the ground, with grass pulled over it, so that it is not visible from above. Commonly 3–5 eggs are laid in October–January. Data on incubation and fledging not available.

CONSERVATION STATUS
Not threatened. Locally common throughout; increasing in numbers and expanding its range in response to the spread of cultivation of rice and other grain crops.

SIGNIFICANCE TO HUMANS
May cause local damage to crops. ◆

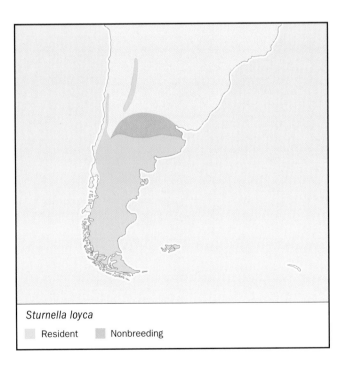

Sturnella loyca

☐ Resident   ■ Nonbreeding

REPRODUCTIVE BIOLOGY
Monogamous. Nest is a domed cup placed on the ground (except in the Falkland Islands where the nest may be above the ground, built on a pedestal of tussock grass). Generally 3–5 eggs are laid in September–January (August–November in Falklands). Data on incubation and fledging not available. May be double brooded.

CONSERVATION STATUS
Not threatened. Common throughout their range.

SIGNIFICANCE TO HUMANS
None known. ◆

Western meadowlark
Sturnella neglecta

TAXONOMY
Sturnella neglecta Audubon, 1844, Old Fort Union, North Dakota.

OTHER COMMON NAMES
French: Sturnelle de l'ouest; German: Wiesenstärling; Spanish: Triguera de Occidente, Pradero Occidental.

PHYSICAL CHARACTERISTICS
8.5–9.5 in (21.5–24 cm); female 3.2 oz (89 g), male 4 oz (115 g). Sexes alike in color. Throat and underparts bright yellow, with a black V on the breast; yellow stripe over the eye; back, wings, and tail mottled with brown; outermost tail feathers white.

DISTRIBUTION
Breeds from central British Columbia and Alberta east to southern Ontario, and south through the Great Plains to cen-

tral Mexico and northern Baja California. Winters from southern British Columbia and the central Plains to central Mexico and Baja California.

HABITAT
Grasslands, pasture, savanna, and cultivated fields.

BEHAVIOR
Territorial during the breeding season. Males defend a territory with songs, fights, and chases. Flight songs are common. In winter, they are found in small groups.

FEEDING ECOLOGY AND DIET
Forage almost entirely on ground, gathering grain, seed, and insects. Feeds mostly on insects during the summer, grain in winter and early spring, and weed seeds in the fall.

REPRODUCTIVE BIOLOGY
Males often are simultaneously mated to two or more females. Females build a domed nest that is placed on the ground. Three to 7, most commonly 5 eggs are laid late April–early August. Incubation 13–14 days; fledging at about 10–12 days. Single brooded, but females renest if their first effort was unsuccessful.

CONSERVATION STATUS
Not threatened. Widespread and common, but numbers declining in many areas. Agricultural practices affect breeding through habitat degradation, destruction of nests by equipment and trampling of nests by livestock.

SIGNIFICANCE TO HUMANS
Western meadowlarks destroy sprouting grain by drilling beside the new shoots and removing the kernels. However, they eat large numbers of insects that can damage crops. ◆

Brown-and-yellow marshbird
Pseudoleistes virescens

TAXONOMY
Agelaius virescens Vieillot, 1819, near Buenos Aires, Argentina.

OTHER COMMON NAMES
French: Troupiale dragon; German: Drachenstärling; Spanish: Dragón, Pechiamarillo Chico.

PHYSICAL CHARACTERISTICS
8–9 in (21–24 cm); female 2.3 oz (64 g), male 3.1 oz (88 g). Sexes similar in color. Body (including flanks), wings, and tail olive brown; breast, belly, and undertail-coverts, and epaulet bright yellow; bill black. Juveniles like adults, but duller, with a yellowish throat, and brown streaking across breast.

DISTRIBUTION
Resident in extreme southern Brazil, Uruguay, and northeastern Argentina.

HABITAT
In and near marshes and adjacent grasslands and agricultural fields.

BEHAVIOR
Territorial during the breeding season. Once mated, the pair travel together. In winter, they tend to be found in flocks, often large flocks.

Sturnella neglecta

☐ Resident ■ Breeding ■ Nonbreeding

Pseudoleistes virescens

◼ Resident

Gnorimopsar chopi

◼ Resident

FEEDING ECOLOGY AND DIET
Brown-and-yellow marshbirds forage almost exclusively on the ground, often at the edge of a marsh or in a plowed field. They thrust their bills into the ground and gape to expose food (invertebrates and seeds).

REPRODUCTIVE BIOLOGY
Monogamous. Nests are placed in marshes, but often in vegetation at the edge of marshes; they also nest in wet roadside ditches. Females build the nest that often is placed in the center of a tuft of pampas grass. Clutch size is 4–5 eggs, which are laid late September–December. Information on incubation and fledging are not available. Often several "helper" adults will help feed the young in a nest; the relationship of these adults to the young is not known.

CONSERVATION STATUS
Not threatened. Common to locally abundant.

SIGNIFICANCE TO HUMANS
None known, although they may be agricultural pests in some areas. ◆

Chopi blackbird
Gnorimopsar chopi

TAXONOMY
Agelaius chopi Vieillot, 1819, Paraguay. Two geographically discrete forms are recognized.

OTHER COMMON NAMES
English: Chopi-grackle; French: Quiscale chopi; German: Chopistärling; Spanish: Turpial Chopí.

PHYSICAL CHARACTERISTICS
9–9.5 in (23–24 cm); female 2.6–3 oz (75–84 g), male 2.8 oz (79.5 g). Sexes similar in color. Entirely black, showing slight blue iridescence, with a black eye; crown and nape feathers are narrow and pointed and a groove on the lower bill is sometimes visible. Juveniles are similar, but lack iridescence and the pointed feathers of the adults.

DISTRIBUTION
Resident from eastern and southern Brazil, northern Bolivia, and southeastern Peru south through Paraguay, Uruguay, and northern Argentina.

HABITAT
Farms, pastures, savannas, and marshy areas, from sea level to 3,300 ft (1,000 m).

BEHAVIOR
Chopi blackbirds are gregarious, and are generally seen in flocks, which may be large. They invariably roost in large groups, commonly in palm trees. They are extremely vocal, and there is continuous vocalization in flocks, even when flying. Singing and courtship tend to take place in trees.

FEEDING ECOLOGY AND DIET
Forages more commonly on the ground than in trees where they eat a variety of plant and animal food.

REPRODUCTIVE BIOLOGY
Monogamous. Chopi blackbirds commonly nest in holes or crevices in trees, fenceposts, banks, other birds' nests, or old buildings. When not nesting in a cavity, the nest is open and cup-shaped, placed in a dense bush or tree. Four to five eggs are laid in September–January. Information on incubation and fledging not available.

CONSERVATION STATUS

Not threatened. They are common and locally abundant, and probably have benefited from the clearing of land for agricultural purposes.

SIGNIFICANCE TO HUMANS

Because of their attractive song, chopi blackbirds are often kept as cage birds in Brazil. ◆

Melodious blackbird

Dives dives

TAXONOMY

Icterus dives W. Deppe, 1830, Oaxaca, Mexico.

OTHER COMMON NAMES

English: Scrub blackbird; French: Quiscale chanteur; German: Trauerstärling, Buschstärling; Spanish: Tordo Cantor.

PHYSICAL CHARACTERISTICS

9–11.5 in (23–29 cm); 2.9–3.6 oz (83.5–102 g); females somewhat smaller than males. Sexes similar in color. Entirely black, showing slight iridescence, with a black eye, and moderately long tail. Juveniles are brownish black, and lack iridescence.

DISTRIBUTION

Resident from central Mexico south to northern Costa Rica.

HABITAT

Tropical lowland forests, secondary forests, pine forests, edge, and settled areas, from sea level to 6,600 ft (2,000 m).

BEHAVIOR

Pairs are territorial, and individuals of both sexes defend their territories with songs and displays (tail-flicking and a fluttering

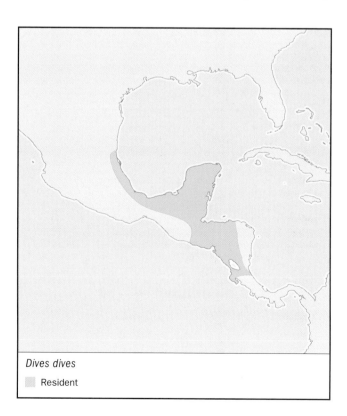

Dives dives
▨ Resident

"bill-up" flight). In winter, pairs may gather into small groups, flocking sometimes with grackles and cowbirds. Sometimes roosts in dense cane with other blackbirds.

FEEDING ECOLOGY AND DIET

Forages mainly on the ground, commonly on lawns and other cleared areas. When foraging in trees, they pick insects and larvae from foliage, and drink nectar from flowers. Their diet consists of seeds, fruits, nectar, and invertebrates (especially insects).

REPRODUCTIVE BIOLOGY

Monogamous. The nest is an open cup placed in a bush or tree; both sexes assist in building the nest. Three to four eggs are laid in April–July. Incubation about 14 days; information on fledging not available. Single brooded.

CONSERVATION STATUS

Not threatened. Common, and expanding its range southward into the Pacific slope of Guatemala, El Salvador, and northern Costa Rica, probably as a consequence of clearing of land for agriculture.

SIGNIFICANCE TO HUMANS

They eat the ripening ears of maize, and are considered to be agricultural pests. ◆

Great-tailed grackle

Quiscalus mexicanus

TAXONOMY

Corvus mexicanus Gmelin, 1788, Veracruz, Mexico. At least eight geographically discrete forms recognized.

OTHER COMMON NAMES

French: Quiscale à longue queue; German: Dohlengrackel; Spanish: Clarinero, Zanate Grande, Zanate Mayor.

PHYSICAL CHARACTERISTICS

10.5–18.5 in (26.5–47 cm); female 3.6–4.7 oz (102–132 g), male 6.2–8.9 oz (175–253 g). Sexually dimorphic in color. Males black glossed with purple on the head, becoming blue on the belly and back, and a yellow eye. They have a long keel-shaped tail. Females brown, with pale brown throat and stripe above the eye; geographically variable. Juveniles resemble females.

DISTRIBUTION

Resident in central California, central Colorado, and eastern Iowa, south to the Gulf coast south through Mexico and Central America to the coast of northwestern Venezuela and south to northwestern Peru. Some individuals from northern populations move south in winter.

HABITAT

Pastures, agricultural lands, second-growth scrub, mangrove forests, and secondary forests to 7,500 ft (2,300 m).

BEHAVIOR

Territorial during the breeding season. Males defend a territory with songs, "rough out" and "bill pointing" displays, and chasing. In winter, found in flocks.

FEEDING ECOLOGY AND DIET

Forage mostly on ground or low in vegetation. Foods taken vary seasonally, but diets consist principally of animal matter (especially grasshoppers and other insects).

Quiscalus mexicanus
◻ Resident

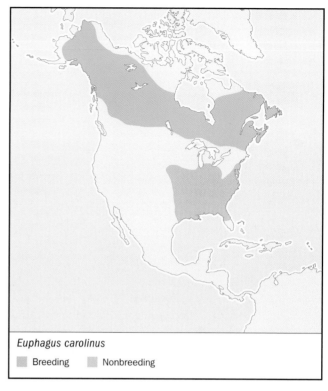

Euphagus carolinus
◻ Breeding   ◻ Nonbreeding

REPRODUCTIVE BIOLOGY

Successful males mate with several different females in a single season. Females build the bulky, cup-shaped nest, which is placed above ground, generally as high as the vegetation permits. Generally 2–4 eggs are laid in January–July. Nesting commences earlier and clutch sizes are smaller in the south than in the north. Incubation takes 13–14 days; fledging about 16–19 days. Two broods may be produced in a single season.

CONSERVATION STATUS

Not threatened. Widespread and common, and its range is rapidly expanding northward in the Great Plains.

SIGNIFICANCE TO HUMANS

They may be considered agricultural pests in some areas; however, they eat many insects that could cause crop damage. ◆

Rusty blackbird

Euphagus carolinus

TAXONOMY

Turdus carolinus Müller, 1776, Carolina. Two geographically discrete forms recognized.

OTHER COMMON NAMES

French: Quiscale rouilleux; German: Roststärling; Spanish: Tordo Canadiense.

PHYSICAL CHARACTERISTICS

8–9 in (20.5–23 cm); female 1.7–2.7 oz (47–76.5 g), male 1.6–2.8 oz (46–80.5 g). Sexes similar in color. Entirely black with a greenish or sometimes purplish gloss, a square-tipped tail, thin bill, and pale yellow eyes. Juveniles brownish, with a paler brown throat and stripe above the eye, and sometimes with the black flecking on the undersides.

DISTRIBUTION

Breeds from western Alaska east across northern Yukon, central Northwest Territories, northern Manitoba and north-central Labrador and Newfoundland south to southern Alaska, northwest British Columbia, central Saskatchewan, southern Ontario, northeastern New York, and western Massachusetts. Winters from the northern states east of the Rockies, south to the Gulf coast and northern Florida, west to eastern Texas.

HABITAT

Moist woodlands (primarily coniferous) and bogs. In winter, found in open, wet woodland, pastures, and cultivated fields.

BEHAVIOR

Territorial during the breeding season. They are usually solitary nesters, but small colonies are not infrequent in Newfoundland and Labrador. Males defend a territory with songs and chasing. In winter, usually found in flocks; when alarmed, they fly to the nearest bush or tree. In winter, they may flock with other species of blackbirds, but flocks of only rusty blackbirds are not uncommon.

FEEDING ECOLOGY AND DIET

Forage mostly on ground or low in vegetation. Their food is principally insects, but they will take small birds, and eat some grain and other seeds.

REPRODUCTIVE BIOLOGY

Monogamous. The nest is a bulky cup of twigs, grass, and other plant matter, which may be placed on the ground or as high as 23 ft (7 m). Generally 4–5 eggs are laid in May–June. Incubation about 14 days; fledging about 12 days. Single brooded.

CONSERVATION STATUS

Not threatened. Widespread and common, but numbers declining.

SIGNIFICANCE TO HUMANS
None known. ◆

Baywing
Agelaioides badius

TAXONOMY
Agelaius badius Vieillot, 1819, Parguay. There are two geographically discrete forms. Many treat the pale baywing (*A. fringillarius*) as a third subspecies.

OTHER COMMON NAMES
English: Bay-winged cowbird; French: Vacher à ailes baies; German: Braunkuhstärling; Spanish: Músico, Tordo Mulato.

PHYSICAL CHARACTERISTICS
7–7.5 in (18–19 cm); 1.4–1.8 oz (41–50 g). Sexes similar in color. Olive-gray, dusky between the bill and the eye, blackish tail, and blackish wings broadly edged with rufous. They have a short conical bill. Juveniles resemble adults.

DISTRIBUTION
Resident in northeastern Brazil, and northern and eastern Bolivia south through western and central Paraguay, Uruguay, to central Argentina.

HABITAT
Open, wooded, or shrubby terrain and adjacent pastures. Found to 9,500 ft (2,880 m) in Bolivia.

BEHAVIOR
This is a social blackbird, typically found in small groups of four to 25 individuals. They roost communally. Baywings sing frequently, but unlike most blackbirds, song is not accompanied by displays. They sing from a perch, and often from near the nest.

FEEDING ECOLOGY AND DIET
They feed both in trees, where they glean for insects, and on the ground, generally near woodland or shrubby cover.

REPRODUCTIVE BIOLOGY
Socially monogamous. Baywings most commonly use the abandoned nest of another species. Covered nests are preferred, such as the nests of ovenbirds (Furnariidae); they will nest in woodpecker holes and in nest boxes, and they do sometimes build their own cup-shaped nests. Clutch size is generally 4–5 eggs, which are laid late November–early January. They are cooperative breeders and one or more "helpers" will help raise the young. They are frequently parasitized by screaming cowbirds (*Molothrus rufoaxillaris*), making it difficult to obtain accurate information on clutch size, and incubation and fledging times because the baywings often suspend nesting to reduce parasitism.

CONSERVATION STATUS
Not threatened. They are common to fairly common.

SIGNIFICANCE TO HUMANS
None known. ◆

Brown-headed cowbird
Molothrus ater

TAXONOMY
Oriolus ater Boddaert, 1783, Carolina. Three geographically discrete forms are recognized.

OTHER COMMON NAMES
French: Vacher à tête brune; German: Braunkopf-Kuhstärling; Spanish: Vaquero Cabecicafé, Tordo Negro.

PHYSICAL CHARACTERISTICS
6.5–7.5 in (16.5–18 cm); female 1.1–1.8 oz (30.5–51 g), male 1.1–2 oz (32.5–58 g). Sexually dimorphic in color. Males black with a brown head; the black is greenish-glossed, and purple-glossed on the neck. Females are entirely brown, with the throat somewhat paler. Juveniles resemble females, but have scaly backs and boldly-streaked underparts. The bill is short and conical.

DISTRIBUTION
Breeds from central and northeastern British Columbia, Alberta, central Saskatchewan and southern Manitoba, east through central Ontario and Quebec to Newfoundland, and south to the Isthmus of Tehuntepec, Mexico. Resident from Nova Scotia, Maine, Illinois, eastern Kansas, west across Oklahoma, Texas, southern New Mexico, and Arizona, and along the Pacific coast north to southwestern British Columbia; does not breed in southern Florida, on the Gulf lowlands of eastern Mexico, or the lowlands of southwestern Mexico. Winters along the Gulf of Mexico coast of Mexico, and on the Pacific coast from Jalisco south to the Isthmus of Tehuntepec, and in southern Florida.

HABITAT
Open woodlands and deciduous forest edge; in migration and winter in open areas, cultivated lands, fields, pastures, and scrub.

Agelaioides badius
Resident

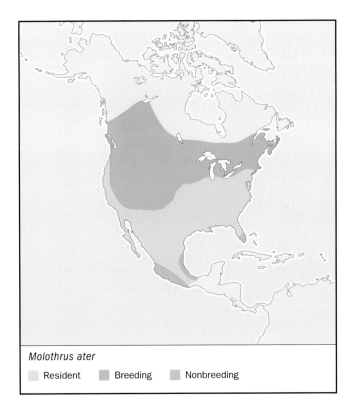

Molothrus ater

■ Resident ■ Breeding ■ Nonbreeding

Bobolink
Dolichonyx oryzivorus

TAXONOMY
Fringilla oryzivora Linnaeus, 1758, South Carolina.

OTHER COMMON NAMES
French: Goglu de pré; German: Bobolink; Spanish: Charlatán, Triste-pia.

PHYSICAL CHARACTERISTICS
6–7 in (15–17.5 cm); female 0.9–1.6 oz (26.5–44.3 g), male 1–2 oz (28.5–56.3 g). Sexually dimorphic in color. Males black face and underparts, light yellow on the back of head and nape, pale grayish white rump, wings black with pale edges to some of the feathers, white scapulars, and a white-tipped black tail. Males in winter, females, and juveniles brown or yellow ochre, with a pale stripe above the eye, streaked and mottled back, pale throat, and varying amounts of streaking on the breast and flanks.

DISTRIBUTION
Breeds from southwestern Newfoundland west to central British Columbia, south to central Colorado, northern

BEHAVIOR
Males display with a full "song-spread" display. In some populations, where they are monogamous, males guard their mates. Females lay their eggs in the nests of other species of birds. They usually do this early in the morning, and remove one of the host's eggs, replacing it with one of theirs. In winter, found in flocks that usually contain several different species of blackbirds as well as European starlings (*Sturnus vulgaris*).

FEEDING ECOLOGY AND DIET
Forage low in vegetation or on the ground, often near the feet of grazing ungulates, where they pick up insects that have been flushed. During the nonbreeding season, they eat primarily grain.

REPRODUCTIVE BIOLOGY
In some areas, males are monogamous; in others they commonly are simultaneously paired to two or more females. No nest is built, but they have been recorded to have parasitized the nests of more than 220 host species (144 of which have been seen to fledge cowbird young). Females do not lay clutches in the usual sense, but one egg is produced each 1–7 days, interrupted by 2 days when no eggs are produced; eggs are laid from March to early-August. Incubation 10–12 days; fledging at 8–13 days.

CONSERVATION STATUS
Not threatened. Common to abundant, and has benefited from the clearing of land for agriculture, and wild bird feeding.

SIGNIFICANCE TO HUMANS
It is considered a pest in feedlots and grain fields. Its brood parasitic habits have caused serious declines in some populations of songbirds. ◆

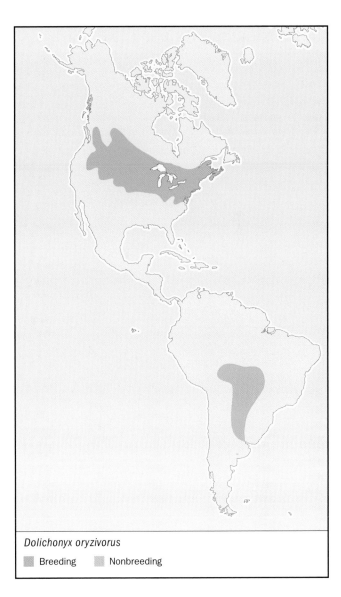

Dolichonyx oryzivorus

■ Breeding ■ Nonbreeding

Missouri, and central Virginia. Winters in the pampas of southwestern Brazil south to central Argentina.

HABITAT
Breeds in moist grassy meadows or old fields. Originally wintered in grasslands and marshes, but now commonly is found in rice fields.

BEHAVIOR
Territorial during the breeding season. Males defend a territory with flight-songs, fights and chases. In migration and winter, often occurs in large flocks.

FEEDING ECOLOGY AND DIET
Forage on the ground. Young are fed insects and other invertebrates. Winter diet is mostly various grains and seeds.

REPRODUCTIVE BIOLOGY
Males are paired simultaneously with several females. Females select the nest site and build the nest, which is an open cup of woven grass and other vegetation, placed on the ground, often at the base of a forb. Three to seven (commonly five) eggs are laid from mid-May to mid-July. Incubation 12–14 days; fledging at 10 or 11 days. Single brooded, but replacement clutches may be produced.

CONSERVATION STATUS
Not threatened. Widespread and locally common, but numbers declining in many areas. Destruction of suitable habitat in both their wintering and breeding range may be affecting numbers.

SIGNIFICANCE TO HUMANS
It is considered an agricultural pest outside of its breeding range. ◆

Resources

Books

American Ornithologists' Union. *Check-list of North American Birds.* 7th ed. Washington, DC: American Ornithologists' Union, 1998.

Collar, N.J., L.P. Gonzaga, N. Krabbe, A. Madroño Nieto, L.G. Naranjo, T.A. Parker III, and D.C. Wege. *Threatened Birds of the Americas.* Cambridge, United Kingdom: International Council for Bird Preservation, 1992.

Jaramillo, A., and P. Burke. *New World Blackbirds: The Icterids.* Princeton: Princeton University Press, 1999.

Orians, G. *Blackbirds of the Americas.* Seattle: University of Washington Press, 1985.

Periodicals

Lanyon, S.M. "Polyphyly of the Blackbird Genus *Agelaius* and the Importance of Assumptions of Monophyly in Comparative Studies." *Evolution* 10 (1994): 679–693.

Nero, R.W. "A Behavior Study of the Red-Winged Blackbird I: Mating and Nesting Activities." *Wilson Bulletin* 68 (1956): 4–37.

Johnson, K.P., and S.M. Lanyon. "Molecular Systematics of the Grackles and Allies, and the Effect of Additional Sequence (Cyt *b* and ND2)." *Auk* 116 (1999): 759–768.

James David Rising, PhD

Finches
(Fringillidae)

Class Aves
Order Passeriformes
Suborder Passeri (Oscines)
Family Fringillidae

Thumbnail description
Small to medium-sized, seed-eating birds with a conical-shaped, pointed beak, a short neck, compact body, and plumage that varies from rather drab to quite colorful, especially in male birds

Size
Body length about 4–10 in (10–25 cm) and weight 0.3–2.1 oz (8–60 g)

Number of genera, species
20 genera; 137 species

Habitat
Forest, shrubland, grassland, agricultural areas, parks, and gardens

Conservation status
Critically Endangered: 1 species; Endangered: 5 species; Vulnerable: 3 species; Near Threatened: 4 species; Conservation Dependent: 1 species; Data Deficient: 2 species; Extinct: 1 species

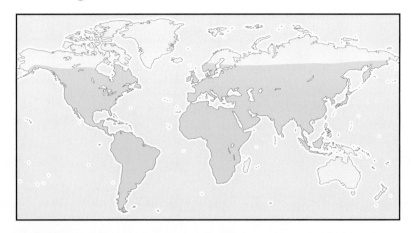

Distribution
Almost global distribution, except for Madagascar, Australasia, many Pacific Islands, and Antarctica. A few species have been introduced beyond their natural range

Evolution and systematics

The family Fringillidae, or the "true" finches, consists of 20 genera and about 137 species. The family is divided into two subfamilies: the Fringillinae, consisting of three species of chaffinches, and the Carduelinae, comprised of numerous species variously known as bullfinches, canaries, citrils, crossbills, goldfinches, grosbeaks, linnets, rosefinches, seedeaters, serins, and siskins, among other common names. The systematics of the Fringillidae is not, however, entirely settled. Some ornithologists group additional families of birds within a greatly expanded Fringillidae, including the tanagers (Thraupidae), sparrows and buntings (Emberizidae), Hawaiian honeycreepers (Drepanididae), and Galapagos finches (Geospizinae).

Physical characteristics

The species of true finches range in body length from 4–10 in (10–25 cm) and in weight from 0.3–2.1 oz (8–60 g). The shape and structure of the beak can vary enormously within the family, but all are conical-shaped, stout, short, and pointed. The beak is well adapted for holding seeds and removing the outer shell (or seed-coat). The true finches also have rather small outer primaries on their wings, and these are entirely concealed by the wing coverts. Species of fringillids also differ from the emberizid finches (Emberizidae) in that the edges of their mandibles fit closely together all along the length of the beak. Some true finches have a par-

ticularly large beak for dealing with relatively large seeds, for example, the evening grosbeak (*Coccothraustes vespertinus*). Other species have a smaller beak with crossed mandibles adapted for extracting seeds from conifer cones, such as the red crossbill (*Loxia curvirostra*). Plumage coloration varies widely among species of fringillids. Species may be brown, yellow, grey, orange, or red, and they may be patterned with spots, patches, or streaks. Most species are dimorphic, with males often being more brightly colored than females.

Distribution

Species of true finches occur extremely widely over the Americas, Eurasia, and Africa, being absent only from Madagascar, Australasia, many Pacific islands, and Antarctica.

Habitat

Species of true finches inhabit a wide range of terrestrial habitats, including many types of forests and woodlands, shrubby places, savannas, grasslands, agricultural areas, and gardens and horticultural parks.

Behavior

Finches may occur as solitary pairs, particularly during the breeding season, but most species are at least seasonally

An American goldfinch (*Carduelis tristis*) pair at their nest. (Photo by Anthony Mercieca. Photo Researchers, Inc. Reproduced by permission.)

gregarious, particularly when not breeding. Finches are strong fliers, and ground-foraging species hop and run well over short distances. Species that breed in regions with a highly seasonal climate, such as the tundra, boreal, or temperate zones, are often migratory during the winter. Some species migrate over relatively long distances to warmer climates, while others form flocks and wander extensively looking for locally abundant food sources. Finches have short, sharp call notes to communicate within flocks and to warn of impending danger. They also have distinctive songs that males use to defend a breeding territory and attract a mate. Young male finches have an innate ability to learn the song of their species, but when young they must hear mature males singing in order to learn how to perfect their own song. Widespread species of finches may have local song dialects. Some finches, such as the canary, are kept as prized pets because of their enthusiastic and musical singing ability.

Feeding ecology and diet

Finches mostly eat seeds, grains, and other vegetable matter, often supplemented by insects and other small invertebrates. Many species forage on the ground, while others feed mainly on tree seeds. Chaffinches are particularly insectivorous when feeding their young, which receive little plant food until they are fledged. All finches have a strong beak used to crush seeds so the edible kernel can be extracted and eaten. To do this, the seed is wedged against a special groove at the side of the palate, and then crushed by raising the lower jaw. The shell is then removed with the aid of the tongue, and the edible kernel is swallowed. The beaks of finches vary greatly, however, depending on the kinds of foods they specialize on. The crossed points of the beak of crossbills enables them to extract seeds from the cones of conifers; they hardly feed on anything else. The beak of goldfinches is long and narrow enough to reach the seeds of the teasel, which lie at the base of a rigid tubular structure. Hawfinches have a particularly stout beak, used to feed on the pits of cherries and rose-hips.

Reproductive biology

Most finches build a cup-shaped nest of grasses and other plant fibers and locate it in a tree or shrub or in a rocky crevice. Most species breed as isolated pairs, but some others are loosely colonial. Once a pair of finches bonds for the breeding season, they are typically monogamous. They lay two to six eggs, which vary in color and markings among species. The eggs may be incubated by the female or by both sexes in turn. Both of the parents share in tending the young and fledglings.

Conservation status

The IUCN lists 17 species in the family Fringillidae as being at various levels of conservation risk. One of them, the Bonin siskin (*Chaunoproctus ferreorostris*), was only known from the Japanese islands of Chichi-jima and Ogasawara-shoto (Peel and Bonin Islands). Unfortunately, the Bonin siskin became extinct in the late nineteenth century, likely because of deforestation and uncontrolled predation by introduced cats and rats. The Sao Tome grosbeak (*Neospiza concolor*) of Sao Tome and Principe is listed as Critically Endangered because of the loss of almost all its natural habitat of primary forest on its tiny island home in the eastern Atlantic just off the west coast of tropical Africa. The red siskin (*Carduelis cucullata*) of Colombia, Puerto Rico, Trinidad and Tobago, and Venezuela is Endangered because of a rapid population decline caused by habitat loss and uncontrolled trapping for the commercial pet trade. The Warsangli linnet (*Carduelis johannis*) is a highly local (or endemic) species of Somalia that has become Endangered because most of its limited habitat is being lost to timber harvesting. The Hispaniolan crossbill (*Loxia megaplaga*) of the Dominican Republic, Haiti, and Jamaica is Endangered because of severe habitat loss and fragmentation. The Azores bullfinch (*Pyrrhula murina*) is Endangered because it only survives in a tiny area of habitat on one of the Azores Islands off northwestern Africa. The Ankober serin (*Serinus ankoberensis*) of Ethiopia is Endangered because of its very small range (it is known from only four locations). The saffron siskin (*Carduelis siemiradzkii*) of Ecuador and Peru is

Asian rosy-finches (*Leucosticte arctoa*) feeding. (Photo by T.J. Ulrich/ VIREO. Reproduced by permission.)

An evening grosbeak (*Coccothraustes vespertinus*) calls from its perch. (Photo by J. Trott/VIREO. Reproduced by permission.)

A house finch (*Carpodacus mexicanus*) feeds on a berry. (Photo by J. Schumacher/VIREO. Reproduced by permission.)

listed as Vulnerable because of extensive habitat loss. The yellow-faced siskin (*Carduelis yarrellii*) of Brazil and Venezuela is Vulnerable because of uncontrolled trapping for the pet trade and habitat loss. The yellow-throated seedeater (*Serinus flavigula*) of Ethiopia is Vulnerable because of habitat loss through agricultural activities. Species listed as Near Threatened include the Vietnam greenfinch (*Carduelis monguilloti*) of Vietnam, the Kipingere seedeater (*Serinus melanochrous*) of Tanzania, the Syrian serin (*Serinus syriacus*) of the Middle East (Egypt, Iraq, Israel, Jordan, Lebanon, and Syria), and the Salvadori's serin (*Serinus xantholaema*) of Ethiopia. The blue chaffinch (*Fringilla teydea*) of the Canary Islands (Spain) is considered Conservation Dependent. (This designation is given to species that would become threatened within five years if conservation programs targeting them were suspended.) Sillem's mountain-finch (*Leucosticte sillemi*) of China and the Scottish crossbill (*Loxia scotica*) of the United Kingdom are listed as Data Deficient, meaning that appropriate data on their abundance and/or distribution are lacking.

Significance to humans

The island canary (*Serinus canaria*) was the first passerine bird to be domesticated and kept as a cage-bird. The Spaniards conquered the Canary Islands in 1478, and they soon brought canaries to Europe in large numbers. A lively trade in this popular cage-bird soon developed. The most common color of the domestic canary is the well-known bright yellow, or "canary yellow," but numerous other color varieties also have been

bred. Red-colored canaries owe their origin, and their reddish coloration, to captive interbreeding of the island canary with the black-capped red siskin (*Carduelis atriceps*). Canaries are still a common pet and are prized all over the world as eager songsters. Selective breeding has produced varieties of canaries with distinctly different songs. A variety of other finches also are kept as cage-birds for their song, lively behavior, and/or attractive plumage. Other than the domesticated canary, all finches have some indirect, local economic importance through ecotourism associated with birding.

A red crossbill (*Loxia curvirostra*) in water. (Photo by S. Holt/VIREO. Reproduced by permission.)

1. Gray-crowned rosy finch (*Leucosticte tephrocotis*); 2. Canary Islands finch (*Fringilla teydea*); 3. Red crossbill (*Loxia curvirostra*); 4. Chaffinch (*Fringilla coelebs*); 5. Common redpoll (*Acanthis flammea*); 6. American goldfinch (*Carduelis tristis*); 7. Brambling (*Fringilla montifringilla*); 8. Greenfinch (*Carduelis chloris*). (Illustration by Barbara Duperron)

1. Hawfinch (*Coccothraustes coccothraustes*); 2. Eurasian siskin (*Carduelis spinus*); 3. Evening grosbeak (*Coccothraustes vespertinus*); 4. Pine grosbeak (*Pinicola enucleator*); 5. European goldfinch (*Carduelis carduelis*); 6. European serin (*Serinus serinus*); 7. Island canary (*Serinus canaria*); 8. Eurasian bullfinch (*Pyrrhula pyrrhula*). (Illustration by Barbara Duperron)

Species accounts

Chaffinch
Fringilla coelebs

SUBFAMILY
Fringillinae

TAXONOMY
Fringilla coelebs Linnaeus, 1758. Seventeen subspecies are recognized.

OTHER COMMON NAMES
English: Common chaffinch; French: Pinson des arbres; German: Buchfink; Spanish: Pinzón Común.

PHYSICAL CHARACTERISTICS
Chaffinches are 5.5–7.1 in (14–18 cm) in body length, have a wingspan of similar length, and weigh 0.7–0.9 oz (20–25 g). They have a white patch on the shoulder, a white wing-bar, and white markings on the tail. Males have a slate-blue back of the head, a pink to deep-red face and breast, and a gray-blue tail. The female is yellow-brown in color, with a lighter belly. However, there is significant geographic variation in the coloration and patterns of streaking of chaffinches, especially in males.

DISTRIBUTION
Chaffinches are widely distributed, occurring in almost all of Europe, across the Middle East, through Ukraine and western Russia to Afghanistan, and in North Africa, the Canary Islands, and the Azores.

HABITAT
Chaffinches occur in a wide variety of woodlands and open forests, urban and suburban parks and gardens, and fields with hedgerows. They tend to occur in more open habitats during the winter.

BEHAVIOR
Chaffinches are migratory in winter, but the sexes do this differently. Their scientific name, *coelebs*, is derived from the Latin word for "without marriage," and acknowledges the preponderance of male chaffinches that winter in northern parts of their range, while females migrate further to the south. Studies of banded birds have shown that more males winter in Scandinavia, Britain, and parts of central Europe, while more females winter in Ireland. The territorial song is a bright series of rattling notes.

FEEDING ECOLOGY AND DIET
Chaffinches forage on the ground and in trees for seeds and fruit, including pine seeds. Unlike other kinds of true finches, the young of chaffinches are mostly fed insect larvae, butterflies, moths, and other invertebrates, which are regurgitated by the parents. When the ground is snow-covered, chaffinches will attend bird feeders, or they may gather in farm yards to eat seed put out for domestic fowl and at barns where seed is stored.

REPRODUCTIVE BIOLOGY
Chaffinches build a well-camouflaged, cup-shaped nest of grasses and lichens. The nest is neatly constructed and sturdy, and is located in a tree or shrub close to the trunk or a large branch. The eggs are incubated for 11–13 days. Only a single brood is raised each year.

CONSERVATION STATUS
Not threatened. The chaffinch is a widespread and abundant species. It probably has benefited from relatively open habitats created when older forests were converted into urbanized and agricultural land-uses, as long as some trees, shrubs, and hedgerows persisted.

SIGNIFICANCE TO HUMANS
Chaffinches are common, much-appreciated birds that enrich residential and agricultural areas with their beauty and song. They have been kept in cages as prized songbirds. ◆

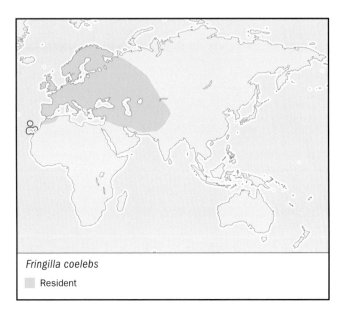

Fringilla coelebs
  Resident

Blue chaffinch
Fringilla teydea

SUBFAMILY
Fringillinae

TAXONOMY
Fringilla teydea Webb Berthelot & Moquin-Tandon, 1841. Two subspecies are recognized.

OTHER COMMON NAMES
English: Canary Islands chaffinch, Teydefinch; French: Pinson bleu; German: Teydefink; Spanish: Pinzón Azul.

PHYSICAL CHARACTERISTICS
The body length is about 5.9 in (15 cm). The male is uniformly slate-blue, darker on the back than on the belly, and whitish beneath the rump. It has faint wing-bars and a whitish eye-ring. The female is more drably gray-blue.

DISTRIBUTION
The blue chaffinch is endemic to the Canary Islands off northwestern Africa. It is restricted to the islands of Tenerife and Gran Canaria.

Fringilla teydea
■ Resident

HABITAT
The blue chaffinch inhabits pine forest almost exclusively, at altitudes ranging from 2,300 to 6,600 ft (700 and 2,000 m). It inhabits both natural pine forest and older planted stands. It prefers areas with an undergrowth of broom (*Chamaecytisus proliferus*).

BEHAVIOR
The blue chaffinch is a non-migratory species. It is a melodic singer and is rather unafraid of humans.

FEEDING ECOLOGY AND DIET
The blue chaffinch feeds on seeds, particularly those of pine. It also eats insects and other arthropods, and feeds its young almost exclusively with this food.

REPRODUCTIVE BIOLOGY
The female blue chaffinch builds a nest of pine needles and branches of broom lined with moss, feathers, grasses, and rabbit hair. It is usually located in a pine tree, but occasionally may be found in heath (*Erica arborea*) or laurel (*Laurus azorica*). The female incubates a clutch of (usually) two eggs for 14–16 days. The chicks are blind and down-covered when hatched and are fed by both the male and female. Fledging takes place in 17–18 days.

CONSERVATION STATUS
The IUCN lists the blue chaffinch as Conservation Dependent. This songbird has declined greatly in abundance. This decline began in the early decades of the nineteenth century because of the destruction and disturbance of their restricted habitat of mountain pine forests, and also because of excessive shooting by naturalists and commercial specimen collectors. In 2001, only about 1,500 breeding pairs were left, and their remaining forest habitat is becoming increasingly lost and fragmented, largely because of inappropriate forestry management. Without effective conservation of this rare species and its habitat, it could soon become endangered.

SIGNIFICANCE TO HUMANS
The blue chaffinch is a rare species, and its sightings are much appreciated by birders and other naturalists that visit its island habitat. This can lead to some local economic benefits through ecotourism. In the past their melodic songs made them a prized possession among sailors, but commercial trade in these rare birds has ceased. ◆

Brambling
Fringilla montifringilla

SUBFAMILY
Fringillinae

TAXONOMY
Fringilla montifringilla Linnaeus, 1758.

OTHER COMMON NAMES
French: Pinson du Nord; German: Bergfink; Spanish: Pinzón Real.

PHYSICAL CHARACTERISTICS
Bramblings are small, stout birds, with a body length of about 5.7 in (14.5 cm). Males have a black head and back, a rich orange throat and breast, wings and tail marked with white and black, and a whitish belly. Females have a similar but much duller coloration. During the winter, males look similar to the females, but they start to molt into their breeding plumage in late winter.

DISTRIBUTION
Bramblings range widely through northern Eurasia, from Scandinavia to eastern Siberia as far as the Kamchatka Peninsula. During the winter, bramblings may wander extensively. In the United Kingdom, for example, it is a sporadic winter visitor, arriving in early October and departing for the more northerly breeding grounds by mid-March.

Fringilla montifringilla
■ Resident

HABITAT

Bramblings breed in subarctic birch and willow groves and shrub tundra of the northern boreal and tundra regions. Its common name, brambling, means "the little bramble bird," implying it occurs in thorny thickets, but its natural habitat is actually northern deciduous woodlands and shrubby tundra. During the winter it may occur in more open habitats.

BEHAVIOR

Bramblings are migratory, wandering extensively during the winter. They often occur in mixed flocks with other species of finches. During a particularly cold and snowy winter in 1946–1947, an estimated eleven million bramblings plus other finches were observed feeding on an abundant crop of beech mast (or beech-nuts) at the village of Porrentruy in Switzerland. Each night these innumerable birds gathered in a particular, small valley to roost communally. Winter irruptions of bramblings typically, however, vary greatly from year to year. They are influenced by both local and large-scale weather and snow conditions over their wintering range. In addition, a lack of suitable food in northern parts of the wintering range may trigger immense out-migrations into more southern regions. As such, bramblings are extremely unpredictable in their migratory routes and wanderings, often appearing in the millions in a region in one winter, but not in other years. Bramblings are territorial during the breeding season. The male has a wheezy song, and the birds also have high-pitched, wheezy "yeep" flight calls during the non-breeding season.

FEEDING ECOLOGY AND DIET

Bramblings eat a wide variety of seeds, including the relatively large nuts (or mast) of beech trees.

REPRODUCTIVE BIOLOGY

Bramblings court and mate in the late winter and breed as territorial pairs. They build a cup-shaped nest in a tree or shrub.

CONSERVATION STATUS

Not threatened. A widespread and abundant species.

SIGNIFICANCE TO HUMANS

Bramblings are popular birds that enrich the lives of many people. They are sought by birders and other naturalists, and this can result in local economic benefits through ecotourism. ◆

Greenfinch

Carduelis chloris

SUBFAMILY

Carduelinae

TAXONOMY

Carduelis chloris Linnaeus, 1758. Nine subspecies are recognized.

OTHER COMMON NAMES

English: European greenfinch; French: Verdier d'Europe; German: Grünling, Grünfink; Spanish: Verderon Común.

PHYSICAL CHARACTERISTICS

Greenfinches are about 5.5 in (14 cm) in body length. They have pink legs and a stout beak. Both sexes are colored overall yellow-brown, with grayish, yellow-edged wings, and black on the tail. Females are duller than males, and juveniles have streaked breasts.

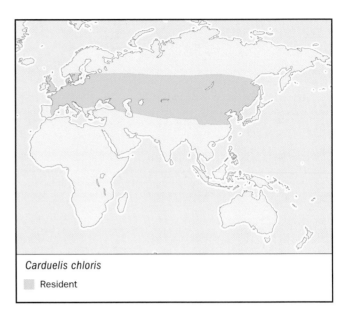

Carduelis chloris
■ Resident

DISTRIBUTION

Greenfinches range widely across Europe and western Asia. They have also been introduced to parts of South America and Australasia, where they persist as wild, non-native songbirds.

HABITAT

Greenfinches inhabit a wide range of forests and woodlands, orchards, parks, gardens, and farmland containing hedgerows.

BEHAVIOR

Greenfinches are migratory birds, breeding in northern parts of their range and spending the winter further to the south. They are social birds, especially during the non-breeding season, and are often found in small flocks. They may also occur in mixed-species flocks with other finches. The territorial song is a nasal, high-pitched call. They also have characteristic, wheezy flight notes.

FEEDING ECOLOGY AND DIET

Greenfinches feed on variety of seeds, including those of trees, shrubs, and herbs. The young are fed partly with insects and spiders, as well as plant matter.

REPRODUCTIVE BIOLOGY

Greenfinches build an unruly nest of sticks lined with feathers. Nests are often grouped together as a loose colony. They typically have two broods each year, with four to five eggs per clutch.

CONSERVATION STATUS

Not threatened. A widespread and abundant species.

SIGNIFICANCE TO HUMANS

None known. ◆

Common redpoll

Acanthis flammea

SUBFAMILY

Carduelinae

TAXONOMY

Acanthis flammea Linnaeus, 1758. Three subspecies are recognized.

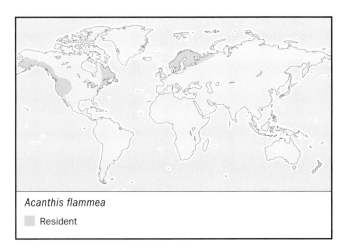

Acanthis flammea
☐ Resident

birds is a scene of feverish activity as they tear dried flower stalks apart and then drop to the ground to pick up the seeds.

REPRODUCTIVE BIOLOGY
Common redpolls are somewhat nomadic in their local breeding. If the local supply of birch seed is abundant they may settle in numbers. After raising their first crop of fledglings, they may move elsewhere to exploit another abundant resource of birch seeds. Their nest is a neat cup of woven grass, moss, and twigs placed in a fork of a willow branch. The clutch is four to six pale green eggs incubated by the female for 10–11 days. The altricial chicks are brooded by the female. They are fed primarily by the female with some male assistance at times. Fledging takes place in 9–14 days.

CONSERVATION STATUS
Not threatened. A widespread and abundant species.

SIGNIFICANCE TO HUMANS
Common redpolls are lively and pleasant birds, and are sought after by birdwatchers, particularly during the winter months. In New Zealand they are sometimes considered an introduced nuisance because of damage caused to fruit trees when their buds are eaten. ◆

OTHER COMMON NAMES
English: Redpoll; French: Sizerin flammé; German: Birkenzeisig; Spanish: Pardillo Sizerin.

PHYSICAL CHARACTERISTICS
The common redpoll is about 4.7–5.5 in (12–14 cm) in body length and weighs about 0.5 oz (14 g). The tail is forked and the beak is sharply pointed and has a black tip. The overall body coloration is gray-brown, with gray wings having light wingbars, a lighter belly streaked with brown, a red crown on the top of the head, and a black patch beneath the lower mandible. The male has orange-red on the face and chest, but the extent of this varies among geographical races of this widespread species.

DISTRIBUTION
The common redpoll is a very wide-ranging species with a circumboreal distribution, occurring in suitable habitat in northern North America as well as in Eurasia. It occurs in Newfoundland, northern Quebec and Labrador, across the rest of northern Canada to Alaska, and through Siberia and northern Russia to northern Europe and Iceland. It is an irregular migrant that may occur as far south in the United States as California, Oklahoma, and the Carolinas, and also through much of southern Europe, Russia, the Caucasus, and central China. It was introduced to New Zealand in the nineteenth century, where it persists as a non-native songbird.

HABITAT
The common redpoll breeds in shrubby tundra, and in the winter occurs in brushy pastures, open forest and thickets, and weedy fields. During the winter they may sleep in snow tunnels to keep warm. They are able to hang upside down chickadee-like to pry birch seeds from hanging catkins. South of the boreal tree-line, the local wanderings and population densities of common redpolls depends on how abundant their winter food is.

BEHAVIOR
Common redpolls are active and mobile birds. Even at rest, much fidgeting and twittering is evident. They are highly social birds, particularly during the non-breeding season when they aggregate into flocks, often with other species of finches. In areas where the ranges of the common redpoll and the hoary redpoll (*Acanthis hornemanni*) overlap, such as northern Norway, the two species may form mixed breeding pairs and produce hybrids of intermediate appearance.

FEEDING ECOLOGY AND DIET
Common redpolls feed on grains and seeds, particularly favoring birch seeds. A stand of winter weeds visited by a flock of these

American goldfinch
Carduelis tristis

SUBFAMILY
Carduelinae

TAXONOMY
Carduelis tristis Linnaeus, 1758.

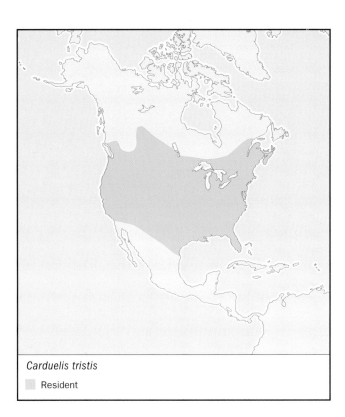

Carduelis tristis
☐ Resident

OTHER COMMON NAMES
English: Goldfinch, wild canary; French: Chardonneret jaune; German: Goldzeisig; Spanish: Dominiquito Canario.

PHYSICAL CHARACTERISTICS
The American goldfinch is about 4.3 in (11 cm) in body length and weighs 0.5 oz (14 g). The male is colored overall a bright canary yellow, with black wings marked with white bars, a black tail, and a black face cap. The female is a more subdued yellow, with dark wings and tail. The juveniles are olive-yellow, with darker wings. During the winter, the male is not so brightly colored.

DISTRIBUTION
The American goldfinch breeds throughout much of southern Canada and the northern half of the United States. It winters in extreme southern Canada, through most of the United States, and northern Mexico.

HABITAT
The American goldfinch breeds in open, mixed-species forests and shrubby places. It winters in shrubby habitats, old fields, and parks and gardens.

BEHAVIOR
American goldfinches have a distinctive, bounding flight. They are migratory birds, breeding in northern parts of their range and spending the winter wandering in the southern reaches. They are highly social birds, particularly during the non-breeding season when they may form large flocks, often with other finches. They often breed in loose colonies. Their courtship and territorial display includes acrobatic aerial maneuvers by the male, which also sings during flight. The song is a high-pitched twittering, and there are also distinctive call notes.

FEEDING ECOLOGY AND DIET
American goldfinches feeds on small seeds and grains, particularly favoring plants in the aster family, including sunflower, lettuce, and thistles.

REPRODUCTIVE BIOLOGY
American goldfinches arrive at the northern parts of their breeding range in April or May, about the time dandelions and some other early-flowering plants begin to set seed. However, the species does not breed until about mid-July. In more southerly locales, such as California, breeding can begin in March and continue through July and even, in exceptional cases, into November. Most pairs probably rear only one brood per year. The reasons for the relatively late-starting breeding of the American goldfinch is not understood, but it may be related to the timing of the maturation of the seeds of thistles, which are a major food for the young birds. The cup-shaped nest is woven of grasses and other plant fibers. It may be placed in a large thistle or other tall weed, or in a shrub or tree. The eggs are colored pale blue, and a typical clutch size is four to five eggs. Incubation is 12–12 days by the female. Both male and female feed the nestlings and fledging occurs in 11–17 days.

CONSERVATION STATUS
Not threatened. The American goldfinch is a widespread and abundant species. There is evidence that its abundance has been decreasing during recent decades, but it is not yet considered to be a species at risk. The population decrease is likely due to habitat loss through urbanization and the intensification of agricultural practices, which result in fewer areas with weed-rich, shrubby habitat.

SIGNIFICANCE TO HUMANS
American goldfinches are well-known and popular birds. They can be tamed and have been kept as caged songbirds, although this is now rarely done. ◆

European goldfinch
Carduelis carduelis

SUBFAMILY
Carduelinae

TAXONOMY
Carduelis carduelis Linnaeus, 1758. Ten subspecies are recognized.

OTHER COMMON NAMES
English: Goldfinch; French: Chardonneret élégant; German: Stieglitz; Spanish: Jilguero Europeo.

PHYSICAL CHARACTERISTICS
European goldfinches are 5–6 in (13–15 cm) in body length. They have a sharply pointed beak, and a forked tail. The back is colored dark olive-brown, the wings are black with a yellow patch, the tail black, the belly whitish, and the face is red, bordered with white and black. Females are olive-brown with yellow highlights and darker wings and tail.

DISTRIBUTION
European goldfinches range over almost all of Europe, the Azores, the Canary Islands, and as far east as western Russia. They have also been introduced to a few places in the United States, Central America, and Australasia, where they survive in urbanized areas.

HABITAT
European goldfinches inhabit open woodlands, shrubby areas, orchards, parks and gardens, and well-vegetated cultivated areas.

BEHAVIOR
The European goldfinch has a distinctive bounding flight. It is a migratory species, breeding in northern parts of its range and

Carduelis carduelis
▨ Resident

spending the winter in southern reaches. It is a highly social bird, particularly during the non-breeding season when it occurs in flocks, often with other finches. The courtship and territorial displays include aerial maneuvers and singing by the male. The song is a high-pitched twittering, and there are also distinctive call notes. They often position and hold food using their toes.

FEEDING ECOLOGY AND DIET
The European goldfinch feeds on small seeds and grains, particularly favoring species in the aster family, such as dandelions, thistles, burdock, lettuce, and sunflowers. The young are fed partly with insects.

REPRODUCTIVE BIOLOGY
The European goldfinch builds a small, neat, cup-shaped nest of woven grass, moss, and lichens. Typically, the nest is positioned at the end of a small branch in an open-grown tree or shrub. They breed twice or sometimes three times each year, with each clutch consisting of four to five eggs. In general, relatively warm, dry summers result in greater reproductive success than those with cool, wet weather.

CONSERVATION STATUS
Not threatened. The European goldfinch suffered widespread population declines over parts of its range during the 1800s because of uncontrolled live-trapping for the commercial pet trade. Their population has since recovered, however, and they are now a widespread and abundant species.

SIGNIFICANCE TO HUMANS
European goldfinches are well-known and popular birds. They can be tamed and have been kept as caged songbirds, although this is now uncommon. In the past, they provided amusement for people, who provided the birds with food tied to the end of a hanging thread. The bird would grasp the thread with its beak, pull up a section, hold it with its foot, and continue to do this until it had pulled up the food. European goldfinches are so good at this trick that for centuries they were kept in special cages constructed so that the birds could only survive by pulling up and holding onto threads, one of which provided seed and the other a thimble containing water. In the sixteenth century, this contraption was so popular in parts of Europe that the birds were commonly known as "dippers." ◆

Eurasian siskin
Carduelis spinus

SUBFAMILY
Carduelinae

TAXONOMY
Carduelis spinus Linnaeus, 1758.

OTHER COMMON NAMES
English: Siskin; French: Tarin des aulnes; German: Erlenzeisig; Spanish: Lúgano.

PHYSICAL CHARACTERISTICS
Eurasian siskins are about 5.1 in (13 cm) in length. They have a thin, pointed beak. The upperparts of the male are colored gray, with yellow bars on the wings, a yellow belly and face, and black on the crown of the head and just beneath the lower mandible. The female is much duller and more heavily streaked, and has a whitish belly and few yellow markings.

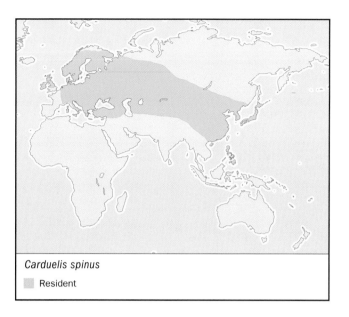

Carduelis spinus

▨ Resident

DISTRIBUTION
Eurasian siskins range widely across Eurasia, from the United Kingdom, through virtually all of Europe, across Asia to eastern Russia, northern China, and Japan. They winter irregularly in more southern regions within their range, sometimes in large irruptions.

HABITAT
Eurasian siskins breed in various kinds of coniferous forest, including boreal and montane types. They often winter in more open kinds of habitats, including gardens.

BEHAVIOR
Eurasian siskins can be rather tame, and may even perch on people when being fed. They have a bounding flight pattern. Eurasian siskins are migratory, breeding in northern parts of their range and spending the winter in southern reaches. They are highly social bird, particularly during the non-breeding season when they occur in large flocks, often with other finches. Their courtship and territorial displays include aerial maneuvers and singing by the male. The song is a high-pitched twittering, and there are also distinctive call notes.

FEEDING ECOLOGY AND DIET
Eurasian siskins feed on grains and tree seeds and buds. They also use bird feeders put out by people during the winter.

REPRODUCTIVE BIOLOGY
Eurasian siskins build a cup-shaped nest of grasses and other plant fibers. The nest is usually placed in a conifer tree. The clutch consists of four or five spotted eggs. They build their nest and lay eggs earlier than usual in years when there is an abundant supply of conifer seeds, even doing while snow is still on the ground. This can allow them to raise additional broods during that nesting season.

CONSERVATION STATUS
Not threatened. They are a widespread and abundant songbird.

SIGNIFICANCE TO HUMANS
None known. ◆

European serin

Serinus serinus

SUBFAMILY
Carduelinae

TAXONOMY
Serinus serinus Linnaeus, 1766.

OTHER COMMON NAMES
English: Serin; French: Serin cini; German: Girlitz; Spanish:
Verdecillo.

PHYSICAL CHARACTERISTICS
European serins are small finches, with a body length of about
4.3 in (11 cm). They have a short, strong, pointed beak and a
slightly forked tail. The male is colored greenish streaked with
black on the back, with dark wings and tail, and a yellow rump,
head, and chest. The female is darker and duller, and much
less yellow.

DISTRIBUTION
The European serin breeds widely in Europe and around the
Mediterranean basin, including parts of coastal North Africa. It
winters in more southern regions of its range.

HABITAT
The European serin inhabits wooded and shrubby hillsides,
and also utilizes well-vegetated agricultural areas, such as vineyards, orchards, and plantations.

BEHAVIOR
The European serin is a migratory species. It is gregarious, especially during the non-breeding season when it occurs in flocks,
often with other finches. The male defends a breeding territory
and attracts a mate by an aerial display and melodic song.

FEEDING ECOLOGY AND DIET
The European serin feeds on grains and tree seeds.

REPRODUCTIVE BIOLOGY
The nest is built in a tree or bush, usually 6.6–10 ft (2–3 m)
above the ground. Eggs are laid from March onward in North
Africa, from April on in southern Europe, and in May in central Europe. There is more than one brood per year.

CONSERVATION STATUS
Not threatened. A widespread and abundant species.

SIGNIFICANCE TO HUMANS
This close relative of the canary is sometimes kept as a cage-
bird. ◆

Hawfinch

Coccothraustes coccothraustes

SUBFAMILY
Carduelinae

TAXONOMY
Coccothraustes coccothrauses Linnaeus, 1758. Three subspecies are
recognized.

OTHER COMMON NAMES
French: Grosbec; German: Kernbeißer; Spanish: Picogordo.

PHYSICAL CHARACTERISTICS
Hawfinches are relatively large, heavy-bodied birds with a massive beak; they have a stocky, top-heavy appearance. Their
body length is about 7 in (18 cm) and they weigh about 1.9 oz
(54 g). The male has a black back, wings, and tail, is chestnut
on the head and belly, and has a black chin and blue beak. The
female is somewhat duller. The massive beak and skull design
allow hawfinches to crack open large, tough seeds, such as
those of cherries and olives.

DISTRIBUTION
Hawfinches are found throughout much of the temperate and
southern boreal regions of Europe and Asia.

Serinus serinus
◻ Resident

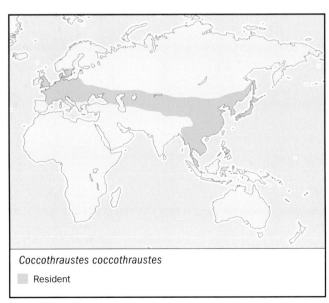

Coccothraustes coccothraustes
◻ Resident

HABITAT
Hawfinches occur in hardwood and mixedwood forests of various kinds, in addition to well-vegetated parks and gardens.

BEHAVIOR
Hawfinches are shy birds that are wary of noise and movement. The male has a rather soft, feeble song that is used to defend its breeding territory.

FEEDING ECOLOGY AND DIET
Hawfinches eat seeds and fruit, including tough nuts and fruit-stones. They sometimes feed on the ground on fallen fruit, but are wary when doing so.

REPRODUCTIVE BIOLOGY
Hawfinches have an elaborate courtship ritual, involving the drooping of a wing to display iridescent purple and green flight feathers. The male also bows deeply to a prospective mate and tucks his beak under his belly, revealing a gray nape patch. Aerial chases between a male and female are also part of pair formation. Hawfinches often breed in loose colonies. Their small, cup-shaped nest is built of roots, twigs, and lichens. It is lined with plant fibers, hair, and rootlets, and placed low in a tree. Three to seven greenish eggs with blackish brown markings are incubated for 9–14 days, mostly by the female. The young are brooded by the female and tended by both sexes. Fledging occurs in 10–14 days.

CONSERVATION STATUS
Not threatened. Hawfinches are a widespread and abundant species, but they are vulnerable to habitat loss due to logging operations.

SIGNIFICANCE TO HUMANS
None known. ◆

Eurasian bullfinch
Pyrrhula pyrrhula

SUBFAMILY
Carduelinae

TAXONOMY
Pyrrhula pyrrhula Linnaeus, 1758. Five subspecies are recognized.

OTHER COMMON NAMES
English: Bullfinch; French: Bouvreuil pivoine; German: Gimpel; Spanish: Camachuelo Común.

PHYSICAL CHARACTERISTICS
The Eurasian bullfinch is a relatively large finch with a body length of about 5.5–6.3 in (14–16 cm) and weighing about 0.8–1.1 oz (22–30 g). They have a short, stout, dark-colored beak. The male is gray on the upper body, with a black head-cap, black wings with a prominent white wing-bar, a reddish belly and sides of head, and white rump. The female and the young are duller, being more brownish pink and lacking the red breast. There is considerable variation in size and coloration among the subspecies.

DISTRIBUTION
The bullfinch ranges widely over Eurasia, occurring in almost all of Europe and most of Asia south of the boreal forest, including the Kamchatka Peninsula and Japan.

Pyrrhula pyrrhula

▨ Resident

HABITAT
The bullfinch inhabits coniferous forest, mountain slopes and ravines, stony edges of deserts, and parks, gardens, and well-vegetated cultivated land.

BEHAVIOR
Bullfinches are shy and wary birds and seldom forage on the ground. They tend to live in family groups, or in small flocks during the non-breeding season. The territorial song is soft, trisyllabic, and creaky.

FEEDING ECOLOGY AND DIET
Bullfinches feed on shoot buds, seeds, and other fruits. Buds are eaten mostly in the winter and spring when the main food of seeds is less abundant.

REPRODUCTIVE BIOLOGY
The female bullfinch constructs a cup-shaped nest of twigs, lichen, and moss in a dense shrub or on a tree limb. The clutch consists of four to six pale-blue eggs marked with reddish brown and is incubated by the female for 12–14 days. The altricial young are brooded by the female and fed by both parents. Fledging occurs in 12–18 days. There are up to two broods per year.

CONSERVATION STATUS
Not threatened. Bullfinches are a widespread species and are abundant over most of their range. However, some local populations and subspecies are threatened, including the Azores race, *Pyrrhula pyrrhula murina*. Populations of bullfinches have declined substantially over much of western Europe since about 1955, likely because of extensive habitat loss through urbanization, deforestation, and the intensification of agricultural practices, including the loss of shrubby hedgerows.

SIGNIFICANCE TO HUMANS
Bullfinches are inconspicuous birds and many people do not realize that they occur nearby. Bullfinches were popular cage-birds in the nineteenth century, but rarely are kept now. ◆

Evening grosbeak

Coccothraustes vespertinus

SUBFAMILY
Carduelinae

TAXONOMY
Coccothraustes vespertinus Cooper, 1825.

OTHER COMMON NAMES
French: Grosbec errant; German: Abendkernbeißer; Spanish: Picogordo Vespertino.

PHYSICAL CHARACTERISTICS
The evening grosbeak has a body length of about 7–8.7 in (18–22 cm) and weighs about 2.1 oz (60 g). It has a rather stout body, a short tail, and a stout yellow beak. The male is bright yellow, with black wings with a large white wing-patch, a black tail, and a black crown on the top of the head. Females are a duller gray and brown pattern, with white wing-patches.

DISTRIBUTION
The evening grosbeak inhabits the southern boreal forest and montane forest regions of North America. The range includes Canada, the western United States, and northern Mexico. They may winter in the breeding range or wander extensively, particularly to the east and south of the breeding range. They periodically irrupt from the usual wintering areas and may then be abundant in areas where they are not commonly seen.

HABITAT
Evening grosbeaks breed in mixed conifer forest, but may winter in more open habitats.

BEHAVIOR
Evening grosbeaks are highly social birds, especially during the non-breeding season when they may occur in large flocks. The territorial song is a repeated chirp-like call.

FEEDING ECOLOGY AND DIET
Evening grosbeaks feed mainly on tree seeds, but also frequent winter feeding sites to get sunflower seeds. Their diet also includes insects, buds, sap, fruits, and berries.

REPRODUCTIVE BIOLOGY
Breeding pairs are monogamous and nest in colonies. The female builds a frail, cup-shaped nest of twigs, grass, moss, roots, and pine needles on a horizontal tree branch far out from the trunk about 20–60 ft (6–18 m) above the ground. She incubates three to five pale blue to bluish eggs spotted with gray, purple, or brown for 11–14 days. The male feeds the incubating female. The altricial young are brooded by the female and fed by both sexes. They fledge in 13–14 days. One to two broods per year.

CONSERVATION STATUS
Not threatened. The evening grosbeak is a widespread and abundant species and may be increasing in abundance and range.

SIGNIFICANCE TO HUMANS
Evening grosbeaks flock to areas infested with spruce budworm to breed and raise their young. (If an evening grosbeak were to get all its daily energy from budworm larvae it would eat 1,000 a day.) Because of its appetite for this destructive pest, the evening grosbeak is considered a beneficial bird. ◆

Red crossbill

Loxia curvirostra

SUBFAMILY
Carduelinae

TAXONOMY
Loxia curvirostra Linnaeus, 1758. Six subspecies are recognized.

OTHER COMMON NAMES
English: Common crossbill; French: Bec-croisé des sapins; German: Fichtenkreuzschnabel; Spanish: Piquituerto Común.

PHYSICAL CHARACTERISTICS
The red crossbill has a body length of about 5.5 in (14 cm) and weighs about 1.4 oz (40 g). It has a rather heavy body, a short forked tail, and a stout beak in which the tips of the upper and lower mandibles cross over as an adaptation to extracting seeds from conifer cones. Males are colored overall brick-red, with blackish wings and tail. Females are a dull yellow-brown with darker wings. Juveniles have weakly crossed mandibles, gray-olive upperparts and whitish underparts both streaked with dark brown, and a buff-yellow rump.

DISTRIBUTION
The red crossbill is an extremely widespread species that inhabits the boreal and montane forest regions of both North America and Eurasia. It occurs from coast to coast in suitable habitats on both continents. It periodically irrupts from its usual wintering regions and may then be abundant in areas where it is not usually seen.

HABITAT
Red crossbills breed and winter in pine-containing conifer forests of various kinds.

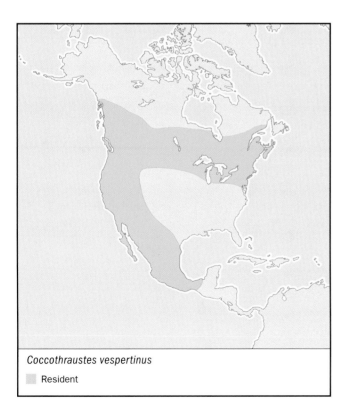

Coccothraustes vespertinus

▨ Resident

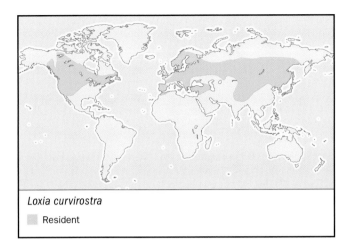

Loxia curvirostra
▨ Resident

OTHER COMMON NAMES
French: Roselin à tête grise; German: Rosenbauch-Schneegimpel; Spanish: Pinzón Rosado de Corona Gris.

PHYSICAL CHARACTERISTICS
The gray-crowned rosy finch has a body length of about 6.5 in (16.5 cm). It has a short, slightly forked tail, and a stout, conical, pointed beak. The male is colored overall red, with pinkish red patches on the wings, a dark tail, a gray head, and black patches on the face. The female has a browner, more subdued body coloration, and lacks a black patch on the chin. Juveniles are gray-brown.

DISTRIBUTION
The gray-crowned rosy finch occurs in the Rocky Mountain region and western Arctic tundra of North America, ranging from Alaska, through western Canada, to the southwestern United States. It breeds in higher-altitude habitats and winters in lower altitudes.

HABITAT
The gray-crowned rosy finch breeds in alpine tundra above the timberline, and also in northern tundra in Alaska and Yukon. It winters in lowlands, including open habitats and conifer forest.

BEHAVIOR
Gray-crowned rosy finches are rather tame and unafraid of humans. They are social birds, especially during the non-breeding season when they occur in flocks. Males are weakly territorial and the song is a repeated series of simple, high-pitched chirps, often given in flight.

FEEDING ECOLOGY AND DIET
Gray-crowned rosy finches feed on seeds of various kinds, supplemented by invertebrates. They forage on ripe herbaceous plants and also on the ground and in snowbanks.

BEHAVIOR
Red crossbills are highly social birds, especially during the non-breeding season when they may occur in large flocks. The territorial song is a repeated series of simple chirps, often given in flight. The male displays to the female by flying above her, vibrating his wings, and delivering an in-flight song.

FEEDING ECOLOGY AND DIET
Red crossbills feed on the seeds of conifers, particularly species of pines. They use their peculiar, crossed bill to force the scales of conifer cones apart and then scoop the seed into their mouths with their tongues. Their diet also includes insects and caterpillars.

REPRODUCTIVE BIOLOGY
Breeding pairs are monogamous and solitary. The female builds a cup-shaped nest of twigs, bark, grass, and rootlets, lined with finer grasses, feathers, fur, hair, and moss. The nest is located on a tree branch far out from the trunk about 6.6–40 ft (2–12 m) above the ground. A clutch of three to four light green or blue eggs spotted with brown and lilac is incubated by the female for 12–18 days. The altricial young are brooded by the female and fed by both parents. They fledge in 15–20 days. Nestlings have straight mandibles that cross gradually after they have been out of the nest for about three weeks. One to two broods per year.

CONSERVATION STATUS
Not threatened. The red crossbill is a widespread and abundant species. Some populations, however, have declined greatly and are considered to be at risk. The subspecies native to the island of Newfoundland, for example, has become extremely rare. Logging operations have destroyed and continue to damage red crossbill habitat.

SIGNIFICANCE TO HUMANS
None known. ◆

Gray-crowned rosy finch
Leucosticte tephrocotis

SUBFAMILY
Carduelinae

TAXONOMY
Leucosticte tephrocotis Swainson, 1832.

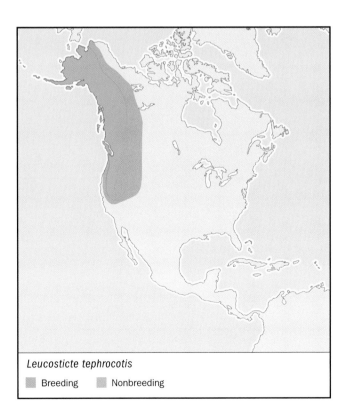

Leucosticte tephrocotis
▨ Breeding   ▨ Nonbreeding

REPRODUCTIVE BIOLOGY
The female chooses the breeding territory and the male follows and defends the female from other males. Breeding pairs are monogamous and loosely colonial. The female builds a cup-shaped nest of grass, rootlets, lichen, and moss, lined with fine grass, plant down, and feathers. The nest is sited in clefts of rock and cliffs, or sometimes in a cave or the eaves of a building. A clutch of four to five whitish eggs occasionally dotted with reddish brown are incubated by the female for 12–14 days. The altricial young are brooded by the female and fed by both sexes. Both males and females develop a gular pouch in the upper throat with an opening in the floor of the mouth so that they can carry larger amounts of seed back to their young. The young fledge in 16–22 days. Birds breeding in the mountains have one brood per year; those living in lower tundra have as many as two.

CONSERVATION STATUS
Not threatened. The gray-crowned rosy finch is a widespread and abundant species.

SIGNIFICANCE TO HUMANS
None known. ◆

Pine grosbeak
Pinicola enucleator

SUBFAMILY
Carduelinae

TAXONOMY
Pinicola enucleator Linnaeus, 1758. Two subspecies are recognized.

OTHER COMMON NAMES
English: Pine rosefinch; French: Durbec des sapins; German: Hakengimpel; Spanish: Camachuelo Picogrueso.

PHYSICAL CHARACTERISTICS
The pine grosbeak is a large, stout-bodied finch with a body length of about 7.9 in (20 cm) and weighing about 2 oz (57 g). It has a short, slightly forked tail and a short, stout, conical beak. The male is colored overall red, with black wings with white wing-bars, a dark tail, and grayish patches on the belly. The female has a yellowish olive head and rump, and gray un-

derparts and back. Juveniles resemble adult females, but are duller with washes of dull yellow on the head, back, and rump.

DISTRIBUTION
The pine grosbeak is an extremely widespread species that inhabits the boreal forest and montane forest regions of both North America and Eurasia. It occurs from coast to coast in suitable habitats on both continents. Pine grosbeaks sporadically irrupt from their usual wintering regions and may then be abundant in areas where they are not commonly seen.

HABITAT
The pine grosbeak breeds in conifer forest in both the northern boreal region and in montane areas in the Rocky Mountains. During the winter they occur more widely in various kinds of forest.

BEHAVIOR
The pine grosbeak is a rather tame species. It is a social bird that often occurs in flocks during the non-breeding season. The territorial song is a series of warbled notes. There is also a variety of simple, high-pitched chirps, often given in flight. Males feed females as part of the courtship ritual.

FEEDING ECOLOGY AND DIET
The pine grosbeak feeds on seeds and small fruits of various kinds. They also eat buds and insects, and mostly forage in trees and shrubs.

REPRODUCTIVE BIOLOGY
Breeding pairs are monogamous and solitary. The female builds a cup-shaped nest of twigs, plant fibers, and rootlets, lined with moss, lichen, fine grass, and rootlets. It is located on the limb of a tree or shrub about 2–25 ft (0.6–7.6 m) above the ground. Two to five blue-green eggs dotted with black, purple, and brown are incubated by the female for 13–15 days. The altricial young are brooded by the female, fed by both parents, and fledge in 13–20 days. Like many finches, both males and females develop gular pouches during the nesting season to carry food to their young. One brood per year.

CONSERVATION STATUS
Not threatened. The pine grosbeak is a widespread and abundant species, but it is vulnerable to habitat loss due to logging.

SIGNIFICANCE TO HUMANS
None known. ◆

Pinicola enucleator
▨ Breeding ▨ Nonbreeding

Island canary
Serinus canaria

SUBFAMILY
Carduelinae

TAXONOMY
Serinus canaria Linnaeus, 1758.

OTHER COMMON NAMES
English: Common canary, canary; French: Serin des Canaries; German: Kanarengirlitz; Spanish: Canario Sylvestre.

PHYSICAL CHARACTERISTICS
The island canary is a small, slender finch with a body length of about 5 in (12.5 cm). It has a rather long, forked tail, and a short, stout, conical, pointed beak. The male is colored overall

Serinus canaria
▨ Resident

DISTRIBUTION
The canary is a highly local (or endemic) species indigenous only to the Azores, Canaries, and Madeira Islands of the eastern temperate Atlantic Ocean. It has been domesticated for centuries, however, and is kept as a caged songbird in many countries.

HABITAT
The island canary inhabits forest, open habitats with shrubs, gardens, and orchards.

BEHAVIOR
The island canary is a non-migratory species. It is a social bird that may occur in small flocks when not breeding. The song is a highly musical series of warbled notes, often given in flight.

FEEDING ECOLOGY AND DIET
The island canary feeds on seeds and small fruits of various kinds.

REPRODUCTIVE BIOLOGY
The island canary weaves a cup-shaped nest of plant fibers and usually locates it in a shrub or low tree. They are often polygamous breeders, meaning a male may mate with several females. The clutch size is typically about five, but can vary from one to ten. Hatching occurs about 14 days after the hen begins to incubate. There may be more than one brood per year.

CONSERVATION STATUS
The island canary is an endemic species of only a few islands in the eastern Atlantic Ocean, but it is locally abundant there. Domesticated varieties are abundant in captivity.

SIGNIFICANCE TO HUMANS
The island canary has been kept as a caged songbird for more than 500 years. It is a highly prized pet because of the loud, enthusiastic, musical song of the male.

olive-brown, with yellow on the face and belly. The female is somewhat duller in color. Domesticated varieties, however, can vary widely in coloration, with as many as several hundred types being recognized. Red canaries are among the extremes of coloration, and are derived from fertile hybrids of the island canary and the black-capped red siskin (*Carduelis atriceps*) of South America.

Resources

Books

Bent, A.C. *Life Histories of North American Cardinals, Grosbeaks, Buntings, Towhees, Finches, Sparrows, and Allies; Order Passeriformes: Family Fringillidae.* Mineola, NY: Dover Publications, 1968.

BirdLife International. *Threatened Birds of the World.* Barcelona, Spain and Cambridge, UK: Lynx Edicions and BirdLife International, 2000.

Clement, P., A. Harris, and J. Davis. *Finches and Sparrows: An Identification Guide.* Princeton, NJ: Princeton University Press, 1993.

Newton, I. *The Finches.* Glasgow: William Collins Sons and Co., Ltd., 1972.

Perrins, C.M., and A.L.A. Middleton, eds. "Fringilline finches." In *The Encyclopedia of Birds.* New York: Facts on File, 1985.

Periodicals

Berger, Cynthia. "Superflight." *National Wildlife* (December–January 1998).

Middleton, A.L.A. "The Annual Cycle of the American Goldfinch." *Condor* 80 (1978): 401–406.

Organizations

BirdLife International. Wellbrook Court, Girton Road, Cambridge, Cambridgeshire CB3 0NA United Kingdom. Phone: +44 1 223 277 318. Fax: +44-1-223-277-200. E-mail: birdlife@birdlife.org.uk Web site: <http://www.birdlife.net>

Department of Ecology and Environmental Biology, Cornell University. E145 Corson Hall, Ithaca, NY 14853-2701 USA. Phone: (607) 254-4201. Web site: <www.es.cornell.edu/winkler/botw/fringillidae.html>

Department of Ecology and Evolutionary Biology, Tulane University. 310 Dinwiddie Hall, New Orleans, LA 70118-5698 USA. Phone: (504) 865-5191. Web site: <www.tulane.edu/eeob/Courses/Heins/Evolution/lecture17.html>

IUCN–The World Conservation Union. Rue Mauverney 28, Gland, 1196 Switzerland. Phone: +41-22-999-0001. Fax: +41-22-999-0025. E-mail: mail@hq.iucn.org Web site: <http://www.iucn.org>

Bill Freedman, PhD
Brian Douglas Hoyle, PhD

Hawaiian honeycreepers

(Drepanididae)

Class Aves

Order Passeriformes

Suborder Passeri (Oscines)

Family Drepanidinae

Thumbnail description
Small to medium, compact, finch-like, often brilliantly and variously colored birds with a wide variation of bill shapes and sizes

Size
3.9–8.3 in (10–21 cm)

Number of genera, species
21 genera; 51 species, including 13 subfossil and 15 extinct historically

Habitat
Forest

Conservation status
Critically Endangered: 6 species; Endangered: 5 species; Vulnerable: 7 species; Near Threatened: 5 species

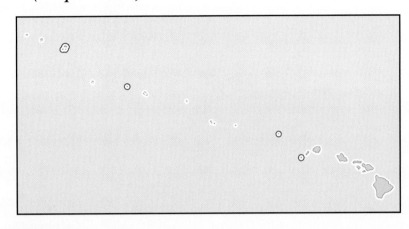

Distribution
Endemic to the Hawaiian Islands

Evolution and systematics

Hawaiian honeycreepers are a splendid living textbook and gazetteer of adaptive radiation, or the evolution of many species, with varied characteristics, from one.

Establishing the relationships among the many genera and species of honeycreepers is ongoing and far from settled. As of 2002, ornithologists consider all honeycreeper species to be monophyletic, i.e., sharing one common ancestral species, probably a cardueline finch species from North America, a small flock of which reached the Hawaiian Islands sometime between 3 and 5 million years ago. Although the Hawaiian honeycreepers are as of 2002 often listed taxonomically as subfamily Drepanidinae within the family Fringillidae, the finches, for the purposes of this account they are treated as a separate family—the Drepanididae.

Once established in Hawaii, the founder species evolved and radiated explosively, filling empty ecological niches and producing a fantastically varied toolbox of bill forms for dealing with virtually every sort of food. Researchers have found fossils of honeycreepers as old as 120,000 years. Fossil ages in Hawaii are limited by the nature of the islands, which are volcanic and relatively transient in geological time.

Physical characteristics

Hawaiian honeycreepers are small to medium-sized, compactly built, finchlike birds, their plumage colors varying widely from dull olive green to brilliant yellow, crimson, and multicolors. The tongue is tubular in most species, with a fringed tip adapted to nectar feeding. Bill shapes and sizes vary enormously, and correlate directly with the foods favored by particular species. Some species retain stubby, cone-shaped, finchlike bills for generalized feeding on insects, seeds, and fruits. Others favor heavier, parrotlike bills for cracking seeds and chopping into especially tough food sources. The various sickle-billed species use their long, thin, downcurved bills to draw nectar from the depths of tubular and bell-shaped flowers. The akepa developed a bill with a crossed lower tip, which it uses like pliers. The Lanai hookbill sported a parrotlike bill with a sizeable gap between the two mandibles, for a function that will probably remain unknown.

Subfossil species had even more outrageous versions of the simple finch bill. Two species of shovelbills (genus *Vangulifer*), had bills that were long and thin but with broad, rounded tips. Unique among honeycreepers, these may have fitted the species out for snagging insects on the wing. The gapers (genus *Aidemedia*) had bills with powerful muscles for opening the bill against pressure. This trait has a parallel with the meadowlark (*Sturnella* sp.), which uses a gaping bill to force open sod for reaching earthworms. The extinct King Kong finch (*Chloridops regiskongi*) had the largest known honeycreeper bill. A reporter used the name "King Kong" as an adjective to convey the massiveness of the bill, and the comparison worked its way into scientific nomenclature.

Nearly all honeycreepers give off a strong, musky odor, which persists even in museum specimens.

Distribution

Honeycreepers are found only on the Hawaiian Islands. As considered here, the Hawaiian Islands include the familiar

A Laysan finch (*Psittirostra cantans*) feeds on a seabird egg that it has broken open with its bill. (Photo by M.J. Rauzon/VIREO. Reproduced by permission.)

major islands, as well as a string of minor islets to the northwest of the main group, which includes Laysan Island, Nihoa Island, and the Pearl and Hermes Reef.

Habitat

Most honeycreepers are forest-dwellers. Exceptions include the Laysan finch (*Psittirostra cantans*) and Nihoa finch, which live on small, treeless islets. Although various species on the main islands originally lived throughout the lowlands as well as in the higher, mountainous areas, habitat loss and the proliferation of mosquitoes carrying avian disease now confines them mainly to mid-level and high tropical mesic (moist) forests and rainforests, 1,968 ft (600 m) above sea level and higher, most of them dominated by ohia trees (*Metrosideros polymorpha*). Some species prefer drier forests or a mix of dry, mesic, and rainforests. Many species depend on nectar from ohia flowers as a major or supplementary food source.

Behavior

The living honeycreeper species, despite their bewildering differences, share behavioral traits passed down from their finch-like ancestors.

All honeycreepers are diurnal. Individuals may be solitary foragers or forage in family groups, while some species show mixed-flock activity when foraging, especially when seeking out patches of ohia blooms, notably amakihis (*Hemignathus spp.*) and iiwis (*Drepanis coccinea*). Individuals of opposite sexes form strong pair bonds, and such pairings are monogamous and long-term for most species.

Honeycreepers produce a range of calls and songs as varied and distinct as the instruments of an orchestra, from ethe-

real music to coarse raspings. Calls and songs may vary even within a species. Some species stake out and defend nesting and feeding areas, other species tolerate intruders of their own or other species. Amakihis stake out breeding territories of 1–2.5 acres (0.4–1 ha), which they aggressively defend. The akohekohe (*Palmeria dolei*) is perhaps the scrappiest of all the honeycreepers in this regard.

Feeding ecology and diet

Honeycreepers eat almost anything edible, and run the entire gamut from generalist, omnivorous feeding, e.g., Laysan finches and amakihis, to specialization on a single plant species. Individual species tend to confine their diets to one or a few major sources. Foods include nectar of ohia (a primary food source for several species), insects, spiders, land snails, slugs, fruit, tree sap, seeds and seedpods, seabird eggs, and carrion. Species with long, curved bills poke them into flowers to reach nectar, while insect-grubbers pull or tear up epiphytic growth and bark to find arthropods. Species may pluck and eat fruit, break open seabird eggs, and cut or tear open seed pods. The po'o-uli (*Melamprosops phaesoma*) is unique among honeycreepers in preferring land snails as its main food, eating them shell and all. The Laysan and Nihoa finches have developed a taste for seabird eggs, which they peck or break open to get at the soupy innards.

An apapane (*Himatione sanguinea*) feeds on flowers. (Photo by P. La Tourrette/VIREO. Reproduced by permission.)

Laysan finches have survived lean times by eating carrion of dead seabirds.

The finch-billed and parrot-billed honeycreepers can be roughly assumed under various genera: the Laysan and Nihoa finches; the palila (*Loxioides*); the Kona, Wahi, and King Kong grosbeaks (*Chloridops*); the greater and lesser koa finches (*Rhodacanthis*), the o'u (*Psittirostra*), the Lanai hookbill (*Dysmorodrepanis*); and the Maui parrotbill (*Pseudonestor*). Bills of these genera and species are short, strong, and finchlike or robust and parrotlike. They may generalize their diets like the Laysan finch, or specialize in getting at and eating tough foodstuffs like seedpods and hard seeds or tearing apart "structurally defended resources," such as arthropods hidden under tree bark or seeds in tough, fibrous pods. Females and young among the finch-bills and parrot-bills tend toward olive green or grayish green plumage, the males of some species add brighter shades of yellow, orange, and red to the basic olive or gray. The Maui parrotbill (*Pseudonestor xanthophrys*) uses its aptly named mouthpiece to rip into branches and stems; pluck and open fruit; and lift bark and lichen clumps to find, snag, and eat caterpillars and the larvae and pupae of beetles. The diet of the Kona grosbeak (*Chloridops kona*) was highly specialized, its main food source being the small, rock-hard seeds of the naio or bastard sandalwood tree (*Myoporum sandwicense*). The bird used its massive, heavily muscle-powered bill to crack open the fruits.

Honeycreepers with long, narrow, downcurved bills are scattered among several genera: the curve-billed honeycreepers (*Hemignathus*), the sicklebills (*Drepanis*); and the alauahios (*Paroreomyza*). The bills become almost needlelike in some of these and can be half the length of the body, all for reaching nectar in the depths of ohia and other blossoms. Honeycreepers in the genus *Hemignathus* (amakihis, akialoas, nukupu'us, and akiapolaau) have a shorter lower mandible. Akiapolaau have a straight, thick lower mandible only half as long as the upper. This bird hammers audibly on tree bark with its lower mandible and plucks insects out of holes with the upper.

The amakihis, which are among the least specialized of the honeycreepers, take an wide food selection: caterpillars, plant lice, and spiders, which they uncover by probing leaves and twigs; nectar from ohia, lobelias, mamane, canna, and pritchardia palm; and non-native bananas and fuchsias.

Reproductive biology

Breeding takes place generally November through July, the males display to females, some species in groups resembling leks, and the sexes form strong, monogamous pair-bonds. Pairs build simple cup nests of twigs, lichens, and other plant materials on tree branches. As with all passerines, Drepanididae hatchlings are altricial, i.e., naked, blind, and helpless. The first young leave the nests anytime from the end of January through July and August. Only the female broods, while the male feeds the brooding female and both parents feed the chicks. The sexes of many honeycreeper species differ subtly or markedly in colors (sexual dichromatism), the males being more vividly and variously colored, the females tending to

A pair of Hawaii amakihi (*Hemignathus virens*) mating. (Photo by A. Walther/VIREO. Reproduced by permission.)

more cryptic coloration, favoring olive-greens and grays, although in some species both sexes have relatively plain rainment. A few species show sexual variation in size (sexual dimorphism), while some species show no sexual variation in size or color.

Conservation status

Fifteen species of Drepanididae have become extinct within historical times, mostly from loss of habitat through deforestation and environmental degradation. By IUCN standards, seven living species are Critical, five are Endangered, and three are Vulnerable. Nevertheless, all species are threatened to various degrees by a combination of loss of habitat, the introduction of mosquito-borne diseases of *Culex quinquefasciatus*, especially avian malaria (*Plasmodium* spp.) and avian pox, along with competition from and predation by invasive species. The o'u, Oahu alauahio (*Paroreomyza maculata*), and Maui nukupuu (*Hemignathus lucidus affinis*) may be extinct, none having been seen for several years. At the very cliff-edge of extinction is the po'o-uli, of which only three individuals remain.

At the opposite end of the scale, the amakihis are the most widespread, populous, and successful honeycreepers; their total population, all species, is estimated at an impressive

870,000. The several subspecies were formerly found on all the major islands. The iiwi, fairly common on the major islands, has a total population estimated at 340,000.

Significance to humans

Ancient Hawaiians made brilliant royal cloaks, mostly yellow and crimson, from thousands, if not millions, of tiny feathers of honeycreepers and other endemic birds of Hawaii. The practice has long been discontinued, but the honeycreepers earn their keep by pollinating ohia and other native plants and keeping the insect populations in check. Besides rendering those ecosystem services, they have become dependable ecotourism magnets. Their main value may be educational, as a visible, instructional living example of adaptive radiation. Finally, as poignant visions of endangered life, honeycreepers have become rallying symbols for conservation.

1. Lanai hookbill (*Dysmorodrepanis munroi*); 2. Anianiau (*Viridonia parvus*); 3. Akohekohe (*Palmeria dolei*); 4. Po'ouli (*Melamprosops phaeosoma*); 5. Apapane (*Himatione sanguinea*); 6. Akikiki (*Oreomystis bairdi*); 7. Akepa (*Loxops coccineus*); 8. Palila (*Loxioides bailleui*); 9. Greater koa finch (*Psittirostra palmeri*); 10. Laysan finch (*Psittirostra cantans*). (Illustration by Patricia Ferrer)

Species accounts

Akepa
Loxops coccineus

TAXONOMY
Loxops coccineus Gmelin, 1789.

OTHER COMMON NAMES
French: Loxopse des Hawaï; German: Akepa; Spanish: Akepa.

PHYSICAL CHARACTERISTICS
5 in (12.5 cm); 4–6 oz (10–12 g). Male mostly bright orange, female gray-green with a paler underside. Bill is distinctive feature; it has a lower tip bent to one side, as in the crossbills (*Loxia* spp).

DISTRIBUTION
Big Island of Hawaii. Extinct on Oahu, probably extinct on Molokai.

HABITAT
Old growth koa and ohia forest, wet and mesic, 3,610–6,890 ft (1,100–2,100 m) above sea level.

BEHAVIOR
Lively and nimble forager. Male's call is high-pitched, descending trill.

FEEDING ECOLOGY AND DIET
Uses its unusual bill to pry apart scales of buds to reach tiny insects inside; caterpillars are main diet.

REPRODUCTIVE BIOLOGY
Monogamous. Pairs form from July through September, start nest building in March, female lays eggs anytime from mid-March to late May, hatchlings fledge by end of June. Both parents feed chicks, but only female builds the cuplike nest. Males undergo a long delay in attaining adult plumage. Second-year males have mostly gray-green coloration like females, and rarely try to breed. Third-year males are partially orange, and small numbers breed; the next year males have full orange plumage and full breeding capacity.

CONSERVATION STATUS
Listed as endangered federally, by state, and by the IUCN.

SIGNIFICANCE TO HUMANS
None known. ◆

Akikiki
Oreomystis bairdi

TAXONOMY
Oreomystis bairdi Stejneger, 1887.

OTHER COMMON NAMES
English: Kauai creeper; French: Grimpeur de Kauai; German: Akikiki; Spanish: Akikiki.

PHYSICAL CHARACTERISTICS
4–4.7 in (10–12 cm); 0.4–0.6 oz (11–17 g). Mostly medium gray with a white belly; short, pink bill; legs and feet are stout and strong.

DISTRIBUTION
Kauai.

Loxops coccineus
Resident

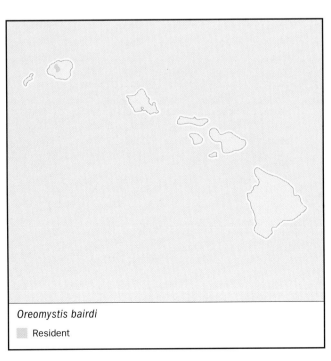

Oreomystis bairdi
Resident

HABITAT
Montane mesic and rainforest above 1,968 ft (600 m).

BEHAVIOR
Lively and active, yet elusive and quiet, moves with a distinctive creeping motion over tree bark while foraging (thus the name "creeper"), reminiscent of nuthatches (family Sittidae); can climb along trunks and branches in any direction. Its unassuming call is the source of its name.

FEEDING ECOLOGY AND DIET
Feeds mainly on insects and their larvae, spiders, millipedes, and slugs; nectar and fruit only rarely. Tongue is short, nontubular, and forked, efficient at snagging and seizing insects from tight niches, unlike the brush-tipped nectar-lapping tongues common among honeycreepers. Birds forage singly, in pairs, or in family groups of up to four individuals. May also form flocks of up to 12 individuals of same species or in mixed flocks with anianiau and akeke'e.

REPRODUCTIVE BIOLOGY
Male and female build cup nest of bark and plant fiber for clutch of two eggs. Previous year's fledglings often help in raising their parents' next brood.

CONSERVATION STATUS
Abundant, but confined to limited area in central Kauai. Listed as Critically Endangered by IUCN, and of special concern by the U.S. Fish and Wildlife Service (USFWS).

SIGNIFICANCE TO HUMANS
None known. ◆

Akohekohe
Palmeria dolei

TAXONOMY
Palmeria dolei S. B. Wilson, 1891.

OTHER COMMON NAMES
English: Crested honeycreeper; French: Palmérie huppée; German: Schopfkleidervogel; Spanish: Akojekoje.

PHYSICAL CHARACTERISTICS
7 in (18 cm); 1 oz (29 g); largest living honeycreeper. Sexes are monochromatic. Unique appearance of black plumage splashed and streaked with brilliant orange-scarlet and silver. Prominent, bushy, whitish crest curves forward over the bill. Crest is often dusted with ohia pollen, thus may serve as a pollinating organ, as well as a display feature.

DISTRIBUTION
Eastern Maui.

HABITAT
Montane mesic and rainforest from 3,800–6,500 ft (1,158–1,981 m) above sea level.

BEHAVIOR
Boisterous and aggressive, with a wide range of rough, throaty calls; native name is derived from one such call. Strong flyer, chases off other species and conspecifics from preferred feeding and nesting sites, even attacking with bill and wings.

FEEDING ECOLOGY AND DIET
Feeds mainly on ohia blossom nectar, but supplements the liquid diet with caterpillars, flies, and spiders.

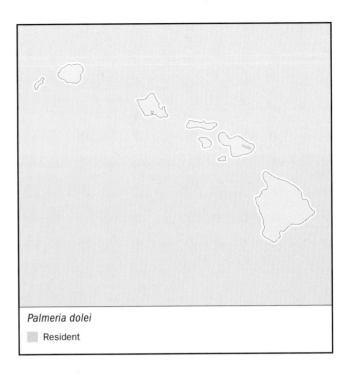

Palmeria dolei
▨ Resident

REPRODUCTIVE BIOLOGY
Monogamous, nesting from November to early June. Females lay two eggs, parents fledge one or both chicks. Females do incubation and brooding, males feed females and chicks.

CONSERVATION STATUS
Listed as Endangered federally and by state, and as Vulnerable by the IUCN. However, the population is stable at 3,800, and the adult survival rate is 95%.

SIGNIFICANCE TO HUMANS
None known. ◆

Anianiau
Viridonia parvus

TAXONOMY
Viridonia parvus Stejneger, 1887.

OTHER COMMON NAMES
English: Lesser amakihi; French: Hémignathe anianiau; German: Anianiau; Spanish: Anianiau.

PHYSICAL CHARACTERISTICS
4.5 in (11.4 cm); 0.32–0.35 oz (9–10 g). Smallest living bird species of Hawaii. Compact build, looks almost spherical when perching. Bill is small, thin, and slightly downturned. Male is bright yellow over most of body, with white rump and slightly darker wings; female duller.

DISTRIBUTION
Widespread in mountain forests of Kauai Island.

HABITAT
Variety of habitats, from dry valleys to rain-soaked Alakai area.

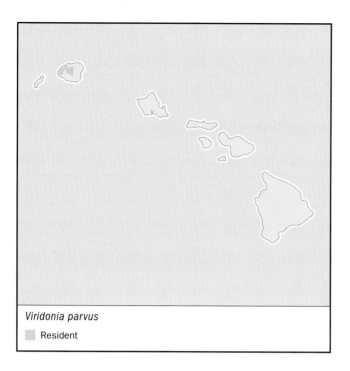

Viridonia parvus

■ Resident

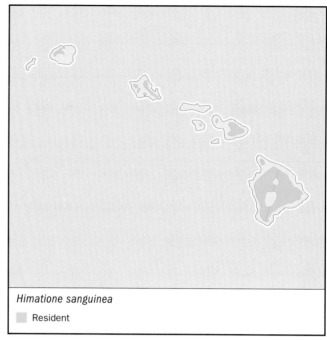

Himatione sanguinea

■ Resident

BEHAVIOR

Lively and active, hops along branches while foraging, seldom or never descends to the ground. Bonded male and female pair will defend a small territory about the nest site, the tree itself, or larger surrounding area up to 25–30 ft (7.6–9 m) across. Males chase off conspecific interlopers, tolerate other species up to a critical distance from the nest. Females join in repelling intruders only if they come too close to the nest. Song is melodious, high-pitched trill rendered "weesee" or "weesity" in sets of four; call is musical, high-pitched "orps-seet."

FEEDING ECOLOGY AND DIET

Prefers to forage above 1,968 ft (600 m) but may range as low as 330 ft (100 m); sips nectar and gleans insects. Forage and feed mostly as individuals, also operate in bonded pairs, family groups, and flocks at favorite nectar sites. May also join in mixed foraging and feeding flocks with akeke'e (*Loxops caeruleirostris*) and akikiki (*Oreomystis bairdi*).

REPRODUCTIVE BIOLOGY

Pairs build nests from February to late May, eggs laid between March and June; young fledge between early April and early July.

CONSERVATION STATUS

Not threatened. Tolerates considerable habitat disturbance and maintains a stable population estimated at 24,000.

SIGNIFICANCE TO HUMANS

None known. ◆

Apapane

Himatione sanguinea

TAXONOMY

Himatione sanguinea Gmelin, 1788.

OTHER COMMON NAMES

French: Picchion cramoisi; German: Apapane; Spanish: Apapane.

PHYSICAL CHARACTERISTICS

5.25 in (13.3 cm); 0.5–0.56 oz (14–16 g). Plumage bright crimson with black wings and tail, white undertail and abdomen; long, downcurved bill.

DISTRIBUTION

Not threatened. Most abundant honeycreeper species, also one of the most conspicuous. Common in forests over 3,300 ft (1,000 m) on Hawaii, Maui, Kauai, and Oahu, rare or extinct on Molokai and Lanai.

HABITAT

Forests over 3,300 ft (1,000 m).

BEHAVIOR

Social, gather into sizeable flocks and range about forests in search of blooming ohia. Calls varied, include squeaks, whistles, raspings, clickings, and musical trillings. Blunt wing tips make a loud and distinctive noise in flight, probably for group cohesion.

FEEDING ECOLOGY AND DIET

Flocks are lively and obvious when feeding on nectar of flowering ohia trees and flying considerable distances between forest patches of seasonal blooms. Flocks can become as dense as 3,000 individuals per 0.4 sq mi (1 sq km) when ohias are blooming. Then groups fly about looking for patches of ohia trees in bloom, and descend to drink nectar and glean insects.

REPRODUCTIVE BIOLOGY

Breed throughout year, mostly during February to June, coinciding with the seasonal availability of ohia nectar.

CONSERVATION STATUS

Not threatened. However, apapane are primary carriers of avian malaria and avian pox, making them major vectors of these diseases, since they fly so extensively.

SIGNIFICANCE TO HUMANS

None known. ◆

Greater koa finch

Psittirostra palmeri

TAXONOMY

Psittirostra palmeri/Rhodacanthis palmeri Rothschild, 1892.

OTHER COMMON NAMES

French: Psittirostre de Palmer; German: Orangebrust-Koagimpel; Spanish: Koa Grande.

PHYSICAL CHARACTERISTICS

9 in (23 cm); weight unknown; was largest known honeycreeper. Sexually dimorphic; male was brilliant scarlet-orange on head, neck, and breast, with lighter orange below, and olive brown with orange touches on back, wings, and tail; female was brownish olive, somewhat lighter below.

DISTRIBUTION

Confined areas on the Big Island (Hawaii). Neighboring species were the closely related lesser koa finch (*Rhodacanthis flaviceps*) and the Kona grosbeak.

HABITAT

Mesic, high-elevation forests.

BEHAVIOR

Active and agile, compared to the lethargic Kona grosbeak. Had little or no fear of humans and would approach a person who imitated their call or song. Male song a whistled series of up to six notes, each successive note held longer; male call a series of two or three whistles, descending in pitch and each like a long "weeek." Female alarm call was a low, deep single note, difficult to locate.

FEEDING ECOLOGY AND DIET

Greater and lesser koa finches and the Kona grosbeak had powerful bills for dealing with tough foods. Koa finches cut open and ate green pods and seeds of their staple food the koa tree (*Acacia koa*), and had large stomachs for processing masses of vegetable matter.

REPRODUCTIVE BIOLOGY

Not known.

CONSERVATION STATUS

Greater and lesser koa finches, and the Kona grosbeak were extinct within 10 years after discovery in the 1890s. All three species had restricted ranges.

SIGNIFICANCE TO HUMANS

None known. ◆

Lanai hookbill

Dysmorodrepanis munroi

TAXONOMY

Dysmorodrepanis munroi Perkins, 1919.

OTHER COMMON NAMES

English: Ou, Lanai finch, akiapolaau; French: Psittirostre de Munro; German: Ou; Spanish: Akiapolaau.

PHYSICAL CHARACTERISTICS

6 in (15 cm); weight unknown. Light gray tinged with green on upperparts and whitish below, light band on wing and light mark over eyes. Most distinctive feature was bill resembling pair of pliers or a paper punch, the upper mandible curving downward, the lower mandible curving upward, leaving a gap between the two. At first glance bill looked deformed, and was once interpreted as a deformity. Some ornithologists classified the one specimen as an aberrant female of the o'u, but extraction of skull in 1986 proved it to be a distinct species.

DISTRIBUTION

Endemic to Lanai, recorded from the southwestern side.

HABITAT

Dry, montane akoko (*Euphorbia lorifolia*) forest at 2,000 ft (610 m) above sea level.

Psittirostra palmeri

  Resident

Dysmorodrepanis munroi

  Resident

BEHAVIOR
Voice described by George C. Munro as "a strange bird chirp." Monroe characterized the species as a quiet, retiring bird, easily overlooked.

FEEDING ECOLOGY AND DIET
Preferred food is unknown. Individual shot by Munro in 1913 had been feeding on fruit of the opuhe (*Urera sandwichensis*), Munro speculated it also fed on the akoko (*Euphorbia lorifolia*). Dissection of the stomach found native berries, but hooked bill and relatively weak jaw musculature suggest it may have fed mostly on land snails.

REPRODUCTIVE BIOLOGY
Not known.

CONSERVATION STATUS
Extinct. First seen alive February 22, 1913, when Munro collected a single specimen in the Kaiholena Valley of Lanai. Munro was the only person to see the species, so nearly everything known about it is in his book *The Birds of Hawaii* (1960). Only existing specimen (Munro's) is in Bishop Museum in Honolulu. After 1913, Munro saw the bird on March 16, 1916, in the Kaiholena Valley, and on August 12, 1918, at Waiakeakua. The 1918 sighting was the last, by that time most of the native akoko forest on Lanai had been replaced by pineapple plantations.

SIGNIFICANCE TO HUMANS
None known. ◆

Laysan finch
Psittirostra cantans

TAXONOMY
Psittirostra cantans Wilson, 1890.

OTHER COMMON NAMES
French: Psittirostre de Laysan; German: Laysangimpel; Spanish: Certiola de Laysan.

PHYSICAL CHARACTERISTICS
6–6.5 in (15–18 cm). Bill large and parrotlike, tip of upper mandible forms slight downward hook. Adult males have bright yellow head, throat, and breast; lower back and rump grayish brown, abdomen whitish, gray collar around neck. Females less gaudy, dark streaks in yellowish crown, some streaking on flanks, gray collar, yellowish throat and breast, dark brown spots along back.

DISTRIBUTION
Laysan Island and Pearl and Hermes Reef, a coral atoll; both sites part of a long string of such islets northwest of main Hawaiian Islands. Shares both sites with nesting seabirds.

HABITAT
Laysan is a low-lying, sandy island about 1,000 acres (405 ha) in area; no trees, but abundant shrubbery and grasses. Pearl and Hermes Reef is a coral atoll containing several small islands.

BEHAVIOR
Lively and gregarious; do not fear humans, will even eat food out of hands of observers.

FEEDING ECOLOGY AND DIET
Omnivorous. Feed on carrion, insects, seeds, roots, sprouts, soft parts of plants and seeds, and interiors of tern eggs, whose shells it punctures with its beak to reach the soupy innards.

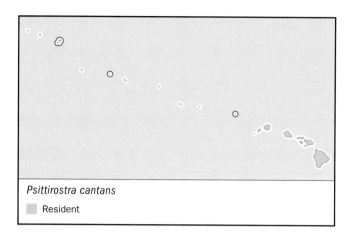

Psittirostra cantans
◻ Resident

REPRODUCTIVE BIOLOGY
At beginning of the breeding season, males gather and display to females in groups resembling leks. As Laysan has no trees, finch secures cup-shaped nest of grass and twigs in clumps of grass or in small bushes.

CONSERVATION STATUS
Listed as Endangered by the USFWS and the State of Hawaii, and as Vulnerable by the IUCN. Habitat is vulnerable to violent storms and the proliferation of alien species. Nearly became extinct in the 1920s. Population had declined to about 100 individuals in 1923, but diet of carrion and seabird eggs helped them to survive.

SIGNIFICANCE TO HUMANS
Significance lies in the successful efforts of wildlife biologists to preserve the species, and in the implications for the study of adaptive evolution. Biologists transferred 108 birds from Laysan to Pearl and Hermes Reef in 1967. In less than 30 years, the beaks shortened in accordance with their new food sources, demonstrating how quickly species can physically change in adapting to a local environment. About 350 birds survive at Pearl and Hermes. ◆

Palila
Loxioides bailleui

TAXONOMY
Loxioides bailleui Oustalet, 1877.

OTHER COMMON NAMES
French: Psittirostre palila; German: Palila; Spanish: Palila.

PHYSICAL CHARACTERISTICS
6–6.5 in (15–16.5 cm); 2 oz (56 g). Fairly large, with large, parrotlike bill. Sexes show little variation in coloring, both have bright yellow crowns, faces and necks, gray backs, white bellies and flanks, and dark beaks; wings are gray edged with yellow. Male has a dark patch surrounding each eye, somewhat muted in the female.

DISTRIBUTION
Western slope of Mauna Kea on Big Island of Hawaii, 6,000–8,000 ft (1,829–2,438 m) above sea level.

HABITAT
Cool, montane, mamane-niao forest.

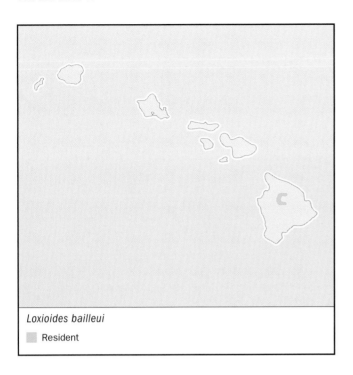

Loxioides bailleui

■ Resident

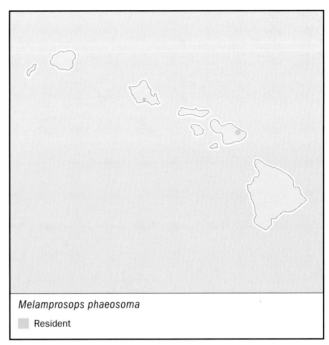

Melamprosops phaeosoma

■ Resident

BEHAVIOR

Tall trees and extensive crown cover among mamane (*Sophora chrysophylla*) and naio (*Myoporum sandwicense*) forests, with a high proportion of native understory plants for foraging and nesting. Call is loud, clear "chee-clee-o."

FEEDING ECOLOGY AND DIET

Feeds mainly on green seed pods of mamane trees, also mamane flowers, buds, naio berries, and caterpillars.

REPRODUCTIVE BIOLOGY

Breeding season March to September; female lays two eggs.

CONSERVATION STATUS

Listed as Endangered federally and by the IUCN. The population is fairly large, with an upper estimate of 5,000, but within a restricted range. A complicating factor with palila is their site tenacity. Most palila translocated by wildlife biologists to other areas, even with adequate food sources, return to their original sites or die of stress.

SIGNIFICANCE TO HUMANS

None known. ◆

Po'o-uli

Melamprosops phaeosoma

TAXONOMY

Melamprosops phaeosoma Casey Jacobi 1974.

OTHER COMMON NAMES

English: Black-faced honeycreeper; French: Po-o-uli masqué; German: Mauigimpel; Spanish: Puli.

PHYSICAL CHARACTERISTICS

5.5 in (15 cm); 0.9 oz (25.5 g). Differs from all other honeycreeper species in odor, tongue structure, and vocalizations,

and its color pattern is unique among all Hawaiian native birds. In 1992 some ornithologists proposed excluding the po'o-uli from the Drepanididae, but this was not carried, and as of 2002, the species was included. Main color brown, male's crown is gray merging into dark brown on nape, belly is buff washed with brown. Most striking and memorable feature is black mask offset by white throat. Has short, black, finchlike bill and long, pale legs. Female is similarly colored, but more cryptically.

DISTRIBUTION

Northeastern slope of Haleakala Crater on Maui.

HABITAT

Lives just below timberline in rainy (up to 550 in [14 m] a year), high-altitude, nearly impassable ohia forest draped in epiphytic mosses, lichens, and ferns.

BEHAVIOR

Spend most of their time foraging in dense forests in small family groups. Seldom vocalize, main calls are single or repetitive "chit."

FEEDING ECOLOGY AND DIET

Hop along tree limbs, tearing apart epiphytes and loose bark with finchlike bills and gleaning leaves and bark in the subcanopy and understory, searching for beetles, spiders, and other invertebrates, especially land snails. Only honeycreeper to prefer land snails as major dietary item.

REPRODUCTIVE BIOLOGY

Breeds February to June; lays one or two eggs in a cup nest.

CONSERVATION STATUS

Critically Endangered. As of 2001, only three individuals (two females and one male), are known. Invasive rats probably main agents of this near extinction.

SIGNIFICANCE TO HUMANS

A rallying symbol for conservation. ◆

Resources

Books

Carlquist, Sherwin. *Hawaii, A Natural History.* New York: Natural History Press, American Museum of Natural History, 1992.

Denny, James. *The Birds of Kauai* Honolulu: University of Hawaii Press, 2001.

Hilton-Taylor, C., comp. *2000 IUCN Red List of Threatened Species.* Species Survival Commission (SSC) Red List Programme. Cambridge, UK: IUCN 2000.

Poole, A., P. Stettenheim, and F. Gill, eds. *Birds of North America: Life Histories for the 21st Century.* (Monograph Series) Po'o-uli, no. 272; Akepa, no. 294; Apapane, no. 296; Maui Parrotbill, no. 311; Anianiau, no. 312; Iiwi, no. 327; Ou and Lanai Hookbill, no. 335–336; Amakihis, no. 360; Akohekohe, no. 400; Kona Grosbeak, Greater, and Lesser Koa-Finches, no. 424; Kakawahie and Oahu Alauahio, no. 503; Greater and Lesser Akialoa, no. 512; Akikiki, no. 552; Akiapala'au and Nukupuu, no. 600; Maui Alauahio, no. 681. Washington, DC: American Ornithologists Union, YEAR.

Pratt, H. Douglas, Phillip L. Bruner, and Delwyn G. Berrett. *The Birds of Hawaii and the Tropical Pacific.* Princeton, NJ: Princeton University Press, 1987.

Scott, J. M., S. Conant, and C. Van Riper III, eds. *Evolution, Ecology, Conservation, and Management of Hawaiian Birds: A Vanishing Avifauna.* Studies in Avian Biology 22. Camarillo, CA: Allen Press, Cooper Ornithological Society, 2001.

Stattersfield, Alison J., and David R. Capper, eds. *Threatened Birds of the World: The Official Source for Birds on the IUCN Red List.* Barcelona: Lynx Ediciones, 2001.

Periodicals

"Hawaii, Showcase of Evolution." *Natural History* 91, no. 12 (December 1982).

Olson, Storrs L., and Helen F. James. "Description of Thirty-Two New Species of Birds From the Hawaiian Islands; Parts I and II, Passeriformes." *Ornithological Monographs* 45&46, bound as one; American Ornithologists' Union, 1991.

Simon, Chris. "Hawaiian Evolutionary Biology: An Introduction;" *Trends in Ecology and Evolution* 2, no. 7 (July 1987): 175–178.

Simon, J. C., T. K. Pratt, K. E. Berlin, and J. R. Kowalsky. "Reproductive Ecology and Demography of the Akohekohe.;" *Condor* 103, no. 4:(Nov 2001): 736–745.

Organizations

The Bishop Museum. 1525 Bernice Street, Honolulu, HI 96817-0916 USA. Phone: (808) 847-3511. E-mail: museum @bishopmuseum.org Web site: <http://www.bishopmuseum .org>

Pacific Island Ecosystems Research Center. 3190 Maile Way, St. John Hall, Room 408, Honolulu, HI 96822 USA. Phone: (808) 956-5691. Fax: (808) 956-5687. E-mail: Bill Steiner@usgs.gov Web site: <http://biology.usgs.gov/pierc/ piercwebsite.htm>

Other

Payne, Robert B. "Bird Families of the World." University of Michigan Museum of Zoology, Bird Division. 12 Jan. 2000 (16 Mar. 2002). <http://www.ummz.lsa.umich.edu/birds/ birddivresources/families.html.>.

Pratt, Thane. "Birds of Hawaii." Pacific Island Ecosystems Research Center. 18 June 2001 (16 Mar. 2002). <http:// biology.usgs.gov/pierc/PLPrattTPage.htm.>.

Kevin F. Fitzgerald, BS

Waxbills and grassfinches
(Estrildidae)

Class Aves

Order Passeriformes

Suborder Passeri (Oscines)

Family Estrildidae

Thumbnail description
Small, often brightly colored, highly social birds with large, conical bills

Size
3.5–6.7 in (9–17 cm)

Number of genera, species
29 genera, 129 species

Habitat
Savanna, forest, and semi-desert

Conservation status
Endangered: 2 species; Vulnerable: 8 species; Near Threatened: 6 species

Distribution
Sub-Saharan Africa, southeastern Asia, Australia, and South Pacific islands. Small populations have been introduced throughout other parts of the world

Evolution and systematics

Weaverfinches, or estrildids, are related to the weavers with which they were formerly placed in the family Ploceidae. They do, however, differ clearly from weavers not only in external appearance, but also in behavior and in a number of digestive tract characteristics. For these reasons, they are now generally considered a distinct family that has diverged far from the common ancestral stock shared with the weavers. Sibley and Monroe, on the basis of genetic studies on DNA-DNA hybridization, have grouped several of the existing families, including Estrildidae and Ploceidae, into the Passeridae family. Although there is presently no consensus among ornithologists about their work, it will certainly play an important role in avian taxonomy in the future.

Physical characteristics

Weaverfinches are relatively quite small in size ranging from the 3.5 in (9 cm) quailfinches (*Ortygospiza* spp.) to the 6.7 in (17 cm) Java sparrow (*Padda oryzivora*). There are about 29 genera with about 129 species. Particularly characteristic of the weaverfinches are the projections or swellings of thickened connective tissue known as tubercles or papillae shown by the young at the edges of the bill and at the gape. These are a striking white, blue, or yellow color, often emphasized by black surroundings. In the Gouldian finch (*Chloebia gouldiae*) and the parrotfinches (*Erythrura* spp.) the tubercles have developed into organs which seem to reflect light and thus show up in the semi-darkness of the nest. A characteristic of the weaverfinches that varies according to the genus and species is the gape pattern of the nestlings. These patterns consist of dark spots or lines on the palate, the tongue, and

the floor of the mouth. In contrast to the colored bulges at the angles of the gape, the patterns in the interior of the mouth are, in many cases, retained for life.

The plumage is sometimes inconspicuous, but often very attractively colored. It is never, as in many weavers, striped in a sparrow or bunting-like fashion. Adult plumage is attained over a period of six to eight weeks without a distinguishable intermediate phase as in the fringillid finches. As in the whydahs, the outermost primary is generally very much shortened. Using the physical characteristics of an individual's plumage, one can usually place it in one of the many Estrildidae groups. For example, the parrotfinches usually display combinations of vibrant greens, blues, and reds in contrast to the munias and mannikins (*Lonchura* spp.), which are usually characterized by various shades of browns and tans. For some of the groups, the name is descriptive and indicates that group's distinguishing feature, such as in the olive-backs (*Nesocharis* spp.), the crimson-wings (*Cryptospiza* spp.), and the bluebills (*Spermophaga* spp.). Sometimes, however, a group's name can be misleading, such as with the firefinches (*Lagonosticta* spp.) whose plumage usually contains colors ranging from pink to crimson red, but never what most would consider a "fiery" red. The patterning of the plumage can also help place estrildids into groups. The twinspots have white spots on their underparts and sides with each feather containing two spots, the characteristic for which they are named. The pytilias (*Pytilia* spp.), on the other hand, have barring in these areas in addition to a bright red face in the males. The firetails (*Emblema* spp.) can have either barring or spots on their sides and underparts, but their distinguishing characteristic is their bright red rump and tail.

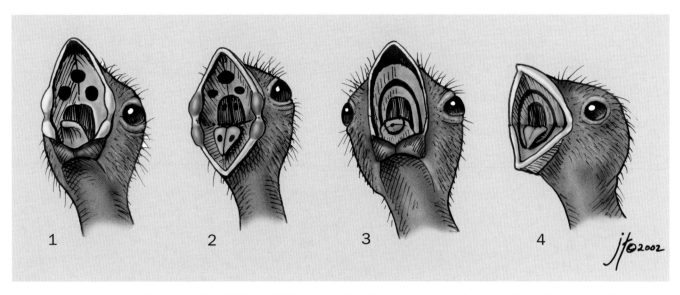

Nestlings of species found within the family Estrildidae exhibit a combination of mouth, tongue, and palate patterns that help identify them as a particular species. Species of whydahs and indigobirds (*Vidua* spp.), brood parasites, have evolved to exhibit mouth patterns similar to their corresponding estrildid host species, thereby increasing the chances that the host parents will accept their new "adopted" chicks. 1. Red-billed firefinch (*Lagonosticta senegala*); 2. Gouldian finch (*Chloebia gouldiae*); 3. White-headed munia (*Lonchura maja*); 4. African silverbill (*Lonchura cantans*). (Illustration by Joseph E. Trumpey)

The bill of the almost exclusively insectivorous species, such as the negro-finches (*Nigrita* spp.) and the flowerpecker weaver-finches (*Parmoptila* spp.), is as slim as that of warblers. In species that eat large seeds, like the bluebills and the seedcrackers (*Pyrenestes* spp.), it is almost as thick and strong as that of hawfinches. The waxbills (*Estrilda* spp.) fall somewhere in the middle of this size range with their often bright red "waxy" bills. Tail size is also quite variable, ranging from the long central tail feathers of the grassfinches (*Poephila* spp.) to the unusually short tail of the quailfinches.

Distribution

The family Estrildidae is an Old World family with a natural distribution around and south of the equator in the Ethiopian, Oriental, and Australasian biogeographic regions. Most species within this family can be found in Africa. The negro-finches and flowerpecker weaver-finches can be found in equatorial east central Africa. The olive-backs have very small, often patchily distributed ranges within central Africa. The other Africa-endemic groups have more diversified ranges. These include the pytilias, the multiple genera of twinspots, the crimson-wings, the seedcrackers, the bluebills, the firefinches, the waxbills, the quailfinches, the cordonbleus (*Uraeginthus* spp.), and the two species in the genus *Amadina*. The munias and mannikins have radiated throughout Africa, including Madagascar, across southern Asia, and into Malaysia, Indonesia, Australia, and surrounding islands. The parrotfinches are concentrated in Southeast Asia throughout its many islands as well as the mainland. Some species have a very limited range sometimes consisting of only a single island, a major disadvantage to the conservation of a species when a threat to its population arises. Australia, like Africa, has its own set of endemic estrildid groups. These in-

clude the grassfinches, the firetails, the two species within *Neochmia*, as well as the monotypic genera *Chloebia* and *Aidemosyne*. Introduced populations of several estrildid species have been established throughout the world from either intentional releases or from escaped captive birds.

Habitat

Most weaverfinches live in grass or bush steppes, savannas, and open dry area forests. A few have penetrated deserts and semi-deserts, particularly in South Africa and central Australia. Others resumed their family's apparent earlier position as true forest dwellers, particularly in western and central Africa, southeastern Asia, and the Indo-Australian insular area. Recently, several species have become closely linked with man and have moved into fields and gardens, some even into city parks. The red-billed firefinch (*Lagonosticta senegala*) comes into native huts in its search for food. Zebra finches (*Taeniopygia guttata*) and crimson finches (*Neochmia phaeton*) breed on and in buildings.

Behavior

In recent years the display behavior of weaverfinches has been studied with particular intensity. The bond between members of a pair, and sometimes among members of a small flock, is usually strong. Unlike in many groups of birds, male weaverfinches do not feed the female as part of a courtship ritual. In most species the male has a "display dance" in which it sings and either hops towards the female or performs characteristic bows or stretching movements while hopping about in front of the female. Many avadavats (*Amandava* spp.) and some Australian grassfinches hold a feather or a grass stalk in the bill during this display, apparently as a nest symbol. The

Bill morphology can often give insight into the diet of a species. Those with slender bills typically consume more insects in their diet, while those with conical bills eat primarily a seed-based diet. The relative size of the bill is often directly proportional to the size of the food that the species can consume. These species illustrate the range of bill shapes and sizes represented within the family Estrildidae: 1. Common waxbill (*Estrilda astrild*); 2. Red-fronted flowerpecker weaver-finch (*Parmoptila rubrifrons*); 3. White-breasted negro-finch (*Nigrita fusconota*); 4. Double barred finch (*Poephila bichenovii*); 5. Green-winged pytilia (*Pytilia melba*); 6. Java sparrow (*Padda oryzivora*); 7. Crimson seedcracker (*Pyrenestes sanguineus*). (Illustration by Joseph E. Trumpey)

female weaverfinch's way of indicating readiness for mating is unique among songbirds. She cowers on a branch and trembles her tail, which is held vertically while the wings are kept still. In contrast, other songbird females tremble their wings and keep the tail quite still.

The song of weaverfinches is often soft and sometimes inaudible to human ears. This may, in part, be due to the fact that weaverfinches do not use songs to indicate aggression or territoriality, and therefore do not need their song to be heard by neighbors. An often unmusical and short song is uttered just loud enough for a nearby female to hear as part of the male's courtship. Unlike other vocalizations, the song is not instinctual. Instead, it is learned during a very narrow window during development of the fledgling (25 to 35 days of age for the zebra finch). Captive birds raised by a different species often learn the song of the male foster parent, but not the other calls. However, they can learn the meaning of the foster parents' calls and will often respond with the corresponding call from their own repertoire.

Feeding ecology and diet

The predominant food of weaverfinches is half-ripe and ripe grass seeds. Particularly in the breeding season many also take arthropods. They are attracted by nuptial swarms of ants and termites at the beginning of the rainy season and they pick the insects up from the ground or sally after them like flycatchers in a short fluttering flight. Some Australian weaverfinches have developed a manner of drinking which is evidently unique among passerine birds and occurs outside this order in only a few groups of birds. They suck in the water like pigeons, immersing the bill almost up to its base. This behavior has evolved independently several times in birds of arid areas probably because the birds are exposed to danger at the water holes. By sucking up the water they can reduce the time required to stock up with fluid.

Reproductive biology

Estrildid nests are always roofed over and are, as a rule, almost spherical with a diameter of about 4–8 in (10–20 cm). Many species attach a long entry tube to the nest, but this, in contrast to many weaver nests, never hangs down vertically. Usually both partners participate in nest building. Males mainly gather the nesting material and the females build with it. Most species use fresh or dry grass stalks and many line the nest cup with feathers or other soft materials. In many cases nests are built outside the breeding season as well and

A zebra waxbill (*Amandava subflava*) sits atop its nest. (Photo by P. Craig-Cooper/VIREO. Reproduced by permission.)

these are used for roosting. In many species a whole group of birds use such sleeping nests together. Nests are usually placed in bushes or low trees. Some species nest on the ground, while others suspend their nests between grass stalks or reeds or nest in tree holes.

The clutch usually consists of four to six eggs, with rare cases of up to nine. Both sexes incubate eggs and brood young. In the daytime they relieve one another at approximately equal intervals of about one and a half hours, while at night both sexes in many species sit together in the nest. The male, when appearing for relief at the nest, often brings as a "present" a bit of grass or feather. The incubation period is 12–16 days. Young receive mainly half-ripe seeds. Parents regurgitate these in small portions from the crop and push the food into the young bird's gape. The nestling with wide open bill grabs the adult's bill about the angles. The begging posture of the young is also unique among passerines. They do not stretch the head and neck towards the parents, but lay the neck flat on the nest floor, turning only the gape upwards. The chick's bill is wide open, displaying the characteristic pattern inside the mouth and the head is moved from side to side and turned in a lively manner. There are no trembling wing movements so characteristic in other food-begging young birds. This begging posture is retained after leaving the nest. The nestling period lasts about three weeks, which is surprisingly long for such small birds. Even after fledging, the young have not definitely left the nest, for the parents guide them back to it for sleeping and, at first, even for feeding. Young remain dependent on the parents for food for one or two weeks after fledging.

Some species of weaverfinches serve as brood hosts for the whydahs and indigobirds (*Vidua* spp.). Each parasitic species has a corresponding estrildid host species with which it shares many similarities including juvenal plumage as well as the characteristic species-specific mouth markings. This result of

convergent evolution is used to fool the host species into accepting their "adopted" charges. The similarities between the groups have even fooled ornithologists in the past to place them within the same family. The whydahs and indigobirds are now rightly placed in family Ploceidae with the weavers, their true relatives.

Conservation status

According to the IUCN, the family Estrildidae contains six species which are considered Near Threatened, eight which are Vulnerable, and two that have reached the status of Endangered. The reason for the decline in most of these species' numbers is multifactorial. Several species, including the green avadavat (*Amandava formosa*), the green-faced parrotfinch (*Erythrura viridifacies*), and the Timor sparrow (*Padda fuscata*), which are all listed as Vulnerable, have suffered from habitat loss and modification in addition to trapping for the pet trade. Early and strict laws governing the exportation of Australian fauna have nearly eliminated the threat of trapping for the pet trade on that continent. However, habitat modifications in the form of fragmentation, overgrazing by cattle, and widespread burning have affected the populations of the star finch (*Neochmia ruficauda*), the diamond firetail (*Stagonopleura guttata*), and the Endangered gouldian finch in Australia. Having a range that is limited only to a small island or two means that a threat to an already decreased population can have major consequences. This has been the case for several parrotfinches including the royal parrotfinch (*Erythrura regia*), considered by some ornithologists to be a subspecies of the red-headed parrotfinch (*Erythrura cyaneovirens*). This species has felt the impact of logging and cattle grazing in its native range of Vanuatu and Banks islands. However, the Java sparrow, another island denizen with a small population, is gradually succumbing to the combined effects of trapping for the pet trade, killing, and egg-robbing because of their effects on

The western bluebill (*Spermophaga haematina*) has a distinctive, colorful bill. (Photo by Doug Wechsler/VIREO. Reproduced by permission.)

rice crops, and hunting for food. Population numbers are not the only factors considered when classifying the conservation status of a species, however. Population trends play a major role as well. This is well illustrated by comparing the Anambra waxbill (*Estrilda poliopareia*) and the pink-billed parrotfinch (*Erythrura kleinschmidti*), both of which have populations fewer than 1,000 individuals. The former's population is stable and it is therefore classified as Vulnerable whereas the latter's population is declining, earning it an Endangered status.

Significance to humans

Weaverfinches, in contrast to many other songbirds, have neither long nor very attractive songs, and only on a few rare circumstances when they are handfed do they become tame. Nevertheless, they have, throughout the history of aviculture, been among the most popular and frequent pet and aviary birds. Several species such as the Java sparrow, the zebra finch, and the gouldian finch have reached the status of domesticated species, a rare title among birds kept as pets. The Ben-galese or society finch has never occurred in the wild. Instead, it is a form of the white-backed munia (*Lonchura striata*) that was domesticated in the Far East during the early 1700s or even earlier. The deficiencies of weaverfinches are balanced in most species by attractive colors and patterns. They are lively, sociable, in most cases peaceable, and not demanding in their maintenance. They are particularly suitable for large aviaries where a mixed group of different species and colors can often be kept harmoniously. If planting within the aviary and feeding are suitable, breeding can usually be expected once the pair becomes established.

In contrast to their pet quality, some estrildids, namely the munias and mannikins, can have a devastating impact on agricultural crops such as rice. This often leads to their widespread persecution. Hunting of the pests becomes a sport and children are sometimes hired to collect eggs from nests. Some species, such as the Java sparrow, are also hunted for food, while others, like the spotted munia (*Lonchura punctulata*), are collected for religious purposes.

1. White-breasted negro-finch (*Nigrita fusconota*); 2. Peter's twinspot (*Hypargos niveoguttatus*); 3. Red-headed finch (*Amadina erythrocephala*); 4. Male (top) and female red-fronted flowerpecker weaver-finch (*Parmoptila rubrifrons*); 5. Green-winged pytilia (*Pytilia melba*); 6. Jameson's fire-finch (*Lagonosticta rhodopareia*); 7. Crimson seedcracker (*Pyrenestes sanguineus*); 8. Common waxbill (*Estrilda astrild*); 9. Male (top) and female red-cheeked cordon-bleu (*Uraeginthus bengalus*); 10. African silverbill (*Lonchura cantans*); 11. African quailfinch (*Ortygospiza atricollis*). (Illustration by Joseph E. Trumpey)

1. Diamond firetail (*Stagonopleura [Emblema] guttata*); 2. Spotted munia (*Lonchura punctulata*); 3. Red avadavat (*Amandava amandava*); 4. Female (left) and male pin-tailed parrotfinch (*Erythrura prasina*); 5. Female (left) and male Gouldian finch (*Chloebia gouldiae*); 6. Pink-billed parrotfinch (*Erythrura kleinschmidti*); 7. Java sparrow (*Padda oryzivora*); 8. Male (left) and female zebra finch (*Poephila guttata*); 9. Double-barred finch (*Poephila bichenovii*). (Illustration by Joseph E. Trumpey)

Species accounts

Red-fronted flowerpecker weaver-finch

Parmoptila rubrifrons

SUBFAMILY
Estrildinae

TAXONOMY
Pholidornis rubrifrons Sharpe and Ussher, 1872

OTHER COMMON NAMES
English: Jameson's antpecker, red-fronted antpecker; French: Parmoptile à front rouge; German: Ameisenpicker; Spanish: Pinzón Hormiguero de Jameson.

PHYSICAL CHARACTERISTICS
3.9–4.3 in (10–11 cm). Similar to warblers, with which they were previously classified. Sexually dimorphic with males having a red forehead and cinnamon-brown underparts; females lack the red forehead and have spotted underparts. Juveniles are similar to adult males but lack the red forehead.

DISTRIBUTION
Two populations: one in Liberia and southwestern Côte d'Ivoire and one in northern Democratic Republic of Congo, eastern Congo, and western Uganda.

HABITAT
Inhabits forest edges and scrub, usually low to the ground.

BEHAVIOR
Found at mid-level or near the ground in pairs, small groups, or sometimes mixed-species groups. The voice of this secretive species has not been recorded.

FEEDING ECOLOGY AND DIET
A longer more slender bill than that of most estrildids reflects this species' more insectivorous diet of mostly ants, including their larvae and pupae. This species, along with the closely-related flowerpecker weaver-finch (*Parmoptila woodhousei*), possesses a brush-like tongue which is believed to be an adaptation to a diet of ants. When searching for food, this species examines both live and dead leaves.

REPRODUCTIVE BIOLOGY
As for negro-finches, the nesting behavior and the nestlings' mouth patterns of *P. woodhousei* are what convinced taxonomists that the flower-peckers are indeed estrildids, albeit aberrant examples. However, the nest and nestlings of *P. rubrifrons* have not been found or described.

CONSERVATION STATUS
CITES: Appendix III. Not considered threatened by the IUCN.

SIGNIFICANCE TO HUMANS
None known. ◆

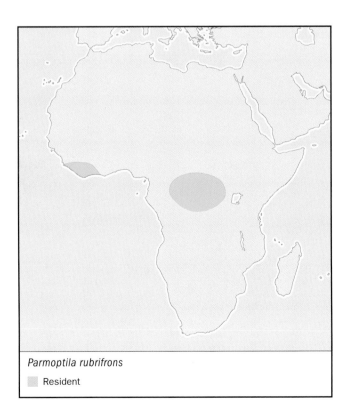

Parmoptila rubrifrons
▨ Resident

White-breasted negro-finch

Nigrita fusconota

SUBFAMILY
Estrildinae

TAXONOMY
Nigrita fusconotus Fraser, 1843.

OTHER COMMON NAMES
French: Nigrette à ventre blanc; German: Weißbrust-mantelschwärzling; Spanish: Negrita Pechiblanca.

PHYSICAL CHARACTERISTICS
3.9 in (10 cm). Sexes similar with females sometimes having slightly paler underparts with less gray. Juveniles are duller than adults and have a dark brown rump and tail compared to the adults' blue-black rump and tail.

DISTRIBUTION
Guinea and Sierra Leone, east to western Kenya, and south to northern Angola.

HABITAT
Occurs in secondary growth, cleared areas, forest edges, and gallery forest.

BEHAVIOR
Found singly, in pairs, or in small groups. The song is described as a descending, trilling "tz-tz-tz-tz-tzeeee" sometimes ending with a few "tsip" or "chip" notes. Males sing from April to October both from a high perch in a tree and while feeding.

FEEDING ECOLOGY AND DIET
Feeds at varying heights of shrubs and trees; rarely seen on the ground. Unlike most estrildids, negro-finches forage by searching leaves in a warbler-like fashion. Their diet consists

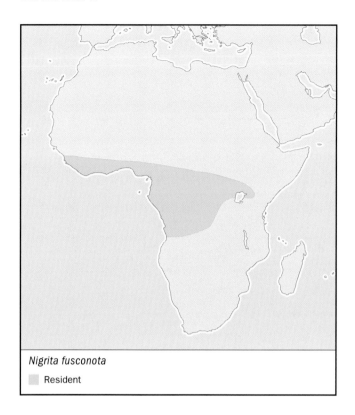

Nigrita fusconota

■ Resident

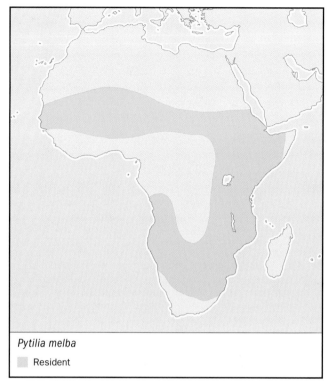

Pytilia melba

■ Resident

of small insects, caterpillars, berries and other small fruits, the oily cases of oil palm nuts, and small seeds. The bill, which is more slender than most estrildids, is probably an adaptation for this species' increased insectivorous portion of the diet.

REPRODUCTIVE BIOLOGY
The mouth patterns of nestlings and the nest, a sphere with a side entrance, are some of the only features that associate this species with the family Estrildidae. Nesting material includes bark strips, leaf fibers, dry grass, and moss. A clutch is typically three to six white eggs.

CONSERVATION STATUS
CITES: Appendix III. Not considered threatened by the IUCN.

SIGNIFICANCE TO HUMANS
Likely due to the lack of bright coloration and to the more insectivorous diet, this species has failed to become popular in the pet trade. ◆

Green-winged pytilia
Pytilia melba

SUBFAMILY
Estrildinae

TAXONOMY
Fringilla melba Linnaeus, 1758.

OTHER COMMON NAMES
English: Melba finch, melba waxbill; French: Beaumarquet melba; German: Buntastrild; Spanish: Pinzón Melba.

PHYSICAL CHARACTERISTICS
4.7–5.1 in (12–13 cm). Sexually dimorphic with females lacking red on the head. Juveniles resemble a duller version of the female.

DISTRIBUTION
Senegal east across northern Nigeria and southern Niger to Ethiopia and Somalia, south through Tanzania to northern South Africa.

HABITAT
Prefers dry, open areas including semi-desert, thorn scrub, acacia woodland, grassland, savanna, and cultivated areas.

BEHAVIOR
Found either singly or in pairs except at watering holes where small flocks might temporarily congregate. The call is a "seeeh," "wick" or "wit" note. The song, which is sometimes lengthy, is a series of whistles and trills interspersed with "kwik" notes.

FEEDING ECOLOGY AND DIET
Feeds on the ground eating mainly grass seeds and termites, although other seeds and insects are probably eaten.

REPRODUCTIVE BIOLOGY
The nest is usually round or dome-shaped and built of grass and lined with feathers. Three to six white eggs are laid and incubated for 12–13 days. The breeding season lasts from November to June, peaking after the heaviest rains. Nests are often parasitized by the paradise whydah (*Vidua paradisaea*).

CONSERVATION STATUS
Not threatened.

SIGNIFICANCE TO HUMANS
Kept in aviculture where it proves to be a challenge to breed, requiring a variety of insects in the diet. In captivity males

defend a territory against conspecific individuals as well as any bird showing red coloration on the head. ◆

Crimson seedcracker
Pyrenestes sanguineus

SUBFAMILY
Estrildinae

TAXONOMY
Pirenestes sanguineus Swainson, 1837.

OTHER COMMON NAMES
French: Pyréneste gros-bec; German: Karmesinastrild; Spanish: Pinzón Casca Nueces Rojo.

PHYSICAL CHARACTERISTICS
5.1–5.5 in (13–14 cm). Sexes differ slightly. Males have a red head, nape, breast, and flanks. Females lack red on the flanks and are duller on the nape and breast. Juveniles lack red except on the rump and tail.

DISTRIBUTION
Southern Côte d'Ivoire to southern Senegal.

HABITAT
Prefers wet habitats including marsh, flooded rice fields, swamps, and undergrowth near water.

BEHAVIOR
This shy species is found in pairs or small groups. The call is a sharp "zeet," while the song is described as a melodious warble, sometimes given during flight.

FEEDING ECOLOGY AND DIET
Feeds on or near the ground. Their diet is poorly understood, but, based in part on bill morphology, is presumed to be mostly seeds, especially those with hard coverings.

REPRODUCTIVE BIOLOGY
Three to four white eggs are incubated for 16 days. The few nests described were composed mostly of reeds.

CONSERVATION STATUS
Not threatened.

SIGNIFICANCE TO HUMANS
In the past, this species could be sporadically found in zoos and aviaries, but it has failed to become established in aviculture. ◆

Peters' twinspot
Hypargos niveoguttatus

SUBFAMILY
Estrildinae

TAXONOMY
Spermophaga niveoguttata Peters, 1868.

OTHER COMMON NAMES
English: Red-throated twinspot, Peters' spotted firefinch; French: Sénégali enflammé; German: Tropfenastrild; Spanish: Pinzón Dos Puntos de Peter.

PHYSICAL CHARACTERISTICS
4.7–5.1 in (12–13 cm). Sexually dimorphic; females lack the male's bright red face and breast. Juveniles similar to females but lack white spots on flanks and underparts.

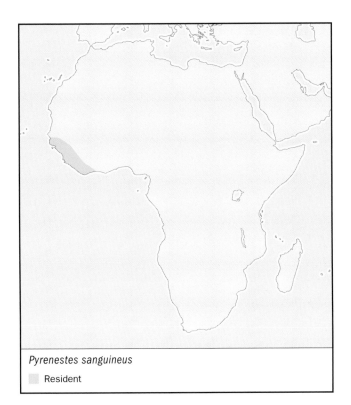

Pyrenestes sanguineus
☐ Resident

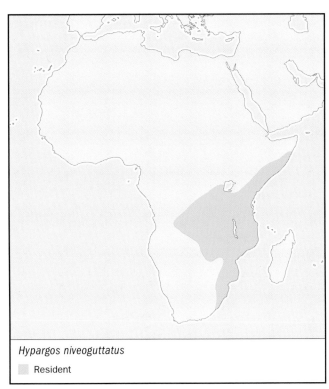

Hypargos niveoguttatus
☐ Resident

DISTRIBUTION
Southern Somalia, south to northeastern South Africa, west to east-central Angola.

HABITAT
Found in grass or the undergrowth of habitats including grassland, evergreen thicket, forest edges, and other brushy cover often near water.

BEHAVIOR
Usually found in pairs or small groups. This species is quiet and shy, but may give a "tseet" note. Its song is an insect-like trill.

FEEDING ECOLOGY AND DIET
Feeds on the ground, usually near cover, on a variety of seeds. A small part of the diet consists of insects.

REPRODUCTIVE BIOLOGY
The spherical nest, usually composed of grass and other plant fibers, is built on the ground or low in a shrub. Three to six white eggs are incubated for 12–13 days.

CONSERVATION STATUS
Not threatened.

SIGNIFICANCE TO HUMANS
This species often becomes tolerant of humans and can frequently be found near human settlements. The ease by which they become adapted to human contact makes them likely aviary subjects. ◆

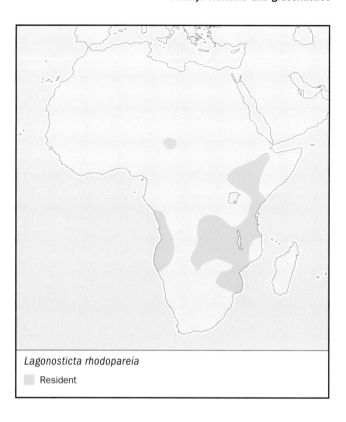

Lagonosticta rhodopareia
▨ Resident

Jameson's firefinch
Lagonosticta rhodopareia

SUBFAMILY
Estrildinae

TAXONOMY
Estrelda rhodopareia Heuglin, 1868.

OTHER COMMON NAMES
English: Pink-backed firefinch; French: Amarante de Jameson; German: Rosenamarant; Spanish: Pinzón Candela de Jameson.

PHYSICAL CHARACTERISTICS
3.9–4.3 in (10–11 cm). Sexually dimorphic; males bright red on head, breast, and underparts; females light brown except for bright red lores. Juveniles similar to females but lack red lores.

DISTRIBUTION
Central Ethiopia south to Mozambique, Zimbabwe, and Zambia. Discontinuous populations in western Angola and in southwestern Chad.

HABITAT
Occurs in areas containing both low cover and ample grass including grassland, thicket, forest edges, bracken-briar, and thorn scrub.

BEHAVIOR
Found in pairs or small groups. The call is a "ti-ti-ti-ti-ti" or a softer "tsit, tsit." The song is a melodious "we-we-we-we-we" or "weet-weet-weet-weet" with a whistling "feeeee" repeated several times.

FEEDING ECOLOGY AND DIET
Feeds on or near the ground on seeds of grasses and other plants, and occasionally on small invertebrates.

REPRODUCTIVE BIOLOGY
The breeding period is at the end of the rainy season and the beginning of the dry season. The round nest is built by the male in small shrubs and made of grasses, rootlets, and other plant fibers. This species is the breeding host of the purple indigobird (*Vidua purpurascens*).

CONSERVATION STATUS
Not threatened.

SIGNIFICANCE TO HUMANS
This species has occasionally been kept in aviculture, but never in great numbers. ◆

Red-cheeked cordon-bleu
Uraeginthus bengalus

SUBFAMILY
Estrildinae

TAXONOMY
Fringilla bengalus Linnaeus, 1766.

OTHER COMMON NAMES
English: Cordon-bleu, red-cheeked blue waxbill; French: Cordonbleu à joues rouges; German: Schmetterlingsastrild; Spanish: Coliazul Bengalí.

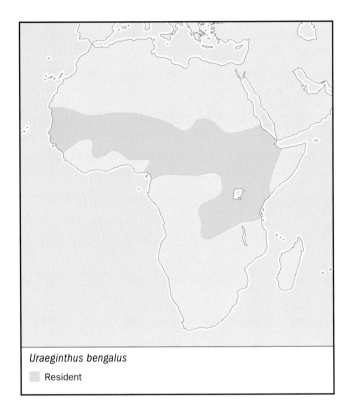

Uraeginthus bengalus

Resident

Common waxbill
Estrilda astrild

SUBFAMILY
Estrildinae

TAXONOMY
Loxia astrild Linnaeus, 1758.

OTHER COMMON NAMES
English: St. Helena waxbill, barred waxbill, brown waxbill, pheasant finch; French: Astrild ondulé; German: Wellenastrild; Spanish: Astrilda Común.

PHYSICAL CHARACTERISTICS
4.3–5.1 in (11–13 cm). Sexes are alike. Juveniles are paler than adults and have fainter barring.

DISTRIBUTION
Found naturally in southern Senegal, east to Ethiopia, south to South Africa, essentially throughout most of sub-Saharan Africa. Has been introduced in Brazil, Portugal, and many islands throughout the world including Hawaii, Tahiti, the Seychelles, Bermuda, and Puerto Rico.

HABITAT
Prefers areas with tall grass including marsh, reed beds, abandoned cultivated areas, gardens, grassy clearings or paths, and farms or plantations.

BEHAVIOR
Being highly gregarious, this species is found in small flocks during the breeding season and larger flocks when not breeding. Calls include a "chip," "tchic," and "pit" while the song is described as a "tcher-tcher-preee," although it can be highly variable.

PHYSICAL CHARACTERISTICS
4.9–5.1 in (12.5–13 cm). Sexually dimorphic; male has a red spot on cheeks, a pink bill, light brown or pink-brown legs; female lacks red spots on cheeks, feathers are paler than for the male. Juveniles lack blue on flanks and have a darker bill.

DISTRIBUTION
Southern Mauritania, east to Ethiopia, south to northern Zambia. An introduced population exists in Hawaii.

HABITAT
Occurs in grassland, savanna, thorn scrub, dry woodland, forest edges and clearings, gardens and villages, roadsides, and cultivated areas.

BEHAVIOR
Found in pairs or small flocks during the breeding season. Otherwise, this species can gather in large sometimes mixed-species flocks. The call is a "tsee-tsee-tsee." The song is a "te tchee-wa-tcheee" or a "ssee-deedelee-deedelee-ssee-see."

FEEDING ECOLOGY AND DIET
Feeds on the ground on a variety of seeds and insects, including termites which are occasionally caught in flight.

REPRODUCTIVE BIOLOGY
Uses old *Ploceus* weaver nests or builds its own round nest of grass several meters off the ground. Three to six white eggs are incubated for 11 days.

CONSERVATION STATUS
CITES: Appendix III. Not considered threatened by the IUCN.

SIGNIFICANCE TO HUMANS
This species becomes accustomed to humans and can be found in villages and gardens. It is also a commonly kept and bred aviary bird. ◆

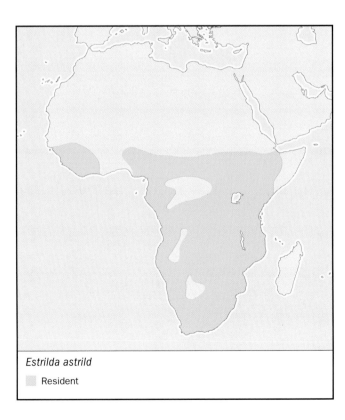

Estrilda astrild

Resident

FEEDING ECOLOGY AND DIET
Feeds mainly on a diversity of seeds taken both from growing plants and off the ground. Swarming termites as well as other insects make up a small portion of the diet.

REPRODUCTIVE BIOLOGY
Builds a pear-shaped nest of grass stems at or near the ground. Four to six white eggs are incubated for 11–12 days. This species is the brood host for the pin-tailed whydah (*Vidua macroura*).

CONSERVATION STATUS
CITES: Appendix III. Not threatened according to IUCN criteria

SIGNIFICANCE TO HUMANS
Commonly kept in captivity, this species has been bred in aviaries and cages and is readily offered for beginning aviculturists. ◆

Red avadavat
Amandava amandava

SUBFAMILY
Estrildinae

TAXONOMY
Fringilla amandava Linnaeus, 1758.

OTHER COMMON NAMES
English: Red munia, avadavat, strawberry finch, tiger finch; French: Bengali rouge; German: Tigerfink; Spanish: Bengalí Rojo.

PHYSICAL CHARACTERISTICS
3.9 in (10 cm). Sexually dimorphic; females have less extensive spotting and red. Juveniles similar to females but have a darker

bill. Males are unique among estrildids due to their non–breeding or eclipse plumage with which they look more like the females.

DISTRIBUTION
Eastern Pakistan through India into Burma and south-central China with populations in southern Thailand, Cambodia and Vietnam and on Java. Introduced populations in Japan, Hawaii, Fiji, the Philippines, Malaysia, Sumatra, Saudi Arabia, and Puerto Rico.

HABITAT
Occurs in grassy areas including marsh, cane fields, jungle clearings, open woodland, reed beds, edges of cultivated or inhabited areas, and gardens.

BEHAVIOR
In pairs or small flocks except in winter when flocks may reach 100 or more birds. The call, given when perched or in flight, is a "tee" or "tsi" but can be quite variable. The song is a weak, but melodious warble. Females also give a shorter version of the song, although they utter it less frequently than do males.

FEEDING ECOLOGY AND DIET
Feeds in vegetation and on the ground on a variety of grass seeds. Observations of captive birds suggest that a small quantity of insects is also consumed.

REPRODUCTIVE BIOLOGY
Nests are placed on or near the ground and are composed of grass blades and stems as well as other plant matter. A nest is lined with soft materials such as feathers, vegetable down, and fine grass. Four to six white eggs are incubated for 11–12 days.

CONSERVATION STATUS
Not threatened.

SIGNIFICANCE TO HUMANS
This species is a popular cage bird that has been kept and bred in captivity for many years. In the pet trade, it is better known as the strawberry finch. ◆

Amandava amandava
▨ Resident

African quailfinch
Ortygospiza atricollis

SUBFAMILY
Estrildinae

TAXONOMY
Fringilla atricollis Vieillot, 1817.

OTHER COMMON NAMES
English: Quailfinch, ground finch, partridge finch; French: Astrild-caille à lunettes; German: Wachtelastrild; Spanish: Astrilda Aperdizada.

PHYSICAL CHARACTERISTICS
3.7–3.9 in (9.5–10 cm). A short tail and lark-like legs are indicative of this species' terrestrial nature. Sexes slightly differ; males have a black face and brown breast. Females are lighter in these areas. Juveniles similar to the female but have fainter barring and a darker bill.

DISTRIBUTION
Senegal east to western Cameroon. Southern Sudan to Angola and south to South Africa.

Ortygospiza atricollis

■ Resident

Stagonopleura guttata

■ Resident

HABITAT
Found in open areas with patchy grass growth, often near water, including sandy grassland, marsh, farms and croplands, and recently mowed areas.

BEHAVIOR
This shy species spends almost all of its time on the ground in pairs or small flocks, being seen only when flushed, one of the few reasons it ever takes flight. The call is a metallic "trillink" or "chwillink" while the song is a series of "click, clack, cluck" notes delivered rapidly and repeatedly.

FEEDING ECOLOGY AND DIET
Feeds on small grass seeds and on occasional spiders or insects.

REPRODUCTIVE BIOLOGY
A dome-shaped nest of grass stems and blades is built on the ground. Four to six white eggs are incubated by both parents.

CONSERVATION STATUS
CITES: Appendix III. Not considered threatened by the IUCN.

SIGNIFICANCE TO HUMANS
Has appeared in aviculture in very low numbers in the past, but is not a popular aviary subject, probably due to its shy and flighty disposition. ◆

Diamond firetail

Stagonopleura guttata or *Emblema guttata*

SUBFAMILY
Poephilinae

TAXONOMY
Loxia guttata Shaw, 1796.

OTHER COMMON NAMES
English: Diamond sparrow, diamond Java sparrow, spotted-sided finch; French: Diamant à gouttelettes; German: Diamantamadine; Spanish: Pinzón Cola de Fuego Diamante.

PHYSICAL CHARACTERISTICS
4.7 in (12 cm). The adult male is brown, above and including wings; rump and upper tail coverts are bright crimson, tail feathers black; forehead, crown of head, and hind neck ashy gray; lores black; throat white; sides of foreneck and flanks black, with some feathers having a subterminal white spot; breast, abdomen, and undertail-coverts white; bill vinous red; legs and feet dark gray; iris red. The adult female is very similar but usually smaller, with a narrower black band on foreneck, and paler lores.

DISTRIBUTION
East-central and southeastern Australia.

HABITAT
Inhabits savanna woodland, eucalypt forests, acacia scrub, mallee, orchards, cultivated areas, and parks and gardens.

BEHAVIOR
Found in pairs or small, often mixed-species flocks. The contact call is a "twooo-heee" while the alarm call is a loud "tay tay tay." The song is described as a series of low-pitched rasping and buzzing notes. Roosts in specially built nests similar to those used for rearing young, but lacking an entrance tube and lining.

FEEDING ECOLOGY AND DIET
Feeds on the ground where it characteristically hops in search of a variety of seeds and insects.

REPRODUCTIVE BIOLOGY
The breeding season is from August to January during which this species builds a grass nest with an entrance tube up to 6 in (15 cm) long. The nest is lined with fine grasses and feathers

in which four to six white eggs are laid. The nest is sometimes built under or near those of raptors.

CONSERVATION STATUS
Near Threatened. The decline in numbers is thought to be due to alterations of the natural habitat.

SIGNIFICANCE TO HUMANS
This species is the only firetail commonly kept in zoos. ◆

Zebra finch
Taeniopygia guttata

SUBFAMILY
Poephilinae

TAXONOMY
Fringilla guttata Vieillot, 1817.

OTHER COMMON NAMES
English: Spotted-sided finch, chestnut-eared finch; French: Diamant mandarin; German: Zebrafink; Spanish: Pinzón Zebra.

PHYSICAL CHARACTERISTICS
3.9 in (10 cm). Sexually dimorphic; females lack the orange cheek patch and the white-spotted chestnut flanks found in adult males. Juveniles similar to females but have a dark bill.

DISTRIBUTION
Throughout most of the interior of Australia and parts of Indonesia.

HABITAT
Inhabits a wide variety of habitats but prefers open areas such as plains, savanna, woodland, mulga scrub, grassland, salt-

marshes, cultivated areas and farmlands, orchards, and inhabited areas and gardens. Water can always be found nearby.

BEHAVIOR
A highly gregarious species, the zebra finch can be found in pairs or, more often, large flocks. The call is a "tya" or "tchee." The song, given by displaying males, is a mixture of trills and nasal notes.

FEEDING ECOLOGY AND DIET
Feeds on the ground on a variety of grass seeds and shoots. The ability to go long periods without water (up to 513 days in one study) and the pigeon-like manner of drinking has allowed this species to survive long periods of drought.

REPRODUCTIVE BIOLOGY
Round nests made of variable materials are either built new or made by renovating roosting nests or other species' nests. Three to eight white eggs are incubated for 11–15 days.

CONSERVATION STATUS
Not threatened.

SIGNIFICANCE TO HUMANS
This species is the most commonly kept and studied estrildid and probably ranks in the top five of most commonly kept birds. Kept since the mid to late 1800s, this species has become domesticated with as many as 30 separate mutations and many combinations thereof developed. Strains developed in Germany and England are several times larger than the wild birds. ◆

Double-barred finch
Taeniopygia bichenovii

SUBFAMILY
Poephilinae

TAXONOMY
Fringilla bichenovii Vigors and Horsfield, 1827.

OTHER COMMON NAMES
English: Bicheno finch, owl finch, owl-faced finch, banded finch, ringed finch, black-ringed finch; French: Diamant de Bichenov; German: Ringelastrild; Spanish: Pinzón de Dos Barras.

PHYSICAL CHARACTERISTICS
3.9–4.3 in (10–11 cm). Sexes alike. Juveniles are a paler version of adults. The double-barred finch is brown with white spots on the upper wings. The underparts and face are cream, with the face surrounded by a black ring. There is another black bar across the lower breast.

DISTRIBUTION
Northern and eastern Australia.

HABITAT
Inhabits dry, open areas including grass plains, open woodland, forest edges, cane fields, inhabited and cultivated areas, and parks and gardens.

BEHAVIOR
Found in small flocks during the breeding season and in larger flocks when not breeding. Roosts communally in specially built nests. The call is a "tat, tat" or a "tiaat, tiaat." The song is a softer version of that of the zebra finch.

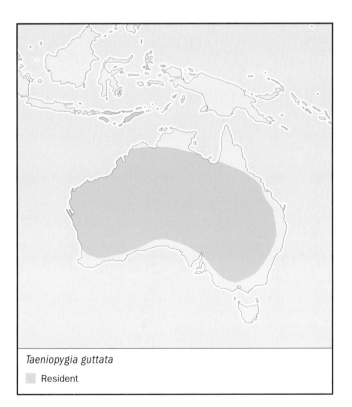

Taeniopygia guttata
■ Resident

Taeniopygia bichenovii

■ Resident

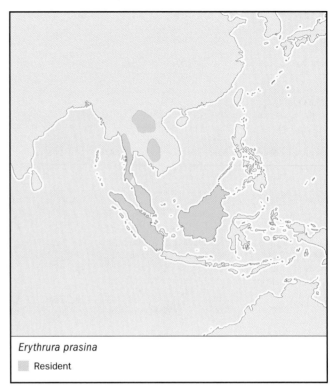

Erythrura prasina

■ Resident

FEEDING ECOLOGY AND DIET
Feeds among grasses and on the ground where it consumes a variety of seeds and undoubtedly a small quantity of insects. This species drinks in a pigeon-like manner.

REPRODUCTIVE BIOLOGY
Breeds year-round with three to six white eggs laid. Builds an almost spherical nest of dry grass stems or uses old nests of other species.

CONSERVATION STATUS
Not threatened.

SIGNIFICANCE TO HUMANS
Commonly kept and bred in captivity where it is known as the owl finch. ◆

Pin-tailed parrotfinch
Erythrura prasina

SUBFAMILY
Erythrurinae

TAXONOMY
Loxia prasina Sparrman, 1788.

OTHER COMMON NAMES
English: Pin-tailed nonpareil, long-tailed munia; French: Diamant quadricolore; German: Lauchgrüne papageiamadine; Spanish: Pinzón Loro de Cola Aguda.

PHYSICAL CHARACTERISTICS
Male: 5.9 in (15 cm); female: 4.5–4.7 in (11.5–12 cm). Sexually dimorphic. Males have a longer tail, a blue face, and red rump,

tail, and belly. A yellow variant exists where the red is replaced with yellow. Females have less blue on the face, shorter tail, and a lack of red on the belly or breast. Juveniles resemble females but have a yellowish lower bill.

DISTRIBUTION
Laos south to Borneo and Sumatra.

HABITAT
Found in forest edges, secondary growth, rice fields, and bamboo.

BEHAVIOR
Usually found in small flocks, but larger flocks can form especially in rice fields. The call is a "tseet-tseet" or "tsit-tsit" while the song is described as a series of clinking, crackling, and chirping.

FEEDING ECOLOGY AND DIET
This species has only been observed feeding on rice or bamboo seed heads in the wild, but in captivity it takes a diversity of seeds as well as leafy green vegetables.

REPRODUCTIVE BIOLOGY
The round nest is built at variable heights and is made of bamboo blades and other plant fibers. Four to six white eggs are incubated 12–14 days.

CONSERVATION STATUS
Not threatened.

SIGNIFICANCE TO HUMANS
In Borneo, this species is a major pest on rice crops. Due to the bright coloration, this and other parrotfinch species are popular aviary subjects. ◆

Pink-billed parrotfinch
Erythrura kleinschmidti

SUBFAMILY
Erythrurinae

TAXONOMY
Amblynura kleinschmidti Finsch, 1878.

OTHER COMMON NAMES
English: Black-faced parrotfinch; French: Diamant à bec rose; German: Schwarzstirn-papageiamadine; Spanish: Pinzón Loro de Pico Rojo.

PHYSICAL CHARACTERISTICS
4.3 in (11 cm). Sexes alike. Unusually shaped bill. They have a strikingly pale 1.5 cm bill. Black head color changing to blue. At the base of the bill the black is replaced by yellow-green up to the ear. The back is green, the upper tail coverts are red and the tail is black. They have pale feet and dark eyes. Juveniles have a dark-tipped bill.

DISTRIBUTION
Found only on Viti Levu, Fiji.

HABITAT
Found in mature rainforests and sometimes in cocoa plantations.

BEHAVIOR
Found alone, in pairs, or in small family flocks, but may join mixed-species flocks when feeding. The call is a "tsee-tsee" or "chee-chee-chee." A formal song is not described but this species also utters a clicking sound.

FEEDING ECOLOGY AND DIET
Feeds on the ground and in trees on figs and other fruit, flower buds, and on diverse invertebrates. The unusually shaped bill is probably an adaptation for its insect-seeking methods whereby it cracks open dead stems of tree-ferns.

REPRODUCTIVE BIOLOGY
Nests are made of bamboo, leaves, and small twigs and lichens.

CONSERVATION STATUS
Endangered. The replacing of native trees in Fiji with those of non-native conifers for forestry is thought to be a major factor in the decline of this species.

SIGNIFICANCE TO HUMANS
None known. ◆

Gouldian finch
Chloebia gouldiae

SUBFAMILY
Erythrurinae

TAXONOMY
Amadina gouldiae Gould, 1844.

OTHER COMMON NAMES
English: Rainbow finch, painted finch, lady Gould, purple-breasted finch; French: Diamant de Gould; German: Gouldamadine; Spanish: Pinzón de Gould.

PHYSICAL CHARACTERISTICS
4.9–5.5 in (12.5–14 cm). The most brightly colored estrildid, this species is sexually dimorphic; females have a shorter tail and a paler breast. Juveniles are a duller, paler version of the adults. Black-headed, red-headed, and yellow-headed varieties exist naturally in the wild. The black-headed is the common morph (75% of the population) while only one in several thousand is of the yellow-headed variety.

DISTRIBUTION
Fragmented areas in north-central Australia.

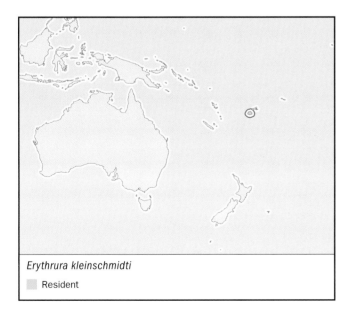

Erythrura kleinschmidti
☐ Resident

Chloebia gouldiae
☐ Resident

HABITAT
Found in dry grassland, plains, areas near water, forest edges, scrubland, and savanna woodland.

BEHAVIOR
Found in flocks of up to several hundred individuals. This shy species tends to avoid areas inhabited by humans. The call is a "sit," "ssit-ssit," or a "sree." The song is a very soft, almost inaudible series of whispers, hisses, whinings, and clicks.

FEEDING ECOLOGY AND DIET
Feeds among grasses on a variety of seeds, especially those of sorghum, and on diverse insects and spiders, especially during the breeding season. Drinks in a pigeon-like manner.

REPRODUCTIVE BIOLOGY
Breeds from November to April during which it builds a globular nest without an entrance tube and sometimes without a roof. Occasionally this species will also use holes in trees or termite mounds as a nesting site. Four to seven white eggs are incubated for 14–15 days.

CONSERVATION STATUS
Endangered. The decrease in numbers is thought to be due to widespread burning of grasses and increase in grazing within its native range.

SIGNIFICANCE TO HUMANS
Commonly kept and bred in captivity where young are often fostered by Bengalese or society finches, a domesticated form of the white-backed munia (*Lonchura striata*). In captivity this species has an unusually high predisposition to the air sac mite (*Sternostoma tracheacolum*). ◆

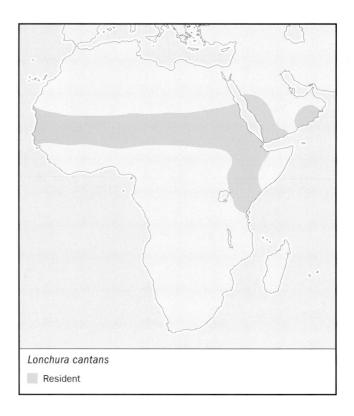

Lonchura cantans
■ Resident

African silverbill
Lonchura cantans

SUBFAMILY
Lonchurinae

TAXONOMY
Loxia cantans Gmelin, 1789.

OTHER COMMON NAMES
English: Warbling silverbill, black-rumped silverbill, silverbill; French: Capucin bec-d'argent; German: Silberschnäbelchen; Spanish: Monjita Pico-de-plata.

PHYSICAL CHARACTERISTICS
3.9–4.5 in (10–11.5 cm). Sexes indistinguishable; brown, belly white, rump and tail black. Juveniles have slightly paler underparts.

DISTRIBUTION
Southern Mauritania, east to Eritrea, south to northeastern Tanzania. Two Asian populations, one in southern Saudi Arabia and western Yemen, and one in southern Oman. Introduced populations are established in Hawaii and Puerto Rico.

HABITAT
Can be found in dry savanna, thorn scrub, acacia woodland, semi-desert, and inhabited or cultivated areas, usually near a water source.

BEHAVIOR
This highly social species can be found in large, often dense, flocks. The call note is a sharp "cheep" or "tseep" while the

song is a series of rising and falling trills for which it is sometimes called the warbling silverbill.

FEEDING ECOLOGY AND DIET
Feeds on grass seeds picked from the growing plant or off the ground. Although it has been reported to eat aphids, insects are not a major part of its diet. This has been supported by captive birds, which rear their young solely on vegetable matter.

REPRODUCTIVE BIOLOGY
Uses old weaver nests or sometimes builds a round nest of grass where three to six white eggs are incubated for 11–13 days.

CONSERVATION STATUS
CITES: Appendix III. Not considered threatened by the IUCN.

SIGNIFICANCE TO HUMANS
Often found near human settlements, sometimes nesting in the eaves of houses. This species is commonly found in aviculture. ◆

Spotted munia
Lonchura punctulata

SUBFAMILY
Lonchurinae

TAXONOMY
Loxia punctulata Linnaeus, 1758.

OTHER COMMON NAMES
English: Scaly-breasted munia/mannikin/finch, barred munia, spice finch/bird, ricebird, nutmeg mannikin/finch, spotted mannikin, common munia; French: Capucin damier; German: Muskatamadine; Spanish: Capuchino Nutmeg.

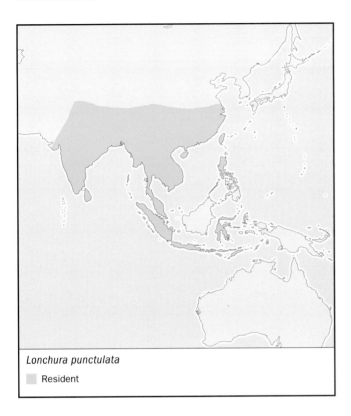

Lonchura punctulata

■ Resident

PHYSICAL CHARACTERISTICS
3.9–4.7 in (10–12 cm). Sexes alike with juveniles a paler brown and lacking the scaled pattern on the underparts.

DISTRIBUTION
India, southern China, and Southeast Asia including parts of Malaysia and Indonesia. Populations introduced in Australia, Hawaii, Puerto Rico, Japan, and the Seychelles.

HABITAT
Inhabits open or semi-open habitats including cultivated and inhabited areas, parks and gardens, grassland, rice fields, and forest edges.

BEHAVIOR
Found in large sometimes mixed-species flocks. The call is a repeated "kitty-kitty-kitty." The soft song is a "klik-klik-klik" followed by a series of whistles and ending with a "weeee," although it can be highly variable.

FEEDING ECOLOGY AND DIET
Feeds on grass seeds, including rice, both on the ground and on live plants. There are several reports of birds feeding on flattened roadkills, possibly as a source of minerals from bones. This species also feeds in human trash dumps taking scraps of bread and other food.

REPRODUCTIVE BIOLOGY
As many as several hundred round nests of grass and bark can comprise a breeding colony. Three to seven white eggs are incubated for 14 days.

CONSERVATION STATUS
Not threatened.

SIGNIFICANCE TO HUMANS
This species is commonly kept and bred in captivity where it is known as the spice finch. Large numbers are caught for the pet

trade, with no noticeable impact on population numbers except in Vietnam and Southeast Asia where the species is also caught in large numbers for human consumption and Buddhist religious purposes. ◆

Java sparrow
Padda oryzivora

SUBFAMILY
Lonchurinae

TAXONOMY
Loxia oryzivora Linnaeus, 1758.

OTHER COMMON NAMES
English: Ricebird, Java temple bird, Java finch, rice munia, paddy bird; French: Padda de Java; German: Reisfink; Spanish: Gorrión de Java.

PHYSICAL CHARACTERISTICS
5.1–6.7 in (13–17 cm). Sexes alike. Upperparts are gray, the head and tail are black, the underparts are rosy, the cheeks are white, and the bill is bright pink to red. Young and immature Javas are dull brown and gray, with bill darker than that of adults.

DISTRIBUTION
Found naturally only on the island of Java. Introduced in many areas throughout the world including numerous south Pacific islands, southeast Asia, Hawaii, Puerto Rico, and Florida.

HABITAT
Found in open woodland, grassland and savanna, but more common in cultivated and inhabited areas.

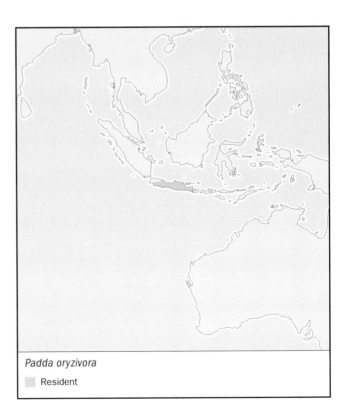

Padda oryzivora

■ Resident

BEHAVIOR

Occurs in pairs or in small flocks, although large flocks usually gather in rice fields. The call is a "tchuk" while the song is a mixture of "diks," "tchuks," "wees," and clicks.

FEEDING ECOLOGY AND DIET

Feeds in vegetation or on the ground on a variety of seeds, fruits, and small insects.

REPRODUCTIVE BIOLOGY

Nest is built in tree holes or crevices in buildings. It is roundish in shape and composed of grass stems. Four to six white eggs are incubated for 13–14 days.

CONSERVATION STATUS

CITES: Appendix II. IUCN considers this species Vulnerable. Causes of population decline in the native range include trapping of live birds for the pet trade, hunting for sport and food, and raiding of eggs from the nest to decrease their numbers and therefore their depredation of rice crops.

SIGNIFICANCE TO HUMANS

This species has long been kept in aviculture where it has been domesticated and where several color mutations have been developed. On its native island of Java, it is hunted for human consumption and persecuted for its impact on rice crops. ◆

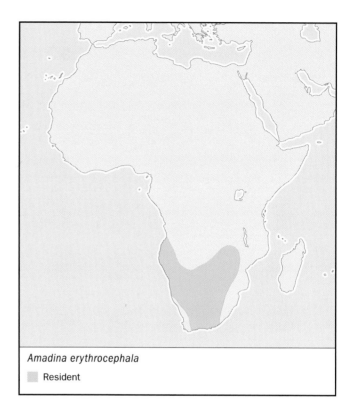

Amadina erythrocephala

▨ Resident

Red-headed finch

Amadina erythrocephala

SUBFAMILY

Lonchurinae

TAXONOMY

Loxia erythrocephala Linnaeus, 1758.

OTHER COMMON NAMES

English: Red-headed weaverfinch; French: Amadine à tête rouge; German: Rotkopfamadine; Spanish: Capuchino de Cabeza Roja.

PHYSICAL CHARACTERISTICS

4.7–5.1 in (12–13 cm). One of the more robust estrildids. The red-headed finch, as its name implies, has a bright, blood red head. Lores are buffish. The lower throat through the belly is marked with black and white scalloping with some chestnut feathers in the middle belly and then off-white in the lower belly areas. Females lack red on the head but may carry a few red feathers. Juveniles are a paler version of the female.

DISTRIBUTION

Southwestern Angola, east to western Mozambique, south to South Africa.

HABITAT

Inhabits semi-desert, savanna, dry grassland, thorn scrub, and the edges of cultivated and inhabited areas.

BEHAVIOR

Found in small flocks. The call is a "chuk, chuk" or, while in flight, a "zree zree." The song is a series of soft buzzing notes.

FEEDING ECOLOGY AND DIET

This species both walks and hops as it feeds on the ground on a variety of seeds and insects.

REPRODUCTIVE BIOLOGY

This species uses old nests of weavers (e.g. sociable weaver [*Philetairus socius*]) or sparrows, or sometimes holes in buildings. Three to eight white eggs are incubated for 12–14 days.

CONSERVATION STATUS

Not threatened.

SIGNIFICANCE TO HUMANS

Occasionally kept in zoos and aviaries but not nearly as commonly as its relative the cut-throat finch (*Amadina fasciata*).

Resources

Books

Clement, P., A. Harris, and J. Davis. *Finches and Sparrows*. Princeton: Princeton University Press, 1993.

Goodwin, D. *Estrildid Finches of the World*. Ithaca, New York: Cornell University Press, 1982.

Pizzey, G. *A Field Guide to the Birds of Australia*. Princeton: Princeton University Press, 1980.

Restall, R. *Munias and Mannikins*. New Haven: Yale University Press, 1997.

Simpson, K. and N. Day. *The Princeton Field Guide to the Birds of Australia*. Princeton: Princeton University Press, 1996.

Zann, R. *The Zebra Finch*. New York: Oxford University Press, 1996.

Resources

Zimmerman, D., D. Turner, and D. Pearson. *Birds of Kenya and Northern Tanzania*. Princeton: Princeton University Press, 1996.

Periodicals

Stripling, R., A. Kruse, and D. Clayton. "Development of Song Responses in the Zebra Finch Caudomedial Neostriatum: Role of Genomic and Electrophysiological Activities." *Journal of Neurobiology* 48 (2001): 163-180.

Other

Birds Australia Nest. Birds Australia. 14 November 2001. <http://www.birdsaustralia.com.au/>.

2000 IUCN Red List of Threatened Species. International Union for Conservation of Nature and Natural Resources. 14 November 2001. <http://www.redlist.org/>.

UNEP-WCMC Database. United Nations Environment Programme World Conservation Monitoring Centre. 20 November 2001. <http://www.unep-wcmc.org/>.

Joseph Allen Smith

▲

Weavers

(Ploceidae)

Class Aves

Order Passeriformes

Suborder Passeri (Oscines)

Family Ploceidae

Thumbnail description

Small to medium-sized passerine birds; bill conical or pointed; plumage plain yellow or black, or these colors in combination with red, brown, or orange, or else sparrowy brown; often there is a seasonal change in plumage, which may include development of greatly elongated tail-feathers; many species highly social, occurring in large flocks

Size

4.3–10 in, up to 28 in with elongated tail (11–25 up to 70 cm); 0.3–2.3 oz (9–65 g)

Number of genera, species

19 genera; 135 species

Habitat

Forest, woodland, swamps, savanna, semi-arid regions

Conservation status

Critically Endangered: 1 species; Endangered: 6 species; Vulnerable: 7 species; Near Threatened: 3 species; Data Deficient: 2 species

Distribution

Sub-Saharan Africa, Arabian Peninsula, South and Southeast Asia, Madagascar, Mauritius, Réunion, Comoros, Seychelles

Evolution and systematics

The fossil record of passerine birds is fragmentary and difficult to interpret. There are no conspicuous skeletal characters in ploceids that distinguish their bones from those of other passerine families, which makes it unlikely that fossils will provide clear evidence of their origins. However, both fossils and molecular data suggest that the passerine birds are an old group, and that many extant families could be as old as 40 million years. Thus the weaver group are likely to have evolved in Africa over a period during which there have been dramatic changes in climate, with the extent of forest cover fluctuating greatly. This would have promoted speciation in both forest and open-country habitats.

Earlier studies based on anatomy and other morphological characters linked the Ploceidae to the Estrildidae, another family of seed-eating birds which is primarily African. This was supported by the DNA-hybridization studies of Charles Sibley, but he placed both these groups as sub-families in a new family (Passeridae), along with the sparrows (Passerinae), wagtails and pipits (Motacillinae), and accentors (Prunellinae). This arrangement remains controversial, and we have followed a more traditional system. The four subfamilies of Ploceidae are most easily defined by their breeding habits: the buffalo weavers, Bubalornithinae (three species), build large nest structures of sticks and have a unique, rigid phalloid or-

gan on the abdomen; the sparrow-weavers, Plocepasserinae (nine species), build nests of straight grass stalks which are not woven but stuck into the nest structure; the "true" weavers Ploceinae (103 species) weave a closed nest, with the entrance either at the side or below; the whydahs, indigobirds, and cuckoo finch Viduinae (20 species) are brood parasites, which lay their eggs in the nests of other birds. The position of the Viduinae is debatable. Behavioral studies by J. Nicolai suggested links to some Ploceinae such as the bishop-birds, whereas skeletal and molecular data imply that the waxbills (Estrildidae) are their closest relatives. The cuckoo finch (*Anomalospiza imberbis*) has usually been classified as a weaver, but both morphological and molecular data show that it belongs in the Viduinae.

The genus *Ploceus* (Ploceinae) is one of the largest bird genera at present with more than 60 species. An examination of skull characteristics suggests that there are several distinct groups within this genus, and new studies may lead to it being broken up into several distinct genera. The relationships between the African and Asian *Ploceus* weavers are not clear, and they have evidently been separated for a long time. The *Foudia* species on the Indian Ocean islands appear to be derived from the African genera *Quelea* or *Euplectes*, whereas the two Madagascar *Ploceus* species could be African or Asian in origin. These conclusions are based on plumage, nest

Weaver nests in Rajasthan, India. (Photo by M.A. Fisher. Bruce Coleman Inc. Reproduced by permission.)

structure, carotenoid pigments, and some skeletal characters; no molecular studies had been published by 2001.

Physical characteristics

The weavers have no defining physical characteristics which are shared by all or even most members of the family. The sexes may be virtually indistinguishable, even in the hand, or highly dimorphic. Tails can be short or extravagantly long. The bill is always straight, not curved, but varies from short and heavy to longer and quite slender. At the sub-family level, there is more consistency. Buffalo weavers are either mainly black or mainly white, with heavy seed-eater bills. Sparrow-weavers are all "sparrowy" brown in appearance, with some black or white plumage areas. There is no obvious seasonal plumage change in either of these groups, and little sexual dimorphism, although males are usually larger. Within the parasitic Viduinae, there is marked sexual dimorphism in plumage during the breeding season, after which males molt into a plumage which resembles that of the females. They can usually be disinguished from other small seed-eating birds by

black stripes on the crown of the head. Male indigobirds are blackish, with pale or reddish bill and legs, in varying combinations. Male whydahs have mainly black or black-and-white breeding plumage with very long central tail feathers, which may be either narrow or broadened. The male cuckoo finch is canary-yellow in breeding plumage.

Among the Ploceinae, there are conspicuous differences between genera. Males are almost always larger than females, while sexual dimorphism in plumage is especially marked in polygynous species. However, even in dimorphic species, the males do not always have a seasonal plumage change. Eye color often changes with age from brown to red, yellowish, or creamy; in many cases only males have a distinctively colored eye. The bill color of male birds may change seasonally from brown to black, in response to increased levels of male sex hormones. The genus *Malimbus* is remarkably uniform. All species are predominantly black with some red, or in one case yellow, plumage; males and females differ in plumage, and juvenile birds have a distinctive plumage, different to both adults. There is no seasonal change in plumage. In contrast the open-country bishops and widows (*Euplectes*) all have sparrowy brown females, while males molt into a breeding plumage which is wholly or partly black, with either red or orange to yellow areas, and in some cases a long, black tail. Young birds resemble females, and males do not usually acquire breeding plumage until at least their second year. The large genus *Ploceus* includes species that are sexually dimorphic with or without a seasonal change in plumage, and species in which the sexes are identical. Black and/or yellow are the predominant plumage colors in males, with some green, brown, or orange, but never red, feathers.

Distribution

Weavers occur throughout sub-Saharan Africa, where all sub-families are represented. Only two genera of Ploceinae are found outside Africa; the fodies (*Foudia*) which are endemic to Madagascar and other Indian Ocean islands, and *Ploceus* with two species on Madagascar and five in Asia. One East African species, Rüppell's weaver (*Ploceus galbula*), also occurs on the Arabian peninsula. Several species are commonly exported as cage birds, and escapes or deliberate releases have led to their establishment, sometimes temporary, in other regions, including Australia, California, Portugal, Hawaii, St. Helena, and some islands in the West Indies. An Asian species, the streaked weaver (*Ploceus manyar*), is now established in the Nile delta in Egypt, and is believed to have escaped from Alexandria Zoo.

Habitat

Many weavers are associated with water, since they breed in wetlands, along rivers, dams, and lakes, nesting in reeds or other waterside vegetation. However, in these cases they often move to grassland or savanna during the non-breeding season. Several species may breed in wetlands, but also in trees far from open water, and have adapted well to man-modified habitats such as farmland. Only members of the sparrow-weavers and buffalo weavers are permanent residents of arid and semi-arid areas. Some species are exclusively forest birds, either in lowland or montane evergreen forest, and may spend much of their time in the canopy 100 ft (30 m) above the ground. All members of the genus *Malimbus* are strictly forest inhabitants.

Behavior

Although many species of weavers move about extensively during the dry season, these are local movements rather than predictable, long-distance migration. The red-billed quelea (*Quelea quelea*) does carry out predictable movements in many regions, and these seem to be correlated with rainfall patterns. This appears to be the only species that could qualify as a migrant throughout its range.

Although they may have a wide range of different calls, few weavers would be considered "songbirds" in the conventional sense. The songs that male weavers use to advertise their territories are often a harsh, repetitive chatter with no tuneful, musical notes. Some forest species do sing short phrases, sometimes as duets, which are more attractive to our ears. The parasitic indigobirds learn elements of the song of their host species while in the nest, and later incorporate these into the songs which they use in courtship.

Feeding ecology and diet

Categorizing weavers as insectivorous or granivorous is misleading. All species will take insects when they are available, and the young are often fed primarily insects, especially in the first days after hatching. There is frequently a seasonal change in diet, with seeds the main or even the only food source in the dry season, and insects more important in the rainy season. The heavy bill of the grosbeak weaver (*Amblyospiza albifrons*) enables the birds to open sunflower seeds, but they have also been seen to catch small frogs. Small lizards are on the menu of several other species in the wild. Fruit and berries are eaten readily, and nectar from plants such as *Aloe* and *Erythrina*. Here weavers are messy feeders, often eating the whole flower and stripping the plants, leaving with their faces caked with pollen. The Cape weaver (*Ploceus capensis*) is probably the main pollinating agent for the endemic South African crane flower *Strelitzia regina*.

Reproductive biology

Social organization in weavers shows clear correlations with habitat and feeding ecology, as J. H. Crook first demonstrated in his innovative comparative studies. Forest weavers are generally insectivorous and remain in pairs throughout the year, whereas seed-eating species of the open savanna associate in flocks, and form colonies for breeding. This, in turn, influences their breeding systems, with monogamy usual in the forest species, while many of the colonial weavers are polygynous, with one male mating in turn with a succession of different females.

Nests and nest construction have attracted most attention in this group of birds. The pioneering work of Nicholas and

A masked weaver (*Ploceus velatus*) with the nest he has built to attract a female. (Photo by M.P. Kahl. Photo Researchers, Inc. Reproduced by permission.)

Elsie Collias—who observed many species in the field, in captivity, and in museum collections—has provided an excellent framework for the evolution of nest-building in the family. In buffalo weavers and sparrow-weavers the technique is simple, with the nests formed as piles of interlocking material. These birds are associated with nests throughout the year, and thus maintain the structures with periodic building at all seasons; both sexes participate to some degree. In the true weavers (Ploceinae) nest-building is seasonal and these are short-lived structures, which mostly do not survive beyond one breeding season. The commonest pattern is for the male to produce a nest frame by weaving and knotting strips of material collected and prepared for this purpose. Once the female has mated and accepted a particular nest, she then adds the lining. However, the female's contribution varies greatly, depending on the mating system; in Jackson's widow (*Euplectes jacksoni*) the female is solely responsible for building and lining the nest, which is not on the male's territory.

Courtship in sparrow-weavers and buffalo weavers involves song and visual displays, generally near the nest structures. In

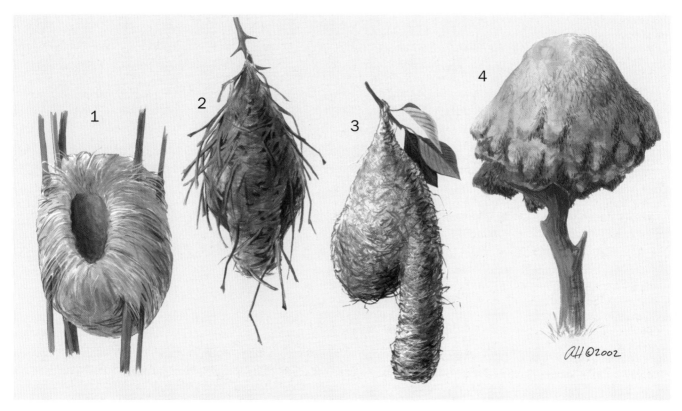

Weaver nests: 1. Thick-billed weaver (*Amblyospiza albifrons*); 2. Red-headed weaver (*Anaplectes rubriceps*); 3. Spectacled weaver (*Ploceus ocularis*); 4. Sociable weaver (*Philetairus socius*). (Illustration by Amanda Humphrey)

the Ploceinae, among the monogamous forest species, courtship frequently takes place away from the nest, even before construction begins. By contrast the colonial species set up territories and build nests before intensive courtship starts, and the male often displays hanging at the nest entrance. Bursts of display activity may sweep through a colony as groups of females arrive, and the males all appear to be vying for their attention. In the polygynous species, each male will build a series of nests, and try to attract as many females as possible. Nest that are not accepted by females, or are no longer occupied, are often demolished. The male may then rebuild at the same site so that a single male masked weaver (*Ploceus velatus*) can build more than 20 nests over a three-month season. There is one exceptional species, Jackson's widow, in which males display at dancing grounds which constitute a lek: a male courtship arena which females visit to select a mate, after which they go off to nest and have no further contact with the male. In the parasitic Viduinae, males set up "song-posts" at which they display, and to which females are attracted. The breeding system is thus a form of lek, where the females visit males only to mate, although the males do not occupy a communal display ground.

Parental care is closely correlated with the mating system. In monogamous species, both incubation and feeding of the young may be shared equally between the partners, whereas in polygynous species the female normally does all the parental duties without assistance. In some cases, polygynous males may feed at the nest late in the season, or occasionally feed the young once they have fledged. Compared to some

other African bird families such as the starlings, cooperative breeding in which several related or unrelated individuals help the parents to rear the young is rare in weavers. It occurs primarily in some of the sparrow-weavers in arid country, where a nesting tree forms a permanent base for the group, which roosts in the nests throughout the year. The situation in the parasitic Viduinae is especially interesting. Whydahs and indigobirds lay their eggs in the nests of waxbills; the eggs of both host and parasite are plain white, and similar in size. Waxbill young have highly distinctive mouth-markings, which are matched by the young parasites. The young are raised together, so although the waxbill parents have extra mouths to feed, they do not lose their whole brood as is often the case for cuckoo hosts. However, the cuckoo finch parasitizes small grassland warblers, and the host young seldom survive.

Conservation status

BirdLife International has produced a review of globally threatened birds, and an account of the Important Bird Areas of Africa. The major threat to weaver species is habitat loss, since some of them have very restricted ranges. Three island fodies are threatened both by habitat loss and introduced predators on Mauritius, Seychelles, and Rodrigues, respectively. *Foudia rubra* may be Critically Endangered, whereas *F. sechellarum* and *F. flavicans* are currently regarded as Vulnerable. The Asian yellow weaver (*Ploceus megarhynchus*) is a grassland species with a restricted range in India. Although the

Asian golden weaver (*Ploceus hypoxanthus*) occurs in several countries, it is uncommon and regarded as Near Threatened.

On mainland Africa, the golden-naped weaver (*P. aureonucha*) and the yellow-footed weaver (*P. flavipes*) are both known only from the Ituri Forest, and have been seen just a few times in the last 30 years. Their canopy habitat and the political problems in this region make it difficult to obtain accurate information. Four localized species in West Africa, Bannerman's weaver (*P. bannermani*), Bates's weaver (*P. batesi*), the Gola malimbe (*Malimbus ballmanni*) and the Ibadan malimbe (*M. ibadanensis*), occur in forest that is disappearing rapidly throughout this region. The situation is most critical for the Ibadan malimbe, which has the smallest range. Two little-known species, the Loango weaver (*P. subpersonatus*) on the coastal strip and the black-chinned weaver (*P. nigrimentum*) in open savanna, range from Gabon southwards towards Angola.

In East Africa, Clarke's weaver (*P. golandi*) is restricted to the Arabuko-Sokoke Forest in Kenya, while the Tanzanian mountain weaver (*P. nicolli*) is found in relict forest patches on the Usambara Mountains and a few other sites. Fortunately both areas are now the site of active conservation programs. Agricultural changes in the highland grasslands of Kenya are a potential threat to Jackson's widow, while Fox's weaver (*P. spekeoides*) is apparently confined to one lake system in central Uganda, but remains unstudied. The Kilombero weaver (*P. burnieri*) was a surprising discovery in Tanzania, described in 1990 and evidently limited to a small area.

Significance to humans

Several colonial weaver species are closely associated with human settlements, nesting in exotic vegetation, and in forested areas, taking advantage of habitat changes to colonize new clearings. Eggs and nestlings may be utilized for food on occasion, but often the relationship is quite harmonious. The long tail feathers of breeding male long-tailed widows (*Euplectes progne*) were once used as elements in traditional head-dresses for warrior tribes in South Africa, but otherwise the colored plumages have not been utilized.

A red-headed weaver (*Anaplectes rubriceps*) works on his nest. (Photo by E.R. Degginger. Photo Researchers, Inc. Reproduced by permission.)

For hundreds of years, grain-eating weavers have been a pest for farmers in Africa. M. Adanson, a French botanist for whom the baobab genus *Adansonia* is named, spent several years in Senegal from 1747, and reported that the inhabitants suffered greatly from the depredations of the weavers. He described several traditional bird-scaring methods which are still in use in Africa today. Since the 1960s the red-billed quelea has been recognized as the major pest of cultivated cereals in Africa. Despite international efforts to reduce its numbers, using aerial spraying and fire-bombs set under roost sites, it remains enormously abundant: in March 2000 the South African department of agriculture reported that an estimated 21 million queleas had been killed in control operations during the past month! It seems that in the past, queleas bred prolifically in good years, and then starved when food supplies declined. Today when wild grass seeds are unavailable, they find crops a very acceptable alternative and consequently agriculture enables them to maintain high population levels. To the interested naturalist, a vast flock of queleas "roller-feeding" (in constant motion, with the birds at the back flying up over those ahead of them to be first at the untouched plants) is one of the great spectacles of Africa, but it is a catastrophe for the small farmer, and there is no simple, effective solution.

1. Red-headed weaver (*Anaplectes rubriceps*); 2. Sociable weaver (*Philetairus socius*); 3. White-browed sparrow weaver (*Plocepasser mahali*); 4. Dark-backed weaver (*Ploceus bicolor*); 5. Thick-billed weaver (*Amblyospiza albifrons*); 6. Village weaver (*Ploceus cucullatus*); 7. Blue-billed malimbe (*Malimbus nitens*); 8. Spectacled weaver (*Ploceus ocularis*); 9. Red-billed buffalo weaver (*Bubalornis niger*). (Illustration by Amanda Humphrey)

1. Sakalava weaver (*Ploceus sakalava*); 2. Red-collared widow (*Euplectes ardens*); 3. Red bishop (*Euplectes orix*); 4. Pin-tailed whydah (*Vidua macroura*); 5. Red-billed quelea (*Quelea quelea*); 6. Cuckoo finch (*Anomalospiza imberbis*); 7. Madagascar fody (*Foudia madagascariensis*); 8. Dusky indigobird (*Vidua funerea*); 9. Jackson's widow (*Euplectes jacksoni*); 10. Baya weaver (*Ploceus philippinus*). (Illustration by Amanda Humphrey)

Species accounts

Red-billed buffalo weaver
Bubalornis niger

SUBFAMILY
Bubalornithinae

TAXONOMY
Bubalornis niger A. Smith, 1836, Kurrichane, South Africa.

OTHER COMMON NAMES
French: Alecto à bec rouge; German: Büffelweber; Spanish: Tejedor Búfalo de Pico Rojo.

PHYSICAL CHARACTERISTICS
8.7 in (22 cm); unsexed 2.7–2.9 oz (78–82 g). Male dark blackish brown, white patch in wings in flight. White bases to body feathers may show when plumage ruffled. Bill and legs red. Female dark brown, variably flecked with white on underparts. Bill and legs brown. Juveniles paler with more white on underparts. Bill orange-yellow.

DISTRIBUTION
Ethiopia and Somalia through eastern Africa to Angola, Zambia, and northern Mozambique, south to northern South Africa.

HABITAT
Dry thornveld with large trees.

BEHAVIOR
In groups or non-breeding flocks up to 50 birds, may associate with other species. Usually present at nest sites throughout the year.

FEEDING ECOLOGY AND DIET
Mainly insects, also seeds and fruit. Most food collected on the ground.

REPRODUCTIVE BIOLOGY
Colonial and polygynous, often polygynandrous (each male mates with several females, and female mates with more than one male), since in most broods genetic studies indicate multiple paternity. Nest is a large mass of thorny twigs, containing up to 13 nest chambers, lined with green vegetation. Male builds main structure and starts lining chambers; female adds further lining before laying. Phalloid organ not inserted during copulation, but stimulation from this structure may be essential for successful mating. Lays three to four eggs in spring and summer. Incubation 11 days, fledging 20–23 days. Female alone incubates and does most feeding of chicks; male feeds young occasionally.

CONSERVATION STATUS
Not threatened; dependent on large trees, but much habitat is sparsely populated.

SIGNIFICANCE TO HUMANS
None known; may use human-made structures as nest sites, or nest near homesteads. ◆

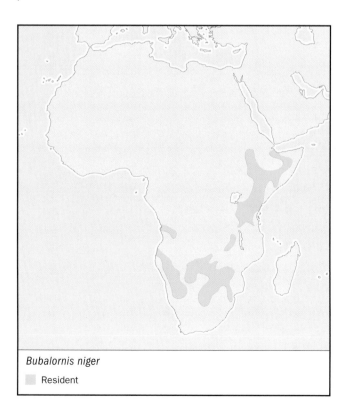

Bubalornis niger
█ Resident

White-browed sparrow-weaver
Plocepasser mahali

SUBFAMILY
Plocepasserinae

TAXONOMY
Plocepasser mahali A. Smith, 1836, Orange River, South Africa.

OTHER COMMON NAMES
French: Mahali à sourcils blancs; German: Augenbrauenmahali; Spanish: Tejedor Gorrión de Cejas Blancas.

PHYSICAL CHARACTERISTICS
6.7 in (17 cm); female and male 1.6–1.8 oz (45–52 g). Brown upperparts with broad white eyebrow and white rump, white underparts. Bill brown to black. Sexes alike, juvenile paler than adult with a pale bill.

DISTRIBUTION
Southern Sudan, Ethiopia, and Somalia south through eastern Africa to Zambia, South Angola, and northern South Africa.

HABITAT
Mopane and acacia savanna in relatively dry country.

BEHAVIOR
Groups of up to 12 birds resident, defend territory of about 55 yd (50 m) in diameter with complex songs and group displays. Strong dominance hierarchy within group, with a single breeding pair. Roost singly in nests.

FEEDING ECOLOGY AND DIET
Insects and seeds, in variable proportions. Most food collected on the ground; birds will dig in soil, roll over small stones, clods, and elephant droppings.

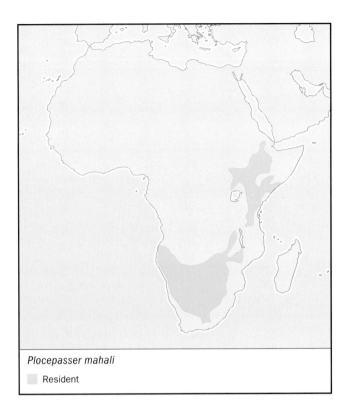

Plocepasser mahali
■ Resident

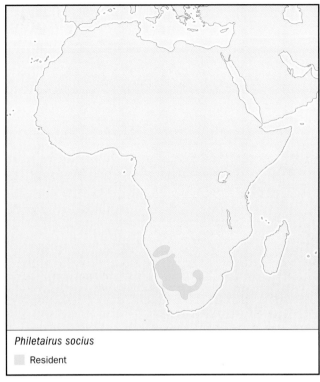

Philetairus socius
■ Resident

REPRODUCTIVE BIOLOGY
Colonial, cooperative breeder. Nest is an elongated retort
made of straight dry grass stems, resting on thin branches, initially with an opening at each end. Both male and female, and
other group members, may contribute to building nest. Second
entrance closed in breeding nests, which are then lined with
feathers. Nest orientation related to prevailing winds. Lays two
to three eggs, mainly in spring and summer but also in other
months. Incubation 14–16 days, fledging 18–23 days. Juveniles
still fed occasionally up to three months after leaving nest. Female alone incubates and feeds young for first three days. Male
seldom feeds young; helpers related to breeding pair do much
of the feeding. Unrelated helpers assist in territorial defense,
but not in raising young.

CONSERVATION STATUS
Not threatened; common in many thinly populated areas.

SIGNIFICANCE TO HUMANS
None known, although may feed on wheat or maize in winter,
usually only grains left behind in fallow lands. ◆

Sociable weaver
Philetairus socius

SUBFAMILY
Plocepasserinae

TAXONOMY
Loxia socia Latham, 1790, Great Namaqualand, South Africa.

OTHER COMMON NAMES
French: Républicain social; German: Siedelweber; Spanish:
Tejedor Sociable.

PHYSICAL CHARACTERISTICS
5.5 in (14 cm); female and male 0.8–1.1 oz (24–32 g). Sandy
brown with black chin and throat, dark feathers with pale
edges on mantle and flanks. Bill blue-gray. Sexes alike. Juvenile
uniform sandy brown, with no darker feathers. Bill pale brown.

DISTRIBUTION
Namibia, southwestern Botswana, northwestern South Africa.

HABITAT
Open, arid regions with scattered trees and bare ground.

BEHAVIOR
Gregarious, resident at nest sites, roosting in chambers
throughout the year. Predators such as cobras may live within
nest structure, also "lodgers" like the pygmy falcon (*Polihierax
semitorquatus*), which is an obligate commensal. Other birds
may occasionally roost or breed in vacant nest chambers.

FEEDING ECOLOGY AND DIET
Seeds and insects, particularly harvester termites. Proportion of
insect and seed food varies seasonally, collected primarily on
the ground. Feed in flocks within 1 mi (1.5 km) radius of nest
site. Seldom drink water.

REPRODUCTIVE BIOLOGY
Colonial, monogamous with cooperative breeding. Communal
nest is a huge mass of dry grass stems, with individual nest
chambers entered from below. Up to 13 ft (4 m) deep and 24
ft (7.2 m) long, supported by large branches; in treeless areas
may use telephone poles. Pair bond may last for only one
breeding attempt, even if both partners resident in same nest
mass; helpers chiefly offspring of pair from earlier broods. Lays
two to six eggs; season entirely dependent on rainfall, and
breeding may start in any month. Incubation 13–14 days,
fledging 21–24 days. Both sexes incubate and feed the young;
up to nine helpers may feed chicks.

CONSERVATION STATUS
Not threatened. Range is thinly populated, includes major conservation areas.

SIGNIFICANCE TO HUMANS
Nest material sometimes used for stock fodder in times of drought. ◆

Blue-billed malimbe
Malimbus nitens

SUBFAMILY
Ploceinae

TAXONOMY
Ploceus nitens J. E. Gray, 1831, Sierra Leone.

OTHER COMMON NAMES
English: Gray's malimbe; French: Malimbe à bec bleu; German: Rotkehlweber; Spanish: Malimbe de Gray.

PHYSICAL CHARACTERISTICS
5.7–6.7 in (14.5–17.0 cm); female 1.0–1.2 oz (29–36 g), male 1.3–1.7 oz (38–47 g). Black with scarlet throat; female less glossy than male, black tinged with brown, and red less intense. Bill blue-gray, eye red. Juvenile sooty brown with throat and breast dull orange-brown. Eye gray-brown.

DISTRIBUTION
Senegal east to extreme western Uganda, south to Democratic Republic of Congo and northern Angola.

HABITAT
Lowland forest, oil palms, swamp forest, and mangroves; occasionally dense savanna woodland.

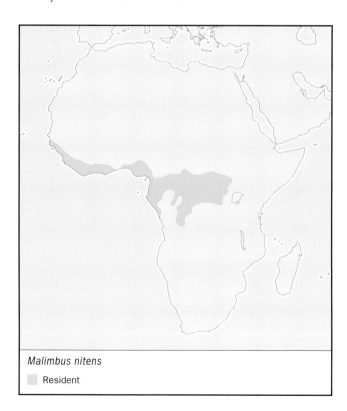

Malimbus nitens
▨ Resident

BEHAVIOR
Usually solitary or in pairs, occasionally groups up to seven birds; regularly joins mixed-species flocks. Very shy except at the nest.

FEEDING ECOLOGY AND DIET
Mainly insects, also spiders and some fruit. Forages on thin twigs, clusters of dry leaves, and vine tangles, mainly at mid-levels of forest.

REPRODUCTIVE BIOLOGY
Solitary and monogamous, although sometimes several pairs nest close together. Male courtship, singing and posturing, occurs away from the nest. Nest is ball-shaped, woven from palm strips, rootlets, or fibers, with canopy over entrance placed low at one side; always overhanging water. In Ghana the birds appear to select nest sites close to crocodile dens. Lays one to two eggs during late summer to autumn. Incubation 14 days, fledging 16 days. Female alone incubates, and broods small chicks; both male and female feed young.

CONSERVATION STATUS
Not threatened, but will not survive without undisturbed forest habitat.

SIGNIFICANCE TO HUMANS
None known. ◆

Spectacled weaver
Ploceus ocularis

SUBFAMILY
Ploceinae

TAXONOMY
Ploceus ocularis A. Smith, 1828, Eastern Cape = Grahamstown, South Africa.

OTHER COMMON NAMES
French: Tisserin à lunettes; German: Brillenweber; Spanish: Tejedor Moteado.

PHYSICAL CHARACTERISTICS
5.9–6.3 in (15–16 cm); female 0.7–1.0 oz (21–30 g), male 0.8–1.1 oz (22–32 g). Greenish yellow weaver with slender, dark bill and dark "spectacle" line through the eye. Eyes pale cream. Male has dark bib on the throat, lacking in female. Juvenile lacks spectacle line or bib, eye brown, bill pale brown.

DISTRIBUTION
Cameroon east to Sudan, Ethiopia, south to northern Namibia, northern Botswana, and eastern South Africa.

HABITAT
Open woodland, forest edge, thickets, and gardens.

BEHAVIOR
Singly or in pairs throughout the year, family groups after breeding. May join mixed-species flocks of insectivorous birds. Territorial, calling regularly, a descending "tee-tee-tee-tee."

FEEDING ECOLOGY AND DIET
Mainly insectivorous, gleaning leaves and branches and probing bark. Also takes berries, small geckos, nectar, and bread and chicken feed in gardens.

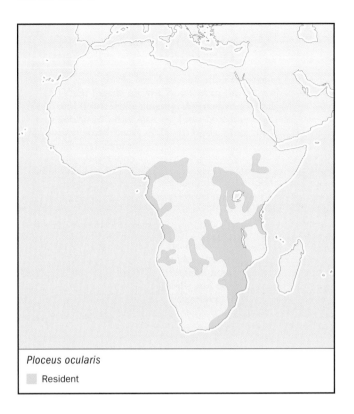

Ploceus ocularis
■ Resident

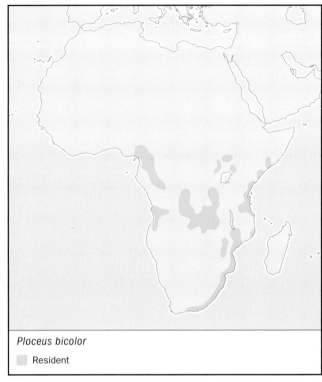

Ploceus bicolor
■ Resident

REPRODUCTIVE BIOLOGY
Nests are finely woven, suspended singly from tip of vegetation, with entrance tunnel 4 in (10 cm) long. Lays one to four eggs, usually two to three, in spring to summer in different regions. Incubation 13–14 days, fledging 15–19 days. Both sexes incubate and feed young. Occasionally parasitized by Diederik cuckoo (*Chrysococcyx caprius*).

CONSERVATION STATUS
Not threatened. Widespread over large area, occurs in man-modified habitats such as suburban gardens.

SIGNIFICANCE TO HUMANS
None known. ◆

Dark-backed weaver
Ploceus bicolor

SUBFAMILY
Ploceinae

TAXONOMY
Ploceus bicolor Vieillot, 1819, 'Senegal' = South Africa.

OTHER COMMON NAMES
English: Forest weaver; French: Tisserin bicolore; German: Waldweber; Spanish: Tejedor Bicolor.

PHYSICAL CHARACTERISTICS
5.5 in (14 cm); female 1.0–1.3 oz (29–36 g), male 1.1–1.6 oz (32–46 g). Upperparts dark, underparts golden yellow. Bill dark with blue-gray rims. Eye brown or red in different

regions. Sexes alike. Juvenile duller than adult, bill light brown.

DISTRIBUTION
Nigeria east to southern Sudan and eastern Africa, south to eastern South Africa but absent from Botswana, Namibia.

HABITAT
Forested areas and dense riverine vegetation, subtropical thicket.

BEHAVIOR
In pairs throughout the year, small family parties after breeding. Joins mixed-species flocks. Song a musical duet, heard at all times of the year.

FEEDING ECOLOGY AND DIET
Insects, mainly gleaned from branches, bark, and tangled vegetation in mid-story of forest. Also fruit and berries, and nectar.

REPRODUCTIVE BIOLOGY
Monogamous, solitary nester. Often returns to same site, so several old nests may be in close proximity. Nest is retort-shaped with broad entrance tunnel pointing downwards; suspended at tip of branch or creeper. Woven of thin vines and creepers, appears rough and always looks old and dry. Lays two to four eggs in summer. Incubation 15–17 days, fledging 22 days. Probably both sexes incubate; both feed young.

CONSERVATION STATUS
Not threatened, but dependent on preservation of well-wooded habitats.

SIGNIFICANCE TO HUMANS
None known. ◆

Village weaver
Ploceus cucullatus

SUBFAMILY
Ploceinae

TAXONOMY
Ploceus cucullatus P. L. S. Müller, 1776, Senegal.

OTHER COMMON NAMES
English: Spotted-backed weaver; black-headed weaver, V-marked weaver; French: Tisserin gendarme; German: Textorweber; Spanish: Tejedor de la Villa.

PHYSICAL CHARACTERISTICS
5.9–6.7 in (15–17 cm); female 1.1–1.5 oz (31–43 g), male 1.1–1.6 oz (32–45 g). Breeding male has head black; forehead yellow in southern birds, and extent of black on throat and breast varies. All populations have upperparts yellow spotted with black, underparts plain yellow. Bill black, eye red. Female and non-breeding male upperparts dull olive, eyebrow, throat, and breast yellow to buff, belly whitish. Bill brown; older females may have red eye. During breeding season, female more yellow on underparts. Juvenile like female, eye brown.

DISTRIBUTION
Senegal east to Somalia, south to northern Namibia, northern Botswana, eastern South Africa. Introduced to Haiti, Dominican Republic, Puerto Rico, Mauritius, and Réunion.

HABITAT
Open wooded areas; in forest zone, in clearings and secondary growth, cultivated areas.

BEHAVIOR
Gregarious, in flocks when foraging and forms large roosts, often with other weavers. May be nomadic in dry season, and possible regular movements in some regions.

FEEDING ECOLOGY AND DIET
Varied diet includes seeds, insects, flowers, nectar. Forages on the ground or gleans on vegetation and tree trunks.

REPRODUCTIVE BIOLOGY
Colonial, polygynous. In central Africa often in large mixed colonies with Vieillot's black weaver (*Ploceus nigerrimus*). Nest is oval with entrance below, may have a short spout. Woven by male, who displays hanging below nest, fluttering wings, and calling. Breeding varies regionally, may continue throughout the year in central Africa. Lays two to five eggs. Incubation 12 days, fledging 17–21 days. Female alone incubates, male may feed nestlings. Often parasitized by Diederik cuckoo.

CONSERVATION STATUS
Not threatened, very widespread, common in human-modified habitats and often abundant.

SIGNIFICANCE TO HUMANS
Familiar commensal throughout central Africa; can be significant crop pest for subsistence farmers. ◆

Sakalava weaver
Ploceus sakalava

SUBFAMILY
Ploceinae

TAXONOMY
Ploceus sakalava Hartlaub, 1867, Madagascar.

OTHER COMMON NAMES
French: Tisserin sakalave; German: Sakalavenweber; Spanish: Fodi Sakalava.

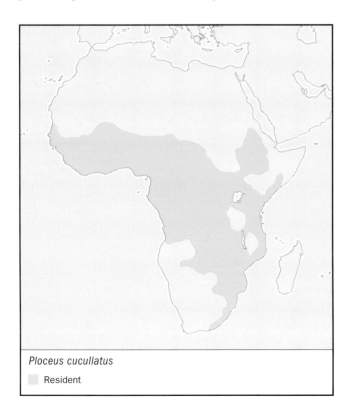

Ploceus cucullatus
▨ Resident

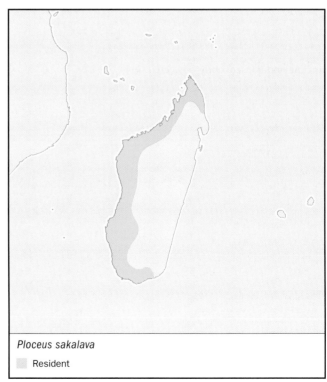

Ploceus sakalava
▨ Resident

PHYSICAL CHARACTERISTICS
5.1–5.9 in (13–15 cm); adult 0.7–0.9 oz (20–27 g). Breeding male has yellow head, gray underparts and back, brown wings and tail. Bare reddish skin around eye, unique in weavers, blue-gray bill. Female is paler below, with whitish throat and distinct brown stripes on side of head, above and below eye; bill pale gray. Non-breeding male like female, but bare pinkish skin around eye. Juvenile like female, but paler with bill horn-colored.

DISTRIBUTION
Western Madagascar.

HABITAT
Open, lowland areas including cultivated land, spiny bush, and deciduous dry forests.

BEHAVIOR
Highly gregarious, typically in flocks of 200 or more.

FEEDING ECOLOGY AND DIET
Feed primarily on the ground, collecting seeds, but also forage in trees and marshes, and young are fed primarily insects.

REPRODUCTIVE BIOLOGY
Colonial, some males may be polygynous, but mating system not studied. Nest is retort-shaped, often suspended on a short woven rope, with entrance tunnel up to 16 in (40 cm) long. Both male and female build nest, and colonies usually in trees, often within villages, and even attached to thatched roofs of huts. Small colonies may be placed under nests of crows or large birds of prey. Breeding season varies with rainfall, especially in dry southwest. Lays two to four eggs; incubation and fledging periods unrecorded. Female alone incubates, but both sexes feed young.

CONSERVATION STATUS
Not threatened, widespread in open and cultivated areas, and seldom molested.

SIGNIFICANCE TO HUMANS
Appear to take only waste rice, and not regarded as agricultural pests. In many areas weaver colonies in villages, especially those nesting close to a house, are considered a sign of good fortune and consequently protected. ◆

Baya weaver
Ploceus philippinus

SUBFAMILY
Ploceinae

TAXONOMY
Loxia philippina Linnaeus, 1766, Philippines = Sri Lanka.

OTHER COMMON NAMES
French: Tisserin baya; German: Bayaweber; Spanish: Tejedor de Baya.

PHYSICAL CHARACTERISTICS
5.1–5.9 in (13–15 cm); female 0.7–1.0 oz (20–28 g), male 0.7–0.9 oz (20–26 g). Breeding male, yellow crown, black face mask, mottled brown upperparts, paler, unstreaked underparts; bill black. Female and non-breeding male, mottled rufous-brown upperparts, some streaking on underparts, bill brown. Juvenile like female.

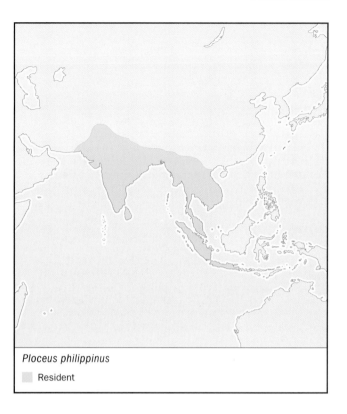

Ploceus philippinus

☐ Resident

DISTRIBUTION
Pakistan, India, and Sri Lanka east to southwestern China, south throughout Malay Peninsula, Sumatra, Java, Bali.

HABITAT
Forest edge, open savanna and scrub, cultivated areas. Appears to prefer agricultural land.

BEHAVIOR
Gregarious in flocks, and forms large communal roosts in reedbeds or sugarcane, together with other weavers, seed-eating birds, starlings, and bulbuls.

FEEDING ECOLOGY AND DIET
Adults mainly seed-eating, including rice, sorghum, millet, and wheat. Also consumes insects, even frogs recorded, and young fed primarily insects.

REPRODUCTIVE BIOLOGY
Colonial, polygynous. Nest is retort-shaped with entrance tunnel of varying length, often suspended over water in trees or bushes. Male builds nest, displays there to attract mates. Blobs of mud, sometimes dung, regularly added to inside of nest. Long-standing but unsubstantiated legend that male embeds fireflies in mud, so that they illuminate the inside of the nest. Lays two to five eggs, breeding from spring through to autumn, depending on timing of monsoon. Incubation 14–15 days, fledging 13–16 days. Female alone incubates, male may assist in feeding young. Nests robbed by snakes and crows.

CONSERVATION STATUS
Not threatened; locally common to abundant, and regarded as a pest in grain-growing areas.

SIGNIFICANCE TO HUMANS
At times damages crops, so that farmers often systematically destroy nests at egg or nestling stage. ◆

Red-billed quelea
Quelea quelea

SUBFAMILY
Ploceinae

TAXONOMY
Emberiza quelea Linnaeus, 1758, 'India' = Senegal.

OTHER COMMON NAMES
English: Red-billed dioch; French: Travailleur à bec rouge; German: Blutschnabelweber; Spanish: Quelea de Pico Rojo.

PHYSICAL CHARACTERISTICS
4.7 in (12 cm); female 0.5–0.9 oz (15–25 g), male 0.6–0.9 oz (16–26 g). Breeding male has face mask, either black or white, with pink or yellowish border; upperparts light brown with dark central streaks, underparts whitish. Bill red, legs pink. Non-breeding male and female lack face mask, gray-brown, streaked upperparts and whitish underparts. Bill red; yellow in breeding females. Juvenile like female.

DISTRIBUTION
Throughout unforested sub-Saharan Africa.

HABITAT
Open grassland and savanna.

BEHAVIOR
Highly gregarious, flocks sometimes numbering millions. Movements highly synchronized in flocks. Huge roosts may break tree branches. Migratory with clear seasonal patterns in some regions.

FEEDING ECOLOGY AND DIET
Primarily small seeds about 0.1 in (2 mm) in diameter, also insects. Drinks regularly, even in arid regions; flocks may sweep over water, drinking on the wing.

REPRODUCTIVE BIOLOGY
Colonial, monogamous. Breeding activities in colony closely synchronized; eggs and chicks may be abandoned when flock moves on. May breed several times in same season, depending on local food supply. Nest built by male, a thin-walled ball with large side entrance. Lays one to five eggs. Incubation 10–12 days, fledging 11–13 days. Both sexes incubate and feed young. Vast colonies with 500 nests per tree attract hundreds of predators, including eagles, vultures, storks, and carnivorous mammals.

CONSERVATION STATUS
Not threatened; considered one of the most abundant bird species. Population can tolerate huge losses, and control efforts have had no noticeable effect on numbers.

SIGNIFICANCE TO HUMANS
Queleas are the major animal pest of cereal crops in Africa, and international programs coordinated by the U.N. Food and Agriculture Organization began in the 1960s. In 1989 losses caused by this bird were estimated at $22 million per annum. However, many other factors contribute to crop losses in Africa. Current research focuses on management rather than attempts to eliminate queleas or reduce their overall numbers. In parts of West Africa, traditional hunters net queleas to pluck, dry, and sell in village markets. ◆

Madagascar fody
Foudia madagascariensis

SUBFAMILY
Ploceinae

TAXONOMY
Loxia madagascariensis Linnaeus, 1766, Madagascar.

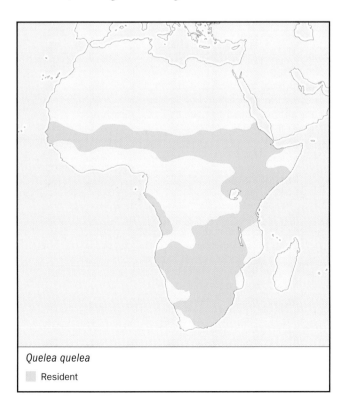

Quelea quelea
▨ Resident

Foudia madagascariensis
▨ Resident

OTHER COMMON NAMES
English: Red fody, Madagascar weaver; French: Foudi rouge,
Foudi de Madagascar; German: Madagaskarweber; Spanish:
Fodi del Madagascar.

PHYSICAL CHARACTERISTICS
4.7–5.1 in (12–13 cm); both sexes 0.5–0.7 oz (14–19 g). Breeding male bright red, black line through eye, olive-brown wings
and tail. Bill black. Female and non-breeding male olive-brown
upperparts, gray-brown underparts. Bill horn-brown. Juvenile
like female but more buffy in appearance.

DISTRIBUTION
Madagascar; introduced to Amirantes, Comoros, Seychelles,
Mauritius, Réunion, St. Helena.

HABITAT
Open savanna, grassland, forest clearings, and cultivated areas;
avoids intact forest.

BEHAVIOR
Gregarious, foraging in flocks and roosting communally in
sugarcane, bamboos, or trees. Solitary and territorial during
breeding season.

FEEDING ECOLOGY AND DIET
Primarily a seed-eater, but also forages for insects in trees, and
takes nectar.

REPRODUCTIVE BIOLOGY
Solitary nests, monogamous. Nest is oval, upright with side entrance near top; woven by male, but female participates from
early stages. Lays two to four eggs, breeding season from
spring through summer to autumn. Female alone incubates,
both sexes feed young. Incubation 11–14 days, fledging 15–16
days.

CONSERVATION STATUS
Not threatened; thrives in human-modified habitats and has
been introduced successfully to other regions.

SIGNIFICANCE TO HUMANS
An important pest in rice fields in Madagascar. Villagers use
traditional cage traps, attempt to scare the birds away from the
crops, and destroy nests. ◆

Southern red bishop
Euplectes orix

SUBFAMILY
Ploceinae

TAXONOMY
Loxia orix Linnaeus, 1758, Angola.

OTHER COMMON NAMES
English: Red bishop-bird, Grenadier weaver; French: Euplecte
ignicolore; German: Oryxweber; Spanish: Obispo Rojo.

PHYSICAL CHARACTERISTICS
5.1 in (13 cm); female 0.6–0.9 oz (17–26 g), male 0.7–1.0 oz
(21–30 g). Breeding male has red and black plumage, with
brown wings and tail. Bill black. Female and non-breeding
male sparrowy brown, pale underparts with some streaking.
Bill brown. Juvenile like female, buffy edges to feathers before

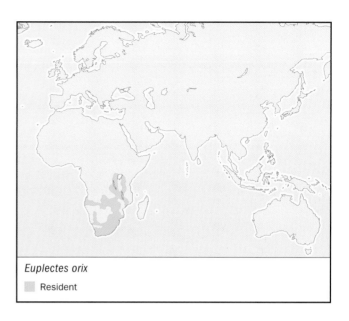

Euplectes orix

☐ Resident

first molt. Males first molt into breeding plumage in second
year.

DISTRIBUTION
Southern Kenya and Uganda south to southern tip of South
Africa.

HABITAT
Tall grassland and cultivation, usually near water.

BEHAVIOR
Gregarious throughout the year, forming large flocks which
feed and roost in association with other seed-eaters. Males return to same breeding localities, often to same territory, in
successive years; return rate of females much lower. Both sexes
may live more than 10 years in the wild.

FEEDING ECOLOGY AND DIET
Mainly seed-eating, also takes many insects, particularly when
feeding young.

REPRODUCTIVE BIOLOGY
Colonial and polygynous, often hundreds of males holding territories in a single reed-bed. Male builds a series of nests, oval
upright structures with side entrances, typically supported by
vertical vegetation. Female lines nest once mated and ready to
lay. Old nests not demolished. In courtship male fluffs out
plumage, resembling black and red bumble-bee, making short
flights towards female. May have up to seven females on territory simultaneously. Breeding season dependent on rainfall; in
winter rainfall region of South Africa, starts in winter, ends in
early summer. Elsewhere during rainy season, usually summer.
Lays one to five eggs (generally three). Incubation 12–13 days,
fledging 11–15 days. Female alone incubates and feeds young.
Nests often subject to heavy predation, and parasitized by
Diederik cuckoo.

CONSERVATION STATUS
Not threatened; abundant in many areas, benefits from farming
activities and building of dams, which provide additional
breeding sites.

SIGNIFICANCE TO HUMANS
Locally an important pest of grain crops; in wheatlands of Western Cape, South Africa, large numbers are killed annually. ◆

Red-collared widow-bird
Euplectes ardens

SUBFAMILY
Ploceinae

TAXONOMY
Fringilla ardens Boddaert, 1783, Cape of Good Hope.

OTHER COMMON NAMES
English: Red-collared widow; French: Veuve noire; German: Schildwida; Spanish: Obispo de Collar Rojo.

PHYSICAL CHARACTERISTICS
5.1 in (13 cm), with long tail 9.8–11.8 in (25–30 cm); female 0.6–0.7 oz (16–21 g), male 0.7–0.9 oz (20–25 g). Breeding male black with long tail, red collar on upper breast, or red on head and breast; some populations wholly black. Bill black. Female brown with dark streaking above, yellowish eyebrow, underparts buffy and unstreaked; bill brown. Non-breeding adult male like female, but retains black wing feathers. Juvenile with feathers of upperparts broadly edged buff.

DISTRIBUTION
Guinea east to Ethiopia, south to Angola and through Zambia to eastern Zimbabwe and eastern South Africa.

HABITAT
Open or bushed grassland, cultivated areas; also highland grasslands from 4,900–9,850 ft (1,500–3,000 m).

BEHAVIOR
Gregarious, forming large roosts even during breeding season, feeding in flocks of 200 birds or more. Often associated with other *Euplectes*.

FEEDING ECOLOGY AND DIET
Takes mainly seeds and insects, which may be hawked in the air; rarely berries, nectar.

REPRODUCTIVE BIOLOGY
Territorial, polygynous, with males well-dispersed. Nest in tall grass, a woven ball with side entrance. Frame started by male, most building done by female. Breeding follows spring or summer rains. Lays two to three eggs; incubation 12–15 days, fledging 14–17 days. Female alone incubates and feeds young.

CONSERVATION STATUS
Not threatened; widespread in lowlands, but distinctive montane populations have restricted range.

SIGNIFICANCE TO HUMANS
None known. ◆

Jackson's widow-bird
Euplectes jacksoni

SUBFAMILY
Ploceinae

TAXONOMY
Drepanoplectes jacksoni Sharpe, 1891, Kikuyu, Kenya.

OTHER COMMON NAMES
English: Jackson's widow, Jackson's dancing whydah; French: Euplecte de Jackson; German: Leierschwanzwida; Spanish: Obispo de Jackson.

PHYSICAL CHARACTERISTICS
5.5 in (14 cm) with long tail 11.4 in (29 cm); female 1.0–1.5 oz (29–42 g), male 1.4–1.7 oz (40–49 g). Breeding male black with

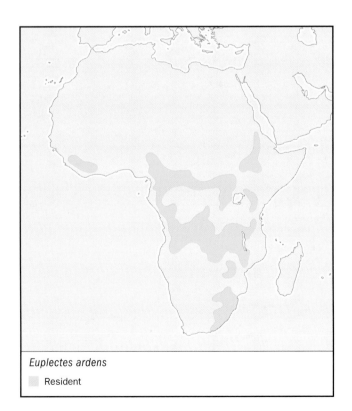

Euplectes ardens
☐ Resident

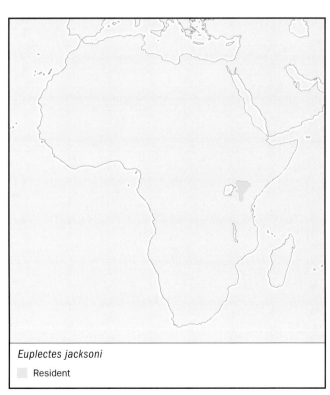

Euplectes jacksoni
☐ Resident

brown wings, curved black tail. Bill steel blue. Female and sub-adult birds, upperparts dark brown with paler edges to feathers, underparts buffy. Bill brown. Non-breeding adult male like female, but bill dark bluish.

DISTRIBUTION
Central Kenya and northern Tanzania.

HABITAT
Highland grasslands, above 4,900 ft (1,500 m).

BEHAVIOR
Gregarious, in flocks when feeding, gathering in communal roosts at night. Breeding areas traditional, and same sites used in successive years.

FEEDING ECOLOGY AND DIET
Mainly grass seeds, also some insects such as winged termites.

REPRODUCTIVE BIOLOGY
Polygynous, with a lek mating system. Male displays at dancing ring, a circle of flattened grass about 24 in (60 cm) in diameter, surrounding a central tuft. Facing tuft, male jumps to various heights, up to 1 yd (1 m) into the air while calling. If female lands in ring, courtship and mating may follow. Female builds nest, a ball of woven grass with side entrance, close to the ground in a grass tuft. Nesting area usually about 330 yd (300 m) from lek. Lays two to four eggs, usually after main rains. Incubation 12–13 days, fledging 17 days. Female alone incubates, and feeds young on regurgitated grass seeds, not insects.

CONSERVATION STATUS
Vulnerable because of limited range, dependence on grasslands which are being altered by agricultural activity.

SIGNIFICANCE TO HUMANS
Flocks may damage grain crops of subsistence farmers. ◆

Thick-billed weaver
Amblyospiza albifrons

SUBFAMILY
Ploceinae

TAXONOMY
Pyrrhula albifrons Vigors, 1831, Algoa Bay, Eastern Cape, South Africa.

OTHER COMMON NAMES
English: Grosbeak weaver; French: Grosbec à front blanc; German: Weißstirnweber; Spanish: Tejedor de Pico Grueso.

PHYSICAL CHARACTERISTICS
6.7–7.5 in (17–19 cm); female 1.1–1.6 oz (31–45 g), male 1.5–2.1 oz (43–60 g). Breeding male chocolate brown with white forehead, white patch in wing conspicuous in flight. Heavy black bill. White on forehead variable, absent in non-breeding plumage. Female has brown upperparts, underparts white heavily streaked with brown. Heavy yellowish bill. Juvenile like female, more rufous above and buffy below. Bill dull brown.

DISTRIBUTION
Sierra Leone east to southern Sudan, western Ethiopia, south to northern Namibia, northern Botswana, eastern Zimbabwe, and eastern South Africa.

HABITAT
Reedbeds, cultivated areas, plantation, and forest.

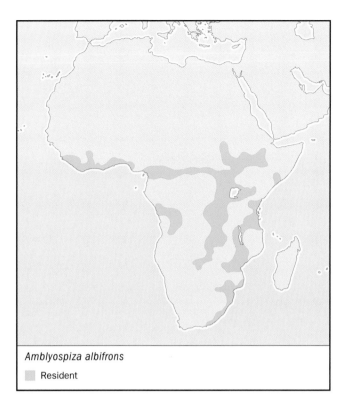

Amblyospiza albifrons
▨ Resident

BEHAVIOR
Gregarious, roosting in reedbeds and breeding there. Flocks move daily up to 19 mi (30 km) to forage when not breeding.

FEEDING ECOLOGY AND DIET
Fruit, seeds, including large, hard-shelled seeds, and insects. In non-breeding season often forages in forest, feeding on fruit in canopy, also on the ground.

REPRODUCTIVE BIOLOGY
Colonial, some males polygynous. Nest highly distinctive, woven of very fine reed strips, slung between upright stems, with large side entrance. Built by male; once female accepts nest, male reduces entrance to a narrow circular hole. Female lines nest. Lays two to five eggs in summer. Incubation 14–16 days, fledging 19–22 days. Female alone incubates, feeds young.

CONSERVATION STATUS
Not threatened; widespread and range expanding in some areas such as Zimbabwe and South Africa.

SIGNIFICANCE TO HUMANS
None known. ◆

Red-headed weaver
Anaplectes rubriceps

SUBFAMILY
Ploceinae

TAXONOMY
Ploceus (Hyphantornis) rubriceps Sundevall, 1850, Upper Caffraria, near the Tropic = Mohapoani, Rustenburg district, South Africa.

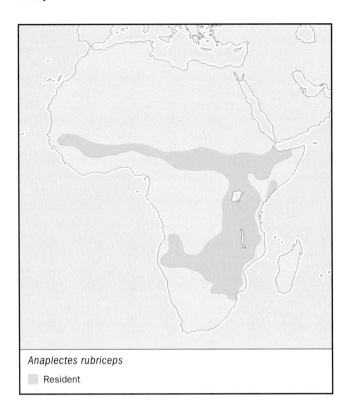

Anaplectes rubriceps
▪ Resident

OTHER COMMON NAMES
French: Tisserin écarlate; German: Scharlachweber; Spanish: Tejedor de Cabeza Roja.

PHYSICAL CHARACTERISTICS
5.9–6.7 in (15–17 cm); female 0.7 oz (20 g); male 0.9 oz (25 g). Breeding male has head, breast, and upper back scarlet, upperparts otherwise gray, underparts white. Wing edged yellow or red in different population; some also have black patch around eye. Bill red, eye red. Female and non-breeding male, scarlet replaced by dull orange on upperparts, below yellow to buff. Bill pink to orange, eye brown. Juvenile like female.

DISTRIBUTION
Senegal east to Somalia, south to Namibia, Botswana, and northeastern South Africa.

HABITAT
Woodland and acacia savanna, gardens.

BEHAVIOR
Solitary or in pairs, joins mixed-species flocks. Local movements during dry season. May stay near, and roost in, old nests.

FEEDING ECOLOGY AND DIET
Mainly insects, spiders, some fruit and seeds. Forages in foliage, on branches and creepers; hawks flying insects.

REPRODUCTIVE BIOLOGY
Often solitary, monogamous or polygynous; sometimes several males in same tree. Nest built by male, at same site in successive seasons, so that several nests may be close together. Retort-shaped structure, woven from twigs and mid-ribs of leaves with rough appearance; long vertical entrance spout. Suspended from tree, often one in which raptor is nesting. Female lines nest. Lays one to four eggs from late winter through

into summer. Incubation 11–13 days, fledging 17 days. Both sexes incubate and feed young, but female contributes more.

CONSERVATION STATUS
Not threatened; extensive range, and well-represented in conservation areas.

SIGNIFICANCE TO HUMANS
None known. ◆

Pin-tailed whydah
Vidua macroura

SUBFAMILY
Viduinae

TAXONOMY
Fringilla macroura Pallas, 1764, 'East Indies' = Angola.

OTHER COMMON NAMES
English: King-of-six; French: Veuve dominicaine; German: Dominikanerwitwe; Spanish: Viuda de Cola Aguda.

PHYSICAL CHARACTERISTICS
4.7–5.1 in (12–13 cm), male with long tail 10.2–13.4 in (26–34 cm); female 0.5–0.6 oz (14–16 g), male 0.5–0.7 oz (14–19 g). Female and non-breeding male, brownish upperparts with broad black stripes on top of head, buff to white underparts. Bill brownish red. Breeding male, black and white with four long, black central tail feathers. Bill bright red. Juvenile plain brown above, buff below.

DISTRIBUTION
Sub-Saharan Africa. Introduced to Hawaii but apparently now extinct.

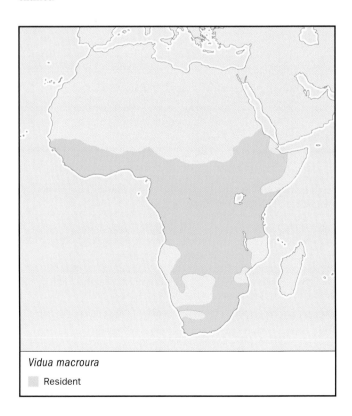

Vidua macroura
▪ Resident

HABITAT
Open savanna and grassland, farmland, gardens.

BEHAVIOR
Male sings from perch, but does not imitate songs of host species. When female arrives, bounces in the air with tail flipping up and down while singing. Aggressive towards other males, but also to other species. Non-breeding birds gregarious, forming small flocks of 20–30 birds, often mixed with other small seed-eaters.

FEEDING ECOLOGY AND DIET
Mainly seeds, also some insects. Collects most food on the ground. Scratches with backward hops to unearth buried seeds.

REPRODUCTIVE BIOLOGY
Brood parasite, polygynous. Lays one to two eggs per nest, removing host egg for each egg added. Incubation about 11 days, fledging about 20 days. Most frequent host is common waxbill (*Estrilda astrild*), also other waxbills and occasionally warblers. Host and parasite young reared together.

CONSERVATION STATUS
Not threatened; widespread and common.

SIGNIFICANCE TO HUMANS
None known; annoys those who put out birdseed, as breeding male pin-tailed whydah will attempt to drive all other birds away from feeding site. ◆

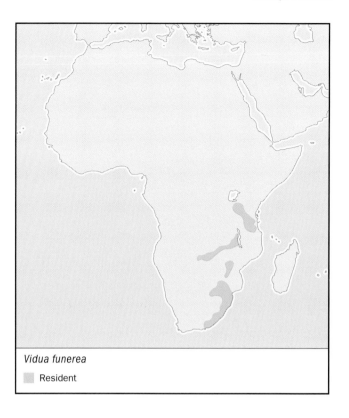

Vidua funerea

Resident

Dusky indigobird
Vidua funerea

SUBFAMILY
Viduinae

TAXONOMY
Fringilla funerea de Tarragon, 1847, Natal.

OTHER COMMON NAMES
English: Black widow finch, variable indigobird; French: Combassou variable; German: Purpuratlaswitwe; Spanish: Viuda Variable.

PHYSICAL CHARACTERISTICS
4.3–4.7 in (11–12 cm), female 0.4–0.5 oz (12–16 g), male 0.5–0.6 oz (14–17 g). Breeding male black, with whitish bill and red legs. Female and non-breeding male buff upperparts with broad black stripes on head, whitish underparts. Juvenile like female but head dark, unstriped.

DISTRIBUTION
Tanzania to eastern South Africa.

HABITAT
Grassy areas including fringes of cultivation, gardens, and roadsides.

BEHAVIOR
Male sings from exposed perch for long periods, including song and calls of African firefinch (*Lagonosticta rubricata*) interspersed with chirping notes. Displays to female in a bobbing flight. In non-breeding season birds associate in flocks, join mixed flocks of other small seed-eaters, and become nomadic.

FEEDING ECOLOGY AND DIET
Seeds, mainly collected on the ground, including buried seeds dug out by scratching backwards with both feet.

REPRODUCTIVE BIOLOGY
Brood parasite, polygynous, with males at display sites which females visit for mating. Lays one egg per host nest. Incubation and fledging periods apparently unrecorded. Host is African firefinch, and young of parasite specifically match mouth markings of this species; male indigobirds learn vocalizations of host while in nest, and later incorporate these elements into their own song.

CONSERVATION STATUS
Not threatened. Locally common, with an extensive range in Africa.

SIGNIFICANCE TO HUMANS
None known. ◆

Cuckoo finch
Anomalospiza imberbis

SUBFAMILY
Viduinae

TAXONOMY
Crithagra imberbis Cabanis, 1868, East Africa = Zanzibar.

OTHER COMMON NAMES
English: Parasitic weaver; French: Anomalospize parasite; German: Kuckucksfink; Spanish: Tejedor Parásito.

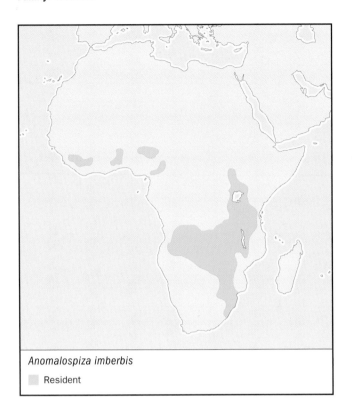

Anomalospiza imberbis

   Resident

PHYSICAL CHARACTERISTICS
5.1 in (13 cm); unsexed birds 0.8–0.9 oz (23–26 g). Breeding male, yellow with some streaking, bill black. Non-breeding male, yellowish head, upperparts olive with heavy streaks, bill brown. Female mainly buffy, heavily streaked on upperparts. Short-tailed, with a stubby bill, deep at base. Juvenile resembles female.

DISTRIBUTION
Local in western and central Africa, through eastern Africa to southern Africa.

HABITAT
Open grassland with scattered trees, wetlands, cultivated lands.

BEHAVIOR
Little-known and probably nomadic; likely to be overlooked in mixed flocks of seedeaters. When breeding, in pairs or small groups. Male has rasping song in display, defends grassland territory. Non-breeding birds form large roosts in reedbeds, sometimes holding more than 500 birds.

FEEDING ECOLOGY AND DIET
Seeds, mostly collected while perching on grasses and weeds.

REPRODUCTIVE BIOLOGY
Brood parasite, mating system not known, but probably polygynous. Lays one to two eggs per nest, removing one or more host eggs. Incubation 14 days, fledging 18 days. Hosts are warblers of the genera *Cisticola* or *Prinia*, host young usually trampled in nest, rarely reared with parasite young. Two cuckoo finch young may be reared together.

CONSERVATION STATUS
Wide range and not considered threatened.

SIGNIFICANCE TO HUMANS
None known. ◆

Resources

Books

Ali, Salim, and S. Dillon Ripley. *Handbook of the Birds of India and Pakistan, Together with Those of Bangladesh, Nepal, Bhutan and Sri Lanka.* Vol. 10, *Flowerpeckers to Buntings.* Delhi: Oxford University Press, 1987.

Bruggers, Richard D., Clive C. H. Elliott. *Quelea quelea Africa's Bird Pest.* Oxford: Oxford University Press, 1999.

Craig, Adrian J. F. K. "Weaving A Story: The Relationships of the Endemic Ploceidae of Madagascar." In *Proceedings of the 22nd International Ornithological Congress,* edited by Nigel J. Adams and Robert H. Slotow. Johannesburg: BirdLife South Africa, 1984: 3063–3070.

Fishpool, Lincoln D. C., and Michael I. Evans. *Important Bird Areas for Africa and Associated Islands: Priority Sites for Conservation.* Newbury and Cambridge, United Kingdom: Pisces Publications and BirdLife International, 2001.

Fry, C. Hilary, Stuart Keith, and Emil K. Urban. *The Birds of Africa.* Vol. VII. London: Academic Press, in press.

Goodman, Steven M., and John P. Benstead. *The Natural History of Madagascar.* Chicago: University of Chicago Press, in press.

Sibley, Charles G., and Jon E. Ahlquist. *Phylogeny and Classification of Birds.* New Haven and London: Yale University Press, 1990.

Stattersfield, Alison J., and David R. Capper. *Threatened Birds of the World: The Official Source for Birds on the IUCN Red List.* Barcelona and Cambridge: BirdLife International/Lynx Edicions, 2000.

Periodicals

Andersson, Staffan. "Bowers on the Savanna: Display Courts and Mate Choice in a Lekking Widowbird." *Behavioral Ecology* 2 (1991): 210–218.

Barnard, Phoebe. "Territoriality and the Determinants of Male Mating Success in Southern African Whydahs (*Vidua*)." *Ostrich* 60 (1989): 103–117.

Brosset, Andre. "Social Organization and Nest Building in the Forest Weaver Birds of the Genus *Malimbus* (Ploceinae)." *Ibis* 120 (1987): 27–37.

Collias, Nicholas E., and Elsie C. Collias. "Evolution of Nest-Building Behavior in the Weaverbirds (Ploceidae)." *University of California Publications in Zoology* 73 (1964): 1–162.

Crook, John H. "The Evolution of Social Organisation and Visual Communication in the Weaverbirds (Ploceinae)." *Behaviour Supplement* 10 (1964): 1–178.

Hudgens, Brian R. "Nest Predation Avoidance by the Blue-Billed Malimbe *Malimbus nitens* (Ploceinae)." *Ibis* 139 (1997): 692–694.

Resources

Nicolai, Jürgen. "Der Brutparasitismus der Viduinae als ethologisches Problem." *Zeitschrift für Tierpsychologie* 21 (1964): 129–204.

Payne, Robert B. "Brood Parasitism in Birds: Strangers in the Nest." *Bioscience* 48 (1998): 377–386.

Winterbottom, M., T. Burke, and T. R. Birkhead. "The Phalloid Organ, Orgasm and Sperm Competition in a Polygynandrous Bird: The Red-Billed Buffalo Weaver (*Bubalornis niger*)." *Behavioural Ecology and Sociobiology* 50 (2001): 474–482.

Adrian Craig, PhD

Sparrows
(Passeridae)

Class Aves
Order Passeriformes
Suborder Passeri (Oscines)
Family Passeridae

Thumbnail description
Small, seed-eating songbirds with stout bills

Size
4.5–7 in (12–17.5 cm)

Number of genera, species
Five genera; 39 species

Habitat
Grassland savanna, inhabited areas

Conservation status
No species threatened, no recent extinctions

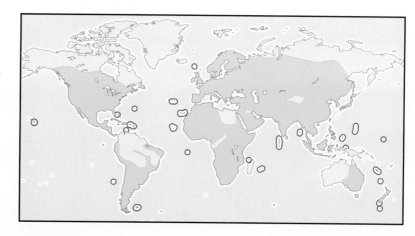

Distribution
Worldwide except for Antarctica, Eurasia north of cultivation, north and west Australia

Evolution and systematics

Sparrows (Passeridae) are seed-eaters, separated by their method of dehusking seeds and their digestive tract. Originally placed with the finches (Fringillidae) and then with the weavers (Ploceidae), sparrows are now recognized as a separate family. It is thought that these birds evolved in the Afrotropical Region during the middle of the Miocene. One group, the snow finches and ground sparrows, probably arose from an early radiation into the Palearctic. The birds in Africa then split into two groups: the rock sparrows and the true sparrows, which subsequently colonized Africa and gave rise to secondary colonizations of Eurasia.

As of 2001, five genera of sparrows are recognized: the snow finches (*Montifringilla*) and the ground sparrows (*Pyrgilauda*) derived from the original Eurasian radiation; the pale rock sparrow (*Carpospiza brachydactyla*) was separated from the rock sparrows *Petronia* and placed in a monotypic genus *Carpospiza*; the remainder are the true sparrows in the genus *Passer* with 26 representatives.

Physical characteristics

The sparrows, with their short, robust bills, have specialized in the seeds of grasses, particularly the cultivated cereals that have been developed from them. Their tongues have a unique skeletal structure that plays a part in dehusking the seeds. These birds are also distinguished by a complete post-juvenile moult. The bills of the males change from horn to black when they become sexually active. Most are comparatively sedentary. The true sparrows and the rock sparrows have short, blunt wings and are not strong fliers, making short, direct flights. The snow finches and the ground sparrows, living in more open country, have proportionately longer wings with varying amounts of white in their plumage

that show prominently in the display flights that are characteristic of open-country birds. The sexes in the snow finches, ground sparrows, and rock sparrows are basically alike, the latter being distinguished by a yellow spot on the throat. In contrast, most of the true sparrows are dimorphic with the males characterized by black bibs and well developed head patterns.

Distribution

From their Afrotropical origins the sparrows now occupy most of Africa and Eurasia. Many species have restricted distributions, but two, the house sparrows (*Passer domesticus*) and tree sparrows (*Passer montanus*), occur widely throughout Eurasia and have increased the range of the family through deliberate introductions from the mid-nineteenth century onwards. The house sparrows have now an almost worldwide distribution, absent only from Antarctica and parts of tropical Africa. The tree sparrows have modest populations in North America and Australia, and are now expanding in the southeast Asian archipelagos and some of the Pacific island groups, partly by introduction, but also by natural spread.

The capacity of the house sparrows for rapid range extension is shown by the way they spread from the Urals to the Pacific coast in the nineteenth and twentieth centuries. Following the building of the trans-Siberian railway, they spread more than 3,000 mi (5,500 km) in a little over 100 years. The rates of spread of this and introduced populations have ranged 9–50 mi per year (15–80 km per year).

Most species are sedentary or disperse nomadically outside the breeding season, though a few have a more defined migration. This particularly applies to high latitude and high altitude populations that withdraw to milder regions in the winter. Pale rock sparrows move south to Arabia and northeast Africa in the winter, and the subspecies of the house sparrow *Passer d. bactrianus* breeds in the Central Asian Republics and Afghanistan and winters in the northern plains of Pakistan. Spanish sparrows (*Passer hispaniolensis*) breed in a Mediterranean-type climate with a short spring flush, migrating to the northeast for successive broods as suitable conditions wax and wane.

Habitat

Sparrows are predominantly birds of open country with scattered trees, extending into the semi-desert and high, arid steppes of Asia (the snow finches and ground sparrows) and light woodland (the African rock sparrows that merit more the epithet bush than rock sparrow). The most significant feature, particularly in the true sparrows, is a close association with humans. Originally through feeding in cultivated areas and then moving into built-up areas, no fewer than 17 of the 26 species of true sparrows nest in holes in occupied buildings. The house and tree sparrows are the supreme examples with some individuals spending most of their time in buildings and the house sparrows even living and breeding in a coal mine in England, 2,100 ft (640 m) below ground level.

Behavior

Most sparrows are gregarious, collecting in large foraging flocks and forming colonial roosts. The majority have a clumped breeding distribution. Colonial nesting is marked in the Spanish sparrows, with nesting colonies of many hundred thousand birds in central Asia. In such colonies, the nests are closely packed together with up to 200 nests in one tree. More usually the nests are more scattered, limited by the availability of suitable sites, but 65,000 nests of golden sparrows (*Passer luteus*) have been estimated in a loose colony spread over 1,500 acres (630 ha). More typically the colonies consist of 20–30 pairs.

The sparrows are one of the few passerines that indulge in both dusting and bathing in water. Both are social activities. The foraging flocks alternate bursts of seed collection with resting spells in good cover. While presumably digesting the hard seeds, the birds remain close together and maintain social contact with soft, conversational calls.

Feeding ecology and diet

The sparrows are essentially granivorous, feeding on the seeds of small plants. In many cases this has led to specialization on the seeds of cultivated cereals and from this to food put out for domestic animals and household scraps. Small berries and tree seeds are also taken. In all cases the young are reared largely on animal food. During the breeding season the adults also take a proportion of invertebrate food, consisting mainly of slow-moving insects, though some are also caught in flight.

Reproductive biology

Pair formation normally takes place by the male calling and displaying at a nest site. The territory is effectively limited to the immediate vicinity of the nest. The open-country snow finches and ground sparrows have more developed songs and display flights that are given near the nest. Nests are domed over. Clutches are typically four to six eggs, the majority of species being multi-brooded. Both sexes take part in breeding activities. The young are born with natal down; flight feathers develop rapidly and the young are fully fledged in 12–20 days. In the sedentary species, like the house and tree sparrows, the birds maintain the nest throughout the year and pair for life. Breeding takes place in the spring and summer in temperate regions, following the end of the rains in the tropics. The birds are nominally monogamous, but DNA studies show that colonial breeding leads to a considerable degree of promiscuity with up to 15% of the chicks not fathered by the attendant male.

Conservation status

Although there has been some retraction in range of rock sparrows (*Petronia petronia*) in Europe and house sparrows are suffering a major decline in western Europe, no species is under serious threat.

Sparrows nesting in a built-up habitat. (Illustration by Amanda Humphrey)

Significance to humans

Those species specializing on the seeds of cultivated grains may reach pest status, but otherwise the birds have little impact on man. Some sparrows take over the nest and nest sites of other species and may affect their numbers, particularly when suitable sites are limited.

1. Rock sparrow (*Petronia petronia*); 2. House sparrow (*Passer domesticus*); 3. Golden sparrow (*Passer luteus*); 4. Tree sparrow (*Passer montanus*); 5. Père David's ground sparrow (*Pyrgilauda davidiana*); 6. Snow finch (*Montifringilla nivalis*); 7. Pale rock sparrow (*Carpospiza brachydactyla*); 8. Southern rufous sparrow (*Passer motitensis*). (Illustration by Amanda Humphrey)

Species accounts

House sparrow

Passer domesticus

TAXONOMY

Fringilla domestica Linnaeus, 1758, Sweden. 11 subspecies.

OTHER COMMON NAMES

English: English sparrow; French: Moineau domestique; German: Haussperling; Spanish: Gorrión Común.

PHYSICAL CHARACTERISTICS

5.5–6.3 in (14–16 cm); 0.7–1.4 oz (20–40 g). Male has a gray crown bordered by chestnut and a small black bib. Female drab brown. Juvenile similar to female but paler.

DISTRIBUTION

North Africa and Eurasia to limit of cultivation, except for Thailand east to Japan. Through introductions from the mid-nineteenth century onwards is now present throughout most of the inhabited world.

HABITAT

Almost entirely associated with humans.

BEHAVIOR

Mainly sedentary, living in small colonies throughout the year.

FEEDING ECOLOGY AND DIET

Seeds and household scraps. Young reared largely on invertebrates.

REPRODUCTIVE BIOLOGY

Preferred nest site is a hole in building or tree, though also builds free-standing domed nest in trees. Up to five clutches of two to five eggs per year. Incubation 11–14 days; fledging 14–16 days. Both sexes take part in breeding activities.

CONSERVATION STATUS

Major decline in western Europe at end of the twentieth century; but not considered threatened by the IUCN as of 2000.

SIGNIFICANCE TO HUMANS

Generally regarded with affection, but can be a pest of cereal cultivation. ◆

Tree sparrow

Passer montanus

TAXONOMY

Fringilla montana Linnaeus, 1758, Southern Italy. Seven subspecies.

OTHER COMMON NAMES

English: Eurasian tree sparrow; French: Moineau friquet; German: Feldsperling; Spanish: Gorrión Molinero.

Passer domesticus
Resident

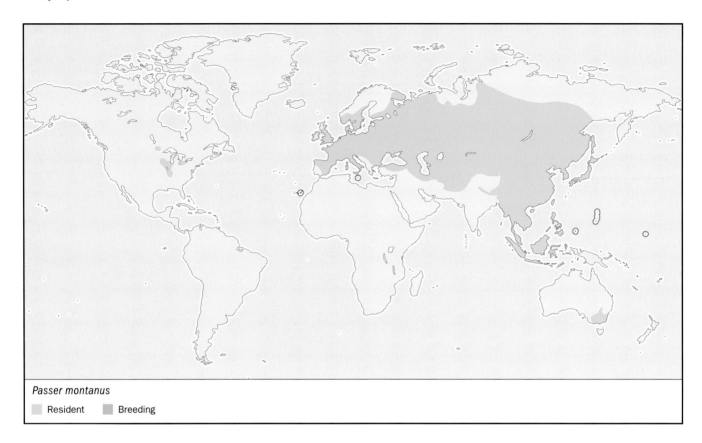

Passer montanus

◻ Resident ◼ Breeding

PHYSICAL CHARACTERISTICS
5.5–6 in (14–15 cm); 0.6–1.0 oz (17–28 g). Small sparrow with particularly neat appearance. Sexes alike. Head chocolate, cheeks white with a prominent black patch, and small black bib. Juvenile paler.

DISTRIBUTION
Widely distributed in Europe and Asia; small introduced populations in North America and Australia.

HABITAT
Built-up areas in the east, more in open country with trees in the west.

BEHAVIOR
A gregarious, familiar bird in the east, becoming shyer and more unobtrusive in the west.

FEEDING ECOLOGY AND DIET
Seeds and a wide range of invertebrates during the breeding season.

REPRODUCTIVE BIOLOGY
Preferred nest site is a hole in building or tree, though also builds free-standing domed nest in trees. Up to five clutches of two to five eggs. Incubation 11–14 days; fledging 14–16 days. Both sexes take part in breeding activities.

CONSERVATION STATUS
Not threatened. Widely distributed and numerous.

SIGNIFICANCE TO HUMANS
Can be a pest in rice paddies. ◆

Golden sparrow
Passer luteus

TAXONOMY
Fringilla lutea Lichtenstein, 1823, Dongola, north Sudan. Formerly placed in genus *Auripasser*.

OTHER COMMON NAMES
French: Moineau doré; German: Gelbsperling; Spanish: Gorrión Aureo.

PHYSICAL CHARACTERISTICS
4.7–5.1 in (12–13 cm); 0.4–0.6 oz (11–17 g). Male golden-yellow apart from a chestnut back and some black on the wings. Female is sandy brown, juvenile paler.

DISTRIBUTION
A narrow zone south of the Sahara from Mauritania to the Red Sea.

HABITAT
Arid sandy areas with a low density of trees.

BEHAVIOR
A gregarious, nomadic bird, forming roosts of up to half a million individuals.

FEEDING ECOLOGY AND DIET
Seeds and a small amount of insects. Nestlings mainly reared on invertebrates.

REPRODUCTIVE BIOLOGY
Breeding occurs when there is a flush of insects following rains. Forms dispersed colonies of up to 50,000 nests. Clutch

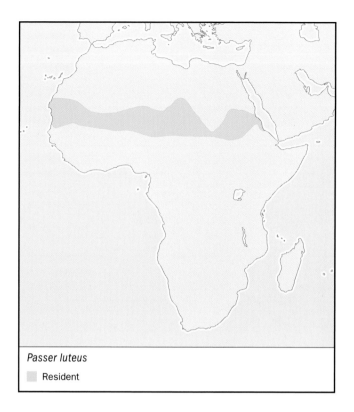

Passer luteus
☐ Resident

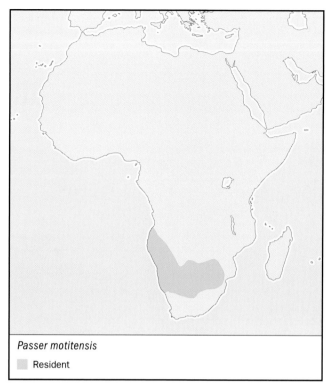

Passer motitensis
☐ Resident

two to three eggs; incubation 10–12 days, fledging 13–14 days. Second brood may be raised in a new location where conditions have become suitable.

CONSERVATION STATUS
Not threatened. A common species.

SIGNIFICANCE TO HUMANS
Can be a pest where large flocks are attracted to ripening cereals. ◆

Southern rufous sparrow
Passer motitensis

TAXONOMY
Pyrgita motitensis A. Smith, 1836, northern Cape Province, South Africa.

OTHER COMMON NAMES
English: Great sparrow; French: Grand moineau; German: Rostsperling; Spanish: Gorrión Grande.

PHYSICAL CHARACTERISTICS
5.9–6.3 in (15–16 cm); 1.0–1.3 oz (28–37 g). Large, robust sparrow. Male has blue-gray crown and a chestnut band circling the rear of the pale cheek. Female is similar, but less well marked. Juvenile paler.

DISTRIBUTION
Widely, but sparsely, distributed in southern Africa.

HABITAT
Dry acacia savanna.

BEHAVIOR
A solitary species.

FEEDING ECOLOGY AND DIET
Mainly grass seeds. Nestlings are reared on insects.

REPRODUCTIVE BIOLOGY
Builds domed nest in thorny tree. Two clutches of three to six eggs. Both sexes take part in breeding activities.

CONSERVATION STATUS
Not threatened.

SIGNIFICANCE TO HUMANS
None known. ◆

Pale rock sparrow
Carpospiza brachydactyla

TAXONOMY
Petronia brachydactyla Bonaparte, 1850, Kunfuda, western Arabia. From its behavior and form of nest has been considered to be a carduline finch, but the horny palate and digestive tract confirm it belongs to Passeridae.

OTHER COMMON NAMES
French: Moineau soulcie pâle; German: Arabian Steinsperling; Spanish: Gorrión Palida.

PHYSICAL CHARACTERISTICS
5.3–5.7 in (13.5–14.5 cm); 1.0–1.4 oz (28–40 g). Generally rather featureless pale brown bird with a short, stout bill. Long, triangular-shaped wings recall lark in flight. Sexes are similar, but juvenile paler, more sandy-colored.

DISTRIBUTION
Southwest Asia, withdrawing south in winter to Arabia and northeast Africa.

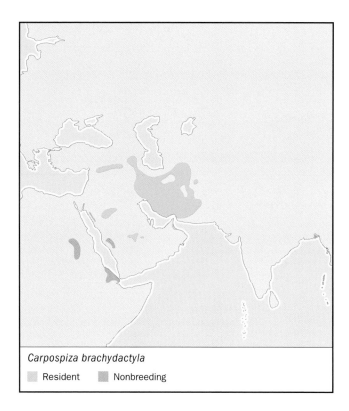

Carpospiza brachydactyla
■ Resident ■ Nonbreeding

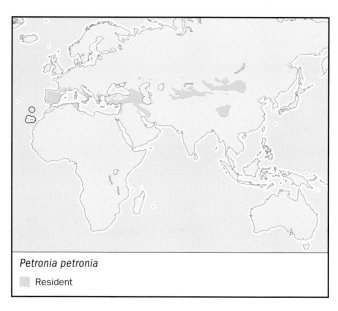

Petronia petronia
■ Resident

HABITAT
Sparsely vegetated regions up to 9,850 ft (3,000 m).

BEHAVIOR
Spends much of the time on the ground, forming flocks of up to several hundreds outside the breeding season that collect near water or fields of ripening grain.

FEEDING ECOLOGY AND DIET
Mainly seeds of grasses, including cultivated cereals, but takes some insects during breeding season. Nestlings are mainly reared on insects.

REPRODUCTIVE BIOLOGY
Nest is open, untidy cup of twigs lined with softer material in bush or tree close to ground. Clutch is four to five eggs, incubation 13–14 days, fledging 11–16 days. Incubation by female only, but both sexes feed the young.

CONSERVATION STATUS
Not threatened. Locally common.

SIGNIFICANCE TO HUMANS
None known. ◆

Rock sparrow
Petronia petronia

TAXONOMY
Petronia petronia Linnaeus 1766, northern Italy. 10 subspecies.

OTHER COMMON NAMES
French: Moineau soulcie; German: Steinsperling; Spanish: Gorrión Pintado.

PHYSICAL CHARACTERISTICS
5.5–6.0 in (14–15 cm); 1.1–1.4 oz (28–40 g). Bulky, grayish brown sparrow, heavily streaked and spotted, head boldly striped. Bill noticeably heavier than those of true sparrows.

DISTRIBUTION
Has a wide but scattered range from the Canary Islands and Madeira, through the Mediterranean region east to western China.

HABITAT
Generally a bird of open, treeless country from semi-desert to rocky slopes up to 17,400 ft (5,300 m). Comes into cultivated areas, particularly large open fields and even small human settlements.

BEHAVIOR
Outside the breeding season collects in compact flocks of up to a few thousand birds, often with finches. Rather wary, flying off when disturbed. Spends much time on the ground where it runs about actively.

FEEDING ECOLOGY AND DIET
Mainly feeds on the seeds of grasses and low herbs, but also on a wide range of invertebrates in the spring and berries in the autumn. Young are mainly fed on insects and insect larvae, taking somewhat larger prey than the true sparrows.

REPRODUCTIVE BIOLOGY
Breeds mostly in loose colonies with the nests in crevices in rocks and trees, also in buildings, occasionally occupied ones. Two clutches of four to seven eggs are laid. Incubation 11–14 days, fledging 18–19 days. Female takes most part, though male occasionally feeds the nestlings.

CONSERVATION STATUS
Not threatened. Some withdrawal from northern parts of range in Europe and Atlantic islands has occurred in twentieth century, but is still a common species.

SIGNIFICANCE TO HUMANS
Though occasionally causes some damage to crops, has little impact on humans. ◆

Snow finch
Montifringilla nivalis

TAXONOMY
Fringilla nivalis Linnaeus 1766, Switzerland. Eight subspecies, reflecting the extensive but discontinuous distribution of a bird that is restricted to high mountain areas.

OTHER COMMON NAMES
English: White-winged snow finch; French: Niverolle des nieges; German: Schneefink; Spanish: Gorrión Alpino.

PHYSICAL CHARACTERISTICS
6.7–6.9 in (17–17.5 cm); 1.0–1.9 oz (28–54 g). A large, plumpish finch-like bird with a blue-gray head and brownish body, showing a lot of white in flight, particularly in display. The sexes are similar, the female paler with less white on the wings.

DISTRIBUTION
A high altitude species that occur on mountains over 6,600 ft (2,000 m), remaining at high altitudes throughout the year.

HABITAT
A bird of barren rocky ground and alps from 6,600–11,500 ft (2,000–3,500 m), frequently occurring near buildings where these occur within its range.

BEHAVIOR
Sociable, forming nomadic groups and even large flocks outside the breeding season. Spends most of its time on the ground where it hops inconspicuously with the wings folded.

FEEDING ECOLOGY AND DIET
Mainly granivorous in the winter, but takes invertebrates at other times. Feeds on scraps around habitations, particularly at ski resorts. The young are fed exclusively on animal food.

REPRODUCTIVE BIOLOGY
Normally in loose colonies of up to five to six pairs. Nests in a rock crevice or hole in building, filling the cavity with grass and moss lined with feathers. Two clutches of three to four

eggs are laid, incubation 13–14 days, fledging 20–21 days. Young are fed by both adults.

CONSERVATION STATUS
A common, even locally abundant bird that has probably benefited through scraps provided at winter resorts.

SIGNIFICANCE TO HUMANS
None known. ◆

Père David's ground sparrow
Pyrgilauda davidiana

SUBFAMILY
none

TAXONOMY
Pyrgilauda davidiana Verreaux, 1871, Suiyuan, Inner Mongolia. Two subspecies occur in disjunct populations.

OTHER COMMON NAMES
English: Père David's snow finch; French: Niverolle du Père David; German: Mongolen Erdsperlin; Spanish: Gorrión de David.

PHYSICAL CHARACTERISTICS
5–6 in (13–15 cm). A small ground sparrow with upperparts mostly fawn-brown and a black bib. The juvenile is dingier and lacks the black bib.

DISTRIBUTION
Separate populations occur in eastern Inner Mongolia and from Mongolia to the Siberian Altai.

Pyrgilauda davidiana
◻ Resident

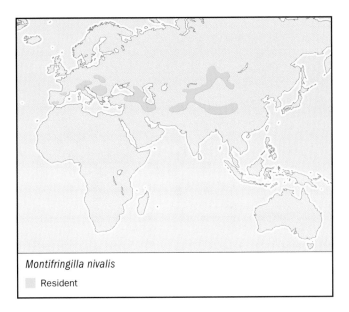

Montifringilla nivalis
◻ Resident

HABITAT
High, sparsely vegetated, semi-desert plains at 6,600–15,750 ft (2,000–4,800 m), usually close to water.

BEHAVIOR
A nomadic species, usually in small flocks. Spends much time hidden in animal burrows when not foraging.

FEEDING ECOLOGY AND DIET
Feeds on grass seeds and insects.

REPRODUCTIVE BIOLOGY
Breeds in burrows of small mammals. Lays five to six eggs. Both sexes feed the young.

CONSERVATION STATUS
Not threatened. Common.

SIGNIFICANCE TO HUMANS
None known. ◆

Resources

Books

Cramp, S., and C. M. Perrins, eds. *The Birds of the Western Palearctic.* Vol. 8, *Crows to Finches.* Oxford: Oxford University Press, 1994.

Ivanitskii, V. V. *The Sparrows and Related Groups of Granivorous Birds: Behaviour, Ecology, Evolution (in Russian).* Moscow: KMK Scientific Press, 1997.

Summers-Smith, J. D. *The Sparrows.* Calton: Poyser, 1989.

Periodicals

Gebauer, A., and M. Kaiser. "Biologie und Verhalten zentral asiatischer Schneefinken (*Montifringilla*) und Erdsperlinge (*Pyrgilauda*)." *Journal für Ornithologie* 135 (1994): 55–71.

Stephan, B. "Die Arten der Familie Passeridae (Gattungen *Montifringilla*, *Petronia*, *Passer*) und ihre phylogenetischen Beziehungen." *Bonner zoologische Beiträge* 49 (2000): 39–70.

J. Denis Summers-Smith, PhD

▲
Starlings and mynas
(Sturnidae)

Class Aves
Order Passeriformes
Suborder Passeri (Oscines)
Family Sturnidae

Thumbnail description
Somewhat stocky, small to medium-sized birds, short-winged, short-tailed, stout-legged, with straight bill. Many are black or dark; some have much white or color, many have iridescent plumage; many have colorful bare facial skin or wattles; some are crested

Size
7–17 in (18–43 cm)

Number of genera, species
25–32 genera, 104–118 species

Habitat
Barren semi-desert to grassland, to dry and moist, evergreen and deciduous forest; agricultural and urban areas

Conservation status
Extinct: 5 species; Critically Endangered: 2 species; Endangered: 2 species; Vulnerable: 5 species; Near Threatened: 8 species; Data Deficient: 1 species

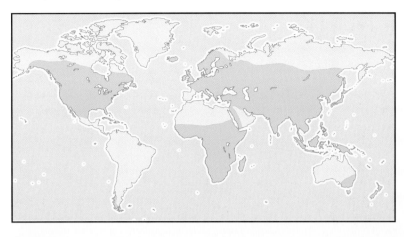

Distribution
Naturally distributed through Eurasia and Africa with one genus reaching Australia; center of diversity is in the Old World tropics. As a result of introductions, there are now breeding sturnids on all continents, except South America and Antarctica and on many oceanic islands

Evolution and systematics

Based on anatomical evidence, the closest living relatives of sturnids have been suggested to be drongos (Dicruridae), Old World orioles (Oriolidae), crows (Corvidae), or mimic thrushes (Mimidae). Based on nest structure, other scientists have suggested affinities with weavers (Ploceidae). Several biochemical analyses, including DNA-DNA hybridization studies, support a close relationship between Old World Sturnidae and New World Mimidae, and Sibley and Monroe (1990) included these two groups as tribes (Sturnini: starlings and mynahs; Mimini: mockingbirds, thrashers, catbirds) within the family Sturnidae.

The family Sturnidae, recognized here as starlings, mynas, and oxpeckers, includes about 27 genera and 111 species divided into two subfamilies, the Sturninae (starlings and mynas: 26 genera, 109 species) and the Buphaginae (oxpeckers: 1 genus, 2 species). The number of genera and species recognized is in a great state of flux as a result of many little-known, closely related, geographically variable forms. Added to the inherent diversity in the group are divergent opinions of scientists as to relationships. New technologies and greater knowledge of the behavioral ecology of starlings promise resolution to many systematic questions.

Physical characteristics

Starlings and mynahs are small to medium-sized birds that vary in length from 7–17 in (18–43 cm). The terms starling

and myna have no significance relative to their relationships with one another, but rather with the common names applied in different regions. The word starling comes from Old English and almost certainly was applied to what is now known as the European starling. The word myna or mynah comes from the Hindi word *maina*, which has its origins in the Sanskrit word *madana* meaning "joyful" or "it bubbles"; this probably refers to the bubbly notes of the hill myna. In practice, the common names of many sturnids have bounced back and forth between being starlings and being mynas. For example, the Bali myna (*Leucopsar rothschildi*) is often called the Bali starling. In general, however, those sturnids that look more like the European starling are referred to as starlings, and those that look a lot like the common hill myna tend to be called mynas.

Most starlings and mynas are stocky with strong legs and a strong, straight bill, a short squared-off tail, and rounded (in resident and forest species) to somewhat long (in migrant and open country species) wings. While sturnid bills are generally straight and often nearly as long as the head, they range from thin and pointed (e.g., European starling) to heavy and somewhat blunt (e.g., white-eyed starling, *Aplonis brunneicapilla*). Mynas often show white wing patches on the primaries. Sturnids often have long, narrow hackle feathers on the neck; those of males are generally most distinctive. Juveniles tend to have darker, duller, sometimes streaked plumage. Starlings have only one molt a year, following breeding, but seasonal differences in appearance are found in some species as a result

A brahminy starling (*Sturnus pagodarum*) pair in India. (Photo by Kim Taylor. Bruce Coleman Inc. Reproduced by permission.)

of wear. The European starling, for example, has white tips to body feathers in fresh plumage, giving a spotted appearance in fall and winter. These tips wear off, leaving the purple-green iridescent black plumage of the breeding season.

Distribution

This family was naturally distributed only in Africa and Eurasia, with a single genus that had reached the South Pacific and northeast Australia. Today, however, the family has breeding populations on every continent, except South America and Antarctica as a result of deliberate introductions and accidental releases of species. The champion among these feathered conquerors is the European starling, which now breeds throughout North America and in Australia, New Zealand, and on many islands.

Habitat

Habitats of sturnids range from barren semi-desert to temperate grasslands, to tropical savanna, to temperate woodlands, to tropical rainforest. The greater numbers of species seem to be associated with forested areas (e.g., *Aplonis* species), tied to them by their need for cavities as nest sites. Those species that are most wide-ranging and most migratory, however, tend to be ground-feeding, grassland species (e.g., *Sturnus* species).

Behavior

Most starlings and mynas are at least somewhat social, often nesting in loose colonies. Some are downright gregarious and aggressive, especially some of the introduced species. This has created serious problems for endemic species that share their foraging or nesting niches, and often for local agriculture that has had to contend with their numbers. Associated with their social nature are voices that are loud and varied, though often raucous, sometimes mechanical, and rarely melodious. Many species are excellent mimics.

A few species such as the brown-winged (*Aplonis grandis*) and Shelley's (*Lamprotornis shelleyi*) starlings are much more solitary in their nesting, but at least the Shelley's starling still gathers in non-breeding flocks. Solitary species tend to be more shy and quiet.

Starlings have exceptional flight abilities. They are swift, yet highly adept at maneuvering, so much so that dense flocks of starlings can twist and turn as one. Temperate-nesting species often migrate to warmer climates for the winter; migratory flights are accomplished at night.

Most tropical species are more or less forest birds, but some, especially in temperate areas, forsake the trees for very open habitats. The only requisite is that there be some substitute for tree cavities as nest sites. Niches in rocky areas, cavities of other birds in dirt banks, nest boxes, and recesses on buildings and bridges often fill that need.

The soft-part colors of sturnids are important in social circumstances and seasonally vary somewhat in intensity, becoming brighter during the breeding season. Examples include the bill color of the European starling, the blue facial color of the Bali myna, and the yellow wattles of several mynas.

Feeding ecology and diet

Several genera (e.g., *Aplonis, Mino, Basilornis, Streptocitta, Ampeliceps*) focus their appetite on fruits and berries. Members of the genus *Sturnus* tend to be more insectivorous, but are very opportunistic. Their northward migrations in spring in the Northern Hemisphere allow them to take advantage of the longer days and the proliferation of insects associated with renewal of plant growth in temperate grasslands. Most species show seasonal shifts in diet, often including an increase in animal food during nesting, thus providing protein for growing nestlings. This is followed by a shift to fruit in late summer, taking advantage of the season's largesse. A few species such as the red-winged starling (*Onychognathus morio*) are essentially omnivorous, feeding on everything from fruits and insects to grain, dead fish, garbage, and nectar.

A characteristic of starlings that aids in their search for food is a unique probing behavior. A starling probes into the substrate and then opens its bill, pushing the substrate aside, to create an open area in which to look for food. With a starling in hand, when fingers are loosely closed over the bill, the bird, seemingly as a reflex, will open its bill, pushing the fingers away. This behavioral adaptation of starlings is also reflected in adaptations of the muscles and the bony parts of the skull to which those muscles are attached.

A yellow-billed oxpecker (*Buphagus africanus*) perches on a giraffe in Kenya. (Photo by P. Davey/VIREO. Reproduced by permission.)

Oxpeckers (subfamily Buphaginae) fill a unique foraging niche, removing ticks and flies from the head and neck of large mammals ranging from giraffes and rhinoceroses to cattle. They also often take some blood and tissue from their host, leading to local campaigns to eradicate the birds.

Reproductive biology

Most sturnids are secondary cavity nesters, often somewhat reliant on woodpeckers and barbets for their nest sites. They compensate for an inability to make their own nest cavities with an aggressiveness that often allows them to appropriate newly excavated ones. Other sturnids such as Tristram's red-winged starling (*Onychognathus tristramii*) make use of niches among rocks, and many have expanded their nest-site preferences to include nest boxes and recesses associated with buildings and other man-made structures.

Once in possession of a cavity, sturnids typically construct a bulky nest of grasses, leaves, fine twigs, and other materials. Sometimes these include man-made objects, and there have been cases of European starlings adding discarded, but lighted, cigarettes to a nest with disastrous results. The amount of material added tends to be whatever it takes to fill the cavity. Red-

billed oxpeckers (*Buphagus erythrorhynchus*) add the dung of ungulates to their nests. Several species add flowers or green leaves to the nest, and it has been suggested that they may select plants laden with chemicals having insecticidal properties. Both sexes are usually involved in nest construction, and nests may be refurbished and cavities used again and again.

Starling eggs are often pale blue, but sometimes white to cream-colored. They may be solid in color (*Acridotheres, Leucopsar, Gracupica, Sturnia, Temenuchus, Pastor, Creatophora, Sturnus*) or have dark spots on them (most species in other genera). The occurrence of colored eggs and eggs with spots has been evidence that sturnid ancestors were open-nesting birds and that cavity nesting is a relatively recent secondary development.

In some cases, only the female incubates; in others, both sexes incubate. Incubation periods are sometimes less than two weeks. Sturnid young at hatching are generally pink, at most with sparse patches of down on top of the head and back, and have their eyes closed for the first few days of life. Both parents contribute to feeding the young and, in some species (e.g., the red-billed oxpecker and possibly the babbling starling, *Neocichla gutteralis*), there are helpers at the

A common (or European) starling (*Sturnus vulgaris*) feeding chicks at its nest hole. (Photo by B & C Calhoun. Bruce Coleman Inc. Reproduced by permission.)

nest who feed the young as well. In such cooperative breeding, the helpers are usually offspring of the same pair from earlier efforts. Nestlings often fledge within three weeks, but young oxpeckers may remain in the nest for nearly a month. Many species can produce two, sometimes three, broods in a year.

Conservation status

At least five starlings have become Extinct within historic times. The Rodriguez starling and Reunion starling, island endemics in the Mascarene Islands, may have disappeared in the eighteenth century. The Kosrae Mountain starling (*Aplonis corvine*), of Kosrae (Kusai) Island in the Caroline Islands, was first and last seen in 1828. The Norfolk Island starling (*Aplonis fusca fusca*) is believed to have disappeared in 1925, and its sister race, the Lord Howe Island starling (*A. f. hulliana*), was last seen in 1919.

The Pohnpei Mountain starling (*Aplonis pelzelni*), found only on Pohnpei (Ponape) in Micronesia, is considered Crit-

ically Endangered and may now be Extinct. The Bali myna is also Critically Endangered, suffering from capture for the cage-bird trade, with only 14 individuals known in the wild in 1998.

The white-eyed starling (*Aplonis brunneicapilla*) of Papua New Guinea and the Solomon Islands is Endangered and suffers from taking of chicks for food, a practice aided by the colonial nesting of the species. The black-winged starling (*Acridotheres melanopterus*) of western Indonesia is also Endangered and suffering from the cage-bird trade.

Vulnerable species include the Rarotonga starling (*Aplonis cinerascens*), atoll starling (*Aplonis feadensis*), mountain starling (*Aplonis santovestris*), Abbott's starling (*Cinnyricinclus femoralis*), and the white-faced starling (*Sturnus albofrontatus*). The Rarotonga starling from Rarotonga in the Cook Islands is rare and probably numbered fewer than 100 individuals in 1987.

Classified Near Threatened are the Tanimbar starling (*Aplonis crassa*), yellow-eyed starling (*Aplonis mystacea*), rusty-winged starling (*Aplonis zelandica*), helmeted myna (*Basilornis galeatus*), Apo myna (*Basilornis miranda*), Sri Lanka myna (*Gracula ptilogenys*), copper-tailed glossy-starling (*Lamprotornis cupreocauda*), and the bare-eyed myna (*Streptocitta albertinae*).

The preponderance of starlings of the genus *Aplonis* on this list of threatened species is primarily a result of the diversity of this group and the cause of that diversity: isolation on small islands of the South Pacific and Indonesia. With limited habitat, bird populations are also limited. As human populations have grown on these islands, forest habitats have come under intense pressure. The colonial nature of many of the birds has also facilitated use of these birds as food or their capture for the cage-bird trade.

A common (or European) starling (*Sturnus vulgaris*) arriving at its nest with food for its young. (Photo by Kim Taylor. Bruce Coleman Inc. Reproduced by permission.)

Significance to humans

Many species are of economic importance as agricultural pests. Some, such as the European starling, occur in such numbers in cities that the uric acid from their droppings damages buildings and monuments. Some are considered hazardous to human health because of their large roosting congregations near or in human cities. Many help control insect pests; others help maintain forest tree-species diversity through dispersal of seeds. The ability of many starlings and mynas to mimic the human voice has made them particularly attractive in the cage-bird trade, but starlings and mynas are also often captured for food. A visit to the bird market in any Indonesian city will reveal many starlings and mynas available for small sums. A visit to even major hotels in Indonesia will also reveal *burung burung* on the menu—birds, including starlings and mynas, are regular fare.

1. Helmeted myna (*Basilornis galeatus*); 2. Bare-eyed myna (*Streptocitta albertinae*); 3. Male golden-crested myna (*Ampeliceps coronatus*); 4. Red-billed oxpecker (*Buphagus erythrorhynchus*); 5. Hill myna (*Gracula religiosa*); 6. Male Bali myna (*Leucopsar rothschildi*); 7. Black-winged starling (*Acridotheres melanopterus*); 8. Common myna (*Acridotheres tristis*); 9. Sri Lanka myna (*Gracula ptilogenys*); 10. Crested myna (*Acridotheres cristatellus*). (Illustration by Emily Damstra)

1. Male magpie starling (*Speculipastor bicolor*); 2. Rarotonga starling (*Aplonis cinerascens*); 3. Male white-eyed starling (*Aplonis brunneicapilla*); 4. Male European starling (*Sturnus vulgaris*) in spring plumage; 5. Male spot-winged starling (*Saroglossa spiloptera*); 6. Copper-tailed glossy-starling (*Lamprotornis cupreocauda*); 7. Male Kenrick's starling (*Poeoptera kenricki*); 8. Male red-winged starling (*Onychognathus morio*); 9. Babbling starling (*Neocichla gutteralis*); 10. Rosy starling (*Sturnus rosea*) in spring plumage. (Illustration by Emily Damstra)

Species accounts

White-eyed starling
Aplonis brunneicapilla

SUBFAMILY
Sturninae

TAXONOMY
Rhinopsar brunneicapilla Danis, 1938.

OTHER COMMON NAMES
French: Stourne aux yeux blancs; German: Weissaugenstar;
Spanish: Estornino de Ojos Blancos.

PHYSICAL CHARACTERISTICS
11.4–12.6 in (29–32 cm); 2.1–2.6 oz (59–73 g). A purple and
greenish glossy bird with whitish eyes, the male has a heavy,
high-arched, black bill, a slight crest of short, almost bristle-
like feathers, and elongate central tail feathers; female is
slightly less iridescent and with a slightly less arched bill. Juvenile has a less robust bill, duller upper plumage, dark eyes and
underparts, with streaking on lower breast and belly.

DISTRIBUTION
Papua New Guinea, and Bougainville, Rendova, Choiseul, and
Guadalcanal in the Solomon Islands.

HABITAT
Upland and lowland forest, forest edge, cultivated areas.

BEHAVIOR
Gregarious both in nesting and feeding.

FEEDING ECOLOGY AND DIET
Diet consists primarily of fruit and berries taken in the canopy.

REPRODUCTIVE BIOLOGY
Colonial cavity nester.

CONSERVATION STATUS
Endangered as a result of fragmentation of populations and
taking of chicks as food.

SIGNIFICANCE TO HUMANS
Cavity trees reportedly cut down to obtain nestlings as food. ◆

Rarotonga starling
Aplonis cinerascens

SUBFAMILY
Sturninae

TAXONOMY
Aplonis cinerascens Hartlaub and Finsch, 1871.

OTHER COMMON NAMES
French: Stourne de Rarotonga; German: Rarotongastar; Spanish: Estornino de Rarotonga.

PHYSICAL CHARACTERISTICS
8.3 in (21 cm). A chunky gray-brown bird with white undertail-
coverts.

DISTRIBUTION
Mountains of Rarotonga in the Cook Islands.

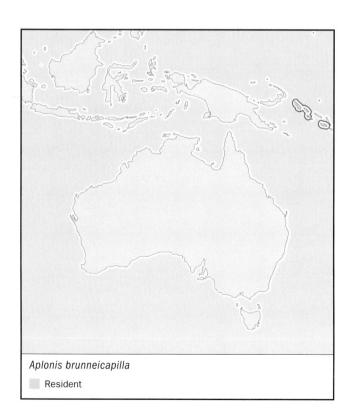

Aplonis brunneicapilla
Resident

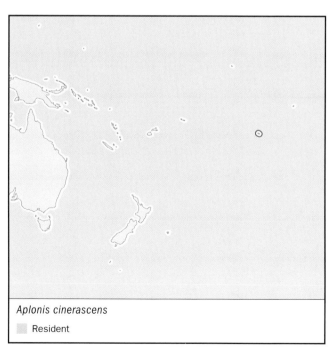

Aplonis cinerascens
Resident

HABITAT
Undisturbed mountain forests.

BEHAVIOR
Quiet, inconspicuous, shy, solitary or in pairs, usually found in the forest canopy.

FEEDING ECOLOGY AND DIET
Sparse data suggest a diet of diverse insects, fruit, and possibly nectar.

REPRODUCTIVE BIOLOGY
Cavity nester. Only two nests known; nest materials are dried leaves and other plant fibers.

CONSERVATION STATUS
Vulnerable; in 1987, the population in the wild was estimated at fewer than 100 birds. Major problems include habitat destruction and predation by introduced black rats.

SIGNIFICANCE TO HUMANS
None known. ◆

Helmeted myna
Basilornis galeatus

SUBFAMILY
Sturninae

TAXONOMY
Basilornis galeatus Meyer, 1894.

OTHER COMMON NAMES
English: Greater crested myna, greater myna, crested myna, Sula myna, king myna, greater king starling, Sula starling; French: Basilorne huppé; German: Helmatzel; Spanish; Estornino Real Grande.

PHYSICAL CHARACTERISTICS
9.4–9.8 in (24–25 cm). Sexes similar; a slightly iridescent black bird with pale yellow bill, large white and buff patches on the side of the neck and breast, and feathers of forehead, crown,

and nape pointed inward with central ones longer to form a distinct iridescent purple-black crest. Juveniles have a shorter, duller crest, a brown chin, and a dark upper bill.

DISTRIBUTION
Banggai east of Sulawesi and the adjacent Sula Islands in the Moluccas.

HABITAT
Favors undisturbed forest; also found in disturbed and cultivated areas and in mangroves.

BEHAVIOR
Usually found in pairs, but also flocks; frequents tall trees.

FEEDING ECOLOGY AND DIET
Only known to take fruit and berries from canopy to mid-levels of trees.

REPRODUCTIVE BIOLOGY
No information.

CONSERVATION STATUS
Near Threatened.

SIGNIFICANCE TO HUMANS
None known. ◆

Bare-eyed myna
Streptocitta albertinae

SUBFAMILY
Sturninae

TAXONOMY
Charitornis albertinae Schlegel, 1866.

OTHER COMMON NAMES
English: Sula myna, Schlegel's myna, Sula magpie, Sula starling, Albertina's starling; French: Streptocitte des Sula; German: Sulaatel; Spanish: Estornino de Sula.

Basilornis galeatus
▢ Resident

Streptocitta albertinae
▢ Resident

PHYSICAL CHARACTERISTICS
16.5–17.7 in (42–45 cm). A striking black-and-white bird; white body, nape, and crown to the bill; black face, throat, wings, and very long tail; yellow bill and legs. Sexes similar. Juvenile has forehead mottled with dark brown.

DISTRIBUTION
Sula Islands, Moluccas; sea level to about 750 ft (228 m).

HABITAT
Cutover and open forest and cultivated areas.

BEHAVIOR
Little known; usually seen alone, in pairs, or in trios; voice descending and a bit like a squeaky gate.

FEEDING ECOLOGY AND DIET
No information.

REPRODUCTIVE BIOLOGY
No information.

CONSERVATION STATUS
Near Threatened, probably due to habitat losses.

SIGNIFICANCE TO HUMANS
One found for sale in a market. ◆

Spot-winged starling
Saroglossa spiloptera

SUBFAMILY
Sturninae

TAXONOMY
Lamprotornis spiloptera Vigors, 1831.

OTHER COMMON NAMES
French: Étourneau à ailes tachetées; German: Marmorstar; Spanish: Estornino de Alas Moteadas.

Saroglossa spiloptera

☐ Resident

PHYSICAL CHARACTERISTICS
7.5 in (19 cm); weight 1.7 oz (48 g). Slender bird with thin, slightly down-curved bill; male has a chestnut throat, rusty flanks, lighter lower breast and belly, and a textured appearance above resulting from gray feathers on head and back that are bordered by blackish brown; female has pale underparts with brown streaking, somewhat mottled brown back and head. In flight, these birds show white spotting in the middle of each primary feather. Juveniles not well known, but apparently vary and are somewhat lighter in color than adults.

DISTRIBUTION
Foothills of the Himalayas of north-central India; winters from eastern India to Burma and Thailand.

HABITAT
Open country, forest edge, agricultural land.

BEHAVIOR
Primarily arboreal, shy, restless. Gives noisy chattering calls reminiscent of mynas. Appears to have an east-west migration within Asia. Flocks sometimes include several hundred birds.

FEEDING ECOLOGY AND DIET
Diet includes insects such as ants and termites, sometimes taken in flight; also takes nectar.

REPRODUCTIVE BIOLOGY
A cavity nester that uses woodpecker and barbet holes in which to build a nest of green leaves and other materials. Nesting is usually April–June. A clutch includes three to four pale gray to blue-green eggs with small reddish brown spots and blotches. No data on incubation or nestling periods.

CONSERVATION STATUS
Near Threatened.

SIGNIFICANCE TO HUMANS
None known. ◆

Golden-crested myna
Ampeliceps coronatus

SUBFAMILY
Sturninae

TAXONOMY
Ampeliceps coronatus Blyth, 1842.

OTHER COMMON NAMES
English: Gold-crested myna; French: Martin couronné; German: Kronenatzel; Spanish: Mainá de Cresta Dorada.

PHYSICAL CHARACTERISTICS
7.5–8.3 in (19–21 cm); weight 2.8–3.5 oz (78–99 g). Glossy black plumage with yellow crown, crest, orbital skin, throat, and base of primaries. Bill yellow and gray; legs orange-brown.

DISTRIBUTION
South Asia from northeastern India to Burma, northern Malaya, Thailand, Laos, Cambodia, Vietnam; birds in southern China may be escapees.

HABITAT
Dense lowland evergreen and moist deciduous forest; open forest and cultivated areas with tall trees.

Ampeliceps coronatus
☐ Resident

Gracula religiosa
☐ Resident

BEHAVIOR
Favors the forest canopy; usually found in pairs or small groups.

FEEDING ECOLOGY AND DIET
Diet seems to include primarily fruit, but also insects obtained in the forest canopy.

REPRODUCTIVE BIOLOGY
A cavity nester; nests found April–May, and juveniles in April–June. The female incubates a clutch of up to four blue-green eggs for about 14–15 days; males may assist in feeding young which fledge at about 25–26 days.

CONSERVATION STATUS
Not threatened.

SIGNIFICANCE TO HUMANS
None known. ◆

Hill myna
Gracula religiosa

SUBFAMILY
Sturninae

TAXONOMY
Gracula religiosa Linnaeus, 1758. Seven races recognized.

OTHER COMMON NAMES
English: Indian hill mynah, common grackle, grackle, talking mynah; French: Mainate religieux; German: Beo, kleinbeo; Spanish: Mainá del Himalaya.

PHYSICAL CHARACTERISTICS
11.0–11.8 in (28–30 cm). Black with a heavy orange-yellow bill, feet, and fleshy wattles below each eye and on the nape;

rounded black wings with a white stripe running midway through the primaries.

DISTRIBUTION
Southern Asia; India east to southern China, Thailand, Malaysia, and Indochina, through the Greater and Lesser Sunda islands, east to Alor and Palawan in the Philippines. Population on St. Helena may be introduced; introduced and breeding on Christmas Island in the Indian Ocean, and in Florida, Hawaii, and Puerto Rico.

HABITAT
Found in areas of lush vegetation and high rainfall, especially at edges of lush forest, but also within dense forest, second growth, and cultivated areas; most common at elevations of 900–6,000 ft (275–1,825 m).

BEHAVIOR
Very arboreal; hop rather than walk; very vocal.

FEEDING ECOLOGY AND DIET
Mainly frugivorous, but also opportunistic, taking nectar, insects, other small animals. Focuses more on animal prey when feeding young. Regurgitates seeds, thus facilitating seed dispersal.

REPRODUCTIVE BIOLOGY
Monogamous; breeding varies geographically from January–July; nests in natural cavities and bird boxes; both sexes aid in constructing a crude nest of small twigs, leaves, and feathers. Clutch size two to three light blue to blue-green eggs with brown to lavender spotting; incubation by both sexes for 13–17 days; parental care by both adults; young fledge at 25–28 days; post-fledging care is minimal. May produce as many as three broods per year in some areas.

CONSERVATION STATUS
Not threatened. Listed on CITES Appendix II. Some concern over excessive capture of birds for the pet trade.

SIGNIFICANCE TO HUMANS
Once commonly eaten, now a very popular cage bird because of their ability to mimic the human voice. May be important pollinators of forest trees. ◆

Sri Lanka myna
Gracula ptilogenys

SUBFAMILY
Sturninae

TAXONOMY
Gracula ptilogenys Blyth, 1846.

OTHER COMMON NAMES
English: Ceylon myna, Sri Lanka hill myna, grackle; French: Mainate de Ceylan; German: Dschungelatzel; Spanish: Mainá de Ceilón.

PHYSICAL CHARACTERISTICS
9.1–9.8 in (23–25 cm). A small, stubby black myna with a white stripe through the primaries, orange bill that is dark at the base, yellow legs, and a bare yellow nape wattle.

DISTRIBUTION
Sri Lanka lowlands to 6,300 ft (1,920 m).

HABITAT
Moist forest; also, cultivated areas with trees.

BEHAVIOR
Gregarious and arboreal, rarely coming to the ground; typically found in pairs in colonies; wings produce a loud humming sound in flight; an excellent mimic of the human voice.

FEEDING ECOLOGY AND DIET
Diet mainly fruit, especially figs, and seeds.

Gracula ptilogenys

 Resident

REPRODUCTIVE BIOLOGY
Cavity nester, nesting February–May and sometimes in August–September. Cavities sometimes used without nest materials, but a nest of grasses and other materials is usual. Clutch normally includes two Prussian-blue eggs mottled with reddish brown. No information available on parental care.

CONSERVATION STATUS
Near Threatened.

SIGNIFICANCE TO HUMANS
Nestlings often taken for the pet trade. ◆

Crested myna
Acridotheres cristatellus

SUBFAMILY
Sturninae

TAXONOMY
Gracula cristatella Linnaeus, 1766. Three races recognized.

OTHER COMMON NAMES
English: Chinese jungle myna, Chinese crested myna, Chinese starling, tufted myna; French: Martin huppé; German: Haubenmaina; Spanish: Mainá China.

PHYSICAL CHARACTERISTICS
8.7–10.2 in (22–26 cm). Sexes similar; black with an ivory bill, orange eye, and longer feathers forming a crest on the lower forehead; juvenile slightly browner, with less of a crest, and blue eyes.

DISTRIBUTION
Lowlands of south Asia; introduced to Malaya, the Philippines (Luzon, Negros), and Vancouver, British Columbia; *A. c. cristatellus*, eastern Burma to southeast and central China; *A. c. brevipennis*, Hainan; *A. c. formosanus*, Taiwan.

HABITAT
Open country, farmlands.

BEHAVIOR
Gregarious, but pair members remain obviously together.

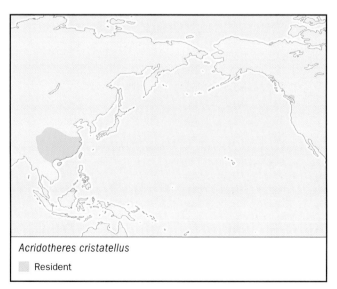

Acridotheres cristatellus

 Resident

FEEDING ECOLOGY AND DIET
Very opportunistic and somewhat omnivorous; feeds mainly on the ground by inserting the bill into substrate and opening it to probe; also chases insects flushed from ground; some fruit included in the diet; sometimes associated with cattle; known to scavenge on beaches.

REPRODUCTIVE BIOLOGY
Colonial or solitary cavity nester that uses woodpecker holes, niches on buildings, bird houses, and similar sites. Clutch of four to seven pale blue or blue-green eggs are incubated about 14 days; chicks fledge at about 21 days.

CONSERVATION STATUS
Not threatened.

SIGNIFICANCE TO HUMANS
Often kept as a cage bird. ◆

Common myna
Acridotheres tristis

SUBFAMILY
Sturninae

TAXONOMY
Paradisea tristis Linnaeus, 1766. Two races recognized; see distribution.

OTHER COMMON NAMES
English: Indian myna, house myna, common mynah, locust starling, myna; French: Martin triste; German: Hirtenmaina; Spanish: Mainá Común.

PHYSICAL CHARACTERISTICS
9.1–9.8 in (23–25 cm); weight 2.9–5.0 oz (82–143 g). Sexes similar; a stocky, brown bird with a glossy black head and throat, yellow bill, bare yellow skin behind its eyes, and yellow legs; juveniles are duller.

DISTRIBUTION
Lowlands to 4,500 ft (1,370 m) in southern Asia from southeastern Iran though Afghanistan, Pakistan, India, Sri Lanka,

Acridotheres tristis
Resident

southern China, and Vietnam. Introduced and established in Arabia, South Africa, Madagascar, Australia, New Zealand, Fiji, Cook Islands, Society Islands, Hivaoa in the Marquesas Islands, and Hawaii (Midway and from Kauai east). An introduced breeding population on Kwajalein in the Marshall Islands died out. *A. t. tristis*, most of range; *A. t. melanosternus*, Sri Lanka.

HABITAT
Open habitats including farmlands, cities.

BEHAVIOR
Tame, bold, and noisy; usually seen in pairs or small flocks.

FEEDING ECOLOGY AND DIET
Feeds mainly on ground on insects, but takes small vertebrates, carrion, occasionally eggs and nestlings of other birds, fruit, and grain.

REPRODUCTIVE BIOLOGY
Builds bulky nest in tree cavities, niches on buildings, and in dense vegetation. Clutch of four to five glossy, pale blue eggs is incubated by both parents for 13–18 days; young leave the nest as early as 22 days, but may not be able to fly for another week or more.

CONSERVATION STATUS
Not threatened.

SIGNIFICANCE TO HUMANS
Considered a pest in Australia where noisy winter roosts of several thousand can occupy city trees and buildings, and also a pest in many areas where it seasonally may take grain or fruit. ◆

Black-winged myna
Acridotheres melanopterus

SUBFAMILY
Sturninae

TAXONOMY
Gracula melanoptera Daudin, 1800.

OTHER COMMON NAMES
English: Black-winged starling, white-breasted starling; French: Étourneau á ailes noires; German: Schwarzflügelstar; Spanish: Estornino de Alas Negras.

PHYSICAL CHARACTERISTICS
8.7–9.4 in (22–24 cm). Sexes alike; a mostly white myna with black wings and tail; bases of primaries are white, showing as a white stripe in the wing of a flying bird; black tail has white-tipped feathers; yellow bill, flesh around eye, and feet. Juvenile has a gray crown, nape, and mantle.

DISTRIBUTION
Java, Bali, and Lombok in western Indonesia.

HABITAT
Open country around human habitations and cultivated areas.

BEHAVIOR
Arboreal in some areas, but also spends a lot of time on the ground; roosts communally in trees.

FEEDING ECOLOGY AND DIET
Diet includes fruit and insects taken from trees, shrubs, and ground.

Acridotheres melanopterus
▨ Resident

REPRODUCTIVE BIOLOGY
A colonial cavity nester naturally nesting in holes among rocks or in trees.

CONSERVATION STATUS
Endangered. Capture for the cage bird trade is the most significant cause of this species' decline; pesticide use also considered a problem. May compete with the critically Endangered Bali myna with which it now coexists.

SIGNIFICANCE TO HUMANS
Commonly sold in markets on Java. ◆

Bali myna
Leucopsar rothschildi

SUBFAMILY
Sturninae

TAXONOMY
Leucopsar rothschildi Stresemann, 1906.

OTHER COMMON NAMES
English: Bali starling, Rothschild's myna, Rothschild's starling, white starling; French: Martin de Rothschild; German: Balistar; Spanish: Mainá de Bali.

PHYSICAL CHARACTERISTICS
9.8 in (25 cm). White body, feathers of lower forehead are bristle-like, feathers of nape and crown long and slender to form a crest that can be raised, especially in males; white tail and primaries have black tips; bright light-blue bare skin around eye; bill gray, yellow towards tip. Juveniles similar to adults, sometimes with gray wash on back and cinnamon tinge on wings.

DISTRIBUTION
Northwest Bali, Indonesia.

HABITAT
Open woods with grassy understory.

Leucopsar rothschildi
▨ Resident

BEHAVIOR
Strongly arboreal, but will come to the ground for food, water, and bathing; can be gregarious outside of breeding season.

FEEDING ECOLOGY AND DIET
Diet includes seeds, fruit, insects and other invertebrates and small vertebrates.

REPRODUCTIVE BIOLOGY
Breeds during the rainy season (January–March); cavity nester that uses old woodpecker holes or nest boxes. Clutch includes two to three pale blue eggs; incubation mostly by female lasts 12–14 days; young fed by both parents and fledge at 21–28 days; parental care continues for about five weeks.

CONSERVATION STATUS
Listed under CITES Appendix I; Critically Endangered under IUCN criteria. In 1925, hundreds could be seen; by 1976, 127 were counted; as of 1998, only 14 birds were known in the wild. Many zoos have this species, and there are strong captive breeding programs, but habitat deterioration and poaching for the pet trade continue to threaten the species.

SIGNIFICANCE TO HUMANS
Valued in the pet trade. ◆

Rosy starling
Sturnus roseus

SUBFAMILY
Sturninae

TAXONOMY
Turdus roseus Linnaeus, 1758, Lapland and Switzerland.

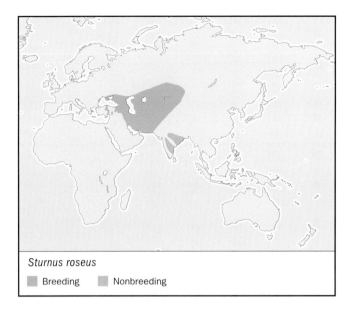

Sturnus roseus

■ Breeding ■ Nonbreeding

OTHER COMMON NAMES
English: Rose-colored starling, rosy pastor; French: Étourneau roselin; German: Rosenstar; Spanish: Estornino Rosado.

PHYSICAL CHARACTERISTICS
7.5–9.1 in (19–23 cm); weight 2.1–3.1 oz (60–88 g). Black head to upper back, chin to base of throat, wing feathers, tail, and vent. Pale rosy back, breast, and abdomen. Bill ivory-yellow.

DISTRIBUTION
South-central Eurasia; generally found where locust swarms are regular; winters primarily in India, but wanders widely.

HABITAT
Found in semi-desert areas with thorn bushes; other barren areas, including industrial sites, agricultural areas, lawns, and pastures.

BEHAVIOR
Migrant; very gregarious. This is often a roadside bird, coming to puddles to bathe and drink as well as to bare areas to feed.

FEEDING ECOLOGY AND DIET
Diet includes primarily insects, especially grasshoppers, crickets, and locusts, but also grapes, mulberries, and other fruit. It often feeds in association with cattle.

REPRODUCTIVE BIOLOGY
Nests colonially in holes among rocks, in walls, or similar sites. In some studies, young were fed almost exclusively crickets and grasshoppers.

CONSERVATION STATUS
Not threatened.

SIGNIFICANCE TO HUMANS
Long recognized for helping to control locust swarms. Sometimes hailed for eating locusts in the spring and hated for taking grapes in the summer. ◆

European starling
Sturnus vulgaris

SUBFAMILY
Sturninae

TAXONOMY
Sturnus vulgaris Linnaeus, 1758. Twelve races recognized.

OTHER COMMON NAMES
English: Common starling, northern starling, English starling, purple-winged starling, starling; French: Étourneau sansonnet; German: Star; Spanish: Estornino Pinto.

PHYSICAL CHARACTERISTICS
8.3–8.7 in (21–22 cm); weight 2.0–3.7 oz (58–105 g). A purple-green iridescent, short-tailed black bird with a long thin bill that changes seasonally from black in winter to yellow during nesting. Following the fall molt, starlings are very spotted with white as a result of white-tipped body feathers. As the winter progresses, the white tips wear off little by little so that, during courtship, the birds show mostly the iridescent black with little spotting. Males have longer, narrower hackle feathers and, during nesting season, a blue base to the bill, while females have a pink base to the bill. Juveniles are gray-brown with a streaked breast and dark bill.

DISTRIBUTION
Most of temperate Eurasia from Iceland east. Introduced and established in South Africa, Polynesia (Fiji, Tonga), Australia, New Zealand, Bermuda, North America (from coast to coast, and southern Alaska into Mexico), Jamaica, and Puerto Rico. It has been seen in Hawaii, but has not become established there.

HABITAT
Open country, open woods, urban and suburban areas.

BEHAVIOR
An aggressive competitor for woodpecker cavities and nest boxes, as well as for niches on buildings; roosts in flocks that sometimes number in the millions. Is a good vocal mimic, often mimicking other birds, but can be taught to mimic the human voice.

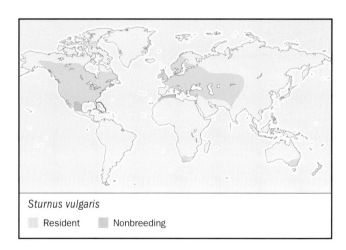

Sturnus vulgaris

■ Resident ■ Nonbreeding

FEEDING ECOLOGY AND DIET
Characteristically feeds on the ground, often in large flocks; takes a diversity of insects, other arthropods, grain, and fruit.

REPRODUCTIVE BIOLOGY
Can be a solitary or loosely colonial nester, nests March–May in Northern Hemisphere, September–December in South Africa. A serious competitor for woodpecker cavities and nest boxes with many cavity nesting birds, especially where it has been introduced. Clutch of three to six pale blue eggs is incubated by the female for 11–15 days. Young are fed by both parents and fledge at 20–21 days.

CONSERVATION STATUS
Not threatened.

SIGNIFICANCE TO HUMANS
Consumes a lot of harmful insects and weed seeds. Its introduction into North America was a result of the desire of a homesick immigrant to the United States deciding to introduce to North America all of the birds mentioned in Shakespeare. He began with the European starling, which is mentioned in Henry IV; the rest is history. Judgment is generally tipped against the starling as a result of its enormous winter flocks, proclivity for close association with humans, building messy nests on buildings, taking grain and fruit, and competing with songbirds and woodpeckers for nest sites. ◆

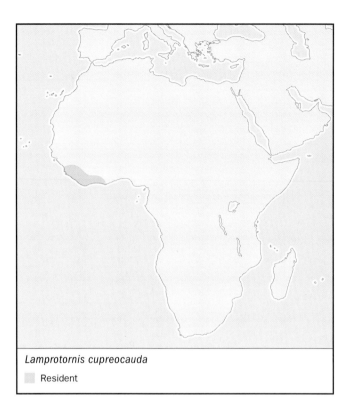

Lamprotornis cupreocauda

◻ Resident

Copper-tailed glossy-starling
Lamprotornis cupreocauda

SUBFAMILY
Sturninae

TAXONOMY
Lamprocolius cupreocauda Hartlaub, 1857. Considered closely related to *Lamprotornis purpureiceps* and some consider these as variants of the same species.

OTHER COMMON NAMES
English: Coppery-tailed glossy-starling; French: Choucador à queu bronzée; German: Kupferglanzstar; Spanish: Estornino de Cola Cobriza.

PHYSICAL CHARACTERISTICS
7.1–8.3 in (18–21 cm); weight 1.9–2.3 oz (53–66 g). Dark blue-violet glossy plumage with copper-brown tail feathers and blackish bill and legs. Juvenile is sooty brown with a slight purple gloss on the crown and a tinge of glossy green on upper parts.

DISTRIBUTION
West Africa from Sierra Leone to Ghana.

HABITAT
Mature lowland forest canopy in both primary and cutover forest; especially riverine forest.

BEHAVIOR
Endemic resident. Seen in flocks of up to 50 birds; sometimes joins mixed-species flocks; flight swift and direct.

FEEDING ECOLOGY AND DIET
Diet includes fruit, especially figs, and insects found in the forest canopy.

REPRODUCTIVE BIOLOGY
Nest and eggs unknown, but adults have been seen at holes in dead trees in October and fledged young seen in December; adult also seen with three fledged young in February.

CONSERVATION STATUS
Near Threatened.

SIGNIFICANCE TO HUMANS
None known. ◆

Magpie starling
Speculipastor bicolor

SUBFAMILY
Sturninae

TAXONOMY
Speculipastor bicolor Reichenow, 1879, Kenya.

OTHER COMMON NAMES
English: Pied starling; French: Spreó pie; German: Spiegelstar; Spanish: Estornino Urraca.

PHYSICAL CHARACTERISTICS
7.5 in (19 cm); weight 2.2–2.4 oz (61–69 g). Blackish head, throat, and upperparts with blue sheen; underparts white. Juvenile dark brown above with paler crown and cheeks, a dark breast band, and buff-white belly.

DISTRIBUTION
Northeast Africa, including Somalia, southern Ethiopia, northeastern Uganda, Kenya.

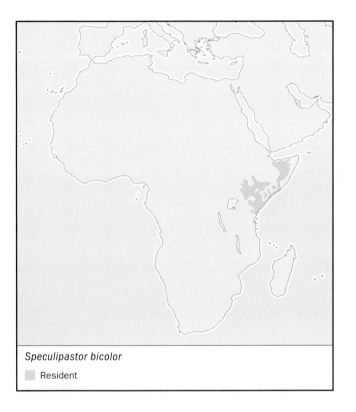

Speculipastor bicolor

■ Resident

Onychognathus morio

■ Resident

HABITAT
Dry savanna and acacia.

BEHAVIOR
Shrill whistling flight call; gather at food trees; apparently migrant, though may be nomadic; some remain year round.

FEEDING ECOLOGY AND DIET
Diet is primarily fruit, especially figs, and insects.

REPRODUCTIVE BIOLOGY
Nests of grass and leaves are placed in cavities in termite mounds or in banks; clutch of three to five blue-green eggs incubated by female for about 18 days; both parents feed nestlings; young fledge at about 21 days.

CONSERVATION STATUS
Not threatened.

SIGNIFICANCE TO HUMANS
None known. ◆

Red-winged starling
Onychognathus morio

SUBFAMILY
Sturninae

TAXONOMY
Turdus morio Linnaeus, 1766, Cape of Good Hope.

OTHER COMMON NAMES
English: Redwing, African red-winged starling; French: Rufipenne morio; German: Rotschwingenstar; Spanish: Estornino de Alas Rojas Africano.

PHYSICAL CHARACTERISTICS
10.6–12.2 in (27–31 cm); weight 4.1–5.6 oz (117–158 g). Dark plumage with dark blue sheen. Reddish brown primaries. Eyes red. Juvenile similar to adult male, but less glossy and with brown eyes and bill.

DISTRIBUTION
East and southeast Africa, from Kenya to South Africa; most common at 3,000–7,500 ft (915–2,285 m).

HABITAT
Rocky hills in savanna, now commonly nests in urban areas.

BEHAVIOR
Endemic resident. Gathers in flocks that can at times exceed 1,000 birds.

FEEDING ECOLOGY AND DIET
Somewhat omnivorous; eats fruits of many trees, but also takes nectar, insects, other arthropods, and small vertebrates, including nestlings of other birds.

REPRODUCTIVE BIOLOGY
Cavity nester. Clutch of two to four blue eggs with red-brown spots is incubated mainly by the female for 13–14 days. Hatchlings are pink with tufts of gray down on top of the head and mid-back. Both adults feed nestlings until they fledge at 22–28 days. Nest sites are often reused.

CONSERVATION STATUS
Not threatened.

SIGNIFICANCE TO HUMANS
None known. ◆

Kenrick's starling
Poeoptera kenricki

SUBFAMILY
Sturninae

TAXONOMY
Paeoptera [sic] kenricki Shelley, 1894, Tanganyika. Sometimes considered conspecific with *Poeoptera stuhlmanni*.

OTHER COMMON NAMES
French: Rufipenne de Kenrick; German: Kenrickstar; Spanish: Estornino de Kenrick.

PHYSICAL CHARACTERISTICS
5.9 in (15 cm); weight 1.3–1.9 oz (38–54 g). Dark plumage with dark brownish wings and olive-gray sheen. Bill and legs black. Eyes yellow. Juvenile is similar to adult female, but duller and more sooty on the body.

DISTRIBUTION
Disjunct populations in mountains of East Africa in central Kenya and northeastern Tanzania.

HABITAT
Forest canopy in high rainfall areas.

BEHAVIOR
Endemic resident. Social, traveling in noisy flocks, but does not join mixed-species flocks. Apparently nomadic, sometimes visits isolated forest areas.

FEEDING ECOLOGY AND DIET
Primarily frugivorous; favors figs (*Ficus*); stays in forest canopy to mid-canopy.

REPRODUCTIVE BIOLOGY
Cavity nester that uses old woodpecker or barbet holes; no details; laying suggested to be mainly in October.

CONSERVATION STATUS
Not threatened.

SIGNIFICANCE TO HUMANS
None known. ◆

Babbling starling
Neocichla gutteralis

SUBFAMILY
Sturninae

TAXONOMY
Crateropus gutturalis Barbosa du Bocage, 1871, Angola. Two races are recognized.

OTHER COMMON NAMES
French: Spréo à gorge noire; German: Weissflügelstar; Spanish: Estornino de Alas Blancas.

PHYSICAL CHARACTERISTICS
8.7–9.4 in (22–24 cm); weight 2.3–2.5 oz (64–72 g). Gray brown head with dark wings and tail; white wing patches are conspicuous in flight. Black throat patch and buff underparts. Eyes yellow.

DISTRIBUTION
Disjunct populations known from western Angola (*N. g. gutturalis*) and Zambia west to northwestern Malawi, and in central and southwestern Tanzania (*N. g. angusta*).

Poeoptera kenricki
▨ Resident

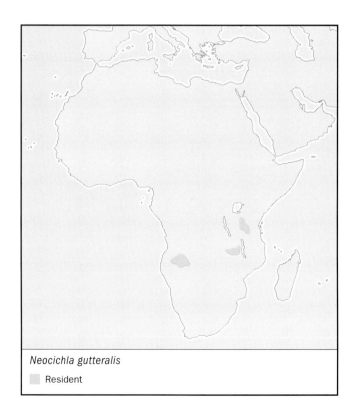

Neocichla gutteralis
▨ Resident

HABITAT
Open *Brachystegia* woodland with open understory.

BEHAVIOR
Found in pairs and small flocks; usually seen on the ground.

FEEDING ECOLOGY AND DIET
Feeds primarily on and close to the ground on a diversity of insects, especially termites and beetles; forages by running and searching; often forages just after sunset.

REPRODUCTIVE BIOLOGY
Apparently a cooperative breeder. Nests in tree cavities lined with lichens, moss, and grass; clutch size from one nest was two smooth white eggs with red-brown spots on a bluish base; nestlings have been found in October–November.

CONSERVATION STATUS
Not threatened.

SIGNIFICANCE TO HUMANS
None known. ◆

Red-billed oxpecker
Buphagus erythrorhynchus

SUBFAMILY
Buphaginae

TAXONOMY
Tanagra erythrorhyncha Stanley, 1814, Ethiopia.

OTHER COMMON NAMES
French: Piqueboeuf à bec rouge; German: Rotschnabel-Madenhacker; Spanish: Picabuey de Pico Rojo.

PHYSICAL CHARACTERISTICS
7.5–8.7 in (19–22 cm); weight 1.5–2.1 oz (42–59 g). Olive-brown above, tan on rump and breast; red bill and eyes with conspicuous yellow flesh around the eyes. Juveniles have a dark bill and eyes and brown area around the eyes.

DISTRIBUTION
East and southeast Africa with a highly fragmented distribution from western Central African Republic, Sudan, Ethiopia, and Somalia, south in Uganda, Kenya, Tanzania, eastern and southern Zaire to northern and eastern South Africa.

HABITAT
Open savanna areas in association with large mammals, including domestic livestock, up to about 9,000 ft (2,745 m).

BEHAVIOR
Endemic resident. Intimately associated with large mammals and, in different areas, different hosts seem to be preferred; usual perch and site of feeding is on the head and neck.

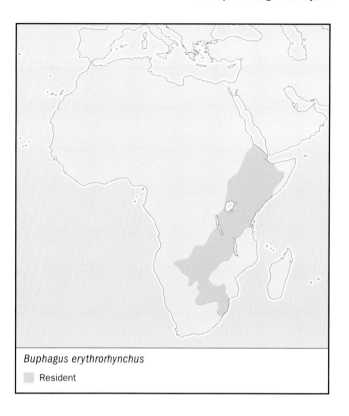

Buphagus erythrorhynchus
▨ Resident

FEEDING ECOLOGY AND DIET
Feeds primarily on ticks and biting flies plucked from host and on host blood and tissues. Apparently, the birds do not make wounds on the animals, but take blood from existing wounds.

REPRODUCTIVE BIOLOGY
Cooperative breeders; courtship often takes place on the backs of host mammals. Breeding occurs at different times in different areas, often associated with beginning of rainy season, and has been reported in all months. Nest of grasses lined with hair and dung is in a natural tree cavity. Clutch size two to five creamy white eggs with brown to lilac speckles; incubation by both parents for 12–13 days; feeding of nestlings is done by parents and helpers; young fledge at about 30 days and are independent about a month later.

CONSERVATION STATUS
Not threatened; dipping of livestock in pesticides to control pests seems to be associated with declines in oxpecker numbers in some areas.

SIGNIFICANCE TO HUMANS
Perform a service for livestock owners, but this has often not been recognized; evidence suggests some negative impacts and oxpecker extermination programs have been carried out in some areas. ◆

Resources

Books
BirdLife International. *Threatened Birds of the World.* Barcelona, Spain: Lynx Edicions and BirdLife International, 2000.

Cramp, S., and C.M. Perrins, eds. *The Birds of the Western Palearctic,* Vol. 8. Oxford, England: Oxford University Press, 1994.

Resources

Feare, C. *The Starling*. Oxford, England: Oxford University Press, 1984.

Feare, C., and A. Craig. *Starlings and Mynas*. Princeton, NJ: Princeton University Press, 1999.

Fry, C.H., S. Keith, and E.K. Urban. *The Birds of Africa*, Vol. 6. New York, NY: Academic Press, 2000.

Grimmett, R., C. Inskipp, and T. Inskipp. *A Guide to the Birds of India, Pakistan, Nepal, Bangladesh, Bhutan, Sri Lanka, and the Maldives*. Princeton, NJ: Princeton University Press, 1999.

Hollom, P.A.D., R.F. Porter, S. Christensen, and I. Willis. *Birds of the Middle East and North Africa*. London, England: Poyser, 1988.

Lever, C. *Naturalised Birds of the World*. London, England: Longmans, 1987.

Sibley, C.G., and J.E. Ahlquist. *Phylogeny and Classification of Birds*. New Haven, CT: Yale University Press, 1990.

Sibley, C.G., and B.L. Monroe, Jr. *Distribution and Taxonomy of Birds of the World*. New Haven, CT: Yale University Press, 1990.

Periodicals

Sibley, C.G., and J.E. Ahlquist. "The Relationships of the Starlings (Sturnidae: Sturnini) and the Mockingbirds (Sturnidae: Mimini)." *Auk* 101 (1984): 230–243.

Jerome A. Jackson, PhD

Old World orioles and figbirds
(Oriolidae)

Class Aves
Order Passeriformes
Suborder Passeri (Oscines)
Family Oriolidae

Thumbnail description
Orioles and figbirds are tree-living song-birds commonly patterned in brilliant yellows and blacks, and streaked ventrally in juveniles and some adults.

Size
7–11.5 in (20–28 cm); 2–5 oz (50–135 g)

Number of genera, species
2 genera; 29-30 species

Habitat
Medium to tall woodland and forest, including rainforest

Conservation status
Endangered: 1 species; Vulnerable: 2 species; Near Threatened: 3 species

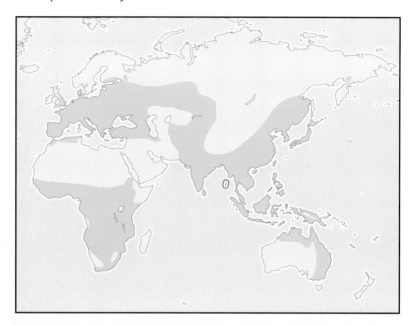

Distribution
Sub-Saharan and far northwest Africa, temperate Eurasia (except central deserts), south and east to India, southeast Asia, and Indonesian archipelagos as far as New Guinea, and north and east Australia

Evolution and systematics

Among songbirds, orioles and figbirds are so typical in appearance and behavior that their origins and relationships have excited little interest. But molecular evidence shows that they are closely allied to Old World cuckoo-shrikes (Campephagidae) and other crow-like songbirds, not starlings (Sturnidae) as traditionally held. Like cuckoo-shrikes, orioles and figbirds are outlying members of a massive evolutionary explosion in crow-like songbirds that originated in Australia in early- to mid-Teriary times, more than 30 million years ago, and then spread north and west through the tropical and temperate Old World. Whether today's orioles and figbirds actually arose in Australasia, or diversified outside to re-enter Australia later within the last few million years, is still in dispute.

Oriolidae comprises three species of figbirds (*Sphecotheres*) and 26-27 of orioles (*Oriolus*). The three figbirds replace one another geographically in different parts of tropical Indo-Australasia east of Wallace's Line. The orioles comprise four, perhaps five, species-groups: golden-headed orioles, a widespread migratory group of four species in west and east Eurasia, India, and central Africa; a black-headed group of seven resident species in Africa, and the black-hooded oriole (*O. xanthornus*) in east Asia; the four sedentary dusky orioles of Southeast Asia and adjacent archipelagos; and the streaked, juvenal-plumaged Australasian orioles, with seven species. The three or four remaining oriole species, all Southeast Asian, are of uncertain affinity.

Physical characteristics

Orioles and figbirds are thrush-like in shape and size, with wings that are long yet rounded, and a square-tipped, 12-feathered tail. The tenth primary flight feather is well developed, while the number of secondary feathers is usually 11 in orioles and only 10 in figbirds. Bills are straight, stout, and notched at the tip of the upper mandible, fitted for grasping and pulling at food; bristles flanking the bill are vestigial and nostrils are narrow slits partially protected by a membrane. Adapted for perching, feet are stout, and the tarsus (leg) coarsely scaled and shorter than the longest toe.

Spectacular yellows and blacks are the badge of oriole plumage; vermillion eyes, brick-red bills, and slate-gray feet complete the picture. Only figbirds and dusky Asian orioles have black or slate-colored bills, sometimes pale eyes, and flesh-colored or black feet. Male orioles are invariably more brightly plumaged that females; juveniles have brown-olive backs with dull bills, eyes, and feet, and are white ventrally with dusky streaks. In the golden-headed group, adult males are almost wholly golden-yellow except for black wings, center tail, and a black mark through the eye, while females are plumaged like dull males or bright juveniles. In black-headed orioles, both males and females have yellow bodies, often gray wings, and black or dark olive heads; in the dusky group, males have dusky or silver bodies with splashes of deep red on the undersurface and tail, and females are browner and sometimes streaked on the belly. In the Indo-Australian streaked group,

An olive-backed oriole (*Oriolus sagittatus*) tends to its nestlings in an Australian forest. (Photo by Wayne Lawler. Photo Researchers, Inc. Reproduced by permission.)

both sexes are dull brown or olive-yellow with varied ventral streaking, as in juveniles. Adult plumage is gained at one to three years of age. Male figbirds have black crowns and bills, olive backs, and yellow, white, olive, or gray breasts, while females and juveniles are plumaged like juvenile orioles. Apart from bulk and a broad head, the only other outward difference in figbirds is a patch of bare skin around the eye, which is enlarged and warty in males, and reddens with pumped blood in excitement, indicating a signal function.

Distribution

The tropics and temperate regions of the Old World east to Australia are the home of orioles and figbirds today. Only two orioles occur in the north temperate region, one in the west, including Europe, and the other in the east, including Japan; both are migratory. To the south, there are three centers of diversity: sub-Saharan Africa, which holds nine species (seven resident black-headed and two golden-headed orioles, one a non-breeding migrant); Southeast Asia south of the Himalayas and central China, west to peninsular India and Sri Lanka, and east to the Greater Sundas and Philippines, which has up to 12 species in three or four species groups, all resident endemics except for two migrant golden-headed orioles; and east and north Australia-New Guinea west to the Moluccas and Lesser Sundas, where the figbirds and streaked orioles are endemic.

Habitat

Living almost exclusively in the crowns of trees, orioles and figbirds are found in rather open broadleaf forests and woodlands in temperate regions, but also in rainforests in the tropics. Most migratory species forage through a range of forest and woodland types, while sedentary species occupy narrower niches. Densities when breeding range from about two to 10 pairs per 0.6 sq mi (1 sq km), depending on habitat productivity and connectivity.

Behavior

Orioles are usually solitary and secretive, while figbirds are communal and extrovert. Figbirds often perch conspicuously on high bare branches in loose flocks of up to 30 or more, giving their simple one- or two-note whistled songs year-round to maintain contact. In contrast, orioles sing a short rolling glottal warble that is repeated monotonously during breeding and carries for almost half a mile (500 m) to advertise territory. The grouping and pitch of the notes may vary among species, but its character remains much the same everywhere. In agitation, both orioles and figbirds utter short, harsh squawks. Flight is direct and undulating, from tree to tree; it is swifter in orioles and ends in a flashing up-swoop on to a perch. Within tree crowns, orioles and figbirds are quiet and measured in their movements, perching still, or hopping about in search of food. They sun-bathe and rain-bathe there, but will also go to ground to water and infrequently feed, drinking mainly by "pumping" (sucking).

Feeding ecology and diet

Orioles eat small, soft fruit and insects opportunistically, whereas figbirds are more strictly frugivorus. By quartering

Presumed mimicry of friarbirds by orioles in east Indonesia. From top to bottom: dusky brown oriole (*Oriolus phaeochromus*) and dusky friarbird (*Philemon fuscicapillus*); Seram oriole (*Oriolus forsteni*) and Seram friarbird (*Philemon subcorniculatus*). (Illustration by Emily Damstra)

Frugivory in a female southern figbird (*Sphecotheres vieilloti*) feeding on a native fig. Figbirds are seasonally nomadic birds (they follow fruiting trees, especially figs) found in Australia. (Photo by Wayne Lawler. Photo Researchers, Inc. Reproduced by permission.)

tracts of forest in groups, figbirds converge on trees and work them over for the last ripe fruit; small fruits are swallowed whole. Often more territorial in their foraging, orioles take the same range of small fruits, but include hairy caterpillars and many other insects in their diet. They beat the larger insects vigorously against a perch to clean and soften them. It has been claimed that, for their protection when feeding and breeding, the streaked Indo-Australasian orioles have come to mimic other more aggressive birds in their appearance. Yet such mimicry may exist only in the eye of the human beholder, given that birds see color in a different way than humans.

Reproductive biology

Breeding occurs erratically year-round in the tropics but is confined to spring and early summer in temperate regions. Orioles and figbirds are essentially monogamous. Oriole males establish and hold territory while females build the nest and incubate with minimal help. Orioles build thick, deep basket-shaped nests of dry plant fiber bound with animal wool, moss, and lichen, which they sling in a horizontal fork in the outer branches of trees, usually well above the ground. Foundation strips are glued with saliva to hold the nest. Figbirds, in contrast, build a transparently flimsy cup of twigs and tendrils in small outer branches. Eggs, two to four per clutch,

are pink white to pale cream-buff in orioles or pale gray olive in figbirds, and sparingly spotted and speckled with black to red-browns. They hatch in 16–18 days. Young have yellow down and are fed by regurgitation by both parents, and occasional male helpers; they fledge in equivalent time. The brood, usually just one a year, is dispersed almost as soon as the fledglings become independent.

Conservation status

Only the Isabela oriole (*O. isabellae*) on isolated mountains in the northern Philippines is considered Endangered. Its habitat is threatened by clearing and logging, even in reserves. There are two Vulnerable species: the silver oriole (*O. mellianus*), which breeds locally in south-central China, and winters south to Thailand and Cambodia, and the Sao Tome oriole (*O. crassirostris*) of remnant primary forests on Sao Tome Island, tropical West Africa. Habitat disruption and, for the latter, pesticides are thought to have contributed to decline. Three species are Near Threatened: the black oriole (*O. hosii*) of montane northeast Borneo, the dark-throated oriole (*O. xanthonotus*) of Southeast Asia, and the Wetar figbird (*Sphecotheres hypoleucus*), which is endemic to Wetar in the Lesser Sundas.

Significance to humans

Despite their color, orioles and figbirds have made little impression on human society. The Eurasian golden oriole (*O. oriolus*) is a harbinger of impending summer in Europe and has sometimes been brought into captivity there, as have other species in Africa and Southeast Asia. No serious attempts have been made at introduction or acclimatization.

Male golden oriole (*Oriolus oriolus*) feeds chicks in the nest. (Photo by H.D. Brandl/Okapia. Photo Researchers, Inc. Reproduced by permission.)

1. Crimson-bellied oriole (*Oriolus cruentus*); 2. Eastern black-headed oriole (*Oriolus larvatus*); 3. Timor figbird (*Sphecotheres viridis*); 4. Olive-backed oriole (*Oriolus sagittatus*); 5. Eurasian golden oriole (*Oriolus oriolus*); 6. Australasian figbird (*Sphecotheres vieilloti*). (Illustration by Emily Damstra)

Species accounts

Australasian figbird
Sphecotheres vieilloti

TAXONOMY
Sphecotheres vieilloti Vigors & Horsfield, 1827, Keppel Bay, Queensland, Australia. Four subspecies: two (*flaviventris, ashbyi*) with yellow-ventrumed males in northern Australia and the Kai Islands, differing in tone of back; one (*salvadorii*) with gray-, yellow-, and white-ventrumed males in southeast New Guinea, and one (*vieilloti*) with gray- and olive-ventrumed males in central east Australia. Taxonomy controversial.

OTHER COMMON NAMES
English: Green figbird, southern figbird; French: Sphécothère de Vieillot. German: Feigenpirol; Spanish: Papahigos Verde.

PHYSICAL CHARACTERISTICS
10–11.5 in (25–29 cm); 4–4.5 oz (110–130 gm), both sexes. A stout, shortish-tailed figbird; upperparts olive-green, except black on head, primaries, and tail. Throat is gray with buff-red bare eye skin.

DISTRIBUTION
Coastal northern and eastern Australia (Kimberley Division to Illawarra district), southeast New Guinea, and Kai Islands (Banda Sea).

HABITAT
Rainforest edge, gallery vine forest, mangroves, and adjacent gardens.

Sphecotheres vieilloti
Resident

BEHAVIOR
Communal, in noisy, loose, locally nomadic flocks of up to 30–50 in tree canopy; often perches high on bare branches and power lines to group and call: loud, single- or double-note whistles.

FEEDING ECOLOGY AND DIET
Feeds in tree crowns primarily on small, soft fruit such as figs and native cherries; ink weed and tobacco bush are staple diet; bananas, guavas, and mulberries are occasionally taken in orchards.

REPRODUCTIVE BIOLOGY
Monogamous in small groups, holding small territories during later austral spring and summer (October–February). Nests are shallow, flimsy, and saucer-like, built of plant fiber and tendrils. Eggs usually three. Both parents share all nesting duties, and additional birds may help feed the young.

CONSERVATION STATUS
No populations are under threat anywhere, having adapted to habitats subject to disturbance and regeneration.

SIGNIFICANCE TO HUMANS
None known, apart from occasional nuisance value to fruit growers. ◆

Timor figbird
Sphecotheres viridis

TAXONOMY
Specoterasic viridis Vieillot, 1816, Australasia = Kupang, Timor.

OTHER COMMON NAMES
English: Green figbird; French: Sphécothère figuier; German: Feigenpirol; Spanish: Papahigos de Timor.

PHYSICAL CHARACTERISTICS
8.7–9.5 in (22–24 cm); 2.4–2.8 oz (75–80 gm), both sexes. A small, slender figbird; males are plain olive ventrally, with gray throat, undertail, and feet. Dark head with rough bare eyeskin colored deep buff but changing to red when excited. Throat to vent greenish yellow.

DISTRIBUTION
Endemic to Timor.

HABITAT
Primary and secondary vine and gallery forest, forest edge, woodland, and mangroves.

BEHAVIOR
Solitary to communal, feeding locally through tree crowns in small loose flocks. Calls from exposed position in tree tops: muted, burred, metallic whistled trills.

FEEDING ECOLOGY AND DIET
Little recorded; observed feeding on small, soft fruit in trees, including figs.

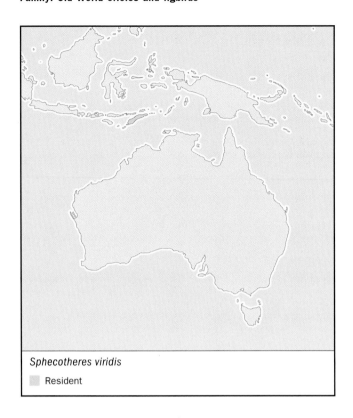

Sphecotheres viridis

▨ Resident

Oriolus sagittatus

▨ Resident

REPRODUCTIVE BIOLOGY
Not known.

CONSERVATION STATUS
Widespread and moderately abundant; copes with habitats affected by disturbance.

SIGNIFICANCE TO HUMANS
None known. ◆

Olive-backed oriole
Oriolus sagittatus

TAXONOMY
Coracias sagittata Latham, 1802, "Nova Wallia Australi" = central coast, New South Wales, Australia.

OTHER COMMON NAMES
English: Green oriole, white-bellied oriole; French: Loriot sagittal; German: Streifenpirol; Spanish: Oropéndola de Lomo Olivo.

PHYSICAL CHARACTERISTICS
10.2–11 in (26–28 cm); 3.5–3.8 oz (90–105 gm), both sexes. Plain olive upperparts; wings and tail dark gray with white edging. Whitish underparts with dark steaks. Eyes red.

DISTRIBUTION
Coastal and subcoastal northern and eastern Australia up to about 500 mi (800 km) inland, between Kimberley Division and Victoria, also dry south New Guinea.

HABITAT
Open eucalypt forest and tall woodland; also paperbark (*Melaleuca*) woodland in north Australia and New Guinea.

BEHAVIOR
Solitary in forest/woodland canopy, pairing only to breed. Gives low carrying glottal warble from set perches throughout year, and incessantly, with mimicry, when breeding. Widely nomadic after breeding.

FEEDING ECOLOGY AND DIET
Forages opportunistically on small soft fruits in trees and tall shrubs and on insects such as leaf beetles, mantids, ants, and caterpillars; mostly captures prey while quietly hopping and gleaning upper and outer branches of trees.

REPRODUCTIVE BIOLOGY
Monogamous; female builds nest (about 14 days) and incubates (17–18 days), but both sexes feed young, fledging them in 15–27 days.

CONSERVATION STATUS
No populations under significant threat anywhere.

SIGNIFICANCE TO HUMANS
None known. ◆

Crimson-bellied oriole
Oriolus cruentus

TAXONOMY
Leptopteryx cruenta Wagler, 1827, Java. Four subspecies, one each in four regions of accurrence.

OTHER COMMON NAMES
English: Black-and-crimson oriole; French: Loriot ensanglanté; German: Rotbrustpirol; Spanish: Oropéndola de Pecho Rojo.

Oriolus cruentus
Resident

PHYSICAL CHARACTERISTICS
8.5–9 in (21.5–23 cm); no weights available. A dusky, gray-billed oriole with crimson or chestnut lower breast and belly and mark in wing.

DISTRIBUTION
Greater Sunda Land: Malay Peninsula, Sumatra, Java, and north Borneo.

HABITAT
Closed, primarily high montane rainforests above 3,300 ft (1,000 m).

BEHAVIOR
Retiring and sedentary, working locally through the inner forest canopy and forest midstage singly or in pairs, although foraging out into cloud forest at high altitudes. Song is a short, mewing, nasal warble of about four syllables, repeated at intervals.

FEEDING ECOLOGY AND DIET
Little known, though hops about actively when foraging and often descends to logs in forest substage. Eats small soft fruits such as figs and berries, and takes a wide range of insects including termites, diverse larvae, and particularly caterpillars, both hairless and haired.

REPRODUCTIVE BIOLOGY
Insufficiently known.

CONSERVATION STATUS
None of the four mountain island populations appear to be under threat at present, although accelerating habitat destruction could impact them in the future.

SIGNIFICANCE TO HUMANS
None known. ◆

Eastern black-headed oriole
Oriolus larvatus

TAXONOMY
Oriolus larvatus Lichtenstein, 1823, "Terr. Caffror" = Cape Province, South Africa. Five or six regional subspecies differing in size and tone of golden plumage.

OTHER COMMON NAMES
English: Black-headed oriole, African black-headed oriole; French: Loriot à tête noire oriental; German: Maskenpirol; Spanish: Oropéndola de Cabeza Negra Africana.

PHYSICAL CHARACTERISTICS
8–8.5 in (20–21.5 cm); 2.2–2.5 oz (60–70 gm). Sexes similar. A black-headed oriole with citrine back, golden ventrum, and gray-green wings and tail; only immatures duller and streaked ventrally.

DISTRIBUTION
Central and southeastern Africa, north to Sudan and southwest Ethiopia, south to Cape Province, and west through much of Angola locally to west coast.

HABITAT
Acacia and broad-leaved woodlands, especially in galleries along streams; also thorn scrub, forest edge, tree plantations, and gardens.

BEHAVIOR
Solitary and rather sedentary, in pairs or small family groups only when breeding or at food concentrations. Calls in considerable variety from tree perches throughout year, mimicking and giving both fluted glottal warbles and figbird-like whistles.

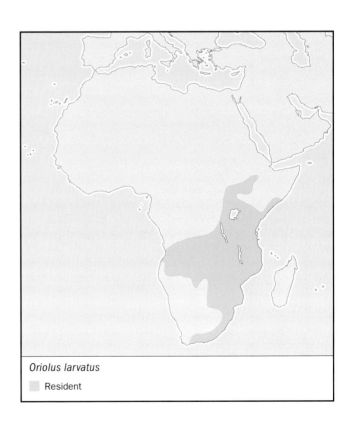

Oriolus larvatus
Resident

FEEDING ECOLOGY AND DIET

Forages mainly in tree crowns, hopping and fluttering, for fruits and caterpillars. Food comprises both small, soft native fruits such as figs, grapes, and mulberries, and insects, particularly large haired and hairless caterpillars, all taken opportunistically.

REPRODUCTIVE BIOLOGY

Monogamous and territorial, pairs holding territories of 12–120 acres (5–50 ha). Breeds erratically year-round in tropics, but limited to spring and summer in temperate southern latitudes. Female builds nest and broods, and male holds territory and assists in rearing young. Nests slung in a three-way fork in horizontal branches at 15–50 ft (5–15 m) above the ground. Eggs in clutches of one to two in tropics and two to three in temperate zone, hatch in 14–15 days; young fledge in another 15–18.

CONSERVATION STATUS

No forms of this common and widespread species are under threat.

SIGNIFICANCE TO HUMANS

None known. ◆

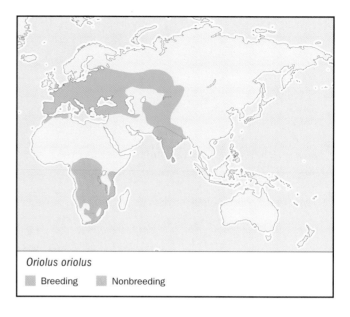

Oriolus oriolus

█ Breeding █ Nonbreeding

Eurasian golden oriole
Oriolus oriolus

TAXONOMY

Oriolus oriolus Linnaeus, 1758, "in Europe, Asia" = Sweden. Two subspecies, one in Europe to west Asia and the other in Central Asia, differing in size, proportions, and black stripe through eye.

OTHER COMMON NAMES

English: Golden oriole, oriole; French: Loriot d'Europe; German: Pirol; Spanish: Oropéndola Europea.

PHYSICAL CHARACTERISTICS

8.3–9.5 in (21–24 cm); 2–2.8 oz (55–80 gm), both sexes. An all-golden oriole with black wings, tail, and stripe through eye; female duller and greenish-backed.

DISTRIBUTION

Europe and far northwest Africa east to Asia Minor, the Caspian Sea, western Siberia, and, in winter, sub-Saharan Africa; Central Asia, from western Siberia south in winter to Afghanistan and Himalayas, peninsular India, and north Sri Lanka.

HABITAT

Woodlands and open forests of primarily broadleaf and deciduous mature trees, including parkland.

BEHAVIOR

Solitary and retiring, keeping within the upper stages of large trees. Song, given year-round by both sexes but more by males during breeding, is a loud, liquid, warbled whistle of three or four syllables, repeated; a grating, drawn-out squalling given in agitation, as well as a range of warbled sub-songs.

FEEDING ECOLOGY AND DIET

Forages by picking mainly in foliage of tree crowns, but also occasionally by hovering and perch-pouncing near to ground. Diet includes both a range of insects, including hairy caterpillars, and a range of small fruits, as well as some seeds, buds, small reptiles, nestling birds, and eggs.

REPRODUCTIVE BIOLOGY

Monogamous, single-brooded, and territorial, males holding dispersed territories. Nest, a shallow cup of plant fiber and stems slung from a thin horizontal fork in high foliage, is built in 6–12 days by female, with initial assistance from male. Eggs, three to four per clutch, are creamy or pink-white with scattered dark brown and blackish spots. Incubated by female with assistance from male, they hatch in 15–18 days. Young, fed by both parents and infrequent trios, fledge in 16–20 days.

CONSERVATION STATUS

Neither subspecies appears to be under threat at present, though status of central Asian race unclear. For western subspecies, declines in central and eastern Europe over past 50 years partly offset by increases in western Europe.

SIGNIFICANCE TO HUMANS

A noted harbinger of spring and summer in northern Europe, both for its golden plumage and fluted song. Occasionally brought into aviculture in Europe as well. ◆

Resources

Books

Cramp, S., and C. M. Perrins, eds. "Flycatchers to Shrikes." *The Birds of the Western Palearctic*, Vol. VII. Oxford: Oxford University Press, 1993.

Du Pont, J. E. *Philippine Birds*. Monograph Series No. 2. Delaware: Delaware Museum of Natural History, 1971.

Fry, C. H., S. Keith, and E. K. Urban, eds. *The Birds of Africa*. Vol. VI. New York: Academic Press, 2000.

Resources

Greenway, J. C. "Oriolidae." *Check-List of Birds of the World.* Vol. 15. Cambridge, Mass.: Museum of Comparative Zoology, 1962.

MacKinnon, J., and K. Phillipps. *A Field Guide to the Birds of Borneo, Sumatra, Java, and Bali.* Oxford: Oxford University Press, 1993.

Schodde, R., and S. C. Tidemann, eds. *Reader's Digest Complete Book of Australian Birds.* 2nd edition. Sydney: Reader's Digest Services, 1986.

Schodde, R., and I. J. Mason. *The Directory of Australian Birds Passerines.* Melbourne: CSIRO Publishing, 1999.

Sibley, C. G., and J. E. Ahlquist. *Phylogeny and Classification of Birds.* New Haven: Yale University Press, 1990.

Sibley, C.G., and B. L. Monroe, Jr. *Distribution and Taxonomy of Birds of the World.* New Haven: Yale University Press, 1990.

Smythies, B. E. *The Birds of Borneo.* 3rd edition. Kota Kinabalu and Kuala Lumpur, Malaysia: The Sabah Society and the Sabah Nature Society, 1981.

UNEP World Conservation Monitoring Centre. *Threatened Animals of the World.* <http://www.unepwcmc.org/species/animals/animal redlist.html. January 2002.

Periodicals

Diamond, J. M. "Mimicry of Friarbirds by Orioles." *Auk* 99 (1982): 187-196.

Organizations

Australian National Wildlife Collection. GPO Box 284, Canberra, ACT 2601 Australia. Phone: +61 2 6242 1600. Fax: +61-2-6242-1688.

Other

Australian Biological Resources Study. Canberra, ACT, Australia.

Richard Schodde, PhD

Drongos
(Dicruridae)

Class Aves

Order Passeriformes

Suborder Passeri (Oscines)

Family Dicruridae

Thumbnail description
Glossy black, fork-tailed, starling-sized song-birds that live in trees and hawk for insect food from vantage perches

Size
7–25 in (18–63 cm), including tail; 0.8–4.8 oz (25–135 g)

Number of genera, species
2 genera; about 22–24 species

Habitat
Tropical and subtropical forest to forest edge, open woodland, and gardens

Conservation status
Endangered: 2 species; Near Threatened: 4 species

Distribution
Old World tropics and subtropics, from sub-Saharan Africa to southern Asia southeast through Indonesian archipelagos to Solomon Islands and Australia

Evolution and systematics

Long thought to be related to starlings and orioles, the drongos have been shown by recent molecular and skeletal research to be nested among the Australasian monarch fly-catchers (family Monarchidae) and related to corvoid song-birds. Unlike starlings but like monarchs, they have a single pneumatized fossa in the head of the humerus; and like monarchs but unlike orioles, they have a heavily ossified nasal cavity and narrow ectethmoidal plate. To manhandle large and hard-cased insect food, they also have an extended bony plate in the roof of the palate, thickened nasal bars, and a large depression in each temple flanked by a long zygomatic process for the attachment of powerful jaw muscles; drongos are strong-billed birds. Outwardly, monarchs also match drongos in their glossy black plumages, arboreal hawking behavior, flimsy nests perched in horizontal tree forks, and reddish-marked eggs.

The drongo-monarch group is a core branch in a massive radiation of crow-like songbirds that appears to have exploded in Australia some 20–30 million years ago, and quickly spread through the Old World tropics. The drongo lineage would have been in the vanguard, reaching Africa and radiating into 11–13 species in Southeast Asia and fringing archipelagos. Left behind in montane New Guinea, signposting the source of radiation as it were, was the pygmy drongo (*Chaetorhynchus papuensis*), the most monarch-like and ancestrally structured drongo of all. The fossil record, limited to the Pleistocene for drongos, preserves little of this information.

Today there are 22–24 species of drongos, one in the genus *Chaetorhynchus* and the rest in the larger but still tight-knit genus *Dicrurus*. Only the taxonomy of the *balicassius-hottentotottus* species-group in the Indo-Australasian archipelagos is seriously controversial, opinions being divided over whether there are anywhere from four to eight species in the 36 taxon complex.

Physical characteristics

Resembling stream-lined, long-tailed starlings, drongos are a picture in black—black in plumage, bill, and feet—except for the gray ashy drongo (*D. leucophaeus*) in Southeast Asia; immatures are duller and sootier, and in some species faintly scalloped, barred, or spotted paler. Eyes, brilliant red in most species (though whitish or brown in some forms), provide the only color contrast; immatures are brown-eyed without exception. Plumage is extensively glossed in green, blue, or purplish sheens, the gloss sometimes spangling hackles on head and breast. Some species are also crested. Crests vary, according to species, from a short tuft of erect feathers on the forehead to bare hair-like plumes or a mane of broadened curled feathers curving back over the head.

Bills are stout, deep, and rather aquiline, well-notched for grasping prey, and clothed with dense, forward-directed bristles at the base. The bristles, which hide slit-like nostrils, are thought to protect the face from retaliation by captured prey; in *Chaetorhynchus* they extend to the bill tip. Feet are short but

Black drongo (*Dicrurus macrocercus*) chicks at their nest. (Photo by T.D. Singh/VIREO. Reproduced by permission.)

strong, and with the toes about as long as the scutellate "leg," are better fitted for perching than movement. Adapted for aerial maneuvers, wings are rather long and pointed, with 10 primaries (tenth well-developed) and nine secondaries plus a remicle. It is the tail that sets drongos apart from other songbirds. Sometimes square-tipped but usually long and forked, it comprises 12 feathers in *Chaetorhynchus* and only 10 in *Dicrurus*, and is often diversely modified by a great lengthening and curling of the outermost pair of feathers. In two species, these plumes are largely bare of webbing except for spatula-like tips; but the function of such modification is not clear. Tails in all species are at least as long as the body, and up to three times as long, even more, in some. Because of this, drongos vary enormously in length, from 7 in to over 23.5 in (18–60 cm).

Distribution

Drongos occur throughout the Old World tropics and subtropics: there are four species in tropical and subtropical sub-Saharan Africa, four in Madagascar and nearby island archipelagos, 11–13 centered in southern and Southeast Asia, from east Iran and India to south Manchuria (Bol Hai), the Philippines and central Indonesian archipelagos, and just four (possibly five) from the east Indonesian Archipelagos to the Solomon Islands and north and east Australia. Most species are sedentary, but populations of those that breed in more temperate latitudes, of the black drongo (*Dicrurus macrocercus*), ashy drongo, and hair-crested drongo (*D. hottentottus*) in China, and the spangled drongo (*D. bracteatus*) in east Australia, are migratory, shifting to the tropics in winter.

Habitat

Tropical and subtropical forests, secondary growth, and forest edge to mangroves, open woodlands, and even urban environs are the habitat of drongos, according to species. Some, such as the New Guinean pygmy drongo, the African shining drongo (*D. atripennis*), and Sulawesi drongo (*D. montanus*) are confined to primary rainforest; but others, notably the widespread Asian black drongo, occupy towns, gardens, and open areas, and are a familiar sight perched on electric lines. Where different species occur together, they co-exist by occupying different habitats. Thus in Southeast Asia, the black, ashy, and greater racket-tailed (*D. paradiseus*) drongos live in more open woods, urban areas, and marshes, while the bronzed (*D. aeneus*), lesser racket-tailed (*D. remifer*), and hair-crested drongos keep to denser forests, often replacing one another altitudinally. In Africa, square-tailed (*D. ludwigii*), velvet-mantled (*D. modestus*), and shining (*D. atripennis*) drongos replace one another in different strata in different structural habitats. Wherever they occur, both resident and migratory populations are well dispersed at densities of about 5–40 birds per mi^2, depending on productivity and connectivity of habitat.

Behavior

Solitary except when paired or in family groups during breeding, drongos are nevertheless showy and noisy birds. Their daily routine is one of perching in the bare middle tiers of trees or their edges, on exposed vantage points from which to sally out in buoyant but agile evolutions on the wing after food, and then return. On the perch they sit upright, long tail hanging down and sporadically waving or twitching from side to side. All movement is on the wing, the birds never moving about on foot. To bathe, they plunge-dive from a perch or flight. Among the first birds to rise before dawn and the last to go to roost, drongos call regularly throughout the day year round, mostly from perches. The calls, of a great variety of grating chatters, creaking hinge notes, discordant chuckles, and melodious whistles, usually have something of a metallic twang; and each song stanza is rarely longer than five or six quick syllables. Some species, perhaps all, are accomplished mimics. Although some species are more retiring than others, all are rather quarrelsome, and are bold and pugnacious in defense of territory. They will attack and chase off birds as large as crows and medium-sized raptors, and mob owls, hornbills, and small predatory mammals.

Feeding ecology and diet

Drongos feed by hawking from perches, either catching prey in the air or picking it from the surface of leaves, branches, or ground. They often join flocks of mixed insectivorous birds quartering their habitat, benefiting from insects disturbed by other species. Drongos of more open habitat also converge opportunistically on fires to snap up insects flushed by flame and smoke. They take an enormous variety of arthropods, many of them large and hard-shelled and most of them flying: grasshoppers, crickets, beetles, mantids, cicadas, moths and butterflies, dragonflies, ants, and even venomous Hymenoptera (wasps and bees), occasional arachnids, and sometimes small birds; they also rifle blossoms for nectar. Termite emergences have a special attraction. Large items are carried in the foot and held by it at a perch, to be torn apart with the bill; moths, butterflies, and dragonflies have their wings torn off before being swallowed.

Reproductive biology

Monogamous and aggressively territorial, drongos breed in dispersed pairs, some species often rearing several broods in a season. In most regions, breeding peaks over spring and summer, particularly in temperate zones north and south of the tropics, where post-breeding molt is also more consistently seasonal. Little is known of courtship, which may comprise little more than head bobbing and bowing, and duetting and counter-singing, as male and female sit together. Both sexes share all nesting duties from nest construction to incubation and rearing of young. Nests are flimsy, shallow saucers of loosely but neatly intertwined rootlets, tendrils and, in some species, leaves, lined with finer material and sometimes wool, often bound on the outside with cobweb and camouflaging lichen, and hung in a horizontal fork at the end of a branch at 6.6–82 ft (2–25 m) above the ground. In clutches of two to five, eggs are salmon-buff to pale cream or pinkish white, covered in freckles and small blotches of brownish red, black, or umber and lilac according to species, with underlying smudges and streaks of pale gray and purple; they measure about 0.8–1.2 by 0.5–0.8 in (20–30 by 15–21 mm). In the African fork-tailed drongo (*D. adsimilis*), eggs hatch in 16–17 days and young fledge in another 17–18 days.

Conservation status

Drongo survival is only threatened when total treed habitat is limited in area. This is the case for species confined to small islands, such as those in the Comoro group off the central east coast of Africa. Both the Comoro drongo (*D. fuscipennis*) on Grand Comoro Island and Mayotte drongo (*D. waldenii*) on Mayotte Island are listed by IUCN as Endangered. Two of the four species in the Near Threatened category also occur on small islands: *D. aldabranus* on Aldabra Island just north of the Comoros, and *D. andamanensis* in the

A fork-tailed drongo (*Dicrurus adsimilis*) visits its nest. (Photo by Clem Haagner. Bruce Coleman Inc. Reproduced by permission.)

Andamans in the Bay of Bengal. Of the other two, the velvet-mantled drongo (*D. modestus*) occurs in west central Africa and the Sumatran drongo (*D. sumatranus*) is endemic to Sumatra, Indonesia. Destruction of forest habitat is their primary threat.

Significance to humans

Despite their extrovert behavior, drongos have made little impact on human society and culture except for the black drongo. This species, a familiar urban commensal across southern Asia, is often cultivated in captivity there. Black drongos from Taiwan were also introduced successfully to Rota in the southern Marianas (Micronesia) in the 1930s, and from there had colonized neighboring Guam by the early 1960s. Its other vernacular name—king crow—celebrates its nerve and pugnacity in driving off predatorial birds much larger than itself.

1. Greater racket-tailed drongo (*Dicrurus paradiseus*); 2. Ashy drongo (*Dicrurus leucophaeus*); 3. Square-tailed drongo (*Dicrurus ludwigii*); 4. Ribbon-tailed drongo (*Dicrurus megarhynchus*); 5. Pygmy drongo (*Chaetorhynchus papuensis*); 6. Black drongo (*Dicrurus macrocercus*). (Illustration by Brian Cressman)

Species accounts

Pygmy drongo
Chaetorhynchus papuensis

TAXONOMY
Chaetorhynchus papuensis A. B. Meyer, 1874, Arfak Mountains, northwest New Guinea. Monotypic. Monarch-like in form and behavior, this species has been confused with the monarch-flycatcher group in the past.

OTHER COMMON NAMES
English: Papuan drongo; French: Drongo papou; German: Rundschwanzdrongo; Spanish: Drogo Papúa.

PHYSICAL CHARACTERISTICS
8–8.5 in (20–22 cm); 1.2–1.6 oz (35–45 g), both sexes. A very small drongo, all black with blue gloss over head and back. It has rictal bristles that extend beyond the bill tip, a short rounded crest over the head, a square-tipped tail of 12 feathers, a concealed white patch on the inner wing coverts under the scapulars in both sexes, and brown eye; immatures are duskier and glossless, and lack the concealed white wing spot.

DISTRIBUTION
Lower slopes of mountain ranges throughout mainland New Guinea, between 1,600 and 5,000 ft (488–1,524 m) above sea level.

HABITAT
Interior of primary and tall secondary hill rainforest.

BEHAVIOR
Solitary or in pairs within lower stages of rainforest where territorial sallying for arthropod food on wing or perching mo-
tionless on bare exposed twigs and branches, sitting near upright with tail hanging down and occasionally twitching from side to side or raised, fantail (*Rhipidura*)-like, on alighting. Sporadically vocal, uttering two types of song of unknown function, one an explosive jumble of nasal and metallic rasps and squeaks (4.5–5.0 seconds), the other a loud melodious mix of whistles, chips, and warbles; other calls comprise a range of metallic clicks, slurs, and squeaks.

FEEDING ECOLOGY AND DIET
Forages by drongo-like sallying in more open mid and lower strata of forest, capturing a range of insects; commonly associates with feeding flocks of other bird species, benefiting from insects disturbed.

REPRODUCTIVE BIOLOGY
Not known.

CONSERVATION STATUS
Not threatened.

SIGNIFICANCE TO HUMANS
None known. ◆

Square-tailed drongo
Dicrurus ludwigii

TAXONOMY
Edolius ludwigii A. Smith, 1834, Port Natal = Durban, South Africa. Five subspecies.

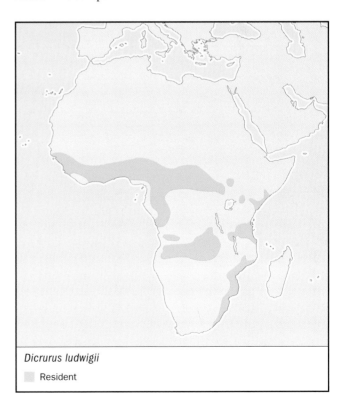

Chaetorhynchus papuensis
▨ Resident

Dicrurus ludwigii
▨ Resident

OTHER COMMON NAMES
French: Drongo de Ludwig; German: Geradschwanzdrongo;
Spanish: Drogo de Cola Cuadrada.

PHYSICAL CHARACTERISTICS
7–7.5 in (18–19 cm); 0.8–1.3 oz (25–35 g). Smallest of drongos,
plain headed, red-eyed, and all black, glossed purplish or
green, with a characteristically short and squarish tail; females
are similar to males or duller, and immatures dull in both sexes
and speckled pale gray on mantle and breast.

DISTRIBUTION
Endemic to Africa where patchy in distribution: subspecies
sharpei occurs in a narrow band through the west and central
tropics from Guinea to Uganda, western Kenya, and other minor races in central Angola, Zambia-Zaire, southeast Somalia,
eastern Tanzania, and coastal Mozambique to the southeast
Cape Province.

HABITAT
Middle and lower strata of gallery forest, moist thicket,
wooded glades, primary rainforest, and denser woodland from
sea-level up to 6,600 ft (2,000 m) above sea level in the tropics.
D. l. sharpei, in particular, keeps to gallery forests and woodlands around the fringe of primary rainforest, being replaced
within by the shining drongo (*D. atripennis*).

BEHAVIOR
Usually permanently territorial, in pairs or family groups.
Though more retiring than other African drongos, it still forages by sallying, sits on exposed vantage perches with tail
drooped and twitched from side to side, defends nest pugna-
ciously, and calls often and rather loudly: repeated single-note
upslurred or down-slurred whistles and buzzes. Song duets are
a quiet and rapidly delivered medley of short whistles and liquid chattery notes.

FEEDING ECOLOGY AND DIET
Aerial insectivore, often accompanying foraging bands of mixed
species of birds. Diet mainly of rather large insects: moths,
grasshoppers, mantids, and beetles; also exploits termite emergences.

REPRODUCTIVE BIOLOGY
Breeds April through November north of equator and September through April in south. Nest a small, neat saucer of lichen
and dry stems bound thickly with cobweb at rim, 2.9 in diameter x 1 in deep (75 mm x 25 mm), suspended by rim in horizontal fork at branchlet extremity 6.6–26 ft (2–8 m) above
ground; eggs two to three per clutch, 0.8–0.9 x 0.6–0.7 in
(20–23 x 15–16.5 mm), white to pale buff, spotted with lilac
and brown, mostly at larger end.

CONSERVATION STATUS
Not threatened.

SIGNIFICANCE TO HUMANS
None known. ◆

Ribbon-tailed drongo

Dicrurus megarhynchus

TAXONOMY
Edolius megarhynchus Quoy and Gaimard, 1830, "Dorérei" =
Port Praslin, New Ireland. Monotypic member of spangled
drongo (*D. bracteatus*) superspecies.

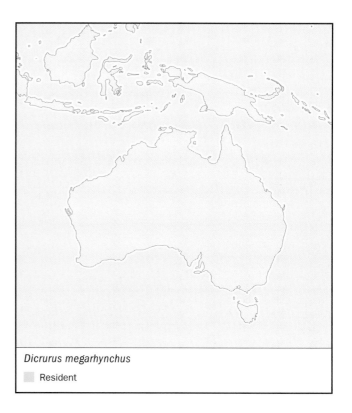

Dicrurus megarhynchus
▨ Resident

OTHER COMMON NAMES
French: Drongo de Nouvelle-Irlande; German: Bandschwanz-
drongo; Spanish: Drogo de Nueva Irlanda.

PHYSICAL CHARACTERISTICS
22–25 in (55–63 cm). Largest of drongos, with strong bill and
enormously elongate, streaming, in-curling outer tail feathers
that are slightly spathulate at the tips. Iris rich red and
plumage all black, with blue gloss on upper surface and wings,
paler blue-glossed spangles on breast, and white tipping on
concealed under-wing coverts; females are smaller and immatures are duller, near glossless.

DISTRIBUTION
Endemic to New Ireland in Bismarck Archipelago, at all altitudes from sea-level to 5,900 ft (1,800 m) above sea level.

HABITAT
Mainly interior of mid and upper strata of primary and tall secondary rainforest.

BEHAVIOR
More retiring than other drongos, but territorial resident year-
round and similarly solitary or in pairs, sitting upright on open
perches within cover, tail dangling. It flares the tail when calling; calls include a loud, liquid medley of whistles (probably
song) unlike the metallic twanging chatter of other drongos.

FEEDING ECOLOGY AND DIET
Aerial insectivore, sallying actively like other drongos, tail trailing, taking a range of large arthropods in flight or picked from
the surface of leaves and branches.

REPRODUCTIVE BIOLOGY
Not known.

CONSERVATION STATUS
Not threatened.

SIGNIFICANCE TO HUMANS
None known. ◆

Ashy drongo
Dicrurus leucophaeus

TAXONOMY
Dicrurus leucophaeus Vieillot, 1817, "Ceylon" = Java. About fourteen subspecies, differing principally in depth of body tone, tail furcation, and presence and extent of a white facial mark. All subspecies intergrade where their breeding ranges meet.

OTHER COMMON NAMES
French: Drongo cendré; German: Graudrongo; Spanish: Drogo Cenzio.

PHYSICAL CHARACTERISTICS
10.5–12.0 in (26–30 cm); 1.4–2.2 oz (40–60 g). The only gray drongo, slim in body and with well-forked tail, varying from pale to dark slaty ashen, paler ventrally and with light green-blue gloss dorsally that disappears in the palest races; sexes alike and rich red-eyed but immatures duller and brown-eyed.

DISTRIBUTION
Southeast Afghanistan through Himalayas, peninsular India and Sri Lanka to all Southeast Asia north through central and east China to south Manchuria and southeast in Greater Sundas and off-shore islands north to Palawan (Philippines) and east to Lombok (Lesser Sundas). More northern populations migrate to more tropical latitudes out of breeding during November through March.

HABITAT
Edges and open interior of taller, intact forest, including rainforest, bamboo forest, and mixed pine-oak forest, as well as shady village groves when resident or breeding, at altitudes from sea level to over 9,900 ft (3,000 m); migrants commonly enter more open woodlands, gardens and plantations at winter quarters.

BEHAVIOR
Usually in pairs or small groups, particularly on migration, ashy drongos work through the upper strata of trees, perching high up and launching in swooping sallies after food, from tree top to near the ground. Calls are a varied assortment of harsh screeches, metallic chatterings, and pleasant musical whistles; mimicry is often included.

FEEDING ECOLOGY AND DIET
Crepuscular aerial insectivore, feeding by sallying from vantage perches from dawn almost until dark. Diet includes a variety of flying insects—locusts, grasshoppers, crickets, moths and butterflies, ants, beetles, and venomous Hymenoptera—as well as occasional small reptiles and birds. Ashy drongos gather in groups of up to 30 or so at termite emergences, and also rifle nectar frequently.

REPRODUCTIVE BIOLOGY
Northern migratory populations breed in the late boreal spring into early summer (April–June), but residents in more tropical latitudes are less seasonal (December–February in Borneo). Nest: a shallow saucer of lichen, dry leaves, and stems secured with or without cobweb in a slender fork at the end of a branchlet at 33–66 ft (10–20 m) above ground. Eggs: two to four per clutch, 0.9–1 x 0.70–0.74 (22.5–25.0 x 17–19 mm), variably pinkish cream to sometimes dark buff, and rather heavily spotted and blotched with reddish brown and black.

CONSERVATION STATUS
Although the species as a whole is not threatened, its forms on small islands off Sumatra—*D. l. periophthalmicus* and *D. l. siberu*—are probably under threat from habitat depletion.

SIGNIFICANCE TO HUMANS
None known. ◆

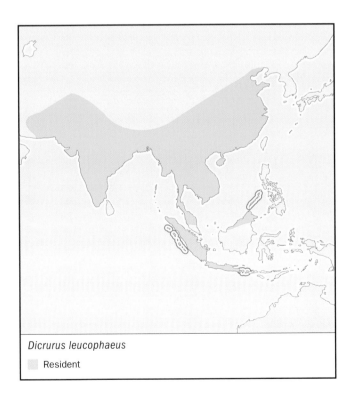

Dicrurus leucophaeus
Resident

Greater racket-tailed drongo
Dicrurus paradiseus

TAXONOMY
Cuculus paradiseus Linnaeus, 1766, "Siam" = region between Ayuthia and Gulf of Thailand. About 14 subspecies, differing principally in size and presence and form of crest and tail rackets.

OTHER COMMON NAMES
English: Large racket-tailed drongo; French: Drongo à raquettes; German: Flaggendrongo; Spanish: Drogo de Cola Raqueta Grande.

PHYSICAL CHARACTERISTICS
13–14 in (33–36 cm), excluding rackets; 2.5–4.4 oz (70–125 g). Most extravagantly plumed drongo, red-eyed and all black glossed bluish to greenish, with scaly-hackled mantle, a large crest varying from forward-facing bristles to, usually, webbed plumes curving back over crown in a mane, and an extraordinary tail with two outer racket-tipped streamers that, of variable form, are often bare of webbing except for short, twisted spoon-shaped vanes at tips; sexes similar but immatures dull and brown-eyed with vestigial crests and shortened tail streamers.

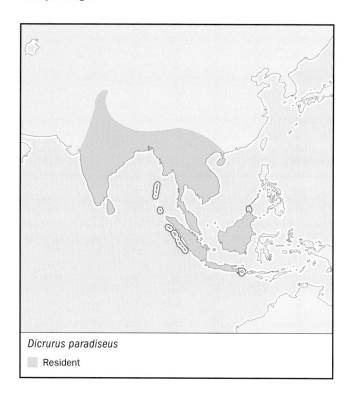

Dicrurus paradiseus
■ Resident

DISTRIBUTION
Lowland Southeast Asia, locally to 4,900 ft (1,500 m) above sea level, from India and Sri Lanka north to south Himalayas, and east to all Southeast Asia south of south China (Yunnan, Hainan Island) and north Vietnam, and throughout Greater Sundas and archipelagos in the Bay of Bengal.

HABITAT
Open interior and edges of evergreen and deciduous forests, including rainforests, and taller close woodlands, from sea-level to 4,900 ft (1,500 m); also plantations, garden edges, and open secondary growth.

BEHAVIOR
Either solitary, in pairs or small family groups, greater racket-tailed drongos are resident and territorial throughout their range. They commonly associate with foraging bands of mixed bird species, perching on exposed branches in and on the edge of forest mid-strata and sallying forth on wing. Flight is buoyant and dipping, of rapid flapping then gliding, the tail streamers making a distinctive humming. Racket-tails are extrovertly noisy, often displaying and counter-singing in small groups, and are one of the first and last birds stirring at dawn and dusk. They have a lusty repertoire of resonant warbles, whistles, and bell-like notes, mixed with typically drongo-like dry chatterings and metallic twanging; mimicry is also a feature.

FEEDING ECOLOGY AND DIET
Hawks, like other drongos, catching prey in mid-air or from vegetation surfaces and often carrying it in its claws to a perch for dismembering. Diet comprises mainly a range of large flying insects (termites, moths, beetles, dragonflies, ants, bees, locusts, and mantids); nectar is an important supplement.

REPRODUCTIVE BIOLOGY
Breeds over summer (June–July) in northern parts of range, but earlier from winter (February) in the low tropics. Nest a deep, loose cup, 5.9 in diameter x 1.9 in deep (15 cm x 5 cm),

of twigs, bark, and tendrils, unlined and with little, if any, binding cobweb, fastened at the rim into a horizontal fork at a branch end, 16–50 ft (5–15 m) above the ground. Eggs, in clutches of 2–4, are 1–1.1 in x 0.7–0.8 in (26–29 mm x 19–21 mm) and creamy white to pale pink, blotched and speckled with deep red-brown and underlying pink-grays.

CONSERVATION STATUS
Although the species as a whole is not threatened, some of its races on small islands are vulnerable to habitat depletion, notably *D. p. banguey*, *D. p. microlophus*, *D. p. lophorinus*, *D. p. otiosus*, *D. p. nicobariensis*, and possibly *D. p. johni*.

SIGNIFICANCE TO HUMANS
None known. ◆

Black drongo
Dicrurus macrocercus

TAXONOMY
Dicrurus macrocercus Vieillot, 1817, "Afrique" = Madras, India. About seven subspecies, differing in size, tone of gloss and wing lining, tail function, and presence of a white facial spot.

OTHER COMMON NAMES
English: King crow; French: Drongo royal; German: Königs-drongol; Spanish: Drogo Real.

PHYSICAL CHARACTERISTICS
11–13 in (26–32 cm); 1.5–2.2 oz (40–60 gm). The archetypal drongo, slender bodied, jet black with blue or green gloss, red eye, uncrested and without hackles but well bristled around bill, and tail deeply forked; sexes are similar, and immatures dull, shorter-tailed, and brown-eyed.

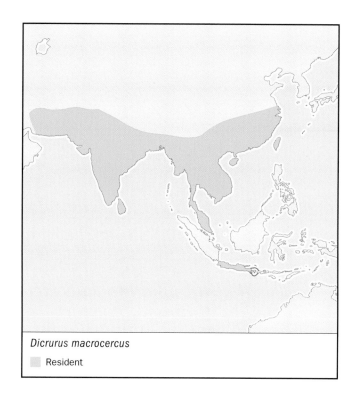

Dicrurus macrocercus
■ Resident

DISTRIBUTION

Southeast Iran and south Afghanistan through south Himalayas to all India, Sri Lanka and mainland Southeast Asia north through central and east China to Formosa and south Manchuria, at altitudes from sea level to 6,600 ft (2,000 m); also an outlying population on Java and Bali.

HABITAT

Savannas, fields, and urban habitats outside forest, in more open environments than occupied by other drongos; over much of its range the black drongo has become a commensal of man.

BEHAVIOR

Thrives in open habitat, seeking exposed vantage perches in isolated trees, fence posts, tops of banks, buildings, and electric wires and poles, and sallying out from them on wing in buoyant, acrobatic evolutions. Tropical populations are resident but opportunistically nomadic out of breeding, and more temperate populations north through China are strongly migratory. This species is more social than other drongos, particularly out of breeding and may roost in flocks, dispersing at dawn to respective feeding territories. It is also especially pugnacious and fearless in defending feeding and breeding territory from larger, predatorial birds. Calls comprise a variety of harsh metallic chatterings; in apparent courtship, pairs or competitive trios perch close or face to face duetting or counter-singing in harsh scolding notes accompanied by violent up-and-down head-bowing and wing fluttering aerial chases.

FEEDING ECOLOGY AND DIET

An opportunistic aerial insectivore, congregating loosely at concentrations of food and environmental disturbances that flush it, such as fire, grazing domestic stock, field clearing, and ploughing. Black drongos even chase other birds piratically for captured prey, and will sometimes settle on the ground to pick up ants and emerging termites. Their staple diet comprises a range of large, hard-cased field insects—locusts and crickets, beetles, and bees—and also some moths and butterflies and, infrequently, small reptiles, birds, and bats. Nectar is an important supplement, and the drongos may play a useful role in plant pollination.

REPRODUCTIVE BIOLOGY

Breeds from February to August, earlier in the tropics and later in more temperate regions, coincident with peaks in insect activity. Nest a flimsy saucer of twigs, grass, and fiber, cemented and bound with cobweb to a horizontal fork at end of a branch about 13–39 ft (4–12 m) above the ground, commonly in an isolated tree with clear view of surroundings. Eggs, in clutches of 2–5, are 0.8–1.1 x 0.7–0.8 in (20–28 x 18–20 mm), whitish to pinkish cream and spotted and blotched with reddish brown and black. Nests are often parasitized by cuckoos, e.g., *Surniculus lugubris* and *Eudynamys scolopacea*.

CONSERVATION STATUS

Not threatened.

SIGNIFICANCE TO HUMANS

None known. ◆

Resources

Books

Ali, S., and S.D. Ripley. *Handbook of the Birds of India and Pakistan*, Vol. 5. Bombay: Oxford University Press, 1972.

Coates, B.J. *The Birds of Papua New Guinea*. Vol. 2. Alderley, Qld: Dove Publications, 1990.

Du Pont, J.E. *Philippine Birds*, Monograph Series No. 2. Wilmington: Delaware Museum of Natural History, 1971.

Fry, C.H, S. Keith, and E.K. Urban, eds. *The Birds of Africa*. Vol. VI. New York: Academic Press, 2000.

MacKinnon, J., and K. Phillipps. *A Field Guide to the Birds of Borneo, Sumatra, Java, and Bali*. Oxford: Oxford University Press, 1993.

Schodde, R., and S.C. Tidemann, consultant eds. *Reader's Digest Complete Book of Australian Birds*. 2nd ed. Sydney: Reader's Digest Services, 1986.

Schodde, R., and I.J. Mason. *The Directory of Australian Birds, Passerines*. Melbourne: CSIRO Publishing, 1999.

Sibley, C.G., and J.E. Ahlquist. *Phylogeny and Classification of Birds*. New Haven: Yale University Press, 1990.

Sibley, C.G., and B.L. Monroe, Jr. *Distribution and Taxonomy of Birds of the World*. New Haven: Yale University Press, 1990.

Smythies, B.E. *The Birds of Borneo*. 3rd ed. Kota Kinabalu and Kuala Lumpur: The Sabah Society and the Sabah Nature Society, 1981.

Vaurie, C. "Dicruridae." *Check-List of Birds of the World*. Vol. 3. Cambridge, MA: Museum of Comparative Zoology, 1962.

Periodicals

Beehler, B.M. "Notes on the Mountain Birds of New Ireland." *Emu* 78 (1978): 65–70.

Vaurie, C. "A Revision of the Bird Family Dicruridae." *Bulletin of the American Museum of Natural History* 93 (1949): 199–342.

Organizations

Australian National Wildlife Collection. GPO Box 284, Canberra, ACT 2601 Australia. Phone: +61 2 6242 1600. Fax: +61-2-6242-1688.

Other

Australian Biological Resources Study. Canberra, ACT, Australia.

UNEP World Conservation Monitoring Center. *Threatened Animals of the World*. <http://www.unepwcmc.org/species/animals/animal redlist.html>. January 2002.

Richard Schodde, PhD

New Zealand wattle birds
(Callaeidae)

Class Aves
Order Passeriformes
Suborder Passeri (Oscines)
Family Callaeidae

Thumbnail description
Medium-sized songbirds with a distinctive, fleshy, bright-colored flap of bare skin, known as a "wattle," on each side of the corners of the beak. They have rounded wings and tail, and strong legs and toes.

Size
10–21 in (25–53 cm)

Number of genera, species
3 genera; 3 species

Habitat
Temperate forest

Conservation status
Extinct: 1 species; Endangered: 1 species; Near Threatened: 1 species

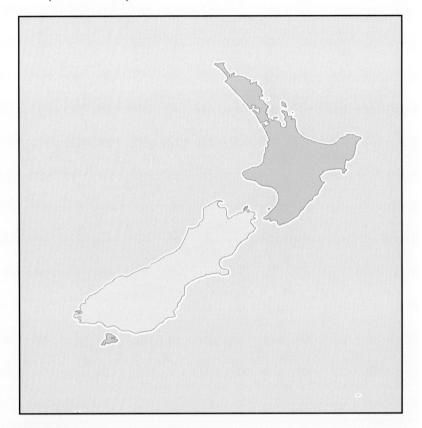

Distribution
New Zealand

Evolution and systematics

The wattle birds are three species of passerine birds, each assigned to its own genus, that comprise the family Callaeidae and live only in New Zealand. They likely evolved from a single, founder species that colonized the primeval forests of New Zealand more than one million years ago.

Physical characteristics

The body length is 10–21 in (25–53 cm), with a moderately long tail, short rounded wings, rather long legs, strong perching feet, and a curved, stout, dark beak. The body coloration is overall black, brown, or blue-gray, and with a blue or orange wattle (this is bare skin behind the gape of the beak). The sexes are dimorphic, differing in wattle color and bill size and shape.

Distribution

New Zealand.

Habitat

Temperate forest

Behavior

Wattle birds walk and hop on the ground but are weak fliers. They generally occur as non-migratory pairs that defend a breeding territory, or as unpaired birds. They have slow and distinct songs consisting of loud, bell-like notes. Pairs may sing duets in the morning.

Feeding ecology and diet

Wattle birds search for food on or close to the ground, where they hop about in long, deliberate jumps. They feed on insects, other invertebrates, fruit, and young leaves.

Reproductive biology

The nest is a flat platform of twigs placed as much as 33 ft (10 m) above the ground. It is loosely built and usually

The huia (*Heteralocha acutirostris*) has been extinct since 1907. (Photo by Tom McHugh. Photo Researchers, Inc. Reproduced by permission.)

roofed over. The clutch consists of two to four eggs that are colored pale gray or pale brown with dark spots.

Conservation status

The huia is recently Extinct, the kokako is Endangered, and the saddleback is dependent on continued conservation efforts for its survival.

Significance to humans

Wattle birds are of cultural significance to the indigenous Maori people of New Zealand. Otherwise, they are of no direct significance to people. They also provide economic benefits associated with bird-watching and ecotourism.

Species accounts

Kokako
Callaeas cinerea

TAXONOMY
Callaeas cinerea Gmelin, 1788. Two subspecies.

OTHER COMMON NAMES
English: Wattled crow; organbird; bellbird; French: Glaucope cendré; German: Graulappenvogel; Spanish: Kakapo.

PHYSICAL CHARACTERISTICS
16–18 in (40–45 cm) and colored bluish gray with a black mask. The North Island kokako (*C. c. wilsoni*) has a blue wattle and the South Island subspecies (*C. c. cinerea*) has an orange or yellow one.

Callaeas cinerea

DISTRIBUTION
New Zealand

HABITAT
Temperate mixed forest, dominated by either coniferous or broadleaf (angiosperm) trees. The North Island kokako is found mainly in mature podocarp-hardwood forest. The South Island kokako utilized similar habitat, but it may now be extinct.

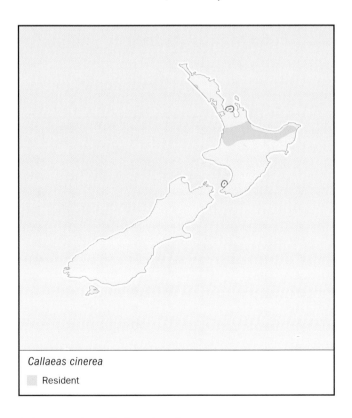

Callaeas cinerea
▨ Resident

BEHAVIOR
Actively walks and hops on the ground and from branch to branch, but is a weak flier. It generally occurs as non-migratory pairs that defend a breeding territory, or as unpaired individuals. It has a mellow, deliberate song consisting of loud bell-like notes. Pairs may sing duets in the morning, and other nearby kokako may also join in as a so-called "bush choir."

FEEDING ECOLOGY AND DIET
Feeds on small fruits, leaves, fern-fronds, flowers, and insects and other arthropods found on the forest floor or on foliage.

REPRODUCTIVE BIOLOGY
Builds a cup-shaped nest of sticks on the ground or low in a shrub or tree. The nest is lined with moss, plant fibers, and feathers. Lays two to three gray eggs marked with brown. The female is responsible for the rather prolonged (about 50 days) incubation and the brooding of the young. This exposes the female to a much higher risk of predation by introduced mammals, resulting in some populations having a great imbalance of male over female birds.

CONSERVATION STATUS
The kokako is listed as Endangered by the IUCN. The total area presently occupied by this rare bird is extremely small and fragmented. The number of breeding pairs is very small because of intense predation by introduced mammals, which has left many populations with an excess of unpaired males. Intensive conservation efforts are improving productivity at some sites. However, unmanaged populations continue to decline and at least one has become extirpated since 1990. Conservation biologists estimate that fewer than 400 pairs of the North Island kokako survive, in several isolated populations on islands off central and northern North Island. Since about 1980, there has been a marked decline in numbers of the North Island kokako, although management is now reversing that trend in some places. The South Island kokako is assumed extinct, although it is possible they survive in tiny numbers in remote parts of South Island or Stewart Island. The goal of the recovery plan for the kokako is to restore its population to about 1,000 pairs by the year 2020. A key element of the recovery program is the re-introduction of birds to predator-free islands having suitable habitat. Birds are being managed in captivity in an attempt to establish a captive-breeding and release program, and to prevent the extinction of local island races.

SIGNIFICANCE TO HUMANS
Economic benefits of ecotourism. ◆

Tieke
Philesturnus carunculatus

TAXONOMY
Philesturnus carunculatus Gmelin, 1789. Two subspecies.

OTHER COMMON NAMES
English: Saddleback; French: Créadion rounoir; German: Sattelvogel; Spanish: Tieke.

PHYSICAL CHARACTERISTICS

10 in (25 cm). Glossy black, with a conspicuous chestnut-brown patch or "saddle" on its back, as well as chestnut on the tip of its tail, a black bill, black legs, and orange wattles on either side of the beak

DISTRIBUTION

New Zealand. There are two subspecies, occurring on North Island (*P. c. rufusater*) and South Island (*P. c. carunculatus*).

Philesturnus carunculatus

HABITAT

Middle and lower layers of native, temperate forest, usually on or near the ground.

BEHAVIOR

Actively walks and hops on the ground and among branches, but is a weak flier. It generally occurs as non-migratory pairs that defend a breeding territory, or as unpaired birds. Males have a repertoire of melodious calls used during mating and to proclaim their territory.

FEEDING ECOLOGY AND DIET

Probes in dead wood and leaf litter to find its prey of insects and other invertebrates, and also eats fruits.

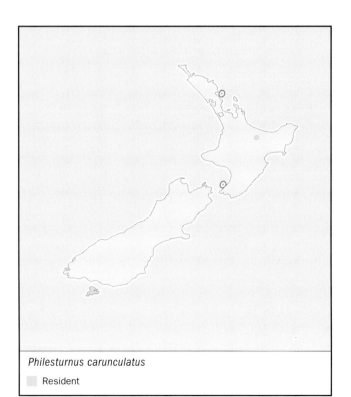

Philesturnus carunculatus

■ Resident

REPRODUCTIVE BIOLOGY

Builds a cup-shaped nest of sticks on the ground or low in a shrub. Lays two to three eggs.

CONSERVATION STATUS

Extinct on the mainland of North and South Island. Until recently, the North Island tieke only survived on Hen Island and the South Island subspecies on three small islands off Stewart Island. However, re-introductions to other suitable, small islands have been made, and it now occurs on about 20 islands. The tieke has suffered from the clearing of it habitat of native forest, and from the debilitating effects of introduced mammalian predators. It is listed Near Threatened by the IUCN. This means that its present non-threatened status can only be maintained through ongoing conservation efforts to enhance its surviving populations and maintain its habitat. According to conservation biologists, the population of the South Island tieke in the year 2000 was only about 650 individuals, and there were 2,000 individuals of the North Island subspecies. Active conservation management includes the establishment of new breeding populations on islands having suitable habitat. As of 2000, the North Island tieke was resident on nine relatively large islands (total of 17,000 acres or 7,000 hectares) and its prospects for survival were relatively favorable. The South Island subspecies was on 11 smaller islands (1,250 acres or 500 ha) and it needs translocation to additional predator-free islands to recover to a safer abundance.

SIGNIFICANCE TO HUMANS

Economic benefits of ecotourism. ◆

Huia

Heteralocha acutirostris

TAXONOMY

Heteralocha acutirostris Gould, 1837.

OTHER COMMON NAMES

French: Huia dimorphe; German: Hopflappenvogel; Spanish: Huia.

PHYSICAL CHARACTERISTICS

18 in (45 cm). Bluish black plumage, white band across the tail end, and bright orange wattles.

DISTRIBUTION

New Zealand.

HABITAT

Temperate forest.

Heteralocha acutirostris

BEHAVIOR

Actively walked and hopped but was a weak flier. Occurred as non-migratory pairs that defended a breeding territory.

FEEDING ECOLOGY AND DIET

Fed on insects and other arthropods found on the forest floor or in foliage and also ate fruits.

Heteralocha acutirostris

■ Resident

REPRODUCTIVE BIOLOGY
Built a cup-shaped nest of sticks on the ground or low in a shrub. Laid two to three eggs.

CONSERVATION STATUS
Extinct.

SIGNIFICANCE TO HUMANS
Revered by the Maori culture, but was hunted for feathers and as food.

Resources

Books

BirdLife International. *Threatened Birds of the World.* Barcelona, Spain and Cambridge, U.K.: Lynx Edicions and BirdLife International, 2000.

Heather, B., and H. Robertson. *Field Guide to the Birds of New Zealand.* Auckland, NZ: Viking Press, 1996.

Oliver, W. R. B. *New Zealand Birds.* Wellington, NZ, 1930.

Turbott, E. G. *Buller's Birds of New Zealand.* Auckland, NZ: Whitcomb & Tombs, 1967.

Organizations

BirdLife International. Wellbrook Court, Girton Road, Cambridge, Cambridgeshire CB3 0NA United Kingdom. Phone: +44 1 223 277 318. Fax: +44-1-223-277-200. E-mail: birdlife@birdlife.org.uk Web site: <http://www.birdlife.net>

IUCN–The World Conservation Union. Rue Mauverney 28, Gland, 1196 Switzerland. Phone: +41-22-999-0001. Fax: +41-22-999-0025. E-mail: mail@hq.iucn.org Web site: <http://www.iucn.org>

Other

Native Animals of New Zealand (website). Department of Conservation, Government of New Zealand. <http://www.doc.govt.nz/Conservation/001>

Plants-and-Animals/001

Native-Animals/index.asp

Bill Freedman, PhD

Mudnest builders
(Grallinidae)

Class Aves

Order Passeriformes

Suborder Passeri (Oscines)

Family Grallinidae

Thumbnail description
Medium to large birds with black and white or gray and brown plumage

Size
8–18 in (20–45 cm)

Number of genera, species
3 genera; 4 species

Habitat
Open woodland

Conservation status
Not threatened

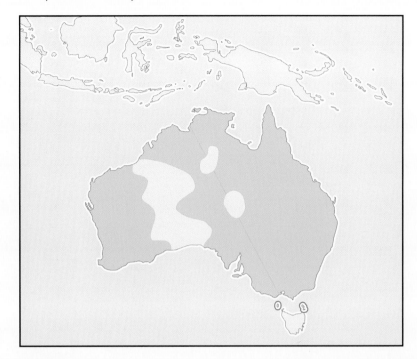

Distribution
Australasia

Evolution and systematics

The four species of mudnesters fall into three subfamilies, the Corcoracinae, the Struthideinae, and the Grallininae. Corcoracinae consists of one genus and one species, the white-winged chough (*Corcorax melanorhamphos*). Struthideinae also has only one genus and one species, the apostlebird (*Struthidea cinerea*). The Grallininae subfamily contains a single genus with two species, the Australian magpie-lark (*Grallina cyanoleuca*) and the torrent-lark (*G. bruijni*) that lives only in the New Guinea highlands. The two species differ in morphology, behavior, and genetics. *Grallina* have been shown to have biochemical similarities with monarch flycatchers and drongos.

Physical characteristics

The *Grallina* are medium sized birds with striking black and white plumage. Corcoracinae are larger, with black and white or gray and brown plumage. Juveniles can be distinguished from adults by eye color and, in some cases, plumage.

Distribution

Grallina occur in Australia and New Guinea, whereas Corcoracinae are specific to Australia.

Habitat

All species except the torrent-lark favor open habitats with some trees and water for nesting, and with open ground for foraging. By contrast, torrent-larks are found in forested hills and mountains, usually near fast-flowing streams.

Behavior

Grallina are usually seen in pairs or small family units, while the Corcoracinae live in large groups, sometimes including more than a dozen birds.

Feeding ecology and diet

Torrent-larks, as their name suggests, forage along fast-flowing mountain streams, taking insects from exposed rocks and the water. The magpie-larks are all predominantly ground foragers. The Corcoracinae forage together in scattered groups, moving forward together as they search the ground for food.

Reproductive biology

Most breeding occurs between August and December or February, though in arid areas birds may breed at any time

Apostlebird (*Struthidea cinerea*) on its nest in east Australia. (Photo by Tom McHugh. Photo Researchers, Inc. Reproduced by permission.)

An Australian magpie-lark (*Grallina cyanoleuca*) feeds its chicks in Queensland, Australia. (Photo by Jen & Des Bartlett. Bruce Coleman Inc. Reproduced by permission.)

of year, taking advantage of rain to build their mud nests. Most group members help with building the bowl of mud reinforced with sticks and grass, and lined with fine grass and fibers. They also take turns incubating eggs and feeding chicks. Clutches of three to five eggs hatch after about 18 days, and chicks leave the nest two to four weeks later. *Grallina* breed in monogamous pairs, while Corcoracinae are cooperative breeders, meaning that more than two birds provide care in rearing the young from one nest.

Conservation status

The torrent-lark is endemic to New Guinea and choughs and apostlebirds are endemic to Australia, but all four species are widespread, and common in suitable habitat.

Significance to humans

None known.

1. White-winged chough (*Corcorax melanorhamphos*); 2. Apostlebird (*Struthidea cinerea*); 3. Australian magpie-lark (*Grallina cyanoleuca*). (Illustration by Wendy Baker)

Species accounts

Australian magpie-lark
Grallina cyanoleuca

SUBFAMILY
Grallininae

TAXONOMY
Grallina cyanoleuca Latham, 1801. Monotypic..

OTHER COMMON NAMES
English; Mudlark, peewee, Murray magpie; French: Gralline pie; German: Drosselstelze; Spanish: Alondra Urraca.

PHYSICAL CHARACTERISTICS
Magpie-larks have striking black-and-white plumage and long legs. Facial plumage patterns differ between males, females and juveniles. Juveniles also differ from adults in having dark eyes and bills, whereas those of adults are white. Males are typically slightly larger than females. Adults are 10–12 in (25–30 cm) long, and weigh 3–4 oz (80–115 g).

DISTRIBUTION
Widespread throughout Australia, except in dry deserts and dense forests. Also in Timor, southern New Guinea, and Lord Howe Island.

HABITAT
Magpie-larks occupy a diverse range of habitats from coastal to semi-arid.

Grallina cyanoleuca
▨ Resident

BEHAVIOR
Usually seen in pairs or small family groups, though young birds and unpaired adults gather into large nomadic flocks. Breeding pairs defend all-purpose territories throughout the year. Partners advertise territory ownership by performing precisely coordinated antiphonal duets from conspicuous perches around their territories, often accompanying the duet by a synchronized wing display. They defend their territories vigorously against intruders, and attack their own reflections in windows.

FEEDING ECOLOGY AND DIET
Magpie-larks feed mostly on insects and other invertebrates such as earthworms and freshwater snails, and are common visitors to Australian backyards, foraging on lawns and enjoying the occasional treat of grated cheese.

REPRODUCTIVE BIOLOGY
Magpie-larks breed in monogamous pairs that tend to stay together for life (though Michelle Hall's study showed the occasional divorce when a better option presented itself!). Males and females share parental care, and may rear more than one clutch over the breeding season. Most juveniles leave their natal territory when they reach independence, though some remain over winter. After leaving their natal territories, juveniles join large semi-nomadic flocks until they form pairs and establish their own territories.

CONSERVATION STATUS
Not threatened.

SIGNIFICANCE TO HUMANS
Magpie-larks have adapted well to agricultural and urban environments and are familiar to most Australians. They are welcome residents in pastoral areas as they feed on freshwater snails that are intermediate hosts for a liver-fluke that parasitizes sheep and cattle. ◆

White-winged chough
Corcorax melanorhamphos

SUBFAMILY
Corcoracinae

TAXONOMY
Corcorax melanorhamphos Vieillot, 1817. Monotypic.

OTHER COMMON NAMES
English: Black jay, black magpie; French: Corbicrave leucoptère; German: Drosselkrähe; Spanish: Chova de Alas Blancas.

PHYSICAL CHARACTERISTICS
Choughs are large sooty black birds with long, curved, black bills, long tails, and large white wing patches that are visible in flight. They have red eyes that become brighter during displays when wings and tail are fanned out and moved up and down. Males and females have similar plumage. Juveniles are fluffier, with brown eyes that change to red over four years. Adults are 17–19 in (43–47 cm) long and weigh 11–12 oz (320–350 g).

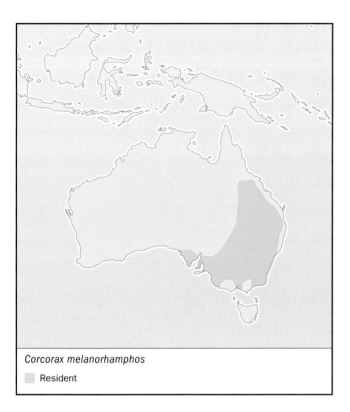

Corcorax melanorhamphos

▨ Resident

DISTRIBUTION
Choughs live in southeastern Australia.

HABITAT
Open woodland with leaf litter and not much understory.

BEHAVIOR
Choughs live in groups of about four to 20 birds. Groups usually consist of a breeding pair with offspring from previous years, though small groups of unrelated birds may come together. Groups sometimes even kidnap juveniles from other groups—the bigger the group the better! Chough are excitable and noisy in their interactions within the group, and the whole group joins in to dive-bomb intruders.

FEEDING ECOLOGY AND DIET
Choughs forage together on the ground, tossing aside leaf litter, probing with their bills and sifting through the soil for insects, worms and other ground-dwelling prey. Rob Heinsohn found that choughs take an unusually long time to learn to forage: juveniles remain dependent on parental care for longer than most other birds, and their foraging skills continue improving over the first four years of life.

REPRODUCTIVE BIOLOGY
Choughs are unable to raise young successfully unless they have help; large groups communally raise all chicks in the brood, while small groups only manage one or two. This cooperative breeding is primarily due to foraging constraints because small groups raise more chicks if they are given supplementary food. For young choughs, whose own foraging skills are still developing, helping at the communal nest is costly. Chris Boland and co-workers found that if no other group members are around when they deliver food to the nest, young birds are likely to swallow it themselves instead of feeding it to the chicks.

CONSERVATION STATUS
Not threatened.

SIGNIFICANCE TO HUMANS
None known. ◆

Apostlebird
Struthidea cinerea

SUBFAMILY
Struthideinae

TAXONOMY
Struthidea cinerea Gould, 1837. A smaller isolated population in the north of Australia may constitute a subspecies with birds that are slightly larger but with smaller bills than the eastern population.

OTHER COMMON NAMES
English; Gray jumper; French: Apôtre gris; German: Gimpelhäher; Spanish: Ave Apóstol.

PHYSICAL CHARACTERISTICS
Mottled ashy gray plumage with brown wings, black tail, and a short, stubby, black bill. Males and females have similar plumage. Juveniles have fluffy plumage and brown eyes which change to gray after their second year. Adults are 11–13 in (29–33 cm) long and weigh 4–5 oz (120–140 g).

DISTRIBUTION
Apostlebirds occur in the inland of eastern Australia, and in a smaller, isolated population in northern Australia.

HABITAT
Dry, open forest with some water.

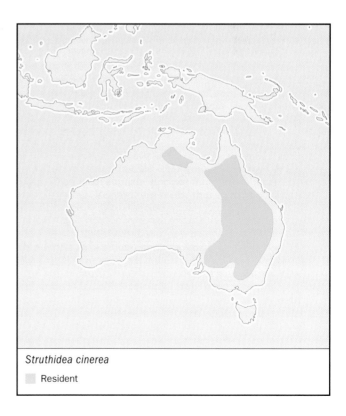

Struthidea cinerea

▨ Resident

BEHAVIOR
Apostlebirds are gregarious birds, foraging, breeding and roosting together. Their name stems from the belief that they live in groups of twelve, though group sizes vary from about three to 20. Apostlebirds are sedentary, defending a territory during the breeding season and wandering further afield after nesting finishes, when groups may aggregate at sources of food or water.

FEEDING ECOLOGY AND DIET
Apostlebirds are predominantly granivorous, collecting seeds from the ground and crushing them with their stout bills. They also eat insects opportunistically.

REPRODUCTIVE BIOLOGY
Like other members of the group, apostlebirds build nests of mud. Graeme Chapman's study showed that family members share nest building, incubation, and feeding of chicks. Large groups may produce several broods in a single breeding season.

CONSERVATION STATUS
Not threatened.

SIGNIFICANCE TO HUMANS
None known. ◆

Resources

Books

Ivison, T. "Magpie-larks" and "Mud-nest builders". In *Finches, Bowerbirds and Other Passerines of Australia*, edited by Robert Strahan. Sydney: Angus and Robertson, 1996.

Morcombe, Michael. *Field Guide to Australian Birds.* Archerfield: Steve Parish Publishing Pty Ltd, 2000.

Periodicals

Boland, C.R.J., R. Heinsohn, and A. Cockburn. "Experimental manipulation of brood reduction and parental care in cooperatively breeding white-winged choughs." *Journal of Animal Ecology* 66 (1997): 683–691.

Boland, C.R.J., R. Heinsohn, and A. Cockburn. "Deception by helpers in cooperatively breeding white-winged choughs

and its experimental manipulation." *Behavioral Ecology and Sociobiology* 41 (1997): 251–256.

Chapman, Graeme. "The social life of the Apostlebird *Struthidea cinerea.*" *Emu* 98 (1998): 178–183.

Hall, Michelle L. "The importance of pair duration and biparental care to reproductive success in the monogamous Australian magpie-lark." *Australian Journal of Zoology* 47 (1999): 439–454.

Hall, Michelle L. "The function of duetting in magpie-larks: conflict, cooperation, or commitment?" *Animal Behaviour* 60 (2000): 667–677.

Heinsohn, Robert G. Slow learning of foraging skills and extended parental care in cooperatively breeding white-winged choughs." *American Naturalist* 137 (1991): 864–881.

Michelle L. Hall, PhD

Woodswallows

(Artamidae)

Class Aves

Order Passeriformes

Suborder Passeri (Oscines)

Family Artamidae

Thumbnail description
Gregarious aerial foragers; mostly some combination of gray, black, and white, with pointed wings and short, stumpy tails

Size
4.7–7.9 in (13–20 cm); 0.5–1.6 oz (13–46 g)

Number of genera, species
1 genus; 11 species

Habitat
Woodswallows are birds of the open forest and woodlands, scrub, and clearings

Conservation status
Not threatened

Distribution
Six species occur in Australia, others through the islands of the South Pacific, across south China to India and Sri Lanka

Evolution and systematics

The taxonomic status of the woodswallows, sometimes called swallow-shrikes, is still under debate. Sibley and Monroe, grouped the woodswallows with the currawongs in the tribe Artamini, subfamily Corvinae, of the family Corvidae, and included several other genera on the basis of DNA-DNA hybridization data. Christidis and Boles in their standard 1994 classification of Australian birds included the woodswallows, butcherbirds, Australian magpies, and currawongs in the family Artamidae. Most recently, Schodde and Mason in their 1999 *The Directory of Australian Birds: Passerines* follow Christidis and Boles except for the order of genera.

Physical characteristics

Woodswallows are swallow-like birds with pointed wings and short tails that forage mainly by soaring and sweeping up flying insects. They are mostly subtle combinations of gray, white, and black, with several species having russet tones as well. Most species have white-tipped tails.

Woodswallows are the only passerines that have powder down patches of feathers that disintegrate into a soft powder that the birds use in grooming feathers in a manner analogous to preen gland secretions in other passerines. They have pointed wings with a vestigial outermost primary, short toes, and weak-grasping feet.

Distribution

Some species are widely distributed, e.g., the ashy woodswallow (*Artamus fuscus*) is found in the lowlands of southern Asia from India through southern China, while others have limited distribution, e.g., the Bismark woodswallow (*Artamus insignis*) is restricted to the Bismark Archipeligo. Six species are found in Australia, four of them endemic.

Habitat

Woodswallows are found in a broad spectrum of habitats throughout their range including mangroves, open areas, orchards, towns, open woodlands, forests, and forest edge.

White-breasted woodswallows (*Artamus leucorhynchus*) perch together on tree branches. (Photo by M.P. Kahl/VIREO. Reproduced by permission.)

A dusky woodswallow (*Artamus cyanopterus*) perches near its nest. (Photo by R. Drummond/VIREO. Reproduced by permission.)

Behavior

Highly gregarious, some woodswallow species are nomadic and travel in flocks of a hundred or more. They sometimes roost communally, with dozens of birds huddling together, perhaps to help with thermoregulation. When not foraging, they may cluster in large numbers on tree branches or wires. In winter they often join mixed species foraging flocks. They utter frequent soft contact calls, and may mob potential predators, while making harsh calls.

Feeding ecology and diet

Although woodswallows are primarily aerial foragers, sweeping flying insects from the canopy and above, they also are proficient ground pouncers, dropping from tree limbs to capture a grasshopper or caterpillar from open ground. These versatile foragers also take nectar and pollen.

Reproductive biology

Many species are opportunistic breeders, well adapted to an unpredictable environment, settling down to nest in loose colonies whenever rains come to arid areas. Nests are usually shallow, flimsy bowls of woven plant fibers including rootlets, twigs, and grass, placed in trees, shrubs, stumps, fence posts, or in rocky crevices. The usual clutch is two to four white eggs spotted or blotched with a variety of colors. Both parents incubate for 12–16 days. Fledging occurs 14-20 days later. Both parents and sometimes a helper feed the young.

Conservation status

Species with broad distributions are not threatened, but species with restricted distributions may be adversely affected by habitat alteration and human disturbance. At present, no woodswallows are considered threatened.

Significance to humans

Woodswallows are highly visible and asthetically pleasing birds, their soft, subtle coloration and dynamic flight making them favorites to many.

1. White-breasted woodswallow (*Artamus leucorynchus*); 2. Male white-browed woodswallow (*Artamus superciliosus*); 3. Little woodswallow (*Artamus minor*); 4. Black-faced woodswallow (*Artamus cinereus*); 5. Dusky woodswallow (*Artamus cyanopterus*). (Illustration by Bruce Worden)

Species accounts

Black-faced woodswallow
Artamus cinereus

TAXONOMY
Artamus cinereus Vieillot, 1817, Timor. Four subspecies.

OTHER COMMON NAMES
French: Langrayen gris; German: Schwarzgesicht-Schwalbenstar; Spanish: Golondrina del Bosque de Cara Negra.

PHYSICAL CHARACTERISTICS
7.1–7.5 in (18–19 cm); 1.1–1.4 oz (32–40 g). *A. c. normani* and *A. c. dealbatus* both have white undertail-coverts, the other subspecies have black. Sexes similar in plumage; gray with black face, silvery underwings, and broad white spots at tip of tail. Juvenile birds are brown and streaked.

DISTRIBUTION
Broadly distributed across Australia. *A. c. normani* and *A. c. dealbatus*, the two white-vented subspecies, are found respectively on the Cape York Peninsula and the northeastern Queensland coastal belt. *A. c. cinereus* is confined to southwestern Western Australia, and *A. c. melanops* is also found in the Lesser Sunda Islands, including Timor.

HABITAT
Occupy a broad range of habitats, often arid and far from water, including open eucalypt woodlands, scrub, and spinifex.

BEHAVIOR
Partly nomadic and an opportunistic breeder; mostly sedentary. Often roost in small flocks tightly clustered. Voice is a sweet chatter.

FEEDING ECOLOGY AND DIET
Primarily aerial feeders, soaring and swooping after flying insect prey. They take ground invertebrates, glean vegetation, and sometimes take flower nectar.

REPRODUCTIVE BIOLOGY
May breed at any time of year after rains when insects become abundant. Courtship displays include wing-waving and tail-rotating, displaying their white patches at the tip of their tails. Nest is a flimsy bowl of plant fibers placed in any available crevice. Clutch is three to four blotched white eggs. Incubation lasts 14–16 days, and fledging occurs 18 days later. They may have helpers at the nest.

CONSERVATION STATUS
Not threatened.

SIGNIFICANCE TO HUMANS
None known. ◆

Little woodswallow
Artamus minor

TAXONOMY
Artamus minor Vieillot, 1817, Shark's Bay, Western Australia. Two subspecies.

OTHER COMMON NAMES
French: Petit Langrayen; German: Zwergschwalbenstar; Spanish: Golondrina del Bosque Pequeña.

PHYSICAL CHARACTERISTICS
4.7–5.1 in (12–13 cm); 0.5–0.6 oz (13–16 g). Chocolate-brown body with dark gray wings and tail; tail spotted white at tip. Juveniles brown and streaked.

DISTRIBUTION
Australian endemic. *A. m. minor* is found in Western Australia, and the central arid regions of northern Southern Australia and southern Northern Territory. *A. m. derbyi* is found in northern Australia from northern Western Australia to northern New South Wales.

HABITAT
Rocky outcrops and gorges, in open arid woodland and acacia scrub.

BEHAVIOR
Sedentary over much of its range. Some eastern populations are nomadic and may be locally migratory. Lives in family groups of up to several dozen birds. Roosts communally. Voice consists of soft contact calls.

FEEDING ECOLOGY AND DIET
Sweep flying insects around cliff faces in flight. Will also forage on ground and glean foliage; sometimes take nectar.

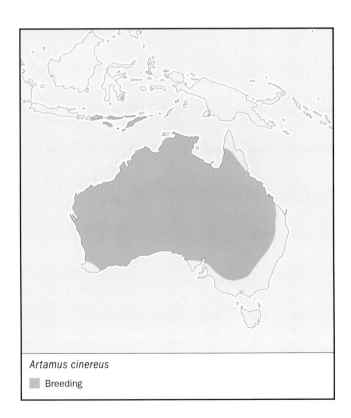

Artamus cinereus

■ Breeding

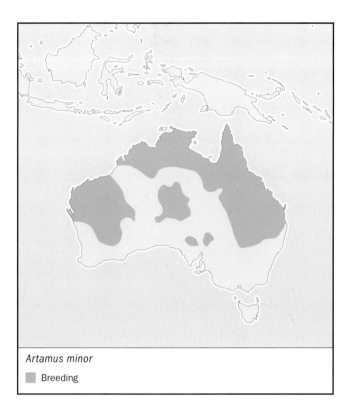

Artamus minor

■ Breeding

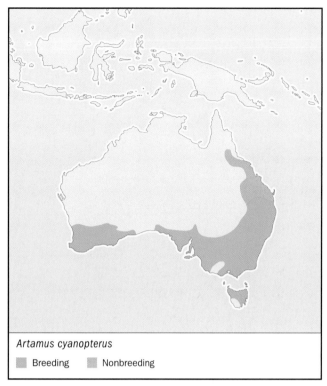

Artamus cyanopterus

■ Breeding ■ Nonbreeding

REPRODUCTIVE BIOLOGY
Nest is built by both parents and is a flimsy cup tucked in a rock crevice or stump. Clutch is two to three splotched white eggs. Both parents care for young.

CONSERVATION STATUS
Not threatened.

SIGNIFICANCE TO HUMANS
None known. ◆

Dusky woodswallow
Artamus cyanopterus

TAXONOMY
Artamus cyanoptera Latham, 1801, Sydney, New South Wales, Australia. Two subspecies.

OTHER COMMON NAMES
French: Langrayen sordide; German: Russchwalbenstar; Spanish: Golondrina del Bosque Ahumada.

PHYSICAL CHARACTERISTICS
6.7–7.1 in (17–18 cm); 1.1–1.6 oz (31–46 g). Smoky-brown body, dark gray wings and tail; wings edged in white, distinctive white spots at end of tail. Underwings silvery.

DISTRIBUTION
Australian endemic. *A. c. perthi* is resident in southwest Australia. *A. c. cyanopterus* is found in eastern South Australia up to southern Queensland, including Tasmania, but migrates north in winter.

HABITAT
Found in eucalypt forests and woodlands, along water courses, and over natural clearings. Tend to prefer rural areas, and wetter habitat than most woodswallows.

BEHAVIOR
Tends to live in small groups of several dozen birds. Social and gregarious, often roosting in a cluster in a tree hollow. Birds often perch close together while resting during the day. Chattering contact call, harsh mobbing call.

FEEDING ECOLOGY AND DIET
Primarily an aerial feeder. Will also glean foliage, take ground insects and occasionally nectar.

REPRODUCTIVE BIOLOGY
Both parents build the flimsy cup nest of plant fibers. Nests are built in loose colonies, with territory around the nest defended. May be cooperative breeders, with helpers at the nest. Clutch is three to four blotched white eggs. Incubation is for 16 days, and fledging occurs 16–20 days after hatching.

CONSERVATION STATUS
Not threatened.

SIGNIFICANCE TO HUMANS
None known. ◆

White-breasted woodswallow
Artamus leucorynchus

TAXONOMY
Artamus leucorynchus Linnaeus, 1771, Manila, Luzon.

Artamus leucorynchus
■ Resident

CONSERVATION STATUS
Not threatened.

SIGNIFICANCE TO HUMANS
None known. ◆

White-browed woodswallow
Artamus superciliosus

TAXONOMY
Artamus superciliosus Gould, 1837, Hunter River, New South Wales, Australia.

OTHER COMMON NAMES
French: Langrayen bridé; German: Weissbrauen-Schwalbenstar; Spanish: Golondrina del Bosque de Cejas Blancas.

PHYSICAL CHARACTERISTICS
7.5–7.9 in (19–20 cm); 1.1–1.3 oz (32–38 g). Male gray above, chestnut below, highlighted by white-eyebrow stripe, and white corners to tail. Female, colors muted; juvenile brown and streaked.

DISTRIBUTION
Nomadic species found throughout much of the eastern half of Australia, primarily away from the coast.

HABITAT
Occupies highly diverse habitat, from eucalypt forests and woodlands to arid spinifex, heathlands, and settled areas including orchards and farmlands.

OTHER COMMON NAMES
French: Langrayen à ventre blanc; German: Weissbauch-Schwalbenstar; Spanish: Golondrina del Bosque de Pecho Blanca.

PHYSICAL CHARACTERISTICS
6.7–7.1 in (17–18 cm); 1.3–1.6 oz (36–46 g). Gray hood, back, and tail, with white below and white rump diagnostic. Lacks white tail spots of most species.

DISTRIBUTION
Widely distributed in Southeast Asia, from Malay Archipelago, and islands of Oceania including Philippines, Moluccas, Greater and Lesser Sundas, New Caledonia, and New Guinea. In Australia, coastal and along rivers from Shark Bay, Western Australia, along northern edge to east coast as far south as Sydney. Usually close to water.

HABITAT
Predominately tropical woodlands, open areas, scrub, mangroves, and settled areas. It occurs in most habitat types.

BEHAVIOR
Largely sedentary, but seasonally nomadic with some winter migratory movements to the north. Very social and gregarious with flocks of several hundred not uncommon. They huddle together on branches or wires. Chattering contact call, harsh warning call.

FEEDING ECOLOGY AND DIET
Forage mostly by soaring through or above canopy. They take ground prey by pouncing, or glean foliage; have been reported taking nectar from flowers.

REPRODUCTIVE BIOLOGY
Lack white tail spots of other woodswallows, and do not use tail in courtship displays. Both parents build flimsy bowl nest of plant fibers, and care for young. Clutch is three to four blotched white eggs, and incubation lasts 13–15 days.

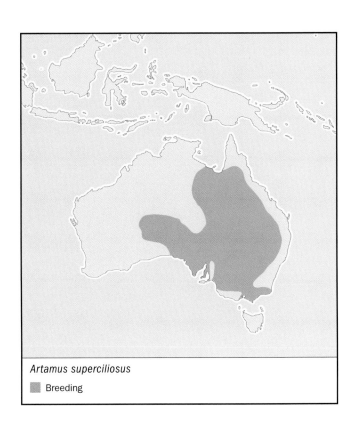

Artamus superciliosus
■ Breeding

BEHAVIOR
Highly nomadic, flocks often contain several woodswallow species. Opportunistic breeder, with flocks settling in areas where insects are abundant or where recent rains have occurred. Chattering contact calls are frequent among flock members, and they frequently cluster together at roost.

FEEDING ECOLOGY AND DIET
A predominately aerial forager, endlessly swooping and soaring, taking flying insects. Will take insects from ground or foliage, and flower nectar.

REPRODUCTIVE BIOLOGY
Usually nest in loose colonies, often opportunistically after rainfall. Both parents build flimsy nest of plant fibers placed in crevice. Clutch is usually two to three blotched white eggs; incubation lasts about 16 days and fledging occurs after about two weeks, although young stay with parents considerably longer.

CONSERVATION STATUS
Not threatened.

SIGNIFICANCE TO HUMANS
None known. ◆

Resources

Books

Christidis, L., and W.E. Boles. *The Taxonomy and Species of Birds of Australia and Its Territories.* Hawthorn East, Victoria, Australia: Royal Australasian Ornithologists Union Monograph 2, 1994.

Pizzey, G., and F. Knight. *The Graham Pizzey and Frank Knight Field Guide to the Birds of Australia.* Sydney: HarperCollins, 1997.

Schodde, R., and I.J. Mason. *The Directory of Australian Birds: Passerines.* Canberra: CSIRO Wildlife and Ecology, 1999.

Schodde, R., and S. C. Tidemann, eds. *Reader's Digest Complete Book of Australian Birds.* 2nd ed. Sydney: Reader's Digest Services, 1986.

Sibley, C.G., and B.L. Monroe, Jr. *Distribution and Taxonomy of Birds of the World.* New Haven: Yale University Press, 1990.

Simpson, K., and N. Day. *Birds of Australia.* Princeton: Princeton University Press, 1999.

Strahan, R., ed. *Finches, Bowerbirds and Other Passerines of Australia.* Sydney: Angus & Robertson, 1996.

William E. Davis, Jr.

Magpie-shrikes
(Cracticidae)

Class Aves
Order Passeriformes
Suborder Passeri (Oscines)
Family Cracticidae

Thumbnail description
Crow- to flycatcher-sized pied or dusky, stout-billed songbirds adapted to diverse niches in trees and on the ground

Size
6.5–22 in (17–55 cm); 1.1–17.5 oz (30–500 g)

Number of genera, species
5 genera; about 14 species

Habitat
Treed habitat on land, including rainforests, savannas, and pastures with trees

Conservation status
Near Threatened: 1 species; Data Deficient: 1 species

Distribution
Australia and New Guinea, with one outlying species in Borneo

Evolution and systematics

No other family of Australasian birds has radiated into so many niches in so few species as the magpie-shrikes. The family's butcherbirds are tree-living, shrike-like predators; its magpies are ground-foraging invertebrate feeders; its currawongs are scavenging generalists; and its peltopses are flycatcher-like, sallying insectivores. Despite such diversity, the group remains similar in skeletal structure, which is characterized by stout, straight, heavily ossified predatorial bills and elongated palates. Due to fusion in the nasal aperture, a massive shelf is formed in the roof of the mouth, an unusual trait in songbirds. The family is a branch of a massive radiation in crow-like birds that arose in Australia in the early to mid-Teriary, 30–40 million years ago. Unlike many others, however, it remained centered on the Australian plate. Only one member occurs outside Australia-New Guinea today, the bristlehead (*Pityriasis*) in Borneo.

Gradual drying in the Australian climate over the last 20 million years drove the diversification of the magpie-shrikes. Scavenging currawongs and ground-feeding magpies, derived from tree-living ancestors that resembled butcherbirds, are centered today in the drier eucalypt woodlands there. Even modern butcherbirds are more diverse in that habitat than in

the remnants of ancestral subtropical rainforest along the east Australian coast and mountain regions of New Guinea.

Physical characteristics

Magpie-shrikes are large songbirds with pied (black-and-white) or blackish plumage and straight, powerful bills. Color is varied with red in only two groups: on the rump, lower belly, and undertail in peltopses and over the head in the bristlehead. Bristleheads are otherwise dusky-gray, stout, and short-tailed, while the small, flycatcher-like peltopses are long-tailed and black with large patches of white on their faces and backs. Currawongs, however, are big but slender crow-sized birds with long, white-tipped tails, rounded wings, startling yellow eyes, and dark gray to blackish plumage broken by patches of white in wings and under the tail. Australian magpies are pied, with black undersurfaces and pied uppersurfaces, and are short in tail, pointed in wing, and long in leg–all adaptations to life in open spaces. Although also basically pied, butcherbirds vary more, from all-black to black-hooded with white patches to wholly white with black, gray, and white patterning. Like magpies, butcherbirds are mid-sized magpie-shrikes with chestnut eyes and bicolored bills; the tip of the bill is blackish and the base whitish or pale bluish gray, while the bills in currawongs, peltopses, and bristleheads are black. Juveniles share the adults' plumage, although it is much duller and grayer, particularly in some butcherbirds. Depending on species or subspecies, fledglings may be rusty-brown, washed olive-yellow, or lack clear head-patterning. They also have uniformly dull gray bills.

Size varies from small (6.7–7.9 in [17–20 cm] long, 1.1–1.2 oz [30–35 g]) in the peltopses to very large in the Tasmanian gray currawong, *Strepera versicolor arguta* (19.7–21.7 in [50–55 cm] long, 16.2–17.6 oz [460–500 g]). As with most other crow-like birds, magpie-shrikes have a single pneumatic depression in the head of the humerus, as well as 12 tail feathers, 10 primary flight feathers (the tenth is usually well-developed), and 10 secondaries (11 in the Australian magpie). Also characteristic are the lack of a gap in the dorsal (back) feather tract and small perforations in the skull above the opening for the nerves of the eye; unusually strong black feet with booted or scaled legs; and powerful bills that range from notched and hooked to pick-shaped at the tip and are swollen at the base of the upper jaw, which lacks bristles and in which the nostrils are sunk in deep bony slits.

Distribution

Except for the bristlehead in the lowlands of Borneo, magpie-shrikes are restricted to Australia and New Guinea and their in-shore islands. Peltopses are endemic to New Guinea, with one species occurring throughout the lowlands and the other in mountain ranges up to 9,900 ft (3,000 m) above sea level. In contrast, currawongs are endemic to Australia, restricted to southern and east coast Bassian regions north to Cape York Peninsula. The Australian magpie is also centered in Australia but, being adapted to open environments, ranges throughout the continent, with outliers in east Tasmania and the central portion of southern New Guinea. Only butcherbirds are represented by species in both Australia and New Guinea.

Habitat

Magpie-shrikes occur in wooded habitats that vary according to the foraging niches of the birds. Bristleheads occupy the mid-strata of mature, lowland alluvial (relating to river and lake systems) and swamp rainforests, while peltopses in New Guinea live on the edge of the canopy in both mature and regrowth rainforest. Currawongs are as much scansorial (adapted for surface-foraging) as arboreal (tree-dwelling) in their feeding and work over foliage, branches, bark, and ground in habitats ranging from dense, tall, wet forests to open, low, eucalypt-dominated woodlands. Australian magpies occupy even more open habitat, centering on savannas and pastures, whereas the wholly arboreal butcherbirds occupy the middle and upper strata of forests and woodlands. Butcherbird species coexist in any one region by partitioning their habitat. The black butcherbird and New Guinean members of the black-hooded groups occur in rainforests, the former at all strata within mature forests and in mangroves, and the latter in the middle and upper strata at forest edges and in secondary growth. In Australia, the three gray-group species occupy drier eucalypt forests and thickets, replacing one another in different parts of the continent. Where it overlaps them, the Australian member of the black-hooded group (*Cracticus nigrogularis*) keeps to more open woodlands.

Behavior

Reflecting adaptations to niche, behaviors are diverse. Tree-living butcherbirds, bristleheads, and New Guinean peltopses are perch-pouncers or salliers, using their feet for little else besides perching, coming to the ground only to snatch prey. All are sedentary, keeping to the same large foraging territories year-round, and are solitary, rarely gathering in groups larger than family parties. In flight, they move between trees in direct flight on rapidly and shallowly flapping wings. Similarly sedentary, the ground-feeding Australian magpie is more gregarious and has evolved a complex social organization; senior pairs or small breeding groups hold permanent territory in optimal habitat, while larger groups of juveniles and subordinate nonbreeders congregate in suboptimal areas. On the ground, magpies walk sedately, and their flight, direct as in butcherbirds, is far swifter. Currawongs are variably social, the gray being solitary at all times, while the other two species gather in large wandering bands when not breeding. At that time, their east coast populations make north to south migratory movements. Currawongs bound about all strata in the forest and hop and run on the ground. These long-tailed birds also fly in loping undulations.

Loud, carrying flutings, glottal (relating to the opening between the vocal chords and larynx) gargles, and bell-like whistles characterize the calls of all magpie-shrikes except the peltopses, which "tick" or "tinckle" mechanically. Australian magpies often chorus together in groups, and flocking currawongs call and answer constantly.

At night, all magpie-shrikes roost in tree foliage at mostly mid-height, Australian magpies and flocking currawongs doing so in loose groups.

Feeding ecology and diet

Magpie-shrikes are predatory passerines that have radiated into a range of foraging niches. Bristleheads are arboreal insect gleaners. Peltopses are arboreal, flycatcher-like salliers of the top and edge of forest canopies, again taking mainly insects. Butcherbirds are more shrike-like, hunting through the mid-strata of forests and woodlands by perching and pouncing, capturing a range of arthropods, small vertebrates, and crustacea in trees and on the ground. They wedge or impale larger prey in crannies and on spikes for tearing with their well-hooked bills before eating. Australian magpies, however, are strict ground-feeders, digging for grubs, worms, and other ground-burrowing invertebrates with their stout bills. Currawongs are opportunistic generalists, particularly the pied currawong (*Strepera graculina*), moving over the ground and through trees and shrubbery to dig, prize, and scavenge for invertebrates, small vertebrates, nestlings, eggs, and berries. Males of most species also have significantly longer, but hardly thicker, bills than females, suggesting some sexual partitioning in feeding.

Reproductive biology

Although breeding may occur erratically year-round in tropical regions, it is restricted from early spring through summer (August to January) in temperate and subtropical Australia. Rarely more than one brood is raised a year. Most species are monogamous, except the polygynous Australian magpie, and breed in pairs; all appear to be strictly territorial. Only the female builds the nest, a rough cup of twigs and rootlets lined with finer fiber, in horizontal to upright forks in the upper branches of trees; and she incubates alone, fed on or off the nest by the male. Clutches are of one to five eggs, cream or pinkish buff to pale green, lined or spotted with red-browns and gray-blacks.

Conservation status

Because it is limited to mature rainforests, particularly dipterocarp-dominated swamp forests that are subject to deforestation for timber, the Bornean bristlehead is classified as Near Threatened. It seems to be the only species under serious threat. Categorized as Data Deficient, the Tagula butcherbird (*Cracticus louisiadensis*) is also listed, presumably because it is restricted to one small island off southeastern New Guinea. Even so, it is a member of the black-hooded butcherbird group that is well-adapted to secondary growth and forest disturbance.

A black butcherbird (*Cracticus quoyi*) in the rainforest of northern Queensland, Australia. (Photo by Michael Fogden/Animals Animals. Reproduced by permission.)

Significance to humans

No magpie-shrikes have made a significant impact on humans except for the Australian magpie. This species, which is widely appreciated in Australia for its fearlessness and caroling song, is cultivated around rural properties and fed in urban areas, and has been introduced successfully to New Zealand, the Solomon Islands (Guadalcanal), Fiji (Taveuni), and several offshore Australian islands. When breeding, rogue males can become a problem, diving to attack children and others. The skulking pied currawong (*Strepera graculina*) has also invaded towns and cities along the east coast of Australia, impacting on smaller birds of all species by robbing nests and taking newly fledged young.

1. Gray butcherbird (*Cracticus torquatus*); 2. Female Bornean bristlehead (*Pityriasis gymnocephala*); 3. Pied currawong (*Strepera graculina*); 4. Clicking peltops (*Peltops blainvillii*); 5. Australian magpie (*Gymnorhina tibicen*). (Illustration by Dan Erickson)

Species accounts

Bornean bristlehead
Pityriasis gymnocephala

SUBFAMILY
Pityriasinae

TAXONOMY
Barita gymnocephala Temminck, 1835, Borneo. Monotypic.

OTHER COMMON NAMES
English: Bald-headed woodshrike; French: Barite chauve; German: Warzenkopf; Spanish: Alcaudón de Borneo.

PHYSICAL CHARACTERISTICS
10–11 in (26–28 cm). Thickset, massive-billed, and stumpy-tailed. Dusky body; black wings, tail, and bill; and red, mostly bare head with a patch of orange-yellow stubble on the crown, another of streaky brown over the ears, and a fringe of scarlet feathering on upper back and breast; lower breast also covered in bristle-like brown and red feathers. Females have a red patch on the flanks. Eyes are chestnut and feet distinctively yellow.

DISTRIBUTION
Lowland Borneo up to altitudes of about 3,900 ft (1,200 m).

HABITAT
Mature lowlands and swamp rainforests.

BEHAVIOR
Patchy throughout range, bristleheads appear to reside in one area, where they work through the mid-strata of forests in noisy groups. Little is known of their behavior. They are ponderous in movement, hopping among branches and crouching and peering into crannies in search of food. Flight is direct, on fast and shallowly beating wings. Calls, presumably given for social cohesion, comprise strange nasal whines, honks, and chortles; members of a group also will chorus, jumbling calls loudly together.

FEEDING ECOLOGY AND DIET
Primarily predatory insectivores, gleaning gregariously among branches and trunk crannies in forest midstage for food. Diet comprises large insects, such as arboreal beetles, grasshoppers, bugs, cockroaches, and larvae. Birds will also gather around recent clearings in search of exposed food.

REPRODUCTIVE BIOLOGY
Little is understood about pair bonding and nest-building duties, nor is the nest and its site described. Eggs are whitish, sparingly blotched with rich brown and slate-gray mostly at the larger end.

CONSERVATION STATUS
Listed as Near Threatened due to occurrence in mature rainforests in areas under threat of extensive deforestation.

SIGNIFICANCE TO HUMANS
None known. ◆

Clicking peltops
Peltops blainvillii

SUBFAMILY
Cracticinae

TAXONOMY
Eurylaimus blainvillii Lesson and Garnot, 1827, Dorey, West Papua. Monotypic. This is the lower altitude member of a pair of sibling species that replace one another altitudinally throughout mainland New Guinea. Despite similarity in appearance and foraging ecology, the two have distinct contact/advertising calls.

OTHER COMMON NAMES
French: Peltopse des plaines; German: Waldpeltops; Spanish: Peltopo del Valle.

PHYSICAL CHARACTERISTICS
6.5–7.3 in (17–18 cm); 1.1–1.2 oz (30–35 g). Stout, black, red-eyed flycatcher-like bird with slender tail; patches of white over the ears and on the upper back; and crimson on belly, undertail, and rump. Sexes are alike but juveniles duller. Differs from its montane (mountain-dwelling) sibling species, *Peltops montanus*, in its slightly smaller size, shorter tail, smaller white patches over ears and on back, and much heavier, broader bill.

DISTRIBUTION
Lowland New Guinea and western Papuan islands up to about 1,640 ft (500 m) above sea level.

Pityriasis gymnocephala
■ Resident

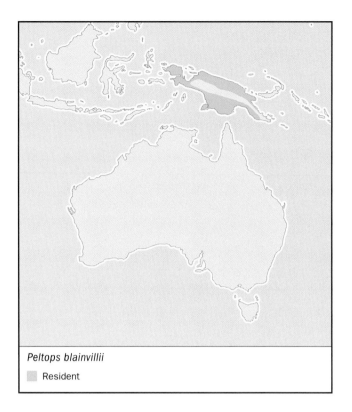

Peltops blainvillii
◻ Resident

HABITAT
Primary and tall secondary rainforests, particularly along edges and around openings such as tree falls, stream edges, and road cuttings. Densities have been estimated at two birds per 25 acres (10 ha) in suitable habitat.

BEHAVIOR
Widely but sparsely distributed year-round residents; solitary, in loose pairs or small family groups of three to five. Live mostly in top of forest canopies, perching upright and motionless for long periods on exposed vantage perches, from which they fly out in extended sallies. From perches and in flight, birds utter distinctive territorial and advertising calls at regular intervals; calls are a series of three or so well-spaced mechanical double *clicks* over four to five seconds. At perches, the singer throws its head violently up and down at each *click*. Other calls include a harsh monarch-like *wheeeit*, possibly in warning or agitation, and a seldom-heard wren-like twittering.

FEEDING ECOLOGY AND DIET
Apparently wholly insectivorous, birds capture food in the air and from the surface of foliage with the bill on sallying flights. Most foraging is done in and above the forest canopy but sometimes extends to lower strata. Food, including dragonflies, beetles, grasshoppers, and other flying insects, is swallowed whole at a perch, without much beating.

REPRODUCTIVE BIOLOGY
Data lacking on timing and duration of events and respective parental contribution. Nests are small, compact cups of twigs and rootlets built in often exposed positions in horizontal tree forks at the ends of branchlets at 20–115 ft (6–35 m) above the ground; and eggs, usually one per clutch, are cream to pale buff, with a ring of black-brown spotting at the larger end.

CONSERVATION STATUS
Not threatened.

SIGNIFICANCE TO HUMANS
None known. ◆

Gray butcherbird
Cracticus torquatus

SUBFAMILY
Cracticinae

TAXONOMY
Lanius torquatus Latham, 1802, Port Jackson (Sydney), Australia. Three, possibly four subspecies; one large and dark with reduced white in wing in Tasmania (*Cracticus torquatus cinereus*), another small and dark with reduced white in wing in coastal southeastern Australia between the New South Wales-Queensland border and Melbourne (nominate *torquatus*), and a third (that may comprise east and west forms) small and paler with extensive white in wing throughout southern and inland Australia from the west coast to coastal Queensland south of Cape York Peninsula (*C. t. leucopterus*). The gray butcherbird forms a superspecies with the silver-backed butcherbird (*C. argenteus*) of northwestern Australia and the black-backed butcherbird (*C. mentalis*) of Cape York Peninsula and dry sectors of southeastern New Guinea.

OTHER COMMON NAMES
French: Cassican à collier; German: Graurücken-Würgatzel; Spanish: Pájaro Matarife Gris.

Cracticus torquatus
◻ Resident

PHYSICAL CHARACTERISTICS

10–12 in (25–30 cm); 2.8–4.0 oz (80–110 g). Medium-sized bull-headed bird with tapered body and black-, gray-, and white-patterned plumage. Head black with white lore spot (between eye and upper bill) and collar, back gray, rump white, tail and wings black with white tips and stripes respectively, and undersurface uniformly grayish white. Eyes dark brown, feet gray, and bills bicolored, with black tip and gray-white base. Females are usually grayer breasted and shorter billed than males. Juveniles dull-gray-billed and dully patterned, with dusky-olive and speckled upperparts and yellowish underparts.

DISTRIBUTION

Most of southern and inland Australia north to 20°S, including Tasmania but excluding treeless deserts.

HABITAT

Closed woodland and open forest of eucalypts and acacias, including mallee (*Eucalyptus*) and mulga (*Acacia*) scrubs, where the space between tree crowns is about the size of the crowns themselves. Densities range from one bird to about 12 to 49 acres (5–20 ha) in suitable habitat.

BEHAVIOR

Retiring and solitary, in pairs or small family groups, gray butcherbirds live in the mid- and upper strata of trees, spending much of their time perching still and coming to ground only to pounce on prey. They are sedentary, with pairs holding the same territory, 20–99 acres (8–40 ha), year-round, with a larger home range. Both male and female duet antiphonally in songs of fluted whistles and ringing caws, which are also given in alarm and aggression.

FEEDING ECOLOGY AND DIET

Gray butcherbirds are raptorial perch-pouncers, watching from tree perches at 6.5–40 ft (2–12 m) up, then swooping down to ground or branches to snap up prey, which is mostly insects but also small birds, nestlings, reptiles, and mice. Fruit also contributes to the diet. Food is carried back to perch, wedged in crannies or forks and torn apart with the bill for eating; the small weak feet are not used for tethering prey.

REPRODUCTIVE BIOLOGY

Monogamous, forming pair bonds in breeding territories reinforced by much duetting during early breeding. Gray butcherbirds breed mostly between July and August and December and January throughout their range. The nest, a rough but tight cup of twigs lined with finer, often reddish fiber, is placed in upright forks in outer foliage at 10–50 ft (3–15 m) up and takes about four weeks to build. Eggs, in clutches of three to five, are 1.20–1.25 x 0.85–0.95 in (30–32 x 22–24 mm), brownish green, and finely freckled in red-browns, often in a zone around the larger end; they hatch in 22–24 days. Female builds nest and incubates unaided, while the male defends territory. He may assist her in feeding young, which fledge around four weeks. Usually, only one brood reared per year.

CONSERVATION STATUS

Not threatened, although many populations have declined locally because of habitat clearing and alienation.

SIGNIFICANCE TO HUMANS

Some local populations frequent camping sites, feeding on scraps and garbage thrown out by campers. They are called commensals in such circumstances because they benefit from a close association with humans. ◆

Australian magpie
Gymnorhina tibicen

SUBFAMILY
Cracticinae

TAXONOMY
Coracias tibicen Latham, 1802, Port Jackson (Sydney), Australia. Polytypic with up to nine subspecies, the limits and identity of which are controversial. Subspecies differ mainly in pattern of back, whether white or black; size; proportions of bill; and width of the black band on the tail tip. Four black-backed subspecies occur across northern Australia south to the Pilbara in the west and the Murray-Darling Basin and south coastal New South Wales in the east; five white-backed subspecies occur in southwestern and southeastern Australia, including Tasmania, with an outlier in central southern New Guinea. Opinion has varied over whether white- and black-backed groups are distinct species, but they hybridize freely wherever they meet. Intriguingly, the white-backed subspecies in southwestern Australia and southern New Guinea have blackish-backed females.

OTHER COMMON NAMES
French: Cassican flûteur; German: Flötenvogel; Spanish: Urraca Canora.

PHYSICAL CHARACTERISTICS
13–17 in (34–43 cm); 8.8–12 oz (250–340 g). Pied, long-legged, short-tailed magpie-shrikes with black heads, white uppersurface with or without a black saddle across the back, white tail with black tip, black wings with white shoulder patch, and entirely black undersurface except for white under-tail coverts; eyes chestnut, feet black, and bills bicolored with black tip and grayish white base. Females have shorter bills

Gymnorhina tibicen
☐ Resident

than males and their dorsal white is clouded with gray; in southwest Australian females, the black feathers of the saddle are edged with white. Juveniles are much duller than adults, dingy gray-billed, and pass through gray-breasted plumage before reaching adulthood in their third or fourth year.

DISTRIBUTION
Throughout Australia, including east Tasmania but excluding central treeless deserts and extreme north (north Kimberley, Arnhem Land, and Cape York Peninsula), with an outlying population in dry sectors of central southern New Guinea. Introduced successfully in New Zealand, Fiji, and Guadalcanal.

HABITAT
Open woodlands, savannas, and rural fields with fringing trees, windbreaks, and copses. A combination of extensive bare or short-pastured ground (for feeding) and scattered groups of trees (for roosting and nesting) is essential.

BEHAVIOR
Australian magpies are bold, gregarious birds in settled areas, adapting to human habitation and benefitting from the clearing of land for rural purposes. Also sedentary and hold territory year-round according to social order; top males occupy and defend optimal territories that include one to several females, while at the bottom rung are loose, locally mobile flocks (of about 10 to 100 or more) of juveniles evicted from parental territories and adults that have not yet gained or have lost territorial status. Groups sing together from perches in rich, organ-like fluted caroling to advertise territory. In group attacks on predators (e.g., raptors) often much larger than themselves, they yell in shrieking yodels, calling in mid-flight and alarming the entire neighborhood. Flight is swift and powerful on rapidly and deeply beating pointed wings. Magpies spend most of the day feeding on the ground but rest and roost on perches in trees or poles, each group sleeping as a loose unit in a single tree or series of adjacent trees.

FEEDING ECOLOGY AND DIET
Ground-foraging invertebrate feeders, Australian magpies feed on bare or short-pastured turf over which they can move easily on long legs. They walk methodically like rooks (*Corvus frugilegus*), head cocked, listening, and watching. Most prey, including grubs, worms, and ground and burrowing insects, is dug out of the ground with their straight, stout bills. Items are dispatched and eaten on the ground at point of capture.

REPRODUCTIVE BIOLOGY
The breeding system is polygynous and territorial, one male mating with one to several females and spending nearly all his time defending them and his territory from other males. Breeding, from nest-building to fledging, extends from early spring to early summer (mid-July through August to December or January), in territories of 7–25 acres (3–10 ha), or more in arid areas. Only one brood is reared per year. All nesting duties are carried out by the mated female, but she may be helped in feeding young by other females in her group. Nests are rough bowls of twigs, lined with finer fiber and placed in the upper forks of trees at 10–50 ft (3–15 m) above the ground. In near-treeless areas, Australian magpies will construct nests with wire and place them on the spars of electricity poles. Eggs, in clutches of three to five, are 1.45–1.53 x 1.02–1.10 in (37–39 x 26–28 mm) and pale green to grayish blue, spotted and/or streaked in earth reds, reddish browns, umber, and dusky. No one clutch resembles another. Eggs hatch in 20–22 days and young fledge in another 28–30.

CONSERVATION STATUS
Not threatened; populations benefit by habitat clearing throughout rural Australia.

SIGNIFICANCE TO HUMANS
A commensal around human habitation. ◆

Pied currawong
Strepera graculina

SUBFAMILY
Cracticinae

TAXONOMY
Corvus graculinus Shaw, 1790, Port Jackson (Sydney), Australia. Three to six subspecies in eastern Australia: one large-billed on eastern Cape York Peninsula (*Strepera graculina magnirostris*), one all-dusky in western Victoria (*S. g. ashbyi*), and the others between which vary regionally in size and tone. One further slender-billed subspecies occurs on Lord Howe Island (*S. g. crissalis*).

OTHER COMMON NAMES
English: Black currawong; French: Grand Réveilleur; German: Dickschnabel-Würgerkrähe; Spanish: Currawong Pálido.

PHYSICAL CHARACTERISTICS
17–20 in (43–50 cm); 10–12.5 oz (280–360 g). Slender, dusky, crow-like birds with hidden white flashes in wings; white tips on long tails; and broad, white bands at base of upper- and undersurface of tail and undertail coverts. Eyes pale yellow and bills and feet black. Females resemble males except for shorter bills; juveniles duller and grayer, with brown eyes and yellow gapes and mouths.

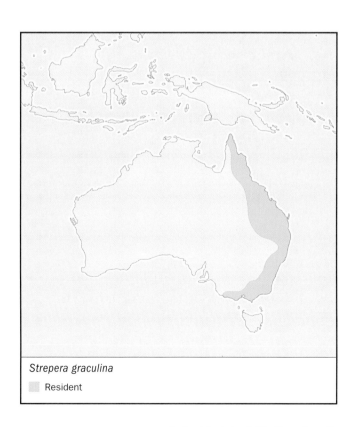

Strepera graculina
☐ Resident

DISTRIBUTION

Coastal and subcoastal eastern mainland Australia between Cape York Peninsula and western Victoria, with outlier on Lord Howe Island.

HABITAT

Closed forests (including rainforests and wet eucalypt forests) to denser, taller wetter woodlands of euclaypts; also urban gardens and parks where tree cover is ample. Cool montane forests and woodlands of eucalypts are core habitat.

BEHAVIOR

Skulking, opportunistic, and piratical predators. Territorial during breeding, pied currawongs congregate in loose foraging flocks of up to a hundred or so at other times; southeast populations move out of mountain ranges to lower altitudes and more northerly regions in early autumn (late March to early April) and return in early spring (September) to breed. Foraging flocks work through all strata of their habitat, running over the ground, bounding about foliage, and flying from tree to tree in slow, loping undulations broken by floppy wing beats. They call constantly to maintain contact or advertise territory with loud whining whistles and clanking glottal chortles. They do not sing or carol like other Australian magpie-shrikes but do call loudly to one another when gathering to roost in tall trees at dusk.

FEEDING ECOLOGY AND DIET

Omnivorous scavengers that search for food anywhere, bounding about shrubbery and branches; poking into foliage and crevices; or walking, running, and hopping over the ground to peck, probe, and jab. Rubbish tips and camping grounds with food waste are favored. Carrion, small birds, nestlings, insects, caterpillars, lizards, snails, food scraps, and berries are all eaten. Over the last 20–30 years, berry-bearing ornamental trees have attracted currawongs to many urban habitats on the east Australian coast, where the birds remain year-round and cause havoc among other bird species during breeding.

REPRODUCTIVE BIOLOGY

Strictly monogamous and territorial. Pairs establish territory of some 12–37 acres (5–15 ha) early in spring (September, or as early as July in the far north) and build from late September to mid-October, so that young fledge from mid-November into December; only one brood reared per year. Although males may assist in gathering nest material and in feeding older young, females bear the brunt of nest construction, incubation, and the brooding and feeding of young. Nests are rough and rather flimsy bowls of twigs, lined with finer fiber and placed high in the outer branchlets of tall trees at about 40–65 ft (12–20 m) above the ground. Eggs, two or usually three per clutch, are 1.58–1.69 x 1.10–1.25 in (40–43 x 28–32 mm) and pinkish buff finely specked and spotted with dark brown and dusky. They hatch in 20–22 days, and young fledge in about another 30.

CONSERVATION STATUS

Not threatened. Because of their predatorial behavior, pied currawongs pose a serious threat themselves to the nesting of many species of Australian birds in human-modified areas.

SIGNIFICANCE TO HUMANS

None known. ◆

Resources

Books

Amadon, D. "Cracticidae." In *Check-List of Birds of the World.* Vol. 15. Cambridge, Mass.: Museum of Comparative Zoology, 1962.

Coates, B.J. *The Birds of Papua New Guinea.* Vol. 2. Alderley, Qld: Dove Publications, 1990.

MacKinnon, J., and K. Phillipps. *A Field Guide to the Birds of Borneo, Sumatra, Java, and Bali.* Oxford: Oxford University Press, 1993.

Schodde, R., and S.C. Tidemann, consultant eds. *Reader's Digest Complete Book of Australian Birds.* 2nd edition. Sydney: Reader's Digest Services, 1986.

Schodde, R., and I.J. Mason. *The Directory of Australian Birds, Passerines.* Melbourne: CSIRO Publishing, 1999.

Sibley, C.G., and J.E. Ahlquist. *Phylogeny and Classification of Birds.* New Haven: Yale University Press, 1990.

Sibley, C.G., and B.L. Monroe Jr. *Distribution and Taxonomy of Birds of the World.* New Haven: Yale University Press, 1990.

Smythies, B.E. *The Birds of Borneo.* 3rd edition. Kota Kinabalu and Kuala Lumpur: The Sabah Society and the Sabah Nature Society, 1981.

Periodicals

Amadon, D. "Taxonomic Notes on the Australian Butcherbirds (Family Cracicidae)." *American Museum Novitates* 1504 (1951): 1–33.

Sibley, C.G. and J.E. Ahlquist. "The Relationships of the Papuan Genus *Peltops.*" *Emu* 84 (1984): 181–183.

Organizations

Australian National Wildlife Collection. GPO Box 284, Canberra, ACT 2601 Australia. Phone: +61 2 6242 1600. Fax: +61-2-6242-1688. E-mail: Richard Schodde@aol.com

Other

Australian Biological Resources Study. Canberra, ACT, Australia.

UNEP World Conservation Monitoring Center. Threatened Animals of the World. <http://www.unep-wcmc.org/species/animals/animal redlist.html> (January 2002).

Richard Schodde, PhD

▲
Bowerbirds
(Ptilonorhynchidae)

Class Aves
Order Passeriformes
Suborder Passeri (Oscines)
Family Ptilonorhynchidae

Thumbnail description
Medium-sized, thrush-like, stocky, strong-footed, and typically stout-billed songbirds. Family includes sexually and cryptically monochromatic to dramatically sexually dichromatic species. Bowerbirds are renowned for the bower building behavior of males of polygynous (one male mated with two or more females) species.

Size
8.7–14.6 in (22–37 cm); 0.18–0.64 lb (80–290 g)

Number of genera, species
8 genera; 20 species

Habitat
Rainforests, moss forests, wet sclerophyll (Australian vegetation with hard, short, and often spiky leaves) forests and woodlands, savanna, rocky wooded gorges, and open woodlands to semi-desert

Conservation status
Vulnerable: 1 species; Near Threatened: 1 species

Distribution
Mainland New Guinea and Australia and offshore islands

Evolution and systematics

Bowerbirds have long been closely associated with birds of paradise (Paradisaeidae) but evidence of a major dichotomy between the two groups in anatomical and biological traits is supported by several molecular studies. Bowerbirds are part of an Australasian radiation thought to have occurred during the past 60 million years. They diverged from lyrebirds (Menuridae) and scrub birds (Atrichornithidae) about 45 million years ago (mya). Results of molecular studies place the separation of bowerbirds (superfamily Menuroidea) from birds of paradise and other corvines (superfamily Corvoidea) at 28 mya and indicate that major lineages within them arose 24 mya.

Satins (*Ptilonorhynchus violaceus*) are the only bowerbirds known from fossil sites; found in Victoria, Australia, two are from the Holocene and one is from the Pleistocene. These fossils are from locations remote from the present wet forest range of the species and attest to a previously more extensive distribution of Australian subtropical rainforests.

The Ptilonorhynchidae comprises 20 species of compact, robust, oscinine songbirds. Three species of socially monogamous and territorial catbirds belong to the genus *Ailuroedus*. The 17 known or presumed polygynous species consist of one *Scenopoeetes*, four *Amblyornis*, one *Archboldia*, one *Prionodura*, four *Sericulus*, one *Ptilonorhynchus*, and five *Chlamydera* species.

Physical characteristics

Bowerbird morphology and anatomy are broadly typical of oscinine passerines with the exception of a few traits. Typical songbirds have 9–10 secondaries (including tertials), but bowerbirds have 11–14. Bowerbirds also have an enlarged lachrymal (part of the skull cranium, near the orbit) that is paralleled only in the Australian lyrebirds (Menuridae). Bowerbirds have high average survivorship, and some individuals live for 20–30 years.

Within the family, great bowerbirds (*Chlamydera nuchalis*) are the largest and golden bowerbirds (*Prionodura newtoniana*) are the smallest. Males are typically, but not always, heavier and are larger in most body measurements than females. Juveniles and immature bowerbirds are generally smaller in wing length and weight than adults. The bill is typically stout and powerful; exceptions are the fine and longer bill of regent bowerbirds (*Sericulus chrysocephalus*) and the falcon-like toothed mandibles of tooth-billed bowerbirds (*Scenopoeetes dentirostris*). Legs and feet are stout, powerful, and scutellate.

The family exhibits 50–60 different plumages. Catbirds are sexually and cryptically monochromatic, and both sexes of the polygynous tooth-billed, Vogelkop (*Amblyornis inornatus*), and *Chlamydera* bowerbirds are nearly identical. The other polygynous species are sexually dichromatic, with adult males adorned with colorful and ornate plumages and females

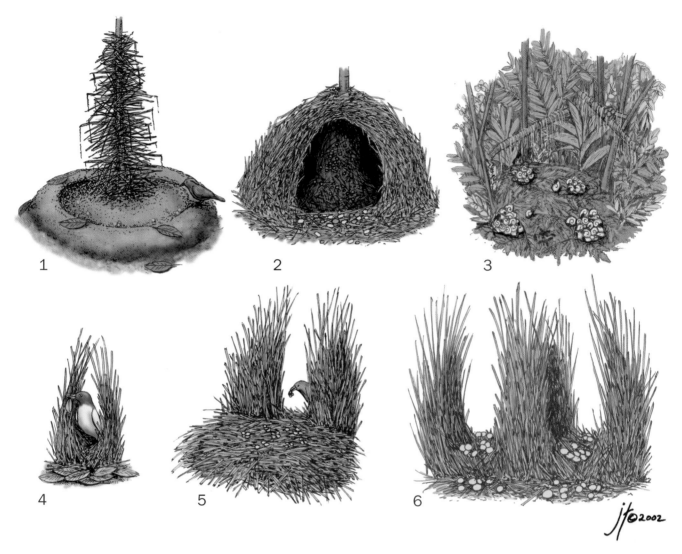

Maypole bowers: 1. Macgregor's bowerbird (*Amblyornis macgregoriae*)—bower decorated with spider's silk and leaves; 2. Streaked bowerbird (*Amblyornis subalaris*)—bower decorated with flowers; 3. Archbold's bowerbird (*Archboldia papuensis*)—bower decorated with snail shells and orchids. Avenue bowers: 4. Flame bowerbird (*Sericulus aureus*); 5. Fawn-breasted bowerbird (*Chlamydera cerviniventris*)—elevated platform decorated with green fruit; 6. Yellow-breasted bowerbird (*Chlamydera lauterbachi*)—avenue bower with central passage, decorated with green fruit. (Illustration by Joseph E. Trumpey)

being drab (some are barred ventrally). Juvenile and immature male plumages are similar to those of adult females. Males take five to seven years to fully acquire adult plumage.

Legs and feet are typically dark brown, olive-brown, olive, blue-gray, or black. Mouth color can be black, pale yellow, or orange-yellow depending on species. The skin of nestlings is pinkish, orange-pink, or pale flesh colored. Bill color is typically dark brown to black but can be pale or sexually dimorphic in some species. Iris color is typically pale to dark brown but is red in adult catbirds, whitish in *Sericulus*, and blue in *Ptilonorhynchus*.

Distribution

Ten species are confined to New Guinea and eight to Australia, and the two remaining species occur on both. Bower-

birds are mainly confined to the tropics and subtropics; only satin bowerbirds extend significantly into and across temperate regions. The Australian *Chlamydera* are mainly lowland dwelling but species in New Guinea occur up to 5,900 ft (1,800 m) altitude. Because of the vast mountain ranges, forest-dwelling New Guinea bowerbirds segregate by altitude. Only white-eared (*Ailuroedus buccoides*) and black-eared (*A. melanotis*) catbirds and great bowerbirds occupy continental islands.

Habitat

Species of *Ailuroudus, Scenopoeetes, Amblyornis,* and *Prionodura* are predominantly confined to rainforests and *Archboldia* to moss forests. Species of *Sericulus* and *Ptilonorhynchus* occur in rainforest but also at rainforest edges, and the latter species within adjacent wet forests and woodlands. The *Chlamydera* bowerbirds are adapted to more open, drier, riverine forests, forest/grassland ecotones, open woodland, savanna, and almost desert.

Behavior

The family includes species with socially monogamous and polygynous mating systems. Monogamous pairs of catbirds defend an all-purpose territory. Males do not assist with nest building, incubation, or brooding of nestlings (which they do feed). The promiscuous males of the 17 polygynous bowerbirds defend only the immediate area of their bowers. A seasonal hyperabundance of fruits permits promiscuous males to spend inordinate amounts of time at their courts, in attracting/ courting females, while also permitting females to nest and provision their offspring unaided.

Uniquely within the avian world, promiscuous males clear court areas and skillfully build complex symmetrical structures of sticks, grasses, or other vegetation, and decorate them. Three types of modified courts are: cleared and leaf-decorated courts, maypole bowers, and avenue bowers. Maypoles consist of branches and/or saplings with accumulations of orchid stems or sticks and an elaborate and decorated discrete mat beneath it. Avenues consist of two parallel walls of sticks or grass stems placed vertically into a foundation that is laid on a ground court that may extend beyond one or both ends of the bower to form a platform.

Male bowerbirds decorate courts and bowers with items such as leaves, flowers, fruits, lichens, beetle wing cases, insect skeletons, tree resin, snail shells, bones, river-worn pebbles, and specific parrot tail feathers and nuptial plumes of adult males of certain birds of paradise. Charcoal, glass, and innumerable other man-made objects may also be used. Males of some species manufacture and apply paint to bowers, even holding a wad of vegetable matter in the bill tip to use as a tool to apply paint. Because of this complex behavior, bowerbirds have been associated with high intelligence and artistic abilities.

Courts and bowers are located on favored topography exhibiting one or more micro-environmental features required by males. Bower sites are occupied for decades, and adult males exhibit long-term (one or more decades) fidelity to them. Immature males spend an apprenticeship of five to six years visiting rudimentary, or practice, courts or bowers of their own construction and bowers of adult males to acquire skills for better bower building, decorating, and displaying to attract females.

Courts and bowers are critical to male reproductive success in the polygynous species. They provide a focal point to which males attract females for courting and mating. Adult males of most species average 50–70% of daylight at their bower sites. Activities at bower sites involve vocalizations (advertisement song and other calls, including mimicry), bower maintenance (building, decorating, painting), display, and chasing unwanted conspecifics away. Rival males damage each others bowers and/or steal favored decorations, in so doing improving their own chances of attracting more potential mates.

Male satin bowerbird (*Ptilonorhynchus violaceus*) in its bower (Australia). (Photo by Tom McHugh. Photo Researchers, Inc. Reproduced by permission.)

Sexual selection, through mate choice by females, is fundamentally important to the evolution of elaborate display traits (ornate plumage and/or bower complexity/decoration) of bowerbirds. In some species colorful and elaborate display plumage has been lost and replaced by, or transferred to, a bower structure and its decorations. Discerning females assess the frequency and intensity of male bower attendance, the quality and/or quantity of bowers and decorations, displays, plumage, and vocalizations before soliciting the male of their choice. It is the older males, those with greater experience, skills, and survival, that are typically selected as mates by females.

Feeding ecology and diet

While typically omnivorous, several bowerbirds are more specialized in having a predominantly fruit diet supplemented by arthropods and other animals such as worms, frogs, skinks, and birds. Flowers, leaves, sap, and few seeds may also be eaten. Unlike birds of paradise, bowerbirds do not use their feet to hold and manipulate food.

Bowerbirds do not digest seeds but act as true seed dispersal agents to the plants on which they feed. The traditional nature of bower sites suggests that a local abundance of food plants might result from the germination of seeds defecated by the birds. Catbirds store or cache fruits about their territories, and males of some polygynous species do so about their bower sites.

Reproductive biology

Courtship of monogamous catbirds is simplistic; a male chases a female through tree foliage to then hop and bounce between perches in front of her before mating. Courtship of most polygynous bowerbirds is far more complex and is typically instigated by the arrival of a female at a bower site, after which the male moves away from the visitor in a ritualized fashion and/or hides from her view while producing a subsong that includes vocal mimicry of other bird calls and

A male golden bower bird (*Prionodura newtoniana*) at his bower that he's decorated with white flowers in Daintree National Park in northern Australia. (Photo by Michael Fogden. Bruce Coleman Inc. Reproduced by permission.)

environmental sounds. Male regent bowerbirds differ in initiating courtship by leading females to their bower from the forest canopy, where they had advertised their location by bright plumage rather than by calls. Courtship display typically commences on the bower court. Females signal their readiness to mate by solicitation posturing. Copulation usually takes place on the bower court, mat, or platform, or within the avenue.

The same nest location, even the specific site, is sometimes used each year by catbird pairs or by the same female of a polygynous species. Nests typically consist of a stick foundation with a nest-cup of dried leaves and twigs atop this and within which a discrete cup lining of finer material holds the clutch.

Elliptical eggs are pale and unmarked in *Ailuroedus*, *Scenopoeetes*, *Amblyornis*, *Archboldia*, and *Prionodura* and colored and vermiculated in *Sericulus*, *Ptilonorhynchus*, and *Chlamydera*. Clutch size is one to three eggs for both monogamous and polygynous bowerbirds. Eggs are laid on alternate days, with incubation usually starting with clutch completion. Renesting occurs following a nest loss, but there is no evidence of two broods being raised in a single season.

Depending on species, incubation lasts 21–27 days and the period lasts 17–30 days. Bowerbirds do not regurgitate meals to nestlings, unlike birds of paradise.

Details of nestling growth and development are known only for *Ailurodus*, *Archboldia*, and *Prionodura*. Nestlings of monogamous parents grow faster than those with only a female parent and also leave the nest when smaller as a proportion of adult size. Nestling bowerbirds fledge well feathered in a plumage similar to that of adult females, with some down remaining on the crown and elsewhere. After leaving the nest, bowerbird offspring depend on their parent(s) for 40–60 days or more. Proportions of successful nests, eggs, and nestlings are greater in monogamous than in polygynous species.

Nesting seasons in New Guinea are poorly known. Bowerbirds nest during the latter part of the dry season (late August through September) in Australian rainforests when temperatures, rainfall, and food resources are increasing. Egg laying peaks during October through December. Fruit and arthropod abundance reach a peak during hotter, wetter, months as females provision nestlings/fledglings (and adults begin their annual molt).

Conservation status

While several Australian species have lost parts of previously more extensive ranges to habitat destruction/degradation, none is rare or endangered as a species. A subspecies of western bowerbirds (*Chlamydera guttata carteri*) may be Near Threatened because of its highly restricted range. Adelbert bowerbirds (*Sericulus bakeri*) are listed as Vulnerable and Archbolds's bowerbird (*Archboldia papuensis*) are listed as Near Threatened. In addition to habitat destruction, the spread of domestic/feral cats and other exotic vertebrates through New Guinea forests may represent a threat to bowerbird populations.

Significance to humans

A few New Guineans and Australian aboriginals have long worn the crests of adult male *Amblyornis* and *Chlamydera* species respectively as personal adornment. Papuan men perceive male bowerbird activities as equivalent to their own efforts in seeking to attract and pay for a bride. Some aboriginal people respect male bowerbirds as avian custodians of ceremonies involving secret business (rites) of their own. They believed that *Chlamydera* species steal the bones of people for their own ceremonial purposes, and so birds and bowers are

Female satin bowerbird (*Ptilonorhynchus violaceus*) visits a male's bower. (Photo by Tom McHugh. Photo Researchers, Inc. Reproduced by permission.)

unmolested. Because birds take items from human middens as decorations they may influence the interpretation of archaeological assemblages. Of the 20 species, eight have been bred in aviaries.

1. Satin bowerbird (*Ptilonorthynchus violaceus*); 2. Regent bowerbird (*Sericulus chrysocephalus*); 3. Tooth-billed bowerbird (*Scenopoeetes dentirostris*); 4.Spotted bowerbird (*Chlamydera maculata*); 5. Archbold's bowerbird (*Archboldia papuensis*); 6. Macgregor's bowerbird (*Amblyornis macgregoriae*); 7. Golden bowerbird (*Prionodura newtoniana*); 8. Green catbird (*Ailuroudus crassirostris*). (Illustration by Joseph E. Trumpey)

Species accounts

Green catbird
Ailuroedus crassirostris

TAXONOMY
Ailuroedus crassirostris Paykull, 1815, Nova Hollandia = Sydney, New South Wales, Australia.

OTHER COMMON NAMES
English: Spotted catbird, large-billed cat bird, Australasian catbird; French: Jardinier vert; German: Grünlaubenvogel; Spanish: Capulinero Verde.

PHYSICAL CHARACTERISTICS
12.2 in (31 cm); female 0.37–0.47 lb (169–211 g), male 0.37–0.64 lb (167–289 g). Brownish head with lime-green upperparts and lighter, streaked coloring underneath; white-tipped wing coverts and tail.

DISTRIBUTION
Subtropical coastal east Australia, from Dawes Range in north to due east of Canberra, at sea level to 3,300 ft (1,000 m) altitude.

HABITAT
Primarily subtropical rainforest, but also adjacent rainforest edges, eucalyptus forests, gardens, and orchards.

BEHAVIOR
Perennial socially monogamous pair bonding within an all-purpose territory. Mean year round home range is up to five acres (two ha) but is smaller during the breeding season. Only females build nest, incubate, and brood. Both sexes feed young. Vocal repertoire is of cat-like wailing territorial song and sharp, high-pitched, tick-like contact notes. No mimicry.

FEEDING ECOLOGY AND DIET
Omnivorous, but predominantly frugivorous; specializes in *Ficus* figs and other fruits. Also eats flowers, buds, leaves, stems, seeds, arthropods, and small vertebrates, including birds. Mostly forages in the canopy but also to the ground.

REPRODUCTIVE BIOLOGY
Breeding mid-September through February/March, egg laying peaks October through December. A large, bulky, open cup nest is mostly built in tree forks, but also found in vine tangles, atop epiphytic ferns, and in tree ferns, at 6.6–60 ft (2–18 m) above ground. Nests are composed of a stick foundation, a cup of large dried leaves and occasional vine stems, a layer of decaying wood and sometimes earthy matter of epiphytic *Asplenium* ferns, and a fine twiglet/vine tendril egg-cup lining. One to three pale buff, unmarked eggs are laid. Incubation last 23–24 days; nestling period is 21 days.

CONSERVATION STATUS
Not threatened. Fairly common and widespread throughout remaining habitat, but rare to absent in rainforest patches of about 6 acres (2.5 ha) and smaller.

SIGNIFICANCE TO HUMANS
Once shot for eating and for sport. Some birds are still killed because they attack cultivated fruit crops. ◆

Tooth-billed bowerbird
Scenopoeetes dentirostris

TAXONOMY
Scenopoeetes dentirostris Ramsay, 1876, Bellenden Ker Range, Queensland, Australia.

Ailuroedus crassirostris
   Resident

Scenopoeetes dentirostris
   Resident

OTHER COMMON NAMES
English: Stagemaker, tooth-billed catbird, leaf turner; French:
Jardinier à bec denté; German: Zahnlaubenvogel; Spanish:
Capulinero de Pico Dentado.

PHYSICAL CHARACTERISTICS
10.6 in (27 cm); female 0.35–0.40 lb (157–182 g), male
0.29–0.44 lb (132–199 g). Medium sized, brownish bird with a
dark notched bill used to cut leaves for decorating court areas.

DISTRIBUTION
Australian wet tropics, north Queensland; from Big Tableland
in the north to Seaview-Paluma Range in south and on Mount
Elliot. Mostly occurs at 1,970–2,950 ft (600–900 m) altitude.

HABITAT
Upland tropical rainforests.

BEHAVIOR
Courts on average are 200 ft (61 m) apart. Court attendance is
during late August through early January, peaks in October
through December. Adult males exhibit
advertisement vocalization, including much vocal mimicry.

FEEDING ECOLOGY AND DIET
Primarily herbivorous, eating mostly fruits and leaves but also
some flowers and arthropods in the canopy; predominantly
folivorous in winter. Fruits and insects, mainly beetles, are
fed to nestlings.

REPRODUCTIVE BIOLOGY
Polygynous, with promiscuous males and exclusively female
nest attendance. Breeding occurs September through January;
egg laying peaks in November and December. Typically nests
in suspended vine tangles 26–88 ft (8–27 m) above ground.
Nests are made of a sparse stick foundation, sometimes with
orchid stems, and an egg-cup lining of fine twigs. Nest
diminutive and sparse. Lays one or two eggs. Incubation and
nestling periods are unknown.

CONSERVATION STATUS
Not threatened. Common throughout remaining, but protected, habitat.

SIGNIFICANCE TO HUMANS
None known. ◆

Macgregor's bowerbird
Amblyornis macgregoriae

TAXONOMY
Amblyornis macgregoriae De Vis, 1890, Musgrave River, Papua
New Guinea. Seven subspecies.

OTHER COMMON NAMES
English: Macgregor's gardenerbird, gardener bowerbird, crested
gardener bird, yellow-crested gardener; French: Jardinier de
Macgregor; German: Goldhaubengärtner; Spanish: Capulinero
de Macgregor.

PHYSICAL CHARACTERISTICS
10.2 in (26 cm); female 0.23–0.31 lb (104–140 g), male
0.22–0.32 lb (100–145 g). Brown with lighter head and underparts; distinctive long red crest.

DISTRIBUTION
Endemic to mountains of eastern and central New Guinea;
widespread on central cordillera, west to Weyland Mountains,

Amblyornis macgregoriae

▨ Resident

Irian Jaya, and on the Adelbert Range, the Huon Peninsula, and
Mount Bosavi. Occurs mostly at 5,250–7,540 ft (1,600–2,300 m)
altitude. *A. m. macgregoriae*: W. Kukukuku and Herzog Range
east to western Owen Stanley Range; *A. m. mayri*: Weyland
Mountains, Irian Jaya, to eastern Star/western Hindenburg
Mountains; *Amblyornis m. lecroyae*: Mount Bosavi; *A. m. kombok*:
Kubor, Hagen, and Bismarck Ranges, probably west to at least
Strickland River or Hindenberg Range and east to Kraetke
Range; *A. m. amati*: Adelbert Mountains; *A. m. germanus*: Huon
Peninsula; *A. m. nubicola*: Simpson-Dayman massifs, eastern
Owen Stanley Range, probably west to Mount Suckling.

HABITAT
Primary tall mixed montane and *Nothofagus* rainforest.

BEHAVIOR
Traditional bower sites are regularly and linearly spaced along
forested ridges. The maypole bower consists of a conical tower
of sticks built about a sapling or tree fern trunk surrounded at
its base by a circular moss mat raised at its circumference into
an elevated rim. Bower may be used for 20 or more years. Decorations include insect frass, charcoal, fungus, tree resin, mammal dung, fruits, and leaves. Bowers maintained for nine to ten
months annually, with peak display during August through December. Advertisement vocalizations include harsh tearing
sounds, growls, thuddings, tappings, whistles, and much vocal
mimicry including human-made sounds.

FEEDING ECOLOGY AND DIET
Primarily frugivorous, taking fruits from numerous trees, shrubs,
and vines. Also eats flower parts and insects.

REPRODUCTIVE BIOLOGY
Polygynous, with promiscuous adult males and exclusively female nest attendance. Breeding season variable across the
species range. Typically builds bulky open cup nest in pandanus tree crown 6.6–10 ft (2–3 m) above ground. Nest is
composed of a sparse stick foundation, a leafy cup, and an egg-
cup lining of supple twiglets/rootlets. Lays a single, pale, unmarked, buff egg. One known incubation period was over 17
days. Nestling period unknown.

CONSERVATION STATUS

Not threatened. Common and widespread throughout range.

SIGNIFICANCE TO HUMANS

Papuans admire the industry/artistry of males at their bowers. By placing a leaf on a bower mat, men and women used the bower mat clearing behavior of males to indicate to them in which direction they might seek a spouse. Crests of adult males may be worn as personal adornment. ◆

Archbold's bowerbird
Archboldia papuensis

TAXONOMY

Archboldia papuensis Rand, 1940, Bele River, Snow Mountains, Irian Jaya. Two subspecies

OTHER COMMON NAMES

English: Sandford's bowerbird, Tomba bowerbird, gold-crested black bowerbird; French: Jardinier d'Archbold; German: Archboldlaubenvogel; Spanish: Capulinero de Archbold.

PHYSICAL CHARACTERISTICS

14.2 in (36 cm); female 0.36–0.41 lb (163–185 g), male 0.40–0.42 lb (180–190 g). Brown with distinctive cropped yellow tuft from forehead to back.

DISTRIBUTION

Patchily distributed along the central New Guinea cordillera, mostly at 7,540–9,500 ft (2,300–2,900 m) altitude. *A. p. papuensis*: Bele River near Lake Habbema, Wissel Lakes, Oranje, Nassau, and Weyland Ranges, Irian Jaya; *A. p. sanfordi*: Mount Hagen and Giluwe, Tari Gap, and southern Karius Range.

HABITAT

Frost-prone moss forests.

BEHAVIOR

The maypole bower consists of a deep terrestrial mat of fern fronds, averaging 10 x 13 ft (3 x 4 m) in size, decorated with discrete piles of snail shells, beetle elytra, tree resin, plumes of adult male King of Saxony birds of paradise (*Pteridophora alberti*), and other objects. Perches above the mat are draped with orchid stems and decorated with fruits and other items. Adult males emit diverse advertisement vocalizations that include mimicry.

FEEDING ECOLOGY AND DIET

Unknown for adults but the nestling diet is mainly of fruit, tree-climbing skinks, beetles, and other arthropods.

REPRODUCTIVE BIOLOGY

Polygynous, with promiscuous adult males and exclusively female nest attendance. Active nests observed during September through February. Large, bulky, open cup nest is typically built in fork of an isolated sapling 10–23 ft (3–7 m) above ground. Nest is made of a stick foundation, a deep substantial cup of large dried leaves (uppermost ones green), and an egg-cup lining of curved twiglets. The single, unmarked, pale buff, egg is incubated for 26–27 days. The nestling period is 30 days.

CONSERVATION STATUS

Considered Near Threatened. Reasonably common and widespread throughout its patchy range.

SIGNIFICANCE TO HUMANS

King of Saxony bird of paradise plumes are highly valued by highland men as personal adornment and are taken from bowers when found. ◆

Golden bowerbird
Prionodura newtoniana

TAXONOMY

Prionodura newtoniana De Vis, 1883, Tully River Scrubs, North Queensland, Australia.

OTHER COMMON NAMES

English: Newton's bowerbird, Queensland gardener; French: Jardinier de Newton; German: Säulengärtner; Spanish: Capulinero de Newton.

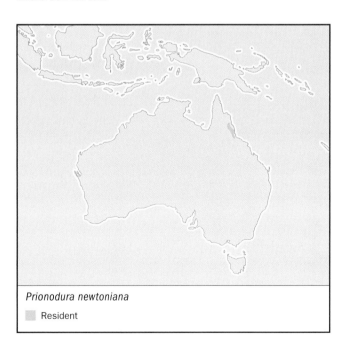

Archboldia papuensis
▨ Resident

Prionodura newtoniana
▨ Resident

PHYSICAL CHARACTERISTICS
9.5 in (24 cm); female 0.13–0.21 lb (62–96 g), male 0.13–0.19
lb (62–86 g). Brown head and wings with bright yellow-gold
underparts, tail, crest, and nape.

DISTRIBUTION
Australian wet tropics, from Thornton Range and Mount
Windsor Tableland in north to Seaview-Paluma Range in
south, mostly at 2,300–3,250 ft (700–990 m) altitude.

HABITAT
Upland tropical rainforests.

BEHAVIOR
Males build bowers to attract females. Traditional bower sites
spatially dispersed throughout suitable topography (flatter terrain and along ridge slopes and ridges), on average 495 ft (151
m) apart. Maypole bowers have one or two towers up to 6.6 ft
(2 m) tall. Bowers are made of sticks around saplings with a
horizontal display perch. Where the perch meets the tower(s),
neatly aligned sticks form a platform(s) upon which grayish
green lichen, creamy-white seed pods, flowers, and fruits are
placed as decorations. Bower structures may remain in use for
20 or more years and traditional sites for much longer. Bowers
are attended during August through December/January, peaking in October through December. Adult males emit rattle-like
advertisement song and medleys of other calls including mimicry. They follow initial display posturing with an extensive
flight and hover display, followed by hiding behind trees while
producing vocal mimicry.

FEEDING ECOLOGY AND DIET
Omnivorous but predominantly frugivorous, eating a variety of
fruits including those of many vines. Also eats flowers, buds,
and arthropods, particularly beetles. Cicadas are important to
the nestling diet. Adults mainly forage in the lower canopy and
subcanopy.

REPRODUCTIVE BIOLOGY
Polygynous, with promiscuous adult males and exclusively female nest attendance. Breeding occurs in late September
through January/February. Egg laying peaks in November and
December. Typically builds its open cup nest within a roofed
tree crevice or crevice-like situation, up to 6.6 ft (2 m) above
ground. Nest is composed of a foundation of stout sticks, a
substantial bowl structure of dead leaves and leaf skeletons, and
an egg-cup lining of fine, supple, springy tendrils. One to three
plain whitish eggs are laid. Incubation lasts 21–23 days and the
nestling period is 17–20 days.

CONSERVATION STATUS
Not threatened. Common and widespread throughout limited
remaining but fully protected habitat.

SIGNIFICANCE TO HUMANS
A small number of traditional bowers are of significance to local tourist industries (to the detriment of several resident males
disturbed by too frequent human visitations). ◆

Regent bowerbird
Sericulus chrysocephalus

TAXONOMY
Sericulus chrysocephalus Lewin, 1808, Patterson's River = Hunter
River, New South Wales, Australia.

Sericulus chrysocephalus

◻ Resident

OTHER COMMON NAMES
English: Regent bird, Australian regent bowerbird, king honey
sucker, golden regent; French: Jardinier prince-régent; German:
Gelbnacken-Laubenvogel; Spanish: Capulinero Governador.

PHYSICAL CHARACTERISTICS
9.8 in (25 cm); female 0.20–0.30 lb (91–138 g), male 0.17–0.24
lb (76–110 g). Black with bright yellow flight feathers and area
from forehead to shoulder; orange-red band from forehead to
nape. Eyes are bright yellow.

DISTRIBUTION
Subtropical coastal zone of central eastern Australia from immediately north of Sydney to the Connors and Clarke Ranges,
Eungella Plateau, inland of Mackay, Queensland, with a gap
in distribution around the Fitzroy River valley inland of Rockhampton. From sea level to 2,950 ft (900 m), but altitude
varies across the range.

HABITAT
Subtropical rainforest, adjacent woodland, and, in winter, more
open habitats, including cultivated country and urban gardens.

BEHAVIOR
Males build bowers to attract females. The sparse and small
avenue bower is well concealed beneath low dense vines/fo-
liage. Traditional bower sites are dispersed through appropriate ridge top habitat. Bowers discovered by rivals are
destroyed, if not by a rival then by the owner. Another bower
is then built at a nearby location. Bower structures last an average of 10 days or less. Decorations include green leaves,
flowers, fruits, snail shells, and cicada ectoskeletons. Seasonal
bower attendance mostly September through January on
Sarabah Range but from July through August elsewhere. Adult
males vocalize by producing harsh grating sounds; when courting they emit a soft complex subsong.

FEEDING ECOLOGY AND DIET
Omnivorous but predominantly frugivorous. Also eats flowers,
nectar, and animals. Females dominate males at feeding trees.

REPRODUCTIVE BIOLOGY
Polygynous, with promiscuous adult males and exclusively female nest attendance. Breeding occurs September through February. Egg laying peaks in November and December. Typically builds relatively frail open cup nest among clumps of vines or mistletoe 6.6–102 ft (2–31 m) above ground. Nests are made of a frail shallow saucer of loose sticks and an egg-cup lining of finer twigs with a few leaves. One to three colored and vermiculated eggs are laid. Incubation in captivity lasted 17–21 days and nestling period at one nest was 17 days.

CONSERVATION STATUS
Not threatened. Common and widespread throughout limited remaining but mostly protected habitat. In some areas numbers are reported as greatly reduced to uncommon because of habitat destruction and degradation.

SIGNIFICANCE TO HUMANS
Adult males were once systematically hunted for mounting as decorative novelties commonly included in cabinets of stuffed birds. They are also a popular cage bird, both within and outside Australia. Birds may be pests to cultivated fruit crops. ◆

Satin bowerbird
Ptilonorhynchus violaceus

TAXONOMY
Ptilonorhynchus violaceus Vieillot, 1816, Nouvelle Hollande = Sydney, New South Wales, Australia. Two subspecies.

OTHER COMMON NAMES
English: Satin bird, satin grackle, purple satin; French: Jardinier satiné; German: Seidenlaubenvogel; Spanish: Capulinero Satinado.

PHYSICAL CHARACTERISTICS
13 in (33 cm); female 0.38–0.57 lb (170–258 g), male 0.38–0.64 lb (173–290 g). Iridescent black plumage with light legs and

Ptilonorhynchus violaceus
◻ Resident

bill. Female slightly smaller with green, gray-green, brown, and buff coloring.

DISTRIBUTION
Eastern and southeastern Australia. *P. v. violaceus*: coastal zone of southeast Australia from Otway Range, immediately east of Melbourne, east and North to Dawes Range just south of the Fitzroy River at Rockhampton; from sea level to 3,600 ft (1,100 m). *P. v. minor*: Australian wet tropics, from Seaview-Paluma Range north to Mount Amos near Cooktown, typically over 1,970 ft (600 m) altitude.

HABITAT
Rainforests, with a strong preference for their edges, and adjacent woodlands with dense sapling understory. Frequents more open habitats when winter flocking, then frequents pastures and urban/suburban areas.

BEHAVIOR
Males build bowers to attract females. Avenue bowers are fairly evenly and linearly dispersed at an average of about 990 ft (300 m) apart along rainforest edges, often further apart in rainforest patches and woodlands. Mostly bluish and greenish yellow items are used as decorations, including flowers, fruits, parrot feathers, snake skin, snail shells, and numerous human-made objects. Seasonal bower attendance commences in late August/September and peaks during October through December. Adult males emit advertisement vocalizations with a clearly-whistled *quoo-eeeew*, various harsh notes, and vocal mimicry.

FEEDING ECOLOGY AND DIET
Omnivorous but predominantly frugivorous. Also eats flowers, leaves, nectar, seeds, and animals including cicadas, beetles, and other arthropods. Forages mostly in the canopy but winter flocks forage on the ground for pasture leaves and herbs.

REPRODUCTIVE BIOLOGY
Polygynous, with promiscuous males and exclusively female nest attendance. Breeding occurs August/September through February. Egg laying peaks in November and December. Typically builds open cup nest in trees or bushes, but also in vine tangles and mistletoe, at 6.6–131 ft (2–40 m) above ground. Nests are composed of a shallow saucer of sticks and twigs and an egg-cup lining of green and dry leaves (mostly of *Eucalyptus* and *Acacia*). One to three colored and blotched, rarely vermiculated, eggs are laid. Incubation is 21–22 days and nestling period is 17–21 days.

CONSERVATION STATUS
Not threatened. A common to reasonably abundant bird in remaining habitat but has lost habitat because of human land use.

SIGNIFICANCE TO HUMANS
Ornithological and popular literature contains numerous stories of males removing jewelry, keys, and other items from homes, vehicles, camps, etc. to decorate bowers ◆

Spotted bowerbird
Chlamydera maculata

TAXONOMY
Chlamydera maculata Gould, 1837, New Holland = Liverpool Plains, Australia, New South Wales, Australia.

Chlamydera maculata

Resident

and just into South Australia along the Murray River system. Occurs from sea level to about 1,640 ft (500 m).

HABITAT
Brigalow and open eucalyptus woodlands, with a preference for riverine woodland.

BEHAVIOR
Grassy avenue bowers are built beneath low bushes or shrubs, 3,300–6,600 ft (1,000–2,000 m) apart. Decorations (to 1,000 or more) include berries, seed pods, pebbles, bones, snail shells, and glass. Adult males emit infrequent, loud, far-carrying advertisement vocalizations of harsh churrings and other notes including vocal mimicry.

FEEDING ECOLOGY AND DIET
Omnivorous. Eats fruits, flowers, leaves, seeds, and arthropods. Grasshoppers are important to the nestling diet.

REPRODUCTIVE BIOLOGY
Polygynous, with promiscuous males and exclusively female nest attendance. Breeding occurs during July through March. Egg laying peaks in October through February. Typically places sparse open cup nest in trees and bushes at 10–40 ft (3–12 m) above ground. Nests are made of a loose bulky foundation of dead twigs and sticks and an egg-cup of fine twiglets, sometimes with dried grass stalks. Incubation period is unknown. Nestling period at one nest was 21 days.

CONSERVATION STATUS
Not threatened, but numbers have historically declined in some areas because of illegal shooting and poisoning, predation by domestic and feral cats and foxes, and widespread clearing and/or modification/fragmentation of habitat. Populations are officially considered Endangered within state of Victoria.

SIGNIFICANCE TO HUMANS
Commonly kept in aviculture, where birds thrive. Bird known to steal items from homes, camps, and vehicles for bower decorations. Birds are killed as pests to gardens and orchards. ◆

OTHER COMMON NAMES
English: Large-frilled bowerbird, cabbage bird; French: Jardinier maculé; German: Fleckenlaubenvogel; Spanish: Capulinero Moteado.

PHYSICAL CHARACTERISTICS
11.4 in (29 cm); female 0.27–0.36 lb (124–162 g), male 0.28–0.33 lb (125–150 g). Mottled brown with lilac bar across the back of the neck.

DISTRIBUTION
Interior of Queensland south of 20 degrees South, except the extreme west and southwest, and interior of west and central New South Wales, except the extreme western border country. Extends a small way into the northwest corner of Victoria

Resources

Books

Frith, C.B., and D. W. Frith. *The Bowerbirds: Ptilonorhynchidae.* Oxford: Oxford University Press, 2002.

Gilliard, E.T. *Birds of Paradise and Bower Birds.* London: Weidenfeld and Nicolson, 1969.

Marshall, A.J. *Bower-birds, Their Displays and Breeding Cycles—A Preliminary Statement.* Oxford: Oxford University Press, 1954.

Schodde, R., and I. J. Mason. *The Directory of Australian Birds, Passerines.* Melbourne: CSIRO, 1999.

Periodicals

Borgia, G. "Why Do Bowerbirds Build Bowers?" *American Scientist* 83 (1995): 541–547.

Dwyer, P., M. Minnegal, and J. Thomson. "Odds and Ends, Bower Birds as Taphonomic Agents." *Australian Archaeology* 21 (1985): 1–10.

Frith, C.B., and D.W. Frith. "Biometrics of the Bowerbirds (Aves: Ptilonorhynchidae): with Observations on Species Limits, Sexual Dimorphism, Intraspecific Variation and Vernacular Nomenclature." *Memoirs of the Queensland Museum* 24 (2001): 512–542.

Madden, J. "Sex, Bowers and Brains." *Proceedings of the Royal Society of London* B268 (2001): 833–838.

Organizations

Birds Australia. 415 Riversdale Road, Hawthorn East, Victoria 3123 Australia. Phone: +61 3 9882 2622. Fax: +61 3 9882 2677. E-mail: mail@birdsaustralia.com.au Web site: <http://www.birdsaustralia.com.au>

Clifford B. Frith, PhD
Dawn W. Frith, PhD

Birds of paradise
(Paradisaeidae)

Class Aves

Order Passeriformes

Suborder Passeri (Oscines)

Family Paradisaeidae

Thumbnail description
Small to very large, powerfully footed, highly animated, and vocal crow-like passerines, most of which are sexually dichromatic. Highly colorful and elaborated, adult male plumages of polygynous species are used in spectacular and complex courtship displays.

Size
6.3–43.3 in (16–110 cm); 0.11–1 lb (50–450 g)

Number of genera, species
17 genera; 42 species

Habitat
Lowland to subalpine rainforests and some associated forests and wet woodland communities

Conservation status
Vulnerable: 4 species; Near Threatened: 8 species

Distribution
Mainland New Guinea and offshore islands, the northern Moluccas of Indonesia, and northeastern and central eastern Australia

Evolution and systematics

Birds of paradise belong to parvorder Corvida, which is considered an ancient lineage of Australo-Papuan passerines derived from Gondwanan stock. Scientists traditionally associated them most closely with bowerbirds (Ptilonorhynchidae), but a major dichotomy between the two groups has been widely accepted. Results of several molecular studies place the separation of birds of paradise and other corvines (superfamily Corvoidea) from bowerbirds (superfamily Menuroidea) at 28 million years ago. The current distribution of birds of paradise strongly supports the thesis that the group radiated in New Guinea. All of the generic radiations are either endemic or largely confined to New Guinea.

The family Paradisaeidae comprises 17 genera and 42 species that are divided into two subfamilies: three species of wide-gaped (Cnemophilinae) and 39 species of typical (Paradisaeinae) birds of paradise. The Cnemophilinae consist of two polygynous *Cnemophilus* and one little known *Loboparadisea* species. The Pardisaeinae comprises seven species in three genera (*Macgregoria, Lycocorax, Manucodia*) known or presumed to be monogamous, and 32 species in 12 genera (*Paradigalla, Astrapia, Parotia, Pteridophora, Ptiloris, Lophorina, Epimachus, Drepanornis, Cicinnurus, Semioptera, Seleucidis, Paradisaea*) known or presumed to be polygynous. Seventy-five subspecies are presently recognized. The two subfamilies are highly distinctive and share few unambiguous derived characters that prove they are of the same lineage.

Physical characteristics

Birds of paradise vary in size from 6 in (15 cm) and 0.11 lb (50 g) for the king bird of paradise (*Cicinnurus regius*) to 17.3 in (44 cm) and 1 lb (450 g) for the curl-crested manucode (*Manucodia comrii*). When tails are added into the size calculations, some birds of paradise exceed 40 in (100 cm) in length. Typically males are larger and heavier than females. As a result of strong morphological radiation, the bill of the various genera varies from short to long, slim to stout, and straight to dramatically curved. The small and finely-tipped weak bill of cnemophilins has a wide gape as an adaptation to an exclusive drupe and berry fruit diet. They also have relatively fine and weak legs and feet. In contrast, paradisaeins have large and powerful legs and feet used to acrobatically cling to substrates and to hold food items.

Several species exhibit areas of pigmented bare skin; for example, Wilson's bird of paradise (*Cicinnurus respublica*) has extensive bright blue bare head skin. Some species have brightly colored wattles or legs and feet. Bare parts tend to be much brighter in males than in females, and they may relate to courtship display and act as species-specific social sig-

A ribbon-tailed astrapia (*Astrapia mayeri*) female feeds a nestling at her nest in the central highlands of Papua New Guinea. (Photo by Frithfoto/Olympus. Bruce Coleman Inc. Reproduced by permission.)

nals. Males of species with a brightly colored mouth interior typically gape the bill widely to present this in courtship calling and display.

Species of *Macgregoria*, *Manucodia*, *Lycocorax*, and *Paradigalla* are generally black and sexually monochromatic with some blue/green iridescence to their feathering. Other species (all polygynous) are sexually dichromatic, with adult males adorned with colorful and often highly elaborated plumage. The remarkably modified and erectile elongate head plumes of *Parotia* and *Pteridophora* are associated with gross cranial modifications to facilitate large muscles required to manipulate the plumes during display. The only crested species is *Cnemophilus macgregorii*, in which both sexes wear a diminutive sagittal crest of a few filamentous sickle-shaped feathers. Females of polygynous species are drably colored in subdued browns and dull yellows (*Paradisaea*) or are brown and/or rufous above and drably paler below with darker barring to give a cryptic appearance. Males take at least five to seven years to fully acquire adult plumage. Limited evidence suggests that females breed at an earlier age than males, probably after their first year or two. Hatchlings have very little down and their skin becomes characteristically dark after several days.

Birds of paradise have ten primaries and twelve tail feathers. In adult males, tail feathers may be highly modified as nuptial display traits. In any given species these may become longer or shorter, even within a single genus, and more ornate with increasing male age. In most genera the wings are rounded. In adult males of several genera some outer primaries are slightly

to highly modified in shape, probably for the production of mechanical sound in flight. The *Manucodia* exhibit a greatly elongated, coiled, trachea that is displaced to sit subcutaneously on top of the pectoral muscles. This structure produces low far-carrying tremulous call notes unique within the group.

Distribution

Thirty-eight species live on the mainland of New Guinea and its adjacent islands, two are peculiar to the northern Moluccas of Indonesia (paradise crows [*Lycocorax pyrrhopterus*] and standardwing birds of paradise [*Semioptera wallacii*]), and two are endemic to areas of eastern Australia (paradise [*Ptiloris paradiseus*] and Victoria's [*P. victoriae*] riflebirds). Magnificent riflebirds (*P. magnificus*) and trumpet manucodes (*Manucodia keraudrenii*) occur on both New Guinea and the extreme northeastern tip of Australia.

Because of the great altitudinal range of New Guinea, forest-dwelling species have segregated and live within different forest types within one or more altitudinal zones. Elevation is perhaps the most important ecological sorting mechanism permitting the adaptive radiation of local avian lineages in birds of paradise, and also offers closely related species the opportunity to avoid competition while establishing limited geographic sympatry (meeting along a narrow altitudinal zone). Thirteen intergeneric and seven intrageneric wild hybrid crosses have been documented where their ranges/favored habitats overlap, dramatically emphasizing the close genetic relationships between the species of the Paradisaeinae.

Habitat

All species depend on closed humid forest over much of their geographical range, and rainforests and/or moss forests are the most typical habitats of the family as a whole. Associated rainforest edges, wet sclerophyll (Australian vegetation with hard, short, and often spiky leaves) forests and woodlands, gardens, savanna, and subalpine woodlands are also used. Only paradise riflebirds range southward across the Tropic of Capricorn to about 32° South within the subtropics. Glossy-mantled manucodes (*Manucodia atra*) are exceptional in inhabiting relatively dry open savanna woodlands in addition to lowland rainforests.

Behavior

The majority of species exhibit polygynous (court- and lek-based type) mating systems with promiscuous males and exclusively female nest attendance. Sexual selection has produced male vocalizations, elborate plumages, and complex courtship choreography. Males of polygynous species occupy a mating area and modify display sites by removing leaves and/or debris to create a visual marker to the site. At their display site, males emit an advertisement song of harsh, loud, crow-, bell-, and bugle-like notes, screeches, and rapid bursts of powerful notes to attract females. Other diverse sounds produced in display include wing beating, bill rattling, primary swishing, and wing snapping, and flight induced noise.

Displays of promiscuous males range from solitary and non-territorial (the most common type) to communal lekking mating systems, with a range of intermediate manifestations. Display sites of some males are not dispersed evenly through habitat but are loosely (exploded leks) or tightly (true leks) clustered. In true lekking species, males display in a tight cluster in the canopy branches of one or more trees, and their leks tend to be distantly spaced, long lived, and traditional. Much display activity apparently maintains a male-male dominance hierarchy that limits the choice of potential mates by females to one or more 'alpha' males occupying the central lek position. A small percentage of males of lekking bird species obtain most of the matings in any single season.

Females of known monogamous species emit identifiable vocalizations, whereas among polygynous species all of the loudly broadcast vocalizations are male advertisement calls and females are virtually mute.

Feeding ecology and diet

Birds of paradise eat a range of food types (omnivorous), but they seem to be primarily fruit and arthropod eaters. Only the Cnemophilinae appear not to eat arthropods. Collectively, birds of paradise are known to eat fruits of several hundred plant species, flowers, nectar, leaves, insects, spiders, frogs, and lizards. Most typical birds of paradise initially feed nestlings arthropods but switch to a predominantly fruit or mixed fruit and arthropod diet after a certain age. Parent birds of paradise regurgitate food items to young.

Birds take insects by bark gleaning, dead wood/foliage probing/tearing, and generalized twig and foliage gleaning.

A male greater bird of paradise (*Paradisaea apoda*) shows off its display. (Photo by C.H. Greenewalt/VIREO. Reproduced by permission.)

The sexes of predominantly insectivorous species show marked sexual dimorphism in bill size and shape, presumably to limit intersexual competition for this resource.

Other than foraging in ones or twos, birds of paradise (mostly brown female-plumaged individuals) commonly join mixed species foraging flocks of typically brown and/or black birds, predominantly in lowland and hill forest.

Reproductive biology

Courtship in monogamous species is simplistic, consisting of little more than chasing, vocalizing, and limited display posturing in canopy foliage. In polygynous species, courtship is far more complex. The elaborate nuptial plumes of adult males are specifically presented to females in a stereotyped courtship. For example, male *Paradisaea* lean forward and downward and lower open wings so that their lacy flank plumes can be raised above their back and head, and males of emperor (*P. guilielmi*) and blue (*P. rudolphi*) birds of paradise hang fully upside down in courtship. Slow and rhythmic leg flexing that enhances the effect of longer plumes is incorporated into some displays.

All species for which nests are known are solitary nesters. Females of polygynous species construct the nest alone; whether both sexes of monogamous species share the task is not known. Nests of *Cnemophilus* species are dense, substantial, roughly spherical, domed structures predominantly composed of slender orchid stems overlaid with fresh mosses and ferns that incorporate a token foundation of a few stout short woody sticks. These nests are extremely cryptic. Nests of members of the Paradisaeinae are typically found in tree branches. Nests of the Paradisaeinae (except *Manucodia* species) are open cup- or bowl-shaped and built of orchid stems, mosses, and fern fronds on branch forks. *Manucodia* construct a sparse and relatively shallow open cup made predominantly of vine tendrils and suspend it between horizontal forking branches.

The female (left) and male king birds of paradise (*Cicinnurus regius*) show the sexual dimorphism typical of this family. (Photo by W. Peckover/VIREO. Reproduced by permission.)

Eggs are typically elliptical, ovate, and pinkish to buff with long, broad, brush-stroke-like markings of browns, grays, and lavender or purplish gray. The clutch consists of one or two (rarely three) eggs. Limited data indicate that most multi-egg clutches are laid on successive days, and egg weights as a proportion of mean adult female body weight average 10–15%. Incubation varies from 14 to 27 days. Nestling periods, which are generally longer in higher-altitude species, vary among species from 14 to more than 30 days. Nestling eyes open at about six days old. The care of dependant young out of the nest is little known. Renesting occurs following a nest loss, but there is no evidence of two broods being successfully raised in a single season.

As a generalization, far more breeding takes place during the months between August and January than during February to July, with March to June being least productive. This broad seasonality appears to coincide with a period of abundance of fruits and arthropod prey. A similar cycle occurs in Australia.

Conservation status

Macgregor's bird of paradise (*Macgregoria pulchra*), black sicklebill (*Epimachus fastuosus*), Wahnes's parotia (*Parotia wahnesi*), and blue bird of paradise are listed as Vulnerable.

The yellow-breasted bird of paradise (*Loboparadisea sericea*), long-tailed paradigalla (*Paradigalla carunculata*), ribbon-tailed astrapia (*Astrapia mayeri*), pale-billed sicklebill (*Drepanornis bruijnii*), and Wilson's, Goldie's (*Paradisaea decora*), red (*P. rubra*), and emperor birds of paradise are considered Near Threatened species.

Significance to humans

Resplendently plumaged adult males of many species have long been killed and their skins preserved for the personal adornment of Papuan men and as highly valued items of trade.

1. Female and 2. Male Lawes's parotia (*Parotia lawesii*). (Illustration by Emily Damstra)

All males: 1. Ribbon-tailed astrapia (*Astrapia mayeri*); 2. Crested bird of paradise (*Cnemophilus macgregorii*); 3. Standardwing bird of paradise (*Semioptera wallacii*); 4. King of Saxony bird of paradise (*Pteridophora alberti*); 5. Crinkle-collared manucode (*Manucodia chalybata*); 6. Greater bird of paradise (*Paradisaea apoda*); 7. Short-tailed paradigalla (*Paradigalla brevicauda*); 8. King bird of paradise (*Cicinnurus regius*); 9. Victoria's riflebird (*Ptiloris victoriae*). (Illustration by Emily Damstra)

Species accounts

Crested bird of paradise
Cnemophilus macgregorii

SUBFAMILY
Cnemophilinae

TAXONOMY
Cnemophilus macgregorii De Vis, 1890, Mount Knutsford, Owen Stanley Range, British New Guinea. Two subspecies.

OTHER COMMON NAMES
English: Sickle-crested bird of paradise, multi-crested bird of paradise; French: Paradisier huppé; German: Furchenvogel; Spanish: Ave del Paraíso de MacGregor.

PHYSICAL CHARACTERISTICS
9.5 in (24 cm); female 0.17–0.28 lb (79–125 g), male 0.20–0.27 lb (90–120 g). Forehead, ear coverts, and upperparts bright flame yellow with orange wash, dulling in color toward rump and tail. Bill, lores, and underparts a dark brownish black. Orange to olive crest, depending on subspecies. Females brownish olive or brownish gray with darker upperparts.

DISTRIBUTION
Occurs patchily in high forests of the central cordillera of New Guinea at altitudes ranging from 6,900 to 11,975 ft (2,100 to 3,650 m). *C. m. macgregorii*: southeast Papua New Guinea northwestward to at least the Ekuti Divide, east of the Watut/Tauri Gap; *C. n. sanguineus*: central and eastern Highlands, Papua New Guinea, west of the range of the nominate subspecies.

HABITAT
Upper montane and subalpine forest and forest edge, including secondary growth and disturbed forest and shrubbery.

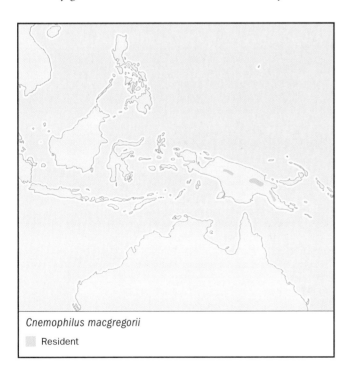

Cnemophilus macgregorii
■ Resident

BEHAVIOR
A less vocal species than most birds of paradise. The flight of adult males produces a loud whirring sound. Display season is likely at least June through November. Mostly frequents middle and lower forest structure.

FEEDING ECOLOGY AND DIET
Notably exceptional in being an obligate frugivore, both as adults and nestlings.

REPRODUCTIVE BIOLOGY
Polygynous, with presumed promiscuous males and exclusively female nest attendance. Breeding at least August through January. Nest is globular, with side entrance, and sits 6.6–13 ft (2–4 m) above ground. Incubation is probably about 26 days or more; nestling period more than 30 days. As only single nestlings have been observed, clutch is probably one egg.

CONSERVATION STATUS
Not threatened. Habitat is more restricted than that of many family members, but the vulnerability of this bird, common in optimal habitat, is low because of its higher altitude.

SIGNIFICANCE TO HUMANS
Rarely, skins of adult males are used as human adornment. ◆

Crinkle-collared manucode
Manucodia chalybata

SUBFAMILY
Paradisaeinae

TAXONOMY
Manucodia chalybata Pennant, 1781, New Guinea (restricted to Arfak Mountains).

OTHER COMMON NAMES
English: Green-breasted manucode, crinkle-breasted manucode, green manucode; French: Paradisier vert; German: Grünmanucodia; Spanish: Manucodia de Cuello Arrugado.

PHYSICAL CHARACTERISTICS
13.0–14.2 in (33–36 cm); female 0.35–0.64 lb (160–289 g), male 0.33–0.58 lb (150–265 g). Male and female have blue-black head and neck with blue-green iridescent feather tips on chin, throat, neck, and nape. Upperparts and uppertail coverts and iridescent violet-black, with dark blue sheen. Underparts also violet-black, but with yellow-green iridescence. Dark bill and legs. Eyes red.

DISTRIBUTION
Hill and lower montane areas of New Guinea mainland and lowlands of Missol Island. Generally at 1,968–4,920 ft (600-1,500 m) but occasionally from sea level to 5,575 ft (1,700 m) altitude.

HABITAT
Hill forests and midmontane rainforest edges.

BEHAVIOR
Courtship displays are simplistic, involving a male chasing a female through foliage via numerous perches. The male gives

Manucodia chalybata

Resident

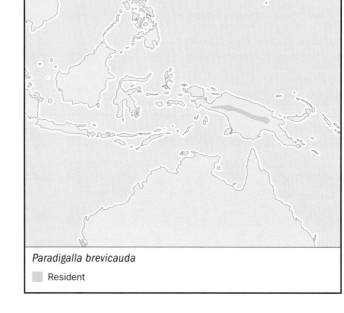

Paradigalla brevicauda

Resident

deep pigeon-like, or deep hollow, calls. Mostly frequents middle to upper forest structure. Flight appears floppy and awkward; wing beats are quick and convulsive, each accompanied by a sharp jerking of the cocked tail and interspersed by short glides.

FEEDING ECOLOGY AND DIET
Predomininatly frugivorous and specializing in fruits of *Ficus* figs. Commonly joins mixed species foraging flocks of predominantly black and/or brown birds.

REPRODUCTIVE BIOLOGY
Socially pair-bonded non-territorial monogamy with both sexes presumed to incubate and provision nestlings, as in other manucodes. Breeding during at least July through September, and in January. Nest is a shallow open cup suspended by its rim from a branch fork. Clutch is one or two eggs.

CONSERVATION STATUS
Not threatened. Apparently common over much of range.

SIGNIFICANCE TO HUMANS
In being monogamous, *Manucodia* provide an interesting contrast to the majority of polygynous family members for socio-ecological study. ◆

Short-tailed paradigalla
Paradigalla brevicauda

SUBFAMILY
Paradisaeinae

TAXONOMY
Paradigalla brevicauda Rothschild and Hartert, 1911, Mount Goliath, central Dutch New Guinea.

OTHER COMMON NAMES
English: Short-tailed wattled bird of paradise; French: Paradisier à queue courte; German: Langschwanz-Paradigalla; Spanish: Paradigalla de Cola Corta.

PHYSICAL CHARACTERISTICS
9.0 in (23 cm); female 0.4–0.38 (155–170 g), male 0.35–0.41 lb (160–184 g). Head, upperparts, and underparts a velvety black. Yellowish green crown to nape and small, light blue wattle at the mandible base. Conspicuous bright yellow foreface. Blackish bill and legs.

DISTRIBUTION
Western and central ranges of New Guinea mainland from the Weyland Mountains eastward to the Bismark Range at altitudes of 4,590–8,460 ft (1,400–2,580 m). May be on the Kratke Range, Papua New Guinea, but unrecorded to date.

HABITAT
Midmontane forests, including beech, forest/garden edges, secondary growth.

BEHAVIOR
Birds give a rising bell-like *zheee* call at about 490 ft (150 m) intervals in moss forest, suggestive of dispersed solitary males advertising from song posts. In flight, wings make an audible rattling or rustling.

FEEDING ECOLOGY AND DIET
Omnivorous, predominantly frugivorous, but little known. Birds acrobatically cling to tree boughs and trunks to tear and probe into epiphytic plant growth for invertebrates and small vertebrates. Nestlings fed a large proportion (65%) of animals, including earthworms, insect larvae, crickets, beetles, mantids, katydids, spiders, frogs, and skinks.

REPRODUCTIVE BIOLOGY
Polygynous, with presumed promiscuous males and exclusively female nest attendance. Breeding on the Tari Valley slopes recorded in all months except March and November. Nest is a substantial, deep, open cup and the clutch is one egg. Incubation lasts more than 19 days and a known nestling period was 25 days.

CONSERVATION STATUS
Not threatened.

SIGNIFICANCE TO HUMANS
None known. ◆

Ribbon-tailed astrapia
Astrapia mayeri

SUBFAMILY
Paradisaeinae

TAXONOMY
Astrapia mayeri Stonor, 1939, Mount Hagen, Papua New Guinea.

OTHER COMMON NAMES
English: Ribbon-tailed bird of paradise, ribbon tail; French: Paradisier à rubans; German: Schmalschwanz-Paradeiselster; Spanish: Ave del Paraíso Cola de Moños.

PHYSICAL CHARACTERISTICS
12.6–13.8 in (32–35 cm), but 20.9–49.2 in (53–125 cm) with adult central rectrices; female 0.23–0.35 lb (102–157 g), male 0.30–0.36 lb (134–164 g). Most easily recognized by the male's long, black-tipped white tail feathers, which are at least three times the length of the bird. Head, throat, and tuft "pom-pom" over the base of the bill is black with intense metallic yellowish green iridescence. Olive-brown upperparts and black breast with coppery-red border. Deep green and copper abdomen to vent. Females duller, with brownish plumage and shorter tail.

DISTRIBUTION
Only an area of the central cordillera of western Papua New Guinea, from Mounts Hagen and Giluwe to Doma Peaks and the southern Karius Range, at altitudes of 5,900–11,320 ft (1,800–3,450 m).

Astrapia mayeri
▨ Resident

HABITAT
Upper montane and subalpine moss forests, forest edges and patches, including disturbed vegetation.

BEHAVIOR
Not yet clear if adult males, which do have traditional display locations and perches, are solitary displaying or do so in twos or larger numbers. Courtship involves males repeatedly hopping between tree perches; displays recorded during June through September. Extensive flight is by shallow undulations consisting of four to six audible, wing beats followed by a brief downward glide with closed wings; horizontally trailing central rectrices of adult males wave and ripple in the air.

FEEDING ECOLOGY AND DIET
Mostly lone individuals, but sometimes two to five birds, forage acrobatically on fruits and arthropods at all levels of the forest structure. Fruit possibly represents more than 50% of the diet. Arthropods and small vertebrate animals are obtained by probing/tearing into foliage, wood, and epiphytic vegetation.

REPRODUCTIVE BIOLOGY
Polygynous with promiscuous (probably lekking) males and exclusively female nest attendance. Breeding known during at least May through March. Often nests in isolated saplings, with no immediately adjacent tree branches or foliage, where the forest canopy is typically lacking directly above. On average, nest is 10–59 ft (3–18 m) above ground. Nest is a deep, substantial, open cup. The clutch is a single egg. In captivity, incubation lasts 21 days and the nestling period is 24 days.

CONSERVATION STATUS
Common to abundant in optimal habitat but geographically highly restricted and habitat is limited. Listed as Near Threatened.

SIGNIFICANCE TO HUMANS
Tail plumes of adult males are highly prized for personal adornment by highland men who trade dried skins. ◆

Lawes's parotia
Parotia lawesii

SUBFAMILY
Paradisaeinae

TAXONOMY
Parotia lawesii Ramsay, 1885, Astrolabe Mountains (subsequently defined as the Aruma Apa-Maguli Mountains, Owen Stanley Range, Papua New Guinea). Two subspecies.

OTHER COMMON NAMES
English: Lawes's six-wired bird of paradise, Lawes's six-wired parotia, Lawes's six-plumed bird of paradise, Helena's parotia; French: Paradisier de Lawes; German: Blaunacken-Paradeisvogel; Spanish: Perotia de Lawes.

PHYSICAL CHARACTERISTICS
9.8 in (25 cm); female 0.27–0.37 lb (122–169 g), male 0.34–0.43 lb (153–195 g). Entirely jet black with short tail. Head decorated with a silvery tuft over base of bill and a frontal crest of dark coppery feathers. Behind each eye are three long, black, wire-like occipital plumes with circular tips. Breast shield of scale-like feathers has highly iridescent yellows, greens, and violets. Female have brown upperparts, black head

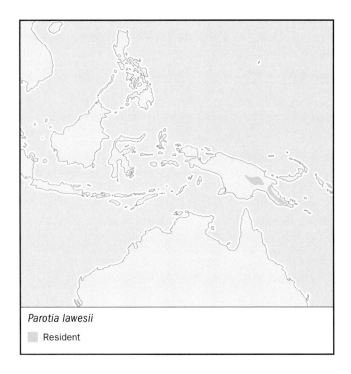

Parotia lawesii

Resident

Pteridophora alberti

Resident

to nape, banded orange and blackish underparts, and lack occipital plumes.

DISTRIBUTION
The eastern third of New Guinea's central cordillera, entirely within Papua New Guinea. *P. l. lawesii*: western and southern highlands of Papua New Guinea southeastward into peninsular Papua New Guinea; *P. l. helenae*: the northern watershed of peninsular Papua New Guinea, from Waria southeast to Milne Bay. Found between altitudes of 1,640–7,540 ft (500–2,300 m).

HABITAT
Midmontane forests including primary mixed oak forest, disturbed forest, secondary growth, and remnant forest patched even within extensive village gardens.

BEHAVIOR
Males clear a terrestrial court to dance upon in courtship display. Courts are typically dispersed to form exploded leks, but some are solitary. Males advertisement-sing from court perches or the forest canopy above but remain mostly silent when interacting with females. Flight swift and buoyant, consisting of four audible wingbeats followed by a short glide.

FEEDING ECOLOGY AND DIET
Omnivorous, but very predominantly frugivorous. Tears epiphytic growth from tree limbs to find arthropods.

REPRODUCTIVE BIOLOGY
Polygynous, with promiscuous, solitary, males and exclusively female nest attendance. Breeds June through January. Nests, built in large trees and vine tangles, consist of a substatial but relatively shallow open cup. Only single egg clutches recorded.

CONSERVATION STATUS
Not threatened. Widespread, common, and tolerant of habitat disturbance.

SIGNIFICANCE TO HUMANS
Plumes of adult males are worn as wig/head dress by highland men. ◆

King of Saxony bird of paradise

Pteridophora alberti

SUBFAMILY
Paradisaeinae

TAXONOMY
Pteridophora alberti Meyer, 1894. Mountains on the Ambernoh (= Mamberamo) River, Irian Jaya, subsequently restricted to the Weyland Mountains.

OTHER COMMON NAMES
English: King of Saxony's bird of paradise, enamelled bird of paradise, enamelled bird; French: Paradisier de Prince Albert; German: Wimpelträger; Spanish: Ave del Paraíso Rey de Sajonia.

PHYSICAL CHARACTERISTICS
8.3 in (21 cm); 19.7 in (50 cm) if head plumes of adult male are included; female 0.15–0.19 lb (68–88 g), male 0.18–0.21 lb (80–95 g). Head, mantle, and back is deep black with dark, bronze-green sheen. Occipital plume behind each eye has a bare central shaft with 40–50 "flags" decorating its outer side; the upperside of each is blue while the underside is brownish. Chin and throat is black with highly iridescent violet tips. Breast, abdomen, and vent dark yellow; brownish rump and tail. Females are brownish gray with black-barred white underparts.

DISTRIBUTION
Western and central two-thirds of the central cordillera of New Guinea, from Weyland Mountains of Irian Jaya east, possibly to the Kratke Range of central Papua New Guinea; at altitudes of 4,590–9,350 ft (1,400–2,850 m).

HABITAT
Mid to upper montane forests and their edges, including lightly disturbed habitat around hunting lodges and tracks.

BEHAVIOR
Adult males typically solitary and territorial, but in some areas they may aggregate into exploded leks. Adult males advertise

vine tendril display perches from perches in the canopy above, giving odd radio static-like advertisement song. A bouncing courtship display is typically subsequently performed upon a suspended vine a few meters from the ground. Singing and display seasonality occurs at least from September to April. Female plumaged birds are far less conspicuous than adult males.

FEEDING ECOLOGY AND DIET
Omnivorous, but predominantly frugivorous. Favors green fruits, with a particular liking for those of false figs (*Timonius*). Forages mostly in the canopy and subcanopy.

REPRODUCTIVE BIOLOGY
Polygynous, with promiscuous males and exclusively female nest attendance. Breeding is occurs at least from July to February. The only observed nesting occurred during late December to the end of January. The nest is a shallow open cup and the only known clutch was a single egg. Incubation is more than 22 days; nestling period is unknown.

CONSERVATION STATUS
Not threatened. In general, widespread and common in most areas of appropriate habitat.

SIGNIFICANCE TO HUMANS
The two elaborate head plumes of adult males are sought by highland men as personal adornment. However, cut-out photocopies of them are sometimes worn, suggesting hunting pressure might be in decline. ◆

Victoria's riflebird
Ptiloris victoriae

SUBFAMILY
Paradisaeinae

TAXONOMY
Ptiloris victoriae Gould, 1850, Barnard Island, North Queensland, Australia.

OTHER COMMON NAMES
English: Queen Victoria's rifle-bird; French: Paradisier de Victoria; German: Victoriaparadeisvogel; Spanish: Ave Fusil de la Reina Victoria.

PHYSICAL CHARACTERISTICS
9.5 in (24 cm); female 0.17–0.21 lb (77–96 g), male 0.20–0.26 lb (91–119 g). Deep black upperparts, chin, cheek, and breast band. Crown and throat a bright metallic greenish blue. Short tail with metallic green central feathers. Lower breast to vent is a darker, iridescent oil-green. Females are red-brown with whitish throat patch and brow stripe.

DISTRIBUTION
The Atherton region of tropical northeast Queensland, Australia, including some off-shore islands; from Big Tableland, south of Cooktown, to Mount Elliot just south of Townsville. Sea level to 3,940 ft (1,200 m) altitude.

HABITAT
Lowland to hill rainforest, adjacent eucalyptus and melaleuca woodlands, and landward edges of mangroves.

BEHAVIOR
Adult males are loudly and frequently vocal in advertising their display sites at the top of vertical broken-off tree stumps, upon

Ptiloris victoriae
☐ Resident

which they perform ritualized courtship postures/movements. Typical advertisement song of adults is an explosive loud *sssssshh* or *yaaaas*. Flight noise produced by adult males is a sharp and dry rattling rustle that probably functions as a social signal to conspecifics. Courtship occurs July through December.

FEEDING ECOLOGY AND DIET
Omnivorous, but arthropods and small vertebrates are taken at least as much as fruits. Nestling diet is mostly animals, including orthopterans, cockroaches, beetles, cicadas, insect larvae, wood lice, spiders, and centipedes. Differences in bill structure between the sexes may reduce competition for animal foods.

REPRODUCTIVE BIOLOGY
Polygynous, with promiscuous solitary males and exclusively female nest attendance. Breeding occurs late August through early January. Nest is a substantial open cup cryptically placed among concealing foliage at 5–66 ft (1.5–20 m) above ground. Clutch is one to two pinkish eggs marked with elongate brush-stroke-like blotches. Incubation is 18–19 days; nestling period is 13-15 days.

CONSERVATION STATUS
Not threatened. Widespread and common throughout remaining and protected habitat.

SIGNIFICANCE TO HUMANS
Once commonly killed and mounted for Victorian bird cabinets as interior decoration. ◆

King bird of paradise
Cicinnurus regius

SUBFAMILY
Paradisaeinae

TAXONOMY
Cicinnurus regius Linnaeus, 1758, East Indies. Two subspecies.

Cicinnurus regius

☐ Resident

OTHER COMMON NAMES
English: Little king bird of paradise; French: Paradisier royal;
German: Königsparadeisvogel; Spanish: Ave del Paraíso Soberbia.

PHYSICAL CHARACTERISTICS
6.3–7.5 in (16–19 cm), but 12.2 in (31 cm) if central rectrices
of adult males included; female 0.08–0.13 lb (38–58 g), male
0.10–0.14 lb (43–65 g). Head, upperparts, and chin to upper
breast is crimson, with an orange wash on the crown and
darker throat. Jet black spot directly over eye. Narrow, dark
green iridescent breast band with whitish lower breast to vent.
Undertail and mantle feathers olive-brown, with iridescent
green tips to the mantle "cape". Long central rectrices with
brownish disks at the ends. Violet legs, bill ivory-yellow. Females have dull olive head and upperparts with yellowish underparts and violet legs.

DISTRIBUTION
Throughout the majority of lowland New Guinea mainland
and Aru, Missol, Salawati, and Yapen islands; from sea level to
3,115 ft (950 m). *C. r. regius*: Aru, Misool, and Salawati Islands,
the Vogelkop, all of south New Guinea, and the northern watershed of southeast Papua New Guinea from Huon Gulf to Milne
Bay; *C. r. coccineifrons*: northern watershed of New Guinea from
the east coast of Geelvink Bay eastward to the Ramu River.
Birds of the north coast of the Huon Peninsula remain to be
subspecifically identified.

HABITAT
Lowland rainforests, gallery forests, forest edges, and disturbed
and tall secondary forests.

BEHAVIOR
An inconspicuous species except for males at their display
trees. Adult males are perhaps more persistent callers than
any other birds of paradise. Courtship involves complex vocalizations, feather manipulations, and body posturing/move-
ments including hanging fully inverted and pendulum-like
swinging.

FEEDING ECOLOGY AND DIET
The diet consists of fruits and animals. Foraging occurs at all
forest levels. Birds often join mixed species foraging flocks to
seek arthropods in the lower forest.

REPRODUCTIVE BIOLOGY
Polygynous, with solitary or lekking, sedentary, promiscuous
adult males dispersed at traditional display tree perches.
Breeding occurs at least during March through October. The
open cup nest is built into a tree cavity (unique within family), within which two eggs are laid. Female builds the nest
and cares for the young without male assistance. In captivity,
incubation lasted 17 days and the nestling period was 14
days.

CONSERVATION STATUS
Not threatened; widespread and abundant.

SIGNIFICANCE TO HUMANS
Plumes of adult males are used for personal adornment but
hunting pressure represents no threat to populations. ◆

Standardwing bird of paradise
Semioptera wallacii

SUBFAMILY
Paradisaeinae

TAXONOMY
Semioptera wallacii Gray, 1859, near Labuha Village, Batchian
(= Bacan Island). Two subspecies.

OTHER COMMON NAMES
English: Wallace's standardwing; French: Paradisier de Wallace;
German: Banderparadeisvogel; Spanish: Ave del Paraíso de
Wallace.

Semioptera wallacii

☐ Resident

PHYSICAL CHARACTERISTICS
9.1–10.2 in (23–26 cm), but 11 in (28 cm) if the standards of adult males be included; female 0.28–0.32 lb (126–143 g), male 0.34–0.38 lb (152-174 g). Light brown head, upperparts, and central tail feathers. Decurved bill and tuft at base of mandible give both sexes a distinctive profile. Chin and upper throat brown, with highly iridescent greenish yellow breast shield. Two white, elongated lesser coverts are often longer than the wing. Legs orange; bill ivory-beige.

DISTRIBUTION
The northern Moluccan Islands of Indonesia. *S. w. wallacii*: Bacan Island, from low hills up to 3,770 ft (1,150 m) altitude, as probably also the population on Kasiruta Island; *S. w. halmaherae*: Halmahera Island, from lower hills at about 820 ft (250 m) to about 3,300 ft (1,000 m) altitude.

HABITAT
Rainforests. Birds apparently absent from flat lowlands and patchy on steeper hilly topography, particularly on limestone. Rare in mature secondary woodland.

BEHAVIOR
Males remove foliage from lek perches. At the latter males are highly vocal and perform very animated courtship plumage manipulations, postures, movements, and aerial flight displays. Advertisement song is typically a single loud nasal upslurred bark. The display season is from about April to December. Birds, shy and inconspicuous except at leks, typically frequent the lower forest canopy and subcanopy.

FEEDING ECOLOGY AND DIET
Typically forage in densely foliaged forest canopies. The diet is fruits and arthropods and probably small vertebrates. May join mixed species foraging flocks.

REPRODUCTIVE BIOLOGY
Polygynous, with densely lekking promiscuous adult males forming aggregations of 30–40 or more at traditional display trees. Breeding during at least May through September. Presumed exclusively female nest attendance. The only nest described was an open cup that included dry leaves and was 33 ft (10 m) above ground; it contained one egg.

CONSERVATION STATUS
Not threatened. Contrary to previous impressions, it was widespread and moderately common on the larger islands in 1999; the status of the Kasiruta Island population is not known.

SIGNIFICANCE TO HUMANS
Historically significant in the context of the field work and discoveries of Alfred Russel Wallace. ◆

Greater bird of paradise
Paradisaea apoda

SUBFAMILY
Paradisaeinae

TAXONOMY
Paradisaea apoda Linnaeus, 1758, India = Aru Islands of Indonesia. Two subspecies.

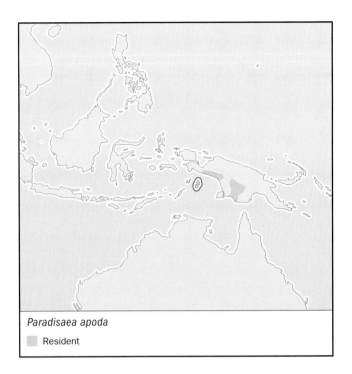

Paradisaea apoda

▪ Resident

OTHER COMMON NAMES
English: Great bird of paradise; French: Paradisier grand-émeraude; German: Göttervogel; Spanish: Ave del Paraíso Grande.

PHYSICAL CHARACTERISTICS
13.8–16.9 in (35–43 cm); female 0.37–0.38 lb (170–173 g), male (no weights avialable). The largest of the plumed birds of paradise. Forehead to nape a pale orange-yellow. Forecrown, lores, ear coverts, chin, and throat have fine, iridescent yellowish green feathers. Dark brown ruffled feathers form a breast cushion; brownish upperparts, tail, and abdomen. Bright yellow elongated flank plumes fade to whitish with beige tips. Bill is pale bluish gray. Females drab brown in color and lack long feathers.

DISTRIBUTION
The south of the western half of the New Guinea mainland, excluding the western peninsulas, and the Aru Islands. Lowlands to about 3,300 ft (1,000 m) altitude. *P. a. apoda*: Aru Islands of Irian Jaya, Indonesia; *P. a. novaeguineae* : southern New Guinea from Timika, Irian Jaya, eastward to the Fly/Strickland Rivers watershed, western Papua New Guinea.

HABITAT
Lowland and hill forests.

BEHAVIOR
Males remove foliage from lek perches. Males are highly vocal and perform very animated flank plumage manipulations, body postures, and movements about lek perches in courtship display. The most common lek advertisement song consists of repeated loud, deep, *wauk* or *wonk* notes.

FEEDING ECOLOGY AND DIET
Actively animated in foraging. Eats fruits and animals (mostly arthropods).

REPRODUCTIVE BIOLOGY
Polygynous, with intensely lekking and promiscuous adult males forming loudly vocal aggregations at traditional display

trees. Breeding during at least August through December, March, and May. Known clutches are all of one egg. Incubation and nestling periods unknown.

CONSERVATION STATUS
Sufficiently widespread and common to be of no concern as a species, but some local populations have declined.

SIGNIFICANCE TO HUMANS
Adult males were of great significance to the plume trade but modern usage is restricted to local head dress adornment. Live and stuffed adult males are, however, still traded throughout Indonesia and beyond, and this illegal exploitation could represent a limited threat to some populations. ◆

Resources

Books
Frith, C.B., and B. M. Beehler. *The Birds of Paradise: Paradisaeidae.* Oxford: Oxford University Press, 1998.

Gilliard, E.T. *Birds of Paradise and Bower Birds.* London: Weidenfeld and Nicolson, 1969.

Periodicals
Beehler, B.M. "The Birds of Paradise." *Scientific American* 261 (1989): 116–123.

Beehler, B.M., and S.G. Pruett-Jones. "Display Dispersion and Diet of Birds of Paradise, a Comparison of Nine Species." *Behavioral Ecology and Sociobiology* 13 (1983): 229–238.

Cracraft, J., and J. Feinstein. "What Is Not a Bird of Paradise? Molecular and Morphological Evidence Places *Macgregoria* in the Meliphagidae and the Cnemophilinae Near the Base of the Corvoid Tree." *Proceedings of the Royal Society of London B.* 267 (2000): 233–241.

Frith, C.B., and D. W. Frith. "Biometrics of Birds of Paradise (Aves: Paradisaeidae) with Observations on Interspecific and Intraspecific Variation and Sexual Dimorphism." *Memoirs of the Queensland Museum* 42 (1997): 159–212.

Organizations
Birds Australia. 415 Riversdale Road, Hawthorn East, Victoria 3123 Australia. Phone: +61 3 9882 2622. Fax: +61 3 9882 2677. E-mail: mail@birdsaustralia.com.au Web site: <http://www.birdsaustralia.com.au>

Clifford B. Frith, PhD
Dawn W. Frith, PhD

Crows and jays

(Corvidae)

Class Aves
Order Passeriformes
Suborder Passeri (Oscines)
Family Corvidae

Thumbnail description
Medium to large-sized birds with large heads and stout, usually slightly hooked beaks, scaly legs, and powerful feet

Size
7.4–26.9 in (19–69 cm); 1.47 oz–3.66 lb (41.6 g–1.56 kg)

Number of genera, species
26 genera; 123 species

Habitat
Widespread family occupying most vegetated land habitats

Conservation status
Critically Endangered: 1 species; Endangered: 4 species; Vulnerable: 8 species; Near Threatened: 11 species

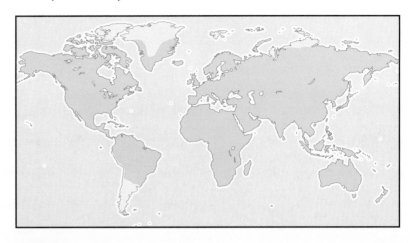

Distribution
All continents except Antarctica; introduced rook (*Corvus frugilegus*) in New Zealand

Evolution and systematics

The earliest fossil corvid, a jay-like "crow" was found in France and dates to the middle Miocene period (about 17 million years ago). The large number of species endemic to eastern Asia (36) and the New World (29) however, suggest that the family may have originated in the Far East, and taken alternative routes a very long time ago.

The ancestors of New World jays spread to the Americas, but no other corvids besides jays are found in South America. Marzluff and Balda conclude that the other genera were relatively recent arrivals in North America—the Bering land bridge that formed between Asia and North America allowed the Old World genera *Corvus*, *Perisoreus*, *Pica*, and *Nucifraga* to cross continents. Scientists are unclear why so few corvid species occur in Africa. The two monotypic genera have features which suggest an ancient origin: most strikingly, the Stresemann's bush-crow (*Zavattariornis stresemanni*), found only in southwest Ethiopia, is shaped like a ground-jay, colored like a nutcracker, and builds a nest like a magpie. Madge speculates that this species may be a relict population of an ancient ancestor to a number of genera.

Taxonomically, the corvids, consisting of 26 genera and 123 species, have fitted historically within the family Corvidae. DNA reclassification work has not changed this essential grouping. Sibley and Ahlquist (1990) widen the family to include such taxa as quail-thrushes, whipbirds, cuckooshrikes, and birds of paradise within a much bigger Corvidae family, placing the corvids together in a Corvinae subfamily. They argue that the enlarged family originated in Australia;

since only members of the *Corvus* genus now live in Australia, they would have been recent re-colonists of their ancestral home.

Physical characteristics

Although corvids show huge variation in size—the largest is more than three times longer than the smallest—a number of features are common to most members of the family. All but three species have their nostrils either partially or completely obscured by bristles or plumes. In the majority of species, the tip of the curved upper mandible overlaps the lower mandible slightly. Some genera show adaptive development of the beak; the long, decurved bills of *Podoces* ground jays and *Pyrrhocorax* choughs are used to probe into narrow holes for food. The feet are proportionally large and powerful, with strong, grasping toes, suited to holding large and awkward items of prey still; the tarsi are scaled at the front and smooth behind.

Corvid wings are generally rounded and rather short in woodland species. Forms that embark on long migrations, such as a number of *Corvus* species, have longer wings. Tails are generally long, exaggeratedly so in some genera. Both the piapiac (*Ptilostomus afer*), which is unique in having 10 rather than 12 tail feathers, and the *Pica* magpies probably use their long tails for maneuverability. In forest-dwelling treepies, the tail helps with balancing. However, elaborate tails such as that of the ratchet-tailed treepie (*Temnurus temnurus*) may also have a display function.

Food-cacheing in corvids: a Clark's nutcracker (*Nucifraga columbiana*) gathers pinenuts in its sublingual pouch, then buries them in trenches to retrieve and eat later. (Illustration by Gillian Harris)

Typical crows of the *Corvus* genus all have predominantly glossy black plumage with a metallic sheen. Species that appear identical in shape, size, and plumage can be sympatric. Australia has five species that offer the birder what Madge calls "the ultimate challenge to his field skills." However, although congeners frequently associate to feed and may overlap or even "share" breeding territories, interspecific pair formation in the wild is unknown. Clearly, the problems of identification rest with humans.

Their black plumage offers corvids considerable advantages. Since this color absorbs solar energy most effectively and radiates less heat, corvids in cold surroundings are better able to regulate body temperature. Dark melanin-pigmented feathers are also tougher than light feathers. In addition, in species such as ravens that are highly sociable and, because of their size as adults, largely untroubled by predators, it is beneficial for them to wear conspicuous black plumage, displaying their presence to conspecifics, even from a distance.

Other genera show a broad and sometimes dazzling range of colorful plumage. Unlike most other passerines, the body plumage is relatively simple, lacking cryptic streaks. In a number of species, the wings and tails are strongly marked, serving both to conceal and reveal. The Eurasian jay (*Garrulus glandarius*) is hard to spot in forests because its white wing patch acts as a disruptive feature to aid concealment when the bird is at rest. Equally, its white rump alerts conspecifics, as it is revealed when the jay leaves its perch. Some species use wing and tail markings in courtship display or in hostile encounters; the yellow-billed magpie (*Pica nuttalli*) flashes its wing patch to warn off rivals. The New World jays reveal a variety of display crests and plumes, from the short, backward-facing crest of the blue jay (*Cyanocitta cristata*), through the nasal tuft of the azure jay (*Cyanocorax caeruleus*), to the aptly named plush-crested jay (*Cyanocorax chrysops*), and the extravagant parakeet-like crest of the white-throated magpie-jay (*Calocitta formosa*). The only Old World species with a sig-

nificant crest or plume is the crested jay (*Platylophus galericulatus*). Since this is thought to be a primitive jay, it seems likely that such display features were simply lost from other Old World species as they evolved.

Juvenal plumage is not generally significantly different to that of adults, although the feathers are softer and duller in color. However, it is common for there to be other physical differences, usually in the eye or bill color, that enable adult conspecifics to rule out juvenals as potential mates or rivals. The first year juvenal Torresian crow (*Corvus orru*), for example, has a brown iris that turns hazel in the second year and white in its third, adult year.

Distribution

In his classic work *Crows of the World*, Derek Goodwin notes that "the amount of adaptive radiation and consequent differentiation found within the Corvidae surpasses that within many Passerine families." With the exception of Antarctica and some remote archipelagos of the Pacific, there is at least one corvid species for most parts of Earth.

In south and Southeast Asia, where Corvids may have originated, arboreal species proliferate. These include the primitive monotype crested jay and black magpie (*Platysmurus leucopterus*) of the Malay Peninsula and Southeast Asian islands, as well as the "tropical" magpies. This group of 18 colorful species consists of the blue *Urocissa* and green *Cissa* magpies, *Dendrocitta* treepies, racket-tailed treepies *Crypsirina*, and ratchet-tailed treepies *Temnurus*. The *Pica* genus of pied magpies has three species, widely distributed throughout the Northern Hemisphere, but geographically distinct.

On the central Asian desert scrub and steppe, stretching from Iran and Turkmenistan in the west to Tibet and Mongolia in the east, live the *Podoces* and *Pseudopodoces* ground-jays. Higher altitudes are occupied by the alpine chough (*Pyrrhocorax graculus*). Both the alpine and the lower-altitude red-billed chough (*Pyrrhocorax pyrrhocorax*) range through central and southern Europe.

The mountainous, coniferous forests of the Old World contain the *Nucifraga* nutcrackers, a genus of three species. The six *Garrulus* and *Perisoreus* jays occupy a wider variety of forest habitats in the Old World, including deciduous, mixed, and coniferous woodland. The Clark's nutcracker (*Nucifraga columbiana*) and gray jay (*Perisoreus canadensis*) are the only members of these genera to colonize the forests of North America.

The New World jays (36 species) are unique to the Americas, found in most forest and scrub south of the Canadian boreal forest. They are all bluish in color, with the exception of the brown jay (*Cyanocorax morio*). The crested *Cyanocitta* genus is found in woodland, parks, and gardens throughout North America; the Steller's jay (*Cyanocitta stelleri*) replacing the blue jay in the west and south. The crestless *Cyanocola* genus of southern United States, Mexico, and Central America includes the three species of scrub-jay. The largest genus in the Americas, *Cyanocorax*, consists of 16 similar species, which range through the forests of Central and South America, with northeast Argentina their southernmost limit.

The mostly blackish *Corvus* genus is the biggest (48 species) and most wide ranging of all genera. It is the only genus found in Africa south of the Sahara, and the only one represented in Australia, where there are five species. The distribution of the most widespread Australian species, the Australian raven (*Corvus coronoides*) in the east, south, and southwest of Australia corresponds with that of sheep. The majority of island endemics belong in this genus. Most are in the islands of the Pacific and Southeast Asia. Oddly, although no corvids other than jays are found in South America, five *Corvus* species have colonized the West Indies.

Wide-ranging *Corvus* species include the northern raven (*Corvus corax*), which has colonized much open country of the Northern Hemisphere, and the pied crow (*Corvus albus*), which adapts well to human habitation and is Africa's most widespread corvid, found in the southern half of the continent. Coexistence with human settlement has enabled other crows to flourish. The American crow (*Corvus brachyrhynchos*) in North America and the carrion crow (*Corvus corone*) of Europe and Asia are found in both rural and urban areas. In the twenty-first century, the house crow (*Corvus splendens*) may prove to be the most successful colonist of all. The native Indian subcontinent population is flourishing, but the bird's readiness to board ships heading for other parts of the world has enabled it colonize and thrive in Malaysia, the Middle East, and Africa. In the South African city of Durban alone, the population reached 12,000 birds just 12 years after the first sighting.

Habitat

Corvids colonize an extraordinarily wide range of habitats. The extremes serve to illustrate the breadth and diversity—alpine choughs have been observed feeding in the Himalayas at 2,700 ft (822 m); the "desert" choughs live in the steppe deserts of central Asia.

Corvids evolved as arboreal species, and some genera have remained wholly dependent on forest habitat. Jays of the *Cyanolyca* genus of Central and South America carry out nearly all of their foraging in the canopy, rarely, if ever, descending to the ground. Genera that have evolved specialized feeding techniques are largely restricted to particular types of forest and, exceptionally, even particular types of tree. Some forms of spotted nutcracker (*Nucifraga caryocatactes*) have unusually slender bills for feeding on the seeds of the Arolla pine (*Pinus cemra*); the two species may have evolved in symbiosis. Other genera are omnivorous, but show a marked preference for certain foods. The *Garrulus* jays favor oak acorns and oak-living invertebrates: their range usually coincides with that of oaks (*Quercus* spp.).

Those genera which rely on trees only for nesting and roosting occupy a much wider variety of habitats. Magpies of the *Pica* genus and some *Corvus* species, such as the Eurasian rook (*Corvus frugilegus*), are able to feed diurnally in a range of open habitats, returning at night to their tree roosts. Some species, most notably the American, carrion, and house crow, have become well adapted to living in areas of dense human settlement. Studies in New York have shown that although urban American crows have smaller territories and nest at

A gray-breasted jay (*Aphelocoma ultramarina*) in flight over Arizona. (Photo by Joe McDonald. Bruce Coleman Inc. Reproduced by permission.)

higher densities than rural crows, nesting success is broadly similar.

An adaptation to nesting on cliffs and on the ground or low bushes has enabled a number of species to colonize open country. The five species of ground-jay are all terrestrial feeders, and the *Pyrrhocorax* choughs feed on invertebrates on grazed pasture. Most raven species nest both in trees and on cliff ledges.

Behavior

When the great naturalist Henry David Thoreau, wrote "if men had wings and bore black feathers, few of them would be wise enough to be crows," he spoke for many, often grudging, corvid admirers. In *The Dictionary of Birds*, published in 1985, this family is thought to "represent the furthest stage so far reached in avian evolution: much in their behavior suggests a highly developed mentality." The numerous behavioral studies carried out since this book was published only serve to confirm a high estimation of corvids. Their huge brains give them a brain to body ratio that is equal to that of dolphins and only slightly smaller than that of humans.

All *Corvus* species are, to a greater or lesser extent, gregarious. The degree of social interaction is dependent on factors that are often complex and open to wide interpretation, but they include season, age, family relationships, defense strategies, and time of day.

Flocking behavior has developed among corvids as a means of exploiting localized food sources, with inexperienced birds gaining from the knowledge of older birds, and the group formation acting to deter predators. Yet the bonding between flock members varies considerably between species. Rooks and western jackdaws (*Corvus monedula*) frequently form mixed feeding flocks, but the only cohesion between individuals in these flocks is between mated pairs; when the flock is

A blue jay (*Cyanocitta cristata*) family at its nest. (Photo by Gregory K. Scott. Photo Researchers, Inc. Reproduced by permission.)

disturbed, the birds disperse randomly, except for paired individuals, which fly off together. Other species, particularly New World jays, show a range of affinities within flocks. Such cohesive flocks may be relatively small: Florida scrub-jays associate in parent-offspring flocks of no more than six birds. Gray-breasted jays (*Aphelocoma ultramarina*) live in year-round extended family flocks of six to 18 birds. In the case of the pinyon jay (*Gymnorhinus cyanocephalus*), the fall "flock" numbering several thousand individuals is actually subdivided into many small clans of related birds.

Gregarious habits can evolve as a reaction to being slightly "afraid." Captive corvids often show agitated behavior towards nightfall. Many of their wild, diurnal compatriots, especially nonbreeders, compensate by roosting communally. These gatherings can reach numbers of epic proportions. A Munich roost has attracted more than one million rooks, jackdaws, and carrion crows, while American crow roosts in Oklahoma and Kansas have been known to draw in around nine million birds. Studies of hooded crows (*Corvus corone cornix*) in Norway show that birds in roosts are able to gain information from others about good feeding sites.

All corvids appear to be territorial, although the extent to which they defend territory is very variable, even within species. At one extreme, the individual pairs among colonial nesting species such as rooks and yellow-billed magpies defend only the area that consists of just a few feet around their nest.

By comparison, a number of species occupy large home ranges, and judging by consistent nest spacing in species such as the northern raven and Eurasian magpie (*Pica pica*), the core

nesting area is defended. The carrion crow is a particularly aggressive nest defender; its biggest nest predator is other carrion crows. Curiously, urban carrion crows are observed to have a territorial "roof," and conspecifics are able to fly above this roof without being challenged. A number of species, including the Eurasian magpie and Eurasian jay, engage in "spring gatherings," where groups of unpaired birds gather at the edge of a territory and apparently "test" the resident pair. If the established birds do not produce a suitably aggressive response, they risk being ousted.

American and northwestern crows (*Corvus caurinus*) take on "helpers." Among New World jays, where at least 12 species employ helpers, research on Florida scrub-jays (*Aphelocoma coerulescens*) has shown that pairs with helpers raise twice as many young as those without. No Old World corvids have been shown to adopt this behavior.

Most corvids are sedentary. In far northern temperate zones, some species migrate, or make cold weather movements. Large-scale irruptions occur in species dependent on an individual food, in years when the food crop fails. The Clark's nutcracker (*Nucifraga columbiana*) irrupts from its conifer habitats to urban backyards, coasts, and even deserts.

For a family that has no song in the recognized sense of the word, corvids are nevertheless remarkably and famously vocal. The song, such as it is, consists of a softly intoned jumble of calls, interspersed with mimicked sounds. Broadly speaking, a corvid's calls can be identified as performing a particular function, whether they are contact, threat, feeding, or territorial calls. For example, the immediacy of a threat will determine the volume and frequency of an alarm call in an American crow.

Vocal mimicry is widespread and can be separated into two categories. Lanceolated jays (*Garrulus lanceolatus*) are among the few species which add the sounds of other species or inanimate objects to their vocal repertoire. Some species do not use mimicry in the wild, but will do so when tamed.

Feeding ecology and diet

In their foraging behavior, corvids are largely restricted to ground feeding and taking food from trees. Use of other feeding techniques, such as aerial feeding and fishing, are rare. Most corvids are omnivorous. Those in tropical zones tend to live on a year-round diet of invertebrates, fruit and berries, reptiles, amphibians, small birds and mammals, and carrion, as well as eggs and young during the breeding season. The range of food items taken by corvids in the Northern Hemisphere, however, is dictated by seasonality. Studies of the Eurasian magpie diet are typical of the *Corvus* genus. They show a preponderance of plant material in the winter and of animal food in summer. Birds in areas of human habitation regularly supplement their diet with food provided by humans.

Very few corvids have specialist diets. Those that do have evolved morphological and behavioral adaptations. The western scrub jay (*Aphelocoma californica*) and Steller's jay both have wide-ranging diets that include insects and other invertebrates, fruits, nuts, berries, as well as the eggs and young of small birds. The seeds of pinyon (*Pinus edulis*) and ponderosa pine (*Pinus ponderosa*) are an important food source for these species too, but not to the extent shown by the pinyon jay, which feeds almost exclusively on pine seeds. Its morphological adaptations include an expandable esophagus that can inflate to hold up to 50 seeds at one time; the scrub jay can manage just five.

Corvids are renown for their intelligence and adaptability in getting food. Examples of optimal foraging include the ability of northwestern crows to weigh whelks to select the biggest, and then to drop them on rocks from a height which experiments have shown is ideal for breaking open large whelk shells. The use of tools was captured on film in the David Attenborough series *The Life of Birds*, in which New Caledonian crows (*Corvus moneduloides*) were shown collecting, trimming, and using twigs to probe into the hollows of logs for invertebrate grubs.

The hiding of food for later consumption is a characteristic of corvids. It confers obvious benefits for species that have a sudden abundance of food and a need to eke out their supplies to cover leaner periods. Studies of spotted nutcrackers in Siberia show that these birds are completely dependent on cached pine seeds from the fall until May. Each individual hides about 100,000 seeds and needs to find about a quarter of these to survive. Research on Eurasian jays indicates that these birds memorize the position of trees and other landmark features, so that they are able to retrieve buried acorns successfully. Although gray jays of boreal forest are unable to find food buried on the forest floor during winter when the ground is covered in deep snow, they solve this problem by gluing insects and pieces of meat onto the bark of trees using their own saliva.

Birds in temperate zones clearly gain from storing food when faced with a glut, but widespread evidence from tropical species too suggests that most, and possibly all corvids, engage in this behavior. In India, rufous treepies (*Dendrocitta vagabunda*) have been observed hanging up pieces of meat in trees. Other largely arboreal species wedge acorns and other seeds in cracks of trees. Some ground-feeders occasionally hide their find under grass or among leaves, but the general tendency is to bury the food underground.

Nevertheless, food storing is a risky strategy, since there is always the possibility that the stored food may be found by a conspecific or other species. Observations have shown that northern ravens and Eurasian jays return to caches that they have stored in the presence of conspecifics, then rebury them when the observers are no longer present. Studies of Florida scrub-jays demonstrate that only those birds that have experience of stealing the caches of other birds will engage in such re-caching.

Reproductive biology

Corvids are largely monogamous. However, the assertion that corvids "mate for life" is open to question, for there is clear evidence that some pair bonds are broken. One study estimated that one-third of black-billed magpie (*Pica hudso-*

nia) pairs "divorced," with one member of the pair seeking a mate in a better territory.

Pair bonds are often formed among flocks at a communal roost, or in noisy group gatherings just prior to the breeding season. In most species, a significant part of courtship behavior involves the male feeding the female—a preliminary to actual feeding when she is on the nest. The female may respond with begging or submissive quivering of the wings or tail, but this reaction is not found in all species. The male of the pair usually shows dominant behavior, such as openly stealing from the female's food store.

The typical corvid nest begins as a stick platform in the fork of a tree. Once the heap of strong sticks is firmly wedged in place, both birds make a bowl-shaped layer of strips of bark and roots, sometimes bound together with mud or animal dung. This is lined with a cup of soft materials such as grass, feathers, fur, and leaves. There are a number of mainly habitat-related adaptations to this basic model. Siberian jays (*Perisoreus infaustus*), one of several tree-nesting species of northern temperate zones, build especially thick nest cups insulated with lichens and reindeer hair. *Pica* magpies, which often nest in open habitat, add a domed roof to reduce the vulnerability of the nest to predators; this is clearly learned behavior, for inexperienced birds are more likely to build open nests. Tropical species usually nest in dense vegetation; their light, flimsy nests show they have no need of extra warmth or added protection.

Some corvids gain extra security or the ability to nest in a wider range of habitats by nesting in holes, cavities, ledges, or on the ground. *Garrulus* jays nest in tree hollows. Western jackdaws evolved the habit of nesting in chimneys and other cavities, possibly to reduce predation by larger carrion crows. *Pyrrhocorax* choughs generally nest in open habitat with few trees. For safety, they lay their eggs on cave ledges or deep cracks in cliffs. Even species that nest on the ground generally site the nest under a protective shelter, either under a bush or on a cliff ledge shielded from the wind. Only the Hume's ground-jay (*Pseudopodoces humilis*) digs a burrow in the ground where it lays its eggs.

Corvids lay between two and seven eggs; these are almost always incubated by the female, with the male feeding and guarding her. Hatching is asynchronous; it is common for the youngest and weakest young to die in the nest. The young are born blind, helpless, and either naked or with a light down. The adults are attentive parents, bringing food back to the nest in their throats. Smaller species fledge after about 20 days, but larger corvids may remain in the nest for five to seven weeks. Among solitary nesters, the young may remain with the parents for up to three months, becoming increasingly independent, before the family group breaks up.

Conservation status

Just over one-tenth of all corvids are classified as under some level of threat according to the IUCN, a reflection of the opportunism and adaptability of this family. However, the 13 species at risk and a further 11 Near Threatened species, are almost all in decline due to factors common to most threatened bird species.

Common crows (*Corvus brachyrhynchos*) raid a cornfield. (Photo by Gary R. Zahm. Bruce Coleman Inc. Reproduced by permission.)

More than any other family, crows are cast as villains of bird conservation, directly accused of causing the declines of other species. Such criticism is, at the very least, simplistic. Generally, crow predation or competition exacerbates a deteriorating situation caused by habitat alteration. Numbers of the endangered Cuban palm crow (*Corvus minutus*) have probably fallen because it is being outcompeted by the Cuban crow (*Corvus nasicus*). Yet such interspecific competition has almost certainly only happened because habitat destruction has resulted in overlapping ranges. The most serious threat to the Amami jay (*Garrulus lidthi*) is predation by other crows and mammals. However, numbers of large-billed crows (*Corvus macrorhynchos*) have recently increased, probably because of increased garbage disposal on the island.

The biggest threat facing endangered corvids, affecting 18 out of the 24 species at risk, is timber extraction, which is often followed by conversion of cleared ground to agriculture or development. When forest habitat is lost, and populations fall as a result, other factors take on greater significance. Fragmentation of habitat has severely depleted populations of the Sri Lanka magpie (*Urocissa ornata*) and the Flores crow (*Corvus florensis*) to the extent that brood-parasitism by the Asian koel (*Eudynamys scolopacea*) has a disproportionately large impact. Similarly, depredations caused by hunting for food by the white-necked crow (*Corvus leucognaphalus*) and Hispaniolan palm crow (*Corvus palmarum*) are now critical factors in their survival.

Habitat loss for development brings additional problems. The case of the Florida scrub-jay is a classic example. Between 1960 and 1980, the human population doubled as housing and the planting of citrus groves destroyed large areas of the bird's scrub habitat. Birds living in the remaining habitat, closer to human habitation, face extra pressure. Human disturbance, predation by feral cats, mortality on roads, and overgrowing of scrub because of fire prevention measures, added to the difficulties. Conservationists are working to ameliorate these problems by establishing controlled, rotational burning of scrub, and setting up a network of protected sites for the birds.

It is hard to see how some problems facing island endemics can be overcome. Added to habitat loss and other pressures are the threats facing small, geographically restricted populations of corvids from invasive species. The BirdLife International statement that "invasive species have entirely or partially caused the majority of all bird extinctions since 1800" appears prophetic for the Hawaiian crow (*Corvus hawaiiensis*) and Mariana crow (*Corvus kubaryi*). Both species have seen their populations eroded by habitat loss, disturbance, and persecution, among other factors. The introduction of the predatory brown tree snake (*Boiga irregularis*) on Guam will probably cause the Mariana crow to disappear from that island, and the population on neighboring Rota is at risk from introduced rats. The position of the Critically Endangered Hawaiian crow is even worse due to predation of nests by introduced rats and the Indian mongoose (*Herpestes auropunctatus*).

Significance to humans

Perhaps more than any other bird, the raven has from ancient times held a notable place in the minds of people in the Northern Hemisphere. This family has been linked inextricably with the lives, religions, and mythology of humans, probably even before ravens appeared on prehistoric cave paintings.

The omnipresence of corvids around human settlements, with crows and ravens often treated synonymously, has undoubtedly been a major factor in their pre-eminence in human mythology. In India, villagers believed the house crow was immortal, for although people died, the crows remained. Attributes given to corvids are legion; the common thread is the magical powers that are invested in these birds. The Innuit and ancient Chinese among others, associated crows with creation. Early Europeans believed that a crow flying around a house calling foretold a death. Viking mythology gives the god Odin two prophetic raven messengers, Hugin (thought) and Munin (memory), and in ancient Greece, the god Apollo had a prophetic raven. The gift of prophecy for both good and evil is powerfully realized among North and Central American aboriginal tribes, where the crow is totemic. Such was the power of crow medicine that the Cheyenne used crow rather than eagle feathers on their warbonnets.

Associations are often linked to observations of bird behavior. As a carrion-feeder, the northern raven waited around battlefields for a feast of human carcasses. On the Welsh and Scottish borderlands, it was named the Corby messenger, foretelling death and slaughter. This author testifies to the power of suggestion. Trapped by a blizzard in the mountains of Glencoe, Scotland, my fate on a cliff ledge appeared settled when, from the silent valley below, came the harsh "krok" of an apparently expectant raven.

Crows appear in the Koran and in the Bible, most memorably in the latter when a raven is sent by Noah to find land. An illustrated Bible from the early twelfth century shows a raven pecking at a drowned corpse. From Shakespeare—who mentioned crows 50 times within his plays—to Edgar Allen Poe, crows have fascinated humans.

Humans discovered early on that these intelligent birds could be tamed. The Roman philosopher Pliny gives instruction on teaching crows to speak. The diminutives in Mag-pie and Jack-daw reveal the familiarity that pet birds would command. Even today, the green magpie (*Cissa chinensis*) is commonly hunted for the cage bird trade.

Some corvid species have been persecuted because of economic damage to farm crops, livestock, poultry, game birds, and waterfowl. In 1938, following recommendations by the U.S. Biological Survey, American crow roosts in Oklahoma were bombed systematically. Seven years earlier, a contest to exterminate the black-billed magpie from British Columbia resulted in 1,033 being killed by just 12 hunters in one season. Persecution dates back much further; bounty hunters in England were given a farthing a crow in the reign of Henry VIII. The English poet William Wordsworth declared of the raven that "this carnivorous fowl is a great enemy to the lambs of these solitudes."

Although the indiscriminate persecution of corvids has now lessened—the Migratory Bird Treaty Act in North America was amended in 1972 to include protection for corvids for example—licensed control of a number of species is still accepted. Federal regulations in North America and national laws throughout Europe allow for killing corvids.

1. Pinyon jay (*Gymnorthinus cyanocephalus*); 2. Gray jay (*Perisoreus canadensis*); 3. Azure-winged magpie (*Cyanopica cyana*); 4. Green magpie (*Cissa chinensis*); 5. Eurasian jay (*Garrulus glandarius*); 6. Rufous treepie (*Dendrocitta vagabunda*); 7. Western scrub-jay (*Aphelocoma california*); 8. Hume's ground-jay (*Pseudopooces humilis*); 9. Eurasian magpie (*Pica pica*); 10. Blue jay (*Cyanocitta cristata*). (Illustration by Gillian Harris)

1. American crow (*Corvus brachyrhynchos*); 2. Spotted nutcracker (*Nucifraga caryocatactes*); 3. Red-billed chough (*Pyrrhocorax pyrrhocorax*); 4. Western jackdaw (*Corvus monedula*); 5. Northern raven (*Corvus corax*); 6. House crow (*Corvus splendens*); 7. Torresian crow (*Corvus orru*); 8. Rook (*Corvus frugilegus*); 9. Carrion crow (*Corvus corone*). (Illustration by Gillian Harris)

Species accounts

Pinyon jay
Gymnorhinus cyanocephalus

SUBFAMILY
Corvinae

TAXONOMY
Gymnorhinus cyanocephalus Wied, 1841, Montana. Monotypic.

OTHER COMMON NAMES
French: Geai des pinèdes; German: Nacktschnabelhäher; Spanish: Chara piñonera.

PHYSICAL CHARACTERISTICS
9.75–10.92 in (25–28 cm); 3.6 oz (103 g). Uniform dull blue plumage is darkest on head, brightest on breast, crown, and forehead, and palest on rump. Throat is whitish. Relatively fine, sharply pointed bill is blackish, as are the legs and feet.

DISTRIBUTION
West-central United States. Central Oregon east to South Dakota; may be found as far south as Baja, California, to western Oklahoma.

HABITAT
Dry mountain slopes of pinyon, juniper, and yellow pine.

BEHAVIOR
Highly gregarious, usually in large flocks, numbering up to 250 individuals. Flight is direct, accompanied by mewing calls.

FEEDING ECOLOGY AND DIET
Heavily dependent on conifer seeds. Will forage widely for invertebrates and other seeds, often visiting backyards.

REPRODUCTIVE BIOLOGY
Cooperative breeder with young adults feeding chicks both before and after fledging. Colonial, well-spaced nests consisting of sticks and vegetation, cup lined with fine plants and wool. Generally three to four eggs laid February through May. Incubation 16 days; fledging 21 days.

CONSERVATION STATUS
Not threatened. Locally common throughout its range.

SIGNIFICANCE TO HUMANS
Known by the Hopi as the bird of war because of its habit of mobbing predators. ◆

Blue jay
Cyanocitta cristata

SUBFAMILY
Corvinae

TAXONOMY
Cyanocitta cristata Linnaeus, 1758, South Carolina. Four subspecies.

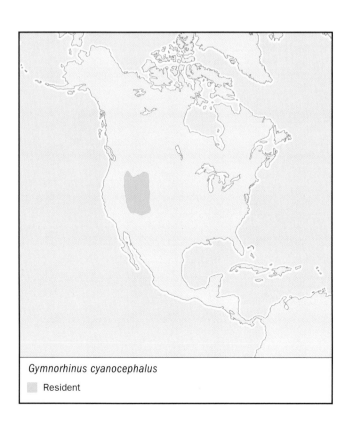

Gymnorhinus cyanocephalus

  Resident

Cyanocitta cristata

  Resident   Breeding

OTHER COMMON NAMES
French: Geai bleu; German: Blauhäher; Spanish: Chara azul.

PHYSICAL CHARACTERISTICS
9.36–11.7 in (24–30 cm); 2.27–3.8 oz (65–109 g). Upperparts
and jaunty crest are purplish blue. Wings and tail are brighter
blue and have prominent black barring; wings also have white
spots. Chin, face, throat and underparts are grayish white.
Long bill, legs, and feet are black.

DISTRIBUTION
Eastern and central North America, with Rocky Mountains
forming broad western limit.

HABITAT
Woodland, parks, and suburbs.

BEHAVIOR
Noisy and bold visitor to many backyards. Generally forages
alone or in pairs.

FEEDING ECOLOGY AND DIET
Highly opportunistic feeder, supplementing a seed and nut diet
with birds, mammals, invertebrates, and human garbage.

REPRODUCTIVE BIOLOGY
Solitary tree nester. Uses mud to hold its twig, root, and
feather nest together. Generally lays four to five eggs late
March through early June. Incubation 16–18 days; fledging
18–21 days. Frequently double brooded.

CONSERVATION STATUS
Not threatened. Very common, with range expanding northwestward.

SIGNIFICANCE TO HUMANS
Iconic status, with the Toronto Blue Jays baseball team, a
record label, and a Beatles song all named after this bird. ◆

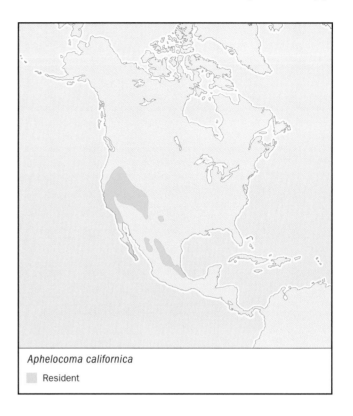

Aphelocoma californica
■ Resident

Western scrub-jay
Aphelocoma californica

SUBFAMILY
Corvinae

TAXONOMY
Aphelocoma californica Vigors, 1839, California. Fifteen subspecies. Isolated Florida scrub-jay in peninsular Florida and the
island scrub-jay (*Aphelocoma insularis*) on Santa Cruz considered
separate species.

OTHER COMMON NAMES
English: California jay; French: Geai buissonnier; German:
Buschhäher; Spanish: Chara Pecho Rayado.

PHYSICAL CHARACTERISTICS
10.53–12.09 in (27–31 cm); 2.8 oz (80 g). Upperparts, head,
wings, and tail are dark blue; mantle and back are grayish
brown. Head features a white "eyebrow" over a dark eye patch.
Underparts are variable whitish, buff, and grayish. White
throat is outlined with a blue necklace and there is a blue band
on the chest. Bill, legs, and feet are blackish.

DISTRIBUTION
Western United States and northwest Mexico.

HABITAT
Mixed shrubs with trees.

BEHAVIOR
Ground-feeder never far from cover, usually in pairs or family
groups. Frequent loud, chattering calls. Pairs are strongly
territorial.

FEEDING ECOLOGY AND DIET
Primarily acorns and pine seeds, often stored for later use. Also
invertebrates, eggs and nestlings, small amphibians, reptiles,
and mammals.

REPRODUCTIVE BIOLOGY
Solitary breeder. Lays two to six eggs late March through early
May in twig nest lined with plant matter. Incubation 16–19
days; fledging about 18 days.

CONSERVATION STATUS
Not threatened; locally common.

SIGNIFICANCE TO HUMANS
Important, if inadvertent role as a planter of trees. ◆

Eurasian jay
Garrulus glandarius

SUBFAMILY
Corvinae

TAXONOMY
Garrulus glandarius Linnaeus, 1758, Sweden. About 33 subspecies.

Garrulus glandarius

☐ Resident

OTHER COMMON NAMES
English: Red-crowned jay (in India); French: Geai des chênes; German: Eichelhaher; Spanish: Arrendejo Común.

PHYSICAL CHARACTERISTICS
12.48–14.04 in (32–36 cm); 4.9–6.55 oz (140–187 g). Subspecies vary significantly in plumage color and pattern, but generally this bird has a pinkish brown body, white wing patches, blue shoulders, and a black tail. The head features a black moustache and crown feathers that appear ruffled when erect. Bill is dark brownish horn, and legs and feet are fleshy-brown.

DISTRIBUTION
Most widespread of all jays, found in all but northernmost parts of Europe, north Africa, Middle East, central and Southeast Asia.

HABITAT
Both deciduous and coniferous woodland.

BEHAVIOR
Normally shy and solitary. Presence given away by loud, screeching call.

FEEDING ECOLOGY AND DIET
Eats and stores enormous quantities of acorns. Also feeds on other seeds, invertebrates, eggs, and nestlings.

REPRODUCTIVE BIOLOGY
Solitary nester, building platform twig nest in fork of tree. Lays three to seven eggs April through May. Incubation 16–19 days; fledging 18–23 days.

CONSERVATION STATUS
Not threatened. Common to abundant in most of its range.

SIGNIFICANCE TO HUMANS
Thought to be primarily responsible for planting of Old World oak forests. Individuals bury thousands of acorns in fall for later consumption. ◆

Gray jay
Perisoreus canadensis

SUBFAMILY
Corvinae

TAXONOMY
Perisoreus canadensis Linnaeus, 1766, Canada. Eight subspecies.

OTHER COMMON NAMES
English: Canada jay, whiskey-jack, venison-hawk; French: Mésangeai du Canada; German: Meisenhäher; Spanish: Chara gris.

PHYSICAL CHARACTERISTICS
9.75–10.92 in (25–28 cm); 2.17–2.5 oz (62–73 g). Upperparts, wings, and tail are dark gray; underparts are lighter gray. Head is pale gray with a black patch on crown and nape; throat is white. Subspecies vary most noticeable in the extent of the black head patch.

DISTRIBUTION
Conifer forests of Canada, Alaska, and northern and western United States.

HABITAT
Coniferous forests away from human habitation.

BEHAVIOR
Usually forages unobtrusively in pairs or family groups.

FEEDING ECOLOGY AND DIET
Invertebrates, small mammals, and birds. Also berries which are "glued" to trees for future consumption, using sticky saliva.

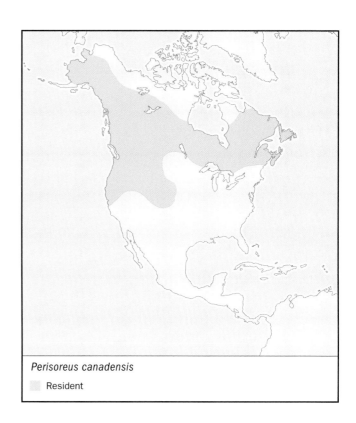

Perisoreus canadensis

☐ Resident

REPRODUCTIVE BIOLOGY
Solitary nester. Lays two to five eggs March through April in twig nest which is well insulated with mosses, lichens, fur, and feathers. Incubation 16–18 days; fledging about 15 days.

CONSERVATION STATUS
Not threatened. Locally common over a wide range.

SIGNIFICANCE TO HUMANS
Bold scavenging from humans has earned it the name "camp robber." ◆

Green magpie
Cissa chinensis

SUBFAMILY
Corvinae

TAXONOMY
Cissa chinensis Boddaert, 1783, Mergui, Tenasserim, Myanmar. Five subspecies.

OTHER COMMON NAMES
English: Green hunting crow, hunting cissa; French: Pirolle verte; German: Jagdelster; Spanish: Urraca Verde.

PHYSICAL CHARACTERISTICS
14.43–15.21 in (37–39 cm); male 4.55–4.65 oz (130–133 g), female 4.2–4.34 oz (120–124 g). A bulky-bodied bright green bird with a wide black mask, chestnut wings, and long, graduated tail. Bill, legs, and feet are bright red.

DISTRIBUTION
Mainland tropical Southeast Asia, also Borneo and Sumatra.

HABITAT
Lowland and hill forest.

BEHAVIOR
Shy bird that gives whistling call from dense undergrowth. Small groups move through forest outside breeding season in company with other bird species.

FEEDING ECOLOGY AND DIET
Hunts low in undergrowth for insects. Also small birds, reptiles, amphibians, fruit, and berries.

REPRODUCTIVE BIOLOGY
Solitary nester on platform nest often built in bamboo or vines. Lays three to seven eggs January through April. No information known about incubation or fledging.

CONSERVATION STATUS
Not threatened, but uncommon throughout its range.

SIGNIFICANCE TO HUMANS
Commonly trapped for the cage bird trade, where it is known as a cissa. ◆

Azure-winged magpie
Cyanopica cyana

SUBFAMILY
Corvinae

TAXONOMY
Cyanopica cyana Pallas, 1776, Dauria. Nine subspecies.

OTHER COMMON NAMES
French: Pie bleue; German: Blauelster; Spanish: Urraca de Rabo Largo.

PHYSICAL CHARACTERISTICS
13.26 in (34 cm); 2.17–2.87 oz (62–82 g). A distinctive bird with gray to pinkish buff upperparts, black hood, white throat, pale blue wings, and long, graduated pale blue tail. Bill, legs, and feet are black.

Cissa chinensis
☐ Resident

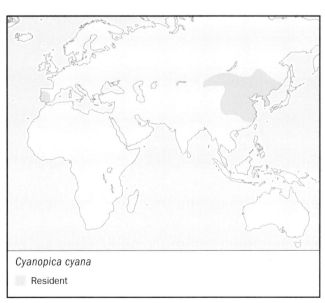

Cyanopica cyana
☐ Resident

DISTRIBUTION
Two separate populations. Western distribution limited to Spain and Portugal. May once have been linked with eastern population, which is distributed widely from Mongolia and eastern Russia through most of China and Korea to Japan.

HABITAT
Woodland and forest edge. Iberian form found in open country.

BEHAVIOR
Highly gregarious magpie, with parties numbering up to 100 outside breeding season. Stays close to cover in most parts of its range, although a bold scavenger in Chinese and Japanese towns and cities.

FEEDING ECOLOGY AND DIET
Insects, fruit, seeds, and berries.

REPRODUCTIVE BIOLOGY
Nests in colonies, with each pair selecting its own tree to build a stick and twig nest, held together with mud. Generally lays five to seven eggs March through June; double-brooded in Far East. Incubation 15 days; fledging about 16 days. Young adults from previous year's brood act as helpers feeding young.

CONSERVATION STATUS
Not threatened. Common and increasing in number over most of its range.

SIGNIFICANCE TO HUMANS
None known. ◆

Dendrocitta vagabunda

▨ Resident

Rufous treepie
Dendrocitta vagabunda

SUBFAMILY
Corvinae

TAXONOMY
Dendrocitta vagabunda Latham, 1790, India. Nine subspecies.

OTHER COMMON NAMES
English: Indian treepie; French: Témia vagabonde; German: Wanderelster; Spanish: Urraca vagabunda.

PHYSICAL CHARACTERISTICS
17.94–19.5 in (46–50 cm); 3.15–4.55 oz (90–130 g). Upperparts are rusty-orange; underparts are paler and buffy. Head, neck, breast, and upper mantle are blackish. Wing coverts and tertials are pale, silvery gray; the rest of the wing is brownish black. Central tail feathers are pale gray ending in a wide, black terminal band; outer tail feathers have more black. Bill is dark gray or blackish; legs and feet are brownish black.

DISTRIBUTION
Widely distributed through Pakistan, India, Burma, and western Thailand. Patchier in Laos, Cambodia, and Vietnam.

HABITAT
Light woodland, open fields with trees, villages, towns, and cities.

BEHAVIOR
Moves in pairs or family parties, uttering loud calls. Largely arboreal, most often seen flying between trees, or perching high in canopy.

FEEDING ECOLOGY AND DIET
Very wide diet, including invertebrates, small birds, mammals, reptiles, and amphibians, nuts, berries, carrion.

REPRODUCTIVE BIOLOGY
Solitary nester usually on lone tree. Generally lays four to five eggs February through May in stick nest. Incubation and fledging periods unknown.

CONSERVATION STATUS
Not threatened. Very common in western part of range, common elsewhere.

SIGNIFICANCE TO HUMANS
None known. ◆

Eurasian magpie
Pica pica

SUBFAMILY
Corvinae

TAXONOMY
Pica pica Linnaeus, 1758, Sweden. Twelve subspecies. Until 2000, the black-billed magpie of northwest North America was considered a subspecies.

OTHER COMMON NAMES
English: Common magpie; French: Pie bavarde; German: Elster; Spanish: Urraca de Pico Negro.

PHYSICAL CHARACTERISTICS
16.77–19.5 in (43–50 cm); 6.3–9.63 oz (180–275 g). Plumage is mostly black with white belly, sides, and scapulars. The black head, neck, mantle, and breast have a weak green and purple

Pica pica

◻ Resident

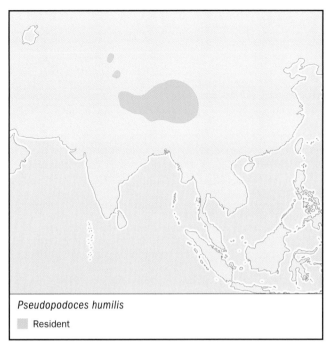

Pseudopodoces humilis

◻ Resident

sheen; the black wings and tail are highly glossed with green/blue/purple iridescence. Bill, legs, and feet are black.

DISTRIBUTION
One of the most widespread of corvids, found throughout Europe and much of Asia, and northwest Africa.

HABITAT
Generally found wherever there are trees, although avoids dense forest.

BEHAVIOR
Presence often betrayed by raucous "chacking" call. Groups of magpies often harass gulls or kites to make them drop food.

FEEDING ECOLOGY AND DIET
Omnivorous diet includes invertebrates, fruit and berries, birds, mammals, carrion including road kills; scavenges human garbage.

REPRODUCTIVE BIOLOGY
Solitary nester. Lays between two to 13, but generally eight to nine eggs March through May in dome-shaped twig nest. Incubation 14–23 days; fledging 10 days; young stay with parents for a short time after leaving nest. Other nonbreeding adults allowed in breeding territory.

CONSERVATION STATUS
Not threatened; common to abundant.

SIGNIFICANCE TO HUMANS
Numbers controlled where perceived as a threat to game birds in Europe. ◆

Hume's ground-jay
Pseudopodoces humilis

SUBFAMILY
Corvinae

TAXONOMY
Pseudopodoces humilis Hume, 1871, Sinkiang. Probably monotypic.

OTHER COMMON NAMES
English: Little ground-jay, Tibetan ground-jay, Hume's groundpecker; French: Podoce de Hume; German: Höhlen-häher; Spanish: Arrandejo Terrestre de Hume.

PHYSICAL CHARACTERISTICS
7.4 in (19 cm); 1.47–1.61 oz (42–46 g). The soft, slightly fluffy body plumage is dull brownish above and off-white below. Wings are darker brown and the tail is white with dark brown central feathers. The black bill is short, thin, and distinctly curved. Legs and feet also are black.

DISTRIBUTION
Tibetan Plateau of China, Nepal, and India.

HABITAT
Grassy, boulder-strewn mountain slopes.

BEHAVIOR
Hops and perches between boulders, flicking wings and tail on landing.

FEEDING ECOLOGY AND DIET
Probes on ground using long bill in search of invertebrates.

REPRODUCTIVE BIOLOGY
Digs long tunnel in bank, wall or building to build grass and moss nest at end. Lays four to six eggs May through June. Incubation and fledging periods unknown.

CONSERVATION STATUS
Not threatened; common.

SIGNIFICANCE TO HUMANS
None known. ◆

Spotted nutcracker
Nucifraga caryocatactes

SUBFAMILY
Corvinae

TAXONOMY
Nucifraga caryocatactes Linnaeus 1758, Sweden. Eight subspecies.

OTHER COMMON NAMES
English: Nutcracker, Eurasian nutcracker, spotted nutcracker; French: Cassenoix moucheté; German: Tannenhäher; Spanish: Cascanueces Moteado.

PHYSICAL CHARACTERISTICS
12.48–13.26 in (32–34 cm); 4.34–7 oz (124–200 g). Body plumage is dark brown profusely spotted with white; lower belly and undertail-coverts are white. Wings are glossy black with white-tipped coverts. Tail is also glossy black with white feather tips, increasing in extent toward the outermost. Bill, legs, and feet are black.

DISTRIBUTION
Coniferous forests of Europe and Asia. North American equivalent is Clark's nutcracker.

HABITAT
Coniferous forests. Widely dispersed during irruptions following failure of seed harvest.

BEHAVIOR
Wary, but perches on conspicuous treetops. Holds year-round territory, keeping several stores of nuts and seeds.

FEEDING ECOLOGY AND DIET
Heavily reliant on conifer seeds and hazel nuts, which are stored as winter supply. Insects and berries eaten seasonally.

REPRODUCTIVE BIOLOGY
Solitary nester. Generally lays three to four eggs March through May in stick nest woven together with plant stems.

Cup lined with moss and grass. Incubation 16–18 days; fledging about 23 days. Young remain with parents for rest of summer.

CONSERVATION STATUS
Not threatened. Abundant throughout its range.

SIGNIFICANCE TO HUMANS
A significant planter of conifers thanks to its habit of storing pine seed underground. ◆

Red-billed chough
Pyrrhocorax pyrrhocorax

SUBFAMILY
Corvinae

TAXONOMY
Pyrrhocorax pyrrhocorax Linnaeus, 1758, England. Eight subspecies.

OTHER COMMON NAMES
English: Chough; French: Crave à bec rouge; German: Alpenkrähe; Spanish: Chova piquirroja.

PHYSICAL CHARACTERISTICS
14.04–15.6 in (36–40 cm); 9.97–13.3 oz (285–380 g). Plumage is velvet-black with a slight bluish purple to greenish gloss on the body; wings and tail are glossier. Slender, curved bill is red, as are legs and feet.

DISTRIBUTION
Widespread in mountainous areas of central and western Asia. Patchy distribution in Europe and northwest Africa.

HABITAT
Rocky, mountainous areas with adjacent animal pasture. Western European populations use sea cliffs.

BEHAVIOR
Territorial during breeding season, but allows third adult in territory. Highly gregarious at other times and roosts colonially in caves or rock crevices.

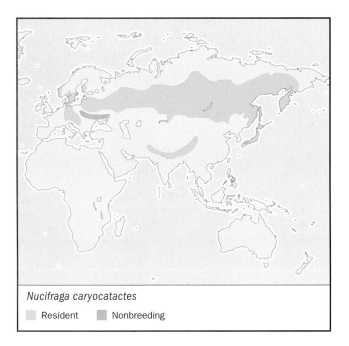

Nucifraga caryocatactes
▨ Resident ▨ Nonbreeding

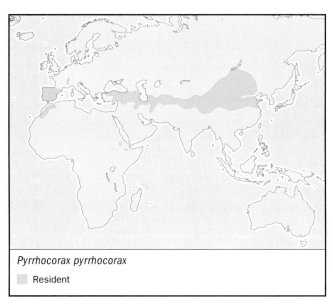

Pyrrhocorax pyrrhocorax
▨ Resident

FEEDING ECOLOGY AND DIET
Soil feeder on ants, beetles, worms, and other invertebrates. Often finds food in animal dung.

REPRODUCTIVE BIOLOGY
Solitary nester, constructing stick and animal hair nest at back of cave. Generally lays three to four eggs March through April. Incubation 17–21 days; fledging 36–41 days.

CONSERVATION STATUS
Not threatened. Common in Asia, populations declining in Europe following changes in land use.

SIGNIFICANCE TO HUMANS
None known. ◆

Western jackdaw
Corvus monedula

SUBFAMILY
Corvinae

TAXONOMY
Corvus monedula Linnaeus, 1758, Sweden. Four subspecies usually recognized.

OTHER COMMON NAMES
French: Choucas des tours; German: Dohle; Spanish: Grajilla Común.

PHYSICAL CHARACTERISTICS
13.26–15.21 in (34–39 cm); 4.86–9.27 oz (139–265 g). Upperparts are grayish black with a slight bluish gloss. Head has a black forecrown and dark gray rear crown, nape, and sides. Wings and tail are black and glossy. Underparts are grayish black. The short bill is black, as are the legs and feet.

DISTRIBUTION
Europe, western Asia, North Africa.

HABITAT
Wide variety of open country with trees, including areas of human habitation.

BEHAVIOR
Sociable groups frequently engage in agile, tumbling flight display, giving loud calls.

FEEDING ECOLOGY AND DIET
Mainly granivorous, except during breeding season when it takes moth caterpillars from tree foliage.

REPRODUCTIVE BIOLOGY
Semicolonial. Stick nests built in tree holes, cliff faces, or artificial structures. Generally four eggs April through May. Incubation 17–19 days; fledging about 30 days.

CONSERVATION STATUS
Not threatened. Abundant; appears to be spreading broadly north and eastward.

SIGNIFICANCE TO HUMANS
Regularly nests in chimneys of inhabited buildings. ◆

House crow
Corvus splendens

SUBFAMILY
Corvinae

TAXONOMY
Corvus splendens Vieillot, 1817, Bengal. Four or five subspecies recognized.

OTHER COMMON NAMES
French: Corbeau familier; German: Glanzkrähe; Spanish: Corneja India.

PHYSICAL CHARACTERISTICS
15.6 in (40 cm); 8.57–12.07 oz (245–371 g). Plumage is black except for the nape, sides of the head, and breast, which are gray. Bill, legs, and feet also black.

DISTRIBUTION
Iran to India, Pakistan and Burma, self-introduced to East Africa, Indian Ocean islands, Malaysia, and South Africa.

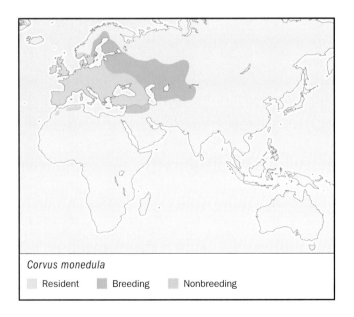

Corvus monedula
▨ Resident   ▨ Breeding   ▨ Nonbreeding

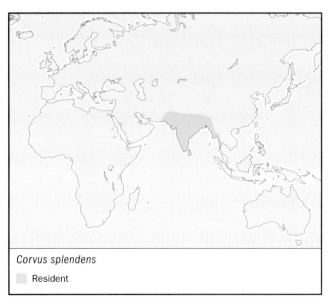

Corvus splendens
▨ Resident

HABITAT
Wholly dependent on human habitation; consequently found in villages, towns, and cities throughout its range.

BEHAVIOR
Highly vocal, gregarious birds, seemingly unafraid of humans. Will attack and chase off any large bird of prey.

FEEDING ECOLOGY AND DIET
Omnivorous. Diet includes seeds, fruit, grain, nectar, berries, bird's eggs, nestlings, mammals, reptiles, amphibians, fish, insects, wide range of carrion.

REPRODUCTIVE BIOLOGY
Solitary nester except in areas of high population density. Will use trees, buildings, or other artificial structures for rough stick nest. Three to four eggs March through July. Incubation 16–17 days; fledging 21–28 days.

CONSERVATION STATUS
Not threatened. Abundant throughout range.

SIGNIFICANCE TO HUMANS
Regarded as a major agricultural and human health pest in self-introduced areas. In South Africa, birds have been reported taking food from school children, killing chicks of domestic fowls, and repeatedly dive-bombing any person near the nest. ◆

Rook
Corvus frugilegus

SUBFAMILY
Corvinae

TAXONOMY
Corvus frugilegus Linnaeus, 1758, Sweden. Two subspecies.

OTHER COMMON NAMES
French: Corbeau freux; German: Saatkrähe; Spanish: Graja Común.

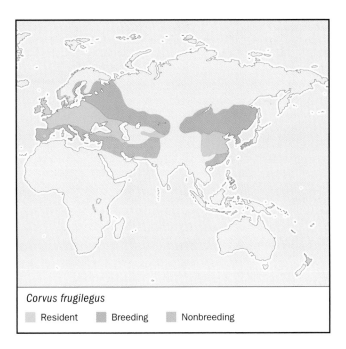

Corvus frugilegus
Resident   Breeding   Nonbreeding

PHYSICAL CHARACTERISTICS
18.33 in (47 cm); 11.8–18.58 oz (337–531 g). Entire plumage is black and highly glossed with metallic blue, green, and purple. Base of the bill, chin, and loral region are bare, giving the forecrown a slightly peaked appearance. Bill is whitish at the base shading to dusky over the terminal half. Legs and feet are black.

DISTRIBUTION
In all but the most mountainous areas of Europe and Asia.

HABITAT
Farmland with scattered trees and wooded steppe.

BEHAVIOR
Significant fall migrant, flying by day from northern Europe and Asia, to western Europe, the Middle East, and southern Asia. In the Seine-et-Marne district of France, the breeding population was estimated at 10,000; the winter population 500,000.

FEEDING ECOLOGY AND DIET
Soil feeder on invertebrates, seed, grain.

REPRODUCTIVE BIOLOGY
Nests colonially in rookeries; there are generally at least 20 nests densely clustered around the tops of several trees. Nests used repeatedly in successive years. Lays two to seven eggs in grass and leaf cup within bulky stick nest. Incubation 16–18 days; fledging 32–33 days.

CONSERVATION STATUS
Not threatened. Abundant to localized in Europe. Evidence of decline in eastern Asia.

SIGNIFICANCE TO HUMANS
Often persecuted because of its tendency to feed on cereal seed. ◆

American crow
Corvus brachyrhynchos

SUBFAMILY
Corvinae

TAXONOMY
Corvus brachyrhynchos Brehm, 1822, Boston, Massachusetts. Four subspecies.

OTHER COMMON NAMES
English: Common crow; French: Corneille d'Amérique; German: Amerikanerkrähe; Spanish: Cuervo Americano.

PHYSICAL CHARACTERISTICS
15.21–19.11 in (39–49 cm); male 1 lb (458 g), female 15.33 oz (438 g). Plumage is entirely black with a light violet-blue gloss. Wings have a more greenish blue gloss. Prominent bristles cover the basal third of the upper mandible. Bill, legs, and feet are black.

DISTRIBUTION
Widespread throughout North America. Canadian birds migrate to central United States and Atlantic seaboard.

HABITAT
Adapted to most habitats with the exception of arid areas and dense forest.

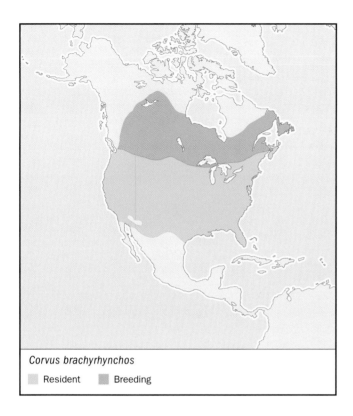

Corvus brachyrhynchos
◻ Resident ◼ Breeding

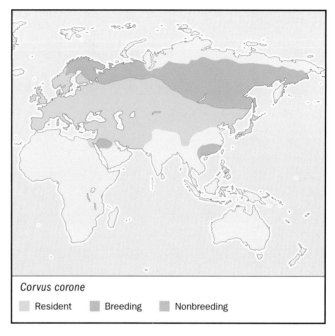

Corvus corone
◻ Resident ◼ Breeding ◼ Nonbreeding

BEHAVIOR
Gathers in huge communal roosts in winter, sometimes containing tens of thousands of birds. Individuals will fly up to 50 mi (80 km) to the roost.

FEEDING ECOLOGY AND DIET
Omnivorous. Main food items include insects, bird nestlings and eggs, and carrion.

REPRODUCTIVE BIOLOGY
Solitary nester. Builds stick nest in fork of tree, bush, or occasionally artificial structure. Generally lays four to five eggs in cup made of roots, grass, and animal hair February through June. Incubation 18 days; fledging 28–35 days.

CONSERVATION STATUS
Not threatened. Abundant throughout its range.

SIGNIFICANCE TO HUMANS
Heavily persecuted as an agricultural pest, largely because its diet includes grain, game birds, and poultry. ◆

Carrion crow
Corvus corone

SUBFAMILY
Corvinae

TAXONOMY
Corvus corone Linnaeus, 1758, England. Two subspecies.

OTHER COMMON NAMES
English: Eurasian crow; French: Corneille noire; German: Aaskrähe; Spanish: Corneja Cenicienta.

PHYSICAL CHARACTERISTICS
18.72–21.84 in (48–56 cm); male 1–1.4 lb (465–650 g), female 1–1.3 lb (450–600 g). Plumage is entirely black with a bluish to purplish sheen. Conspicuous bristles cover base of culmen and basal third of upper mandible. Bill, legs, and feet are black.

DISTRIBUTION
Widely distributed throughout Europe and Asia. The black nominate form is replaced in central Europe and western Asia by the gray and black race known as the hooded crow.

HABITAT
Huge diversity of habitats, ranging from the centers of cities to sea cliffs.

BEHAVIOR
Some territories held year-round. Most defended only during breeding season, especially aggressively when nest-building. Considerable territorial rivalry between corvids; crows and black-billed magpies will destroy each other's nests and predate eggs and chicks.

FEEDING ECOLOGY AND DIET
Mainly carnivorous. Diet includes invertebrates, mollusks, amphibians, fish, birds, and mammals. Also carrion and human garbage.

REPRODUCTIVE BIOLOGY
Solitary nester. Stick nest in tree or bush, cliff ledges, or buildings. Generally lays four to five eggs April through May in cup made of moss, wool, animal hair, and roots. Incubation 17–19 days; fledging 32–36 days.

CONSERVATION STATUS
Not threatened. Common throughout its range.

SIGNIFICANCE TO HUMANS
Perceived as a threat to livestock and game birds throughout its range and heavily persecuted. ◆

Torresian crow

Corvus orru

SUBFAMILY
Corvinae

TAXONOMY
Corvus orru Bonaparte, 1851, New Guinea. Four subspecies.

OTHER COMMON NAMES
English: Australian crow; French: Corbeau de Torres; German: Salvadorikrähe; Spanish: Cuervo Australiano.

PHYSICAL CHARACTERISTICS
19.5–21.45 in (50–55 cm); 15.05–23.45 oz (430–670 g). Plumage is entirely black with a strong purple or bluish purple gloss. Bill, legs, and feet are also black.

DISTRIBUTION
Australia, New Guinea, and neighboring islands of Indonesia.

HABITAT
Edges of rainforest, open forest, woodlands and tall scrub, coastal margins, ranges and gorges of arid areas, human settlements.

BEHAVIOR
The only one of five very similar Australian crow species to adopt a curious post-alighting behavior. The Torresian crow lands, then promptly begins shuffling its wings up and down.

FEEDING ECOLOGY AND DIET
Omnivorous, feeding mainly on insects, fruit, seeds, and carrion.

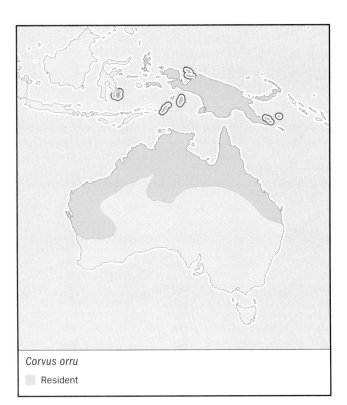

Corvus orru

Resident

REPRODUCTIVE BIOLOGY
Solitary tree nester. Lays two to five eggs. Incubation is 16–20 days; fledging is 34–38 days.

CONSERVATION STATUS
Listed as Near Threatened in the *Action Plan for Australian Birds 2000*.

SIGNIFICANCE TO HUMANS
Considered an agricultural pest, as it feeds on grain, peanuts, and fruit. ◆

Northern raven

Corvus corax

SUBFAMILY
Corvinae

TAXONOMY
Corvus corax Linnaeus, 1758, Sweden. Eight subspecies may be recognized.

OTHER COMMON NAMES
French: Grand corbeau; German: Kolkrabe; Spanish: Cuervo Común.

PHYSICAL CHARACTERISTICS
22.62–26.91 in (58–69 cm); 2.02–3.43 lb (92–156 g). Plumage is glossy black. Prominent nasal bristles cover basal third of upper mandible. Long, heavy bill is black and the distal part of the culmen is strongly decurved. Legs and feet are also black.

DISTRIBUTION
Most widespread corvid found in most of the Northern Hemisphere.

HABITAT
A wide variety of habitats, generally open and away from human habitation, including treeless tundra.

BEHAVIOR
Established pairs remain on large, usually well-spaced territories. The familiar guttural "krok" call is the most recognizable of a highly varied vocal repertoire.

FEEDING ECOLOGY AND DIET
Mainly carrion feeder, but also eats live prey and plant material. Powerful beak used to tear carrion apart and kill live animals. Often robs other predators of food. Seeds and berries seasonally important.

REPRODUCTIVE BIOLOGY
Solitary nester, constructing enormous stick nest in tree or bush, on cliff ledge, or rarely on the ground. Lays three to seven eggs early spring. Incubation 18–21 days; fledging 35–42 days.

CONSERVATION STATUS
Thinly distributed throughout range, but not at risk, despite often heavy persecution.

SIGNIFICANCE TO HUMANS
Mistakenly thought to kill healthy young livestock and consequently persecuted. ◆

Corvus corax

▨ Resident

Resources

Books

Bent, A. C. *Life Histories of North American Jays, Crows, and Titmice*. Washington, DC: Smithsonian Institution, 1946.

Birkhead, T. *The Magpies: The Ecology and Behavior of Black-Billed and Yellow-Billed Magpies*. New York: Academic Press, 1991.

Coombs, F. *The Crows: A Study of the Corvids of Europe*. London: Batsford, 1978.

Cramp, S., and C. M. Perrins. "A Handbook of the Birds of Europe, the Middle East and North Africa." In *The Birds of the Western Palearctic*, Vol. 8, *Crows to Finches*. Oxford, UK: Oxford University Press, 1994.

Goodwin, D. *Crows of the World*. London: British Museum (Natural History), 1986.

Heinrich, B. *Ravens in Winter*. New York: Summit, 1989.

Madge, S., and H. Burn. *Crows and Jays: A Guide to the Crows, Jays and Magpies of the World*. London: Christopher Helm, 1995.

Marzluff, J. M., and R. F. Balda. *The Pinyon Jay*. Orlando, FL: Academic Press, 1992.

Palmer, J. D. *Animal Wisdom: The Definitive Guide to the Myth, Folklore and Medicine Power of Animals*. London: Thorsons, 2001.

Ratcliffe, D. *The Raven*. Orlando, FL: Academic Press, 1997.

Savage, C. *Bird Brains: The Intelligence of Crows, Ravens, Magpies and Jays*. San Francisco: Sierra Club Books, 1997.

Sibley, C. G., and J. E. Ahlquist. *Phylogeny and Classification of Birds: A Study in Molecular Evolution*. New Haven and London: Yale University Press, 1990.

Yapp, B. *Birds in Medieval Manuscripts*. London: British Library, 1981.

Periodicals

Emery, N. J., and N. S. Clayton. "Effects of Experience and Social Context on Prospective Caching Strategies by Scrub Jays." *Nature* 414 (November 22, 2001): 443–446.

Sonerud, G.A, C. A. Smedshaug, and O. G. Brathen. "Ignorant Hooded Crows Follow Knowledgeable Roost-Mates to Food: Support for the Information Centre Hypothesis." *Proceedings of the Royal Society of London* B 268, no. 1469 (2001): 827–831.

Organizations

Roberts VII Project, Percy FitzPatrick Institute of African Ornithology, University of Cape Town. Rondebosch 7701, South Africa. Fax: (021) 650 3295. E-mail: roberts@botzoo.uct.ac.za Web site: <http://www.uct.ac.za/depts/fitzpatrick/docs/r549.html>

Other

BirdLife International. March 18, 2002. <http://www.birdlife.net>.

McGowan, K. "Demographic and Behavioral Comparisons of Urban and Rural American Crows in New York: Abstracts from the Symposium at the 117th Stated Meeting of the American Ornithologists' Union, Ithaca, NY, 14 August 1999." *Abstracts of Presented Papers Given By Kevin J. Mcgowan.* March 18, 2002 <http://www.birds.cornell.edu/crows/aou99.htm.>

Derek William Niemann, BA

For further reading

Ali, S. and Ripley, S. D. *Handbook of the Birds of India and Pakistan.* 2nd edition. 10 Vols. New York: Oxford University Press, 1978-1999.

American Ornithologists' Union. *Check-list of North American Birds: the Species of Birds of North America from the Arctic through Panama, including the West Indies and Hawaiian Islands.* 7th ed. Washington, DC: The American Ornithologists' Union, 1998.

Bennett, Peter M. and I. P. F. Owens. *Evolutionary Ecology of Birds: Life Histories, Mating Systems, and Extinction.* Oxford Series in Ecology and Evolution. Oxford: Oxford University Press, 2002.

Berthold, P. *Bird Migration: A General Survey.* Translated by H.-G. Bauer and V. Westhead. 2nd edition. Oxford Ornithology Series, no. 12. Oxford: Oxford University Press, 2001.

Boles, W. E. *The Robins and Flycatchers of Australia.* Sydney: Angus & Robertson, 1989.

Borrow, Nik, and Ron Demey. *Birds of Western Africa: An Identification Guide.* London: Christopher Helm, 2001.

Brewer, David, and B. K. Mackay. *Wrens, Dippers, and Thrashers.* New Haven: Yale University Press. 2001.

Brown, L. H., E. K. Urban and K. Newman, eds. *The Birds of Africa.* Vol. 1, *Ostriches to Falcons.* London: Academic Press, 1982.

Bruggers, R. L., and C. C. H. Elliott. *Quelea quelea: Africa's Bird Pest.* Oxford: Oxford University Press, 1990.

Burger, J., and Olla, B. I., eds. *Shorebirds: Breeding Behavior and Populations.* New York: Plenum Press, 1984.

Burton, J.A., Ed. *Owls of the World: Their Evolution, Structure and Ecology.* 2nd edition. London: Peter Lowe, 1992.

Byers, Clive, Jon Curson, and Urban Olsson. *Sparrows and Buntings: A Guide to the Sparrows and Buntings of North America and the World.* Boston: Houghton Mifflin Company, 1995.

Castro, I., and A. A. Phillips. *A Guide to the Birds of the Galápagos Islands.* Princeton: Princeton University Press, 1997.

Chantler, P., and G. Driessens. *Swifts: A Guide to the Swifts and Treeswifts of the World.* Sussex: Pica Press, 1995.

Cheke, R. A., and C. Mann. *Sunbirds: A Guide to the Sunbirds, Flowerpeckers, Spiderhunters and Sugarbirds of the World.* Christopher Helm, 2001.

Cleere, N., and D. Nurney. *Nightjars: A Guide to the Nightjars and Related Nightbirds.* Sussex: Pica Press, 1998.

Clement, P. *Finches and Sparrows: An Identification Guide.* Princeton: Princeton University Press, 1993.

Clement, P., et al. *Thrushes.* London: Christopher Helm, 2000.

Clements, J. F., and N. Shany. *A Field Guide to the Birds of Peru.* Temecula, California: Ibis Pub. 2001.

Coates, B. J. *The Birds of Papua New Guinea: Including the Bismarck Archipelago and Bougainville.* 2 vols. Alderley, Queensland: Dove Publications, 1985, 1990.

Coates, B. J., and K. D. Bishop. *A Guide to the Birds of Wallacea: Sulawesi, The Moluccas and Lesser Sunda Islands, Indonesia.* Alderley, Queensland: Dove Publications, 1997.

Cooke, Fred, Robert F. Rockwell, and David B. Lank. *The Snow Geese of La Pérouse Bay: Natural Selection in the Wild.* Oxford Ornithology Series, no. 4. Oxford: Oxford University Press, 1995.

Cooper, W. T., and J. M. Forshaw. *The Birds of Paradise and Bowerbirds.* Sydney: Collins, 1977.

Cramp, S., ed. *Handbook of the Birds of Europe the Middle East and North Africa. The Birds of the Western Palearctic.* Vol. 1, *Ostrich to Ducks.* Oxford: Oxford University Press, 1977.

Cramp, S., ed. *Handbook of the Birds of Europe the Middle East and North Africa. The Birds of the Western Palearctic.* Vol. 2, *Hawks to Bustards.* Oxford: Oxford University Press, .

Cramp, S., ed. *Handbook of the Birds of Europe the Middle East and North Africa. The Birds of the Western Palearctic.* Vol. 3, *Waders to Gulls.* Oxford: Oxford University Press, 1983.

Cramp, S., ed. *Handbook of the Birds of Europe the Middle East and North Africa. The Birds of the Western Palearctic.* Vol. 4, *Terns to Woodpeckers.* Oxford: Oxford University Press, 1985.

Cramp, S., ed. *Handbook of the Birds of Europe the Middle East and North Africa. The Birds of the Western Palearctic.* Vol. 6, *Warblers.* Oxford: Oxford University Press, 1992.

Cramp, S., ed. *Handbook of the Birds of Europe the Middle East and North Africa. The Birds of the Western Palearctic.* Vol. 7, *Flycatchers to Shrikes.* Oxford: Oxford University Press, 1993.

Cramp, S., ed. *Handbook of the Birds of Europe the Middle East and North Africa. The Birds of the Western Palearctic.* Vol. 8, *Crows to Finches.* Oxford: Oxford University Press, 1994.

Cramp, S., ed. *Handbook of the Birds of Europe the Middle East and North Africa. The Birds of the Western Palearctic.* Vol. 9, *Buntings and New World Warblers.* Oxford: Oxford University Press, 1994.

Davies, N. B. *Dunnock Behaviour and Social Evolution.* Oxford: Oxford University Press, 1992.

Davies, N. B. *Cuckoos, Cowbirds and Other Cheats.* London: T. & A. D. Poyser, 2000.

Davis, L. S., and J. Darby, eds. *Penguin Biology.* New York: Academic Press, 1990.

Deeming, D. C., ed. *Avian Incubation: Behaviour, Environment, and Evolution.* Oxford Ornithology Series, no. 13. Oxford: Oxford University Press, 2002.

del Hoyo, J., A. Elliott and J. Sargatal, eds. *Handbook of the Birds of the World.* Vol. 1, *Ostrich to Ducks.* Barcelona: Lynx Edicions, 1992.

del Hoyo, J., A. Elliott and J. Sargatal, eds. *Handbook of the Birds of the World.* Vol. 2, *New World Vultures to Guineafowl.* Barcelona: Lynx Edicions, 1994.

del Hoyo, J., A. Elliott and J. Sargatal, eds. *Handbook of the Birds of the World.* Vol. 3, *Hoatzin to Auks.* Barcelona: Lynx Edicions, 1996.

del Hoyo, J., A. Elliott and J. Sargatal, eds. *Handbook of the Birds of the World.* Vol. 4, *Sandgrouse to Cuckoos.* Barcelona: Lynx Edicions, 1997.

del Hoyo, J., A. Elliott and J. Sargatal, eds. *Handbook of the Birds of the World.* Vol. 5, *Barn-owls to Hummingbirds.* Barcelona: Lynx Edicions, 1999.

del Hoyo, J., A. Elliott and J. Sargatal, eds. *Handbook of the Birds of the World.* Vol. 6, *Mousebirds to Hornbills.* Barcelona: Lynx Edicions, 2001.

del Hoyo, J., A. Elliott and J. Sargatal, eds. *Handbook of the Birds of the World.* Vol. 7, *Jacamars to Woodpeckers.* Barcelona: Lynx Edicions, 2002.

Delacour, J., and D. Amadon. *Currasows and Related Birds.* New York: American Museum of Natural History, 1973.

Diamond, J., and A. B. Bond. *Kea, Bird of Paradox: The Evolution and Behavior of a New Zealand Parrot.* Berkeley: University of California Press, 1999.

Erritzoe, J. *Pittas of the World: a Monograph on the Pitta Family.* Cambridge: Lutterworth, 1998.

Feare, C., and A. Craig. *Starlings and Mynahs.* Princeton: Princeton University Press, 1999.

Feduccia, A. *The Origin and Evolution of Birds.* 2nd edition. New Haven: Yale University Press, 2001.

Ferguson-Lees, J., and D. A. Christie. *Raptors of the World.* Boston: Houghton Mifflin, 2001.

Fjeldsa, J., and N. Krabbe. *Birds of the High Andes.* Svendborg, Denmark: Apollo Books, 1990.

Forshaw, J. M. *Parrots of the World.* 3rd rev. edition. Melbourne: Lansdowne Editions, 1989.

Forshaw, J., ed. *Encyclopedia of Birds.* 2nd edition. McMahons Point, N.S.W.: Weldon Owen, 1998.

Frith, C. B., and B. M. Beehler. *The Birds of Paradise: Paradisaeidae.* Bird Families of the World, no. 6. Oxford University Press, 1998.

Fry, C. H., and K. Fry. *Kingfishers, Bee-eaters and Rollers: A Handbook.* Princeton: Princeton University Press, 1992.

Fry, C. H., S. Keith, and E. K. Urban, eds. *The Birds of Africa.* Vol. 3, *Parrots to Woodpeckers.* London: Academic Press, 1988.

Fry, C H., and S. Keith, eds. *The Birds of Africa.* Vol. 6. *Picathartes to Oxpeckers.* London: Academic Press, 2000.

Fuller, Errol. *Extinct Birds.* Rev. ed. Ithaca, N.Y.: Comstock Pub., 2001.

Gehlbach, Frederick R. *The Eastern Screech Owl: Life History, Ecology, and Behavior in the Suburbs and Countryside.* College Station: Texas A & M University Press, 1994.

Gibbs, D., E. Barnes, and J. Cox. *Pigeons and Doves: A Guide to the Pigeons and Doves of the World.* Robertsbridge: Pica Press, 2001.

Gill, F. B. *Ornithology.* 2nd edition. New York: W. H. Freeman, 1995.

Grant, P. R. *Ecology and Evolution of Darwin's Finches.* Princeton: Princeton University Press, 1986.

Hagemeijer, Ward J. M., and Michael J. Blair, eds. *The EBCC Atlas of European Breeding Birds: Their Distribution and Abundance.* London: T. & A. D. Poyser, 1997.

Hancock, J. A., J. A. Kushlan, and M. P. Kahl. *Storks, Ibises, and Spoonbills of the World.* New York: Academic Press, 1992.

Hancock, J. A., and J. A. Kushlan. *The Herons Handbook.* New York: Harper and Row, 1984.

Harris, Tony, and Kim Franklin. *Shrikes and Bush-shrikes: Including Wood-shrikes, Helmet-shrikes, Flycatcher-shrikes, Philentomas, Batises and Wattle-eyes.* London: Christopher Helm, 2000.

Harrap, S., and D. Quinn. *Chickadees, Tits, Nuthatches, and Treecreepers.* Princeton: Princeton University Press, 1995.

Heinrich, B. *Ravens in Winter.* New York: Summit Books, 1989.

Higgins, P. J., and S. J. J. F. Davies, eds. *The Handbook of Australian, New Zealand and Antarctic Birds.* Vol. 3, *Snipe to Pigeons.* Melbourne: Oxford University Press, 1996.

Higgins, P. J., ed. *The Handbook of Australian, New Zealand and Antarctic Birds.* Vol. 4, *Parrots to Dollarbirds.* Melbourne: Oxford University Press, 1999.

Hilty, S. L., and W. L. Brown. *A Guide to the Birds of Colombia.* Princeton, N. J.: Princeton University Press, 1986.

Holyoak, D.T. *Nightjars and Their Allies: The Caprimulgiformes. Bird Families of the World,* no. 7. Oxford: Oxford University Press, 2001.

Howard, R., and A. Moore. *A Complete Checklist of the Birds of the World.* 2nd edition. London: Macmillan, 1991.

Howell, S. N. G., and S. Webb. *A Guide to the Birds of Mexico and Northern Central America.* Oxford: Oxford University Press, 1995.

Howell, S. N. G. *Hummingbirds of North America: The Photographic Guide.* San Diego: Academic Press, 2002.

Isler, M. L., and P. R. Isler. *The Tanagers: Natural History, Distribution, and Identification.* Washington, D.C.: Smithsonian Institution Press, 1987.

Jaramillo, A., and P. Burke. *New World Blackbirds: The Icterids.* Princeton: Princeton University Press, 1999.

Jehl, Joseph R., Jr. *Biology of the Eared Grebe and Wilson's Phalarope in the Nonbreeding Season: A study of Adaptations to Saline Lakes. Studies in Avian Biology,* no. 12. San Diego: Cooper Ornithological Society, 1988.

Johnsgard, Paul A. *Bustards, Hemipodes, and Sandgrouses, Birds of Dry Places.* Oxford: Oxford University Press, 1991.

Johnsgard, Paul A. *Cormorants, Darters, and Pelicans of the World.* Washington, D.C.: Smithsonian Institution Press, 1993.

Johnsgard, Paul A. *Cranes of the World.* Bloomington: Indiana University Press, 1983.

Johnsgard, Paul A. *Diving Birds of North America.* Lincoln: University of Nebraska Press, 1987.

Johnsgard, Paul A. *The Hummingbirds of North America.* 2nd ed. Washington, D.C.: Smithsonian Institution Press, 1997.

Johnsgard, P. A. *The Plovers, Sandpipers, and Snipes of the World.* Lincoln: University of Nebraska Press, 1981.

Johnsgard, Paul A. *Trogons and Quetzals of the World.* Washington, D.C.: Smithsonian Institution Press, 2000.

Johnsgard, Paul A., and Montserrat Carbonell. *Ruddy Ducks and other Stifftails: Their Behavior and Biology.* Norman: University of Oklahoma Press, 1996.

Jones, D. N., R. W. R. J. Dekker, and C. S. Roselaar. *The Megapodes. Bird Families of the World,* no. 3. Oxford: Oxford University Press, 1995.

Juniper, Tony, and Mike Parr. *Parrots: A Guide to the Parrots of the World.* Sussex: Pica Press, 1998.

Kear, J., and N. Düplaix-Hall, eds. *Flamingos.* Berkhamsted: T. & A. D. Poyser, 1975.

Keith, S., E. K. Urban, and C. H. Fry, eds. *The Birds of Africa.* Vol. 4, *Broadbills to Chats.* London: Academic Press, 1992.

Kemp, A. *The Hornbills. Bird Families of the World,* no. 1. Oxford: Oxford University Press, 1995.

Kennedy, Robert S., et al. *A Guide to the Birds of the Philippines.* Oxford: Oxford University Press, 2000.

Lambert, F., and M. Woodcock. *Pittas, Broadbills and Asities.* Sussex: Pica Press, 1996.

Lefranc, Norbert, and Tim Worfolk. *Shrikes: A Guide to the Shrikes of the World.* Robertsbridge: Pica Press, 1997.

Lenz, Norbert. *Evolutionary Ecology of the Regent Bowerbird Sericulus chrysocephalus.* Special Issue of *Ecology of Birds,* Vol. 22, Supplement. Ludwigsburg, 1999.

MacKinnon, John R., and Karen Phillipps. *A Field Guide to the Birds of China.* Oxford Ornithology Series. Oxford; New York: Oxford University Press, 2000.

Madge, S. *Crows and Jays: A Guide to the Crows, Jays, and Magpies of the World.* New York: Houghton Mifflin, 1994.

Madge, Steve, and Phil McGowan. *Pheasants, Partridges and Grouse: A Guide to the Pheasants, Partridges, Quails, Grouse, Guineafowl, Buttonquails and Sandgrouse of the World.* London: Christopher Helm, 2002.

Marchant, S., and P. Higgins, eds. *The Handbook of Australian, New Zealand and Antarctic Birds.* Vol. 1, parts A , B, *Ratites to Ducks.* Melbourne: Oxford University Press, 1990.

Marchant, S., and P. Higgins, eds. *The Handbook of Australian, New Zealand and Antarctic Birds.* Vol. 2, *Raptors to Lapwings.* Melbourne: Oxford University Press, 1993

Higgins, P. J., J. M. Peter, and W. K. Steele. *The Handbook of Australian, New Zealand and Antarctic Birds.* Vol. 5, *Tyrant-Flycatcher to Chats.* Melbourne: Oxford University Press, 2001.

Marzluff, J. M., R. Bowman, and R. Donnelly, eds. *Avian Ecology and Conservation in an Urbanizing World.* Boston: Kluwer Academic Publishers, 2001.

Matthysen, E. The *Nuthatches.* London: T. A. & D. Poyser, 1998.

Mayfield, H. *The Kirtland's Warbler.* Bloomfield Hills, Michigan: Cranbrook Institute of Science, 1960.

Mayr, E. *The Birds of Northern Melanesia: Speciation, Ecology, and Biogeography.* Oxford: Oxford University Press, 2001.

McCabe, Robert A. *The Little Green Bird: Ecology of the Willow Flycatcher.* Madison, Wis.: Rusty Rock Press, 1991.

Mindell, D. P., ed. *Avian Molecular Evolution and Systematics.* New York: Academic Press, 1997.

Morse, D. H. *American Warblers, An Ecological and Behavioral Perspective.* Cambridge: Harvard University Press, 1989.

Mundy, P., D., et al. *The Vultures of Africa.* London: Academic Press, 1992.

Nelson, J. B. *The Sulidae: Gannets and Boobies.* Oxford: Oxford University Press, 1978.

Nelson, Bryan. *The Atlantic Gannet.* 2nd ed. Great Yarmouth: Fenix, 2002.

Olsen, Klaus Malling, and Hans Larsson. *Skuas and Jaegers: A Guide to the Skuas and Jaegers of the World.* New Haven: Yale University Press, 1997.

Olsen, Klaus Malling, and Hans Larsson. *Gulls of Europe, Asia and North America.* London: Christopher Helm, 2001.

Ortega, Catherine P. *Cowbirds and Other Brood Parasites.* Tucson: University of Arizona Press, 1998.

Padian, K., ed. *The Origin of Birds and the Evolution of Flight.* Memoirs of the California Academy of Sciences, no. 8. San Francisco: California Academy of Sciences, 1986.

Paul, Gregory S. *Dinosaurs of the Air: the Evolution and Loss of Flight in Dinosaurs and Birds.* Baltimore: Johns Hopkins University Press, 2001.

Poole, Alan F., P. Stettenheim, and Frank B. Gill, eds. *The Birds of North America.* 15 vols. to date. Philadelphia: The Academy of Natural Sciences; Washington, D.C.: American Ornithologists' Union, 1992-.

Raffaele, H., et al. *Birds of the West Indies.* London: Christopher Helm, A. & C. Black, 1998.

Ratcliffe, D. *The Peregrine Falcon.* 2nd ed. London: T. & A. D. Poyser, 1993.

Restall, R. *Munias and Mannikins.* Sussex: Pica Press, 1996.

Ridgely, R. S., and G. A. Gwynne, Jr. *A Guide to the Birds of Panama: with Costa Rica, Nicaragua, and Honduras.* 2nd ed. Princeton: Princeton University Press. 1989.

Ridgely, R. S., and G. Tudor. *The Birds of South America.* Vol. 1, *The Oscine Passerines.* Austin: University of Texas Press, 1989.

Ridgely, R. S., and G. Tudor. *The Birds of South America.* Vol. 2, *The Suboscine Passerines.* Austin: University of Texas Press, 1994.

Ridgely, Robert S., and Paul J. Greenfield. *The Birds of Ecuador.* Vol. 1, *Status, Distribution, and Taxonomy.* Ithaca, NY: Comstock Pub., 2001.

Ridgely, Robert S., and Paul J. Greenfield. *The Birds of Ecuador.* Vol. 2, *Field Guide.* Ithaca, NY: Comstock Pub., 2001.

Rising, J. D. *Guide to the Identification and Natural History of the Sparrows of the United States and Canada.* New York: Academic Press, 1996.

Rowley, I., and E. Russell. *Fairy-Wrens and Grasswrens. Bird Families of the World*, no. 4. Oxford: Oxford University Press, 1997.

Scott, J. M., S. Conant, and C. van Riper, III, eds. *Evolution, Ecology, Conservation, and Management of Hawaiian Birds: A Vanishing Avifauna. Studies in Avian Biology*, no. 22. Camarillo, CA : Cooper Ornithological Society, 2001.

Searcy, W. A., and K. Yasukawa. *Polygyny and Sexual Selection in Red-winged Blackbirds. Monographs in Behavior and Ecology.* Princeton: Princeton University Press, 1995.

Shirihai, H., G. Gargallo, and A. J. Helbig. *Sylvia Warblers.* Edited by G. M. Kirwan and L. Svensson. Princeton: Princeton University Press, 2001.

Short, L. L. *Woodpeckers of the World.* Greenville, Del.: Delaware Natural History Museum, 1984.

Short, L. L., and J. F.M. Horne. *Toucans, Barbets and Honeyguides: Ramphastidae, Capitonidae, and Indicatoridae. Bird Families of the World*, no. 8. Oxford: New York: Oxford University Press, 2001.

Sibley, C. G., and J. E. Ahlquist. *Phylogeny and Classification of Birds: A Study in Molecular Evolution.* New Haven: Yale University Press, 1990.

Sibley, C. G., and B. L. Monroe, Jr. *Distribution and Taxonomy of Birds of the World.* New Haven: Yale University Press, 1990.

Sibley, C. G., and B. L. Monroe, Jr. *A Supplement to Distribution and Taxonomy of Birds of the World.* New Haven: Yale University Press, 1993.

Sibley, David Allen. *The Sibley Guide to Birds.* New York: Knopf, 2000.

Sick, H. *Birds in Brazil, a Natural History.* Princeton University Press, 1993.

Simmons, Robert Edward. *Harriers of the World: Their Behaviour and Ecology. Oxford Ornithology Series*, no. 11. Oxford: Oxford University Press, 2000.

Sinclair, I., and O. Langrand. *Birds of the Indian Ocean Islands: Madagascar, Mauritius, Reunion, Rodrigues, Seychelles and the Comoros.* New Holland. 1998.

Skutch, A. F. *Antbirds and Ovenbirds: Their Lives and Homes.* Austin: University of Texas Press, 1996.

Skutch, A. F. *Orioles, Blackbirds, and Their Kin: A Natural History.* Tucson: University of Arizona Press, 1996.

Smith, S. M. *The Black-capped Chickadee: Behavioral Ecology and Natural History.* Ithaca: Cornell University Press, 1991.

Snow, D. W. *The Cotingas, Bellbirds, Umbrella-birds and Their Allies in Tropical America.* London: Brit. Museum (Nat. Hist.), 1982.

Snyder, Noel F. R., and Helen Snyder. *The California Condor: A Saga of Natural History and Conservation*. San Diego, Calif.: Academic Press, 2000.

Stacey, P. B., and W. D. Koenig, eds. *Cooperative Breeding in Birds*. Cambridge: Cambridge University Press, 1989.

Stattersfield, Alison J., et al. eds. *Threatened Birds of the World: The Official Source for Birds on the IUCN Red List*. Cambridge: BirdLife International, 2000.

Stiles, F. G., and A. F. Skutch. *A Guide to the Birds of Costa Rica*. Ithaca: Comstock, 1989.

Stokes, Donald W., and Lillian Q. Stokes. *A Guide to Bird Behavior*. 3 vols. Boston: Little, Brown, 1979-1989.

Stolz, D.F., J. W. Fitzpatrick, T. A. Parker III, and D. K. Moskovits. *Neotropical Birds, Ecology and Conservation*. Chicago: University of Chicago Press, 1996.

Summers-Smith, J. D. *The Sparrows: A Study of the Genus Passer*. Calton: T. & A. D. Poyser, 1988.

Taylor, B., and B. van Perlo. *Rails: A Guide to the Rails, Crakes, Gallinules and Coots of the World*. Sussex: Pica Press, 1998.

Terres, John K. *The Audubon Society Encyclopedia of North American Birds*. New York: Knopf, 1980.

Tickell, W. L. N. *Albatrosses*. Sussex: Pica Press, 2000.

Todd, F. S. *Natural History of the Waterfowl*. Ibis Publishing Co., San Diego Natural History Museum, 1996.

Turner, A., and C. Rose. *A Handbook to the Swallows and Martins of the World*. London: Christopher Helm, 1989.

Tyler, Stephanie J., and Stephen J. *The Dippers*. London: T. & A. D. Poyser, 1994.

Urban, E. K., C. H. Fry, and S. Keith, eds. *The Birds of Africa*. Vol. 2, *Gamebirds to Pigeons*. London: Academic Press, 1986.

Urban, E. K., C. H. Fry, and S. Keith, eds. *The Birds of Africa*. Vol. 5, *Thrushes to Puffback Flycatchers*. London: Academic Press, 1997.

van Rhijn, J. G. *The Ruff: Individuality in a Gregarious Wading Bird*. London: T. & A. D. Poyser, 1991.

Voous, Karel H. *Owls of the Northern Hemisphere*. Cambridge, Mass.: MIT Press, 1988.

Warham, J. *Behaviour and Population Ecology of the Petrels*. New York: Academic Press,1996.

Williams, T. D. *The Penguins. Bird Families of the World*, no. 2. Oxford: Oxford University Press, 1995.

Winkler, H., D. A. Christie, and D. Nurney. *Woodpeckers: A Guide to the Woodpeckers, Piculets and Wrynecks of the World*. Sussex: Pica Press, 1995.

Woolfenden, G. E., and J. W. Fitzpatrick. *The Florida Scrub Jay: Demography of a Cooperative-breeding Bird. Monographs in Population Biology*, no. 20. Princeton: Princeton University Press, 1984.

Zimmerman, D. A., D. A. Turner, and D. J. Pearson. *Birds of Kenya and Northern Tanzania*. London: Christopher Helm, 1996.

Compiled by Janet Hinshaw, Bird Division Collection Manager, University of Michigan Museum of Zoology

Organizations

African Bird Club
  Wellbrook Court, Girton Road
  Cambridge, Cambridgeshire CB3 0NA
  United Kingdom
  Phone: +44 1 223 277 318
  Fax: +44-1-223-277-200
  <http://www.africanbirdclub.org>

African Gamebird Research, Education and Development
(AGRED)
  P.O. Box 1191
  Hilton, KwaZulu-Natal 3245
  South Africa
  Phone: +27-33-343-3784

African-Eurasian Migratory Waterbird Agreement (AEWA)
  UN Premises in Bonn, Martin Luther-King St.
  Bonn D-53175 Germany
  <http://www.wcmc.org.uk/AEWA>

American Ornithologists' Union
  Suite 402, 1313 Dolley Madison Blvd
  McLean, VA 22101
  USA
  <http://www.aou.org>

American Zoo and Aquarium Association
  8403 Colesville Road
  Suite 710
  Silver Spring, Maryland 20910
  <http://www.aza.org>

Association for BioDiversity Information
  1101 Wilson Blvd., 15th Floor
  Arlington, VA 22209
  USA
  <http://www.infonatura.org/>

Association for Parrot Conservation
  Centro de Calidad Ambiental ITESM Sucursal de Correos
  J., C.P. 64849
  Monterrey, N.L.
  Mexico

Australasian Raptor Association
  415 Riversdale Road
  Hawthorn East, Victoria 3123
  Australia
  Phone: +61 3 9882 2622
  Fax: +61 3 9882 2677
  <http://www.tasweb.com.au/ara/index.htm>

Australian National Wildlife Collection
  GPO Box 284
  Canberra, ACT 2601
  Australia
  Phone: +61 2 6242 1600
  Fax: +61-2-6242-1688

The Bird Conservation Society of Thailand
  69/12 Rarm Intra 24
  Jarakhebua Lat Phrao, Bangkok 10230
  Thailand
  Phone: 943-5965
  <http://www.geocities.com/TheTropics/Harbor/7503/
  ruang_nok/princess_bird.html>

BirdLife International
  Wellbrook Court, Girton Road
  Cambridge, Cambridgeshire CB3 0NA
  United Kingdom
  Phone: +44 1 223 277 318
  Fax: +44-1-223-277-200
  <http://www.birdlife.net>

BirdLife International Indonesia Programme
  P. O. Box 310/Boo
  Bogor
  Indonesia
  Phone: +62 251 357222
  Fax: +62 251 357961
  <http://www.birdlife-indonesia.org>

BirdLife International, Panamerican Office
  Casilla 17-17-717
  Quito
  Ecuador
  Phone: +593 2 244 3261
  Fax: +593 2 244 3261
  <http://www.latinsynergy.org/birdlife.html>

BirdLife South Africa
  P. O. Box 515
  Randburg 2125
  South Africa
  Phone: +27-11-7895188
  <http://www.birdlife.org.za>

Birds Australia
  415 Riversdale Road
  Hawthorn East, Victoria 3123
  Australia
  Phone: +61 3 9882 2622
  Fax: +61 3 9882 2677
  <http://www.birdsaustralia.com.au>

Birds Australia Parrot Association, Birds Australia
415 Riversdale Road
Hawthorn East, Victoria 3123
Australia
Phone: +61 3 9882 2622
Fax: +61 3 9882 2677
<http://www.birdsaustralia.com.au>

The Bishop Museum
1525 Bernice Street
Honolulu, HI 96817-0916
Phone: (808) 847-3511
<http://www.bishopmuseum.org>

British Trust for Ornithology
The Nunnery
Thetford, Norfolk IP24 2PU
United Kingdom
Phone: +44 (0) 1842 750050
Fax: +44 (0) 1842 750030
<http://www.bto.org>

Center for Biological Diversity
P.O. Box 710
Tucson, AZ 85702-0701
USA
Phone: (520) 623-5252
Fax: (520) 623-9797
<http://www.biologicaldiversity.org/swcbd/index.html>

Coraciiformes Taxon Advisory Group
<http://www.coraciiformestag.com>

The Cracid Specialist Group
PO Box 132038
Houston, TX 77219-2038
USA
Phone: (713) 639-4776
<http://www.angelfire.com/ca6/cracid>

Department of Ecology and Environmental Biology, Cornell
University
E145 Corson Hall
Ithaca, NY 14853-2701
USA
Phone: (607) 254-4201
<http://www.es.cornell.edu/winkler/botw/fringillidae.html>

Department of Ecology and Evolutionary Biology, Tulane
University
310 Dinwiddie Hall
New Orleans, LA 70118-5698
USA
Phone: (504) 865-5191
<http://www.tulane.edu/
eeob/Courses/Heins/Evolution/lecture17.html>

Department of Zoology, University of Toronto
25 Harbord Street
Toronto, Ontario M5S 3G5
Canada
Phone: (416) 978-3482
Fax: (416) 978-8532
<http://www.zoo.utoronto.ca>

Ducks Unlimited, Inc
One Waterfowl Way
Memphis, TN 38120
USA

Phone: (800) 453-8257
Fax: (901) 758-3850
<http://www.ducks.org>

Emu Farmers Federation of Australia
P.O Box 57
Wagin, Western Australia 6315
Australia
Phone: +61 8 9861 1136

Game Conservancy Trust
Fordingbridge, Hampshire SP6 1EF
United Kingdom
Phone: +44 1425 652381
Fax: +44 1425 651026
<http://www.gct.org.uk>

Gamebird Research Programme, Percy FitzPatrick Institute,
University of Cape Town
Private Bag
Rondebosch, Western Cape 7701
South Africa
Phone: +27 21 6503290
Fax: +27 21 6503295
<http://www.uct.ac.za/depts/fitzpatrick>

Haribon Foundation for the Conservation of Natural Resources
9A Malingap Cot, Malumanay Streets, Teachers' Village,
1101 Diliman
Quezon City
Philippines
Phone: +63 2 9253332
<http://www.haribon.org.ph>

The Hawk and Owl Trust
11 St Marys Close
Newton Abbot, Abbotskerswell, Devon TQ12 5QF
United Kingdom
Phone: +44 (0)1626 334864
Fax: +44 (0)1626 334864
<http://www.hawkandowltrust.org>

Herons Specialist Group
Station Biologique de la Tour du Valat
Le Sambuc, Arles 13200
France
Phone: +33-4-90-97-20-13
Fax: 33-4-90-97-29-19
<http://www.tour-du-valat.com>

Hornbill Research Foundation
c/o Department of Microbiology, Faculty of Science,
Mahidol University, Rama 6 Rd
Bangkok 10400
Thailand
Phone: +66 22 460 063, ext. 4006

International Crane Foundation
P.O. Box 447
Baraboo, WI 53913-0447
USA
Phone: (608) 356-9462
Fax: (608) 356-9465
<http://www.savingcranes.org>

International Shrike Working Group
"Het Speihuis," Speistraat, 17
Sint-Lievens-Esse (Herzele), B-9550

Belgium
Phone: +32 54 503 789

International Species Inventory System
<http://www.isis.org>

International Touraco Society
Brackenhurst, Grange Wood
Netherseal, Nr Swadlincote, Derbyshire DE12 8BE
United Kingdom
Phone: +44 (0)1283 760541

International Waterbird Census
<http://www.wetlands.org>

IUCN Species Survival Commission
219c Huntingdon Road
Cambridge, Cambridgeshire CB3 0DL
United Kingdom
<http://www.iucn.org/themes/ssc>

IUCN–The World Conservation Union
Rue Mauverney 28
Gland 1196
Switzerland
Phone: +41-22-999-0001
Fax: +41-22-999-0025
<http://www.iucn.org>

IUCN–World Conservation Union, USA Multilateral Office
1630 Connecticut Avenue
Washington, DC 20009
USA
Phone: (202) 387-4826
<http://www.iucn.org/places/usa/inter.html>

IUCN/SSC Grebes Specialist Group
Copenhagen DK 2100
Denmark
Phone: +45 3 532 1323
Fax: +45-35321010
<http://www.iucn.org>

Japanese Association for Wild Geese Protection
Minamimachi 16
Wakayangi 989-5502
Japan
Phone: +81 228 32 2004
Fax: +81 228 32 2004
<http://www.japwgp.org>

Ligue pour la Protection des Oiseaux
La Corderie Royale, B.P. 263
17305 Rochefort cedex
France
Phone: +33 546 821 234
Fax: +33 546 839 586
<http://www.lpo-birdlife.asso.fr>

Loro Parque Fundación
Loro Parque S.A. 38400 Puerto de la Cruz
Tenerife, Canary Islands
Spain

National Audubon Society
700 Broadway
New York, NY 10003
USA

Phone: (212) 979-3000
Fax: (212) 978-3188
<http://www.Audubon.org>

National Audubon Society Population & Habitat Program
1901 Pennsylvania Ave. NW, Suite 1100
Washington, DC 20006
USA
Phone: (202) 861-2242
<http://www.audubonpopulation.org>

Neotropical Bird Club
c/o The Lodge
Sandy, Bedfordshire SG19 2DL
United Kingdom

Oriental Bird Club, American Office
4 Vestal Street
Nantucket, MA 02554
USA
Phone: (508) 228-1782
<http://www.orientalbirdclub.org>

Ornithological Society of New Zealand
P.O. Box 12397
Wellington, North Island
New Zealand
<http://osnz.org.nz>

Pacific Island Ecosystems Research Center
3190 Maile Way, St. John Hall, Room 408
Honolulu, HI 96822
USA
Phone: (808) 956-5691
Fax: (808) 956-5687
<http://biology.usgs.gov/pierc/piercwebsite.htm>

ProAves Peru
P.O. Box 07
Piura
Peru

Raptor Conservation Group, Endangered Wildlife Trust
Private Bag X11
Parkview, Gauteng 2122
South Africa
Phone: +27-11-486-1102
Fax: +27-11-486-1506
<http://www.ewt.org.za>

Raptor Research Foundation
P.O. Box 1897, 810 E. 10th Street
Lawrence, KS 66044-88973
USA
<http://biology.biosestate.edu/raptor>

Research Centre for African Parrot Conservation Zoology and Entomology Department
Private Bag X01
Scottsville 3201
Natal Republic of South Africa

Roberts VII Project, Percy FitzPatrick Institute of African Ornithology, University of Cape Town
Rondebosch 7701
South Africa
Fax: (021) 650 3295
<http://www.uct.ac.za/depts/fitzpatrick/docs/r549.html>

Royal Society for the Protection of Birds
Admail 975 Freepost ANG 6335, The Lodge
Sandy, Bedfordshire SG19 2TN
United Kingdom
<http://www.rspb.org.uk>

Ruffed Grouse Society
451 McCormick Rd
Coraopolis, PA 15108
Phone: (888) 564-6747
Fax: (412) 262-9207
<http://www.ruffedgrousesociety.org>

Smithsonian Migratory Bird Center, Smithsonian National
Zoological Park
3001 Connecticut Avenue, NW
Washington, DC 20008
USA
Phone: (202) 673-4800
<http://www.natzoo.si.edu>

The Songbird Foundation
2367 Eastlake Ave. East
Seattle, WA 98102
USA
Phone: (206) 374-3674
Fax: (206) 374-3674
<http://www.songbird.org>

University of Michigan
3019 Museum of Zoology, 1109 Geddes Ave
Ann Arbor, MI 48109-1079
USA
Phone: (734) 647-2208
Fax: (734) 763-4080
<http://www.ummz.lsa.umich.edu/birds/index.html>

Wader Specialist Group, Mr. David Stroud
Monkstone House, City Road
Peterborough PE1 1JY
United Kingdom
Phone: +44 1733 866/810
Fax: +44 1733 555/448

Wader Study Group, The National Centre for Ornithology
The Nunnery
Thetford, Norfolk JP24 2PU
United Kingdom

Waterbird Society
National Museum of Natural History, Smithsonian
Institution
Washington, DC 20560
USA
<http://www.nmnh.si.edu/BIRDNET/cws>

Western Hemisphere Shorebird Reserve Network (WHSRN)
Manomet Center for Conservation Science, P O Box 1770
Manomet, MA 02345
USA
Phone: (508) 224-6521
Fax: (508) 224-9220
<http://www.manomet.org/WHSRN/index.html>

Wetlands International
Droevendaalsesteeg 3A

Wageningen 6700 CA
The Netherlands
Phone: +31 317 478884
Fax: +31 317 478885
<http://www.wetlands.agro.nl>

Wetlands International (the Americas)
7 Hinton Avenue North, Suite 200
Ottawa, Ontario K1Y 4P1
Canada
Phone: (613) 722-2090
<http://www.wetlands.org>

Wetlands International/Survival Service Commission Flamingo
Specialist Group
c/o Station Biologique de la Tour du Valat
Le Sambuc, Arles 13200
France

Wildfowl and Wetlands Trust
Slimbridge, Glos GL2 7BT
United Kingdom
Phone: +44 01453 891900

Woodcock and Snipe Specialist Group
Director, European Wildlife Research Institute
Bonndorf, Glashuette D-79848
Germany
Phone: 949 7653 1891
Fax: 949 7653 9269

Woodhoopoe Research Project, FitzPatrick Institute of African
Ornithology, University of Cape Town
P.O. Rondebosch
Cape Town, Western Cape 7700
South Africa
Phone: +27 (0)21 650-3290
Fax: +27-21-650-3295
<http://www.fitztitute.uct.ac.za>

Working Group on Birds in Madagascar and the Indian Ocean
Islands. World Wide Fund for Nature
Antananarivo 101 BP 738
Madagascar
Phone: +261 3207 80806

Working Group on International Wader and Waterfowl
Research (WIWO)
Stichting WIWO, c/o P O Box 925
Zeist 3700 AX
The Netherlands
<http://www.wiwo-international.org>

World Center for Birds of Prey, The Peregrine Fund
566 West Flying Hawk Lane
Boise, ID 83709
USA
Phone: (208) 362-3716
Fax: (208) 362-2376
<http://www.peregrinefund.org>

World Parrot Trust
Glanmor House
Hayle, Cornwall TR27 4HB
United Kingdom
<http://www.worldparrottrust.org>

World Pheasant Association
P. O. Box 5, Lower Basildon St
Reading RG8 9PF
United Kingdom
Phone: +44 1 189 845 140
Fax: +44 118 984 3369
<http://www.pheasant.org.uk>

World Working Group on Birds of Prey and Owl
P.O. Box 52
Towcester NN12 7ZW
United Kingdom
Phone: +44 1 604 862 331
Fax: +44 1 604 862 331
<http://www.Raptors-International.de>

WPA/BirdLife/SSC Megapode Specialist Group
c/o Department of Ornithology, National Museum of
Natural History, P.O. Box 9517
Leiden 2300 RA
The Netherlands

WPA/BirdLife/SSC Partridge, Quail, and Francolin Specialist
Group
c/o World Pheasant Association, PO Box 5
Lower Basildon, Reading RG8 9PF
United Kingdom
Phone: +44 1 189 845 140
Fax: +118 9843369
<http:/www.pheasant.org.uk>
PQF: <http://www.gct.org.uk/pqf>

Dr. Fritz Dieterlen
Zoological Research Institute, A.
Koenig Museum
Bonn, Germany

Dr. Rolf Dircksen
Professor, Pedagogical Institute
Bielefeld, Germany

Josef Donner
Instructor of Biology
Katzelsdorf, Austria

Dr. Jean Dorst
Professor, National Museum of
Natural History
Paris, France

Dr. Gerti Dücker
Professor and Chief Curator,
Zoological Institute, University of
Münster
Münster, Germany

Dr. Michael Dzwillo
Zoological Institute and Museum,
University of Hamburg
Hamburg, Germany

Dr. Irenäus Eibl-Eibesfeldt
Professor and Director, Institute of
Human Ethology, Max Planck
Institute for Behavioral Physiology
Percha/Starnberg, Germany

Dr. Martin Eisentraut
Professor and Director, Zoological
Research Institute and A. Koenig
Museum
Bonn, Germany

Dr. Eberhard Ernst
Swiss Tropical Institute
Basel, Switzerland

R. D. Etchecopar
Director, National Museum of
Natural History
Paris, France

Dr. R. A. Falla
Director, Dominion Museum
Wellington, New Zealand

Dr. Hubert Fechter
Curator, Lower Animals, Zoological
Collection of the State of Bavaria
Munich, Germany

Dr. Walter Fiedler
Docent, University of Vienna, and
Director, Schönbrunn Zoo
Vienna, Austria

Wolfgang Fischer
Inspector of Animals, Animal Park
Berlin, Germany

Dr. C. A. Fleming
Geological Survey Department of
Scientific and Industrial Research
Lower Hutt, New Zealand

Dr. Hans Frädrich
Zoological Garden
Berlin, Germany

Dr. Hans-Albrecht Freye
Professor and Director, Biological
Institute of the Medical School
Halle a.d.S., Germany

Günther E. Freytag
Former Director, Reptile and
Amphibian Collection, Museum of
Cultural History in Magdeburg
Berlin, Germany

Dr. Herbert Friedmann
Director, Los Angeles County
Museum of Natural History
Los Angeles, California, U.S.A.

Dr. H. Friedrich
Professor, Overseas Museum
Bremen, Germany

Dr. Jan Frijlink
Zoological Laboratory, University of
Amsterdam
Amsterdam, The Netherlands

Dr. H.C. Karl Von Frisch
Professor Emeritus and former
Director, Zoological Institute,
University of Munich
Munich, Germany

Dr. H. J. Frith
C.S.I.R.O. Research Institute
Canberra, Australia

Dr. Ion E. Fuhn
Academy of the Roumanian Socialist
Republic, Trajan Savulescu Institute of
Biology
Bucharest, Romania

Dr. Carl Gans
Professor, Department of Biology,
State University of New York at
Buffalo
Buffalo, New York, U.S.A.

Dr. Rudolf Geigy
Professor and Director, Swiss Tropical
Institute
Basel, Switzerland

Dr. Jacques Gery
St. Genies, France

Dr. Wolfgang Gewalt
Director, Animal Park
Duisburg, Germany

Dr. H.C. Dr. H.C. Viktor Goerttler
Professor Emeritus, University of Jena
Jena, Germany

Dr. Friedrich Goethe
Director, Institute of Ornithology,
Heligoland Ornithological Station
Wilhelmshaven, Germany

Dr. Ulrich F. Gruber
Herpetological Section, Zoological
Research Institute and A. Koenig
Museum
Bonn, Germany

Dr. H. R. Haefelfinger
Museum of Natural History
Basel, Switzerland

Dr. Theodor Haltenorth
Director, Mammalology, Zoological
Collection of the State of Bavaria
Munich, Germany

Barbara Harrisson
Sarawak Museum, Kuching, Borneo
Ithaca, New York, U.S.A.

Dr. Francois Haverschmidt
President, High Court (retired)
Paramaribo, Suriname

Dr. Heinz Heck
Director, Catskill Game Farm
Catskill, New York, U.S.A.

Dr. Lutz Heck
Professor (retired), and Director,
Zoological Garden, Berlin
Wiesbaden, Germany

Dr. H.C. Heini Hediger
Director, Zoological Garder
Zurich, Switzerland

Dr. Dietrich Heinemann
Director, Zoological Garden, Münster
Dörnigheim, Germany

Dr. Helmut Hemmer
Institute for Physiological Zoology,
University of Mainz
Mainz, Germany

Dr. W. G. Heptner
Professor, Zoological Museum,
University of Moscow
Moscow, Russia

Dr. Konrad Herter
Professor Emeritus and Director
(retired), Zoological Institute, Free
University of Berlin
Berlin, Germany

Dr. Hans Rudolf Heusser
Zoological Museum, University of
Zurich
Zurich, Switzerland

Dr. Emil Otto Höhn
Associate Professor of Physiology,
University of Alberta
Edmonton, Canada

Dr. W. Hohorst
Professor and Director,
Parasitological Institute, Farbwerke
Hoechst A.G.
Frankfurt-Höchst, Germany

Dr. Folkhart Hückinghaus
Director, Senckenbergische Anatomy,
University of Frankfurt a.M.
Frankfurt a.M., Germany

Francois Hüe
National Museum of Natural
History
Paris, France

Dr. K. Immelmann
Professor, Zoological Institute,
Technical University of Braunschweig
Braunschweig, Germany

Dr. Junichiro Itani
Kyoto University
Kyoto, Japan

Dr. Richard F. Johnston
Professor of Zoology, University of
Kansas
Lawrence, Kansas, U.S.A.

Otto Jost
Oberstudienrat, Freiherr-vom-Stein
Gymnasium
Fulda, Germany

Dr. Paul Kähsbauer
Curator, Fishes, Museum of Natural
History
Vienna, Austria

Dr. Ludwig Karbe
Zoological State Institute and
Museum
Hamburg, Germany

Dr. N. N. Kartaschew
Docent, Department of Biology,
Lomonossow State University
Moscow, Russia

Dr. Werner Kästle
Oberstudienrat, Gisela Gymnasium
Munich, Germany

Dr. Reinhard Kaufmann
Field Station of the Tropical Institute,
Justus Liebig University, Giessen,
Germany
Santa Marta, Colombia

Dr. Masao Kawai
Primate Research Institute, Kyoto
University
Kyoto, Japan

Dr. Ernst F. Kilian
Professor, Giessen University and
Catedratico Universidad Austral,
Valdivia-Chile
Giessen, Germany

Dr. Ragnar Kinzelbach
Institute for General Zoology,
University of Mainz
Mainz, Germany

Dr. Heinrich Kirchner
Landwirtschaftsrat (retired)
Bad Oldesloe, Germany

Dr. Rosl Kirchshofer
Zoological Garden, University of
Frankfort a.M.
Frankfurt a.M., Germany

Dr. Wolfgang Klausewitz
Curator, Senckenberg Nature
Museum and Research Institute
Frankfurt a.M., Germany

Dr. Konrad Klemmer
Curator, Senckenberg Nature
Museum and Research Institute
Frankfurt a.M., Germany

Dr. Erich Klinghammer
Laboratory of Ethology, Purdue
University
Lafayette, Indiana, U.S.A.

Dr. Heinz-Georg Klös
Professor and Director, Zoological
Garden
Berlin, Germany

Ursula Klös
Zoological Garden
Berlin, Germany

Dr. Otto Koehler
Professor Emeritus, Zoological
Institute, University of Freiburg
Freiburg i. BR., Germany

Dr. Kurt Kolar
Institute of Ethology, Austrian
Academy of Sciences
Vienna, Austria

Dr. Claus König
State Ornithological Station of Baden-
Württemberg
Ludwigsburg, Germany

Dr. Adriaan Kortlandt
Zoological Laboratory, University of
Amsterdam
Amsterdam, The Netherlands

Dr. Helmut Kraft
Professor and Scientific Councillor,
Medical Animal Clinic, University of
Munich
Munich, Germany

Dr. Helmut Kramer
Zoological Research Institute and A.
Koenig Museum
Bonn, Germany

Dr. Franz Krapp
Zoological Institute, University of
Freiburg
Freiburg, Switzerland

Dr. Otto Kraus
Professor, University of Hamburg,
and Director, Zoological Institute and
Museum
Hamburg, Germany

Dr. Hans Krieg
Professor and First Director (retired),
Scientific Collections of the State of
Bavaria
Munich, Germany

Dr. Heinrich Kühl
Federal Research Institute for
Fisheries, Cuxhaven Laboratory
Cuxhaven, Germany

Dr. Oskar Kuhn
Professor, formerly University
Halle/Saale
Munich, Germany

Dr. Hans Kumerloeve
First Director (retired), State
Scientific Museum, Vienna
Munich, Germany

Dr. Nagamichi Kuroda
Yamashina Ornithological Institute,
Shibuya-Ku
Tokyo, Japan

Dr. Fred Kurt
Zoological Museum of Zurich
University, Smithsonian Elephant
Survey
Colombo, Ceylon

Dr. Werner Ladiges
Professor and Chief Curator,
Zoological Institute and Museum,
University of Hamburg
Hamburg, Germany

Leslie Laidlaw
Department of Animal Sciences,
Purdue University
Lafayette, Indiana, U.S.A.

Dr. Ernst M. Lang
Director, Zoological Garden
Basel, Switzerland

Dr. Alfredo Langguth
Department of Zoology, Faculty of
Humanities and Sciences, University
of the Republic
Montevideo, Uruguay

Leo Lehtonen
Science Writer
Helsinki, Finland

Bernd Leisler
Second Zoological Institute, University
of Vienna
Vienna, Austria

Dr. Kurt Lillelund
Professor and Director, Institute for
Hydrobiology and Fishery Sciences,
University of Hamburg
Hamburg, Germany

R. Liversidge
Alexander MacGregor Memorial
Museum
Kimberley, South Africa

Dr. Konrad Lorenz
Professor and Director, Max Planck
Institute for Behavioral Physiology
Seewiesen/Obb., Germany

Dr. Martin Lühmann
Federal Research Institute for the
Breeding of Small Animals
Celle, Germany

Dr. Johannes Lüttschwager
Oberstudienrat (retired)
Heidelberg, Germany

Dr. Wolfgang Makatsch
Bautzen, Germany

Dr. Hubert Markl
Professor and Director, Zoological
Institute, Technical University of
Darmstadt
Darmstadt, Germany

Basil J. Marlow, B.SC. (Hons)
Curator, Australian Museum
Sydney, Australia

Dr. Theodor Mebs
Instructor of Biology
Weissenhaus/Ostsee, Germany

Dr. Gerlof Fokko Mees
Curator of Birds, Rijks Museum of
Natural History
Leiden, The Netherlands

Hermann Meinken
Director, Fish Identification Institute,
V.D.A.
Bremen, Germany

Dr. Wilhelm Meise
Chief Curator, Zoological Institute
and Museum, University of Hamburg
Hamburg, Germany

Dr. Joachim Messtorff
Field Station of the Federal Fisheries
Research Institute
Bremerhaven, Germany

Dr. Marian Mlynarski
Professor, Polish Academy of
Sciences, Institute for Systematic and
Experimental Zoology
Cracow, Poland

Dr. Walburga Moeller
Nature Museum
Hamburg, Germany

Dr. H.C. Erna Mohr
Curator (retired), Zoological State
Institute and Museum
Hamburg, Germany

Dr. Karl-Heinz Moll
Waren/Müritz, Germany

Dr. Detlev Müller-Using
Professor, Institute for Game
Management, University of Göttingen
Hannoversch-Münden, Germany

Werner Münster
Instructor of Biology
Ebersbach, Germany

Dr. Joachim Münzing
Altona Museum
Hamburg, Germany
Dr. Wilbert Neugebauer
Wilhelma Zoo
Stuttgart-Bad Cannstatt, Germany

Dr. Ian Newton
Senior Scientific Officer, The Nature
Conservancy
Edinburgh, Scotland

Dr. Jürgen Nicolai
Max Planck Institute for Behavioral
Physiology
Seewiesen/Obb., Germany

Dr. Günther Niethammer
Professor, Zoological Research
Institute and A. Koenig Museum
Bonn, Germany

Dr. Bernhard Nievergelt
Zoological Museum, University of
Zurich
Zurich, Switzerland

Dr. C. C. Olrog
Institut Miguel Lillo San Miguel de
Tucuman
Tucuman, Argentina

Alwin Pedersen
Mammal Research and aRctic Explorer
Holte, Denmark

Dr. Dieter Stefan Peters
Nature Museum and Senckenberg
Research Institute
Frankfurt a.M., Germany

Dr. Nicolaus Peters
Scientific Councillor and Docent,
Institute of Hydrobiology and
Fisheries, University of Hamburg
Hamburg, Germany

Dr. Hans-Günter Petzold
Assistant Director, Zoological Garden
Berlin, Germany

Dr. Rudolf Piechocki
Docent, Zoological Institute,
University of Halle
Halle a.d.S., Germany

Dr. Ivo Poglayen-Neuwall
Director, Zoological Garden
Louisville, Kentucky, U.S.A.

Dr. Egon Popp
Zoological Collection of the State of
Bavaria
Munich, Germany

Dr. H.C. Adolf Portmann
Professor Emeritus, Zoological
Institute, University of Basel
Basel, Switzerland

Hans Psenner
Professor and Director, Alpine Zoo
Innsbruck, Austria

Dr. Heinz-Siburd Raethel
Oberveterinärrat
Berlin, Germany

Dr. Urs H. Rahm
Professor, Museum of Natural History
Basel, Switzerland

Dr. Werner Rathmayer
Biology Institute, University of
Konstanz
Konstanz, Germany

Walter Reinhard
Biologist
Baden-Baden, Germany

Dr. H. H. Reinsch
Federal Fisheries Research Institute
Bremerhaven, Germany

Dr. Bernhard Rensch
Professor Emeritus, Zoological
Institute, University of Münster
Münster, Germany

Dr. Vernon Reynolds
Docent, Department of Sociology,
University of Bristol
Bristol, England

Dr. Rupert Riedl
Professor, Department of Zoology,
University of North Carolina
Chapel Hill, North Carolina, U.S.A.

Dr. Peter Rietschel
Professor (retired), Zoological
Institute, University of Frankfurt a.M.
Frankfurt a.M., Germany

Dr. Siegfried Rietschel
Docent, University of Frankfurt;
Curator, Nature Museum and
Research Institute Senckenberg
Frankfurt a.M., Germany

Herbert Ringleben
Institute of Ornithology, Heligoland
Ornithological Station
Wilhelmshaven, Germany

Dr. K. Rohde
Institute for General Zoology, Ruhr
University
Bochum, Germany

Dr. Peter Röben
Academic Councillor, Zoological
Institute, Heidelberg University
Heidelberg, Germany

Dr. Anton E. M. De Roo
Royal Museum of Central Africa
Tervuren, South Africa

Dr. Hubert Saint Girons
Research Director, Center for
National Scientific Research
Brunoy (Essonne), France

Dr. Luitfried Von Salvini-Plawen
First Zoological Institute, University
of Vienna
Vienna, Austria

Dr. Kurt Sanft
Oberstudienrat, Diesterweg-Gymnasium
Berlin, Germany

Dr. E. G. Franz Sauer
Professor, Zoological Research
Institute and A. Koenig Museum,
University of Bonn
Bonn, Germany

Dr. Eleonore M. Sauer
Zoological Research Institute and A.
Koenig Museum, University of Bonn
Bonn, Germany

Dr. Ernst Schäfer
Curator, State Museum of Lower
Saxony
Hannover, Germany

Dr. Friedrich Schaller
Professor and Chairman, First
Zoological Institute, University of
Vienna
Vienna, Austria

Dr. George B. Schaller
Serengeti Research Institute, Michael
Grzimek Laboratory
Seronera, Tanzania

Dr. Georg Scheer
Chief Curator and Director,
Zoological Institute, State Museum of
Hesse
Darmstadt, Germany

Dr. Christoph Scherpner
Zoological Garden
Frankfurt a.M., Germany

Dr. Herbert Schifter
Bird Collection, Museum of Natural
History
Vienna, Austria

Dr. Marco Schnitter
Zoological Museum, Zurich
University
Zurich, Switzerland

Dr. Kurt Schubert
Federal Fisheries Research Institute
Hamburg, Germany

Eugen Schuhmacher
Director, Animals Films, I.U.C.N.
Munich, Germany

Dr. Thomas Schultze-Westrum
Zoological Institute, University of
Munich
Munich, Germany

Dr. Ernst Schüt
Professor and Director (retired), State
Museum of Natural History
Stuttgart, Germany

Dr. Lester L. Short Jr.
Associate Curator, American Museum
of Natural History
New York, New York, U.S.A.

Dr. Helmut Sick
National Museum
Rio de Janeiro, Brazil

Dr. Alexander F. Skutch
Professor of Ornithology, University
of Costa Rica
San Isidro del General, Costa Rica

Dr. Everhard J. Slijper
Professor, Zoological Laboratory,
University of Amsterdam
Amsterdam, The Netherlands

Bertram E. Smythies
Curator (retired), Division of Forestry
Management, Sarawak-Malaysia
Estepona, Spain

Dr. Kenneth E. Stager
Chief Curator, Los Angeles County
Museum of Natural History
Los Angeles, California, U.S.A.

Dr. H.C. Georg H.W. Stein
Professor, Curator of Mammals,
Institute of Zoology and Zoological
Museum, Humboldt University
Berlin, Germany

Dr. Joachim Steinbacher
Curator, Nature Museum and
Senckenberg Research Institute
Frankfurt a.M., Germany

Dr. Bernard Stonehouse
Canterbury University
Christchurch, New Zealand

Dr. Richard Zur Strassen
Curator, Nature Museum and
Senckenberg Research Institute
Frankfurt a.M., Germany

Dr. Adelheid Studer-Thiersch
Zoological Garden
Basel, Switzerland

Dr. Ernst Sutter
Museum of Natural History
Basel, Switzerland

Dr. Fritz Terofal
Director, Fish Collection, Zoological
Collection of the State of Bavaria
Munich, Germany

Dr. G. F. Van Tets
Wildlife Research
Canberra, Australia

Ellen Thaler-Kottek
Institute of Zoology, University of
Innsbruck
Innsbruck, Austria

Dr. Erich Thenius
Professor and Director, Institute of
Paleontolgy, University of Vienna
Vienna, Austria

Dr. Niko Tinbergen
Professor of Animal Behavior,
Department of Zoology, Oxford
University
Oxford, England

Alexander Tsurikov
Lecturer, University of Munich
Munich, Germany

Dr. Wolfgang Villwock
Zoological Institute and Museum,
University of Hamburg
Hamburg, Germany

Zdenek Vogel
Director, Suchdol Herpetological
Station
Prague, Czechoslovakia

Dieter Vogt
Schorndorf, Germany

Dr. Jiri Volf
Zoological Garden
Prague, Czechoslovakia
Otto Wadewitz
Leipzig, Germany

Dr. Helmut O. Wagner
Director (retired), Overseas Museum,
Bremen
Mexico City, Mexico

Dr. Fritz Walther
Professor, Texas A & M University
College Station, Texas, U.S.A.

John Warham
Zoology Department, Canterbury
University
Christchurch, New Zealand

Dr. Sherwood L. Washburn
University of California at Berkeley
Berkeley, California, U.S.A.

Eberhard Wawra
First Zoological Institute, University
of Vienna
Vienna, Austria

Dr. Ingrid Weigel
Zoological Collection of the State of
Bavaria
Munich, Germany

Dr. B. Weischer
Institute of Nematode Research,
Federal Biological Institute
Münster/Westfalen, Germany

Herbert Wendt
Author, Natural History
Baden-Baden, Germany

Dr. Heinz Wermuth
Chief Curator, State Nature Museum,
Stuttgart
Ludwigsburg, Germany

Dr. Wolfgang Von Westernhagen
Preetz/Holstein, Germany

Dr. Alexander Wetmore
United States National Museum,
Smithsonian Institution
Washington, D.C., U.S.A.

Dr. Dietrich E. Wilcke
Röttgen, Germany

Dr. Helmut Wilkens
Professor and Director, Institute of
Anatomy, School of Veterinary
Medicine
Hannover, Germany

Dr. Michael L. Wolfe
Utah, U.S.A.

Hans Edmund Wolters
Zoological Research Institute and A.
Koenig Museum
Bonn, Germany

Dr. Arnfrid Wünschmann
Research Associate, Zoological Garden
Berlin, Germany

Dr. Walter Wüst
Instructor, Wilhelms Gymnasium
Munich, Germany

Dr. Heinz Wundt
Zoological Collection of the State of
Bavaria
Munich, Germany

Dr. Claus-Dieter Zander
Zoological Institute and Museum,
University of Hamburg
Hamburg, Germany

Dr. Fritz Zumpt
Director, Entomology and
Parasitology, South African Institute
for Medical Research
Johannesburg, South Africa

Dr. Richard L. Zusi
Curator of Birds, United States
National Museum, Smithsonian
Institution
Washington, D.C., U.S.A.

Glossary

The following glossary is not intended to be exhaustive, but rather includes primarily terms that (1) have some specific importance to our understanding of birds, (2) have been used in these volumes, (3) might have varying definitions relative to birds as opposed to common usage, or (4) are often misunderstood.

Accipiter—This is the genus name for a group of bird-eating hawks (Accipitridae; e.g., sharp-shinned hawk, Cooper's hawk). These birds show similar behavior and appearance and extreme sexual dimorphism. Females are much larger than males and the female of the sharp-shinned hawk often seems as large as the male of the Cooper's hawk, leading to some confusion on the part of birders. In the face of uncertainty, these birds are often just referred to as "Accipiters" and the name is now firmly ensconced in "birding" terminology.

Adaptive radiation—Diversification of a species or single ancestral type into several forms that are each adaptively specialized to a specific niche.

Aftershaft—A second rachis (= shaft) arising near the base of a contour feather, creating a feather that "branches." Aftershafts can be found in many birds (e.g., pheasants) but in most the aftershaft is much smaller than the main shaft of the feather. In ratites (ostrich-like birds), the aftershaft is about the same size as the main shaft. Sometimes the term "aftershaft" is restricted to the rachis that extends from the main rachis and the whole secondary structure is referred to as the "afterfeather."

Agonistic—Behavioral patterns that are aggressive in context. Most aggressive behavior in birds is expressed as song (in songbirds) or other vocal or mechanical sound (e.g. see Drumming). The next level of intensity is display, and only in extreme circumstances do birds resort to physical aggression.

Air sac—Thin-walled, extensions of the lungs, lying in the abdomen and thorax, and extending even into some bones of birds. Air sacs allow an increased respiratory capacity of birds and the removal of oxygen both as air passes in through the lungs and also as it passes back through the lungs as the bird exhales. The flow of air through the air sacs also helps dissipate the heat produced through muscle activity and increases a bird's volume while only minimally increasing weight—thus effectively making birds lighter relative to their size and more efficient in flying. Air sacs are best developed in the strongest flying birds and least developed in some groups that are flightless.

Alcid—Referring to a member of the family Alcidae; including puffins, auks, auklets, murres, razorbills, and guillemots.

Allopatric—Occurring in separate, nonoverlapping geographic areas.

Allopreening—Mutual preening; preening of the feathers of one bird by another; often a part of courtship or pair bond maintenance.

Alpha breeder—The reproductively dominant member of a social unit.

Alternate plumage—The breeding plumage of passerines, ducks, and many other groups; typically acquired through a partial molt prior to the beginning of courtship.

Altricial—An adjective referring to a bird that hatches with little, if any, down, is unable to feed itself, and initially has poor sensory and thermoregulatory abilities.

Alula—Small feathers at the leading edge of the wing and attached to the thumb; also called bastard wing; functions in controlling air flow over the surface of the wing, thus allowing a bird to land at a relatively slow speed.

Anatid—A collective term referring to members of the family Anatidae; ducks, geese, and swans.

Anisodactyl— An adjective that describes a bird's foot in which three toes point forward and one points backwards, a characteristic of songbirds.

Anserine—Goose-like.

Anting—A behavior of birds that involves rubbing live ants on the feathers, presumably to kill skin parasites.

Antiphonal duet—Vocalizations by two birds delivered alternately in response to one another; also known as responsive singing.

AOU—American Ornithologists' Union; the premier professional ornithological organization in North America; the organizational arbiter of scientific and standardized common names of North American birds as given in the periodically revised Check-list of North American Birds.

Arena—See Lek.

Aspect ratio—Length of a wing divided by width of the wing; High aspect wings are long and narrow. These are characteristic of dynamic soaring seabirds such as albatrosses. These birds have tremendous abilities to soar over the open ocean, but poor ability to maneuver in a small area. In contrast, low aspect ratio wings are short and broad, characteristic of many forest birds, and provide great ability to quickly maneuver in a small space.

Asynchronous—Not simultaneous; in ornithology often used with respect to the hatching of eggs in a clutch in which hatching occurs over two or more days, typically a result of initiation of incubation prior to laying of the last egg.

Auricular—An adjective referring to the region of the ear in birds, often to a particular plumage pattern over the ear.

Austral—May refer to "southern regions," typically meaning Southern Hemisphere. May also refer to the geographical region included within the Transition, Upper Austral, and Lower Austral Life Zones as defined by C. Hart Merriam in 1892–1898. These zones are often characterized by specific plant and animal communities and were originally defined by temperature gradients especially in the mountains of southwestern North America.

Autochthonous—An adjective that indicates that a species originated in the region where it now resides.

Barb—One of the hair-like extensions from the rachis of a feather. Barbs with barbules and other microstructures can adhere to one another, forming the strong, yet flexible vane needed for flight and protection and streamlining of body surfaces.

Barbules—A structural component of the barbs of many feathers; minute often interlocking filaments in a row at each side of a barb. As a result of their microstructure, barbules adhere to one another much like "Velcro®" thus assuring that feathers provide a stiff, yet flexible vane.

Basic plumage—The plumage an adult bird acquires as a result of its complete (or near complete) annual molt.

Bergmann's rule—Within a species or among closely related species of mammals and birds, those individuals in colder environments often are larger in body size. Bergmann's rule is a generalization that reflects the ability of warm-blooded animals to more easily retain body heat (in cold climates) if they have a high body surface to body volume ratio, and to more easily dissipate excess body heat (in hot environments) if they have a low body surface to body volume ratio.

Bioacoustics—The study of biological sounds such as the sounds produced by birds.

Biogeographic region—One of several major divisions of the earth defined by a distinctive assemblage of animals and plants. Sometimes referred to as "zoogeographic regions or realms" (for animals) or "phytogeographic regions or realms" (for plants). Such terminology dates from the late nineteenth century and varies considerably. Major biogeographic regions each have a somewhat distinctive flora and fauna. Those generally recognized include Nearctic, Neotropical, Palearctic, Ethiopian, Oriental, and Australian.

Biomagnification—Sometimes referred to as "bioaccumulation." Some toxic elements and chemical compounds are not readily excreted by animals and instead are stored in fatty tissues, removing them from active metabolic pathways. Birds that are low in a food chain (e.g., sparrows that eat seeds) accumulate these chemicals in their fatty tissues. When a bird that is higher in the food chain (e.g., a predator like a falcon) eats its prey (e.g., sparrows), it accumulates these chemicals from the fatty tissue of each prey individual, thus magnifying the level of the chemical in its own tissues. When the predator then comes under stress and all of these chemicals are released from its fat into its system, the effect can be lethal. Chemicals capable of such biomagnification include heavy metals such as lead and mercury, and such manmade compounds as organochlorine pesticides and polychlorinated biphenyls (PCBs).

Booming ground—See Lek.

Booted—An adjective describing a bird tarsus (leg) that has a smooth, generally undivided, rather than scaly (= scutellate) appearance. The extent of the smooth or scaly appearance of a bird tarsus varies among taxonomic groups and there are many different, more specific, patterns of tarsal appearance that are recognized.

Boreal—Often used as an adjective meaning "northern"; also may refer to the northern climatic zone immediately south of the Arctic; may also include the Arctic, Hudsonian, and Canadian Life Zones described by C. Hart Merriam.

Bristle—In ornithology, a feather with a thick, tapered rachis and no vane except for a remnant sometimes found near the bristle base.

Brood—As a noun: the young produced by a pair of birds during one reproductive effort. As a verb: to provide warmth and shelter to chicks by gathering them under the protection of breast and/or wings.

Brood parasitism—Reproductive strategy where one species of bird (the parasite) lays its eggs in the nests of another species (the host). An acceptable host will incubate the eggs and rear the chicks of the brood parasite, often to the detriment or loss of the host's own offspring.

Brood patch—A bare area of skin on the belly of a bird, the brood patch is enlarged beyond the normal apterium (bare area) as a result of loss of feathers. It becomes highly vascularized (many blood vessels just under the surface). The brood patch is very warm to the touch and the bird uses it to cover and warm its chicks. In terms of structure, the brood patch is the same as the incubation patch and the two terms are often used synonymously. Technically the brood patch and incubation patch differ in function: the incubation patch is used in incubating eggs, the brood patch is used to brood the young after the eggs hatch.

Brood reduction—Reduction in the number of young in the nest. Viewed from an evolutionary perspective, mechanisms that allow for brood reduction may assure that at least some offspring survive during stressful times and that during times of abundant resources all young may survive. Asynchronous hatching results in young of different ages and sizes in a nest and is a mechanism that facilitates brood reduction: the smallest chick often dies if there is a shortage of food. The barn owl (*Tyto alba*; Tytonidae) depends on food resources that vary greatly in availability from year to year and it often experiences brood reduction.

Buteo—This is the genus name for a group of hawks that have broad wings and soar. These hawks are often seen at a distance and are easily recognized as "Buteos" although they may not be identifiable as species. Hence the genus name has come into common English usage.

Caecum (pl. caeca)—Blindly-ending branch extending from the junction of the small and large intestine. Most birds have two caeca, but the number and their development in birds is highly variable. Caeca seem to be most highly developed and functional in facilitating microbial digestion of food in those birds that eat primarily plant materials.

Caruncle—An exposed, often brightly colored, fleshy protuberance or wrinkled facial skin of some birds.

Casque—An enlargement at the front of the head (e.g., on cassowaries, Casuaridae) or sometimes of the bill (e.g., on hornbills, Bucerotidae) of a bird. A casque may be bony, cartilaginous, or composed of feathers (e.g., Pri-onopidae). A casque is often sexual ornamentation, but may protect the head of a cassowary crashing through underbrush, may be used for vocal amplification, or may serve a physiological function.

Cavity nester—A species that nests in some sort of a cavity. Primary cavity nesters (e.g., woodpeckers, Picidae; kingfishers, Alcedinidae; some swallows, Hirundinidae) are capable of excavating their own cavities; secondary cavity nesters (e.g., starlings, Sturnidae; House Sparrows, Passeridae; bluebirds, Turdidae) are not capable of excavating their own cavities.

Cere—The soft, sometimes enlarged, and often differently colored basal covering of the upper bill (maxilla) of many hawks (Falconiformes), parrots (Psitaciformes), and owls (Strigiformes). The nostrils are often within or at the edge of the cere. In parrots the cere is sometimes feathered.

Cladistic—Evolutionary relationships suggested as "tree" branches to indicate lines of common ancestry.

Cleidoic eggs—Cleidoic eggs are simply ones that are contained, hence protected, inside of a somewhat impervious shell—such as the eggs of birds. The presence of a shell around an egg freed the amphibian ancestors of reptiles from the need to return to the water to lay eggs and provided greater protection from dying.

Cline—A gradient in a measurable character, such as size and color, showing geographic differentiation. Various patterns of geographic variation are reflected as clines or clinal variation, and have been described as "ecogeographic rules."

Clutch—The set of eggs laid by a female bird during one reproductive effort. In most species, a female will lay one egg per day until the clutch is complete; in some species, particularly larger ones (e.g., New World vultures, Cathartidae), the interval between eggs may be more than one day.

Colony—A group of birds nesting in close proximity, interacting, and usually aiding in early warning of the presence of predators and in group defense.

Commensal—A relationship between species in which one benefits and the other is neither benefited nor harmed.

Congeneric—Descriptive of two or more species that belong to the same genus.

Conspecific—Descriptive of two or more individuals or populations that belong to the same species.

Conspecific colony—A colony of birds that includes only members of one species.

Contact call—Simple vocalization used to maintain communication or physical proximity among members of a social unit.

Contour feather—One of those feathers covering the body, head, neck, and limbs of a bird and giving rise to the shape (contours) of the bird.

Convergent evolution—When two evolutionarily unrelated groups of organisms develop similar characteristics due to adaptation to similar aspects of their environment or niche. The sharply pointed and curved talons of hawks and owls are convergent adaptations for their predatory lifestyle.

Cooperative breeding—A breeding system in which birds other than the genetic parents share in the care of eggs and young. There are many variants of cooperative breeding. The birds that assist with the care are usually referred to as "helpers" and these are often offspring of the same breeding pair, thus genetically related to the chicks they are tending. Cooperative breeding is most common among tropical birds and seems most common in situations where nest sites or breeding territories are very limited. Several studies have demonstrated that "helping" increases reproductive success. By helping a helper is often assuring survival of genes shared with the related offspring. The helper also may gain important experience and ultimately gain access to a breeding site.

Coracoid—A bone in birds and some other vertebrates extending from the scapula and clavicle to the sternum; the coracoid serves as a strut supporting the chest of the bird during powerful muscle movements associated with flapping flight.

Cosmopolitan—Adjective describing the distribution pattern of a bird found around the world in suitable habitats.

Countershading—A color pattern in which a bird or other animal is darker above and lighter below. The adaptive value of the pattern is its ability to help conceal the animal: a predator looking down from above sees the darker back against the dark ground; a predator looking up from below sees the lighter breast against the light sky; a predator looking from the side sees the dark back made lighter by the light from above and the light breast made darker by shading.

Covert—A feather that covers the gap at the base between flight feathers of the wing and tail; coverts help create smooth wing and tail contours that make flight more efficient.

Covey—A group of birds, often comprised of family members that remain together for periods of time; usually applied to game birds such as quail (Odontophorinae).

Crepuscular—Active at dawn and at dusk.

Crèche—An aggregation of young of many colonially-nesting birds (e.g., penguins, Spheniscidae; terns, Laridae). There is greater safety from predators in a crèche.

Crissum—The undertail coverts of a bird; often distinctively colored.

Critically Endangered—A technical category used by IUCN for a species that is at an extremely high risk of extinction in the wild in the immediate future.

Cryptic—Hidden or concealed; i.e., well-camouflaged patterning.

Dichromic—Occurring in two distinct color patterns (e.g., the bright red of male and dull red-brown of female northern cardinals, *Cardinalis cardinalis*)

Diurnal—Active during the day.

Dimorphic—Occurring in two distinct forms (e.g., in reference to the differences in tail length of male and female boat-tailed grackles, *Cassidix major*).

Disjunct—A distribution pattern characterized by populations that are geographically separated from one another.

Dispersal—Broadly defined: movement from an area; narrowly defined: movement from place of hatching to place of first breeding.

Dispersion—The pattern of spatial arrangement of individuals, populations, or other groups; no movement is implied.

Disruptive color—A color pattern such as the breast bands on a killdeer (*Charadrius vociferus*) that breaks up the outline of the bird, making it less visible to a potential predator, when viewed from a distance

DNA-DNA hybridization—A technique whereby the genetic similarity of different bird groups is determined based on the extent to which short stretches of their DNA, when mixed together in solution in the laboratory, are able to join with each other.

Dominance hierarchy—"Peck order"; the social status of individuals in a group; each animal can usually dominate those animals below it in a hierarchy.

Dummy nest—Sometimes called a "cock nest." An "extra" nest, often incomplete, sometimes used for roosting, built by aggressive males of polygynous birds. Dummy nests may aid in the attraction of additional mates, help define a male's territory, or confuse potential predators.

Dump nest—A nest in which more than one female lays eggs. Dump nesting is a phenomenon often linked to young, inexperienced females or habitats in which nest sites are scarce. The eggs in dump nest are usually not incubated. Dump nesting may occur within a species or between species.

Dynamic soaring—A type of soaring characteristic of oceanic birds such as albatrosses (Diomedeidae) in which the bird takes advantage of adjacent wind currents that are of different speeds in order to gain altitude and effortlessly stay aloft.

Echolocation—A method of navigation used by some swifts (Apodidae) and oilbirds (Steatornithidae) to move in darkness, such as through caves to nesting sites. The birds emit audible "clicks" and determine pathways by using the echo of the sound from structures in the area.

Eclipse plumage—A dull, female-like plumage of males of Northern Hemisphere ducks (Anatidae) and other birds such as house sparrows (*Passer domesticus*) typically attained in late summer prior to the annual fall molt. Ducks are flightless at this time and the eclipse plumage aids in their concealment at a time when they would be especially vulnerable to predators.

Ecotourism—Travel for the primary purpose of viewing nature. Ecotourism is now "big business" and is used as a non-consumptive but financially rewarding way to protect important areas for conservation.

Ectoparasites—Relative to birds, these are parasites such as feather lice and ticks that typically make their home on the skin or feathers.

Emarginate—Adjective referring to the tail of a bird that it notched or forked or otherwise has an irregular margin as a result of tail feathers (rectrices) being of different lengths. Sometimes refers to individual flight feather that is particularly narrowed at the tip.

Endangered—A term used by IUCN and also under the Endangered Species Act of 1973 in the United States in reference to a species that is threatened with imminent extinction or extirpation over all or a significant portion of its range.

Endemic—Native to only one specific area.

Eocene—Geological time period; subdivision of the Tertiary, from about 55.5 to 33.7 million years ago.

Erythrocytes—Red blood cells; in birds, unlike mammals, these retain a nucleus and are longer lived. Songbirds tend to have smaller, more numerous (per volume) erythrocytes that are richer in hemoglobin than are the erythrocytes of more primitive birds.

Ethology—The study of animal behavior.

Exotic—Not native.

Extant—Still in existence; not destroyed, lost, or extinct.

Extinct—Refers to a species that no longer survives anywhere.

Extirpated—Referring to a local extinction of a species that can still be found elsewhere.

Extra-pair copulation—In a monogamous species, refers to any mating that occurs between unpaired males and females.

Facial disc—Concave arrangement of feathers on the face of an owl. The facial discs on an owl serve as sound parabolas, focusing sound into the ears around which the facial discs are centered, thus enhancing their hearing.

Fecal sac—Nestling songbirds (Passeriformes) and closely related groups void their excrement in "packages"—enclosed in thin membranes—allowing parents to remove the material from the nest. Removal of fecal material likely reduces the potential for attraction of predators.

Feminization—A process, often resulting from exposure to environmental contaminants, in which males produce a higher levels of female hormones (or lower male hormone levels), and exhibit female behavioral or physiological traits.

Feral—Gone wild; i.e., human-aided establishment of non-native species.

Fledge—The act of a juvenile making its first flight; sometimes generally used to refer to a juvenile becoming independent.

Fledgling—A juvenile that has recently fledged. An emphasis should be placed on "recently." A fledgling generally lacks in motor skills and knowledge of its habitat and fledglings are very vulnerable, hence under considerable parental care. Within a matter of a few days, however, they gain skills and knowledge and less parental care is needed.

Flight feathers—The major feathers of the wing and tail that are crucial to flight. (See Primary, Secondary, Tertial, Alula, Remex, Rectrix)

Flyway—A major pathway used by a group of birds during migration. The flyway concept was developed primarily with regard to North American waterfowl (Anatidae) and has been used by government agencies in waterfowl management. Major flyways described include the Atlantic, Mississippi, Central, and Pacific flyways. While the flyway concept is often used in discussions of other groups of birds, even for waterfowl the concept is an oversimplification. The patterns of movements of migrant waterfowl and other birds vary greatly among species.

Frugivorous—Feeds on fruit.

Galliform—Chicken-like, a member of the Galliformes.

Gape—The opening of the mouth of a bird; the act of opening the mouth, as in begging.

Gizzard—The conspicuous, muscular portion of the stomach of a bird. Birds may swallow grit or retain bits of bone or hard parts of arthropods in the gizzard and these function in a manner analogous to teeth as the strong muscles of the gizzard contract, thus breaking food into smaller particles. The gizzard is best developed in birds that eat seeds and other plant parts; in some fruit-eating birds the gizzard is very poorly developed.

Glareolid—A member of the family Glareolidae.

Gloger's rule—Gloger's rule is an ecogeographic generalization that suggests that within a species or closely related group of birds there is more melanin (a dark pigment) in feathers in warm humid parts of the species' or groups' range, and less melanin in feathers in dry or cooler parts of the range.

Gorget—Colorful throat patch or bib (e.g., of many hummingbirds, Trochilidae).

Graduated—An adjective used to describe the tail of a bird in which the central rectrices are longest and those to the outside are increasingly shorter.

Granivorous—Feeding on seeds.

Gregarious—Occuring in large groups.

Gular—The throat region.

Hallux—The innermost digit of a hind or lower limb.

Hawk—Noun: a member of the family Accipitridae. Verb: catching insects by flying around with the mouth open (e.g. swallows, Hirundinidae; nightjars, Caprimulgidae).

Heterospecific colony—A colony of birds with two or more species.

Heterothermy—In birds, the ability to go into a state of torpor or even hibernation, lowering body temperature through reduced metabolic activity and thus conserving energy resources during periods of inclement weather or low food.

Hibernation—A deep state of reduced metabolic activity and lowered body temperature that may last for weeks; attained by few birds, resulting from reduced food supplies and cool or cold weather.

Holarctic—The Palearctic and Nearctic bigeographic regions combined.

Homeothermy—In birds the metabolic ability to maintain a constant body temperature. The lack of development of homeothermy in new-hatched chicks is the underlying need for brooding behavior.

Hover-dip—A method of foraging involving hovering low over the water, and then dipping forward to pick up prey from the surface (e.g., many herons, Ardeidae).

Hybrid—The offspring resulting from a cross between two different species (or sometimes between distinctive subspecies).

Imprinting—A process that begins with an innate response of a chick to its parent or some other animal (or object!) that displays the appropriate stimulus to elicit the chick's response. The process continues with the chick rapidly learning to recognize its parents. Imprinting typically occurs within a few hours (often 13–16 hours) after hatching. Imprinting then leads to learning behavioral characteristics that facilitate its survival, including such things as choice of foraging sites and foods, shelter, recognition of danger, and identification of a potential mate. The most elaborate (and best studied) imprinting is associated with precocial chicks such as waterfowl (Anatidae).

Incubation patch—See Brood patch.

Indigenous—See Endemic.

Innate—An inherited characteristic; e.g., see Imprinting.

Insectivorous—In ornithology technically refers to a bird that eats insects; generally refers to in birds that feed primarily on insects and other arthropods.

Introduced species—An animal or plant that has been introduced to an area where it normally does not occur.

Iridescent—Showing a rainbow-like play of color caused by differential refraction of light waves that change as the angle of view changes. The iridescence of bird feathers is a result of a thinly laminated structure in the barbules of those feathers. Iridescent feathers are made more brilliant by pigments that underlie this structure, but the pigments do not cause the iridescence.

Irruptive—A species of bird that is characterized by irregular long-distance movements, often in response to a fluctuating food supply (e.g., red crossbill, *Loxia curvirostra*, Fringillidae; snowy owl, *Nyctea scandiaca*, Strigidae).

IUCN—The World Conservation Union; formerly the International Union for the Conservation of Nature, hence IUCN. It is the largest consortium of governmental and nongovernmental organizations focused on conservation issues.

Juvenal—In ornithology (contrary to most dictionaries), restricted to use as an adjective referring to a characteristic (usually the plumage) of a juvenile bird.

Juvenile—A young bird, typically one that has left the nest.

Kleptoparasitism—Behavior in which one individual takes ("steals") food, nest materials, or a nest site from another.

Lachrymal—Part of the skull cranium, near the orbit; lachrymal and Harderian glands in this region lubricate and protect the surface of the eye.

Lamellae—Transverse tooth-like or comb-like ridges inside the cutting edge of the bill of birds such as ducks (Anatidae) and flamingos (Phoenicopteridae). Lamellae serve as a sieve during feeding: the bird takes material into its mouth, then uses its tongue to force water out through the lamellae, while retaining food particles.

Lek—A loose to tight association of several males vying for females through elaborate display; lek also refers to the specific site where these males gather to display. Lek species include such birds as prairie chickens (Phasianidae) and manakins (Pipridae).

Lobed feet—Feet that have toes with stiff scale-covered flaps that extend to provide a surface analogous to webbing on a duck as an aid in swimming.

Lore—The space between the eye and bill in a bird. The loral region often differs in color from adjacent areas of a bird's face. In some species the area is darker, thus helping to reduce glare, serving the same function as the dark pigment some football players apply beneath each eye. In predatory birds, a dark line may extend from the eye to the bill, perhaps decreasing glare, but also serving as a sight to better aim its bill. The color and pattern of plumage and skin in the loral region is species-specific and often of use in helping birders identify a bird.

Malar—Referring to the region of the face extending from near the bill to below the eye; markings in the region are often referred to as "moustache" stripes.

Mandible—Technically the lower half of a bird's bill. The plural, mandibles, is used to refer to both the upper and lower bill. The upper half of a bird's bill is technically the maxilla, but often called the "upper mandible."

Mantle—Noun: The plumage of the back of the bird, including wing coverts evident in the back region on top of the folded wing (especially used in describing hawks (Accipitridae) and gulls (Laridae). Verb: The behavior in which a raptor (typically on the ground) shields its acquired prey to protect it from other predators.

Mesoptile—On chicks, the second down feathers; these grow attached to the initial down, or protoptile.

Metabolic rate—The rate of chemical processes in living organisms, resulting in energy expenditure and growth. Hummingbirds (Trochilidae), for example, have a very high metabolic rate. Metabolic rate decreases when a bird is resting and increases during activity.

Miocene—The geological time period that lasted from about 23.8 to 5.6 million years ago.

Migration—A two-way movement in birds, often dramatically seasonal. Typically latitudinal, though in some species is altitudinal or longitudinal. May be short-distance or long-distance. (See Dispersal)

Mitochondrial DNA—Genetic material located in the mitochondria (a cellular organelle outside of the nucleus). During fertilization of an egg, only the DNA from the nucleus of a sperm combines with the DNA from the nucleus of an egg. The mitochondrial DNA of each offspring is inherited only from its mother. Changes in mitochondrial DNA occur quickly through mutation and studying differences in mitochondrial DNA helps scientists better understand relationships among groups.

Mobbing—A defensive behavior in which one or more birds of the same or different species fly toward a potential predator, such as a hawk, owl, snake, or a mammal, swooping toward it repeatedly in a threatening manner, usually without actually striking the predator. Most predators depend on the element of surprise in capturing their prey and avoid the expenditure of energy associated with a chase. Mobbing alerts all in the neighborhood that a potential predator is at hand and the predator often moves on. Rarely, a predator will capture a bird that is mobbing it.

Molecular phylogenetics—The use of molecular (usually genetic) techniques to study evolutionary relationships between or among different groups of organisms.

Molt—The systematic and periodic loss and replacement of feathers. Once grown, feathers are dead structures that continually wear. Birds typically undergo a complete or near-complete molt each year and during this molt feathers are usually lost and replaced with synchrony between right and left sides of the body, and gradually, so that the bird retains the ability to fly. Some species, such as northern hemisphere ducks, molt all of their flight feathers at once, thus become flightless for a short time. Partial molts, typically involving only contour feathers, may occur prior to the breeding season.

Monophyletic—A group (or clade) that shares a common ancestor.

Monotypic—A taxonomic category that includes only one form (e.g., a genus that includes only one species; a species that includes no subspecies).

Montane—Of or inhabiting the biogeographic zone of relatively moist, cool upland slopes below timberline dominated by large coniferous trees.

Morphology—The form and structure of animals and plants.

Mutualism—Ecological relationship between two species in which both gain benefit.

Nail—The horny tip on the leathery bill of ducks, geese, and swans (Anatidae).

Nectarivore—A nectar-eater (e.g., hummingbirds, Trochilidae; Hawaiian honeycreepers, Drepaniidae).

Near Threatened—A category defined by the IUCN suggesting possible risk of extinction in the medium term future.

Nearctic—The biogeographic region that includes temperate North America faunal region.

Neotropical—The biogeographic region that includes South and Central America, the West Indies, and tropical Mexico.

Nestling—A young bird that stays in the nest and needs care from parents.

New World—A general descriptive term encompassing the Nearctic and Neotropical biogeographic regions.

Niche—The role of an organism in its environment; multidimensional, with habitat and behavioral components.

Nictitating membrane—The third eyelid of birds; may be transparent or opaque; lies under the upper and lower eyelids. When not in use, the nictitating membrane is held at the corner of the eye closest to the bill; in use it moves horizontally or diagonally across the eye. In flight it keeps the bird's eyes from drying out; some aquatic birds have a lens-like window in the nictitating membrane, facilitating vision underwater.

Nidicolous—An adjective describing young that remain in the nest after hatching until grown or nearly grown.

Nidifugous—An adjective describing young birds that leave the nest soon after hatching.

Nocturnal—Active at night.

Nominate subspecies—The subspecies described to represent its species, the first described, bearing the specific name.

Nuclear DNA—Genetic material from the nucleus of a cell from any part of a bird's body other than its reproductive cells (eggs or sperm).

Nuptial displays—Behavioral displays associated with courtship.

Oligocene—The geologic time period occurring from about 33.7 to 23.8 million years ago.

Old World—A general term that usually describes a species or group as being from Eurasia or Africa.

Omnivorous—Feeding on a broad range of foods, both plant and animal matter.

Oscine—A songbird that is in the suborder Passeri, order Passeriformes; their several distinct pairs of muscles within the syrinx allow these birds to produce the diversity of sounds that give meaning to the term "songbird."

Osteological—Pertaining to the bony skeleton.

Palearctic—A biogeographic region that includes temperate Eurasia and Africa north of the Sahara.

Paleocene—Geological period, subdivision of the Tertiary, from 65 to 55.5 million years ago.

Pamprodactyl—The arrangement of toes on a bird's foot in which all four toes are pointed forward; characteristic of swifts (Apodidae).

Parallaxis—Comparing the difference in timing and intensity of sounds reaching each ear (in owls).

Passerine—A songbird; a member of the order Passeriformes.

Pecten—A comb-like structure in the eye of birds and reptiles, consisting of a network of blood vessels projecting inwards from the retina. The main function of the pecten seems to be to provide oxygen to the tissues of the eye.

Pectinate—Having a toothed edge like that of a comb. A pectinate claw on the middle toe is a characteristic of nightjars, herons, and barn owls. Also known as a "feather comb" since the pectinate claw is used in preening.

Pelagic—An adjective used to indicate a relationship to the open sea.

Phalloid organ—Penis-like structure on the belly of buffalo weavers; a solid rod, not connected to reproductive or excretory system.

Philopatry—Literally "love of homeland"; a bird that is philopatric is one that typically returns to nest in the same area in which it was hatched. Strongly philopatric species (e.g., hairy woodpecker, *Picoides borealis*) tend to accumulate genetic characteristics that adapt them to local conditions, hence come to show considerable geographic variation; those species that show little philopatry tend to show little geographic variation.

Phylogenetics—The study of racial evolution.

Phylogeny—A grouping of taxa based on evolutionary history.

Picid—A member of the family Picide (woodpeckers, wrynecks, piculets).

Piscivorous—Fish-eating.

Pleistocene—In general, the time of the great ice ages; geological period variously considered to include the last 1 to 1.8 million years.

Pliocene—The geological period preceding the Pleistocence; the last subdivision of what is known as the Tertiary; lasted from 5.5 to 1.8 million years ago.

Plumage—The complete set of feathers that a bird has.

Plunge-diving—A method of foraging whereby the bird plunges from at least several feet up, head-first into the water, seizes its prey, and quickly takes to the wing (e.g., terns, Laridae; gannets, Sulidae).

Polygamy—A breeding system in which either or both male and female may have two or more mates.

Polyandry—A breeding system in which one female bird mates with two or more males. Polyandry is relatively rare among birds.

Polygyny—A breeding system in which one male bird mates with two or more females.

Polyphyletic—A taxonomic group that is believed to have originated from more than one group of ancestors.

Powder down—Specialized feathers that grow continuously and break down into a fine powder. In some groups (e.g., herons, Ardeidae) powder downs occur in discrete patches (on the breast and flanks); in others (e.g., parrots, Psitacidae) they are scattered throughout the plumage. Usually used to waterproof the other feathers (especially in birds with few or no oil glands).

Precocial—An adjective used to describe chicks that hatch in an advanced state of development such that they generally can leave the nest quickly and obtain their own food, although they are often led to food, guarded, and brooded by a parent (e.g., plovers, Charadriidae; chicken-like birds, Galliformes).

Preen—A verb used to describe the behavior of a bird when it cleans and straightens its feathers, generally with the bill.

Primaries—Unusually strong feathers, usually numbering nine or ten, attached to the fused bones of the hand at the tip of a bird's wing.

Protoptile—The initial down on chicks.

Pterylosis—The arrangement of feathers on a bird.

Quaternary—The geological period, from 1.8 million years ago to the present, usually including two subdivisions: the Pleistocene, and the Holocene.

Quill—An old term that generally refers to a primary feather.

Rachis—The shaft of a feather.

Radiation—The diversification of an ancestral species into many distinct species as they adapt to different environments.

Ratite—Any of the ostrich-like birds; characteristically lack a keel on the sternum (breastbone).

Rectrix (pl. rectrices)—A tail feather of a bird; the rectrices are attached to the fused vertebrae that form a bird's bony tail.

Remex (pl. remiges)—A flight feather of the wing; remiges include the primaries, secondaries, tertials, and alula).

Reproductive longevity—The length of a bird's life over which it is capable of reproduction.

Resident—Nonmigratory.

Rhampotheca—The horny covering of a bird's bill.

Rictal bristle— A specialized tactile, stiff, hairlike feather with elongated, tapering shaft, sometimes with short barbs at the base. Rictal bristles prominently surround the mouth of birds such as many nightjars (Caprimulgidae), New World flycatchers (Tyrannidae), swallows (Hirundinidae), hawks (Accipitridae) and owls (Strigidae). They are occasionally, but less precisely referred to as "vibrissae," a term more appropriate to the "whiskers" on a mammal.

Rookery—Originally a place where rooks nest; now a term often used to refer to a breeding colony of gregarious birds.

Sally—A feeding technique that involves a short flight from a perch or from the ground to catch a prey item before returning to a perch.

Salt gland—Also nasal gland because of their association with the nostrils; a gland capable of concentrating and excreting salt, thus allowing birds to drink saltwater. These glands are best developed in marine birds.

Scapulars—Feathers at sides of shoulders.

Schemochrome—A structural color such as blue or iridescence; such colors result from the structure of the feather rather than from the presence of a pigment.

Scutellation—An arrangement or a covering of scales, as that on a bird's leg.

Secondaries—Major flight feathers of the wing that are attached to the ulna.

Sexual dichromatism—Male and female differ in color pattern (e.g., male hairy woodpecker [*Picoides villosus*, Picidae] has a red band on the back of the head, female has no red).

Sexual dimorphism—Male and female differ in morphology, such as size, feather size or shape, or bill size or shape.

Sibling species—Two or more species that are very closely related, presumably having differentiated from a common ancestor in the recent past; often difficult to distinguish, often interspecifically territorial.

Skimming—A method of foraging whereby the skimmers (Rhynchopidae) fly low over the water with the bottom bill slicing through the water and the tip of the bill above. When the bird hits a fish, the top bill snaps shut.

Slotting—Abrupt narrowing of the inner vane at the tip of some outer primaries on birds that soar; slotting breaks up wing-tip turbulence, thus facilitating soaring.

Sonagram—A graphic representation of sound.

Speciation—The evolution of new species.

Speculum—Colored patch on the wing, typically the secondaries, of many ducks (Anatidae).

Spur—A horny projection with a bony core found on the tarsometatarsus.

Sternum—Breastbone.

Structural color—See Schemochrome.

Suboscine— A songbird in the suborder Passeri, order Passeriformes, whose songs are thought to be innate, rather than learned.

Sympatric—Inhabiting the same range.

Syndactyl—Describes a condition of the foot of birds in which two toes are fused near the base for part of their length (e.g., kingfishers, Alcedinidae; hornbills, Bucerotidae).

Synsacrum—The expanded and elongated pelvis of birds that is fused with the lower vertebrae.

Syrinx (pl. syringes)—The "voice box" of a bird; a structure of cartilage and muscle located at the junction of the trachea and bronchi, lower on the trachea than the larynx of mammals. The number and complexity of muscles in the syrinx vary among groups of birds and have been of value in determining relationships among groups.

Systematist—A specialist in the classification of organisms; systematists strive to classify organisms on the basis of their evolutionary relationships.

Tarsus—In ornithology also sometimes called Tarsometatarsus or Metatarsus; the straight part of a bird's foot immediately above its toes. To the non-biologist, this seems to be the "leg" bone—leading to the notion that a bird's "knee" bends backwards. It does not. The joint at the top of the Tarsometatarsus is the "heel" joint, where the Tarsometatarsus meets the Tibiotarsus. The "knee" joint is between the Tibiotarsus and Femur.

Taxon (pl. taxa)—Any unit of scientific classification (e.g., species, genus, family, order).

Taxonomist— A specialist in the naming and classification of organisms. (See also Systematist. Taxonomy is the older science of naming things; identification of evolutionary relationships has not always been the goal of taxonomists. The modern science of Systematics generally incorporates taxonomy with the search for evolutionary relationships.)

Taxonomy—The science of identifying, naming, and classifying organisms into groups.

Teleoptiles—Juvenal feathers.

Territory—Any defended area. Typically birds defend a territory with sound such as song or drumming. Territorial defense is typically male against male, female against female, and within a species or between sibling species. Area defended varies greatly among taxa, seasons, and habitats. A territory may include the entire home range, only the area immediately around a nest, or only a feeding or roosting area.

Tertiary—The geological period including most of the Cenozoic; from about 65 to 1.8 million years ago.

Tertial—A flight feather of the wing that is loosely associated with the humerus; tertials fill the gap between the secondary feathers and the body.

Thermoregulation—The ability to regulate body temperature; can be either behavioral or physiological. Birds can regulate body temperature by sunning or moving to shade or water, but also generally regulate their body temperature through metabolic processes. Baby birds initially have poor thermoregulatory abilities and thus must be brooded.

Threatened—A category defined by IUCN and by the Endangered Species Act of 1973 in the United States to refer to a species that is at risk of becoming endangered.

Tomium (pl. tomia)—The cutting edges of a bird's bill.

Torpor—A period of reduced metabolic activity and lowered body temperature; often results from reduced availability of food or inclement weather; generally lasts for only a few hours (e.g., hummingbirds, Trochilidae; swifts, Apodidae).

Totipalmate—All toes joined by webs, a characteristic that identifies members of the order Pelecaniformes.

Tribe—A unit of classification below the subfamily and above the genus.

Tubercle—A knob- or wart-like projection.

Urohydrosis—A behavior characteristic of storks and New World vultures (Ciconiiformes) wherein these birds excrete on their legs and make use of the evaporation of the water from the excrement as an evaporative cooling mechanism.

Uropygial gland—A large gland resting atop the last fused vertebrae of birds at the base of a bird's tail; also known as oil gland or preen gland; secretes an oil used in preening.

Vane—The combined barbs that form a strong, yet flexible surface extending from the rachis of a feather.

Vaned feather—Any feather with vanes.

Viable population—A population that is capable of maintaining itself over a period of time. One of the major conservation issues of the twenty-first century is determining what is a minimum viable population size. Population geneticists have generally come up with estimates of about 500 breeding pairs.

Vibrissae—See Rictal bristle.

Vulnerable—A category defined by IUCN as a species that is not Critically Endangered or Endangered, but is still facing a threat of extinction.

Wallacea—The area of Indonesia transition between the Oriental and Australian biogeographical realms, named after Alfred Russell Wallace, who intensively studied this area.

Wattles—Sexual ornamentation that usually consists of flaps of skin on or near the base of the bill.

Zoogeographic region—See Biogeographic region.

Zygodactyl—Adjective referring to the arrangement of toes on a bird in which two toes project forward and two to the back.

Compiled by Jerome A. Jackson, PhD

Aves species list

Struthioniformes [Order]
Struthionidae [Family]
Struthio [Genus]
S. camelus [Species]

Rheidae [Family]
Rhea [Genus]
R. Americana [Species]
Pterocnemia [Genus]
P. pennata [Species]

Casuaridae [Family]
Casuarius [Genus]
C. bennetti [Species]
C. casuarius
C. unappendiculatus

Dromaiidae [Family]
Dromaius [Genus]
D. novaehollandiae [Species]
D. diemenianus

Apterygidae [Family]
Apteryx [Genus]
A. australis [Species]
A. owenii
A. haastii

Tinamiiformes [Order]
Tinamidae [Family]
Tinamus [Genus]
T. tao [Species]
T. solitarius
T. osgoodi
T. major
T. guttatus
Nothocercus [Genus]
N. bonapartei [Species]
N. julius
N. nigrocapillus
Crypturellus [Genus]
C. berlepschi [Species]
C. cinereus
C. soui
C. ptaritepui
C. obsoletus
C. undulatus
C. transfasciatus

C. strigulosus
C. duidae
C. erythropus
C. noctivagus
C. atrocapillus
C. cinnamomeus
C. boucardi
C. kerriae
C. variegatus
C. brevirostris
C. bartletti
C. parvirostris
C. casiquiare
C. tataupa
Rhynchotus [Genus]
R. rufescens [Species]
Nothoprocta [Genus]
N. taczanowski [Species]
N. kalinowskii
N. omata
N. perdicaria
N. cinerascens
N. pentlandii
N. curvirostris
Nothura [Genus]
N. boraquira [Species]
N. minor
N. darwinii
N. maculosa
Taoniscus [Genus]
T. nanus [Species]
Eudromia [Genus]
E. elegans [Species]
E. formosa
Tinamotis [Genus]
T. pentlandii [Species]
T. ingoufi

Procellariiformes [Order]
Diomedidae [Family]
Diomedea [Genus]
D. exulans [Species]
D. epomophora
D. irrorata
D. albatrus
D. nigripes
D. immutabilis

D. melanophrys
D. cauta
D. chrysostoma
D. chlororhynchos
D. bulleri
Phoebetria [Genus]
P. fusca [Species]
P. palpebrata
Macronectes [Genus]
M. giganteus [Species]
M. halli
Fulmarus [Genus]
F. glacialoides [Species]
F. glacialis
Thalassoica [Genus]
T. antarctica [Species]
Daption [Genus]
D. capense [Species]
Pagodroma [Genus]
P. nivea [Species]
Pterodroma [Genus]
P. macroptera [Species]
P. lessonii
P. incerta
P. solandri
P. magentae
P. rostrata
P. macgillivrayi
P. neglecta
P. arminjoniana
P. alba
P. ultima
P. brevirostris
P. mollis
P. inexpectata
P. cahow
P. hasitata
P. externa
P. baraui
P. phaeopygia
P. hypoleuca
P. nigripennis
P. axillaris
P. cookii
P. defilippiana
P. longirostris
P. leucoptera

Halobaena [Genus]
  H. caerulea
Pachyptila [Genus]
  P. vittata [Species]
  P. desolata
  P. belcheri
  P. turtur
  P. crassirostris
Bulweria [Genus]
  B. bulwerii [Species]
  B. fallax
Procellaria [Genus]
  P. aequinoctialis [Species]
  P. westlandica
  P. parkinsoni
  P. cinerea
Calonectris [Genus]
  C. diomedea [Species]
  C. leucomelas
Puffinus [Genus]
  P. pacificus [Species]
  P. bulleri
  P. carneipes
  P. creatopus
  P. gravis
  P. griseus
  P. tenuirostris
  P. nativitatis
  P. puffinus
  P. gavia
  P. huttoni
  P. lherminieri
  P. assimilis
Oceanites [Genus]
  O. oceanicus [Species]
  O. gracilis
Garrodia [Genus]
  G. nereis [Species]
Pelagodroma [Genus]
  P. marina [Species]
Fregetta [Genus]
  F. tropica [Species]
  F. grallaria
Nesofregetta [Genus]
  N. fuliginosa [Species]
Hydrobates [Genus]
  H. pelagicus [Species]
Halocyptena [Genus]
  H. microsoma [Species]
Oceanodroma [Genus]
  O. tethys [Species]
  O. castro
  O. monorhis
  O. leucorhoa
  O. macrodactyla
  O. markhami
  O. tristami
  O. melania
  O. matsudairae
  O. homochroa

  O. hornbyi
  O. furcata
Pelecanoides [Genus]
  P. garnotii [Species]
  P. magellani
  P. georgicus
  P. urinator

Sphenisciformes [Order]
  Spheniscidae [Family]
    Aptenodytes [Genus]
      A. patagonicus [Species]
      A. forsteri
    Pygoscelis [Genus]
      P. papua [Species]
      P. adeliae
      P. antarctica
    Eudyptes [Genus]
      E. chrysocome [Species]
      E. pachyrhynchus
      E. robustus
      E. sclateri
      E. chryoslophus
    Megadyptes [Genus]
      M. antipodes [Species]
    Eudyptula [Genus]
      E. minor [Species]
    Spheniscus [Genus]
      S. demersus [Species]
      S. humboldti
      S. magellanicus
      S. mendiculus

Gaviiformes [Order]

  Gaviidae [Family]
    Gavia [Genus]
      G. stellata [Species]
      G. arctica
      G. immer
      G. adamsii

Podicipediformes [Order]

  Podicipedidae [Family]
    Rollandia [Genus]
      R. rolland [Species]
      R. microptera
    Tachybaptus [Genus]
      T. novaehollandiae [Species]
      T. ruficollis
      T. rufolavatus
      T. pelzelnii
      T. dominicus
    Podilymbus [Genus]
      P. podiceps [Species]
      P. gigas
    Poliocephalus [Genus]
      P. poliocephalus [Species]
      P. rufopectus
    Podiceps [Genus]
      P. major [Species]

      P. auritus
      P. grisegena
      P. cristatus
      P. nigricollis
      P. occipitalis
      P. taczanowskii
      P. gallardoi
    Aechmophorus [Genus]
      A. occidentalis [Species]

Pelecaniformes [Order]
  Phaethontidae [Family]
    Phaethon [Genus]
      P. aethereus [Species]
      P. rubricauda
      P. lepturus

  Fregatidae [Family]
    Fregata [Genus]
      F. magnificens [Species]
      F. minor
      F. ariel
      F. andrewsi

  Phalacrocoracidae [Family]
    Phalacrocorax [Genus]
      P. carbo [Species]
      P. capillatus
      P. nigrogularis
      P. varius
      P. harrisi
      P. auritus
      P. olivaceous
      P. fuscicollis
      P. sulcirostris
      P. penicillatus
      P. capensis
      P. neglectus
      P. punctatus
      P. aristotelis
      P. perspicillatus
      P. urile
      P. pelagicus
      P. gaimardi
      P. magellanicus
      P. bouganvillii
      P. atriceps
      P. albiventer
      P. carunculatus
      P. campbelli
      P. fuscescens
      P. melanoleucos
      P. niger
      P. pygmaeus
      P. africanus
    Anhinga [Genus]
      A. anhinga [Species]
      A. melanogaster

  Sulidae [Family]
    Sula [Genus]
      S. bassana [Species]

S. capensis
S. serrator
S. nebouxii
S. variegata
S. dactylatra
S. sula
S. leucogaster
S. abbotti

Pelecanidae [Family]
Pelecanus [Genus]
P. onocrotalus [Species]
P. rufescens
P. philippensis
P. conspicillatus
P. erythrorhynchos
P. occidentalis

Ciconiiformes [Order]
Ardeidae [Family]
Syrigma [Genus]
S. sibilatrix [Species]
Pilherodius [Genus]
P. pileatus [Species]
Ardea [Genus]
A. cinerea [Species]
A. herodias
A. cocoi
A. pacifica
A. melanocephala
A. hombloti
A. imperialis
A. sumatrana
A. goliath
A. purpurea
A. alba
Egretta [Genus]
E. rufescens [Species]
E. picata
E. vinaceigula
E. ardesiaca
E. tricolor
E. intermedia
E. ibis
E. novaehollandiae
E. caerulea
E. thula
E. garzetta
E. gularis
E. dimorpha
E. eulophotes
E. sacra
Ardeola [Genus]
A. ralloides [Species]
A. grayii
A. bacchus
A. speciosa
A. idae
A. rufiventris
A. striata

Agamia [Genus]
A. agami [Species]
Nyctanassa [Genus]
N. violacea [Species]
Nycticorax [Genus]
N. nycticorax [Species]
N. caledonicus
N. leuconotus
N. magnificus
N. goisagi
N. melanolophus
Cochlearius [Genus]
C. cochlearius [Species]
Tigrisoma [Genus]
T. mexicanum [Species]
T. fasciatum
T. lineatum
Zonerdius [Genus]
Z. heliosylus [Species]
Tigriornis [Genus]
T. leucolophus [Species]
Zebrilus [Genus]
Z. undulatus [Species]
Ixobrychus [Genus]
I. involucris [Species]
I. exilis
I. minutus
I. sinensis
I. eurhythmus
I. cinnamomeus
I. sturmii
I. flavicollis
Botaurus [Genus]
B. pinnatus [Species]
B. lentiginosus
B. stellaris
B. poiciloptilus

Scopidae [Family]
Scopus [Genus]
S. umbretta [Species]

Ciconiidae [Family]
Mycteria [Genus]
M. americana [Species]
M. cinerea
M. ibis
M. leucocephala
Anastomus [Genus]
A. oscitans [Species]
A. lamelligerus
Ciconia [Genus]
C. nigra [Species]
C. abdimii
C. episcopus
C. maguari
C. ciconia
Ephippiorhynchus [Genus]
E. asiaticus [Species]
E. senegalensis
Jabiru [Genus]
J. mycteria [Species]

Leptoptilos [Genus]
L. javanicus [Species]
L. dubius
L. crumeniferus

Balaenicipitidae [Family]
Balaeniceps [Genus]
B. rex [Species]

Threskiornithidae [Family]
Eudocimus [Genus]
E. albus [Species]
E. ruber
Phimosus [Genus]
P. infuscatus [Species]
Plegadis [Genus]
P. falcinellus [Species]
P. chihi
P. ridgwayi
Cercibis [Genus]
C. oxycerca [Species]
Theristicus [Genus]
T. caerulescens [Species]
T. caudatus
T. melanopsis
Mesembrinibis [Genus]
M. cayennensis [Species]
Bostrychia [Genus]
B. hagedash [Species]
B. carunculata
B. olivacea
B. rara
Lophotibis [Genus]
L. cristata [Species]
Threskiornis [Genus]
T. aethiopicus [Species]
T. spinicollis
Geronticus [Genus]
G. eremita [Species]
G. calvus
Pseudibis [Genus]
P. papillosa [Species]
P. gigantea
Nipponia [Genus]
N. nippon [Species]
Platalea [Genus]
P. leucocorodia [Species]
P. minor
P. alba
P. flavipes
P. ajaja

Phoenicopteriformes [Order]
Phoenicopteridae [Family]
Phoenicopterus [Genus]
P. ruber [Species]
P. chilensis
Phoeniconaias [Genus]
P. minor [Species]
Phoenicoparrus [Genus]
P. andinus [Species]
P. jamesii

Falconiformes [Order]
Cathartidae [Family]
  Coragyps [Genus]
    C. atratus [Species]
  Cathartes [Genus]
    C. burrovianus [Species]
    C. melambrotus
  Gymnogyps [Genus]
    G. californianus [Species]
  Vultur [Genus]
    V. gryphus [Species]
  Sarcoramphus [Genus]
    S. papa [Species]

Accipitridae [Family]
  Pandion [Genus]
    P. haliaetus [Species]
  Aviceda [Genus]
    A. cuculoides [Species]
    A. madagascariensis
    A. jerdoni
    A. subcristata
    A. leuphotes
  Leptodon [Genus]
    L. cayanensis [Species]
  Chondrohierax [Genus]
    C. uncinatus [Species]
  Henicopernis [Genus]
    H. longicauda [Species]
    H. infuscata
  Pernis [Genus]
    P. aviporus [Species]
    P. ptilorhynchus
    P. celebensis
  Elanoides [Genus]
    E. forficatus [Species]
  Macheiramphus [Genus]
    M. alcinus [Species]
  Gampsonyx [Genus]
    G. swainsonii [Species]
  Elanus [Genus]
    E. leucurus [Species]
    E. caeruleus
    E. notatus
    E. scriptus
  Chelictinia [Genus]
    C. riocourii [Species]
  Rostrhamus [Genus]
    R. sociabilis [Species]
    R. hamatus
  Harpagus [Genus]
    H. bidentatus [Species]
    H. diodon
  Ictinia [Genus]
    I. plumbea [Species]
    I. misisippiensis
  Lophoictinia [Genus]
    L. isura [Species]
  Hamirostra [Genus]
    H. melanosternon [Species]

Milvus [Genus]
  M. milvus [Species]
  M. migrans
Haliastur [Genus]
  H. sphenurus [Species]
  H. indus
Haliaeetus [Genus]
  H. leucogaster [Species]
  H. sanfordi
  H. vocifer
  H. vociferoides
  H. leucoryphus
  H. albicilla
  H. leucocephalus
  H. pelagicus
Ichthyophaga [Genus]
  I. humilis [Species]
  I. ichthyaetus
Gypohierax [Genus]
  G. angolensis [Species]
Gypaetus [Genus]
  G. barbatus [Species]
Neophron [Genus]
  N. percnopterus [Species]
Necrosyrtes [Genus]
  N. monachus [Species]
Gyps [Genus]
  G. bengalensis [Species]
  G. africanus
  G. indicus
  G. rueppellii
  G. himalayensis
  G. fulvus
Aegypius [Genus]
  A. monachus [Species]
  A. tracheliotus
  A. occipitalis
  A. calvus
Circaetus [Genus]
  C. gallicus [Species]
  C. cinereus
  C. fasciolatus
  C. cinerascens
Terathopius [Genus]
  T. ecaudatus [Species]
Spilornis [Genus]
  S. cheela [Species]
  S. elgini
Dryotriorchis [Genus]
  D. spectabilis [Species]
Eutriorchis [Genus]
  E. astur [Species]
Polyboroides [Genus]
  P. typus [Species]
  P. radiatus
Circus [Genus]
  C. assimilis [Species]
  C. maurus
  C. cyaneus
  C. cinereus

C. macrourus
C. melanoleucos
C. pygargus
C. ranivorus
C. aeruginosus
C. spilonotus
C. approximans
C. maillardi
C. buffoni
Melierax [Genus]
  M. gabar [Species]
  M. metabates
  M. canorus
Accipiter [Genus]
  A. poliogaster [Species]
  A. trivirgatus
  A. griseiceps
  A. tachiro
  A. castanilius
  A. badius
  A. brevipes
  A. butleri
  A. soloensis
  A. francesii
  A. trinotatus
  A. fasciatus
  A. novaehollandiae
  A. melanochlamys
  A. albogularis
  A. rufitorques
  A. haplochrous
  A. henicogrammus
  A. luteoschistaceus
  A. imitator
  A. poliocephalus
  A. princeps
  A. superciliosus
  A. collaris
  A. erythropus
  A. minullus
  A. gularis
  A. virgatus
  A. nanus
  A. cirrhocephalus
  A. brachyurus
  A. erythrauchen
  A. rhodogaster
  A. ovampensis
  A. madagascariensis
  A. nisus
  A. rufiventris
  A. striatus
  A. bicolor
  A. cooperii
  A. gundlachi
  A. melanoleucus
  A. henstii
  A. gentilis
  A. meyerianus
  A. buergersi

A. radiatus
A. doriae
Urotriorchis [Genus]
   U. macrourus [Species]
Butastur [Genus]
   B. rufipennis [Species]
   B. liventer
   B. teesa
   B. indicus
Kaupifalco [Genus]
   K. monogrammicus [Species]
Geranospiza [Genus]
   G. caerulescens [Species]
Leucopternis [Genus]
   L. schistacea [Species]
   L. plumbea
   L. princeps
   L. melanops
   L. kuhli
   L. lacernulata
   L. semiplumbea
   L. albicollis
   L. polionota
Asturina [Genus]
   A. nitida [Species]
Buteogallus [Genus]
   B. aequinoctialis [Species]
   B. subtilis
   B. anthracinus
   B. urubitinga
   B. meridionalis
Parabuteo [Genus]
   P. unicinctus [Species]
Busarellus [Genus]
   B. nigricollis [Species]
Geranoaetus [Genus]
   G. melanoleucus [Species]
Harpyhaliaetus [Genus]
   H. solitarius [Species]
   H. coronatus
Buteo [Genus]
   B. magnirostris [Species]
   B. leucorrhous
   B. ridgwayi
   B. lineatus
   B. platypterus
   B. brachyurus
   B. swainsoni
   B. galapagoensis
   B. albicaudatus
   B. polyosoma
   B. poecilochrous
   B. albonotatus
   B. solitarius
   B. ventralis
   B. jamaicensis
   B. buteo
   B. oreophilus
   B. brachypterus
   B. rufinus

B. hemilasius
B. regalis
B. lagopus
B. auguralis
B. rufofuscus
Morphnus [Genus]
   M. guianensis [Species]
Harpia [Genus]
   H. harpyja [Species]
Pithecophaga [Genus]
   P. jeffreyi [Species]
Ictinaetus [Genus]
   I. malayensis [Species]
Aquila [Genus]
   A. pomarina [Species]
   A. clanga
   A. rapax
   A. heliaca
   A. wahlbergi
   A. gurneyi
   A. chrysaetos
   A. audax
   A. verreauxii
Hieraaetus [Genus]
   H. fasciatus [Species]
   H. spilogaster
   H. pennatus
   H. morphnoides
   H. dubius
   H. kienerii
Spizastur [Genus]
   S. melanoleucus [Species]
Lophaetus [Genus]
   L. occipitalis [Species]
Spizaetus [Genus]
   S. africanus [Species]
   S. cirrhatus
   S. nipalensis
   S. bertelsi
   S. lanceolatus
   S. philippensis
   S. alboniger
   S. nanus
   S. tyrannus
   S. ornatus
Stephanoaetus [Genus]
   S. coronatus [Species]
Oroaetus [Genus]
   O. isidori [Species]
Polemaetus [Genus]
   P. bellicosus [Species]

Sagittariidae [Family]
   Sagittarius [Genus]
      S. serpentarius [Species]

Falconidae [Family]
   Daptrius [Genus]
      D. ater [Species]
      D. americanus

Phalcoboenus [Genus]
   P. megalopterus [Species]
   P. australis
Polyborus [Genus]
   P. plancus [Species]
Milvago [Genus]
   M. chimachima [Species]
   M. chimango
Herpetotheres [Genus]
   H. cachinnans [Species]
Micrastur [Genus]
   M. ruficollis [Species]
   M. gilvicollis
   M. mirandollei
   M. semitorquatus
   M. buckleyi
Spiziapteryx [Genus]
   S. circumcinctus [Species]
Polihierax [Genus]
   P. semitorquatus [Species]
   P. insignis
Microhierax [Genus]
   M. caerulescens [Species]
   M fringillarius
   M. latifrons
   M. erythrogerys
   M. melanoleucus
Falco [Genus]
   F. berigora [Species]
   F. naumanni
   F. sparverius
   F. tinnunculus
   F. newtoni
   F. punctatus
   F. araea
   F. moluccensis
   F. cenchroides
   F. rupicoloides
   F. alopex
   F. ardosiaceus
   F. dickinsoni
   F. zoniventris
   F. chicquera
   F. vespertinus
   F. amurensis
   F. eleonorae
   F. concolor
   F. femoralis
   F. columbarius
   F. rufigularis
   F. subbuteo
   F. cuvieri
   F. severus
   F. longipennis
   F. novaeseelandiae
   F. hypoleucos
   F. subniger
   F. mexicanus
   F. jugger
   F. biarmicus

F. cherrug
F. rusticolus
F. kreyenborgi
F. peregrinus
F. deiroleucus
F. fasciinucha

Anseriformes [Order]
Anatidae [Family]
Anseranas [Genus]
  A. semipalmata [Species]
Dendrocygna [Genus]
  D. guttata [Species]
  D. eytoni
  D. bicolor
  D. arcuata
  D. javanica
  D. viduata
  D. arborea
  D. autumnalis
Thalassornis [Genus]
  T. leuconotus [Species]
Cygnus [Genus]
  C. olor [Species]
  C. atratus
  C. melanocoryphus
  C. buccinator
  C. cygnus
  C. bewickii
  C. columbianus
Coscoroba [Genus]
  C. coscoroba [Species]
Anser [Genus]
  A. cygnoides [Species]
  A. fabalis
  A. albifrons
  A. erythropus
  A. anser
  A. indicus
  A. caerulescens
  A. rossii
  A. canagicus
Branta [Genus]
  B. sandvicensis [Species]
  B. canadensis
  B. leucopsis
  B. bernicla
  B. ruficollis
Cereopsis [Genus]
  C. novaehollandiae [Species]
Stictonetta [Genus]
  S. naevosa [Species]
Cyanochen [Genus]
  C. cyanopterus [Species]
Chloephaga [Genus]
  C. melanoptera [Species]
  C. picta
  C. hybrida
  C. poliocephala
  C. rubidiceps

Neochen [Genus]
  N. jubata [Species]
Alopochen [Genus]
  A. aegyptiaca [Species]
Tadorna [Genus]
  T. ferruginea [Species]
  T. cana
  T. variegata
  T. cristata
  T. tadornoides
  T. tadorna
  T. radjah
Tachyeres [Genus]
  T. pteneres [Species]
  T. brachypterus
  T. patachonicus
Plectropterus [Genus]
  P. gambensis [Species]
Cairina [Genus]
  C. moschata [Species]
  C. scutulata
Pteronetta [Genus]
  P. hartlaubii
Sarkidiornis [Genus]
  S. melanotos [Species]
Nettapus [Genus]
  N. pulchellus [Species]
  N. coromandelianus
  N. auritus
Callonetta [Genus]
  C. leucophrys [Species]
Aix [Genus]
  A. sponsa [Species]
  A. galericulata
Chenonetta [Genus]
  C. jubata [Species]
Amazonetta [Genus]
  A. brasiliensis [Species]
Merganetta [Genus]
  M. armata [Species]
Hymenolaimus [Genus]
  H. malacorhynchos [Species]
Anas [Genus]
  A. waigiuensis [Species]
  A. penelope
  A. americana
  A. sibilatrix
  A. falcata
  A. strepera
  A. formosa
  A. crecca
  A. flavirostris
  A. capensis
  A. gibberifrons
  A. bernieri
  A. castanea
  A. aucklandica
  A. platyrhynchos
  A. rubripes
  A. undulata

A. melleri
A. poecilorhyncha
A. superciliosa
A. luzonica
A. sparsa
A. specularioides
A. specularis
A. acuta
A. georgica
A. bahamensis
A. erythrorhyncha
A. versicolor
A. hottentota
A. querquedula
A. discors
A. cyanoptera
A. platalea
A. smithii
A. rhynchotis
A. clypeata
Malacorhynchus [Genus]
  M. membranaceus [Species]
Marmaronetta [Genus]
  M. angustirostris [Species]
Rhodonessa [Genus]
  R. caryophyllacea [Species]
Netta [Genus]
  N. rufina [Species]
  N. peposaca
  N. erythrophthalma
Aythya [Genus]
  A. valisineria [Species]
  A. ferina
  A. americana
  A. collaris
  A. australis
  A. baeri
  A. nyroca
  A. innotata
  A. novaeseelandiae
  A. fuligula
  A. marila
  A. affinis
Somateria [Genus]
  S. mollissima [Species]
  S. spectabilis
  S. fischeri
Polysticta [Genus]
  P. stelleri [Species]
Camptorhynchus [Genus]
  C. labradorius [Species]
Histrionicus [Genus]
  H. histrionicus [Species]
Clangula [Genus]
  C. hyemalis [Species]
Melanitta [Genus]
  M. nigra [Species]
  M. perspicillata
  M. fusca

Bucephala [Genus]
  B. clangula [Species]
  B. islandica
  B. albeola
Mergus [Genus]
  M. albellus [Species]
  M. cucullatus
  M. octosetaceous
  M. serrator
  M. squamatus
  M. merganser
  M. australis
Heteronetta [Genus]
  H. atricapilla [Species]
Oxyura [Genus]
  O. dominica [Species]
  O. jamaicensis
  O. leucocephala
  O. maccoa
  O. vittata
  O. australis
Biziura [Genus]
  B. lobata [Species]

Anhimidae [Family]
Anhima [Genus]
  A. cornuta [Species]
Chauna [Genus]
  C. chavaria [Species]
  C. torquata

Galliformes [Order]
Megapodiidae [Family]
Megapodius [Genus]
  M. nicobariens [Species]
  M. tenimberens
  M. reinwardt
  M. affinis
  M. eremita
  M. freycinet
  M. laperouse
  M. layardi
  M. pritchardii
Eulipoa [Genus]
  E. wallacei [Species]
Leipoa [Genus]
  L. ocellata [Species]
Alectura [Genus]
  A. lathami [Species]
Talegalla [Genus]
  T. cuvieri [Species]
  T. fuscirostris
  T. jobiensis
Aepypodius [Genus]
  A. arfakianus [Species]
  A. bruijnii
Macrocephalon [Genus]
  M. maleo [Species]

Cracidae [Family]
Nothocrax [Genus]
  N. urumutum [Species]

Mitu [Genus]
  M. tomentosa [Species]
  M. salvini
  M. mitu
Pauxi [Genus]
  P. pauxi [Species]
Crax [Genus]
  C. nigra [Species]
  C. alberti
  C. fasciolata
  C. pinima
  C. globulosa
  C. blumenbachii
  C. rubra
Penelope [Genus]
  P. purpurascens [Species]
  P. ortoni
  P. albipennis
  P. marail
  P. montagnii
  P. obscura
  P. superciliaris
  P. jacu-caca
  P. ochrogaster
  P. pileata
  P. argyrotis
Ortalis [Genus]
  O. motmot [Species]
  O. spixi
  O. araucuan
  O. superciliaris
  O. guttata
  O. columbiana
  O. wagleri
  O. vetula
  O. ruficrissa
  O. ruficauda
  O. garrula
  O. canicollis
  O. erythroptera
Penelopina [Genus]
  P. nigra [Species]
Chamaepetes [Genus]
  C. goudotii [Species]
  C. unicolor
Pipile [Genus]
  P. pipile [Species]
  P. cumanensis
  P. jacutinga
Aburria [Genus]
  A. aburri [Species]
Oreophasis [Genus]
  O. derbianus [Species]

Tetraonidae [Family]
Tetrao [Genus]
  T. urogallus [Species]
  T. parvirostris
Lyrurus [Genus]
  L. tetrix [Species]
  L. mlokosiewiczi

Dendragapus [Genus]
  D. obscurus [Species]
Lagopus [Genus]
  L. scoticus [Species]
  L. lagopus
  L. mutus
  L. leucurus
Canachites [Genus]
  C. canadensis [Species]
  C. franklinii
Falcipennis [Genus]
  F. falcipennis [Species]
Tetrastes [Genus]
  T. bonasia [Species]
  T. sewerzowi
Bonasa [Genus]
  B. umbellus [Species]
Pedioecetes [Genus]
  P. phasianellus [Species]
Tympanuchus [Genus]
  T. cupido [Species]
  T. palladicinctus
Centrocercus [Genus]
  C. urophasianus [Species]

Phasianidae [Family]
Dendrortyx [Genus]
  D. barbatus [Species]
  D. macroura
  D. leucophrys
  D. hypospodius
Oreortyx [Genus]
  O. picta [Species]
Callipepla [Genus]
  C. squamota [Species]
Lophortyx [Genus]
  L. californica [Species]
  L. gambelli
  L. leucoprosopon
  L. douglasii
Philortyx [Genus]
  P. fasciatus [Species]
Colinus [Genus]
  C. virginianus [Species]
  C. nigrogularis
  C. leucopogon
  C. cristatus
Odontophorus [Genus]
  O. gujanensis [Species]
  O. capueira
  O. erythrops
  O. hyperythrus
  O. melanonotus
  O. speciosus
  O. loricatus
  O. parambae
  O. strophium
  O. atrifrons
  O. leucolaemus
  O. columbianus
  O. soderstromii

O. balliviani
O. stellatus
O. guttatus
Dactylortyx [Genus]
   D. thoracicus [Species]
Cyrtonyx [Genus]
   C. montezumae [Species]
   C. sallei
   C. ocellatus
Rhynchortyx [Genus]
   R. cinctus [Species]
Lerwa [Genus]
   L. lerwa [Species]
Ammoperdix [Genus]
   A. griseogularis [Species]
   A. heyi
Tetraogallus [Genus]
   T. caucasicus [Species]
   T. caspius
   T. tibetanus
   T. altaicus
   T. himalayensis
Tetraophasis [Genus]
   T. obscurus [Species]
   T. szechenyii
Alectoris [Genus]
   A. graeca [Species]
   A. rufa
   A. barbara
   A. melanocephala
Anurophasis [Genus]
   A. monorthonyx [Species]
Francolinus [Genus]
   F. francolinus [Species]
   F. pictus
   F. pintadeanus
   F. pondicerianus
   F. gularis
   F. lathami
   F. nahani
   F. streptophorus
   F. coqui
   F. albogularis
   F. sephaena
   F. africanus
   F. shelleyi
   F. levaillantii
   F. finschi
   F. gariepensis
   F. adspersus
   F. capensis
   F. natalensis
   F. harwoodi
   F. bicalcaratus
   F. icterorhynchus
   F. clappertoni
   F. hartlaubi
   F. swierstrai
   F. hildebrandti
   F. squamatus

F. ahantensis
F. griseostriatus
F. camerunensis
F. nobilis
F. jacksoni
F. castaneicollis
F. atrifrons
F. erckelii
Pternistis [Genus]
   P. rufopictus [Species]
   P. afer
   P. swainsonii
   P. leucoscepus
Perdix [Genus]
   P. perdix [Species]
   P. barbata
   P. hodgsoniae
Rhizothera [Genus]
   R. longirostris [Species]
Margaroperdix [Genus]
   M. madagarensis [Species]
Melanoperdix [Genus]
   M. nigra [Species]
Coternix [Genus]
   C. coturnix [Species]
   C. coromandelica
   C. delegorguei
   C. pectoralis
   C. novaezelandiae
Synoicus [Genus]
   S. ypsilophorus [Species]
Excalfactoria [Genus]
   E. adansonii [Species]
   E. chinensis
Perdicula [Genus]
   P. asiatica [Species]
Cryptoplectron [Genus]
   C. erythrorhynchum [Species]
   C. manipurensis
Arborophila [Genus]
   A. torqueola [Species]
   A. rufogularis
   A. atrogularis
   A. crudigularis
   A. mandellii
   A. brunneopectus
   A. rufipectus
   A. gingica
   A. davidi
   A. cambodiana
   A. orientalis
   A. javanica
   A. rubrirostris
   A. hyperythra
   A. ardens
Tropicoperdix [Genus]
   T. charltonii [Species]
   T. chloropus
   T. merlini
Caloperdix [Genus]
   C. oculea [Species]

Haematortyx [Genus]
   H. sanguiniceps [Species]
Rollulus [Genus]
   R. roulroul [Species]
Ptilopachus [Genus]
   P. petrosus [Species]
Bambusicola [Genus]
   B. fytchii [Species]
   B. thoracica
Galloperdix [Genus]
   G. spadicea [Species]
   G. lunulata
   G. bicalcarata
Ophrysia [Genus]
   O. superciliosa [Species]
Ithaginis [Genus]
   I. cruentus [Species]
Tragopan [Genus]
   T. melanocephalus [Species]
   T. satyra
   T. blythii
   T. temminckii
   T. caboti
Lophophorus [Genus]
   L. impejanus [Species]
   L. sclateri
   L. lhuysii
Crossoptilon [Genus]
   C. mantchuricum [Species]
   C. auritum
   C. crossoptilon
Gennaeus [Genus]
   G. leucomelanos [Species]
   G. horsfieldii
   G. lineatus
   G. nycthemerus
Hierophasis [Genus]
   H. swinhoii [Species]
   H. imperialis
   H. edwardsi
Houppifer [Genus]
   H. erythrophthalmus [Species]
   H. inornatus
Lophura [Genus]
   L. rufa [Species]
   L. ignita
Diardigallus [Genus]
   D. diardi [Species]
Lobiophasis [Genus]
   L. bulweri [Species]
Gallus [Genus]
   G. gallus [Species]
   G. lafayetii
   G. sonneratii
   G. varius
Pucrasia [Genus]
   P. macrolopha [Species]
Catreus [Genus]
   C. wallichii [Species]
Phasianus [Genus]
   P. colchicus [Species]

Syrmaticus [Genus]
 S. reevesii [Species]
 S. soemmerringii
 S. humiae
 S. ellioti
 S. mikado
Chrysolophus [Genus]
 C. pictus [Species]
 C. amherstiae
Chalcurus [Genus]
 C. inopinatus [Species]
 C. chalcurus
Polyplecton [Genus]
 P. bicalcaratum [Species]
 P. germaini
 P. malacensis
 P. schleiermacheri
 P. emphanum
Rheinardia [Genus]
 R. ocellata [Species]
Argusianus [Genus]
 A. argus [Species]
Pavo [Genus]
 P. cristatus [Species]
 P. muticus

Numididae [Family]
Phasidus [Genus]
 P. niger [Species]
Agelastes [Genus]
 A. meleagrides [Species]
Numida [Genus]
 N. meleagris [Species]
Guttera [Genus]
 G. plumifera [Species]
 G. edouardi
 G. pucherani
Acryllium [Genus]
 A. vulturinum [Species]

Meleagridae [Family]
Meleagris [Genus]
 M. gallopavo [Species]
Agriocharis [Genus]
 A. ocellata [Species]

Opisthocomidae [Family]
Opisthocomus [Genus]
 O. hoazin [Species]

Gruiformes [Order]
Mesoenatidae [Family]
Mesoenas [Genus]
 M. variegata [Species]
 M. unicolor
Monias [Genus]
 M. benschi [Species]

Turnicidae [Family]
Turnix [Genus]
 T. sylvatica [Species]
 T. worcesteri

T. nana
T. hottentotta
T. tanki
T. suscitator
T. nigricollis
T. ocellata
T. melanogaster
T. varia
T. castanota
T. pyrrhothorax
T. velox
Ortyxelos [Genus]
 O. meiffrenii [Species]
Pedionomus [Genus]
 P. torquatus [Species]

Gruidae [Family]
Grus [Genus]
 G. grus [Species]
 G. nigricollis
 G. monacha
 G. canadensis
 G. japonensis
 G. americana
 G. vipio
 G. antigone
 G. rubicunda
 G. leucogeranus
Bugeranus [Genus]
 B. carunculatus [Species]
Anthropoides [Genus]
 A. virgo [Species]
 A. paradisea
Balearica [Genus]
 B. pavonina [Species]

Aramidae [Family]
Aramus [Genus]
 A. scolopaceus [Species]

Psophiidae [Family]
Psophia [Genus]
 P. crepitans [Species]
 P. leucoptera
 P. viridis

Rallidae [Family]
Rallus [Genus]
 R. longirostris [Species]
 R. elegans
 R. limicola
 R. semiplumbeus
 R. aquaticus
 R. caerulescens
 R. madagascariensis
 R. pectoralis
 R. muelleri
 R. striatus
 R. philippensis
 R. ecaudata
 R. torquatus
 R. owstoni
 R. wakensis

Nesolimnas [Genus]
 N. dieffenbachii [Species]
Cabalus [Genus]
 C. modestus [Species]
Atlantisia [Genus]
 A. rogersi [Species]
Tricholimnas [Genus]
 T. conditicius [Species]
 T. lafresnayanus
 T. sylvestris
Ortygonax [Genus]
 O. rytirhynchos [Species]
 O. nigricans
Pardirallus [Genus]
 P. maculatus [Species]
Dryolimnas [Genus]
 D. cuvieri [Species]
Rougetius [Genus]
 R. rougetii [Species]
Amaurolimnas [Genus]
 A. concolor [Species]
Rallina [Genus]
 R. fasciata [Species]
 R. eurizonoides
 R. canningi
 R. tricolor
Rallicula [Genus]
 R. rubra [Species]
 R. leucospila
Cyanolimnas [Genus]
 C. cerverai [Species]
Aramides [Genus]
 A. mangle [Species]
 A. cajanea
 A. wolfi
 A. gutturalis
 A. ypecaha
 A. axillaris
 A. calopterus
 A. saracura
Aramidopsis [Genus]
 A. plateni [Species]
Nesoclopeus [Genus]
 N. poeciloptera [Species]
 N. woodfordi
Gymnocrex [Genus]
 G. rosenbergii [Species]
 G. plumbeiventris
Gallirallus [Genus]
 G. australis [Species]
 G. troglodytes
Habropteryx [Genus]
 H. insignis [Species]
Habroptila [Genus]
 H. wallacii [Species]
Megacrex [Genus]
 M. inepta [Species]
Eulabeornis [Genus]
 E. castaneoventris [Species]
Himantornis [Genus]
 H. haematopus [Species]

AVES SPECIES LIST

Canirallus [Genus]
   C. oculeus [Species]
Mentocrex [Genus]
   M. kioloides [Species]
Crecopsis [Genus]
   C. egregria [Species]
Crex [Genus]
   C. crex [Species]
Anurolimnas [Genus]
   A. castaneiceps [Species]
Limnocorax [Genus]
   L. flavirostra [Species]
Porzana [Genus]
   P. parva [Species]
   P. pusilla
   P. porzana
   P. fluminea
   P. carolina
   P. spiloptera
   P. flaviventer
   P. albicollis
   P. fusca
   P. paykullii
   P. olivieri
   P. bicolor
   P. tabuensis
Porzanula [Genus]
   P. palmeri [Species]
Pennula [Genus]
   P. millsi [Species]
   P. sandwichensis
Nesophylax [Genus]
   N. ater [Species]
Aphanolimnas [Genus]
   A. monasa [Species]
Laterallus [Genus]
   L. jamaicensis [Species]
   L. spilonotus
   L. exilis
   L. albigularis
   L. melanophaius
   L. ruber
   L. levraudi
   L. viridis
   L. hauxwelli
   L. leucopyrrhus
Micropygia [Genus]
   M. schomburgkii [Species]
Coturnicops [Genus]
   C. exquisita [Species]
   C. noveboracensis
   C. notata
   C. ayresi
Neocrex [Genus]
   N. erythrops [Species]
Sarothura [Genus]
   S. rufa [Species]
   S. lugeus
   S. pulchra
   S. elegans

S. bohmi
S. antonii
S. lineata
S. insularis
S. watersi
Aenigmatolimnas [Genus]
   A. marginalis [Species]
Poliolimnas [Genus]
   P. cinereus [Species]
Porphyriops [Genus]
   P. melanops [Species]
Tribonyx [Genus]
   T. ventralis [Species]
   T. mortierii
Amaurornis [Genus]
   A. akool [Species]
   A. olivacea
   A. isabellina
   A. phoenicurus
Gallicrex [Genus]
   G. cinerea [Species]
Gallinula [Genus]
   G. tenebrosa [Species]
   G. chloropus
   G. angulata
Porphyriornis [Genus]
   P. nesiotis [Species]
   P. comeri
Pareudiastes [Genus]
   P. pacificus [Species]
Porphyrula [Genus]
   P. alleni [Species]
   P. martinica
   P. parva
Porphyrio [Genus]
   P. porphyrio [Species]
   P. madagascariensis
   P. poliocephalus
   P. albus
   P. pulverulentus
Notornis [Genus]
   N. mantelli [Species]
Fulica [Genus]
   F. atra [Species]
   F. cristata
   F. americana
   F. ardesiaca
   F. armillata
   F. caribaea
   F. leucoptera
   F. rufrifrons
   F. gigantea
   F. cornuta

Heliornithidae [Family]
   Podica [Genus]
      P. senegalensis [Species]
   Heliopais [Genus]
      H. personata [Species]
   Heliornis [Genus]
      H. fulica [Species]

Rhynochetidae [Family]
   Rhynochetos [Genus]
      R. jubatus [Species]

Eurypygidae [Family]
   Eurypyga [Genus]
      E. helias [Species]

Cariamidae [Family]
   Cariama [Genus]
      C. cristata [Species]
   Chunga [Genus]
      C. burmeisteri [Species]

Otidae [Family]
   Tetrax [Genus]
      T. tetrax [Species]
   Otis [Genus]
      O. tarda [Species]
   Neotis [Genus]
      N. cafra [Species]
      N. ludwigii
      N. burchellii
      N. Nuba
      N. heuglinii
   Choriotius [Genus]
      C. arabs [Species]
      C. kori
      C. nigriceps
      C. australis
   Chlamydotis [Genus]
      C. undulata [Species]
   Lophotis [Genus]
      L. savilei [Species]
      L. ruficrista
   Afrotis [Genus]
      A. atra [Species]
   Eupodotis [Genus]
      E. vigorsii [Species]
      E. ruppellii
      E. humilis
      E. senegalensis
      E. caerulescens
   Lissotis [Genus]
      L. melanogaster [Species]
      L. hartlaubii
   Houbaropsis [Genus]
      H. bengalensis [Species]
   Sypheotides [Genus]
      S. indica [Species]

Charadriiformes [Order]
Jacanidae [Family]
   Microparra [Genus]
      M. capensis [Species]
   Actophilornis [Genus]
      A. africana [Species]
      A. albinucha
   Irediparra [Genus]
      I. gallinacea [Species]
   Hydrophasianus [Genus]
      H. chirurgus [Species]

Metopidius [Genus]
    M. indicus [Species]
Jacana [Genus]
    J. spinosa [Species]

Rostratulidae [Family]
    Rostratula [Genus]
        R. benghalensis [Species]
    Nycticryphes [Genus]
        N. semicollaris [Species]

Haematopodidae [Family]
    Haematopus [Genus]
        H. ostralegus [Species]
        H. leucopodus
        H. fuliginosus
        H. ater
    Chettusia [Genus]
        C. leucura [Species]
        C. gregaria

Charadriidae [Family]
    Vanellus [Genus]
        V. vanellus [Species]
    Belonopterus [Genus]
        B. chilensis [Species]
    Hemiparra [Genus]
        H. crassirostris [Species]
    Tylibyx [Genus]
        T. melanocephalus [Species]
    Microsarcops [Genus]
        M. cinereus [Species]
    Lobivanellus [Genus]
        L. indicus [Species]
    Xiphidiopterus [Genus]
        X. albiceps [Species]
    Rogibyx [Genus]
        R. tricolor [Species]
    Lobibyx [Genus]
        L. novaehollandiae [Species]
        L. miles
    Afribyx [Genus]
        A. senegallus [Species]
    Stephanibyx [Genus]
        S. lugubris [Species]
        S. melanopterus
        S. coronatus
    Hoplopterus [Genus]
        H. spinosus [Species]
        H. armatus
        H. duvaucelii
    Hoploxypterus [Genus]
        H. cayanus [Species]
    Ptilocelys [Genus]
        P. resplendens [Species]
    Zonifer [Genus]
        Z. tricolor [Species]
    Anomalophrys [Genus]
        A. superciliosus [Species]
    Lobipluvia [Genus]
        L. malabarica [Species]

Sarciophorus [Genus]
    S. tectus [Species]
Squatarola [Genus]
    S. squatarola [Species]
Pluvialis [Genus]
    P. apricaria [Species]
    P. dominica
Pluviorhynchus [Genus]
    P. obscurus [Species]
Charadrius [Genus]
    C. rubricollis [Species]
    C. hiaticula
    C. melodus
    C. dubius
    C. alexandrinus
    C. venustus
    C. falklandicus
    C. alticola
    C. bicinctus
    C. peronii
    C. collaris
    C. pecuarius
    C. sanctaehelenae
    C. thoracicus
    C. placidus
    C. vociferus
    C. tricollaris
    C. mongolus
    C. wilsonia
    C. leschenaultii
Elseyornis [Genus]
    E. melanops [Species]
Eupoda [Genus]
    E. asiatica [Species]
    E. veredus
    E. montana
Oreopholus [Genus]
    O. ruficollis [Species]
Erythrogonys [Genus]
    E. cinctus [Species]
Eudromias [Genus]
    E. morinellus [Species]
Zonibyx [Genus]
    Z. modestus [Species]
Thinornis [Genus]
    T. novaeseelandiae [Species]
Anarhynchus [Genus]
    A. frontalis [Species]
Pluvianellus [Genus]
    P. socialis [Species]
Phegornis [Genus]
    P. mitchellii [Species]

Scopacidae [Family]
    Aechmorhynchus [Genus]
        A. cancellatus [Species]
        A. parvirostris
    Prosobonia [Genus]
        P. leucoptera [Species]
    Bartramia [Genus]
        B. longicauda [Species]

Numenius [Genus]
    N. minutus [Species]
    N. borealis
    N. phaeopus
    N. tahitiensis
    N. tenuirostris
    N. arquata
    N. madagascariensis
    N. americanus
Limosa [Genus]
    L. limosa [Species]
    L. haemastica
    L. lapponica
    L. fedoa
Tringa [Genus]
    T. erythropus [Species]
    T. totanus
    T. flavipes
    T. stagnatilis
    T. nebularia
    T. melanoleuca
    T. ocrophus
    T. solitaria
    T. glareola
Pseudototanus [Genus]
    P. guttifer [Species]
Xenus [Genus]
    X. cinereus [Species]
Actitis [Genus]
    A. hypoleucos [Species]
    A. macularia
Catoptrophorus [Genus]
    C. semipalmatus [Species]
Heteroscelus [Genus]
    H. brevipes [Species]
    H. incanus
Aphriza [Genus]
    A. virgata [Species]
Arenaria [Genus]
    A. interpres [Species]
    A. melanocephala
Limnodromus [Genus]
    L. griseus [Species]
    L. semipalmatus
Coenocorypha [Genus]
    C. aucklandica [Species]
Capella [Genus]
    C. solitaria [Species]
    C. hardwickii
    C. nemoricola
    C. stenura
    C. megala
    C. nigripennis
    C. macrodactyla
    C. media
    C. gallinago
    C. delicata
    C. paraguaiae
    C. nobilis
    C. undulata

Chubbia [Genus]
   C. imperialis [Species]
   C. jamesoni
   C. stricklandii
Scolopax [Genus]
   S. rusticola [Species]
   S. saturata
   S. celebensis
   S. rochussenii
Philohela [Genus]
   P. minor [Species]
Lymnocryptes [Genus]
   L. minima [Species]
Calidris [Genus]
   C. canutus [Species]
   C. tenuirostris
Crocethia [Genus]
   C. alba [Species]
Ereunetes [Genus]
   E. pusillus [Species]
   E. mauri
Eurynorhynchus [Genus]
   E. pygmeus [Species]
Erolia [Genus]
   E. ruficollis [Species]
   E. minuta
   E. temminckii
   E. subminuta
   E. minutilla
   E. fuscicollis
   E. bairdii
   E. melanotos
   E. acuminata
   E. maritima
   E. ptilocnemis
   E. alpina
   E. testacea
Limicola [Genus]
   L. falcinellus [Species]
Micropalama [Genus]
   M. himantopus [Species]
Tryngites [Genus]
   T. subruficollis [Species]
Philomachus [Genus]
   P. pugnax [Species]

Recurvostridae [Family]
   Ibidorhyncha [Genus]
     I. struthersii [Species]
   Himantopus [Genus]
     H. himantopus [Species]
   Cladorhynchus [Genus]
     C. leucocephala [Species]
   Recurvirostra [Genus]
     R. avosetta [Species]
     R. americana
     R. novaehollandiae
     R. andina

Phalaropodidae [Family]
   Phalaropus [Genus]
     P. fulicarius [Species]

Steganopus [Genus]
   S. tricolor [Species]
Lobipes [Genus]
   L. lobatus [Species]

Dromadidae [Family]
   Dromas [Genus]
     D. ardeola [Species]

Burhinidae [Family]
   Burhinus [Genus]
     B. oedicnemus [Species]
     B. senegalensis
     B. vermiculatus
     B. capensis
     B. bistriatus
     B. superciliaris
     B. magnirostris
   Esacus [Genus]
     E. recurvirostris [Species]
   Orthoramphus [Genus]
     O. magnirostris [Species]

Glareolidae [Family]
   Pluvianus [Genus]
     P. aegyptius [Species]
   Cursorius [Genus]
     C. cursor [Species]
     C. temminckii
     C. coromandelicus
   Rhinoptilus [Genus]
     R. africanus [Species]
     R. cinctus
     R. chalcopterus
     R. bitorquatus
   Peltohyas [Genus]
     P. australis [Species]
   Stiltia [Genus]
     S. isabella [Species]
   Glareola [Genus]
     G. pratincola [Species]
     G. maldivarum
     G. nordmanni
     G. ocularis
     G. nuchalis
     G. cinerea
     G. lactea
   Attagis [Genus]
     A. gayi [Species]
     A. malouinus
   Thinocorus [Genus]
     T. orbignyianus [Species]
     T. rumicivorus

Chionididae [Family]
   Chionis [Genus]
     C. alba [Species]
     C. minor

Stercorariidae [Family]
   Catharacta [Genus]
     C. skua [Species]

Stercorarius [Genus]
   S. pomarinus [Species]
   S. parasiticus
   S. longicaudus

Laridae [Family]
   Gabianus [Genus]
     G. pacificus [Species]
     G. scoresbii
   Pagophila [Genus]
     P. eburnea [Species]
   Larus [Genus]
     L. fuliginosus [Species]
     L. modestus
     L. heermanni
     L. leucophthalmus
     L. hemprichii
     L. belcheri
     L. crassirostris
     L. audouinii
     L. delawarensis
     L. canus
     L. argentatus
     L. fuscus
     L. californicus
     L. occidentalis
     L. dominicanus
     L. schistisagus
     L. marinus
     L. glaucescens
     L. hyperboreus
     L. leucopterus
     L. ichthyaetus
     L. atricilla
     L. brunnicephalus
     L. cirrocephalus
     L. serranus
     L. pipixcan
     L. novaehollandiae
     L. melanocephalus
     L. bulleri
     L. maculipennis
     L. ridibundus
     L. genei
     L. philadelphia
     L. minutus
     L. saundersi
   Rhodostethia [Genus]
     R. rosea [Species]
   Rissa [Genus]
     R. tridactyla [Species]
     R. brevirostris
   Creagrus [Genus]
     C. furcatus [Species]
   Xema [Genus]
     X. sabini [Species]
   Chlidonias [Genus]
     C. hybrida [Species]
     C. leucoptera
     C. nigra

Phaetusa [Genus]
  P. simplex [Species]
Gelochelidon [Genus]
  G. nilotica [Species]
Hydroprogne [Genus]
  H. tschegrava [Species]
Sterna [Genus]
  S. aurantia [Species]
  S. hirundinacea
  S. hirundo
  S. paradisaea
  S. vittata
  S. virgata
  S. forsteri
  S. trudeaui
  S. dougallii
  S. striata
  S. repressa
  S. sumatrana
  S. melanogaster
  S. aleutica
  S. lunata
  S. anaethetus
  S. fuscata
  S. nereis
  S. albistriata
  S. superciliaris
  S. balaenarum
  S. iorata
  S. albifrons
Thalasseus [Genus]
  T. bergii [Species]
  T. maximus
  T. bengalensis
  T. zimmermanni
  T. eurygnatha
  T. elegans
  T. sandvicensis
Larosterna [Genus]
  L. inca [Species]
Procelsterna [Genus]
  P. cerulea [Species]
Anous [Genus]
  A. stolidus [Species]
  A. tenuirostris
  A. minutus
Gygis [Genus]
  G. alba [Species]

Rynchopidae [Family]
Rynchops [Genus]
  R. nigra [Species]
  R. flavirostris
  R. albicollis

Alcidae [Family]
Plautus [Genus]
  P. alle [Species]
Pinguinis [Genus]
  P. impennis [Species]
Alca [Genus]
  A. torda [Species]

Uria [Genus]
  U. lomvia [Species]
  U. aalge
Cepphus [Genus]
  C. grylle [Species]
  C. columba
  C. carbo
Brachyramphus [Genus]
  B. marmoratus [Species]
  B. brevirostris
  B. hypoleucus
  B. craveri
Synthliboramphus [Genus]
  S. antiquus [Species]
  S. wumizusume
Ptychoramphus [Genus]
  P. aleuticus [Species]
Cyclorrhynchus [Genus]
  C. psittacula [Species]
Aethia [Genus]
  A. cristatella [Species]
  A. pusilla
  A. pygmaea
Cercorhinca [Genus]
  C. monocerata [Species]
Fratercula [Genus]
  F. arctica [Species]
  F. corniculata
Lunda [Genus]
  L. cirrhata [Species]

Columbiformes [Order]
Pteroclididae [Family]
Syrrhaptes [Genus]
  S. tibetanus [Species]
  S. paradoxus
Pterocles [Genus]
  P. alchata [Species]
  P. namaqua
  P. exustus
  P. senegallus
  P. orientalis
  P. coronatus
  P. gutturalis
  P. burchelli
  P. personatus
  P. decoratus
  P. lichtensteinii
  P. bicinctus
  P. indicus
  P. quadricinctus

Raphidae [Family]
Raphus [Genus]
  R. cucullatus [Species]
  R. solitarius
Pezophaps [Genus]
  P. solitaria [Species]

Columbidae [Family]
Sphenurus [Genus]
  S. apicauda [Species]

  S. seimundi
  S. oxyura
  S. sphenurus
  S. korthalsi
  S. sieboldii
  S. farmosae
Butreron [Genus]
  B. capellei [Species]
Treron [Genus]
  T. curvirostra [Species]
  T. pompadora
  T. fulvicollis
  T. olax
  T. vernans
  T. bicincta
  T. s. thomae
  T. australis
  T. calva
  T. delalandii
  T. waalia
  T. phoenicoptera
Phapitreron [Genus]
  P. leucotis [Species]
  P. amethystina
Leucotreron [Genus]
  L. occipitalis [Species]
  L. fischeri
  L. merrilli
  L. marchei
  L. subgularis
  L. leclancheri
  L. cincta
  L. dohertyi
  L. porphyrea
Ptilinopus [Genus]
  P. dupetithouarsii [Species]
  P. regina
  P. mercierii
  P. purpuratus
  P. coralensis
  P. insularis
  P. rarotongensis
  P. huttoni
  P. porphyraceus
  P. greyii
  P. richardsii
  P. ponapensis
  P. pelewensis
  P. roseicapilla
  P. perousii
  P. superbus
  P. pulchellus
  P. coronulatus
  P. monacha
  P. iozonus
  P. insolitus
  P. rivoli
  P. miquelli
  P. bellus
  P. solomonensis

P. viridis
P. eugeniae
P. geelvinkiana
P. pectoralis
P. naina
P. hyogastra
P. granulifrons
P. melanospila
P. jambu
P. wallacii
P. aurantiifrons
P. ornatus
P. perlatus
P. tannensis
Chrysoena [Genus]
    C. victor [Species]
    C. viridis
    C. luteovirens
Alectroenas [Genus]
    A. pulcherrima [Species]
    A. sganzini
    A. madagascariensis
    A. nitidissima
Drepanoptila [Genus]
    D. holosericea [Species]
Megaloprepia [Genus]
    M. magnifica [Species]
    M. formosa
Ducula [Genus]
    D. galeata [Species]
    D. aurorae
    D. oceanica
    D. pacifica
    D. rubricera
    D. myristicivora
    D. concinna
    D. aenea
    D. oenothorax
    D. pistrinaria
    D. whartoni
    D. rosacea
    D. perspicillata
    D. pickeringii
    D. latrans
    D. bakeri
    D. brenchleyi
    D. goliath
    D. bicolor
    D. luctuosa
    D. melanura
    D. spilorrhoa
    D. cineracea
    D. lacernulata
    D. badia
    D. mullerii
    D. pinon
    D. melanochroa
    D. poliocephala
    D. forsteni
    D. mindorensis

D. radiata
D. rufigaster
D. finschii
D. chalconota
D. zoeae
D. carola
Cryptophaps [Genus]
    C. poecilorrhoa [Species]
Hemiphaga [Genus]
    H. novaeseelandiae [Species]
Lopholaimus [Genus]
    L. antarcticus [Species]
Gymnophaps [Genus]
    G. albertisii [Species]
    G. solomonensis
    G. mada
Columba [Genus]
    C. leuconota [Species]
    C. rupestris
    C. livia
    C. oenas
    C. eversmanni
    C. oliviae
    C. albitorques
    C. palumbus
    C. trocaz
    C. junoniae
    C. leucocephala
    C. picazuro
    C. gymnophtalmos
    C. squamosa
    C. maculosa
    C. unicincta
    C. guinea
    C. hodgsonii
    C. arquatrix
    C. thomensis
    C. albinucha
    C. flavirostris
    C. oenops
    C. inornata
    C. caribaea
    C. rufina
    C. fasciata
    C. albilinea
    C. araucana
    C. elphinstonii
    C. torringtoni
    C. pulchricollis
    C. punicea
    C. palumboides
    C. janthina
    C. versicolor
    C. jouyi
    C. vitiensis
    C. pallidiceps
    C. norfolciensis
    C. argentina
    C. pollenii
    C. speciosa

C. nigriristris
C. goodsoni
C. subvinacea
C. plumbea
C. chiriquensis
C. purpureotincta
C. delegorguei
C. iriditorques
C. malherbii
Nesoenas [Genus]
    . mayeri [Species]
Turacoena [Genus]
    T. manadensis [Species]
    T. modesta
Macropygia [Genus]
    M. unchall [Species]
    M. amboinensis
    M. ruficeps
    M. magna
    M. phasianella
    M. rufipennis
    M. nigrirostris
    M. mackinlayi
Reinwardtoena [Genus]
    R. reinwardtsi [Species]
    R. browni
Coryphoenas [Genus]
    C. crassirostris [Species]
Ectopistes [Genus]
    E. migratoria [Species]
Zenaidura [Genus]
    Z. macroura [Species]
    Z. graysoni
    Z. auriculata
Zenaida [Genus]
    Z. aurita [Species]
    Z. asiatica
Nesopelia [Genus]
    N. galapagoensis [Species]
Streptopelia [Genus]
    S. turtur [Species]
    S. orientalis
    S. lugens
    S. picturata
    S. decaocto
    S. roseogrisea
    S. semitorquata
    S. decipiens
    S. capicola
    S. vinacea
    S. reichenowi
    S. fulvopectoralis
    S. bitorquata
    S. tranquebarica
    S. chinensis
    S. senegalensis
Geopelia [Genus]
    G. humeralis [Species]
    G. striata
    G. cuneata

Metriopelia [Genus]
  M. ceciliae [Species]
  M. morenoi
  M. melanoptera
  M. aymara
Scardafella [Genus]
  S. inca [Species]
  S. squammata
Uropelia [Genus]
  U. campestris [Species]
Columbina [Genus]
  C. picui [Species]
Columbigallina [Genus]
  C. passerina [Species]
  C. talpacoti
  C. minuta
  C. buckleyi
  C. cruziana
Oxypelia [Genus]
  O. cyanopis [Species]
Claravis [Genus]
  C. pretiosa [Species]
  C. mondetoura
  C. godefrida
Oena [Genus]
  O. capensis [Species]
Tympanistria [Genus]
  T. tympanistria [Species]
Turtur [Genus]
  T. afer [Species]
  T. abyssinicus
  T. chalcospilos
  T. brehmeri
Chalcophaps [Genus]
  C. indica [Species]
  C. stephani
Henicophaps [Genus]
  H. albifrons [Species]
  H. foersteri
Petrophassa [Genus]
  P. albipennis [Species]
  P. rufipennis
Phaps [Genus]
  P. chalcoptera [Species]
  P. elegans
Ocyphaps [Genus]
  O. lophotes [Species]
Lophophaps [Genus]
  L. plumifera [Species]
  L. ferruginea
Geophaps [Genus]
  G. scripta [Species]
  G. smithii
Histriophaps [Genus]
  H. histrionica [Species]
Aplopelia [Genus]
  A. larvata [Species]
  A. simplex
Leptotila [Genus]
  L. verreauxi [Species]

L. megalura
L. jamaicensis
L. plumbeiceps
L. rufaxilla
L. wellsi
L. cassini
L. ochraceiventris
Osculatia [Genus]
  O. saphirina [Species]
Oreopeleia [Genus]
  O. veraguensis [Species]
  O. lawrencii
  O. goldmani
  O. costaricensis
  O. chrysia
  O. mystacea
  O. martinica
  O. violacea
  O. montana
  O. caniceps
  O. albifacies
  O. chiriquensis
  O. linearis
  O. bourcieri
  O. erythropareia
Geotrygon [Genus]
  G. versicolor [Species]
Gallicolumba [Genus]
  G. luzonica [Species]
  G. platenae
  G. keayi
  G. criniger
  G. menagei
  G. rufigula
  G. tristigmata
  G. beccarii
  G. salamonis
  G. sanctaecrucis
  G. stairi
  G. canifrons
  G. xanthonura
  G. kubaryi
  G. jobiensis
  G. erythroptera
  G. rubescens
  G. hoedtii
Leucosarcia [Genus]
  L. melanoleuca [Species]
Trugon [Genus]
  T. terrestris [Species]
Microgoura [Genus]
  M. meeki [Species]
Starnoenas [Genus]
  S. cyanocephala [Species]
Otidiphaps [Genus]
  O. nobilis [Species]
Caloenas [Genus]
  C. nicobarica [Species]
Goura [Genus]
  G. cristata [Species]

  G. scheepmakeri
  G. victoria
Didunculus [Genus]
  D. strigirostris [Species]

Psittaciformes [Order]
Psittacidae [Family]
Strigops [Genus]
  S. habroptilus [Species]
Nestor [Genus]
  N. meridionalis [Species]
  N. notabilis
  N. productus
Chalcopsitta [Genus]
  C. atra [Species]
  C. insignis
  C. sintillata
  C. duivenbodei
  C. cardinalis
Eos [Genus]
  E. cyanogenia [Species]
  E. reticulata
  E. squamata
  E. histrio
  E. bornea
  E. semilarvata
  E. goodfellowi
Trichoglossus [Genus]
  T. ornatus [Species]
  T. haematod
  T. rubiginosus
  T. chlorolepidotus
  T. euteles
Psitteuteles [Genus]
  P. flavoviridis [Species]
  P. johnstoniae
  P. goldiei
  P. versicolor
  P. iris
Pseudeos [Genus]
  P. fuscata [Species]
Domicella [Genus]
  D. hypoinochroa [Species]
  D. amabilis
  D. lory
  D. domicella
  D. tibialis
  D. chlorocercus
  D. albidinucha
  D. garrula
Phigys [Genus]
  P. solitarius [Species]
Vini [Genus]
  V. australis [Species]
  V. kuhlii
  V. stepheni
  V. peruviana
  V. ultramarina
Glossopsitta [Genus]
  G. concinna [Species]
  G. porphyrocephala
  G. pusilla

Charmosyna [Genus]
    C. palmarum [Species]
    C. meeki
    C. rubrigularis
    C. aureicincta
    C. diadema
    C. toxopei
    C. placentis
    C. rubronotata
    C. multistriata
    C. wilhelminae
    C. pulchella
    C. margarethae
    C. josefinae
    C. papou
Oreopsittacus [Genus]
    O. arfaki [Species]
Neopsittacus [Genus]
    N. musschenbroekii [Species]
    N. pullicauda
Psittaculirostris [Genus]
    P. desmaresti [Species]
    P. salvadorii
Opopsitta [Genus]
    P. gulielmitertii [Species]
    P. diophthalma
Lathamus [Genus]
    L. discolor [Species]
Micropsitta [Genus]
    M. bruijnii [Species]
    M. keiensis
    M. geelvinkiana
    M. pusio
    M. meeki
    M. finschii
Probosciger [Genus]
    P. aterrimus [Species]
Calyptorhynchus [Genus]
    C. baudinii [Species]
    C. funereus
    C. magnificus
    C. lathami
Callocephalon [Genus]
    C. fimbriatum [Species]
Kakatoe [Genus]
    K. galerita [Species]
    K. sulphurea
    K. alba
    K. moluccensis
    K. Haematuropygia
    K. leadbeateri
    K. ducrops
    K. sanguinea
    K. tenuirostris
    K. roseicapilla
Nymphicus [Genus]
    N. hollandicus [Species]
Anodorhynchus [Genus]
    A. hyacinthinus [Species]
    A. glaucus
    A. leari

Ara [Genus]
    A. ararauna [Species]
    A. caninde
    A. militaris
    A. ambigua
    A. macao
    A. chloroptera
    A. tricolor
    A. rubrogenys
    A. auricollis
    A. severa
    A. spixii
    A. manilata
    A. maracana
    A. couloni
    A. nobilis
Aratinga [Genus]
    A. acuticaudata [Species]
    A. guarouba
    A. holochlora
    A. strenua
    A. finschi
    A. wagleri
    A. mitrata
    A. erythrogenys
    A. leucophthalmus
    A. chloroptera
    A. euops
    A. auricapillus
    A. jandaya
    A. solstitialis
    A. weddellii
    A. astec
    A. nana
    A. canicularis
    A. pertinax
    A. cactorum
    A. aurea
Nandayus [Genus]
    N. nenday [Species]
Leptosittaca [Genus]
    L. branickii [Species]
Conuropsis [Genus]
    C. carolinensis [Species]
Rhynchopsitta [Genus]
    R. pachyrhyncha [Species]
Cyanoliseus [Genus]
    C. patagonus [Species]
    C. whitleyi
Ognorhynchus [Genus]
    O. icterotis [Species]
Pyrrhura [Genus]
    P. cruentata [Species]
    P. devillei
    P. frontalis
    P. perlata
    P. rhodogaster
    P. molinae
    P. hypoxantha
    P. hoematotis

    P. leucotis
    P. picta
    P. viridicata
    P. egregria
    P. melanura
    P. berlepschi
    P. rupicola
    P. albipectus
    P. calliptera
    P. rhodocephala
    P. hoffmanni
Microsittace [Genus]
    M. ferruginea [Species]
Enicognathus [Genus]
    E. leptorhynchus [Species]
Myiopsitta [Genus]
    M. monachus [Species]
Amoropsittaca [Genus]
    A. aymara [Species]
Psilopsaigon [Genus]
    P. aurifrons [Species]
Bolborhynchus [Genus]
    B. lineola [Species]
    B. ferrugineifrons
    B. andicolus
Forpus [Genus]
    F. cyanopygius [Species]
    F. passerinus
    F. conspicillatus
    F. sclateri
    F. coelestis
Brotogeris [Genus]
    B. tirica [Species]
    B. versicolurus
    B. pyrrhopterus
    B. jugularis
    B. gustavi
    B. chrysopterus
    B. sanctithomae
Nannopsittaca [Genus]
    N. panychlora [Species]
Touit [Genus]
    T. batavica [Species]
    T. purpurata
    T. melanonotus
    T. huetii
    T. dilectissima
    T. surda
    T. stictoptera
    T. emmae
Pionites [Genus]
    P. melanocephala [Species]
    P. leucogaster
Pionopsitta [Genus]
    P. pileata [Species]
    P. haematotis
    P. caica
    P. barrabandi
    P. pyrilia
Hapalopsittaca [Genus]
    H. melanotis [Species]

H. fuertesi
H. amazonina
H. pyrrhops
Gypopsitta [Genus]
   G. vulturina [Species]
Graydidascalus [Genus]
   G. brachyurus [Species]
Pionus [Genus]
   P. menstruus [Species]
   P. sordidus
   P. maximiliani
   P. tumultuosus
   P. seniloides
   P. senilis
   P. chalcopterus
   P. fuscus
Amazona [Genus]
   A. collaria [Species]
   A. leucocephala
   A. ventralis
   A. xantholora
   A. albifrons
   A. agilis
   A. vittata
   A. pretrei
   A. viridigenalis
   A. finschi
   A. autumnalis
   A. dufresniana
   A. brasiliensis
   A. arausiaca
   A. festiva
   A. xanthops
   A. barbadensis
   A. aestiva
   A. ochrocephala
   A. amazonica
   A. mercenaria
   A. farinosa
   A. vinacea
   A. guildingii
   A. versicolor
   A. imperialis
Deroptyus [Genus]
   D. accipitrinus [Species]
Triclaria [Genus]
   T. malachitacea [Species]
Poicephalus [Genus]
   P. robustus [Species]
   P. gulielmi
   P. flavifrons
   P. cryptoxanthus
   P. senegalus
   P. meyeri
   P. rufiventris
   P. ruppellii
Psittacus [Genus]
   P. erithacus [Species]
Coracopsis [Genus]
   C. vasa [Species]
   C. nigra

Psittrichas [Genus]
   P. fulgidus [Species]
Lorius [Genus]
   L. roratus [Species]
Geoffroyus [Genus]
   G. geoffroyi [Species]
   G. simplex
   G. heteroclitus
Prioniturus [Genus]
   P. luconensis [Species]
   P. discurus
   P. flavicans
   P. platurus
   P. mada
Tanygnathus [Genus]
   T. lucionensis [Species]
   T. mulleri
   T. gramineus
   T. heterurus
   T. megalorynchos
Mascarinus [Genus]
   M. mascarin [Species]
Psittacula [Genus]
   P. eupatria [Species]
   P. krameri
   P. alexandri
   P. caniceps
   P. exsul
   P. derbyana
   P. longicauda
   P. cyanocephala
   P. intermedia
   P. himalayana
   P. calthorpae
   P. columboides
Polytelis [Genus]
   P. swainsonii [Species]
   P. anthopeplus
   P. alexandrae
Aprosmictus [Genus]
   A. jonquillaceus [Species]
   A. erythropterus
Alisterus [Genus]
   A. amboinensis [Species]
   A. chloropterus
   A. scapularis
Prosopeia [Genus]
   P. tabuensis [Species]
   P. personata
Psittacella [Genus]
   P. brehmii [Species]
   P. picta
   P. modesta
Bolbopsittacus [Genus]
   B. lunulatus [Species]
Psittinus [Genus]
   P. cyanurus [Species]
Agapornis [Genus]
   A. cana [Species]
   A. pullaria
   A. roseicollis

A. taranta
A. swinderniana
A. fischeri
A. personata
A. lilianae
A. nigrigenis
Loriculus [Genus]
   L. vernalis [Species]
   L. beryllinus
   L. pusillus
   L. philippensis
   L. amabilis
   L. stigmatus
   L. galgulus
   L. exilis
   L. flosculus
   L. aurantiifrons
Platycercus [Genus]
   P. elegans [Species]
   P. caledonicus
   P. eximius
   P. icterotis
   P. adscitus
   P. venustus
   P. zonarius
Purpureicephalus [Genus]
   P. spurius [Species]
Northiella [Genus]
   N. haematogaster [Species]
Psephotus [Genus]
   P. haematonotus [Species]
   P. varius
   P. pulcherrimus
   P. chrysopterygius
Neophema [Genus]
   N. elegans [Species]
   N. chrysostomus
   N. chrysogaster
   N. petrophila
   N. pulchella
   N. splendida
   N. bourkii
Eunymphicus [Genus]
   E. cornutus [Species]
Cyanoramphus [Genus]
   C. unicolor [Species]
   C. novaezelandiae
   C. zealandicus
   C. auriceps
   C. malherbi
   C. ulietanus
Melopsittacus [Genus]
   M. undulatus [Species]
Pezoporus [Genus]
   P. wallicus [Species]
Geopsittacus [Genus]
   G. occidentalis [Species]

Cuculiformes [Order]
Musophagidae [Family]
   Tauraco [Genus]
     T. persa [Species]

T. livingstonii
T. corythaix
T. schuttii
T. fischeri
T. erythrolophus
T. bannermani
T. ruspolii
T. leucotis
T. macrorhynchus
T. hartlaubi
T. leucolophus
Gallirex [Genus]
  G. porphyreolophus [Species]
Ruwenzorornis [Genus]
  R. johnstoni [Species]
Musophaga [Genus]
  M. violacea [Species]
Corythaeola [Genus]
  C. cristata [Species]
Crinifer [Genus]
  C. leucogaster [Species]
  C. africanus
  C. concolor
  C. personata

Cuculidae [Family]
  Clamator [Genus]
    C. glandarius [Species]
    C. coromandus
    C. serratus
    C. jacobinus
    C. cafer
  Pachycoccyx [Genus]
    P. audeberti [Species]
  Cuculus [Genus]
    C. crassirostris [Species]
    C. sparverioides
    C. varius
    C. vagans
    C. fugax
    C. solitarius
    C. clamosus
    C. micropterus
    C. canorus
    C. saturatus
    C. poliocephalus
    C. pallidus
  Cercococcyx [Genus]
    C. mechowi [Species]
    C. olivinus
    C. montanus
  Penthoceryx [Genus]
    P. sonneratii [Species]
  Cacomantis [Genus]
    C. merulinus [Species]
    C. variolosus
    C. castaneiventris
    C. heinrichi
    C. pyrrophanus
  Rhamphomantis [Genus]
    R. megarhynchus [Species]

Misocalius [Genus]
  M. osculans [Species]
Chrysococcyx [Genus]
  C. cupreus [Species]
  C. flavigularis
  C. klaas
  C. caprius
Chalcites [Genus]
  C. maculatus [Species]
  C. xanthorhynchus
  C. basalis
  C. lucidus
  C. malayanus
  C. crassirostris
  C. ruficollis
  C. meyeri
Caliechthrus [Genus]
  C. leucolophus [Species]
Surniculus [Genus]
  S. lugubris [Species]
Microdynamis [Genus]
  M. parva [Species]
Eudynamys [Genus]
  E. scolopacea [Species]
Urodynamis [Genus]
  U. taitensis [Species]
Scythrops [Genus]
  S. novaehollandiae [Species]
Coccyzus [Genus]
  C. pumilus [Species]
  C. cinereus
  C. erythropthalmus
  C. americanus
  C. euleri
  C. minor
  C. melacoryphus
  C. lansbergi
Piaya [Genus]
  P. rufigularis [Species]
  P. pluvialis
  P. cayana
  P. melanogaster
  P. minuta
Saurothera [Genus]
  S. merlini [Species]
  S. vetula
Ceuthmochares [Genus]
  C. aereus [Species]
Rhopodytes [Genus]
  R. diardi [Species]
  R. sumatranus
  R. tristis
  R. viridirostris
Taccocua [Genus]
  T. leschenaulti [Species]
Rhinortha [Genus]
  R. chlorophaea [Species]
Zanclostomus [Genus]
  Z. javanicus [Species]
Rhamphococcyx [Genus]

R. calyorhynchus [Species]
R. curvirostris
Phaenicophaeus [Genus]
  P. pyrrhocephalus [Species]
Dasylophus [Genus]
  D. superciliosus [Species]
Lepidogrammus [Genus]
  L. cumingi [Species]
Crotophaga [Genus]
  C. major [Species]
  C. ani
  C. sulcirostris
Guira [Genus]
  G. guira [Species]
Tapera [Genus]
  T. naevia [Species]
Morococcyx [Genus]
  M. erythropygus [Species]
Dromococcyx [Genus]
  D. phasianellus [Species]
  D. pavoninus
Geococcyx [Genus]
  G. californiana [Species]
  G. velox
Neomorphus [Genus]
  N. geoffroyi [Species]
  N. squaminger
  N. radiolosus
  N. rufipennis
  N. pucheranii
Carpococcyx [Genus]
  C. radiceus [Species]
  C. renauldi
Coua [Genus]
  C. delalandei [Species]
  C. gigas
  C. coquereli
  C. serriana
  C. reynaudii
  C. cursor
  C. ruficeps
  C. cristata
  C. verreauxi
  C. caerulea
Centropus [Genus]
  C. milo [Species]
  C. goliath
  C. violaceus
  C. menbeki
  C. ateralbus
  C. chalybeus
  C. phasianinus
  C. spilopterus
  C. bernsteini
  C. chlororhynchus
  C. rectunguis
  C. steerii
  C. sinensis
  C. andamanensis
  C. nigrorufus
  C. viridis

C. toulou
C. bengalensis
C. grillii
C. epomidis
C. leucogaster
C. anselli
C. monachus
C. senegalensis
C. superciliosus
C. melanops
C. celebensis
C. unirufus

Strigiformes [Order]
Tytonidae [Family]
Tyto [Genus]
T. soumagnei [Species]
T. alba
T. rosenbergii
T. inexpectata
T. novaehollandiae
T. aurantia
T. tenebricosa
T. capensis
T. longimembris
Phodilus [Genus]
P. badius [Species]

Strigidae [Family]
Otus [Genus]
O. sagittatus [Species]
O. rufescens
O. icterorhynchus
O. spilocephalus
O. vandewateri
O. balli
O. alfredi
O. brucei
O. scops
O. umbra
O. senegalensis
O. flammeolus
O. brookii
O. rutilus
O. manadensis
O. beccarii
O. silvicola
O. whiteheadi
O. insularis
O. bakkamoena
O. asio
O. trichopsis
O. barbarus
O. guatemalae
O. roboratus
O. cooperi
O. choliba
O. atricapillus
O. ingens
O. watsonii
O. nudipes

O. clarkii
O. albogularis
O. minimus
O. leucotis
O. hartlaubi
Pyrroglaux [Genus]
P. podargina [Species]
Mimizuku [Genus]
M. gurneyi [Species]
Jubula [Genus]
J. lettii [Species]
Lophostrix [Genus]
L. cristata [Species]
Bubo [Genus]
B. virginianus [Species]
B. bubo
B. capensis
B. africanus
B. poensis
B. nipalensis
B. sumatrana
B. shelleyi
B. lacteus
B. coromandus
B. leucostictus
Pseudoptynx [Genus]
P. philippensis [Species]
Ketupa [Genus]
K. blakstoni [Species]
K. zeylonensis
K. flavipes
K. ketupu
Scotopelia [Genus]
S. peli [Species]
S. ussheri
S. bouvieri
Pulsatrix [Genus]
P. perspicillata [Species]
P. koeniswaldiana
P. melanota
Nyctea [Genus]
N. scandiaca [Species]
Surnia [Genus]
S. ulula [Species]
Glaucidium [Genus]
G. passerinum [Species]
G. gnoma
G. siju
G. minutissimum
G. jardinii
G. brasilianum
G. perlatum
G. tephronotum
G. capense
G. brodiei
G. radiatum
G. cuculoides
G. sjostedti
Micrathene [Genus]
M. whitneyi [Species]

Uroglaux [Genus]
U. dimorpha [Species]
Ninox [Genus]
N. rufa [Species]
N. strenua
N. connivens
N. novaeseelandiae
N. scutulata
N. affinis
N. superciliaris
N. philippensis
N. spilonota
N. spilocephala
N. perversa
N. squamipila
N. theomacha
N. punctulata
N. meeki
N. solomonis
N. odiosa
N. jacquinoti
Gymnoglaux [Genus]
G. lawrencii [Species]
Sceloglaux [Genus]
S. albifacies [Species]
Athene [Genus]
A. noctua [Species]
A. brama
A. blewitti
Speotyto [Genus]
S. cunicularia [Species]
Ciccaba [Genus]
C. virgata [Species]
C. nigrolineata
C. huhula
C. albitarsus
C. woodfordii
Strix [Genus]
S. butleri [Species]
S. seloputo
S. ocellata
S. leptogrammica
S. aluco
S. occidentalis
S. varia
S. hylophila
S. rufipes
S. uralensis
S. davidi
S. nebulosa
Rhinoptynx [Genus]
R. clamator [Species]
Asio [Genus]
A. otus [Species]
A. stygius
A. abyssinicus
A. madagascariensis
A. flammeus
A. capensis
Pseudoscops [Genus]
P. grammicus [Species]

AVES SPECIES LIST

Nesasio [Genus]
    N. solomonensis [Species]
Aegolius [Genus]
    A. funereus [Species]
    A. acadicus
    A. ridgwayi
    A. harrisii

Caprimulgiformes [Order]
Steatornithidae [Family]
    Steatornis [Genus]
        S. caripensis [Species]

Podargidae [Family]
    Podargus [Genus]
        P. strigoides [Species]
        P. papuensis
        P. ocellatus
    Batrachostomus [Genus]
        B. auritus [Species]
        B. harteri
        B. septimus
        B. stellatus
        B. moniliger
        B. hodgsoni
        B. poliolophus
        B. javensis
        B. affinis

Nyctibiidae [Family]
    Nyctibius [Genus]
        N. grandis [Species]
        N. aethereus
        N. griseus
        N. leucopterus
        N. bracteatus

Aegothelidae [Family]
    Aegotheles [Genus]
        A. crinifrons [Species]
        A. insignis
        A. cristatus
        A. savesi
        A. bennettii
        A. wallacii
        A. albertisi

Caprimulgidae [Family]
    Lurocalis [Genus]
        L. semitorquatus [Species]
    Chordeiles [Genus]
        C. pusillus [Species]
        C. rupestris
        C. acutipennis
        C. minor
    Nyctiprogne [Genus]
        N. leucopyga [Species]
    Podager [Genus]
        P. nacunda [Species]
    Eurostopodus [Genus]
        E. guttatus [Species]
        E. albogularis

E. diabolicus
E. papuensis
E. archboldi
E. temminckii
E. macrotis
Veles [Genus]
    V. binotatus [Species]
Nyctidromus [Genus]
    N. albicollis [Species]
Phalaenoptilus [Genus]
    P. nuttallii [Species]
Siphonorhis [Genus]
    S. americanus [Species]
Otophanes [Genus]
    O. mcleodii [Species]
    O. yucatanicus
Nyctiphrynus [Genus]
    N. ocellatus [Species]
Caprimulgus [Genus]
    C. carolinensis [Species]
    C. rufus
    C. cubanensis
    C. sericocaudatus
    C. ridgwayi
    C. vociferus
    C. saturatus
    C. longirostris
    C. cayennensis
    C. maculicaudus
    C. parvulus
    C. maculosus
    C. nigrescens
    C. hirundinaceus
    C. ruficollis
    C. indicus
    C. europaeus
    C. aegyptius
    C. mahrattensis
    C. nubicus
    C. eximius
    C. madagascariensis
    C. macrurus
    C. pectoralis
    C. rufigena
    C. donaldsoni
    C. poliocephalus
    C. asiaticus
    C. natalensis
    C. inornatus
    C. stellatus
    C. ludovicianus
    C. monticolus
    C. affinis
    C. tristigma
    C. concretus
    C. pulchellus
    C. enarratus
    C. batesi
Scotornis [Genus]
    S. fossii [Species]
    S. climacurus

Macrodipteryx [Genus]
    M. longipennis [Species]
Semeiophorus [Genus]
    S. vexillarius [Species]
Hydropsalis [Genus]
    H. climacocerca [Species]
    H. brasiliana
Uropsalis [Genus]
    U. segmentata [Species]
    U. lyra
Macropsalis [Genus]
    M. creagra [Species]
Eleothreptus [Genus]
    E. anomalus [Species]

Apodiformes [Order]

Apodidae [Family]
    Collocalia [Genus]
        C. gigas [Species]
        C. whiteheadi
        C. lowi
        C. fuciphaga
        C. brevirostris
        C. francica
        C. inexpectata
        C. inquieta
        C. vanikorensis
        C. leucophaea
        C. vestita
        C. spodiopygia
        C. hirundinacea
        C. troglodytes
        C. marginata
        C. esculenta
    Hirundapus [Genus]
        H. caudacutus [Species]
        H. giganteus
        H. ernsti
    Streptoprocne [Genus]
        S. zonaris [Species]
        S. biscutata
    Aerornis [Genus]
        A. senex [Species]
        A. semicollaris
    Chaetura [Genus]
        C. chapmani [Species]
        C. pelagica
        C. vauxi
        C. richmondi
        C. gaumeri
        C. leucopygialis
        C. sabini
        C. thomensis
        C. sylvatica
        C. nubicola
        C. cinereiventris
        C. spinicauda
        C. martinica
        C. rutila
        C. ussheri

C. andrei
C. melanopygia
C. brachyura
Zoonavena [Genus]
   Z. grandidieri [Species]
Mearnsia [Genus]
   M. picina [Species]
   M. novaeguineae
   M. cassini
   M. bohmi
Cypseloides [Genus]
   C. cherriei [Species]
   C. fumigatus
   C. major
Nephoecetes [Genus]
   N. niger [Species]
Apus [Genus]
   A. melba [Species]
   A. aequatorialis
   A. reichenowi
   A. apus
   A. sladeniae
   A. toulsoni
   A. pallidus
   A. acuticaudus
   A. pacificus
   A. unicolor
   A. myoptilus
   A. batesi
   A. caffer
   A. horus
   A. affinis
   A. andecolus
Aeronautes [Genus]
   A. saxatalis [Species]
   A. montivagus
Panyptila [Genus]
   P. sanctihieronymi [Species]
   P. cayennensis
Tachornis [Genus]
   T. phoenicobia [Species]
Micropanyptila [Genus]
   M. furcata [Species]
Reinarda [Genus]
   R. squamata [Species]
Cypsiurus [Genus]
   C. parvus [Species]

Hemiprocnidae [Family]
Hemiprocne [Genus]
   H. longipennis [Species]
   H. mystacea
   H. comata

Trochilidae [Family]
Doryfera [Genus]
   D. johannae [Species]
   D. ludovicae
Androdon [Genus]
   A. aequatorialis [Species]
Ramphodon [Genus]

R. naevius [Species]
R. dohrnii
Glaucis [Genus]
   G. hirsuta [Species]
Threnetes [Genus]
   T. niger [Species]
   T. leucurus
   T. ruckeri
Phaethornis [Genus]
   P. yaruqui [Species]
   P. guy
   P. syrmatophorus
   P. superciliosus
   P. malaris
   P. eurynome
   P. hispidus
   P. anthophilus
   P. bourcieri
   P. philippii
   P. squalidus
   P. augusti
   P. pretrei
   P. subochraceus
   P. nattereri
   P. gounellei
   P. rupurumii
   P. porcullae
   P. ruber
   P. griseogularis
   P. longuemareus
   P. zonura
Eutoxeres [Genus]
   E. aquila [Species]
   E. condamini
Phaeochroa [Genus]
   P. cuvierii [Species]
Campylopterus [Genus]
   C. curvipennis [Species]
   C. largipennis
   C. rufus
   C. hyperythrus
   C. hemileucurus
   C. ensipennis
   C. falcatus
   C. phainopeplus
   C. villaviscensio
Eupetomana [Genus]
   E. macroura [Species]
Florisuga [Genus]
   F. mellivora [Species]
Melanotrochilus [Genus]
   M. fuscus [Species]
Colibri [Genus]
   C. delphinae [Species]
   C. thalassinus
   C. coruscans
   C. serrirostris
Anthracothorax [Genus]
   A. viridigula [Species]
   A. prevostii

A. nigricollis
A. veraguensis
A. dominicus
A. viridis
A. mango
Avocettula [Genus]
   A. recurvirostris [Species]
Eulampis [Genus]
   E. jugularis [Species]
Sericotes [Genus]
   S. holosericeus [Species]
Chrysolampis [Genus]
   C. mosquitus [Species]
Orthorhyncus [Genus]
   O. cristatus [Species]
Klais [Genus]
   K. guimeti [Species]
Abeillia [Genus]
   A. albeillei [Species]
Stephanoxis [Genus]
   S. lalandi [Species]
Lophornis [Genus]
   L. ornata [Species]
   L. gouldii
   L. magnifica
   L. delattrei
   L. stictolopha
   L. melaniae
Polemistria [Genus]
   P. chalybea [Species]
   P. pavonina
Lithiophanes [Genus]
   L. insignibarbis [Species]
Paphosia [Genus]
   P. helenae [Species]
   P. adorabilis
Popelairia [Genus]
   P. popelairii [Species]
   P. langsdorffi
   P. letitiae
   P. conversii
Discosura [Genus]
   D. longicauda [Species]
Chlorestes [Genus]
   C. notatus [Species]
Chlorostilbon [Genus]
   C. prasinus [Species]
   C. vitticeps
   C. aureoventris
   C. canivetti
   C. ricordii
   C. swainsonii
   C. maugaeus
   C. russatus
   C. gibsoni
   C. inexpectatus
   C. stenura
   C. alice
   C. poortmani
   C. euchloris
   C. auratus

Cynanthus [Genus]
  C. sordidus [Species]
  C. latirostris
Ptochoptera [Genus]
  P. iolaima [Species]
Cyanophaia [Genus]
  C. bicolor [Species]
Thalurania [Genus]
  T. furcata [Species]
  T. watertonii
  T. glaucopis
  T. lerchi
Neolesbia [Genus]
  N. nehrkorni [Species]
Panterpe [Genus]
  P. insignis [Species]
Damophila [Genus]
  D. julie [Species]
Lepidopyga [Genus]
  L. coeruleogularis [Species]
  L. goudoti
  L. luminosa
Hylocharis [Genus]
  H. xantusii [Species]
  H. leucotis
  H. eliciae
  H. sapphirina
  H. cyanus
  H. chrysura
  H. grayi
Chrysuronia [Genus]
  C. oenone [Species]
Goldmania [Genus]
  G. violiceps [Species]
Goethalsia [Genus]
  G. bella [Species]
Trochilus [Genus]
  T. polytmus [Species]
Leucochloris [Genus]
  L. albicollis [Species]
Polytmus [Genus]
  P. guainumbi [Species]
Waldronia [Genus]
  W. milleri [Species]
Smaragdites [Genus]
  S. theresiae [Species]
Leucippus [Genus]
  L. fallax [Species]
  L. baeri
  L. chionogaster
  L. viridicauda
Talaphorus [Genus]
  T. hypostictus [Species]
  T. taczanowskii
  T. chlorocercus
Amazilia [Genus]
  A. candida [Species]
  A. chionopectus
  A. versicolor
  A. hollandi

A. luciae
A. fimbriata
A. lactea
A. amabilis
A. cyaneotincta
A. rosenbergi
A. boucardi
A. franciae
A. veneta
A. leucogaster
A. cyanocephala
A. microrhyncha
A. cyanifrons
A. beryllina
A. cyanura
A. saucerrottei
A. tobaci
A. viridigaster
A. edward
A. rutila
A. yucatanensis
A. tzacatl
A. castaneiventris
A. amazilia
A. violiceps
Eupherusa [Genus]
  E. eximia [Species]
  E. nigriventris
Elvira [Genus]
  E. chionura [Species]
  E. cupreiceps
Microchera [Genus]
  M. albocoronata [Species]
Chalybura [Genus]
  C. buffonii [Species]
  C. urochrysia
Aphantochroa [Genus]
  A. cirrochloris [Species]
Lampornis [Genus]
  L. clemenciae [Species]
  L. amethystinus
  L. viridipallens
  L. hemileucus
  L. castaneoventris
  L. cinereicauda
Lamprolaima [Genus]
  L. rhami [Species]
Adelomyia [Genus]
  A. melanogenys [Species]
Anthocephala [Genus]
  A. floriceps [Species]
Urosticte [Genus]
  U. ruficrissa [Species]
  U. benjamini
Phlogophilus [Genus]
  P. hemileucurus [Species]
  P. harterti
Clytolaema [Genus]
  C. rubricauda [Species]
Polyplancta [Genus]
  P. aurescens [Species]

Heliodoxa [Genus]
  H. rubinoides [Species]
  H. leadbeateri
  H. jacula
  H. xanthogonys
Ionolaima [Genus]
  I. schreibersii [Species]
Agapeta [Genus]
  A. gularis [Species]
Lampraster [Genus]
  L. branickii [Species]
Eugenia [Genus]
  E. imperatrix [Species]
Eugenes [Genus]
  E. fulgens [Species]
Hylonympha [Genus]
  H. macrocerca [Species]
Sternoclyta [Genus]
  S. cyanopectus [Species]
Topaza [Genus]
  T. pella [Species]
  T. pyra
Oreotrochilus [Genus]
  O. chimborazo [Species]
  O. stolzmanni
  O. melanogaster
  O. estella
  O. bolivianus
  O. leucopleurus
  O. adela
Urochroa [Genus]
  U. bougueri [Species]
Patagona [Genus]
  P. gigas [Species]
Aglaeactis [Genus]
  A. cupripennis [Species]
  A. aliciae
  A. castelnaudii
  A. pamela
Lafresnaya [Genus]
  L. lafresnayi [Species]
Pterophanes [Genus]
  P. cyanopterus [Species]
Coeligena [Genus]
  C. coeligena [Species]
  C. wilsoni
  C. prunellei
  C. torquata
  C. phalerata
  C. eos
  C. bonapartei
  C. helianthea
  C. lutetiae
  C. violifer
  C. iris
Ensifera [Genus]
  E. ensifera [Species]
Sephanoides [Genus]
  S. sephanoides [Species]
  S. fernandensis

Boissoneaua [Genus]
    B. flavescens [Species]
    B. matthewsii
    B. jardini
Heliangelus [Genus]
    H. mavors [Species]
    H. clarisse
    H. amethysticollis
    H. strophianus
    H. exortis
    H. viola
    H. micraster
    H. squamigularis
    H. speciosa
    H. rothschildi
    H. luminosus
Eriocnemis [Genus]
    E. nigrivestis [Species]
    E. soderstromi
    E. vestitus
    E. godini
    E. cupreoventris
    E. luciani
    E. isaacsonii
    E. mosquera
    E. glaucopoides
    E. alinae
    E. derbyi
Haplophaedia [Genus]
    H. aureliae [Species]
    H. lugens
Ocreatus [Genus]
    O. underwoodii [Species]
Lesbia [Genus]
    L. victoriae [Species]
    L. nuna
Sappho [Genus]
    S. sparganura [Species]
Polyonymus [Genus]
    P. caroli [Species]
Zodalia [Genus]
    Z. glyceria [Species]
Ramphomicron [Genus]
    R. microrhynchum [Species]
    R. dorsale
Metallura [Genus]
    M. phoebe [Species]
    M. theresiae
    M. purpureicauda
    M. aeneocauda
    M. melagae
    M. eupogon
    M. williami
    M. tyrianthina
    M. ruficeps
Chalcostigma [Genus]
    C. olivaceum [Species]
    C. stanleyi
    C. heteropogon
    C. herrani

Oxypogon [Genus]
    O. guerinii [Species]
Opisthoprora [Genus]
    O. euryptera [Species]
Taphrolesbia [Genus]
    T. griseiventris [Species]
Aglaiocercus [Genus]
    A. kingi [Species]
    A. emmae
    A. coelestis
Oreonympha [Genus]
    O. nobilis [Species]
Augastes [Genus]
    A. scutatus [Species]
    A. lumachellus
Schistes [Genus]
    S. geoffroyi [Species]
Heliothryx [Genus]
    H. barroti [Species]
    H. aurita
Heliactin [Genus]
    H. cornuta [Species]
Loddigesia [Genus]
    L. mirabilis [Species]
Heliomaster [Genus]
    H. constantii [Species]
    H. longirostris
    H. squamosus
    H. furcifer
Rhodopis [Genus]
    R. vesper [Species]
Thaumastura [Genus]
    T. cora [Species]
Philodice [Genus]
    P. evelynae [Species]
    P. bryantae
    P. mitchellii
Doricha [Genus]
    D. enicura [Species]
    D. eliza
Tilmatura [Genus]
    T. dupontii [Species]
Microstilbon [Genus]
    M. burmeisteri [Species]
Calothorax [Genus]
    C. lucifer [Species]
    C. pulcher
Archilochus [Genus]
    A. colubris [Species]
    A. alexandri
Calliphlox [Genus]
    C. amethystina [Species]
Mellisuga [Genus]
    M. minima [Species]
Calypte [Genus]
    C. anna [Species]
    C. costae
    C. helenae
Stellula [Genus]
    S. calliope [Species]

Atthis [Genus]
    A. heloisa [Species]
Myrtis [Genus]
    M. fanny [Species]
Eulidia [Genus]
    E. yarrellii [Species]
Myrmia [Genus]
    M. micrura [Species]
Acestrura [Genus]
    A. mulsanti [Species]
    A. decorata
    A. bombus
    A. heliodor
    A. berlepschi
    A. harteri
Chaetocercus [Genus]
    C. jourdanii [Species]
Selasphorus [Genus]
    S. platycercus [Species]
    S. rufus
    S. sasin
    S. flammula
    S. torridus
    S. simoni
    S. ardens
    S. scintilla

Coliiformes [Order]
Coliidae [Family]
    Colius [Genus]
        C. striatus [Species]
        C. castanotus
        C. colius
        C. leucocephalus
        C. indicus
        C. macrourus

Trogoniformes [Order]
Trogonidae [Family]
    Pharomachrus [Genus]
        P. mocinno [Species]
        P. fulgidus
        P. pavoninus
    Euptilotis [Genus]
        E. neoxenus [Species]
    Priotelus [Genus]
        P. temnurus [Species]
    Temnotrogon [Genus]
        T. roseigaster [Species]
    Trogon [Genus]
        T. massena [Species]
        T. clathratus
        T. melanurus
        T. strigilatus
        T. citreolus
        T. mexicanus
        T. elegans
        T. collaris
        T. aurantiiventris
        T. personatus
        T. rufus

T. surrucura
T. curucui
T. violaceus
Apaloderma [Genus]
A. narina [Species]
A. aequatoriale
Heterotrogon [Genus]
H. vittatus [Species]
Harpactes [Genus]
H. reinwardtii [Species]
H. fasciatus
H. kasumba
H. diardii
H. ardens
H. whiteheadi
H. orrhophaeus
H. duvaucelii
H. oreskios
H. erythrocephalus
H. wardi

Coraciiformes [Order]
Alcedinidae [Family]
Ceryle [Genus]
C. lugubris [Species]
C. maxima
C. torquata
C. alcyon
C. rudis
Chloroceryle [Genus]
C. amazona [Species]
C. americana
C. inda
C. aenea
Alcedo [Genus]
A. hercules [Species]
A. atthis
A. semitorquata
A. meninting
A. quadribrachys
A. euryzona
A. coerulescens
A. cristata
A. leucogaster
Myioceyx [Genus]
M. lecontei [Species]
Ispidina [Genus]
I. picta [Species]
I. madagascariensis
Ceyx [Genus]
C. cyanopectus [Species]
C. argentatus
C. goodfellowi
C. lepidus
C. azureus
C. websteri
C. pusillus
C. erithacus
C. rufidorsum
C. melanurus
C. fallax

Pelargopsis [Genus]
P. amauroptera [Species]
P. capensis
P. melanorhyncha
Lacedo [Genus]
L. pulchella [Species]
Dacelo [Genus]
D. novaeguineae [Species]
D. leachii
D. tyro
D. gaudichaud
Clytoceyx [Genus]
C. rex [Species]
Melidora [Genus]
M. macrorrhina [Species]
Cittura [Genus]
C. cyanotis [Species]
Halcyon [Genus]
H. coromanda [Species]
H. badia
H. smyrnensis
H. pileata
H. cyanoventris
H. leucocephala
H. senegalensis
H. senegaloides
H. malimbica
H. albiventris
H. chelicuti
H. nigrocyanea
H. winchelli
H. diops
H. macleayii
H. albonotata
H. leucopygia
H. farquhari
H. pyrrhopygia
H. torotoro
H. megarhyncha
H. australasia
H. sancta
H. cinnamomina
H. funebris
H. chloris
H. saurophaga
H. recurvirostris
H. venerata
H. tuta
H. gambieri
H. godeffroyi
H. miyakoensis
H. bougainvillei
H. concreta
H. lindsayi
H. fulgida
H. monacha
H. princeps
Tanysiptera [Genus]
T. hydrocharis [Species]
T. galatea

T. riedelii
T. carolinae
T. ellioti
T. nympha
T. danae
T. sylvia

Todidae [Family]
Todus [Genus]
T. multicolor [Species]
T. angustirostris
T. todus
T. mexicanus
T. subulatus

Momotidae [Family]
Hylomanes [Genus]
H. momotula [Species]
Aspatha [Genus]
A. gularis [Species]
Electron [Genus]
E. platyrhynchum [Species]
E. carinatum
Eumomota [Genus]
E. superciliosa [Species]
Baryphthengus [Genus]
B. ruficapillus [Species]
Momotus [Genus]
M. mexicanus [Species]
M. momota

Meropidae [Family]
Dicrocercus [Genus]
D. hirundineus [Species]
Melittophagus [Genus]
M. revoilii [Species]
M. pusillus
M. variegatus
M. lafresnayii
M. bullockoides
M. bulocki
M. gularis
M. mulleri
Aerops [Genus]
A. albicollis [Species]
A. boehmi
Merops [Genus]
M. leschenaulti [Species]
M. apiaster
M. superciliosus
M. ornatus
M. orientalis
M. viridis
M. malimbicus
M. nubicus
M. nubicoides
Bombylonax [Genus]
B. breweri [Species]
Nyctyornis [Genus]
N. amicta [Species]
N. athertoni

Meropogon [Genus]
    M. forsteni [Species]

Leptosomatidae [Family]
  Leptosomus [Genus]
    L. discolor [Species]

Coraciidae [Family]
  Brachypteracias [Genus]
    B. leptosomus [Species]
    B. squamigera
  Atelornis [Genus]
    A. pittoides [Species]
    A. crossleyi
  Uratelornis [Genus]
    U. chimaera [Species]
  Coracias [Genus]
    C. garrulus [Species]
    C. abyssinica
    C. caudata
    C. spatulata
    C. noevia
    C. benghalensis
    C. temminckii
    C. cyanogaster
  Eurystomus [Genus]
    E. glaucurus [Species]
    E. gularis
    E. orientalis

Upupidae [Family]
  Upupa [Genus]
    U. epops [Species]

Phoeniculidae [Family]
  Phoeniculus [Genus]
    P. purpureus [Species]
    P. bollei
    P. castaneiceps
    P. aterrimus
  Rhinopomastus [Genus]
    R. minor [Species]
    R. cyanomelas
Bucerotidae [Family]
  Tockus [Genus]
    T. birostris [Species]
    T. fasciatus
    T. alboterminatus
    T. bradfieldi
    T. pallidirostris
    T. nasutus
    T. hemprichii
    T. monteiri
    T. griseus
    T. hartlaubi
    T. camurus
    T. erythrorhynchus
    T. flavirostris
    T. deckeni
    T. jacksoni
  Berenicornis [Genus]
    B. comatus [Species]
    B. albocristatus

Ptiloaemus [Genus]
    P. tickelli [Species]
Anorrhinus [Genus]
    A. galeritus [Species]
Penelopides [Genus]
    P. panini [Species]
    P. exarhatus
Aceros [Genus]
    A. nipalensis [Species]
    A. corrugatus
    A. leucocephalus
    A. cassidix
    A. undulatus
    A. plicatus
    A. everetti
    A. narcondami
Anthracoceros [Genus]
    A. malayanus [Species]
    A. malabaricus
    A. coronatus
    A. montani
    A. marchei
Bycanistes [Genus]
    B. bucinator [Species]
    B. cylindricus
    B. subcylindricus
    B. brevis
Ceratogymna [Genus]
    C. atrata [Species]
    C. elata
Buceros [Genus]
    B. rhinoceros [Species]
    B. bicornis
    B. hydrocorax
Rhinoplax [Genus]
    R. vigil [Species]
Bucorvus [Genus]
    B. abyssinicus [Species]
    B. leadbeateri

Piciformes [Order]
Galbulidae [Family]
  Galbalcyrhynchus [Genus]
    G. leucotis [Species]
  Brachygalba [Genus]
    B. lugubris [Species]
    B. phaeonota
    B. goeringi
    B. salmoni
    B. albogularis
  Jacamaralcyon [Genus]
    J. tridactyla [Species]
  Galbula [Genus]
    G. albirostris [Species]
    G. galbula
    G. tombacea
    G. cyanescens
    G. pastazae
    G. ruficauda
    G. leucogastra
    G. dea

Jacamerops [Genus]
    J. aurea [Species]

Bucconidae [Family]
  Notharchus [Genus]
    N. macrorhynchos [Species]
    N. pectoralis
    N. ordii
    N. tectus
  Bucco [Genus]
    B. macrodactylus [Species]
    B. tamatia
    B. noanamae
    B. capensis
  Nystalus [Genus]
    N. radiatus [Species]
    N. chacuru
    N. striolatus
    N. maculatus
  Hypnelus [Genus]
    H. ruficollis [Species]
    H. bicinctus
  Malacoptila [Genus]
    M. striata [Species]
    M. fusca
    M. fulvogularis
    M. rufa
    M. panamensis
    M. mystacalis
  Micromonacha [Genus]
    M. lanceolata [Species]
  Nonnula [Genus]
    N. rubecula [Species]
    N. sclateri
    N. brunnea
    N. frontalis
    N. ruficapilla
    N. amaurocephala
  Hapaloptila [Genus]
    H. castanea [Species]
  Monasa [Genus]
    M. atra [Species]
    M. nigrifrons
    M. morphoeus
    M. flavirostris
  Chelidoptera [Genus]
    C. tenebrosa [Species]

Capitonidae [Family]
  Capito [Genus]
    C. aurovirens [Species]
    C. maculicoronatus
    C. squamatus
    C. hypoleucus
    C. dayi
    C. quinticolor
    C. niger
  Eubucco [Genus]
    E. richardsoni [Species]
    E. bourcierii
    E. versicolor

Semnornis [Genus]
　S. frantzii [Species]
　S. ramphastinus
Psilopogon [Genus]
　P. pyrolophus [Species]
Megalaima [Genus]
　M. virens [Species]
　M. lagrandieri
　M. zeylanica
　M. viridis
　M. faiostricta
　M. corvina
　M. chrysopogon
　M. rafflesii
　M. mystacophanos
　M. javensis
　M. flavifrons
　M. franklinii
　M. oorti
　M. asiatica
　M. incognita
　M. henricii
　M. armillaris
　M. pulcherrima
　M. robustirostris
　M. australis
　M. eximia
　M. rubricapilla
　M. haemacephala
Calorhamphus [Genus]
　C. fuliginosus [Species]
Gymnobucco [Genus]
　G. calvus [Species]
　G. peli
　G. sladeni
　G. bonapartei
Smilorhis [Genus]
　S. leucotis [Species]
Stactolaema [Genus]
　S. olivacea [Species]
　S. anchietae
　S. whytii
Pogoniulus [Genus]
　P. duchaillui [Species]
　P. scolopaceus
　P. leucomystax
　P. simplex
　P. coryphaeus
　P. pusillus
　P. chrysoconus
　P. bilineatus
　P. subsulphureus
　P. atroflavus
Tricholaema [Genus]
　T. lacrymosum [Species]
　T. leucomelan
　T. diadematum
　T. melanocephalum
　T. flavibuccale
　T. hirsutum

Lybius [Genus]
　L. undatus [Species]
　L. vieilloti
　L. torquatus
　L. guifsobalito
　L. rubrifacies
　L. chaplini
　L. leucocephalus
　L. minor
　L. melanopterus
　L. bidentatus
　L. dubius
　L. rolleti
Trachyphonus [Genus]
　T. purpuratus [Species]
　T. vaillantii
　T. erythrocephalus
　T. darnaudii
　T. margaritatus

Indicatoridae [Family]
　Prodotiscus [Genus]
　　P. insignis [Species]
　　P. regulus
　Melignomon [Genus]
　　M. zenkeri [Species]
　indicator [Genus]
　　I. exilis [Species]
　　I. propinquus
　　I. minor
　　I. conirostris
　　I. variegatus
　　I. maculatus
　　I. archipelagicus
　　I. indicator
　　I. xanthonotus
　Melichneutes [Genus]
　　M. robustus [Species]

Ramphastidae [Family]
　Aulacorhynchus [Genus]
　　A. sulcatus [Species]
　　A. calorhynchus
　　A. derbianus
　　A. prasinus
　　A. haematopygus
　　A. coeruleicinctis
　　A. huallagae
　Pteroglossus [Genus]
　　P. torquatus [Species]
　　P. sanguineus
　　P. erythropygius
　　P. castanotis
　　P. aracari
　　P. pluricinctus
　　P. viridis
　　P. bitorquatus
　　P. olallae
　　P. flavirostris
　　P. mariae
　　P. beauharnaesii

Selenidera [Genus]
　S. spectabilis [Species]
　S. culik
　S. reinwardtii
　S. langsdorffi
　S. nattereri
　S. maculirostris
Andigena [Genus]
　A. bailloni [Species]
　A. laminirostris
　A. hypoglauca
　A. cucullata
　A. nigrirostris
Ramphastos [Genus]
　R. vitellinus [Species]
　R. dicolorus
　R. citreolaemus
　R. sulfuratus
　R. swainsonii
　R. ambiguus
　R. aurantiirostris
　R. tucanus
　R. cuvieri
　R. inca
　R. toco

Picidae [Family]
　Jynx [Genus]
　　J. torquilla [Species]
　　J. ruficollis
　Picumnus [Genus]
　　P. cinnamomeus [Species]
　　P. rufiventris
　　P. fuscus
　　P. castelnau
　　P. leucogaster
　　P. limae
　　P. olivaceus
　　P. granadensis
　　P. nebulosus
　　P. exilis
　　P. borbae
　　P. aurifrons
　　P. temminckii
　　P. cirratus
　　P. sclateri
　　P. steindachneri
　　P. squamulatus
　　P. minutissimus
　　P. pallidus
　　P. albosquamatus
　　P. guttifer
　　P. varzeae
　　P. pygmaeus
　　P. asterias
　　P. pumilus
　　P. innominatus
　Nesoctites [Genus]
　　N. micromegas [Species]
　Verreauxia [Genus]
　　V. africana [Species]

Sasia [Genus]
  S. ochracea [Species]
  S. abnormis
Geocolaptes [Genus]
  G. olivaceus [Species]
Colaptes [Genus]
  C. cafer [Species]
  C. auratus
  C. chrysoides
  C. rupicola
  C. pitius
  C. campestris
Nesoceleus [Genus]
  N. fernandinae [Species]
Chrysoptilus [Genus]
  C. melanochloros [Species]
  C. punctigula
  C. atricollis
Piculus [Genus]
  P. rivolii [Species]
  P. auricularis
  P. aeruginosus
  P. rubiginosus
  P. simplex
  P. flavigula
  P. leucolaemus
  P. aurulentus
  P. chrysochloros
Campethera [Genus]
  C. punctuligera [Species]
  C. nubica
  C. bennettii
  C. cailliautii
  C. notata
  C. abingoni
  C. taeniolaema
  C. tullbergi
  C. maculosa
  C. permista
  C. caroli
  C. nivosa
Celeus [Genus]
  C. flavescens [Species]
  C. spectabilis
  C. castaneus
  C. immaculatus
  C. elegans
  C. jumana
  C. grammicus
  C. loricatus
  C. undatus
  C. flavus
  C. torquatus
Micropternus [Genus]
  M. brachyurus [Species]
Picus [Genus]
  P. viridis [Species]
  P. vaillantii
  P. awokera
  P. squamatus

P. viridanus
P. vittatus
P. xanthopygaeus
P. canus
P. rabieri
P. erythropygius
P. flavinucha
P. puniceus
P. chlorolophus
P. mentalis
P. mineaceus
Dinopium [Genus]
  D. benghalense [Species]
  D. shorii
  D. javanense
  D. rafflesii
Gecinulus [Genus]
  G. grantia [Species]
  G. viridis
Meiglyptes [Genus]
  M. tristis [Species]
  M. jugularis
  M. tukki
Mulleripicus [Genus]
  M. pulverulentus [Species]
  M. funebris
  M. fuliginosus
  M. fulvus
Dryocopus [Genus]
  D. martius [Species]
  D. javensis
  D. pileatus
  D. lineatus
  D. erythrops
  D. schulzi
  D. galeatus
Asyndesmus [Genus]
  A. lewis [Species]
Melanerpes [Genus]
  M. erythrocephalus [Species]
  M. portoricensis
  M. herminieri
  M. formicivorus
  M. hypopolius
  M. carolinus
  M. aurifrons
  M. chrysogenys
  M. superciliaris
  M. caymanensis
  M. radiolatus
  M. striatus
  M. rubricapillus
  M. pucherani
  M. chrysauchen
  M. flavifrons
  M. cruentatus
  M. rubrifrons
Leuconerpes [Genus]
  L. candidus [Species]
Sphyrapicus [Genus]

S. varius [Species]
S. thyroideus
Trichopicus [Genus]
  T. cactorum [Species]
Veniliornis [Genus]
  V. fumigatus [Species]
  V. spilogaster
  V. passerinus
  V. frontalis
  V. maculifrons
  V. cassini
  V. affinis
  V. kirkii
  V. callonotus
  V. sanguineus
  V. dignus
  V. nigriceps
Dendropicos [Genus]
  D. fuscescens [Species]
  D. stierlingi
  D. elachus
  D. abyssinicus
  D. poecilolaemus
  D. gabonensis
  D. lugubris
Dendrocopos [Genus]
  D. major [Species]
  D. leucopterus
  D. syriacus
  D. assimilis
  D. himalayensis
  D. darjellensis
  D. medius
  D. leucotos
  D. cathpharius
  D. hyperythrus
  D. auriceps
  D. atratus
  D. macei
  D. mahrattensis
  D. minor
  D. canicapillus
  D. wattersi
  D. kizuki
  D. moluccensis
  D. maculatus
  D. temminckii
  D. obsoletus
  D. dorae
  D. albolarvatus
  D. villosus
  D. pubescens
  D. borealis
  D. nuttallii
  D. scalaris
  D. arizonae
  D. stricklandi
  D. mixtus
  D. lignarius
Picoides [Genus]

P. tridactylus [Species]
P. arcticus
Sapheopipo [Genus]
   S. noguchii [Species]
Xiphidiopicus [Genus]
   X. percussus [Species]
Polipicus [Genus]
   P. johnstoni [Species]
   P. elliotii
Mesopicos [Genus]
   M. goertae [Species]
   M. griseocephalus
Thripias [Genus]
   T. namaquus [Species]
   T. xantholophus
   T. pyrrhogaster
Hemicircus [Genus]
   H. concretus [Species]
   H. canente
Blythipicus [Genus]
   B. pyrrhotis [Species]
   B. rubiginosus
Chrysocolaptes [Genus]
   C. validus [Species]
   C. festivus
   C. lucidus
Phloeoceastes [Genus]
   P. guatemalensis [Species]
   P. melanoleucos
   P. leucopogon
   P. rubricollis
   P. robustus
   P. pollens
   P. haematogaster
Campephilus [Genus]
   C. principalis [Species]
   C. imperialis
   C. magellanicus

Passeriformes [Order]
Eurylaimidae [Family]
Smithornis [Genus]
   S. capensis [Species]
   S. rufolateralis
   S. sharpei
Pseudocalyptomena [Genus]
   P. graueri [Species]
Corydon [Genus]
   C. sumatranus [Species]
Cymbirhynchus [Genus]
   C. macrorhynchos [Species]
Eurylaimus [Genus]
   E. javanicus [Species]
   E. ochromalus
   E. steerii
Serilophus [Genus]
   S. lunatus [Species]
Psarisomus [Genus]
   P. dalhousiae [Species]
Calyptomena [Genus]

C. viridis [Species]
C. hosii
C. whiteheadi

Dendrocolaptidae [Family]
Dendrocincla [Genus]
   D. tyrannina [Species]
   D. macrorhyncha
   D. fuliginosa
   D. anabatina
   D. merula
   D. homochroa
Deconychura [Genus]
   D. longicauda [Species]
   D. stictolaema
Sittasomus [Genus]
   S. griseicapillus [Species]
Glyphorynchus [Genus]
   G. spirurus [Species]
Drymornis [Genus]
   D. bridgesii [Species]
Nasica [Genus]
   N. longirostris [Species]
Dendrexetastes [Genus]
   D. rufigula [Species]
Hylexetastes [Genus]
   H. perrotii [Species]
   H. stresemanni
Xiphocolaptes [Genus]
   X. promeropirhynchus [Species]
   X. albicollis
   X. falcirostris
   X. franciscanus
   X. major
Dendrocolaptes [Genus]
   D. certhia [Species]
   D. concolor
   D. hoffmannsi
   D. picumnus
   D. platyrostris
Xiphorhynchus [Genus]
   X. picus [Species]
   X. necopinus
   X. obsoletus
   X. ocellatus
   X. spixii
   X. elegans
   X. pardalotus
   X. guttatus
   X. flavigaster
   X. striatigularis
   X. lachrymosus
   X. erythropygius
   X. triangularis
Lepidocolaptes [Genus]
   L. leucogaster [Species]
   L. souleyetii
   L. angustirostris
   L. affinis
   L. squamatus
   L. fuscus
   L. albolineatus

Campylorhamphus [Genus]
   C. pucherani [Species]
   C. trochilirostris
   C. pusillus
   C. procurvoides

Furnariidae [Family]
Geobates [Genus]
   G. poecilopterus [Species]
Geositta [Genus]
   G. maritima [Species]
   G. peruviana
   G. saxicolina
   G. isabellina
   G. rufipennis
   G. punensis
   G. cunicularia
   G. antarctica
   G. tenuirostris
   G. crassirostris
Upucerthia [Genus]
   U. dumetaria [Species]
   U. albigula
   U. validirostris
   U. serrana
   U. andaecola
Ochetorhynchus [Genus]
   O. ruficaudus [Species]
   O. certhioides
   O. harteri
Eremobius [Genus]
   E. phoenicurus [Species]
Chilia [Genus]
   C. melanura [Species]
Cinclodes [Genus]
   C. antarcticus [Species]
   C. patagonicus
   C. oustaleti
   C. fuscus
   C. comechingonus
   C. atacamensis
   C. palliatus
   C. taczanowskii
   C. nigrofumosus
   C. excelsior
Clibanornis [Genus]
   C. dendrocolaptoides [Species]
Furnarius [Genus]
   F. rufus [Species]
   F. leucopus
   F. torridus
   F. minor
   F. figulus
   F. cristatus
Limnornis [Genus]
   L. curvirostris [Species]
Sylviorthorhynchus [Genus]
   S. desmursii [Species]
Aphrastura [Genus]
   A. spinicauda [Species]
   A. masafuerae

Phleocryptes [Genus]
  P. melanops [Species]
Leptasthenura [Genus]
  L. andicola [Species]
  L. striata
  L. pileata
  L. xenothorax
  L. striolata
  L. aegithaloides
  L. platensis
  L. fuliginiceps
  L. yanacensis
  L. setaria
Spartonoica [Genus]
  S. maluroides [Species]
Schizoeaca [Genus]
  S. coryi [Species]
  S. fuliginosa
  S. griseomurina
  S. palpebralis
  S. helleri
  S. harterti
Schoeniophylax [Genus]
  S. phryganophila [Species]
Oreophylax [Genus]
  O. moreirae [Species]
Synallaxis [Genus]
  S. ruficapilla [Species]
  S. superciliosa
  S. poliophrys
  S. azarae
  S. frontalis
  S. moesta
  S. cabanisi
  S. spixi
  S. hypospodia
  S. subpudica
  S. albescens
  S. brachyura
  S. albigularis
  S. gujanensis
  S. propinqua
  S. cinerascens
  S. tithys
  S. cinnamomea
  S. fuscorufa
  S. unirufa
  S. rutilans
  S. erythrothorax
  S. cherriei
  S. stictothorax
Hellmayrea [Genus]
  H. gularis [Species]
Gyalophylax [Genus]
  G. hellmayri [Species]
Certhiaxis [Genus]
  C. cinnamomea [Species]
  C. mustelina
Limnoctites [Genus]
  L. rectirostris [Species]

Poecilurus [Genus]
  P. candei [Species]
  P. kollari
  P. scutatus
Cranioleuca [Genus]
  C. sulphurifera [Species]
  C. semicinerea
  C. obsoleta
  C. pyrrhophia
  C. subcristata
  C. hellmayri
  C. curtata
  C. furcata
  C. demissa
  C. erythrops
  C. vulpina
  C. pallida
  C. antisiensis
  C. marcapatae
  C. albiceps
  C. baroni
  C. albicapilla
  C. mulleri
  C. gutturata
Siptornopsis [Genus]
  S. hypochondriacus [Species]
Asthenes [Genus]
  A. pyrrholeuca [Species]
  A. dorbignyi
  A. berlepschi
  A. baeri
  A. patagonica
  A. steinbachi
  A. humicola
  A. modesta
  A. pudibunda
  A. ottonis
  A. heterura
  A. wyatti
  A. humilis
  A. anthoides
  A. sclateri
  A. hudsoni
  A. virgata
  A. maculicauda
  A. flammulata
  A. urubambensis
Thripophaga [Genus]
  T. macroura [Species]
  T. cherriei
  T. fusciceps
  T. berlepschi
Phacellodomus [Genus]
  P. sibilatrix [Species]
  P. rufifrons
  P. striaticeps
  P. erythrophthalmus
  P. ruber
  P. striaticollis
  P. dorsalis

Coryphistera [Genus]
  C. alaudina [Species]
Anumbius [Genus]
  A. annumbi [Species]
Siptornis [Genus]
  S. striaticollis [Species]
Xenerpestes [Genus]
  X. minlosi [Species]
  X. singularis
Metopothrix [Genus]
  M. aurantiacus [Species]
Roraimia [Genus]
  R. adusta [Species]
Margarornis [Genus]
  M. squamiger [Species]
  M. bellulus
  M. rubiginosus
  M. stellatus
Premnornis [Genus]
  P. guttuligera [Species]
Premnoplex [Genus]
  P. brunnescens [Species]
Pseudocolaptes [Genus]
  P. lawrencii [Species]
  P. boissonneautii
Berlepschia [Genus]
  B. rikeri [Species]
Pseudoseisura [Genus]
  P. cristata [Species]
  P. lophotes
  P. gutturalis
Hyloctistes [Genus]
  H. subulatus [Species]
Ancistrops [Genus]
  A. strigilatus [Species]
Anabazenops [Genus]
  A. fuscus [Species]
Syndactyla [Genus]
  S. rufosuperciliata [Species]
  S. subalaris
  S. guttulata
  S. mirandae
Simoxenops [Genus]
  S. ucayalae [Species]
  S. striatus
Anabacerthia [Genus]
  A. striaticollis [Species]
  A. temporalis
  A. amaurotis
Philydor [Genus]
  P. atricapillus [Species]
  P. erythrocercus
  P. pyrrhodes
  P. dimidiatus
  P. baeri
  P. lichtensteini
  P. rufus
  P. erythropterus
  P. ruficaudatus
Automolus [Genus]

A. leucophthalmus [Species]
A. infuscatus
A. dorsalis
A. rubiginosus
A. albigularis
A. ochrolaemus
A. rufipileatus
A. ruficollis
A. melanopezus
Hylocryptus [Genus]
  H. erythrocephalus [Species]
  H. rectirostris
Cichlocolaptes [Genus]
  C. leucophrus [Species]
Heliobletus [Genus]
  H. contaminatus [Species]
Thripadectes [Genus]
  T. flammulatus [Species]
  T. holostictus
  T. melanorhynchus
  T. rufobrunneus
  T. virgaticeps
  T. scrutator
  T. ignobilis
Xenops [Genus]
  X. milleri [Species]
  X. tenuirostris
  X. rutilans
  X. minutus
Megaxenops [Genus]
  M. parnaguae [Species]
Pygarrhichas [Genus]
  P. albogularis [Species]
Sclerurus [Genus]
  S. scansor [Species]
  S. albigularis
  S. mexicanus
  S. rufigularis
  S. caudacutus
  S. guatemalensis
Lochmias [Genus]
  L. nematura [Species]

Formicariidae [Family]
  Cymbilaimus [Genus]
  C. lineatus [Species]
  Hypoedaleus [Genus]
  H. guttatus [Species]
  Batara [Genus]
  B. cinerea [Species]
  Mackenziaena [Genus]
  M. leachii [Species]
  M. severa
  Frederickena [Genus]
  F. viridis [Species]
  U. unduligera
  Taraba [Genus]
  T. major [Species]
  Sakesphorus [Genus]
  S. canadensis [Species]
  S. cristatus

S. bernardi
S. melanonotus
S. melanothorax
S. luctuosus
Biatas [Genus]
  B. nigropectus [Species]
Thamnophilus [Genus]
  T. doliatus [Species]
  T. multistriatus
  T. palliatus
  T. bridgesi
  T. nigriceps
  T. praecox
  T. nigrocinereus
  T. aethiops
  T. unicolor
  T. schistaceus
  T. murinus
  T. aroyae
  T. punctatus
  T. amazonicus
  T. insignis
  T. caerulescens
  T. torquatus
  T. ruficapillus
Pygiptila [Genus]
  P. stellaris [Species]
Megastictus [Genus]
  M. margaritatus [Species]
Neoctantes [Genus]
  N. niger [Species]
Clytoctantes [Genus]
  C. alixii [Species]
Xenornis [Genus]
  X. setifrons [Species]
Thamnistes [Genus]
  T. anabatinus [Species]
Dysithamnus [Genus]
  D. stictothorax [Species]
  D. mentalis
  D. striaticeps
  D. puncticeps
  D. xanthopterus
  D. ardesiacus
  D. saturninus
  D. occidentalis
  D. plumbeus
Thamnomanes [Genus]
  T. caesius [Species]
Myrmotherula [Genus]
  M. brachyura [Species]
  M. obscura
  M. sclateri
  M. klagesi
  M. surinamensis
  M. ambigua
  M. cherriei
  M. guttata
  M. longicauda
  M. hauxwelli

M. gularis
M. gutturalis
M. fulviventris
M. leucophthalma
M. haematonota
M. ornata
M. erythrura
M. erythronotos
M. axillaris
M. schisticolor
M. sunensis
M. longipennis
M. minor
M. iheringi
M. grisea
M. unicolor
M. behni
M. urosticta
M. menetriesii
M. assimilis
Dichrozona [Genus]
  D. cincta [Species]
Myrmorchilus [Genus]
  M. strigilatus [Species]
Herpsilochmus [Genus]
  H. pileatus [Species]
  H. sticturus
  H. stictocephalus
  H. dorsimaculatus
  H. roraimae
  H. pectoralis
  H. longirostris
  H. axillaris
  H. rufimarginatus
Microrhopias [Genus]
  M. quixensis [Species]
Formicivora [Genus]
  F. iheringi [Species]
  F. grisea
  F. serrana
  F. melanogaster
  F. rufa
Drymophila [Genus]
  D. ferruginea [Species]
  D. genei
  D. ochropyga
  D. devillei
  D. caudata
  D. malura
  D. squamata
Terenura [Genus]
  T. maculata [Species]
  T. callinota
  T. humeralis
  T. sharpei
  T. spodioptila
Cercomacra [Genus]
  C. cinerascens [Species]
  C. brasiliana
  C. tyrannina

C. nigriscens
C. serva
C. nigricans
C. carbonaria
C. melanaria
C. ferdinandi
Sipia [Genus]
S. berlepschi [Species]
S. rosenbergi
Pyriglena [Genus]
P. leuconota [Species]
P. atra
P. leucoptera
Rhopornis [Genus]
R. ardesiaca [Species]
Myrmoborus [Genus]
M. leucophrys [Species]
M. lugubris
M. myotherinus
M. melanurus
Hypocnemis [Genus]
H. cantator [Species]
H. hypoxantha
Hypocnemoides [Genus]
H. melanopogon [Species]
H. maculicauda
Myrmochanes [Genus]
M. hemileucus [Species]
Gymnocichla [Genus]
G. nudiceps [Species]
Sclateria [Genus]
S. naevia [Species]
Percnostola [Genus]
P. rufifrons [Species]
P. schistacea
P. leucostigma
P. caurensis
P. lophotes
Myrmeciza [Genus]
M. longipes [Species]
M. exsul
M. ferruginea
M. ruficauda
M. laemosticta
M. disjuncta
M. pelzelni
M. hemimelaena
M. hyperythra
M. goeldii
M. melanoceps
M. fortis
M. immaculata
M. griseiceps
Myrmoderus [Genus]
M. loricatus [Species]
M. squamosus
Myrmophylax [Genus]
M. atrothorax [Species]
M. stictothorax
Formicarius [Genus]

F. colma [Species]
F. analis
F. nigricapillus
F. rufipectus
Chamaeza [Genus]
C. campanisona [Species]
C. nobilis
C. ruficauda
C. mollissima
Pithys [Genus]
P. albifrons [Species]
P. castanea
Gymnopithys [Genus]
G. rufigula [Species]
G. salvini
G. lunulata
G. leucaspis
Rhegmatorhina [Genus]
R. gymnops [Species]
R. berlepschi
R. cristata
R. hoffmannsi
R. melanosticta
Hylophylax [Genus]
H. naevioides [Species]
H. naevia
H. punctulata
H. poecilonota
Phlegopsis [Genus]
P. nigromaculata [Species]
P. erythroptera
P. borbae
Phaenostictus [Genus]
P. mcleannani [Species]
Myrmornis [Genus]
M. torquata [Species]
Pittasoma [Genus]
P. michleri [Species]
P. rufopileatum
Grallaricula [Genus]
G. flavirostris [Species]
G. ferrugineipectus
G. nana
G. loricata
G. peruviana
G. lineifrons
G. cucullata
Myrmothera [Genus]
M. campanisona [Species]
M. simplex
Thamnocharis [Genus]
T. dignissima [Species]
Grallaria [Genus]
G. squamigera [Species]
G. excelsa
G. gigantea
G. guatimalensis
G. varia
G. alleni
G. haplonota

G. milleri
G. bangsi
G. quitensis
G. erythrotis
G. hypoleuca
G. przewalskii
G. capitalis
G. nuchalis
G. albigula
G. ruficapilla
G. erythroleuca
G. rufocinerea
G. griseonucha
G. rufula
G. andicola
G. macularia
G. fulviventris
G. berlepschi
G. perspicillata
G. ochroleuca

Conopophagidae [Family]
Conopophaga [Genus]
C. lineata [Species]
C. cearae
C. aurita
C. roberti
C. peruviana
C. ardesiaca
C. castaneiceps
C. melanops
C. melanogaster
Corythopis [Genus]
C. delalandi [Species]
C. torquata

Rhinocryptidae [Family]
Pteroptochos [Genus]
P. castaneus [Species]
P. tarnii
P. megapodius
Scelorchilus [Genus]
S. albicollis [Species]
S. rubecula
Rhinocrypta [Genus]
R. lanceolata [Species]
Teledromas [Genus]
T. fuscus [Species]
Liosceles [Genus]
L. thoracicus [Species]
Merulaxis [Genus]
M. ater [Species]
Melanopareia [Genus]
M. torquata [Species]
M. maximiliani
M. maranonicus
M. elegans
Scytalopus [Genus]
S. unicolor [Species]
S. speluncae
S. macropus

S. femoralis
S. argentifrons
S. chiriquensis
S. panamensis
S. latebricola
S. indigoticus
S. magellanicus
Psilorhamphus [Genus]
  P. guttatus [Species]
Myornis [Genus]
  M. senilis [Species]
Eugralla [Genus]
  E. paradoxa [Species]
Acropternis [Genus]
  A. orthonyx [Species]

Tyrannidae [Family]
  Phyllomyias [Genus]
    P. fasciatus [Species]
    P. burmeisteri
    P. virescens
    P. sclateri
    P. griseocapilla
    P. griseiceps
    P. plumbeiceps
    P. nigrocapillus
    P. cinereiceps
    P. uropygialis
  Zimmerius [Genus]
    Z. vilissimus [Species]
    Z. bolivianus
    Z. cinereicapillus
    Z. gracilipes
    Z. viridiflavus
  Ornithion [Genus]
    O. inerme [Species]
    O. semiflavum
    O. brunneicapillum
  Camptostoma [Genus]
    C. imberbe [Species]
    C. obsoletum
  Phaeomyias [Genus]
    P. murina [Species]
  Sublegatus [Genus]
    S. modestus [Species]
    S. obscurior
  Suiriri [Genus]
    S. suiriri [Species]
  Tyrannulus [Genus]
    T. elatus [Species]
  Myiopagis [Genus]
    M. gaimardii [Species]
    M. caniceps
    M. subplacens
    M. flavivertex
    M. cotta
    M. viridicata
    M. leucospodia
  Elaenia [Genus]
    E. martinica [Species]
    E. flavogaster

E. spectabilis
E. albiceps
E. parvirostris
E. mesoleuca
E. strepera
E. gigas
E. pelzelni
E. cristata
E. ruficeps
E. chiriquensis
E. frantzii
E. obscura
E. dayi
E. pallatangae
E. fallax
Mecocerculus [Genus]
  M. leucophrys [Species]
  M. poecilocercus
  M. hellmayri
  M. calopterus
  M. minor
  M. stictopterus
Serpophaga [Genus]
  S. cinerea [Species]
  S. hypoleuca
  S. nigricans
  S. araguayae
  S. subcristata
Inezia [Genus]
  I. inornata [Species]
  I. tenuirostris
  I. subflava
Stigmatura [Genus]
  S. napensis [Species]
  S. budytoides
Anairetes [Genus]
  A. alpinus [Species]
  A. agraphia
  A. agilis
  A. reguloides
  A. flavirostris
  A. fernandezianus
  A. parulus
Tachuris [Genus]
  T. rubrigastra [Species]
Culicivora [Genus]
  C. caudacuta [Species]
Polystictus [Genus]
  P. pectoralis [Species]
  P. superciliaris
Pseudocolopteryx [Genus]
  P. sclateri [Species]
  P. dinellianus
  P. acutipennis
  P. flaviventris
Euscarthmus [Genus]
  E. meloryphus [Species]
  E. rufomarginatus
Mionectes [Genus]
  M. striaticollis [Species]

M. oliveceus
M. oleagineus
M. macconnelli
M. rufiventris
Leptopogon [Genus]
  L. rufipectus [Species]
  L. taczanowskii
  L. amaurocephalus
  L. superciliaris
Phylloscartes [Genus]
  P. nigrifrons [Species]
  P. poecilotis
  P. chapmani
  P. ophthalmicus
  P. eximius
  P. gualaquizae
  P. flaviventris
  P. venezuelanus
  P. orbitalis
  P. flaveolus
  P. roquettei
  P. ventralis
  P. paulistus
  P. oustaleti
  P. difficilis
  P. flavovirens
  P. virescens
  P. superciliaris
  P. sylviolus
Pseudotriccus [Genus]
  P. pelzelni [Species]
  P. simplex
  P. ruficeps
Myiornis [Genus]
  M. auricularis [Species]
  M. albiventris
  M. ecaudatus
Lophotriccus [Genus]
  L. pileatus [Species]
  L. eulophotes
  L. vitiosus
  L. galeatus
Atalotriccus [Genus]
  A. pilaris [Species]
Poecilotriccus [Genus]
  P. ruficeps [Species]
  P. capitale
  P. tricolor
  P. andrei
Oncostoma [Genus]
  O. cinereigulare [Species]
  O. olivaceum
Hemitriccus [Genus]
  H. minor [Species]
  H. josephinae
  H. diops
  H. obsoletus
  H. flammulatus
  H. zosterops
  H. aenigma

H. orbitatus
H. iohannis
H. striaticollis
H. nidipendulus
H. spodiops
H. margaritaceiventer
H. inoratus
H. granadensis
H. mirandae
H. kaempferi
H. rufigularis
H. furcatus
Todirostrum [Genus]
  T. senex [Species]
  T. russatum
  T. plumbeiceps
  T. fumifrons
  T. latirostre
  T. sylvia
  T. maculatum
  T. poliocephalum
  T. cinereum
  T. pictum
  T. chrysocrotaphum
  T. nigriceps
  T. calopterum
Cnipodectes [Genus]
  C. subbrunneus [Species]
Ramphotrigon [Genus]
  R. megacephala [Species]
  R. fuscicauda
  R. ruficauda
Rhynchocyclus [Genus]
  R. brevirostris [Species]
  R. olivaceus
  R. fulvipectus
Tolmomyias [Genus]
  T. sulphurescens [Species]
  T. assimilis
  T. poliocephalus
  T. flaviventris
Platyrinchus [Genus]
  P. saturatus [Species]
  P. cancrominus
  P. mystaceus
  P. coronatus
  P. flavigularis
  P. platyrhynchos
  P. leucoryphus
Onychorhynchus [Genus]
  O. coronatus [Species]
Myiotriccus [Genus]
  M. ornatus [Species]
Terenotriccus [Genus]
  T. erythrurus [Species]
Myiobius [Genus]
  M. villosus [Species]
  M. barbatus
  M. atricaudus
Myiophobus [Genus]

M. flavicans [Species]
M. phoenicomitra
M. inornatus
M. roraimae
M. lintoni
M. pulcher
M. ochraceiventris
M. cryptoxanthus
M. fasciatus
Aphanotriccus [Genus]
  A. capitalis [Species]
  A. audax
Xenotriccus [Genus]
  X. callizonus [Species]
  X. mexicanus
Pyrrhomyias [Genus]
  P. cinnamomea [Species]
Mitrephanes [Genus]
  M. phaeocercus [Species]
  M. olivaceus
Contopus [Genus]
  C. borealis [Species]
  C. fumigatus
  C. ochraceus
  C. sordidulus
  C. virens
  C. cinereus
  C. nigrescens
  C. albogularis
  C. caribaeus
  C. latirostris
Empidonax [Genus]
  E. flaviventris [Species]
  E. virescens
  E. alnorum
  E. traillii
  E. albigularis
  E. euleri
  E. griseipectus
  E. minimus
  E. hammondii
  E. wrightii
  E. oberholseri
  E. affinis
  E. difficilis
  E. flavescens
  E. fulvifrons
  E. atriceps
Nesotriccus [Genus]
  N. ridgwayi [Species]
Cnemotriccus [Genus]
  C. fuscatus [Species]
Sayornis [Genus]
  S. phoebe [Species]
  S. saya
  S. nigricans
Pyrocephalus [Genus]
  P. rubinus [Species]
Ochthoeca [Genus]
  O. cinnamomeiventris [Species]
  O. diadema

O. frontalis
O. pulchella
O. rufipectoralis
O. fumicolor
O. oenanthoides
O. parvirostris
O. leucophrys
O. piurae
O. littoralis
Myiotheretes [Genus]
  M. striaticollis [Species]
  M. erythropygius
  M. rufipennis
  M. pernix
  M. fumigatus
  M. fuscorufus
Xolmis [Genus]
  X. pyrope [Species]
  X. cinerea
  X. coronata
  X. velata
  X. dominicana
  X. irupero
Neoxolmis [Genus]
  N. rubetra [Species]
  N. ruficentris
Agriornis [Genus]
  A. montana [Species]
  A. andicola
  A. livida
  A. microptera
  A. murina
Muscisaxicola [Genus]
  M. maculirostris [Species]
  M. fluviatilis
  M. macloviana
  M. capistrata
  M. rufivertex
  M. juninensis
  M. albilora
  M. alpina
  M. cinerea
  M. albifrons
  M. flavinucha
  M. frontalis
Lessonia [Genus]
  L. oreas [Species]
  L. rufa
Knipolegus [Genus]
  K. striaticeps [Species]
  K. hudsoni
  K. poecilocercus
  K. signatus
  K. cyanirostris
  K. poecilurus
  K. orenocensis
  K. aterrimus
  K. nigerrimus
  K. lophotes
Hymenops [Genus]
  H. perspicillata [Species]

Fluvicola [Genus]
    F. pica [Species]
    F. nengeta
    F. leucocephala
Colonia [Genus]
    C. colonus [Species]
Alectrurus [Genus]
    A. tricolor [Species]
    A. risora
Gubernetes [Genus]
    G. yetapa [Species]
Satrapa [Genus]
    S. icterophrys [Species]
Tumbezia [Genus]
    T. salvini [Species]
Muscigralla [Genus]
    M. brevicauda [Species]
Hirundinea [Genus]
    H. ferruginea [Species]
Machetornis [Genus]
    M. rixosus [Species]
Muscipipra [Genus]
    M. vetula [Species]
Attila [Genus]
    A. phoenicurus [Species]
    A. cinnamomeus
    A. torridus
    A. citriniventris
    A. bolivianus
    A. rufus
    A. spadiceus
Casiornis [Genus]
    C. rufa [Species]
    C. fusca
Rhytipterna [Genus]
    R. simplex [Species]
    R. holerythra
    R. immunda
Laniocera [Genus]
    L. hypopyrrha [Species]
    L. rufescens
Sirystes [Genus]
    S. sibilator [Species]
Myiarchus [Genus]
    M. semirufus [Species]
    M. yucatanensis
    M. barbirostris
    M. tuberculifer
    M. swainsoni
    M. venezuelensis
    M. panamensis
    M. ferox
    M. cephalotes
    M. phaeocephalus
    M. apicalis
    M. cinerascens
    M. nuttingi
    M. crinitus
    M. tyrannulus
    M. magnirostris

M. nugator
M. validus
M. sagrae
M. stolidus
M. antillarum
M. oberi
Deltarhynchus [Genus]
    D. flammulatus [Species]
Pitangus [Genus]
    P. lictor [Species]
    P. sulphuratus
Megarhynchus [Genus]
    M. pitangua [Species]
Myiozetetes [Genus]
    M. cayanensis [Species]
    M. similis
    M. granadensis
    M. luteiventris
Conopias [Genus]
    C. inornatus [Species]
    C. parva
    C. trivirgata
    C. cinchoneti
Myiodynastes [Genus]
    M. hemichrysus [Species]
    M. chrysocephalus
    M. bairdii
    M. maculatus
    M. luteiventris
Legatus [Genus]
    L. leucophaius [Species]
Empidonomus [Genus]
    E. varius [Species]
    E. aurantioatrocristatus
Tyrannopsis [Genus]
    T. sulphurea [Species]
Tyrannus [Genus]
    T. niveigularis [Species]
    T. albogularis
    T. melancholicus
    T. couchii
    T. vociferans
    T. crassirostris
    T. verticalis
    T. forficata
    T. savana
    T. tyrannus
    T. dominicensis
    T. caudifasciatus
    T. cubensis
Xenopsaris [Genus]
    X. albinucha [Species]
Pachyramphus [Genus]
    P. viridis [Species]
    P. versicolor
    P. spodiurus
    P. rufus
    P. castaneus
    P. cinnamomeus
    P. polychopterus

P. marginatus
P. albogriseus
P. major
P. surinamus
P. aglaiae
P. homochrous
P. minor
P. validus
P. niger
Tityra [Genus]
    T. cayana [Species]
    T. semifasciata
    T. inquisitor
    T. leucura

Pipridae [Family]
    Schiffornis [Genus]
        S. major [Species]
        S. turdinus
        S. virescens
    Sapayoa [Genus]
        S. aenigma [Species]
    Piprites [Genus]
        P. griseiceps [Species]
        P. chloris
        P. pileatus
    Neopipo [Genus]
        N. cinnamomea [Species]
    Chloropipo [Genus]
        C. flavicapilla [Species]
        C. holochlora
        C. uniformis
        C. unicolor
    Xenopipo [Genus]
        X. atronitens [Species]
    Antilophia [Genus]
        A. galeata [Species]
    Tyranneutes [Genus]
        T. stolzmanni [Species]
        T. virescens
    Neopelma [Genus]
        N. chrysocephalum [Species]
        N. pallescens
        N. aurifrons
        N. sulphureiventer
    Heterocercus [Genus]
        H. flavivertex [Species]
        H. aurantiivertex
        H. lineatus
    Machaeropterus [Genus]
        M. regulus [Species]
        M. pyrocephalus
        M. deliciosus
    Manacus [Genus]
        M. manacus [Species]
    Corapipo [Genus]
        C. leucorrhoa [Species]
        C. gutturalis
    Ilicura [Genus]
        I. militaris [Species]
    Masius [Genus]

M. chrysopterus [Species]
Chiroxiphia [Genus]
  C. linearis [Species]
  C. lanceolata
  C. pareola
  C. caudata
Pipra [Genus]
  P. pipra [Species]
  P. coronata
  P. isidorei
  P. coeruleocapilla
  P. nattereri
  P. vilasboasi
  P. iris
  P. serena
  P. aureola
  P. fasciicauda
  P. filicauda
  P. mentalis
  P. erythrocephala
  P. rubrocapilla
  P. chloromeros
  P. cornuta

Cotingidae [Family]
  Phoenicircus [Genus]
    P. carnifex [Species]
    P. nigricollis
  Laniisoma [Genus]
    L. elegans [Species]
  Phibalura [Genus]
    P. flavirostris [Species]
  Tijuca [Genus]
    T. atra [Species]
  Carpornis [Genus]
    C. cucullatus [Species]
    C. melanocephalus
  Ampelion [Genus]
    A. rubrocristatus [Species]
    A. rufaxilla
    A. sclateri
    A. stresemanni
  Pipreola [Genus]
    P. riefferii [Species]
    P. intermedia
    P. arcuata
    P. auroeopectus
    P. frontalis
    P. chlorolepidota
    P. formosa
    P. whitelyi
  Ampelioides [Genus]
    A. tschudii [Species]
  Iodopleura [Genus]
    I. pipra [Species]
    I. fusca
    I. isabellae
  Calyptura [Genus]
    C. cristata [Species]
  Lipaugus [Genus]
    L. subalaris [Species]

    L. cryptolophus
    L. fuscocinereus
    L. vociferans
    L. unirufus
    L. lanioides
    L. streptophorus
  Chirocylla [Genus]
    C. uropygialis [Species]
  Porphyrolaema [Genus]
    P. porphyrolaema [Species]
  Cotinga [Genus]
    C. amabilis [Species]
    C. ridgwayi
    C. nattererii
    C. maynana
    C. cotinga
    C. maculata
    C. cayana
  Xipholena [Genus]
    X. punicea [Species]
    X. lamellipennis
    X. atropurpurea
  Carpodectes [Genus]
    C. nitidus [Species]
    C. antoniae
    C. hopkei
  Conioptilon [Genus]
    C. mcilhennyi [Species]
  Gymnoderus [Genus]
    G. foetidus [Species]
  Haematoderus [Genus]
    H. militaris [Species]
  Querula [Genus]
    Q. purpurata [Species]
  Pyroderus [Genus]
    P. scutatus [Species]
  Cephalopterus [Genus]
    C. glabricollis [Species]
    C. penduliger
    C. ornatus
  Perissocephalus [Genus]
    P. tricolor [Species]
  Procnias [Genus]
    P. tricarunculata [Species]
    P. alba
    P. averano
    P. nudicollis
  Rupicola [Genus]
    R. rupicola [Species]
    R. peruviana

Oxyruncidae [Family]
  Oxyruncus [Genus]
    O. cristatus [Species]

Phytotomidae [Family]
  Phytotoma [Genus]
    P. raimondii [Species]
    P. rara
    P. rutila

Pittidae [Family]
  Pitta [Genus]
    P. phayrei [Species]
    P. nipalensis
    P. soror
    P. oatesi
    P. schneideri
    P. caerulea
    P. cyanea
    P. elliotii
    P. guajana
    P. gurneyi
    P. kochi
    P. erythrogaster
    P. arcuata
    P. granatina
    P. venusta
    P. baudii
    P. sordida
    P. brachyura
    P. nympha
    P. angolensis
    P. superba
    P. maxima
    P. steerii
    P. moluccensis
    P. versicolor
    P. anerythra

Philepittidae [Family]
  Philepitta [Genus]
    P. castanea [Species]
    P. schlegeli
  Neodrepanis [Genus]
    N. coruscans [Species]
    N. hypoxantha

Acanthisittidae [Family]
  Acanthisitta [Genus]
    A. chloris [Species]
  Xenicus [Genus]
    X. longipes [Species]
    X. gilviventris
    X. lyalli

Menuridae [Family]
  Menura [Genus]
    M. novaehollandiae [Species]
    M. alberti

Atrichornithidae [Family]
  Atrichornis [Genus]
    A. clamosus [Species]
    A. rufescens

Alaudidae [Family]
  Mirafra [Genus]
    M. javanica [Species]
    M. hova
    M. cordofanica
    M. williamsi
    M. cheniana

M. albicauda
M. passerina
M. candida
M. pulpa
M. hypermetra
M. somalica
M. africana
M. chuana
M. angolensis
M. rufocinnamomea
M. apiata
M. africanoides
M. collaris
M. assamica
M. rufa
M. gilleti
M. poecilosterna
M. sabota
M. erythroptera
M. nigricans
Heteromirafra [Genus]
H. ruddi [Species]
Certhilauda [Genus]
C. curvirostris [Species]
C. albescens
C. albofasciata
Eremopterix [Genus]
E. australis [Species]
E. leucotis
E. signata
E. verticalis
E. nigriceps
E. grisea
E. leucopareia
Ammomanes [Genus]
A. cincturus [Species]
A. phoenicurus
A. deserti
A. dunni
A. grayi
A. burrus
Alaemon [Genus]
A. alaudipes [Species]
A. hamertoni
Ramphocoris [Genus]
R. clotbey [Species]
Melanocorypha [Genus]
M. calandra [Species]
M. bimaculata
M. maxima
M. mongolica
M. leucoptera
M. yeltoniensis
Calandrella [Genus]
C. cinerea [Species]
C. blanfordi
C. acutirostris
C. raytal
C. rufescens
C. razae

C. conirostris
C. starki
C. sclateri
C. fringillaris
C. obbiensis
C. personata
Chersophilus [Genus]
C. duponti [Species]
Pseudalaemon [Genus]
P. fremantlii [Species]
Galerida [Genus]
G. cristata [Species]
G. theklae
G. malabarica
G. deva
G. modesta
G. magnirostris
Lullula [Genus]
L. arborea [Species]
Alauda [Genus]
A. arvensis [Species]
A. gulgula
Eremophila [Genus]
E. alpestris [Species]
E. bilopha

Hirundinidae [Family]
Pseudochelidon [Genus]
P. eurystomina [Species]
Tachycineta [Genus]
T. bicolor [Species]
T. albilinea
T. albiventer
T. leucorrhoa
T. leucopyga
T. thalassina
Callichelidon [Genus]
C. cyaneoviridis [Species]
Kalochelidon [Genus]
K. euchrysea [Species]
Progne [Genus]
P. tapera [Species]
P. subis
P. dominicensis
P. chalybea
P. modesta
Notiochelidon [Genus]
N. murina [Species]
N. cyanoleuca
N. flavipes
N. pileata
Atticora [Genus]
A. fasciata [Species]
A. melanoleuca
Neochelidon [Genus]
N. tibialis [Species]
Alopochelidon [Genus]
A. fucata [Species]
Stelgidopteryx [Genus]
S. ruficollis [Species]
Cheramoeca [Genus]

C. leucosternum [Species]
Pseudhirundo [Genus]
P. griseopyga [Species]
Riparia [Genus]
R. paludicola [Species]
R. congica
R. riparia
R. cincta
Phedina [Genus]
P. borbonica [Species]
P. brazzae
Ptyonoprogne [Genus]
P. rupestris [Species]
P. obsoleta
P. fuligula
P. concolor
Hirundo [Genus]
H. rustica [Species]
H. lucida
H. angolensis
H. tahitica
H. albigularis
H. aethiopica
H. smithii
H. atrocaerulea
H. nigrita
H. leucosoma
H. megaensis
H. nigrorufa
H. dimidiata
Cecropis [Genus]
C. cucullata [Species]
C. abyssinica
C. semirufa
C. senegalensis
C. daurica
C. striolata
Petrochelidon [Genus]
P. rufigula [Species]
P. preussi
P. andecola
P. nigricans
P. spilodera
P. pyrrhonota
P. fulva
P. fluvicola
P. ariel
P. fuliginosa
Delichon [Genus]
D. urbica [Species]
D. dasypus
D. nipalensis
Psalidoprocne [Genus]
P. nitens [Species]
P. fuliginosa
P. albiceps
P. pristoptera
P. oleaginea
P. antinorii
P. petiti

P. holomelaena
P. orientalis
P. mangebettorum
P. chalybea
P. obscura

Motacillidae [Family]
  Dendronanthus [Genus]
    D. indicus [Species]
  Motacilla [Genus]
    M. flava [Species]
    M. citreola
    M. cinerea
    M. alba
    M. grandis
    M. madaraspatensis
    M. aguimp
    M. clara
    M. capensis
    M. flaviventris
  Tmetothylacus [Genus]
    T. tenellus [Species]
  Macronyx [Genus]
    M. capensis [Species]
    M. croceus
    M. fullebornii
    M. sharpei
    M. flavicollis
    M. aurantiigula
    M. ameliae
    M. grimwoodi
  Anthus [Genus]
    A. novaeseelandiae [Species]
    A. leucophrys
    A. vaalensis
    A. pallidiventris
    A. melindae
    A. campestris
    A. godlewskii
    A. berthelotii
    A. similis
    A. brachyurus
    A. caffer
    A. trivialis
    A. nilghiriensis
    A. hodgsoni
    A. gustavi
    A. pratensis
    A. cervinus
    A. roseatus
    A. spinoletta
    A. sylvanus
    A. spragueii
    A. furcatus
    A. hellmayri
    A. chacoensis
    A. lutescens
    A. correndera
    A. nattereri
    A. bogotensis
    A. antarcticus

A. gutturalis
A. sokokensis
A. crenatus
A. lineiventris
A. chloris

Campephagidae [Family]
  Pteropodocys [Genus]
    P. maxima [Species]
  Coracina [Genus]
    C. novaehollandiae [Species]
    C. fortis
    C. atriceps
    C. pollens
    C. schistacea
    C. caledonica
    C. caeruleogrisea
    C. temminckii
    C. larvata
    C. striata
    C. bicolor
    C. lineata
    C. boyeri
    C. leucopygia
    C. papuensis
    C. robusta
    C. longicauda
    C. parvula
    C. abbotti
    C. analis
    C. caesia
    C. pectoralis
    C. graueri
    C. cinerea
    C. azurea
    C. typica
    C. newtoni
    C. coerulescens
    C. dohertyi
    C. tenuirostris
    C. morio
    C. schisticeps
    C. melaena
    C. montana
    C. holopolia
    C. mcgregori
    C. panayensis
    C. polioptera
    C. melaschistos
    C. fimbriata
    C. melanoptera
  Campochaera [Genus]
    C. sloetii [Species]
  Chlamydochaera [Genus]
    C. jefferyi [Species]
  Lalage [Genus]
    L. melanoleuca [Species]
    L. nigra
    L. sueurii
    L. aurea
    L. atrovirens

L. leucomela
L. maculosa
L. sharpei
L. leucopygia
Campephaga [Genus]
  C. phoenicea [Species]
  C. quiscalina
  C. lobata
Pericrocotus [Genus]
  P. roseus [Species]
  P. divaricus
  P. cinnamomeus
  P. lansbergei
  P. erythropygius
  P. solaris
  P. ethologus
  P. brevirostris
  P. miniatus
  P. flammeus
Hemipus [Genus]
  H. picatus [Species]
  H. hirundinaceus
Tephrodornis [Genus]
  T. gularis [Species]
  T. pondicerianus

Pycnonotidae [Family]
  Spizixos [Genus]
    S. canifrons [Species]
    S. semitorques
  Pycnonotus [Genus]
    P. zeylanicus [Species]
    P. striatus
    P. leucogrammicus
    P. tympanistrigus
    P. melanoleucos
    P. priocephalus
    P. atriceps
    P. melanicterus
    P. squamatus
    P. cyaniventris
    P. jocosus
    P. xanthorrhous
    P. sinensis
    P. taivanus
    P. leucogenys
    P. cafer
    P. aurigaster
    P. xanthopygos
    P. nigricans
    P. capensis
    P. barbatus
    P. eutilotus
    P. nieuwenhuisii
    P. urostictus
    P. bimaculatus
    P. finlaysoni
    P. xantholaemus
    P. penicillatus
    P. flavescens
    P. goiavier

P. luteolus
P. plumosus
P. blanfordi
P. simplex
P. brunneus
P. erythropthalmos
P. masukuensis
P. montanus
P. virens
P. gracilis
P. ansorgei
P. curvirostris
P. importunus
P. latirostris
P. gracilirostris
P. tephrolaemus
P. milanjensis
Calyptocichla [Genus]
C. serina [Species]
Baeopogon [Genus]
B. indicator [Species]
B. clamans
Ixonotus [Genus]
I. guttatus [Species]
Chlorocichla [Genus]
C. falkensteini [Species]
C. simplex
C. flavicollis
C. flaviventris
C. laetissima
Thescelocichla [Genus]
T. leucopleura [Species]
Phyllastrephus [Genus]
P. scandens [Species]
P. terrestris
P. strepitans
P. cerviniventris
P. fulviventris
P. poensis
P. hypochloris
P. baumanni
P. poliocephalus
P. flavostriatus
P. debilis
P. lorenzi
P. albigularis
P. fischeri
P. orostruthus
P. icterinus
P. xavieri
P. madagascariensis
P. zosterops
P. tenebrosus
P. xanthophrys
P. cinereiceps
Bleda [Genus]
B. syndactyla [Species]
B. eximia
B. canicapilla
Nicator [Genus]

N. chloris [Species]
N. gularis
N. vireo
Criniger [Genus]
C. barbatus [Species]
C. calurus
C. ndussumensis
C. olivaceus
C. finschii
C. flaveolus
C. pallidus
C. ochraceus
C. bres
C. phaeocephalus
Setornis [Genus]
S. criniger [Species]
Hypsipetes [Genus]
H. viridescens [Species]
H. propinquus
H. charlottae
H. palawanensis
H. criniger
H. philippinus
H. siquijorensis
H. everetti
H. affinis
H. indicus
H. mcclellandii
H. malaccensis
H. virescens
H. flavala
H. amaurotis
H. crassirostris
H. borbonicus
H. madagascariensis
H. nicobariensis
H. thompsoni
Neolestes [Genus]
N. torquatus [Species]
Tylas [Genus]
T. eduardi [Species]

Irenidae [Family]
Aegithina [Genus]
A. tiphia [Species]
A. nigrolutea
A. viridissima
A. lafresnayei
Chloropsis [Genus]
C. flavipennis [Species]
C. palawanensis
C. sonnerati
C. cyanopogon
C. cochinchinensis
C. aurifrons
C. hardwickei
C. venusta
Irena [Genus]
I. puella [Species]
I. cyanogaster
Eurocephalus [Genus]

E. ruppelli [Species]
E. anguitimens

Laniidae [Family]
Prionops [Genus]
P. plumata [Species]
P. poliolopha
P. caniceps
P. alberti
P. retzii
P. gabela
P. scopifrons
Lanioturdus [Genus]
L. torquatus [Species]
Nilaus [Genus]
N. afer [Species]
Dryoscopus [Genus]
D. pringlii [Species]
D. gambensis
D. cubla
D. senegalensis
D. angolensis
D. sabini
Tchagra [Genus]
T. minuta [Species]
T. senegala
T. tchagra
T. australis
T. jamesi
T. cruenta
Laniarius [Genus]
L. ruficeps [Species]
L. luhderi
L. ferrugineus
L. barbarus
L. mufumbiri
L. atrococcineus
L. atroflavus
L. fulleborni
L. funebris
L. leucorhynchus
Telophorus [Genus]
T. bocagei [Species]
T. sulfureopectus
T. olivaceus
T. nigrifrons
T. multicolor
T. kupeensis
T. zeylonus
T. viridis
T. quadricolor
T. dohertyi
Malaconotus [Genus]
M. cruentus [Species]
M. lagdeni
M. gladiator
M. blanchoti
M. alius
Corvinella [Genus]
C. corvina [Species]
C. melanoleuca

Lanius [Genus]
    L. tigrinus [Species]
    L. souzae
    L. bucephalus
    L. cristatus
    L. collurio
    L. collueioides
    L. gubernator
    L. vittatus
    L. schach
    L. validirostris
    L. mackinnoni
    L. minor
    L. ludovicianus
    L. excubitor
    L. excubitoroides
    L. sphenocercus
    L. cabanisi
    L. dorsalis
    L. somalicus
    L. collaris
    L. newtoni
    L. senator
    L. nubicus
Pityriasis [Genus]
    P. gymnocephala [Species]

Vangidae [Family]
  Calicalicus [Genus]
    C. madagascariensis [Species]
  Schetba [Genus]
    S. rufa [Species]
  Vanga [Genus]
    V. curvirostris [Species]
  Xenopirostris [Genus]
    X. xenopirostris [Species]
    X. damii
    X. polleni
  Falculea [Genus]
    F. palliata [Species]
  Leptopterus [Genus]
    L. viridis [Species]
    L. chabert
    L. madagascarinus
  Oriolia [Genus]
    O. bernieri [Species]
  Euryceros [Genus]
    E. prevostrii [Species]

Bombycillidae [Family]
  Bombycilla [Genus]
    B. garrulus [Species]
    B. japonica
    B. cedrorum
  Ptilogonys [Genus]
    P. cinereus [Species]
    P. caudatus
  Phainopepla [Genus]
    P. nitens [Species]
  Phainoptila [Genus]
    P. melanoxantha [Species]

Hypocolius [Genus]
    H. ampelinus [Species]

Dulidae [Family]
  Dulus [Genus]
    D. dominicus [Species]

Cinclidae [Family]
  Cinclus [Genus]
    C. cinclus [Species]
    C. pallasii
    C. mexicanus
    C. leucocephalus

Troglodytidae [Family]
  Campylorhynchus [Genus]
    C. jocosus [Species]
    C. gularis
    C. yucatanicus
    C. brunneicapillus
    C. griseus
    C. rufinucha
    C. turdinus
    C. nuchalis
    C. fasciatus
    C. zonatus
    C. megalopterus
  Odontorchilus [Genus]
    O. cinereus [Species]
    O. branickii
  Salpinctes [Genus]
    S. obsoletus [Species]
    S. mexicanus
  Hylorchilus [Genus]
    H. sumichrasti [Species]
  Cinnycerthia [Genus]
    C. unirufa [Species]
    C. peruana
  Cistothorus [Genus]
    C. platensis [Species]
    C. meridae
    C. apolinari
    C. palustris
  Thryomanes [Genus]
    T. bewickii [Species]
    T. sissonii
  Ferminia [Genus]
    F. cerverai [Species]
  Thryothorus [Genus]
    T. atrogularis [Species]
    T. fasciatoventris
    T. euophrys
    T. genibarbis
    T. coraya
    T. felix
    T. maculipectus
    T. rutilus
    T. nigricapillus
    T. thoracicus
    T. pleurostictus
    T. ludovicianus

    T. rufalbus
    T. nicefori
    T. sinaloa
    T. modestus
    T. leucotis
    T. superciliaris
    T. guarayanus
    T. longirostris
    T. griseus
  Troglodytes [Genus]
    T. troglodytes [Species]
    T. aedon
    T. solstitialis
    T. rufulus
    T. browni
  Uropsila [Genus]
    U. leucogastra [Species]
  Henicorhina [Genus]
    H. leucosticta [Species]
    H. leucophrys
  Microcerculus [Genus]
    M. marginatus [Species]
    M. ustulatus
    M. bambla
  Cyphorhinus [Genus]
    C. thoracicus [Species]
    C. aradus

Mimidae [Family]
  Dumetalla [Genus]
    D. carolinensis [Species]
  Melanoptila [Genus]
    M. glabrirostris [Species]
  Melanotis [Genus]
    M. caerulescens [Species]
    M. hypoleucus
  Mimus [Genus]
    M. polyglottos [Species]
    M. gilvus
    M. gundlachii
    M. thenca
    M. longicaudatus
    M. saturninus
    M. patagonicus
    M. triurus
    M. dorsalis
  Nesomimus [Genus]
    N. trifasciatus [Species]
  Mimodes [Genus]
    M. graysoni [Species]
  Oreoscoptes [Genus]
    O. montanus [Species]
  Toxostoma [Genus]
    T. rufum [Species]
    T. longirostre
    T. guttatum
    T. cinereum
    T. bendirei
    T. ocellatum
    T. curvirostre
    T. lecontei

T. redivivum
T. dorsale
Cinclocerthia [Genus]
C. ruficauda [Species]
Ramphocinclus [Genus]
R. brachyurus [Species]
Donacobius [Genus]
D. atricapillus [Species]
Allenia [Genus]
A. fusca [Species]
Margarops [Genus]
M. fuscatus [Species]

Prunellidae [Family]
Prunella [Genus]
P. collaris [Species]
P. himalayana
P. rubeculoides
P. strophiata
P. montanella
P. fulvescens
P. ocularis
P. atrogularis
P. koslowi
P. modularis
P. rubida
P. immaculata

Turdidae [Family]
Brachypteryx [Genus]
B. stellata [Species]
B. hyperythra
B. major
B. calligyna
B. leucophrys
B. montana
Zeledonia [Genus]
Z. coronata [Species]
Erythropygia [Genus]
E. coryphaeus [Species]
E. leucophrys
E. hartlaubi
E. galactotes
E. paena
E. leucosticta
E. quadrivirgata
E. barbata
E. signata
Namibornis [Genus]
N. herero [Species]
Cercotrichas [Genus]
C. podobe [Species]
Pinarornis [Genus]
P. plumosus [Species]
Chaetops [Genus]
C. frenatus [Species]
Drymodes [Genus]
D. brunneopygia [Species]
D. superciliaris
Pogonocichla [Genus]
P. stellata [Species]

P. swynnertoni
Erithacus [Genus]
E. gabela [Species]
E. cyornithopsis
E. aequatorialis
E. erythrothorax
E. sharpei
E. gunningi
E. rubecula
E. akahige
E. komadori
E. sibilans
E. luscinia
E. megarhynchos
E. calliope
E. svecicus
E. pectoralis
E. ruficeps
E. obscurus
E. pectardens
E. brunneus
E. cyane
E. cyanurus
E. chrysaeus
E. indicus
E. hyperythrus
E. johnstoniae
Cossypha [Genus]
C. roberti [Species]
C. bocagei
C. polioptera
C. archeri
C. isabellae
C. natalensis
C. dichroa
C. semirufa
C. heuglini
C. cyanocampter
C. caffra
C. anomala
C. humeralis
C. ansorgei
C. niveicapilla
C. heinrichi
C. albicapilla
Modulatrix [Genus]
M. stictigula [Species]
Cichladusa [Genus]
C. guttata [Species]
C. arquata
C. ruficauda
Alethe [Genus]
A. diademata [Species]
A. poliophrys
A. fuelleborni
A. montana
A. lowei
A. poliocephala
A. choloensis
Copsychus [Genus]

C. saularis [Species]
C. sechellarum
C. albospecularis
C. malabaricus
C. stricklandii
C. luzoniensis
C. niger
C. pyrropygus
Irania [Genus]
I. gutturalis [Species]
Phoenicurus [Genus]
P. alaschanicus [Species]
P. erythronotus
P. caeruleocephalus
P. ochruros
P. phoenicurus
P. hodgsoni
P. frontalis
P. schisticeps
P. auroreus
P. moussieri
P. erythrogaster
Rhyacornis [Genus]
R. bicolor [Species]
R. fuliginosus
Hodgsonius [Genus]
H. phaenicuroides [Species]
Cinclidium [Genus]
C. leucurum [Species]
C. diana
C. frontale
Grandala [Genus]
G. coelicolor [Species]
Sialia [Genus]
S. sialis [Species]
S. mexicana
S. currucoides
Enicurus [Genus]
E. scouleri [Species]
E. velatus
E. ruficapillus
E. immaculatus
E. schistaceus
E. leschenaulti
E. maculatus
Cochoa [Genus]
C. purpurea [Species]
C. viridis
C. azurea
Myadestes [Genus]
M. townsendi [Species]
M. obscurus
M. elisabeth
M. genibarbis
M. ralloides
M. unicolor
M. leucogenys
Entomodestes [Genus]
E. leucotis [Species]
E. coracinus

Stizorbina [Genus]
　S. fraseri [Species]
　S. finschii
Neocossyphus [Genus]
　N. rufus [Species]
　N. poensis
Cercomela [Genus]
　C. sinuata [Species]
　C. familiaris
　C. tractrac
　C. schlegelii
　C. fusca
　C. dubia
　C. melanura
　C. scotocerca
　C. sordida
Saxicola [Genus]
　S. rubetra [Species]
　S. macrorbyncha
　S. insignis
　S. dacotiae
　S. torquata
　S. leucura
　S. caprata
　S. jerdoni
　S. ferrea
　S. gutturalis
Myrmecocichla [Genus]
　M. tholloni [Species]
　M. aethiops
　M. formicivora
　M. nigra
　M. arnotti
　M. albifrons
　M. melaena
Thamnolaea [Genus]
　T. cinnamomeiventris [Species]
　T. coronata
　T. semirufa
Oenanthe [Genus]
　O. bifasciata [Species]
　O. isabellina
　O. bottae
　O. xanthoprymna
　O. oenanthe
　O. deserti
　O. hispanica
　O. finschii
　O. picata
　O. lugens
　O. monacha
　O. alboniger
　O. pleschanka
　O. leucopyga
　O. leucura
　O. monticola
　O. moesta
　O. pileata
Chaimarrornis [Genus]
　C. leucocephalus [Species]

Saxicoloides [Genus]
　S. fulicata [Species]
Pseudocossyphus [Genus]
　P. imerinus [Species]
Monticola [Genus]
　M. rupestris [Species]
　M. explorator
　M. brevipes
　M. rufocinereus
　M. angolensis
　M. saxatilis
　M. cinclorhynchus
　M. rufiventris
　M. solitarius
Myophonus [Genus]
　M. blighi [Species]
　M. melanurus
　M. glaucinus
　M. robinsoni
　M. horsfieldii
　M. insularis
　M. caeruleus
Geomalia [Genus]
　G. heinrichi [Species]
Zoothera [Genus]
　Z. schistacea [Species]
　Z. dumasi
　Z. interpres
　Z. erythronota
　Z. wardii
　Z. cinerea
　Z. peronii
　Z. citrina
　Z. everetti
　Z. sibirica
　Z. naevia
　Z. pinicola
　Z. piaggiae
　Z. oberlaenderi
　Z. gurneyi
　Z. cameronensis
　Z. princei
　Z. crossleyi
　Z. guttata
　Z. spiloptera
　Z. andromedae
　Z. mollissima
　Z. dixoni
　Z. dauma
　Z. talaseae
　Z. margaretae
　Z. monticola
　Z. marginata
　Z. terrestris
Amalocichla [Genus]
　A. sclateriana [Species]
　A. incerta
Cataponera [Genus]
　C. turdoides [Species]
Nesocichla [Genus]

　N. eremita [Species]
Cichlherminia [Genus]
　C. lherminieri [Species]
Phaeornis [Genus]
　P. obscurus [Species]
　P. palmeri
Catharus [Genus]
　C. gracilirostris [Species]
　C. aurantiirostris
　C. fuscater
　C. occidentalis
　C. mexicanus
　C. dryas
　C. fuscescens
　C. minimus
　C. ustulatus
　C. guttatus
Hylocichla [Genus]
　H. mustelina [Species]
Platycichla [Genus]
　P. flavipes [Species]
　P. leucops
Turdus [Genus]
　T. bewsheri [Species]
　T. olivaceofuscus
　T. olivaceus
　T. abyssinicus
　T. helleri
　T. libonyanus
　T. tephronotus
　T. menachensis
　T. ludoviciae
　T. litsipsirupa
　T. dissimilis
　T. unicolor
　T. cardis
　T. albocinctus
　T. torquatus
　T. boulboul
　T. merula
　T. poliocephalus
　T. chrysolaus
　T. celaenops
　T. rubrocanus
　T. kessleri
　T. feae
　T. pallidus
　T. obscurus
　T. ruficollis
　T. naumanni
　T. pilaris
　T. iliacus
　T. philomelos
　T. mupinensis
　T. viscivorus
　T. aurantius
　T. ravidus
　T. plumbeus
　T. chiguanco
　T. nigriscens

T. fuscater
T. serranus
T. nigriceps
T. reevei
T. olivater
T. maranonicus
T. fulviventris
T. rufiventris
T. falcklandii
T. leucomelas
T. amaurochalinus
T. plebejus
T. ignobilis
T. lawrencii
T. fumigatus
T. hauxwelli
T. haplochrous
T. grayi
T. nudigenis
T. jamaicensis
T. albicollis
T. rufopalliatus
T. swalesi
T. rufitorques
T. migratorius

Orthonychidae [Family]
  Orthonyx [Genus]
    O. temminckii [Species]
    O. spaldingii
  Androphobus [Genus]
    A. viridis [Species]
  Psophodes [Genus]
    P. olivaceus [Species]
    P. nigrogularis
  Sphenostoma [Genus]
    S. cristatum [Species]
  Cinclostoma [Genus]
    C. punctatum [Species]
    C. castanotum
    C. cinnamomeum
    C. ajax
  Ptilorrhoa [Genus]
    P. leucosticta [Species]
    P. caerulescens
    P. castanonota
  Eupetes [Genus]
    E. macrocercus [Species]
  Melampitta [Genus]
    M. lugubris [Species]
    M. gigantea
  Ifrita [Genus]
    I. kowaldi [Species]

Timaliidae [Family]
  Pellorneum [Genus]
    P. ruficeps [Species]
    P. palustre
    P. fuscocapillum
    P. capistratum
    P. albiventre

Trichastoma [Genus]
  T. tickelli [Species]
  T. pyrrogenys
  T. malaccense
  T. cinereiceps
  T. rostratum
  T. bicolor
  T. separium
  T. celebense
  T. abbotti
  T. perspicillatum
  T. vanderbilti
  T. pyrrhopterum
  T. cleaveri
  T. albipectus
  T. rufescens
  T. rufipenne
  T. fulvescens
  T. puveli
  T. poliothorax
Leonardina [Genus]
  L. woodi [Species]
Ptyrticus [Genus]
  P. turdinus [Species]
Malacopteron [Genus]
  M. magnirostre [Species]
  M. affine
  M. cinereum
  M. magnum
  M. palawanense
  M. albogulare
Pomatorhinus [Genus]
  P. hypoleucos [Species]
  P. erythrogenys
  P. horsfieldii
  P. schisticeps
  P. montanus
  P. ruficollis
  P. ochraceiceps
  P. ferruginosus
Garritornis [Genus]
  G. isidorei [Species]
Pomatostomus [Genus]
  P. temporalis [Species]
  P. superciliosus
  P. ruficeps
Xiphirhynchus [Genus]
  X. superciliaris [Species]
Jabouilleia [Genus]
  J. danjoui [Species]
Rimator [Genus]
  R. malacoptilus [Species]
Ptilocichla [Genus]
  P. leucogrammica [Species]
  P. mindanensis
  P. falcata
Kenopia [Genus]
  K. striata [Species]
Napothera [Genus]
  N. rufipectus [Species]

N. atrigularis
N. macrodactyla
N. marmorata
N. crispifrons
N. brevicaudata
N. crassa
N. rabori
N. epilepidota
Pnoepyga [Genus]
  P. albiventer [Species]
  P. pusilla
Spelaeornis [Genus]
  S. caudatus [Species]
  S. troglodytoides
  S. formosus
  S. chocolatinus
  S. longicaudatus
Sphenocichla [Genus]
  S. humei [Species]
Neomixis [Genus]
  N. tenella [Species]
  N. viridis
  N. striatigula
  N. flavoviridis
Stachyris [Genus]
  S. rodolphei [Species]
  S. rufifrons
  S. ambigua
  S. ruficeps
  S. pyrrhops
  S. chrysaea
  S. plateni
  S. capitalis
  S. speciosa
  S. whiteheadi
  S. striata
  S. nigrorum
  S. hypogrammica
  S. grammiceps
  S. herberti
  S. nigriceps
  S. poliocephala
  S. striolata
  S. oglei
  S. maculata
  S. leucotis
  S. nigricollis
  S. thoracica
  S. erythroptera
  S. melanothorax
Dumetia [Genus]
  D. hyperythra [Species]
Rhopocichla [Genus]
  R. atriceps [Species]
Macronous [Genus]
  M. flavicollis [Species]
  M. gularis
  M. kelleyi
  M. striaticeps
  M. ptilosus

Micromacronus [Genus]
   M. leytensis [Species]
Timalia [Genus]
   T. pileata [Species]
Chrysomma [Genus]
   C. sinense [Species]
Moupinia [Genus]
   M. altirostris [Species]
   M. poecilotis
Chamaea [Genus]
   C. fasciata [Species]
Turdoides [Genus]
   T. nipalensis [Species]
   T. altirostris
   T. caudatus
   T. earlei
   T. gularis
   T. longirostris
   T. malcolmi
   T. squamiceps
   T. fulvus
   T. aylmeri
   T. rubiginosus
   T. subrufus
   T. striatus
   T. affinis
   T. melanops
   T. tenebrosus
   T. reinwardtii
   T. plebejus
   T. jardineii
   T. squamulatus
   T. leucopygius
   T. hindei
   T. hypoleucus
   T. bicolor
   T. gymnogenys
Babax [Genus]
   B. lanceolatus [Species]
   B. waddelli
   B. koslowi
Garrulax [Genus]
   G. cinereifrons [Species]
   G. palliatus
   G. rufifrons
   G. perspicillatus
   G. albogularis
   G. leucolophus
   G. monileger
   G. pectoralis
   G. lugubris
   G. striatus
   G. strepitans
   G. milleti
   G. maesi
   G. chinensis
   G. vassali
   G. galbanus
   G. delesserti
   G. variegatus

G. davidi
G. sukatschewi
G. cineraceus
G. rufogularis
G. lunulatus
G. maximus
G. ocellatus
G. caerulatus
G. mitratus
G. ruficollis
G. merulinus
G. canorus
G. sannio
G. cachinnans
G. lineatus
G. virgatus
G. austeni
G. squamatus
G. subunicolor
G. elliotii
G. henrici
G. affinis
G. erythrocephalus
G. yersini
G. formosus
G. milnei
Liocichla [Genus]
   L. phoenicea [Species]
   L. steerii
Leiothrix [Genus]
   L. argentauris [Species]
   L. lutea
Cutia [Genus]
   C. nipalensis [Species]
Pteruthius [Genus]
   P. rufiventer [Species]
   P. flaviscapis
   P. xanthochlorus
   P. melanotis
   P. aenobarbus
Gampsorhynchus [Genus]
   G. rufulus [Species]
Actinodura [Genus]
   A. egertoni [Species]
   A. ramsayi
   A. nipalensis
   A. waldeni
   A. souliei
   A. morrisoniana
Minla [Genus]
   M. cyanouroptera [Species]
   M. strigula
   M. ignotincta
Alcippe [Genus]
   A. chrysotis [Species]
   A. variegaticeps
   A. cinerea
   A. castaneceps
   A. vinipectus
   A. striaticollis

A. ruficapilla
A. cinereiceps
A. rufogularis
A. brunnea
A. brunneicauda
A. poioicephala
A. pyrrhoptera
A. peracensis
A. morrisonia
A. nipalensis
A. abyssinica
A. atriceps
Lioptilus [Genus]
   L. nigricapillus [Species]
   L. gilberti
   L. rufocinctus
   L. chapini
Parophasma [Genus]
   P. galinieri [Species]
Phyllanthus [Genus]
   P. atripennis [Species]
Crocias [Genus]
   C. langbianis [Species]
   C. albonotatus
Heterophasia [Genus]
   H. annectens [Species]
   H. capistrata
   H. gracilis
   H. melanoleuca
   H. auricularis
   H. pulchella
   H. picaoides
Yuhina [Genus]
   Y. castaniceps [Species]
   Y. bakeri
   Y. flavicollis
   Y. gularis
   Y. diademata
   Y. occipitalis
   Y. brunneiceps
   Y. nigrimenta
   Y. zantholeuca
Malia [Genus]
   M. grata [Species]
Myzornis [Genus]
   M. pyrrhoura [Species]
Horizorhinus [Genus]
   H. dohrni [Species]
Oxylabes [Genus]
   O. madagascariensis [Species]
Mystacornis [Genus]
   M. crossleyi [Species]

Panuridae [Family]
Panurus [Genus]
   P. biarmicus [Species]
Conostoma [Genus]
   C. oemodium [Species]
Paradoxornis [Genus]
   P. paradoxus [Species]
   P. unicolor

P. flavirostris
P. guttaticollis
P. conspicillatus
P. ricketti
P. webbianus
P. alphonsianus
P. zappeyi
P. przewalskii
P. fulvifrons
P. nipalensis
P. davidianus
P. atrosuperciliaris
P. ruficeps
P. gularis
P. heudei

Picathartidae [Family]
Picathartes [Genus]
P. gymnocephalus [Species]
P. oreas

Polioptilidae [Family]
Microbates [Genus]
M. collaris [Species]
M. cinereiventris
Ramphocaenus [Genus]
R. melanurus [Species]
Polioptila [Genus]
P. caerulea [Species]
P. melanura
P. lembeyei
P. albiloris
P. plumbea
P. lactea
P. guianensis
P. schistaceigula
P. dumicola

Sylviidae [Family]
Oligura [Genus]
O. castaneocoronata [Species]
Tesia [Genus]
T. superciliaris [Species]
T. olivea
T. cyaniventer
Urosphena [Genus]
U. subulata [Species]
U. whiteheadi
U. squameiceps
U. pallidipes
Cettia [Genus]
C. diphone [Species]
C. annae
C. parens
C. ruficapilla
C. fortipes
C. vulcania
C. major
C. flavolivacea
C. robustipes
C. brunnifrons
C. cetti

Bradypterus [Genus]
B. baboecala [Species]
B. graueri
B. grandis
B. carpalis
B. alfredi
B. sylvaticus
B. barratti
B. victorini
B. cinnamomeus
B. thoracicus
B. major
B. tacsanowskius
B. luteoventris
B. palliseri
B. seebohmi
B. caudatus
B. accentor
B. castaneus
Bathmocercus [Genus]
B. cerviniventris [Species]
B. rufus
B. winifredae
Dromaeocercus [Genus]
D. brunneus [Species]
D. seeboehmi
Nesillas [Genus]
N. typica [Species]
N. aldabranus
N. mariae
Thamnornis [Genus]
T. chloropetoides [Species]
Melocichla [Genus]
M. mentalis [Species]
Achaetops [Genus]
A. pycnopygius [Species]
Sphenoeacus [Genus]
S. afer [Species]
Megalurus [Genus]
M. pryeri [Species]
M. timoriensis
M. palustris
M. albolimbatus
M. gramineus
M. punctatus
Cincloramphus [Genus]
C. cruralis [Species]
C. mathewsi
Eremiornis [Genus]
E. carteri [Species]
Megalurulus [Genus]
M. bivittata [Species]
M. mariei
Cichlornis [Genus]
C. whitneyi [Species]
C. llaneae
C. grosvenori
Ortygocichla [Genus]
O. rubiginosa [Species]
O. rufa

Chaetornis [Genus]
C. striatus [Species]
Graminicola [Genus]
G. bengalensis [Species]
Schoenicola [Genus]
S. platyura [Species]
Locustella [Genus]
L. lanceolata [Species]
L. naevia
L. certhiola
L. ochotensis
L. pleskei
L. fluvialtilis
L. luscinioides
L. fasciolata
L. amnicola
Acrocephalus [Genus]
A. melanopogon [Species]
A. paludicola
A. schoenobaenus
A. sorghophilus
A. bistrigiceps
A. agricola
A. concinens
A. scirpaceus
A. cinnamomeus
A. baeticatus
A. palustris
A. dumetorum
A. arundinaceus
A. stentoreus
A. orinus
A. orientalis
A. luscinia
A. familiaris
A. aequinoctialis
A. caffer
A. atyphus
A. vaughani
A. rufescens
A. brevipennis
A. gracilirostris
A. newtoni
A. aedon
Bebrornis [Genus]
B. rodericanus [Species]
B. sechellensis
Hippolais [Genus]
H. caligata [Species]
H. pallida
H. languida
H. olivetorum
H. polyglotta
H. icterina
Chloropeta [Genus]
C. natalensis [Species]
C. similis
C. gracilirostris
Cisticola [Genus]
C. erythrops [Species]
C. lepe

C. cantans
C. lateralis
C. woosnami
C. anonyma
C. bulliens
C. chubbi
C. hunteri
C. nigriloris
C. aberrans
C. bodessa
C. chiniana
C. cinereola
C. ruficeps
C. rufilata
C. subruficapilla
C. lais
C. restricta
C. njombe
C. galactotes
C. pipiens
C. carruthersi
C. tinniens
C. robusta
C. aberdare
C. natalensis
C. fulvicapilla
C. angusticauda
C. melanura
C. brachyptera
C. rufa
C. troglodytes
C. nana
C. incana
C. juncidis
C. cherina
C. haesitata
C. aridula
C. textrix
C. eximia
C. dambo
C. brunnescens
C. ayresii
C. exilis
Scotocerca [Genus]
S. inquieta [Species]
Rhopophilus [Genus]
R. pekinensis [Species]
Prinia [Genus]
P. burnesi [Species]
P. criniger
P. polychroa
P. atrogularis
P. cinereocapilla
P. buchanani
P. rufescens
P. hodgsoni
P. gracilis
P. sylvatica
P. familiaris
P. flaviventris

P. socialis
P. subflava
P. somalica
P. fluviatilis
P. maculosa
P. flavicans
P. substriata
P. molleri
P. robertsi
P. leucopogon
P. leontica
P. bairdii
P. erythroptera
P. pectoralis
Drymocichla [Genus]
D. incana [Species]
Urolais [Genus]
U. epichlora [Species]
Spiloptila [Genus]
S. clamans [Species]
Apalis [Genus]
A. thoracica [Species]
A. pulchra
A. ruwenzori
A. nigriceps
A. jacksoni
A. chariessa
A. binotata
A. flavida
A. ruddi
A. rufogularis
A. sharpii
A. goslingi
A. bamendae
A. porphyrolaema
A. melanocephala
A. chirindensis
A. cinerea
A. alticola
A. karamojae
A. rufifrons
Stenostira [Genus]
S. scita [Species]
Phyllolais [Genus]
P. pulchella [Species]
Orthotomus [Genus]
O. metopias [Species]
O. moreaui
O. cucullatus
O. sutorius
O. atrogularis
O. derbianus
O. sericeus
O. ruficeps
O. sepium
O. cinereiceps
O. nigriceps
O. samarensis
Camaroptera [Genus]
C. brachyura [Species]

C. brevicauda
C. harterti
C. superciliaris
C. chloronota
Calamonastes [Genus]
C. simplex [Species]
C. stierlingi
C. fasciolatus
Euryptila [Genus]
E. subcinnamomea [Species]
Poliolais [Genus]
P. lopesi [Species]
Graueria [Genus]
G. vittata [Species]
Eremomela [Genus]
E. icteropygialis [Species]
E. flavocrissalis
E. scotops
E. pusilla
E. canescens
E. gregalis
E. badiceps
E. turneri
E. atricollis
E. usticollis
Randia [Genus]
R. pseudozosterops [Species]
Newtonia [Genus]
N. brunneicauda [Species]
N. amphichroa
N. archboldi
N. fanovanae
Sylvietta [Genus]
S. virens [Species]
S. denti
S. leucophrys
S. brachyura
S. philippae
S. whytii
S. ruficapilla
S. rufescens
S. isabellina
Hemitesia [Genus]
H. neumanni [Species]
Macrosphenus [Genus]
M. kempi [Species]
M. flavicans
M. concolor
M. pulitzeri
M. kretschmeri
Amaurocichla [Genus]
A. bocagei [Species]
Hypergerus [Genus]
H. atriceps [Species]
H. lepidus
Hyliota [Genus]
H. flavigaster [Species]
H. australis
H. violacea
Hylia [Genus]

H. prasina [Species]
Phylloscopus [Genus]
    P. ruficapilla [Species]
    P. laurae
    P. laetus
    P. herberti
    P. budongoensis
    P. umbrovirens
    P. trochilus
    P. collybita
    P. sindianus
    P. neglectus
    P. bonelli
    P. sibilatrix
    P. fuscatus
    P. fuligiventer
    P. affinis
    P. griseolus
    P. armandii
    P. schwarzi
    P. pulcher
    P. maculipennis
    P. proregulus
    P. subviridis
    P. inornatus
    P. borealis
    P. trochiloides
    P. nitidus
    P. plumbeitarsus
    P. tenellipes
    P. magnirostris
    P. tytleri
    P. occipitalis
    P. coronatus
    P. ijimae
    P. reguloides
    P. davisoni
    P. cantator
    P. ricketti
    P. olivaceus
    P. cebuensis
    P. trivirgatus
    P. sarasinorum
    P. presbytes
    P. poliocephalus
    P. makirensis
    P. amoenus
Seicercus [Genus]
    S. burkii [Species]
    S. xanthoschistos
    S. affinis
    S. poliogenys
    S. castaniceps
    S. montis
    S. grammiceps
Tickellia [Genus]
    T. hodgsoni [Species]
Abroscopus [Genus]
    A. albogularis [Species]
    A. schisticeps
    A. superciliaris

Parisoma [Genus]
    P. buryi [Species]
    P. lugens
    P. boehmi
    P. layardi
    P. subcaeruleum
Sylvia [Genus]
    S. atricapilla [Species]
    S. borin
    S. communis
    S. curruca
    S. nana
    S. nisoria
    S. hortensis
    S. leucomelaena
    S. rueppelli
    S. melanocephala
    S. melanothorax
    S. mystacea
    S. cantillans
    S. conspicillata
    S. deserticola
    S. undata
    S. sarda
Regulus [Genus]
    R. ignicapillus [Species]
    R. regulus
    R. goodfellowi
    R. satrapa
    R. calendula
Leptopoecile [Genus]
    L. sophiae [Species]
    L. elegans

Muscicapidae [Family]
  Melaenornis [Genus]
    M. semipartitus [Species]
    M. pallidus
    M. infuscatus
    M. mariquensis
    M. microrhynchus
    M. chocolatinus
    M. fischeri
    M. brunneus
    M. edolioides
    M. pammelaina
    M. ardesiacus
    M. annamarulae
    M. ocreatus
    M. cinerascens
    M. silens
  Rhinomyias [Genus]
    R. addita [Species]
    R. oscillans
    R. brunneata
    R. olivacea
    R. umbratilis
    R. ruficauda
    R. colonus
    R. gularis
    R. insignis
    R. goodfellowi

Muscicapa [Genus]
    M. striata [Species]
    M. gambagae
    M. griseisticta
    M. sibirica
    M. dauurica
    M. ruficauda
    M. muttui
    M. ferruginea
    M. sordida
    M. thalassina
    M. panayensis
    M. albicaudata
    M. indigo
    M. infuscata
    M. ussheri
    M. boehmi
    M. aquatica
    M. olivascens
    M. lendu
    M. adusta
    M. epulata
    M. sethsmithii
    M. comitata
    M. tessmanni
    M. cassini
    M. caerulescens
    M. griseigularis
Myioparus [Genus]
    M. plumbeus [Species]
Humblotia [Genus]
    H. flavirostris [Species]
Ficedula [Genus]
    F. hypoleuca [Species]
    F. albicollis
    F. zanthopygia
    F. narcissina
    F. mugimaki
    F. hodgsonii
    F. dumetoria
    F. strophiata
    F. parva
    F. subruba
    F. monileger
    F. solitaris
    F. hyperythra
    F. basilanica
    F. rufigula
    F. buruensis
    F. henrici
    F. harterti
    F. platenae
    F. bonthaina
    F. westermanni
    F. superciliaris
    F. tricolor
    F. sapphira
    F. nigrorufa
    F. timorensis
    F. cyanomelana

Niltava [Genus]
   N. grandis [Species]
   N. macgrigoriae
   N. davidi
   N. sundara
   N. sumatrana
   N. vivida
   N. hyacinthina
   N. hoevelli
   N. sanfordi
   N. concreta
   N. ruecki
   N. herioti
   N. hainana
   N. pallipes
   N. poliogenys
   N. unicolor
   N. rubeculoides
   N. banyumas
   N. superba
   N. caerulata
   N. turcosa
   N. tickelliae
   N. rufigastra
   N. hodgsoni
Culicicapa [Genus]
   C. ceylonensis [Species]
   C. helianthea

Platysteiridae [Family]
  Bias [Genus]
   B. flammulatus [Species]
   B. musicus
  Pseudobias [Genus]
   P. wardi [Species]
  Batis [Genus]
   B. diops [Species]
   B. margaritae
   B. mixta
   B. dimorpha
   B. capensis
   B. fratrum
   B. molitor
   B. soror
   B. pririt
   B. senegalensis
   B. orientalis
   B. minor
   B. perkeo
   B. minulla
   B. minima
   B. ituriensis
   B. poensis
  Platysteira [Genus]
   P. cyanea [Species]
   P. albifrons
   P. peltata
   P. laticincta
   P. castanea
   P. tonsa
   P. blissetti

P. chalybea
P. jamesoni
P. concreta

Maluridae [Family]
  Clytomyias [Genus]
   C. insignis [Species]
  malurus [Genus]
   M. wallacii [Species]
   M. grayi
   M. alboscapulatus
   M. melanocephalus
   M. leucopterus
   M. cyaneus
   M. splendens
   M. lamberti
   M. amabilis
   M. pulcherrimus
   M. elegans
   M. coronatus
   M. cyanocephalus
  Stipiturus [Genus]
   S. malachurus [Species]
   S. mallee
   M. ruficeps
  Amytornis [Genus]
   A. textilis [Species]
   A. purnelli
   A. housei
   A. woodwardi
   A. dorotheae
   A. striatus
   A. barbatus
   A. goyderi

Acanthizidae [Family]
  Dasyornis [Genus]
   D. brachypterus [Species]
   D. broadbenti
  Pycnoptilus [Genus]
   P. floccosus [Species]
  Origma [Genus]
   O. solitaria [Species]
  Crateroscelis [Genus]
   C. gutturalis [Species]
   C. murina
   C. nigrorufa
   C. robusta
  Sericornis [Genus]
   S. citreogularis [Species]
   S. maculatus
   S. humilis
   S. frontalis
   S. beccarii
   S. nouhuysi
   S. magnirostris
   S. keri
   S. spilodera
   S. perspicillatus
   S. rufescens
   S. papuensis

S. arfakianus
S. magnus
Pyrrholaemus [Genus]
   P. brunneus [Species]
Chthonicola [Genus]
   C. sagittatus [Species]
Calamanthus [Genus]
   C. fuliginosus [Species]
   C. campestris
Hylacola [Genus]
   H. pyrrhopygius [Species]
   H. cautus
Acanthiza [Genus]
   A. murina [Species]
   A. inornata
   A. reguloides
   A. iredalei
   A. katherina
   A. pusilla
   A. apicalis
   A. ewingii
   A. chrysorrhoa
   A. uropygialis
   A. robustirostris
   A. nana
   A. lineata
Smicrornis [Genus]
   S. brevirostris [Species]
Gerygone [Genus]
   G. cinerea [Species]
   G. chloronota
   G. palpebrosa
   G. olivacea
   G. dorsalis
   G. chrysogaster
   G. ruficauda
   G. magnirostris
   G. sulphurea
   G. inornata
   G. ruficollis
   G. fusca
   G. tenebrosa
   G. laevigaster
   G. flavolateralis
   G. insularis
   G. mouki
   G. modesta
   G. igata
   G. albofrontata
Aphelocephala [Genus]
   A. leucopsis [Species]
   A. pectoralis
   A. nigricincta
Mohoua [Genus]
   M. ochrocephala [Species]
Finschia [Genus]
   F. novaeseelandiae [Species]
Epthianura [Genus]
   E. albifrons [Species]
   E. tricolor

E. *aurifrons*
E. *crocea*
Ashbyia [Genus]
  A. *lovensis* [Species]

Monarchidae [Family]
  Erythrocercus [Genus]
    E. *mccallii* [Species]
    E. *holochlorus*
    E. *livingstonei*
  Elminia [Genus]
    E. *longicauda* [Species]
    E. *albicauda*
  Trochocercus [Genus]
    T. *nigromitratus* [Species]
    T. *albiventris*
    T. *albonotatus*
    T. *cyanomelas*
    T. *nitens*
  Philentoma [Genus]
    P. *pyrhopterum* [Species]
    P. *velatum*
  Hypothymis [Genus]
    H. *azurea* [Species]
    H. *helenae*
    H. *coelestris*
  Eutrichomyias [Genus]
    E. *rowleyi* [Species]
  Terpsiphone [Genus]
    T. *rufiventer* [Species]
    T. *bedfordi*
    T. *rufocinerea*
    T. *viridis*
    T. *paradisi*
    T. *atrocaudata*
    T. *cyanescens*
    T. *cinnamomea*
    T. *atrochalybeia*
    T. *mutata*
    T. *corvina*
    T. *bourbonnensis*
  Chasiempis [Genus]
    C. *sandwichensis* [Species]
  Pomarea [Genus]
    P. *dimidiata* [Species]
    P. *nigra*
    P. *mendozae*
    P. *iphis*
    P. *whitneyi*
  Mayrornis [Genus]
    M. *versicolor* [Species]
    M. *lessoni*
    M. *schistaceus*
  Neolalage [Genus]
    N. *banksiana* [Species]
  Clytorhynchus [Genus]
    C. *pachycephaloides* [Species]
    C. *vitiensis*
    C. *nigrogularis*
    C. *hamlini*
  Metabolus [Genus]

    M. *rugensis* [Species]
  Monarcha [Genus]
    M. *axillaris* [Species]
    M. *rubiensis*
    M. *cinerascens*
    M. *melanopsis*
    M. *frater*
    M. *erythrostictus*
    M. *castaneiventris*
    M. *richardsii*
    M. *leucotis*
    M. *guttulus*
    M. *mundus*
    M. *sacerdotum*
    M. *trivirgatus*
    M. *leucurus*
    M. *julianae*
    M. *manadensis*
    M. *brehmii*
    M. *infelix*
    M. *menckei*
    M. *verticalis*
    M. *barbatus*
    M. *browni*
    M. *viduus*
    M. *godeffroyi*
    M. *takatsukasae*
    M. *chrysomela*
  Arses [Genus]
    A. *insularis* [Species]
    A. *telescophthalmus*
    A. *kaupi*
  Myiagra [Genus]
    M. *oceanica* [Species]
    M. *galeata*
    M. *atra*
    M. *rubecula*
    M. *ferrocyanea*
    M. *cervinicauda*
    M. *caledonica*
    M. *vanikorensis*
    M. *albiventris*
    M. *azureocapilla*
    M. *ruficollis*
    M. *cyanoleuca*
    M. *alecto*
    M. *hebetior*
    M. *inquieta*
  Lamprolia [Genus]
    L. *victoriae* [Species]
  Machaerirhynchus [Genus]
    M. *flaviventer* [Species]
    M. *nigripectus*
  Peltops [Genus]
    P. *blainvillii* [Species]
    P. *montanus*
  Rhipidura [Genus]
    R. *hypoxantha* [Species]
    R. *superciliaris*
    R. *cyaniceps*

    R. *phoenicura*
    R. *nigrocinnamomea*
    R. *albicollis*
    R. *euryura*
    R. *aureola*
    R. *javanica*
    R. *perlata*
    R. *leucophrys*
    R. *rufiventris*
    R. *cockerelli*
    R. *albolimbata*
    R. *hyperythra*
    R. *threnothorax*
    R. *maculipectus*
    R. *leucothorax*
    R. *atra*
    R. *fuliginosa*
    R. *drownei*
    R. *tenebrosa*
    R. *rennelliana*
    R. *spilodera*
    R. *nebulosa*
    R. *brachyrhyncha*
    R. *personata*
    R. *dedemi*
    R. *superflua*
    R. *teysmanni*
    R. *lepida*
    R. *opistherythra*
    R. *rufidorsa*
    R. *dahli*
    R. *matthiae*
    R. *malaitae*
    R. *rufifrons*

Eopsaltriidae [Family]
  Monachella [Genus]
    M. *muelleriana* [Species]
  Microeca [Genus]
    M. *leucophaea* [Species]
    M. *flavigaster*
    M. *hemixantha*
    M. *griseoceps*
    M. *flavovirescens*
    M. *papuana*
  Eugerygone [Genus]
    E. *rubra* [Species]
  Petroica [Genus]
    P. *bivittata* [Species]
    P. *archboldi*
    P. *multicolor*
    P. *goodenovii*
    P. *phoenicea*
    P. *rosea*
    P. *rodinogaster*
    P. *cucullata*
    P. *vittata*
    P. *macrocephala*
    P. *australis*
    P. *traversi*
  Tregellasia [Genus]

T. capito [Species]
T. leucops
Eopsaltria [Genus]
   E. australis [Species]
   E. flaviventris
   E. georgiana
Peneoenanthe [Genus]
   P. pulverulenta [Species]
Peocilodryas [Genus]
   P. brachyura [Species]
   P. hypoleuca
   P. placens
   P. albonotata
   P. superciliosa
Peneothello [Genus]
   P. sigillatus [Species]
   P. cryptoleucus
   P. cyanus
   P. bimaculatus
Heteromyias [Genus]
   H. cinereifrons [Species]
   H. albispecularis
Pachycephalopsis [Genus]
   P. hattamensis [Species]
   P. poliosoma

Pachycephalidae [Family]
Eulacestoma [Genus]
   E. nigropectus [Species]
Falcunculus [Genus]
   F. frontatus [Species]
Oreoica [Genus]
   O. gutturalis [Species]
Pachycare [Genus]
   P. flavogrisea [Species]
Rhagologus [Genus]
   R. leucostigma [Species]
Hylocitrea [Genus]
   H. bonensis [Species]
Pachycephala [Genus]
   P. raveni [Species]
   P. rufinucha
   P. tenebrosa
   P. olivacea
   P. rufogularis
   P. inornata
   P. hypoxantha
   P. cinerea
   P. phaionota
   P. hyperythra
   P. modesta
   P. philippensis
   P. sulfuriventer
   P. meyeri
   P. soror
   P. simplex
   P. orpheus
   P. pectoralis
   P. flavifrons
   P. caledonica
   P. implicata

P. nudigula
P. lorentzi
P. schlegelii
P. aurea
P. rufiventris
P. lanioides
Colluricincla [Genus]
   C. megarhyncha [Species]
   C. parvula
   C. boweri
   C. harmonica
   C. woodwardi
Pitohui [Genus]
   P. kirhocephalus [Species]
   P. dichrous
   P. incertus
   P. ferrugineus
   P. cristatus
   P. nigrescens
   P. tenebrosus
Turnagra [Genus]
   T. capensis [Species]

Aegithalidae [Family]
Aegithalos [Genus]
   A. caudatus [Species]
   A. leucogenys
   A. concinnus
   A. iouschistos
   A. fuliginosus
Psaltria [Genus]
   P. exilis [Species]
Psaltriparus [Genus]
   P. minimus [Species]
   P. melanotis

Remizidae [Family]
Remiz [Genus]
   R. pendulinus [Species]
Anthoscopus [Genus]
   A. punctifrons [Species]
   A. parvulus
   A. musculus
   A. flavifrons
   A. caroli
   A. sylviella
   A. minutus
Auriparus [Genus]
   A. flaviceps [Species]
Cephalopyrus [Genus]
   C. flammiceps [Species]

Paridae [Family]
Parus [Genus]
   P. palustris [Species]
   P. lugubris
   P. montanus
   P. atricapillus
   P. carolinensis
   P. sclateri
   P. gambeli

P. superciliosus
P. davidi
P. cinctus
P. hudsonicus
P. rufescens
P. wollweberi
P. rubidiventris
P. melanolophus
P. ater
P. venustulus
P. elegans
P. amabilis
P. cristatus
P. dichrous
P. afer
P. griseiventris
P. niger
P. leucomelas
P. albiventris
P. leuconotus
P. funereus
P. fasciiventer
P. fringillinus
P. rufiventris
P. major
P. bokharensis
P. monticolus
P. nuchalis
P. xanthogenys
P. spilonotus
P. holsti
P. caeruleus
P. cyanus
P. varius
P. semilarvatus
P. inornatus
P. bicolor
Melanochlora [Genus]
   M. sultanea [Species]
Sylviparus [Genus]
   S. modestus [Species]
Hypositta [Genus]
   H. corallirostris [Species]

Sittidae [Family]
Sitta [Genus]
   S. europaea [Species]
   S. nagaensis
   S. castanea
   S. himalayensis
   S. victoriae
   S. pygmaea
   S. pusilla
   S. whiteheadi
   S. yunnanensis
   S. canadensis
   S. villosa
   S. leucopsis
   S. carolinensis
   S. krueperi
   S. neumayer

S. tephronota
S. frontalis
S. solangiae
S. azurea
S. magna
S. formosa
Neositta [Genus]
  N. chrysoptera [Species]
  N. papuensis
Daphoenositta [Genus]
  D. miranda [Species]
Tichodroma [Genus]
  T. muraria [Species]

Certhiidae [Family]
  Certhia [Genus]
    F. familiaris [Species]
    F. brachydactyla
    F. himalayana
    F. nipalensis
    F. discolor
  Salpornis [Genus]
    S. spilonotus [Species]

Rhabdornithidae [Family]
  Rhabdornis [Genus]
    R. mysticalis [Species]
    R. inornatus

Climacteridae [Family]
  Climacteris [Genus]
    C. erythrops [Species]
    C. affinis
    C. picumnus
    C. rufa
    C. melanura
    C. leucophaea

Dicaeidae [Family]
  Melanocharis [Genus]
    M. arfakiana [Species]
    M. nigra
    M. longicauda
    M. versteri
    M. striativentris
  Rhamphocharis [Genus]
    R. crassirostris [Species]
  Prionochilus [Genus]
    P. olivaceus [Species]
    P. maculatus
    P. percussus
    P. plateni
    P. xanthopygius
    P. thoracicus
  Dicaeum [Genus]
    D. annae [Species]
    D. agile
    D. everetti
    D. aeruginosum
    D. proprium
    D. chrysorrheum
    D. melanoxanthum

D. vincens
D. aureolimbatum
D. nigrilore
D. anthonyi
D. bicolor
D. quadricolor
D. australe
D. retrocinctum
D. trigonostigma
D. hypoleucum
D. erythrorhynchos
D. concolor
D. pygmaeum
D. nehrkorni
D. vulneratum
D. erythrothorax
D. pectorale
D. eximium
D. aeneum
D. tristrami
D. igniferum
D. maugei
D. sanguinolentum
D. hirundinaceum
D. celebicum
D. monticolum
D. ignipectus
D. cruentatum
D. trochileum
Oreocharis [Genus]
  O. arfaki [Species]
Paramythia [Genus]
  P. montium [Species]
Pardalotus [Genus]
  P. quadragintus [Species]
  P. punctatus
  P. xanthopygus
  P. rubricatus
  P. striatus
  P. ornatus
  P. substriatus
  P. melanocephalus

Nectariniidae [Family]
  Anthreptes [Genus]
    A. gabonicus [Species]
    A. fraseri
    A. reichenowi
    A. anchietae
    A. simplex
    A. malacensis
    A. rhodolaema
    A. singalensis
    A. longuemarei
    A. orientalis
    A. neglectus
    A. aurantium
    A. pallidogaster
    A. pujoli
    A. rectirostris
    A. collaris
    A. platurus

Hypogramma [Genus]
  H. hypogrammicum [Species]
Nectarina [Genus]
  N. seimundi [Species]
  N. batesi
  N. olivacea
  N. ursulae
  N. veroxii
  N. balfouri
  N. reichenbachii
  N. hartlaubii
  N. newtonii
  N. thomensis
  N. oritis
  N. alinae
  N. bannermani
  N. verticalis
  N. cyanolaema
  N. fuliginosa
  N. rubescens
  N. amethystina
  N. senegalensis
  N. adelberti
  N. zeylonica
  N. minima
  N. sperata
  N. sericea
  N. calcostetha
  N. dussumeiri
  N. lotenia
  N. jugularis
  N. buettikoferi
  N. solaris
  N. asiatica
  N. souimanga
  N. humbloti
  N. comorensis
  N. coquerellii
  N. venusta
  N. talatala
  N. oustaleti
  N. fusca
  N. chalybea
  N. afra
  N. mediocris
  N. preussi
  N. neergaardi
  N. chloropygia
  N. minulla
  N. regia
  N. loveridgei
  N. rockefelleri
  N. violacea
  N. habessinica
  N. bouvieri
  N. osea
  N. cuprea
  N. tacazze
  N. bocagii
  N. purpureiventris

N. shelleyi
N. mariquensis
N. bifasciata
N. pembae
N. chalcomelas
N. coccinigastra
N. erythrocerca
N. congensis
N. pulchella
N. nectarinioides
N. famosa
N. johnstoni
N. notata
N. johannae
N. superba
N. kilimensis
N. reichenowi
Aethopyga [Genus]
A. primigenius [Species]
A. boltoni
A. flagrans
A. pulcherrima
A. duyvenbodei
A. shelleyi
A. gouldiae
A. nipalensis
A. eximia
A. christinae
A. saturata
A. siparaja
A. mysticalis
A. ignicauda
Arachnothera [Genus]
A. longirostra [Species]
A. crassirostris
A. robusta
A. flavigaster
A. chrysogenys
A. clarae
A. affinis
A. magna
A. everetti
A. juliae

Zosteropidae [Family]
Zosterops [Genus]
Z. erythropleura [Species]
Z. japonica
Z. palpebrosa
Z. ceylonensis
Z. conspicillata
Z. salvadorii
Z. atricapilla
Z. everetti
Z. nigrorum
Z. montana
Z. wallacei
Z. flava
Z. chloris
Z. consibrinorum
Z. grayi

Z. uropygialis
Z. anomala
Z. atriceps
Z. atrifrons
Z. mysorensis
Z. fuscicapilla
Z. buruensis
Z. kuehni
Z. novaeguineae
Z. metcalfi
Z. natalis
Z. lutea
Z. griseotincta
Z. rennelliana
Z. vellalavella
Z. luteirostris
Z. rendovae
Z. murphyi
Z. ugiensis
Z. stresemanni
Z. sanctaecrucis
Z. samoensis
Z. explorator
Z. flavifrons
Z. minuta
Z. xanthochroa
Z. lateralis
Z. strenua
Z. tenuirostris
Z. albogularis
Z. inornata
Z. cinerea
Z. abyssinica
Z. pallida
Z. senegalensis
Z. virens
Z. borbonica
Z. ficedulina
Z. griseovirescens
Z. maderaspatana
Z. mayottensis
Z. modesta
Z. mouroniensis
Z. olivacea
Z. vaughani
Woodfordia [Genus]
W. superciliosa [Species]
W. lacertosa
Rukia [Genus]
R. palauensis [Species]
R. oleaginea
R. ruki
R. longirostra
Tephrozosterops [Genus]
T. stalkeri [Species]
Madanga [Genus]
M. ruficollis [Species]
Lophozosterops [Genus]
L. pinaiae [Species]
L. goodfellowi

L. squamiceps
L. javanica
L. superciliaris
L. dohertyi
Oculocincta [Genus]
O. squamifrons [Species]
Heleia [Genus]
H. muelleri [Species]
H. crassirostris
Chlorocharis [Genus]
C. emiliae [Species]
Hypocryptadius [Genus]
H. cinnamomeus [Species]
Speirops [Genus]
S. brunnea [Species]
S. leucophoeca
S. lugubris

Meliphagidae [Family]
Timeliopsis [Genus]
T. fulvigula [Species]
T. griseigula
Melilestes [Genus]
M. megarhynchus [Species]
M. bouganvillei
Toxorhamphus [Genus]
T. novaeguineae [Species]
T. poliopterus
Oedistoma [Genus]
O. iliolophum [Species]
O. pygmaeum
Glycichaera [Genus]
G. fallax [Species]
Lichmera [Genus]
L. lombokia [Species]
L. argentauris
L. indistincta
L. incana
L. alboauricularis
L. squamata
L. deningeri
L. monticola
L. flavicans
L. notabilis
L. cockerelli
Myzomela [Genus]
M. blasii [Species]
M. albigula
M. cineracea
M. eques
M. obscura
M. cruentata
M. nigrita
M. pulchella
M. kuehni
M. erythrocephala
M. adolphinae
M. sanguinolenta
M. cardinalis
M. chermesina
M. sclateri

M. lafargei
M. melanocephala
M. eichhorni
M. malaitae
M. tristrami
M. jugularis
M. erythromelas
M. vulnerata
M. rosenbergii
Certhionyx [Genus]
   C. niger [Species]
   C. variegatus
Meliphaga [Genus]
   M. mimikae [Species]
   M. montana
   M. orientalis
   M. albonotata
   M. aruensis
   M. analoga
   M. vicina
   M. gracilis
   M. notata
   M. flavirictus
   M. lewinii
   M. flava
   M. albilineata
   M. virescens
   M. versicolor
   M. fasciogularis
   M. inexpectata
   M. fusca
   M. plumula
   M. chrysops
   M. cratitia
   M. keartlandi
   M. penicillata
   M. ornata
   M. reticulata
   M. leucotis
   M. flavicollis
   M. melanops
   M. cassidix
   M. unicolor
   M. flaviventer
   M. polygramma
   M. macleayana
   M. frenata
   M. subfrenata
   M. obscura
Oreornis [Genus]
   O. chrysogenys [Species]
Foulehaio [Genus]
   F. carunculata [Species]
   F. provocator
Cleptornis [Genus]
   C. marchei [Species]
Apalopteron [Genus]
   A. familiare [Species]
Melithreptus [Genus]
   M. brevirostris [Species]

M. lunatus
M. albogularis
M. affinis
M. gularis
M. laetior
M. validirostris
Entomyzon [Genus]
   E. cyanotis [Species]
Notiomystis [Genus]
   N. cincta [Species]
Pycnopygius [Genus]
   P. ixoides [Species]
   P. cinereus
   P. stictocephalus
Philemon [Genus]
   P. meyeri [Species]
   P. brassi
   P. citreogularis
   P. inornatus
   P. gilolensis
   P. fuscicapillus
   P. subcorniculatus
   P. moluccensis
   P. buceroides
   P. novaeguineae
   P. cockerelli
   P. eichhorni
   P. albitorques
   P. argenticeps
   P. corniculatus
   P. diemenensis
Ptiloprora [Genus]
   P. plumbea [Species]
   P. meekiana
   P. erythropleura
   P. guisei
   P. perstriata
Melidectes [Genus]
   M. fuscus [Species]
   M. princeps
   M. nouhuysi
   M. ochromelas
   M. leucostephes
   M. belfordi
   M. torquatus
Melipotes [Genus]
   M. gymnops [Species]
   M. fumigatus
   M. ater
Vosea [Genus]
   V. whitemanensis [Species]
Myza [Genus]
   M. celebensis [Species]
   M. sarasinorum
Meliarchus [Genus]
   M. sclateri [Species]
Gymnomyza [Genus]
   G. viridis [Species]
   G. samoensis
   G. aubryana

Moho [Genus]
   M. braccatus [Species]
   M. bishopi
   M. apicalis
   M. nobilis
Chaetoptila [Genus]
   C. angustipluma [Species]
Phylidonyris [Genus]
   P. pyrrhoptera [Species]
   P. novaehollandiae
   P. nigra
   P. albifrons
   P. melanops
   P. undulata
   P. notabilis
Ramsayornis [Genus]
   R. fasciatus [Species]
   R. modestus
Plectorhyncha [Genus]
   P. lanceolata [Species]
Conopophila [Genus]
   C. whitei [Species]
   C. albogularis
   C. rufogularis
   C. picta
Xanthomyza [Genus]
   X. phrygia [Species]
Cissomela [Genus]
   C. pectoralis [Species]
Acanthorhynchus [Genus]
   A. tenuirostris [Species]
   A. superciliosus
Manorina [Genus]
   M. melanophrys [Species]
   M. melanocephala
   M. flavigula
   M. melanotis
Anthornis [Genus]
   A. melanura [Species]
Anthochaera [Genus]
   A. rufogularis [Species]
   A. chrysoptera
   A. carunculata
   A. paradoxa
Prosthemadera [Genus]
   P. novaeseelandiae [Species]
Promerops [Genus]
   P. cafer [Species]
   P. gurneyi

Emberizidae [Family]
Melophus [Genus]
   M. lathami [Species]
Latoucheornis [Genus]
   L. siemsseni [Species]
Emberiza [Genus]
   E. calandra [Species]
   E. citrinella
   E. leucocephala
   E. cia
   E. cioides

E. jankowskii
E. buchanani
E. stewarti
E. cineracea
E. hortulana
E. caesia
E. cirlus
E. striolata
E. impetuani
E. tahapisi
E. socotrana
E. capensis
E. yessoensis
E. tristami
E. fucata
E. pusilla
E. chrysophrys
E. rustica
E. elegans
E. aureola
E. poliopleura
E. flaviventris
E. affinis
E. cabanisi
E. rutila
E. koslowi
E. melanocephala
E. bruniceps
E. sulphurata
E. spodocephala
E. variabilis
E. pallasi
E. schoeniclus
Calcarius [Genus]
C. mccownii [Species]
C. lapponicus
C. pictus
C. ornatus
Plectrophenax [Genus]
P. nivalis [Species]
Calamospiza [Genus]
C. melanocorys [Species]
Zonotrichia [Genus]
Z. iliaca [Species]
Z. melodia
Z. lincolnii
Z. georgiana
Z. capensis
Z. querula
Z. leucophrys
Z. albicollis
Z. atricapilla
Junco [Genus]
J. vulcani [Species]
J. hyemalis
J. phaeonotus
Ammodramus [Genus]
A. sandwichensis [Species]
A. maritimus
A. caudacutus

A. leconteii
A. bairdii
A. baileyi
A. henslowii
A. savannarum
A. humeralis
A. aurifrons
Spizella [Genus]
S. arborea [Species]
S. passerina
S. pusilla
S. atrogularis
S. pallida
S. breweri
Pooecetes [Genus]
P. gramineus [Species]
Chondestes [Genus]
C. grammacus [Species]
Amphispiza [Genus]
A. bilineata [Species]
A. belli
Aimophila [Genus]
A. mystacalis [Species]
A. humeralis
A. ruficauda
A. sumichrasti
A. stolzmanni
A. strigiceps
A. aestivalis
A. botterii
A. cassinii
A. quinquestriata
A. carpalis
A. ruficeps
A. notosticta
A. rufescens
Torreornis [Genus]
T. inexpectata [Species]
Orirurus [Genus]
O. superciliosus [Species]
Phrygilus [Genus]
P. atriceps [Species]
P. gayi
P. patagonicus
P. fruticeti
P. unicolor
P. dorsalis
P. erythronotus
P. plebejus
P. carbonarius
P. alaudinus
Melanodera [Genus]
M. melanodera [Species]
M. xanthogramma
Haplospiza [Genus]
H. rustica [Species]
H. unicolor
Acanthidops [Genus]
A. bairdii [Species]
Lophospingus [Genus]

L. pusillus [Species]
L. griseocristatus
Donacospiza [Genus]
D. albifrons [Species]
Rowettia [Genus]
R. goughensis [Species]
Nesospiza [Genus]
N. acunhae [Species]
N. wilkinsi
Diuca [Genus]
D. speculifera [Species]
D. diuca
Idiopsar [Genus]
I. brachyurus [Species]
Piezorhina [Genus]
P. cinerea [Species]
Xenospingus [Genus]
X. concolor [Species]
Incaspiza [Genus]
I. pulchra [Species]
I. ortizi
I. laeta
I. watkinsi
Poospiza [Genus]
P. thoracica [Species]
P. boliviana
P. alticola
P. hypochondria
P. erythrophrys
P. ornata
P. nigrorufa
P. lateralis
P. rubecula
P. garleppi
P. baeri
P. caesar
P. hispaniolensis
P. torquata
P. cinerea
Sicalis [Genus]
S. citrina [Species]
S. lutea
S. uropygialis
S. luteocephala
S. auriventris
S. olivascens
S. columbiana
S. flaveola
S. luteola
S. raimondii
S. taczanowskii
Emberizoides [Genus]
E. herbicola [Species]
Embernagra [Genus]
E. platensis [Species]
E. longicauda
Volatinia [Genus]
V. jacarina [Species]
Sporophila [Genus]
S. frontalis [Species]
S. falcirostris

S. schistacea
S. intermedia
S. plumbea
S. americana
S. torqueola
S. collaris
S. lineola
S. luctuosa
S. nigricollis
S. ardesiaca
S. melanops
S. obscura
S. caerulescens
S. albogularis
S. leucoptera
S. peruviana
S. simplex
S. nigrorufa
S. bouvreuil
S. insulata
S. minuta
S. hypoxantha
S. hypochroma
S. ruficollis
S. palustris
S. castaneiventris
S. cinnamomea
S. melanogaster
S. telasco
Oryzoborus [Genus]
  O. crassirostris [Species]
  O. angolensis
Amaurospiza [Genus]
  A. concolor [Species]
  A. moesta
Melopyrrha [Genus]
  M. nigra [Species]
Dolospingus [Genus]
  D. fringilloides [Species]
Catamenia [Genus]
  C. analis [Species]
  C. inornata
  C. homochroa
  C. oreophila
Tiaris [Genus]
  T. canora [Species]
  T. olivacea
  T. bicolor
  T. fuliginosa
Loxipasser [Genus]
  L. anoxanthus [Species]
Loxigilla [Genus]
  L. portorocensis [Species]
  L. violacea
  L. noctis
Melanospiza [Genus]
  M. richardsoni [Species]
Geospiza [Genus]
  G. magnirostris [Species]
  G. fortis

G. fuliginosa
G. difficilis
G. scandens
G. conirostris
Camarhynchus [Genus]
  C. crassirostris [Species]
  C. psittacula
  C. pauper
  C. parvulus
  C. pallidus
  C. heliobates
Certhidea [Genus]
  C. olivacea [Species]
Pinaroloxias [Genus]
  P. inornata [Species]
Pipilo [Genus]
  P. chlorurus [Species]
  P. ocai
  P. erythrophthalmus
  P. socorroensis
  P. fuscus
  P. aberti
  P. albicollis
Melozone [Genus]
  M. kieneri [Species]
  M. biarcuatum
  M. leucotis
Arremon [Genus]
  A. taciturnus [Species]
  A. flavirostris
  A. aurantiirostris
  A. schlegeli
  A. abeillei
Arremonops [Genus]
  A. rufivirgatus [Species]
  A. tocuyensis
  A. chlorinotus
  A. conirostris
Atlapetes [Genus]
  A. albinucha [Species]
  A. pallidinucha
  A. rufinucha
  A. leucopis
  A. pileatus
  A. melanocephalus
  A. flaviceps
  A. fuscoolivaceus
  A. tricolor
  A. albofrenatus
  A. schistaceus
  A. nationi
  A. leucopterus
  A. albiceps
  A. pallidiceps
  A. rufigenis
  A. semirufus
  A. personatus
  A. fulviceps
  A. citrinellus
  A. brunneinucha
  A. torquatus

Pezopetes [Genus]
  P. capitalis [Species]
Oreothraupis [Genus]
  O. arremonops [Species]
Pselliophorus [Genus]
  P. tibialis [Species]
  P. luteoviridis
Lysurus [Genus]
  L. castaneiceps [Species]
Urothraupis [Genus]
  U. stolzmanni [Species]
Charitospiza [Genus]
  C. eucosma [Species]
Coryphaspiza [Genus]
  C. melanotis [Species]
Saltatricula [Genus]
  S. multicolor [Species]
Gubernatrix [Genus]
  G. cristata [Species]
Coryphospingus [Genus]
  C. pileatus [Species]
  C. cucullatus
Rhodospingus [Genus]
  R. cruentus [Species]
Paroaria [Genus]
  P. coronata [Species]
  P. dominicana
  P. gularis
  P. baeri
  P. capitata
Catamblyrhynchus [Genus]
  C. diadema [Species]
Spiza [Genus]
  S. americana [Species]
Pheucticus [Genus]
  P. chrysopeplus [Species]
  P. aureoventris
  P. ludovicianus
  P. melanocephalus
Cardinalis [Genus]
  C. cardinalis [Species]
  C. phoeniceus
  C. sinuatus
Caryothraustes [Genus]
  C. canadensis [Species]
  C. humeralis
Rhodothraupis [Genus]
  R. celaeno [Species]
Periporphyrus [Genus]
  P. erythromelas [Species]
Pitylus [Genus]
  P. grossus [Species]
Saltator [Genus]
  S. atriceps [Species]
  S. maximus
  S. atripennis
  S. similis
  S. coerulescens
  S. orenocensis
  S. maxillosus

S. aurantiirostris
S. cinctus
S. atricollis
S. rufiventris
S. albicollis
Passerina [Genus]
   P. glaucocaerulea [Species]
   P. cyanoides
   P. brissonii
   P. parellina
   P. caerulea
   P. cyanea
   P. amoena
   P. versicolor
   P. ciris
   P. rositae
   P. leclancherii
   P. caerulescens
Orchesticus [Genus]
   O. albeillei [Species]
Schistochlamys [Genus]
   S. ruficapillus [Species]
   S. melanopis
Neothraupis [Genus]
   N. fasciata [Species]
Cypsnagra [Genus]
   C. hirundinacea [Species]
Conothraupis [Genus]
   C. speculigera [Species]
   C. mesoleuca
Lamprospiza [Genus]
   L. melanoleuca [Species]
Cissopis [Genus]
   C. leveriana [Species]
Chlorornis [Genus]
   C. reifferii [Species]
Compsothraupis [Genus]
   C. loricata [Species]
Sericossypha [Genus]
   S. albocristata [Species]
Nesospingus [Genus]
   N. speculiferus [Species]
Chlorospingus [Genus]
   C. ophthalmicus [Species]
   C. tacarcunae
   C. inornatus
   C. punctulatus
   C. semifuscus
   C. zeledoni
   C. pileatus
   C. parvirostris
   C. flavigularis
   C. flavovirens
   C. canigularis
Cnemoscopus [Genus]
   C. rubrirostris [Species]
Hemispingus [Genus]
   H. atropileus [Species]
   H. superciliaris
   H. reyi

H. frontalis
H. melanotis
H. goeringi
H. verticalis
H. xanthophthalmus
H. trifasciatus
Pyrrhocoma [Genus]
   P. ruficeps [Species]
Thlypopsis [Genus]
   T. fulviceps [Species]
   T. ornata
   T. pectoralis
   T. sordida
   T. inornata
   T. ruficeps
Hemithraupis [Genus]
   H. guira [Species]
   H. ruficapilla
   H. flavicollis
Chrysothlypis [Genus]
   C. chrysomelas [Species]
   C. salmoni
Nemosia [Genus]
   N. pileata [Species]
   N. rourei
Phaenicophilus [Genus]
   P. palmarum [Species]
   P. poliocephalus
Calyptophilus [Genus]
   C. frugivorus [Species]
Rhodinocichla [Genus]
   R. rosea [Species]
Mitrospingus [Genus]
   M. cassinii [Species]
   M. oleagineus
Chlorothraupis [Genus]
   C. carmioli [Species]
   C. olivacea
   C. stolzmanni
Orthogonys [Genus]
   O. chloricterus [Species]
Eucometis [Genus]
   E. penicillata [Species]
Lanio [Genus]
   L. fulvus [Species]
   L. versicolor
   L. aurantius
   L. leucothorax
Creurgops [Genus]
   C. verticalis [Species]
   C. dentata
Heterospingus [Genus]
   H. xanthopygius [Species]
Tachyphonus [Genus]
   T. cristatus [Species]
   T. rufiventer
   T. surinamus
   T. luctuosus
   T. delatrii
   T. coronatus

T. rufus
T. phoenicius
Trichothraupis [Genus]
   T. melanops [Species]
Habia [Genus]
   H. rubica [Species]
   H. fuscicauda
   H. atrimaxillaris
   H. gutturalis
   H. cristata
Piranga [Genus]
   P. bidentata [Species]
   P. flava
   P. rubra
   P. roseogularis
   P. olivacea
   P. ludoviciana
   P. leucoptera
   P. erythrocephala
   P. rubriceps
Calochaetes [Genus]
   C. coccineus [Species]
Ramphocelus [Genus]
   R. sanguinolentus [Species]
   R. nigrogularis
   R. dimidiatus
   R. melanogaster
   R. carbo
   R. bresilius
   R. passerinii
   R. flammigerus
Spindalis [Genus]
   S. zena [Species]
Thraupis [Genus]
   T. episcopus [Species]
   T. sayaca
   T. cyanoptera
   T. ornata
   T. abbas
   T. palmarum
   T. cyanocephala
   T. bonariensis
Cyanicterus [Genus]
   C. cyanicterus [Species]
Buthraupis [Genus]
   B. arcaei [Species]
   B. melanochlamys
   B. rothschildi
   B. edwardsi
   B. aureocincta
   B. montana
   B. eximia
   B. wetmorei
Wetmorethraupis [Genus]
   W. sterrhopteron [Species]
Anisognathus [Genus]
   A. lacrymosus [Species]
   A. igniventris
   A. flavinuchus
   A. notabilis

Stephanophorus [Genus]
  S. diadematus [Species]
Iridosornis [Genus]
  I. porphyrocephala [Species]
  I. analis
  I. jelskii
  I. rufivertex
Dubusia [Genus]
  D. taeniata [Species]
Delothraupis [Genus]
  D. castaneoventris [Species]
Pipraeidea [Genus]
  P. melanonota [Species]
Euphonia [Genus]
  E. jamaica [Species]
  E. plumbea
  E. affinis
  E. luteicapilla
  E. chlorotica
  E. trinitatis
  E. concinna
  E. saturata
  E. finschi
  E. violacea
  E. laniirostris
  E. hirundinacea
  E. chalybea
  E. musica
  E. fulvicrissa
  E. imitans
  E. gouldi
  E. chrysopasta
  E. mesochrysa
  E. minuta
  E. anneae
  E. xanthogaster
  E. rufiventris
  E. pectoralis
  E. cayennensis
Chlorophonia [Genus]
  C. flavirostris [Species]
  C. cyanea
  C. pyrrhophrys
  C. occipitalis
Chlorochrysa [Genus]
  C. phoenicotis [Species]
  C. calliparaea
  C. nitidissima
Tangara [Genus]
  T. inornata [Species]
  T. cabanisi
  T. palmeri
  T. mexicana
  T. chilensis
  T. fastuosa
  T. seledon
  T. cyanocephala
  T. desmaresti
  T. cyanoventris
  T. johannae

T. schrankii
T. florida
T. arthus
T. icterocephala
T. xanthocephala
T. chrysotis
T. parzudakii
T. xanthogastra
T. punctata
T. guttata
T. varia
T. rufigula
T. gyrola
T. lavinia
T. cayana
T. cucullata
T. peruviana
T. preciosa
T. vitriolina
T. rufigenis
T. ruficervix
T. labradorides
T. cyanotis
T. cyanicollis
T. larvata
T. nigrocincta
T. dowii
T. nigroviridis
T. vassorii
T. heinei
T. viridicollis
T. argyrofenges
T. cyanoptera
T. pulcherrima
T. velia
T. callophrys
Dacnis [Genus]
  D. albiventris [Species]
  D. lineata
  D. flaviventer
  D. hartlaubi
  D. nigripes
  D. venusta
  D. cayana
  D. viguieri
  D. berlepschi
Chlorophanes [Genus]
  C. spiza [Species]
Cyanerpes [Genus]
  C. nitidus [Species]
  C. lucidus
  C. caeruleus
  C. cyaneus
Xenodacnis [Genus]
  X. parina [Species]
Oreomanes [Genus]
  O. fraseri [Species]
Diglossa [Genus]
  D. baritula [Species]
  D. lafresnayii

D. carbonaria
D. venezuelensis
D. albilatera
D. duidae
D. major
D. indigotica
D. glauca
D. caerulescens
D. cyanea
Euneornis [Genus]
  E. campestris [Species]
Tersina [Genus]
  T. viridis [Species]

Parulidae [Family]
  Mniotilta [Genus]
    M. varia [Species]
  Vermivora [Genus]
    V. bachmanii [Species]
    V. chrysoptera
    V. pinus
    V. peregrina
    V. celata
    V. ruficapilla
    V. virginiae
    V. crissalis
    V. luciae
    V. gutturalis
    V. superciliosa
  Parula [Genus]
    P. americana [Species]
    P. pitiayumi
  Dendroica [Genus]
    D. petechia [Species]
    D. pensylvanica
    D. cerulea
    D. caerulescens
    D. plumbea
    D. pharetra
    D. pinus
    D. graciae
    D. adelaidae
    D. pityophila
    D. dominica
    D. nigrescens
    D. townsendi
    D. occidentalis
    D. chrysoparia
    D. virens
    D. discolor
    D. vitellina
    D. tigrina
    D. fusca
    D. magnolia
    D. coronata
    D. palmarum
    D. kirtlandii
    D. striata
    D. castanea
  Catharopeza [Genus]
    C. bishopi [Species]

Setophaga [Genus]
  S. ruticilla [Species]
Seiurus [Genus]
  S. aurocapillus [Species]
  S. noveboracensis
  S. motacilla
Limnothlypis [Genus]
  L. swainsonii [Species]
Helmitheros [Genus]
  H. vermivorus [Species]
Protonotaria [Genus]
  P. citrea [Species]
Geothlypis [Genus]
  G. trichas [Species]
  G. beldingi
  G. flavovelata
  G. rostrata
  G. semiflava
  G. speciosa
  G. nelsoni
  G. chiriquensis
  G. aequinoctialis
  G. poliocephala
  G. formosa
  G. agilis
  G. philadelphia
  G. tolmiei
Microligea [Genus]
  M. palustris [Species]
Teretistris [Genus]
  T. fernandinae [Species]
  T. fornsi
Leucopeza [Genus]
  L. semperi [Species]
Wilsonia [Genus]
  W. citrinia [Species]
  W. pusilla
  W. canadensis
Cardellina [Genus]
  C. rubrifrons [Species]
Ergaticus [Genus]
  E. ruber [Species]
  E. versicolor
Myioborus [Genus]
  M. pictus [Species]
  M. miniatus
  M. brunniceps
  M. pariae
  M. cardonai
  M. torquatus
  M. ornatus
  M. melanocephalus
  M. albifrons
  M. flavivertex
  M. albifacies
Euthlypis [Genus]
  E. lachrymosa [Species]
Basileuterus [Genus]
  B. fraseri [Species]
  B. bivittatus

B. chrysogaster
B. flaveolus
B. luteoviridis
B. signatus
B. nigrocristatus
B. griseiceps
B. basilicus
B. cinereicollis
B. conspicillatus
B. coronatus
B. culicivorus
B. rufifrons
B. belli
B. melanogenys
B. tristriatus
B. trifasciatus
B. hypoleucus
B. leucoblepharus
B. leucophrys
Phaeothlypis [Genus]
  P. fulvicauda [Species]
  P. rivularis
Peucedramus [Genus]
  P. taeniatus [Species]
Xenoligea [Genus]
  X. montana [Species]
Granatellus [Genus]
  G. venustus [Species]
  G. sallaei
  G. pelzelni
Icteria [Genus]
  I. virens [Species]
Conirostrum [Genus]
  C. speciosum [Species]
  C. leucogenys
  C. bicolor
  C. margaritae
  C. cinereum
  C. ferrugineiventre
  C. rufum
  C. sitticolor
  C. albifrons
Coereba [Genus]
  C. flaveola [Species]

Drepanididae [Family]
  Himatione [Genus]
    H. sanguinea [Species]
  Palmeria [Genus]
    P. dolei [Species]
  Vestiaria [Genus]
    V. coccinea [Species]
  Drepanis [Genus]
    D. funerea [Species]
    D. pacifica
  Ciridops [Genus]
    C. anna [Species]
  Viridonia [Genus]
    V. virens [Species]
    V. parva
    V. sagittirostris

Hemignathus [Genus]
  H. obscurus [Species]
  H. lucidus
  H. wilsoni
Loxops [Genus]
  L. coccinea [Species]
Paroreomyza [Genus]
  P. maculata [Species]
Pseudonester [Genus]
  P. xanthophrys [Species]
Psittirostra [Genus]
  P. psittacea [Species]
Loxioides [Genus]
  L. cantans [Species]
  L. palmeri
  L. flaviceps
  L. bailleui
  L. kona

Vireonidae [Family]
  Cyclarhis [Genus]
    C. gujanensis [Species]
    C. nigrirostris
  Vireolanius [Genus]
    V. melitophrys [Species]
    V. pulchellus
    V. leucotis
  Vireo [Genus]
    V. brevipennis [Species]
    V. huttoni
    V. atricapillus
    V. griseus
    V. pallens
    V. caribaeus
    V. bairdi
    V. gundlachii
    V. crassirostris
    V. bellii
    V. vicinior
    V. nelsoni
    V. hypochryseus
    V. modestus
    V. nanus
    V. latimeri
    V. osburni
    V. carmioli
    V. solitarius
    V. flavifrons
    V. philadelphicus
    V. olivaceus
    V. magister
    V. altiloquus
    V. gilvus
  Hylophilus [Genus]
    H. poicilotis [Species]
    H. thoracicus
    H. semicinereus
    H. pectoralis
    H. sclateri
    H. muscicapinus
    H. brunneiceps

H. *semibrunneus*
H. *aurantifrons*
H. *hypoxanthus*
H. *flavipes*
H. *ochraceiceps*
H. *decurtatus*

Icteridae [Family]
Psarocolius [Genus]
P. *oseryi* [Species]
P. *latirostris*
P. *decumanus*
P. *viridis*
P. *atrovirens*
P. *angustifrons*
P. *wagleri*
P. *montezuma*
P. *cassini*
P. *bifasciatus*
P. *guatimozinus*
P. *yuracares*
Cacicus [Genus]
C. *cela* [Species]
C. *uropygialis*
C. *chrysopterus*
C. *koepckeae*
C. *leucoramphus*
C. *chrysonotus*
C. *sclateri*
C. *solitarius*
C. *melanicterus*
C. *holosericeus*
Icterus [Genus]
I. *cayanensis* [Species]
I. *chrysater*
I. *nigrogularis*
I. *leucopteryx*
I. *auratus*
I. *mesomelas*
I. *auricapillus*
I. *graceannae*
I. *xantholemus*
I. *pectoralis*
I. *gularis*
I. *pustulatus*
I. *cucullatus*
I. *icterus*
I. *galbula*
I. *spurius*
I. *dominicensis*
I. *wagleri*
I. *laudabilis*
I. *bonana*
I. *oberi*
I. *graduacauda*
I. *maculialatus*
I. *parisorum*
Nesopsar [Genus]
N. *nigerrimus* [Species]
Xanthopsar [Genus]
X. *flavus* [Species]

Gymnomystax [Genus]
G. *mexicanus* [Species]
Xanthocephalus [Genus]
X. *xanthocephalus* [Species]
Agelaius [Genus]
A. *thilius* [Species]
A. *phoeniceus*
A. *tricolor*
A. *icterocephalus*
A. *humeralis*
A. *xanthomus*
A. *cyanopus*
A. *ruficapillus*
Leistes [Genus]
L. *militaris* [Species]
Pezites [Genus]
P. *militaris* [Species]
Sturnella [Genus]
S. *magna* [Species]
S. *neglecta*
Pseudoleistes [Genus]
P. *guirahuro* [Species]
P. *virescens*
Amblyramphus [Genus]
A. *holosericeus* [Species]
Hypopyrrhus [Genus]
H. *pyrohypogaster* [Species]
Curaeus [Genus]
C. *curaeus* [Species]
C. *forbesi*
Gnorimopsar [Genus]
G. *chopi* [Species]
Oreopsar [Genus]
O. *bolivianus* [Species]
Lampropsar [Genus]
L. *tanagrinus* [Species]
Macroagelaius [Genus]
M. *subalaris* [Species]
Dives [Genus]
D. *atroviolacea* [Species]
D. *dives*
Quiscalus [Genus]
Q. *mexicanus* [Species]
Q. *major*
Q. *palustris*
Q. *nicaraguensis*
Q. *quiscula*
Q. *niger*
Q. *lugubris*
Euphagus [Genus]
E. *carolinus* [Species]
E. *cyanocephalus*
Molothrus [Genus]
M. *badius* [Species]
M. *rufoaxillaris*
M. *bonariensis*
M. *aeneus*
M. *ater*
Scaphidura [Genus]
S. *oryzivorus* [Species]

Fringillidae [Family]
Fringilla [Genus]
F. *coelebs* [Species]
F. *teydea*
F. *montifringilla*
Serinus [Genus]
S. *pusillus* [Species]
S. *serinus*
S. *syriacus*
S. *canaria*
S. *citrinella*
S. *thibetanus*
S. *canicollis*
S. *nigriceps*
S. *citrinelloides*
S. *frontalis*
S. *capistratus*
S. *koliensis*
S. *scotops*
S. *leucopygius*
S. *atrogularis*
S. *citrinipectus*
S. *mozambicus*
S. *donaldsoni*
S. *flaviventris*
S. *sulphuratus*
S. *albogularis*
S. *gularis*
S. *mennelli*
S. *tristriatus*
S. *menschensis*
S. *striolatus*
S. *burtoni*
S. *rufobrunneus*
S. *leucopterus*
S. *totta*
S. *alario*
S. *estherae*
Neospiza [Genus]
N. *concolor* [Species]
Linurgus [Genus]
L. *olivaceus* [Species]
Rhynchostruthus [Genus]
R. *socotranus* [Species]
Carduelis [Genus]
C. *chloris* [Species]
C. *sinica*
C. *spinoides*
C. *ambigua*
C. *spinus*
C. *pinus*
C. *atriceps*
C. *spinescens*
C. *yarrellii*
C. *cucullata*
C. *crassirostris*
C. *magellanica*
C. *dominicensis*
C. *siemiradzkii*
C. *olivacea*

C. notata
C. xanthogastra
C. atrata
C. uropygialis
C. barbata
C. tristis
C. psaltria
C. lawrencei
C. carduelis
Acanthis [Genus]
   A. flammea [Species]
   A. hornemanni
   A. flavirostris
   A. cannabina
   A. yemenensis
   A. johannis
Leucosticte [Genus]
   L. nemoricola [Species]
   L. brandti
   L. arctoa
Callacanthis [Genus]
   C. burtoni [Species]
Rhodopechys [Genus]
   R. sanguinea [Species]
   R. githaginea
   R. mongolica
   R. obsoleta
Uragus [Genus]
   U. sibiricus [Species]
Urocynchramus [Genus]
   U. pylzowi [Species]
Carpodacus [Genus]
   C. rubescens [Species]
   C. nipalensis
   C. erythrinus
   C. purpureus
   C. cassinii
   C. mexicanus
   C. pulcherrimus
   C. eos
   C. rhodochrous
   C. vinaceus
   C. edwardsii
   C. synoicus
   C. roseus
   C. trifasciatus
   C. rhodopeplus
   C. thura
   C. rhodochlamys
   C. rubicilloides
   C. rubicilla
   C. puniceus
   C. roborowskii
Chaunoproctus [Genus]
   C. ferreorostris [Species]
Pinicola [Genus]
   P. enucleator [Species]
   P. subhimachalus
Haematospiza [Genus]
   H. sipahi [Species]

Loxia [Genus]
   L. pytyopsittacus [Species]
   L. curvirostra
   L. leucoptera
Pyrrhula [Genus]
   P. nipalensis [Species]
   P. leucogenys
   P. aurantiaca
   P. erythrocephala
   P. erythaca
   P. pyrrhula
Coccothraustes [Genus]
   C. coccothraustes [Species]
   C. migratorius
   C. personatus
   C. icterioides
   C. affinis
   C. melanozanthos
   C. carnipes
   C. vespertinus
   C. abeillei
Pyrrhoplectes [Genus]
   P. epauletta [Species]

Estrildidae [Family]
Parmoptila [Genus]
   P. woodhousei [Species]
Nigrita [Genus]
   N. fusconota [Species]
   N. bicolor
   N. luteifrons
   N. canicapilla
Nesocharis [Genus]
   N. shelleyi [Species]
   N. ansorgei
   N. capistrata
Pytilia [Genus]
   P. phoenicoptera [Species]
   P. hypogrammica
   P. afra
   P. melba
Mandingoa [Genus]
   M. nitidula [Species]
Cryptospiza [Genus]
   C. reichenovii [Species]
   C. salvadorii
   C. jacksoni
   C. shelleyi
Pyrenestes [Genus]
   P. sanguineus [Species]
   P. ostrinus
   P. minor
Spermophaga [Genus]
   P. poliogenys [Species]
   P. haematina
   P. ruficapilla
Clytospiza [Genus]
   C. monteiri [Species]
Hypargos [Genus]
   H. margaritatus [Species]
   H. niveoguttatus

Euschistospiza [Genus]
   E. dybowskii [Species]
   E. cinereovinacea
Lagonosticta [Genus]
   L. rara [Species]
   L. rufopicta
   L. nitidula
   L. senegala
   L. rubricata
   L. landanae
   L. rhodopareia
   L. larvata
Uraeginthus [Genus]
   U. angolensis [Species]
   U. bengalus
   U. cyanocephala
   U. granatina
   U. ianthinogaster
Estrilda [Genus]
   E. caerulescens [Species]
   E. perreini
   E. thomensis
   E. melanotis
   E. paludicola
   E. melpoda
   E. rhodopyga
   E. rufibarba
   E. troglodytes
   E. astrild
   E. nigriloris
   E. nonnula
   E. atricapilla
   E. erythronotos
   E. charmosyna
Amandava [Genus]
   A. amandava [Species]
   A. formosa
   A. subflava
Ortygospiza [Genus]
   O. atricollis [Species]
   O. gabonensis
   O. locustella
Aegintha [Genus]
   A. temporalis [Species]
Emblema [Genus]
   E. picta [Species]
   E. bella
   E. oculata
   E. guttata
Oreostruthus [Genus]
   O. fuliginosus [Species]
Neochmia [Genus]
   N. phaeton [Species]
   N. ruficauda
Poephila [Genus]
   P. guttata [Species]
   P. bichenovii
   P. personata
   P. acuticauda
   P. cincta

Erythrura [Genus]
    E. hyperythra [Species]
    E. prasina
    E. viridifacies
    E. tricolor
    E. coloria
    E. trichroa
    E. papuana
    E. psittacea
    E. cyaneovirens
    E. kleinschmidti
Chloebia [Genus]
    C. gouldiae [Species]
Aidemosyne [Genus]
    A. modesta [Species]
Lonchura [Genus]
    L. malabarica [Species]
    L. griseicapilla
    L. nana
    L. cucullata
    L. bicolor
    L. fringilloides
    L. striata
    L. leucogastroides
    L. fuscans
    L. molucca
    L. punctulata
    L. kelaarti
    L. leucogastra
    L. tristissima
    L. leucosticta
    L. quinticolor
    L. malacca
    L. maja
    L. pallida
    L. grandis
    L. vana
    L. caniceps
    L. nevermanni
    L. spectabilis
    L. forbesi
    L. hunsteini
    L. flaviprymna
    L. castaneothorax
    L. stygia
    L. teerinki
    L. monticola
    L. montana
    L. melaena
    L. pectoralis
Padda [Genus]
    P. fuscata [Species]
    P. oryzivora
Amadina [Genus]
    A. erythrocephala [Species]
    A. fasciata
Pholidornis [Genus]
    P. rushiae [Species]

Ploceidae [Family]
    Vidua [Genus]

V. chalybeata [Species]
V. funerea
V. wilsoni
V. hypocherina
V. fischeri
V. regia
V. macroura
V. paradisaea
V. orientalis
Bubalornis [Genus]
    B. albirostris [Species]
Dinemellia [Genus]
    D. dinemelli [Species]
Plocepasser [Genus]
    P. mahali [Species]
    P. superciliosus
    P. donaldsoni
    P. rufoscapulatus
Histurgops [Genus]
    H. ruficauda [Species]
Pseudonigrita [Genus]
    P. arnaudi [Species]
    P. cabanisi
Philetairus [Genus]
    P. socius [Species]
Passer [Genus]
    P. ammodendri [Species]
    P. domesticus
    P. hispaniolensis
    P. pyrrhonotus
    P. castanopterus
    P. rutilans
    P. flaveolus
    P. moabiticus
    P. iagoensis
    P. melanurus
    P. griseus
    P. simplex
    P. montanus
    P. luteus
    P. eminibey
Petronia [Genus]
    P. brachydactyla [Species]
    P. xanthocollis
    P. petronia
    P. superciliaris
    P. dentata
Montifringilla [Genus]
    M. nivalis [Species]
    M. adamsi
    M. taczanowskii
    M. davidiana
    M. ruficollis
    M. blanfordi
    M. theresae
Sporopipes [Genus]
    S. squamifrons [Species]
    S. frontalis
Amblyospiza [Genus]
    A. albifrons [Species]

Ploceus [Genus]
    P. baglafecht [Species]
    P. bannermani
    P. batesi
    P. nigrimentum
    P. bertrandi
    P. pelzelni
    P. subpersonatus
    P. luteolus
    P. ocularis
    P. nigricollis
    P. alienus
    P. melanogaster
    P. capensis
    P. subaureus
    P. xanthops
    P. aurantius
    P. heuglini
    P. bojeri
    P. castaneiceps
    P. princeps
    P. xanthopterus
    P. castanops
    P. galbula
    P. taeniopterus
    P. intermedius
    P. velatus
    P. spekei
    P. spekeoides
    P. cucullatus
    P. grandis
    P. nigerrimus
    P. weynsi
    P. golandi
    P. dicrocephalus
    P. melanocephalus
    P. jacksoni
    P. badius
    P. rubiginosus
    P. aureonucha
    P. tricolor
    P. albinucha
    P. nelicourvi
    P. sakalava
    P. hypoxanthus
    P. superciliosus
    P. benghalensis
    P. manyar
    P. philippinus
    P. megarhynchus
    P. bicolor
    P. flavipes
    P. preussi
    P. dorsomaculatus
    P. olivaceiceps
    P. insignis
    P. angolensis
    P. sanctithomae
Malimbus [Genus]
    M. coronatus [Species]
    M. cassini

M. scutatus
M. racheliae
M. ibadanensis
M. nitens
M. rubricollis
M. erythrogaster
M. malimbicus
M. rubriceps
Quelea [Genus]
Q. cardinalis [Species]
Q. erythrops
Q. quelea
Foudia [Genus]
F. madagascariensis [Species]
F. eminentissima
F. rubra
F. bruante
F. sechellarum
F. flavicans
Euplectes [Genus]
E. anomalus [Species]
E. afer
E. diadematus
E. gierowii
E. nigroventris
E. hordeaceus
E. orix
E. aureus
E. capensis
E. axillaris
E. macrourus
E. hartlaubi
E. albonotatus
E. ardens
E. progne
E. jacksoni
Anomalospiza [Genus]
A. imberbis [Species]

Sturnidae [Family]
Aplonis [Genus]
A. zelandica [Species]
A. santovestris
A. pelzelni
A. atrifusca
A. corvina
A. mavornata
A. cinerascens
A. tabuensis
A. striata
A. fusca
A. opaca
A. cantoroides
A. crassa
A. feadensis
A. insularis
A. dichroa
A. mysolensis
A. magna
A. minor
A. panayensis

A. metallica
A. mystacea
A. brunneicapilla
Poeoptera [Genus]
P. kenricki [Species]
P. stuhlmanni
P. lugubris
Grafisia [Genus]
G. torquata [Species]
Onychognathus [Genus]
O. walleri [Species]
O. nabouroup
O. morio
O. blythii
O. frater
O. tristramii
O. fulgidus
O. tenuirostris
O. albirostris
O. salvadorii
Lamprotornis [Genus]
L. iris [Species]
L. cupreocauda
L. purpureiceps
L. curruscus
L. purpureus
L. nitens
L. chalcurus
L. chalybaeus
L. chloropterus
L. acuticaudus
L. splendidus
L. ornatus
L. australis
L. mevesii
L. purpuropterus
L. caudatus
Cinnyricinclus [Genus]
C. femoralis [Species]
C. sharpii
C. leucogaster
Speculipastor [Genus]
S. bicolor [Species]
Neocichla [Genus]
N. gutturalis [Species]
Spreo [Genus]
S. fischeri [Species]
S. bicolor
S. albicapillus
S. superbus
S. pulcher
S. hildebrandti
Cosmopsarus [Genus]
C. regius [Species]
C. unicolor
Saroglossa [Genus]
S. aurata [Species]
S. spiloptera
Creatophora [Genus]
C. cinerea [Species]

Necropsar [Genus]
N. leguati [Species]
Fregilupus [Genus]
F. varius [Species]
Sturnus [Genus]
S. senex [Species]
S. malabaricus
S. erythropygius
S. pagodarum
S. sericeus
S. philippensis
S. sturninus
S. roseus
S. vulgaris
S. unicolor
S. cinerascens
S. contra
S. nigricollis
S. burmannicus
S. melanopterus
S. sinensis
Leucopsar [Genus]
L. rothschildi [Species]
Acridotheres [Genus]
A. tristis [Species]
A. ginginianus
A. fuscus
A. grandis
A. albocinctus
A. cristatellus
Ampeliceps [Genus]
A. coronatus [Species]
Mino [Genus]
M. anais [Species]
M. dumontii
Basilornis [Genus]
B. celebensis [Species]
B. galeatus
B. corythaix
B. miranda
Streptocitta [Genus]
S. albicollis [Species]
S. albertinae
Sarcops [Genus]
S. calvus [Species]
Gracula [Genus]
G. ptilogenys [Species]
G. religiosa
Enodes [Genus]
E. erythrophris [Species]
Scissirostrum [Genus]
S. dubium [Species]
Buphagus [Genus]
B. africanus [Species]
B. erythrorhynchus

Oriolidae [Family]
Oriolus [Genus]
O. szalayi [Species]
O. phaeochromus
O. forsteni

O. bouroensis
O. viridifuscus
O. sagittatus
O. flavocinctus
O. xanthonotus
O. albiloris
O. isabellae
O. oriolus
O. auratus
O. chinensis
O. chlorocephalus
O. crassirostris
O. brachyrhynchus
O. monacha
O. larvatus
O. nigripennis
O. xanthornus
O. hosii
O. crentus
O. traillii
O. mellianus
Sphecotheres [Genus]
S. vieilloti [Species]
S. flaviventris
S. viridis
S. hypoleucus

Dicruridae [Family]
Chaetorhynchus [Genus]
C. papuensis [Species]
Dicrurus [Genus]
D. ludwigii [Species]
D. atripennis
D. adsimilis
D. fuscipennis
D. aldabranus
D. forficatus
D. waldenii
D. macrocercus
D. leucophaeus
D. caerulescens
D. annectans
D. aeneus
D. remifer
D. balicassius
D. hottentottus
D. megarhynchus
D. montanus
D. andamanensis
D. paradiseus

Callaeidae [Family]
Callaeas [Genus]
C. cinerea [Species]
Creadion [Genus]
C. carunculatus [Species]
Heterolocha [Genus]
H. acutirostris [Species]

Grallinidae [Family]
Grallina [Genus]

G. cyanoleuca [Species]
G. brujini
Corcorax [Genus]
C. melanorhamphos [Species]
Struthidea [Genus]
S. cinerea [Species]

Artamidae [Family]
Artamus [Genus]
A. fuscus [Species]
A. leucorhynchus
A. monachus
A. maximus
A. insignis
A. personatus
A. superciliosus
A. cinereus
A. cyanopterus
A. minor

Cracticidae [Family]
Cracticus [Genus]
C. mentalis [Species]
C. torquatus
C. cassicus
C. louisiadensis
C. nigrogularis
C. quoyi
Gymnorhina [Genus]
G. tibicen [Species]
Strepera [Genus]
S. graculina [Species]
S. fuliginosa
S. versicolor

Ptilonorhynchidae [Family]
Ailuroedus [Genus]
A. buccoides [Species]
A. crassirostris
Scenopoeetes [Genus]
S. dentirostris [Species]
Archboldia [Genus]
A. papuensis [Species]
Amblyornis [Genus]
A. inornatus [Species]
A. macgregoriae
A. subalaris
A. flavifrons
Prionodura [Genus]
P. newtoniana [Species]
Sericulus [Genus]
S. aureus [Species]
S. bakeri
S. chrysocephalus
Ptilonorhynchus [Genus]
P. violaceus [Species]
Chlamydera [Genus]
C. maculata [Species]
C. nuchalis
C. lauterbachi
C. cerviniventris

Paradisaeidae [Family]
Loria [Genus]
L. loriae [Species]
Loboparadisea [Genus]
L. sericea [Species]
Cnemophilus [Genus]
C. macgregorii [Species]
Macgregoria [Genus]
M. pulchra [Species]
Lycocorax [Genus]
L. pyrrhopterus [Species]
Manucodia [Genus]
M. ater [Species]
M. jobiensis
M. chalybatus
M. comrii
Phonygammus [Genus]
P. keraudrenii [Species]
Ptiloris [Genus]
P. paradiseus [Species]
P. victoriae
P. magnificus
Semioptera [Genus]
S. wallacei [Species]
Seleucidis [Genus]
S. melanuleuca [Species]
Paradigalla [Genus]
P. carunculata [Species]
Drepanornis [Genus]
D. albertisi [Species]
D. brujini
Epimachus [Genus]
E. fastuosus [Species]
E. meyeri
Astrapia [Genus]
A. nigra [Species]
A. splendidissima
A. mayeri
A. stephaniae
A. rothschildi
Lophorina [Genus]
L. superba [Species]
Parotia [Genus]
P. sefilata [Species]
P. carolae
P. lawesii
P. wahnesi
Pteridophora [Genus]
P. alberti [Species]
Cicinnurus [Genus]
C. regius [Species]
Diphyllodes [Genus]
D. magnificus [Species]
D. respublica
Paradisaea [Genus]
P. apoda [Species]
P. minor
P. decora
P. rubra
P. guilielmi
P. rudolphi

Corvidae [Family]
   Platylophus [Genus]
      P. galericulatus [Species]
   Platysmurus [Genus]
      P. leucopterus [Species]
   Gymnorhinus [Genus]
      G. cyanocephala [Species]
   Cyanocitta [Genus]
      C. cristata [Species]
      C. stelleri
   Aphelocoma [Genus]
      A. coerulescens [Species]
      A. ultramarina
      A. unicolor
   Cyanolyca [Genus]
      C. viridicyana [Species]
      C. pulchra
      C. cucullata
      C. pumilo
      C. nana
      C. mirabilis
      C. argentigula
   Cissilopha [Genus]
      C. melanocyanea [Species]
      C. sanblasiana
      C. beecheii
   Cyanocorax [Genus]
      C. caeruleus [Species]
      C. cyanomelas
      C. violaceus
      C. cristatellus
      C. heilprini
      C. cayanus
      C. affinis
      C. chrysops
      C. mysticalis
      C. dickeyi
      C. yncas
   Psilorhinus [Genus]
      P. morio [Species]
   Calocitta [Genus]
      C. formosa [Species]
   Garrulus [Genus]

      G. glandarius [Species]
      G. lanceolatus
      G. lidthi
   Perisoreus [Genus]
      P. canadensis [Species]
      P. infaustus
      P. internigrans
   Urocissa [Genus]
      U. ornata [Species]
      U. caerulea
      U. flavirostris
      U. erythrorhyncha
      U. whiteheadi
   Cissa [Genus]
      C. chinensis [Species]
      C. thalassina
   Cyanopica [Genus]
      C. cyana [Species]
   Dendrocitta [Genus]
      D. vagabunda [Species]
      D. occipitalis
      D. formosae
      D. leucogastra
      D. frontalis
      D. baileyi
   Crypsirina [Genus]
      C. temia [Species]
      C. cucullata
   Temnurus [Genus]
      T. temnurus [Species]
   Pica [Genus]
      P. pica [Species]
      P. nuttali
   Zavattariornis [Genus]
      Z. stresemanni [Species]
   Podoces [Genus]
      P. hendersoni [Species]
      P. biddulphi
      P. panderi
      P. pleskei
   Pseudopodoces [Genus]
      P. humilis [Species]
   Nucifraga [Genus]

      N. columbiana [Species]
      N. caryocatactes
   Pyrrhocorax [Genus]
      P. pyrrhocorax [Species]
      P. graculus
   Ptilostomus [Genus]
      P. afer [Species]
   Corvus [Genus]
      C. monedula [Species]
      C. dauuricus
      C. splendens
      C. moneduloides
      C. enca
      C. typicus
      C. florensis
      C. kubaryi
      C. validus
      C. woodfordi
      C. fuscicapillus
      C. tristis
      C. capensis
      C. frugilegus
      C. brachyrhynchos
      C. caurinus
      C. imparatus
      C. ossifragus
      C. palmarum
      C. jamaicensis
      C. nasicus
      C. leucognaphalus
      C. corone
      C. macrorhynchos
      C. orru
      C. bennetti
      C. coronoides
      C. torquatus
      C. albus
      C. tropicus
      C. cryptoleucus
      C. ruficollis
      C. corax
      C. rhipidurus
      C. albicollis
      C. crassirostris

A brief geologic history of animal life

A note about geologic time scales: A cursory look will reveal that the timing of various geological periods differs among textbooks. Is one right and the others wrong? Not necessarily. Scientists use different methods to estimate geological time—methods with a precision sometimes measured in tens of millions of years. There is, however, a general agreement on the magnitude and relative timing associated with modern time scales. The closer in geological time one comes to the present, the more accurate science can be—and sometimes the more disagreement there seems to be. The following account was compiled using the more widely accepted boundaries from a diverse selection of reputable scientific resources.

Geologic time scale

Era	Period	Epoch	Dates	Life forms
Proterozoic			2,500-544 mya*	First single-celled organisms, simple plants, and invertebrates (such as algae, amoebas, and jellyfish)
Paleozoic	Cambrian		544-490 mya	First crustaceans, mollusks, sponges, nautiloids, and annelids (worms)
	Ordovician		490-438 mya	Trilobites dominant. Also first fungi, jawless vertebrates, starfish, sea scorpions, and urchins
	Silurian		438-408 mya	First terrestrial plants, sharks, and bony fish
	Devonian		408-360 mya	First insects, arachnids (scorpions), and tetrapods
	Carboniferous	Mississippian	360-325 mya	Amphibians abundant. Also first spiders, land snails
		Pennsylvanian	325-286 mya	First reptiles and synapsids
	Permian		286-248 mya	Reptiles abundant. Extinction of trilobytes
Mesozoic	Triassic		248-205 mya	Diversification of reptiles: turtles, crocodiles, therapsids (mammal-like reptiles), first dinosaurs
	Jurassic		205-145 mya	Insects abundant, dinosaurs dominant in later stage. First mammals, lizards, frogs, and birds
	Cretaceous		145-65 mya	First snakes and modern fish. Extinction of dinosaurs, rise and fall of toothed birds
Cenozoic	Tertiary	Paleocene	65-55.5 mya	Diversification of mammals
		Eocene	55.5-33.7 mya	First horses, whales, and monkeys
		Oligocene	33.7-23.8 mya	Diversification of birds. First anthropoids (higher primates)
		Miocene	23.8-5.6 mya	First hominids
		Pliocene	5.6-1.8 mya	First australopithecines
	Quaternary	Pleistocene	1.8 mya-8,000 ya	Mammoths, mastodons, and Neanderthals
		Holocene	8,000 ya-present	First modern humans

*Millions of years ago (mya)

Index

Bold page numbers indicate the primary discussion of a topic; page numbers in italics indicate illustrations.

INDEX

INDEX

INDEX

INDEX

INDEX

INDEX

INDEX

F

INDEX

INDEX

INDEX

INDEX

INDEX

INDEX